Zeolite Catalysis for Low-Carbon Molecule Conversion

Zeolite Catalysis for Low-Carbon Molecule Conversion

Synthesis, Characterization, and Applications

Edited by
Xingang Li and Noritatsu Tsubaki

WILEY-VCH

Editors

Prof. Xingang Li
School of Chemical Engineering & Technology
Tianjin University
Tianjin 300350
P. R. China

Prof. Noritatsu Tsubaki
University of Toyama
Toyama 930-8555
Japan

Cover Design: Wiley
Cover Image: Yingtian Zhang

Typesetting: Lumina Datamatics, Madurai, India

Library of Congress Card No.: applied for

British Library Cataloguing-in-Publication Data
A catalogue record for this book is available from the British Library.

Bibliographic information published by the Deutsche Nationalbibliothek The Deutsche Nationalbibliothek lists this publication in the Deutsche Nationalbibliografie; detailed bibliographic data are available on the Internet at <http://dnb.d-nb.de>.

Print ISBN: 9783527352036
ePDF ISBN: 9783527841813
ePub ISBN: 9783527841820
oBook ISBN: 9783527841837

Printed and bound by
CPI Group (UK) Ltd, Croydon, CR0 4YY

C9783527352036_060726

Contents

Preface

With the 2025 Nobel Prize in Chemistry rewarding the pioneers of metal-organic frameworks (MOFs), porous materials have attracted unprecedented worldwide attention. As another important class of porous materials, **zeolites** have long been recognized for their unique crystalline structures and remarkable chemical and physical properties, making them highly versatile catalysts. First discovered in the eighteenth century by the Swedish mineralogist Axel Cronstedt, zeolites are distinguished by their high surface area, uniform pore structures, and tunable acidic properties. These features have led to their widespread use in catalytic processes, particularly in petroleum refining and petrochemical production. In recent decades, zeolites have also attracted significant attention in emerging fields such as sustainable energy and environmental protection. Advances in synthesis techniques have greatly expanded the diversity of zeolite frameworks, enabling the design of tailored catalysts with enhanced performance. The potential of zeolites in next-generation catalytic applications is vast, with research continually seeking to improve their efficiency and broaden their scope.

Low-carbon molecules—such as methane, carbon monoxide (CO), carbon dioxide (CO_2), and methanol—are key building blocks for sustainable chemical processes. Defined by their renewable origins, these molecules represent critical building blocks for the production of value-added chemicals and fuels. Transforming low-carbon feedstocks into higher-value products, such as olefins and aromatics, offers a pathway to reduce reliance on fossil resources, lower greenhouse gas emissions, and support the transition toward a circular, renewable-carbon economy. Catalytically converting these molecules efficiently and selectively is therefore of great importance for sustainable chemical manufacturing and clean energy production.

The conversion of low-carbon molecules holds tremendous promise for tackling critical environmental challenges, including climate change and resource depletion. Achieving this transformation efficiently demands catalytic processes that minimize energy input while maximizing selectivity toward desired products. Zeolites have emerged as outstanding catalysts for such conversions owing to their high surface area, microporous frameworks, and tunable acidity, which together enable precise control of reaction pathways. They have been successfully applied in processes such as methanol-to-olefins (MTO), and CO_2 conversion to value-added chemicals—key building blocks in the chemical industry. Recent advances in zeolite design have delivered catalysts with superior activity, selectivity, and stability, further establishing them as attractive candidates for sustainable chemical processes.

This book, *Zeolite Catalysis for Low-Carbon Molecule Conversion: Synthesis, Characterization, and Applications*, aims to provide a comprehensive overview of the latest progress in this field. The opening section outlines the challenges and opportunities associated with zeolite-catalyzed conversion of low-carbon molecules. It then introduces the structure and properties of zeolites, emphasizing the characteristics that underpin their catalytic functions. Subsequent chapters examine the conversion of key low-carbon molecules—including CO, CO_2, methane, and methanol—with a focus on the role of zeolites in these processes and the challenges associated with their optimization. The applications of zeolites in hydrocarbon dehydrogenation and hydrogenation are also discussed. By gathering state-of-the-art research and insights, this book seeks to deepen understanding of zeolite catalysis and to inspire the development of more efficient, sustainable catalytic processes. We hope that this book will serve as a valuable resource, as well as a platform for researchers, practitioners, and industrial stakeholders working to advance the field of zeolite catalysis and contribute to the global pursuit of sustainable chemical technologies.

We sincerely thank all individuals who contributed to the compilation, proofreading, and related work for this book. We extend special gratitude to Professor Song Song and PhD candidate Kaijie Han for their efforts in standardizing the chapters and formatting the content. Additionally, we would like to thank Dr. Yingtian Zhang for his contribution to the cover design.

By *Xingang Li* and *Noritatsu Tsubaki*
Tianjin University and
University of Toyama
October 2025

Part I

Introduction

1

Challenges and Opportunities for Zeolites in the Conversion of Low-carbon Molecules

Zaiku Xie, Chuanming Wang, Jing Shi*

State Key Laboratory of Green Chemical Engineering and Industrial Catalysis, Sinopec Shanghai Research Institute of Petrochemical Technology Co., Ltd., Shanghai, China

*Corresponding author: xzk@sinopec.com

1.1 Introduction

Zeolites, i.e., molecular sieves, are inorganic solid porous crystallites with well-defined channel/pore structures, and contribute highly to the development of modern chemical industries in the fields of adsorption, separation, and catalysis. Last several decades have witnessed the wide applications of zeolites in a variety of important industrial catalytic processes, like fluid catalytic cracking, toluene disproportionation, aromatic alkylation, alkane isomerization, olefin catalytic cracking, and epoxidation [1]. With the increasing requirements for the development of a more sustainable and environmentally benign chemical industry, it is essential to utilize other resources like syngas, shale gas, CO_2, and biomass to produce diverse chemicals and fuels. Zeolites play a considerable role in these kinds of low-carbon hydrocarbon catalytic conversions [2, 3]. The unique catalytic properties are highly related to both confinement environments and acidity properties of zeolites. Despite the great progress of zeolite-catalyzed low-carbon hydrocarbon conversions, there still exist several challenges and opportunities attracting widespread interests from both academic and industrial communities.

1.2 Challenges in the Synthesis of Zeolites

Though it is theoretically possible to construct millions of hypothetical zeolite structures, only around 260 zeolites with distinct framework topologies until now have been successfully obtained, and only limited zeolites (e.g., FAU, MFI, MWW, BEA, MOR, CHA, AEI, etc.) have found widespread applications in the industrial practice of chemical processes. Targeted synthesis of zeolites with the desired active sites and framework topologies is a great challenge [4]. Zeolites are usually synthesized via the hydrothermal method, which mainly consists of two stages, i.e., the homogenization and aging of the mixture at lower temperatures (usually <100 °C), and the heating of the mixture at elevated temperatures (usually >100 °C) to form crystals [5]. During the process of premixing raw materials into gel, reactive silicate and aluminate species quickly condense to form a gel structure, making the synthesis system highly viscous and causing it to easily delaminate. During the high-temperature heating process, aluminosilicate species are hydrolyzed, condensed, and rearranged, consequently leading to the onset of the nucleation and crystallization procedures. These transformations can occur in the liquid or solid phase, and more likely at the interface. The complex bonding and nonbonding interactions among multiple components complicate the understanding of the role of organic structure-directing agents (OSDAs), which could be space-filling, charge-balancing, structure-directing, or templating. The precise match between the rates of material transfer/mixing and intrinsic reaction in the complex synthetic system at the molecular scale plays a key role in controlling the crystallization time, zeolite morphology, and distribution of active sites. It is thus necessary to understand the crystallization mechanism and kinetics in order to prepare zeolites with specific topology, framework composition, and pore architecture (size, dimension, orientations, etc.).

1.3 Challenges in Zeolite-catalyzed Low-carbon Hydrocarbon Conversions

Zeolites also play indispensable roles in the low-carbon hydrocarbon conversions, e.g., methanol-to-olefins/aromatics (MTO/MTA) conversion, direct syngas/CO_2-to-olefins/aromatics (STO/STA, CTO/CTA) conversion within bifunctional catalysts, selective oxidation of methane to methanol (MTM), methane dehydroaromatization (MDA), propane dehydrogenation (PDH), etc. The confinement effect imposed by zeolite frameworks and the well-tuned active sites endow zeolites with high catalytic activity, selectivity, and stability toward low-carbon hydrocarbon conversions. However, systematic understanding of the underlying reaction mechanism and the relationships between zeolite structures and catalytic performances is the prerequisite toward the rational design and practical application of zeolite catalysts, which remains a great challenge.

Considering MTO conversion as an example, which has been successfully industrialized (DMTO, S-MTO, UOP-MTO, etc.) in this century using small-pore eight-membered ring SAPO-34 zeolite as the catalyst, the detailed reaction mechanism is still in great dispute [6, 7]. It is generally recognized that the active site for MTO conversion consists of inorganic Brønsted acid center and organic hydrocarbon pool species. While the characteristics of the intrinsic hydrocarbon pool species are under great debate, it is suggested that aromatics like polymethylbenzenes and/or olefins themselves are the dominating hydrocarbon pool species. As different hydrocarbon pool species follow different reaction cycles and dedicate to distinct MTO catalytic performances, it is particularly important to address the distribution and relative contribution to MTO conversion of the hydrocarbon pool species as a function of zeolite structures and reaction conditions. It is also particularly necessary to identify and quantify the dynamic characteristics of the hydrocarbon pool species in zeolites under working conditions. However, the dynamic features of the hydrocarbon pool species formed in situ result in severe challenges in deciphering the structures and following the evolution of the dominating hydrocarbon pool species in detail during the reaction, let alone the complexities of the involved possible reaction pathways. Significant experimental and computational efforts have been dedicated to several challenging issues related to the MTO conversion, like the first C—C bond coupling in the induction period, the nature and evolution of organic hydrocarbon pool species, and the characteristics and effects of coking species formed in situ. It is highly desirable to decrease coking and methanol consumption and to tailor the selectivity of ethene and propene for the development of next-generation MTO zeolite catalysts.

The selective oxidation of methane to methanol is usually considered as one of the "holy grail" reactions in heterogeneous catalysis [8–10]. The traditional methane-to-methanol conversion usually involves two processes, i.e., the reforming of methane to syngas, and the syngas conversion to methanol; however, such an indirect process is quite energy-consuming. It is thus essential to develop a direct methane-selective oxidation process. Inspired by the natural methane monooxygenase (MMO) that can catalyze the selective oxidation of methane by oxygen molecule under ambient conditions, the Cu-based and Fe-based zeolites have shown potential prospect toward the selective oxidation of methane to methanol. Both stepwise and continuous catalytic processes have been suggested and exploited in a variety of zeolite matrixes. However, the conversion rate and the methanol yield are particularly low, far less than the catalytic performance as expected for possible industrialization. Several different Cu active centers confined with zeolites (e.g., $CuOH^+$, Cu_2O^{2+}, $Cu_2O_2^{2+}$, and $Cu_3O_3^{2+}$) were identified and proposed, while the intrinsic dominating active site for MTM reaction is still in dispute. It is likely that these motifs can be interconverted under working conditions. In addition, the critical role of water for reaction needs to be deciphered to enhance catalytic performance. It is also quite important and challenging to address the origin of oxygen atom in methanol and unveil the characteristics of oxidant (H_2O versus O_2). More attention thus should be paid to the dynamic evolution of active sites under working conditions.

Very recently, the bifunctional catalysts pioneered by Bao and other researchers received increasing interests for the direct conversion of syngas and CO_2 to light olefins and aromatics [11, 12]. Such bifunctional catalysts consist of oxide component for the conversion of syngas/CO_2 to bridging oxygenate, and the zeolite component for the conversion of bridging oxygenate to olefins and aromatics. Simpler assembling mode of two components and higher selectivity toward olefins and aromatics have evoked great interests on bifunctional catalysts for several conversion processes. A variety of oxides including binary spinel (e.g., $ZnCrO_x$, $ZnAlO_x$) and several zeolites with different ring openings (e.g., SAPO-34, SAPO-18, and ZSM-5) were taken into account to assemble fascinating bifunctional catalysts and establish relay catalytic processes for the direct conversion of syngas or CO_2 to olefins, aromatics, or fuels. Despite the impressive catalytic selectivity for direct conversion, it is essential to improve the activity of oxides toward CO/CO_2, and balance the catalytic performance of oxides and zeolites under working conditions in order to achieve possible industrial practices. It is noteworthy that the characteristics of bridging oxygenate are still under debate; moreover, the similarities and differences for the conversion of methanol and ketene should be disclosed. It is also necessary to identify the intrinsic active sites of oxide surface (e.g., spinel $ZnCr_2O_4$ oxide) and understand the effects of surface structures and reducibility on syngas conversion. An impressive catalytic performance of more than 80% in both CO conversion and olefin selectivity was reported for the STO conversion using heteroatom-substituted Ge-AlPO-18 zeolite component [13], making it possible to industrialize.

1.4 Challenges in the Construction and Characterization of Active Sites in Zeolites

A typical zeolite-catalyzed reaction usually proceeds on Brønsted site, Lewis site, and/or confined metal cluster within the zeolite matrix. The catalytic performance highly depends on both the reaction and diffusion kinetics. It is thus essential to tailor the confinement effect and shape-selectivity performance of zeolites to balance both reaction and diffusion kinetics. For example, using eight-membered ring zeolites like SAPO-34 is the prerequisite to achieve high ethene and propene selectivity for MTO conversion as the eight-membered ring imposes strong confinement effect on the diffusion of propene and higher olefins. On the other hand, it is possible to construct mesopores in zeolites to decrease coking formation and prolong the lifetime of catalysts [14]. Modifying the hydrophilic or hydrophobic feature of zeolite channel provides another strategy to tailor the diffusion kinetics of zeolites. However, it is a great challenge to precisely construct well-defined structures of zeolites in terms of both isolated active sites and pore structures for achieving the balance in both reaction and diffusion kinetics. The challenge, on the one hand, lies in tracing the dynamic evolution of active sites of zeolite-catalyzed reactions; it also remains a great challenge, to establish both precise macroscopic and microscopic measurements and decipher the anisotropy of the diffusion characteristics in zeolites under working conditions (e.g., reaction–diffusion coupling, multiple components, and high temperature) [15].

1.5 Opportunities and Outlook

We are now stepping into a new era of green and carbon neutrality for sustainable development of chemical industry. The development of advanced in situ/operando characterizations, together with the first-principles-based computational approaches including density functional theory (DFT) calculations, microkinetic modeling (MKM), and ab-initio molecular dynamics (AIMD) simulations, is playing an indispensable role in identifying the intrinsic active sites, uncovering the underlying reaction network, and establishing the relations between zeolite structures and catalytic performances [16–19]. The emergence of artificial intelligence (AI)-driven approach and high-throughput screening (HTS) techniques makes it possible to tackle complex systems involving complicated structures and reaction networks. It is anticipated that the next-generation catalytic processes can be progressed by taking advantage of the merits of zeolites.

References

1 Corma, A. (1997). From microporous to mesoporous molecular sieve materials and their use in catalysis. *Chem. Rev.* 97: 2373–2420.
2 Zhang, Q., Yu, J., and Corma, A. (2020). Applications of zeolites to C1 chemistry: recent advances, challenges, and opportunities. *Adv. Mater.* 32: 2002927.
3 Serrano, D.P., Centi, G., Diddams, P.A. et al. (2024). Outlooks for zeolite catalysts in a low-carbon scenario. *Catal. Today* 426: 114365.
4 Mallette, A.J., Shilpa, K., and Rimer, J.D. (2024). The current understanding of mechanistic pathways in zeolite crystallization. *Chem. Rev.* 124: 3416–3493.
5 Cundy, C.S. and Cox, P.A. (2003). The hydrothermal synthesis of zeolites: history and development from the earliest days to the present time. *Chem. Rev.* 103: 663–702.
6 Tian, P., Wei, Y., Ye, M. et al. (2015). Methanol to olefins (MTO): from fundamentals to commercialization. *ACS Catal.* 5: 1922–1938.
7 Yarulina, I., Chowdhury, A.D., Meirer, F. et al. (2018). Recent trends and fundamental insights in the methanol-to-hydrocarbons process. *Nat. Catal.* 1: 398–411.
8 Newton, M.A., Knorpp, A.J., Sushkevich, V.L. et al. (2020). Active sites and mechanisms in the direct conversion of methane to methanol using Cu in zeolitic hosts: a critical examination. *Chem. Soc. Rev.* 49: 1449–1486.
9 Dummer, N.F., Willock, D.J., He, Q. et al. (2023). Methane oxidation to methanol. *Chem. Rev.* 123: 6359–6411.
10 Zhang, H., Li, J., Wang, D. et al. (2024). A review on the active sites for direct oxidation of methane to methanol by copper-zeolites: coordination structure, formation and activity. *Coord. Chem. Rev.* 503: 215637.
11 Jiao, F., Li, J., Pan, X. et al. (2016). Selective conversion of syngas to light olefins. *Science* 351: 1065–1068.
12 Pan, X., Jiao, F., Miao, D. et al. (2021). Oxide–zeolite-based composite catalyst concept that enables syngas chemistry beyond Fischer–Tropsch synthesis. *Chem. Rev.* 121: 6588–6609.
13 Jiao, F., Bai, B., Li, G. et al. (2023). Disentangling the activity-selectivity trade-off in catalytic conversion of syngas to light olefins. *Science* 380: 727–730.
14 Chen, L.-H., Sun, M.-H., Wang, Z. et al. (2020). Hierarchically structured zeolites: from design to application. *Chem. Rev.* 120: 11194–11294.
15 Liu, X., Wang, C., Zhou, J. et al. (2022). Molecular transport in zeolite catalysts: depicting an integrated picture from macroscopic to microscopic scales. *Chem. Soc. Rev.* 51: 8174–8200.

16 Groppo, E., Rojas-Buzo, S., and Bordiga, S. (2023). The role of in situ/operando IR spectroscopy in unraveling adsorbate-induced structural changes in heterogeneous catalysis. *Chem. Rev.* 123: 12135–12169.

17 Van Speybroeck, V., Hemelsoet, K., Joos, L. et al. (2015). Advances in theory and their application within the field of zeolite chemistry. *Chem. Soc. Rev.* 44: 7044–7111.

18 Xie, W., Xu, J., Chen, J. et al. (2022). Achieving theory–experiment parity for activity and selectivity in heterogeneous catalysis using microkinetic modeling. *Acc. Chem. Res.* 55: 1237–1248.

19 Schwalbe-Koda, D., Kwon, S., Paris, C. et al. (2021). A priori control of zeolite phase competition and intergrowth with high-throughput simulations. *Science* 374: 308–315.

Part II

Structures and Properties of Zeolites

2

Synthesis of Zeolite-encapsulated Metal Catalysts for Catalytic Conversion of Low-carbon Molecules

Kai Zhang[1], Chengxu Li[1], Qiming Sun[1,], Jihong Yu[2]*

[1] *Innovation Center for Chemical Science, College of Chemistry, Chemical Engineering and Materials Science and Jiangsu Key Laboratory of Advanced Negative Carbon Technologies, Soochow University, Suzhou, Jiangsu, China*
[2] *State Key Laboratory of Inorganic Synthesis and Preparative Chemistry, College of Chemistry and International Center of Future Science, Jilin University, Changchun, Jilin, China*

*Corresponding author: sunqiming@suda.edu.cn
K. Zhang and C. Li contributed equally to this work.

2.1 Introduction

Zeolites are crystalline porous materials characterized by high surface area, well-defined microporous networks, outstanding thermal stability, and tunable acid–base properties. These features make them highly valuable in industrial processes such as fluid catalytic cracking, methanol-to-olefin conversion, and toluene disproportionation, where they exhibit remarkable catalytic efficiency and selectivity toward desired products [1–8]. In recent years, zeolites have also emerged as excellent hosts for stabilizing metal species—including nanoparticles (NPs), clusters, single atoms, and isolated ions—enabling high dispersion and efficient utilization of active metals [9–18]. Compared to conventional supported metal catalysts, metal species confined within zeolites offer several distinct advantages, including the generation of ultrasmall, highly active sites, exceptional stability under harsh thermal and hydrothermal conditions, and inherent shape selectivity during catalytic transformations.

In general, the incorporation of metal species into zeolites can be classified into four types based on the nature of the metal entities and their spatial arrangement relative to the zeolite framework: (1) single atoms anchored within micropores or structural defects; (2) sub-nanometer metal clusters confined in the microporous channels; (3) metal NPs embedded within mesopores; and (4) NPs encapsulated in the internal cavities of hollow zeolite crystals. Owing to the high dispersion of active sites, the confinement effect provided by the zeolite matrix, and the synergistic interactions between the framework and the metals, such encapsulated systems

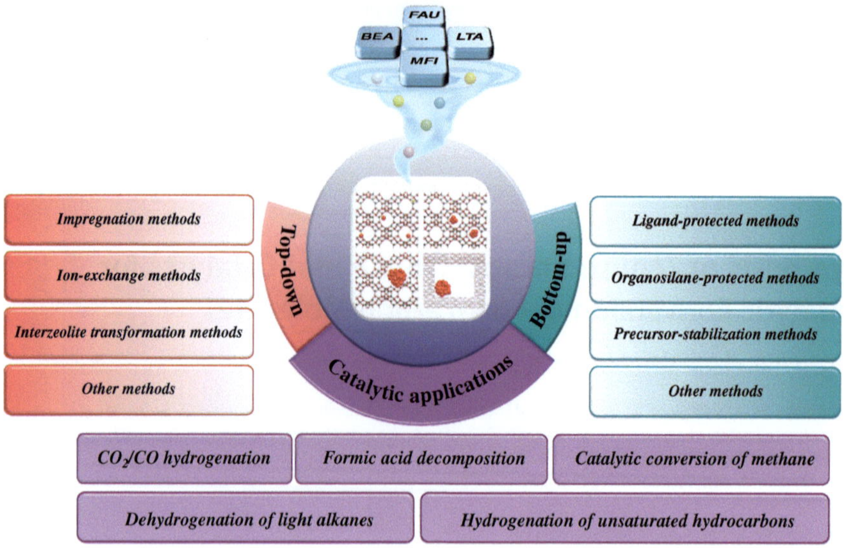

Figure 2.1 Schematic illustration of the construction routes for metal species confined within zeolites and their utilization in catalytic reactions involving low-carbon substrates.

have become highly attractive platforms for a wide range of heterogeneous catalytic processes [19–27].

Given the rapid and substantial advancements in the synthesis of zeolite-encapsulated metal catalysts, along with their promising applications in the crucial catalytic reactions, this chapter aims to provide a comprehensive overview of the cutting-edge synthetic strategies developed over the past decade, including top-down synthetic methods (e.g., impregnation method, ion-exchange method, and interzeolite transformation method) and bottom-up synthetic methods (e.g., ligand-protected method, organosilane-protected method, and precursor-stabilization method) (see Figure 2.1). We also present the important applications in catalytic conversion of low-carbon molecules of the representative zeolite-encapsulated metal catalysts, such as CO/CO_2 hydrogenation, catalytic conversion of methane, dehydrogenation of formic acid, dehydrogenation of light alkanes, and selective hydrogenation of unsaturated hydrocarbons (see Table 2.1). In the end, the future opportunities and challenges for this subject are pointed out.

2.2 Synthetic Methods for Zeolite-encapsulated Metal Catalysts Preparation

2.2.1 Top-down Synthetic Methods

2.2.1.1 Impregnation Methods

The impregnation technique is among the most widely employed methods for preparing metal-loaded zeolite catalysts. Nevertheless, conventional impregnation often results in metal species being deposited on the external surfaces of zeolite

Table 2.1 Summary of the structural information and catalytic applications in low-carbon molecule conversions of the representative zeolite-encapsulated metal catalysts prepared by various synthetic methods.

Catalyst	Metal species	Zeotype(s)	Metal loading (wt.%)	Metal size (nm)	Catalytic application	Source
Impregnation methods						
Rh–ZSM-5_{washed}	Rh	MFI	0.50	SA[a]	Oxidation of methane to acetic acid	[28]
Pt–Zn@DeAlBEA	Pt, Zn	BEA	0.80	SA	PDH	[20]
Pt@Beta	Pt	BEA	0.31	1	Selective hydrogenation	[29]
Ru@zeolite	Ru	CHA, MFI, BEA, and FAU	0.44	1.5	Ammonia borane hydrolysis	[30]
Rh–Ru@SP-MFI	Rh, Ru	MFI	0.44	0.7	Ammonia borane hydrolysis; hydrogenation of nitroarenes	[21]
$Au(CH_3)_2$@NaY	Au	FAU	1	SA	CO oxidation	[31]
Fe@zeolite	Fe	BEA, MFI, MOR, and FER	0.30	SA	Methane oxidation	[32]
M@Y	Ru, La	FAU	1	1	Hydrode oxygenation	[33]
Pd@ZSM-5	Pd	MFI	0.01	SA	Methane catalytic conversion	[34]
Rh@ZSM-5	Rh	MFI	0.10	SA	Methane catalytic conversion	[35]
Pt–REE@mz-deGa	Pt, La, Y, and Ce	MFI	Pt (1); REE (1)	3	PDH	[36]
Ion-exchange methods						
PtZn@MFI	Pt	MFI	1	1.1	Selective hydrogenation	[37]
Pd@H–MOR	Pd	MOR	1	1.3	Methane oxidation	[38]
M@zeolite	Pd, Pt, Co, Cu, Ag, and Au	MOR, MFI, MWW, FAU, and BEA	0.9–5.72	2.8	Dimethyl ether carbonylation; methane coupling	[39]
Co@S-1	Co	MFI	0.45	SA	EDH	[40]

(Continued)

Table 2.1 (Continued)

Catalyst	Metal species	Zeotype(s)	Metal loading (wt.%)	Metal size (nm)	Catalytic application	Source
Interzeolite transformation methods						
M@MFI	Pt, Ru, and Rh	MFI	0.80–1.55	1–1.7	Selective hydrogenation	[41]
Pt@MCM-22	Pt	MWW	0.11	SA-1	PDH	[42]
Other methods						
Au@S-1	Au	MFI	0.08–1.20	2.9–7.7	—	[43]
M@MFI	Co, Ni, and Cu	MFI	1–8	2.7–15	Toluene and mesitylene hydrogenation	[44]
Fe_2O_3–Pt@ hier-ZSM-5	Fe, Pt	MFI	Fe (3.40); Pt (0.50)	3.2–4.4	Phenol degradation	[45]
Pt@MFI	Pt	MFI	0.80–1.20	3	CO oxidation; selective hydrogenation	[46]
Ligand-protected methods						
M@zeolite	Pt, Pd, Ru, and Rh	SOD, GIS	0.03–0.86	1.4–1.7	Selective hydrogenation/ dehydrogenation	[47]
M@LTA	Pt, Pd, Rh, Ir, Re, and Ag	LTA	0.35–1.40	1.1–1.8	Alkanol oxidative dehydrogenation; alkene hydrogenation	[48]
Co@LTA	Co	LTA	0.40–4.50	1.5	CO/NO oxidation; ethanol oxidation	[49]
M@zeolite	Co, Ni, and Fe	LTA, FAU, and MFI	1.20–2.70	1.5–2.3	Ethanol oxidation	[50]
Pd@S-1	Pd	MFI	0.64	1.5–1.8	Formic acid dehydrogenation	[22]
Pd–$M(OH)_2$ @S-1	Pd, Ni, and Co	MFI	0.64	1.7	Formic acid dehydrogenation	[51]
Rh@MFI	Rh	MFI	0.28–0.71	SA	Ammonia borane hydrolysis; hydrogenation of nitroarenes	[52]

Catalyst	Metal species	Zeotype(s)	Metal loading (wt.%)	Metal size (nm)	Catalytic application	Source
M@Y	Pt, Pd, Ru, Rh, Co, Ni, and Cu	FAU	0.09–0.65	SA	EDH; isomerization	[53]
Pd@SOD	Pd	SOD	0.1	0.8	Acetylene-selective hydrogenation	[54]
PdMn@S-1	Pd, Mn	MFI	Pd (0.59–0.63); Mn (0.07–0.25)	1	CO_2 hydrogenation; formic acid dehydrogenation	[55]
PtZn4@S-1	Pt, Zn	MFI	Pt (0.72); Zn (1)	0.78	PDH	[19]
PtSn@MFI	Pt, Sn	MFI	Pt (0.40); Sn (0.90)	0.5–0.6	PDH	[23]
K-PtSn@MFI	Pt, Sn	MFI	Pt (0.42); Sn (0.9)	0.5	PDH	[56]
Ir@MWW	Ir	MWW	0.07–0.25	SA-0.8	PDH	[57]
Co@MFI	Co	MFI	1.12	SA	PDH	[58]
PdCu@S-1	Pd, Cu	MFI	0.11	3.2	Acetylene semihydrogenation	[59]
Na-Ni@CHA	Ni	CHA	3.5	SA	Acetylene semihydrogenation	[60]
Rh@ZSM-5	Rh	MFI	1	SA	Methanol carbonylation	[61]
FeS-1-EDTA	Fe	MFI	0.80	SA	EDH	[62]
Organosilane-protected methods						
M@NaA	Pt, Pd, Ir, Rh, and Ag	LTA	1	1.2–1.7	Ethene hydrogenation	[63]
Pt@CHA	Pt	CHA	0.17–0.45	SA-1.7	Hydrogenation of light olefins	[64]
Pd@MFI	Pd	MFI	0.71	1.7	Hydroconversion of furfural	[65]
Ni@FAU	Ni	FAU	4.50	SA	Chemoselective alkyne/olefin separations	[24]
Pt@Y	Pt	FAU	0.60	SA	Selective hydrogenation	[66]

(Continued)

Table 2.1 (Continued)

Catalyst	Metal species	Zeotype(s)	Metal loading (wt.%)	Metal size (nm)	Catalytic application	Source
Precursor-stabilization methods						
M@zeolite	Pd, Pt, and Ru	BEA, MOR	0.39–0.54	0.9–3.3	Selective hydrogenation	[67]
M@zeolite	Pt, Rh, Ag, Pd, and Fe	BEA, MOR, and MFI	0.39–1.01	0.8–3.6	CO_2 hydrogenation; water–gas shift reaction; oxidative reforming of methane	[26]
Ni@HZSM-5	Ni	MFI	1.02	10	Dry reforming of methane	[27]
Pd@S-1	Pd	MFI	0.34	3.7–11.7	Biomass hydrogenation	[68]
AuPd @ZSM-5	Au, Pd	MFI	Au (3.24); Pd (1.76)	6	Methane oxidation to methanol	[25]
Ru@NaY	Ru	FAU	2.3	4.9	CO hydrogenation	[69]
Fe@NaY	Fe	FAU	5.9	3.5	CO hydrogenation	[70]
Rh@MFI	Rh	MFI	0.42–0.52	4.2–5	CO_2 hydrogenation	[71]
RhMn @S-1	Rh, Mn	MFI	Rh (0.76); Mn (0.85)	2.7	CO hydrogenation	[72]

a. SA means single atom.

crystals, where they are prone to sintering during thermal treatments or catalytic operation. To achieve encapsulation of metal species within the zeolite framework, various improved impregnation strategies have been explored. For example, Flytzani-Stephanopoulos and co-workers introduced a modified incipient wetness impregnation (IWI) approach, which successfully incorporated single-atom Rh species into the micropores of ZSM-5 (MFI-type) zeolite by washing the impregnated samples [28]. C_s-corrected high-angle annular dark-field scanning transmission electron microscopy (HAADF-STEM) measurements revealed that the Rh species were isolated and homogeneously dispersed within zeolite channels in the Rh–ZSM-5_{washed} sample subjected to the water washing and H_2 reduction at 550 °C. In contrast, the large-sized Rh NPs can be observed on the outer surface of zeolites in the non-washed Rh–ZSM-5 sample. In situ CO diffuse reflectance infrared Fourier transform spectroscopy (CO-DRIFTS) tests showed that both the linear adsorption of CO molecules on isolated Rh atoms and the bridged adsorption of CO molecules on Rh NPs could be observed in non-washed Rh–ZSM-5 samples. In contrast, the Rh–ZSM-5_{washed} sample exhibited only two IR absorption bands at 2116 and 2049 cm^{-1}, corresponding to the symmetric and asymmetric CO stretching vibrations of isolated mononuclear $Rh^{I}(CO)_2$ complexes. This suggests that the water-washing step played a crucial role in facilitating the encapsulation of atomically dispersed Rh species within the zeolite matrixes. The Rh ions attached to the outside of zeolite crystals after the IWI process could be removed effectively by the post-washing treatment, which avoided the aggregation of Rh species on the outer surface of zeolites during the high-temperature reductions.

Bell and co-workers developed a silanol nest-stabilized method to synthesize zeolite-encapsulated isolated Pt atoms by a step-by-step impregnation method [20]. Typically, the Zn^{2+} ions were first introduced into the dealuminated Beta zeolites (BEA zeotype, DeAlBEA) with abundant silanol nests by impregnation method to form the ≡ SiOZn—OH group. Subsequently, the $Pt(NH_3)_4(NO_3)_2$ precursor was introduced into the Zn–DeAlBEA support by the impregnation method. Finally, the isolated Pt species was obtained by self-reduction upon calcination in air. Extended X-ray absorption fine structure (EXAFS) results showed that the coordination numbers (CNs) of Zn—(O)—Si bonds in the 0.36Zn–DeAlBEA and 0.04Pt–0.36Zn–DeAlBEA samples were 0.8 ± 0.6 and 0.5 ± 0.1 and the CNs of Zn—Zn bonds of samples were 0.8 ± 0.8 and 0.6 ± 0.1, respectively. The results revealed the Zn species predominantly located in ≡ SiOZn—OH nests, with a portion of Zn atoms situated in close proximity within these sites. C_s-corrected STEM analysis further confirmed that isolated Pt atoms were uniformly distributed across the Beta zeolite and anchored within the ≡ SiOZn—OH nests, forming a well-defined $(\equiv \text{Si—O—Zn})_{4\text{–}6}$ Pt coordination environment.

Gates and co-workers developed a physical adsorption of organometallic compounds method to encapsulate a series of atomically dispersed metal species into zeolites. For example, the calcined NaY zeolites (FAU zeotype) were first vacuumed to remove adsorbed water and gas, and then adsorbed the $Au(CH_3)_2(acac)$ precursor (acac = acetylacetone) from *n*-pentane solution [31]. Thanks to the high dispersion of $Au(CH_3)_2(acac)$ in *n*-pentane and strong interaction between $Au(CH_3)_2(acac)$ precursors and zeolites, the Au species (1 wt.%) were encapsulated within the NaY zeolites when the solvent was removed by evacuation for one day. The EXAFS results showed that only Au—O, Au—C, and Au—Al bonds existed in the catalyst, while no Au—Au bond was observed, indicating the Au species in NaY zeolites were in the form of Au complexes and connected with the Al atoms on zeolite frameworks through O atoms. HAADF-STEM measurements further confirmed the Au species were isolated and confined within the NaY zeolites, and no aggregated Au cluster or NPs could be observed in/on zeolites. Density functional theory (DFT) calculations further demonstrated that the isolated Au atoms preferentially occupy the T5 and T6 crystallographic sites in close proximity to Al centers within the FAU supercage.

Recently, Liu and co-workers introduced an innovative strategy that combines template-guidance synthesis with impregnation to construct metal catalysts confined within zeolite (see Figure 2.2a) [29]. The Beta zeolites were first treated with acetic acid to remove part of organic templates, creating a transport pathway for metal ions to enter the zeolites. Then the acid-treated zeolites were dispersed in an H_2PtCl_6 solution. Thanks to the electrostatic attraction between the negatively charged $PtCl_6^{2-}$ species and positively charged template agents (tetraethylammonium cations), the highly dispersed Pt species were adsorbed and encapsulated within zeolite matrixes. The sub-nanometric Pt clusters were finally formed and confined in Beta zeolites subjected to the H_2 reduction. TEM measurements revealed that the Pt species in Pt@Beta sample were approximately 1 nm in size and were uniformly dispersed throughout the zeolite crystals (see Figure 2.2b). In contrast, the Pt NPs with a mean size of 3.63 nm in the Pt/Beta sample synthesized by the conventional IWI method were mainly located on the outer surface of zeolite crystals (see Figure 2.2c).

In addition to the impregnation approach, the morphology of zeolites plays a crucial role in immobilizing ultrasmall metal species within their micropores. Yu and co-workers demonstrated the incorporation of a series of sub-nanometric RhRu bimetallic clusters onto and into ultrathin self-pillared MFI zeolite nanosheets via a straightforward IWI method (see Figure 2.2d) [21]. These nanosheet-structured zeolites featured high surface areas and numerous isolated Si–OH defect sites, which promoted strong interactions with metal precursors and effectively stabilized the metal clusters after H_2 reduction. C_s-corrected HAADF-STEM measurements revealed that the vast majority of metal species were confined within the micropores of self-pillared MFI zeolite nanosheets (less than 10 nm in thickness) and the average metal sizes of representative $Rh_{0.8}Ru_{0.2}$@SP-S-1 and $Rh_{0.8}Ru_{0.2}$@SP-ZSM-5-100 samples were about 0.7 nm (see Figure 2.2e–j). In contrast, when using the same IWI method on conventional MFI nanocrystals (200–300 nm), the

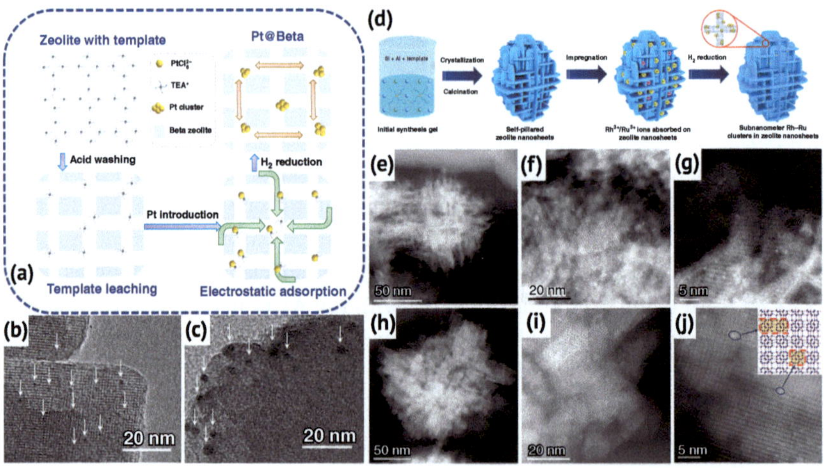

Figure 2.2 (a) Schematic illustration of the synthesis of Beta zeolite-encapsulated Pt catalysts by using template guidance coupled with the impregnation method. *Source*: Reproduced from [29] / with permission of John Wiley & Sons. The TEM images of (b) Pt@Beta and (c) Pt/Beta. *Source*: Reproduced from [29] / with permission of John Wiley & Sons. (d) Schematic representation of the formation process for sub-nanometric Rh–Ru bimetallic clusters confined within self-pillared MFI zeolite nanosheets. *Source*: Reproduced from [21] / with permission of American Chemical Society. C_s-corrected STEM images of (e–g) $Rh_{0.8}Ru_{0.2}$@SP-S-1 sample and (h–j) $Rh_{0.8}Ru_{0.2}$@SP-ZSM-5-100 sample. *Source*: Reproduced from [21] / with permission of American Chemical Society.

resulting metal particles were approximately 4 nm in size and predominantly distributed on the external surfaces of the zeolite with poor uniformity. Benefiting from both the confinement effect of the nanosheet zeolite architecture and the anchoring interactions provided by Si–OH groups, the metal catalysts encapsulated within MFI zeolite nanosheets exhibited remarkable thermal stability under oxidative and reductive conditions at 600 °C and significantly surpassing the performance of conventional metal catalysts supported on nanosized MFI crystals.

2.2.1.2 Ion-exchange Methods

It is well known that aluminosilicate and silicoaluminophosphate zeolites have negatively charged frameworks, as well as the alkali and alkaline earth metal cations (e.g., Na^+, K^+, and Ca^{2+}) or protons are located in the pores/cages of these zeolites to balance the electrical charge. When the zeolites are ion-exchanged by aqueous metal-containing solutions, the alkali cations in zeolites can be replaced by target metal cations. After subsequent calcination and/or a reduction procedure, the metal species will be supported onto/into the zeolite matrixes.

Román-Leshkov and co-workers developed an ion-exchange method coupled with the demetallation of zinc-containing MFI frameworks to encapsulate highly dispersed bimetallic Pt–Zn clusters into MFI zeolites [37]. During the synthesis, $(NH_3)_4Pt(NO_3)_2$ was initially introduced into zincosilicate zeolite (Zn@MFI) via ion exchange, forming Pt^{2+}/Zn@MFI as a result of the framework charge imbalance caused by Zn incorporation (see Figure 2.3a). Subsequent calcination in air followed by H_2 reduction at 400 °C led to the demetallation of Zn from the zeolite lattice and its alloying with Pt, ultimately yielding finely dispersed Pt–Zn bimetallic NPs confined within the MFI zeolite. High-resolution TEM (HRTEM) revealed that the Pt–Zn NPs, with an average size of approximately 1.1 nm, were uniformly embedded within the zeolite crystals, with no detectable presence on the external surfaces (see Figure 2.3b). Complementary differential pair distribution function (PDF) analysis confirmed the formation of a homogeneous Pt–Zn alloy within the zeolite, indicating intimate mixing of the two-metal species without any phase separation. The PtZn@MFI featured excellent shape- and substrate-selective features during the catalytic hydrogenation reactions, giving superb selectivity and production of *p*-chloroaniline by competitive hydrogenation of *p*-chloronitrobenzene and 1,3-dimethyl-5-nitrobenzene. The above shape-selective activity further revealed that the metal species in PtZn@MFI were confined within zeolite matrixes.

In the ion-exchange synthesis of metal-encapsulated zeolite catalysts, the nature of the original cations in the parent zeolite significantly influences the dispersion and distribution of the introduced metal species. Recently, Tsubaki and co-workers developed a universal ammonium pool-assisted ion-exchange method to encapsulate various small metal species (e.g., Cu, Co, Pt, Pd, Au, and Ag) into diverse zeolite structures, such as MOR, MFI, MWW, FAU, and BEA [39]. As compared with H-typed zeolites, filling the zeolites with ammonium ions gave the NH_4^+–zeolite a self-reducing feature. Ammonia released from the ammonia pools during calcination can not only serve as a mild in situ reducing agent for supported metal oxides,

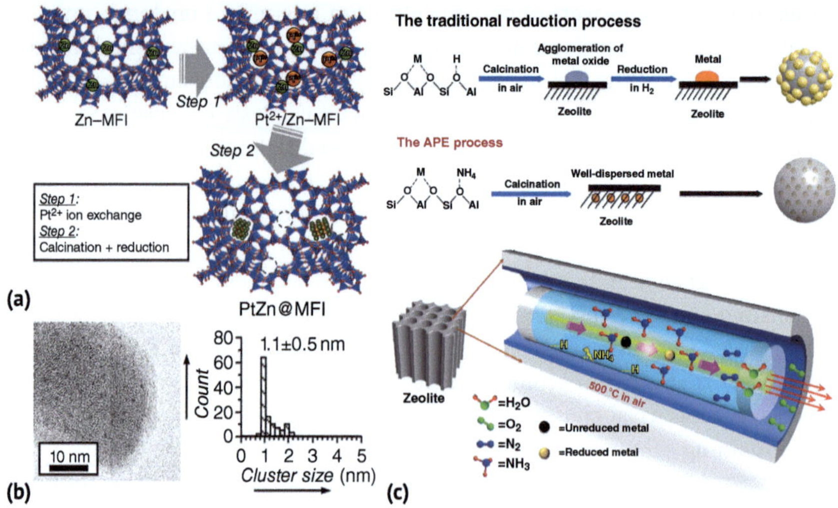

Figure 2.3 (a) Schematic illustration of the preparation for the bimetallic PtZn@MFI sample. *Source*: Reproduced from [37] / with permission of John Wiley & Sons. (b) HRTEM image of PtZn@MFI and the corresponding size distribution. *Source*: Reproduced from [37] / with permission of John Wiley & Sons. (c) Differences between conventional and APE reduction mechanisms for the synthesis of metal–zeolite catalysts and the scheme of the APE process. *Source*: [39] / Springer Nature / CC BY 4.0.

but also help preserve the small particle size and high dispersion of the metal species. Simultaneously, the NH_3 product from the ammonia pools' decomposition and reduction process filled the whole pores of zeolites, which pushed the water molecules out from the zeolite channels. In addition, the NH_3 product also prevented outside air from entering the zeolite crystals, leading to a slight oxidation of metal species (see Figure 2.3c). The ammonium pool effect (APE) process is performed smoothly, and the atmosphere-confined-self-regulation mechanism allowed the highly dispersed metal species to be anchored inside zeolites. Moreover, this APE is independent of the zeolite topology and is effective across a wide range of reducible metal species.

2.2.1.3 Interzeolite Transformation Methods

Interzeolite transformation offers a viable top-down strategy for embedding metal species into zeolites by reconstructing the framework of metal-containing precursor zeolites. For example, Iglesia and co-workers used zeolites (BEA and FAU)-supported metal NP (Pt, Ru, and Rh) samples as raw materials to synthesize the MFI zeolite-encapsulated metal NPs via the interzeolite transition under the hydrothermal condition (see Figure 2.4a) [41]. During the transformation process, the parent 12-membered ring (MR) zeolites (such as BEA and FAU) can be recrystallized into the 10-MR MFI zeolites with a higher framework density, and the encapsulated metal clusters in parent zeolites could be simultaneously transferred into the daughter zeolites. On the basis of the X-ray diffraction measurements, there was no characteristic peak of the parent zeolite in the transformed zeolite, indicating the purity of obtained MFI zeolite-encapsulated metal samples. TEM analysis showed that the metal NPs (1.3–1.7 nm) in the parent BEA and FAU zeolites retained their size after the interzeolite transformation into MFI zeolites. The encapsulation efficiency was assessed by comparing the hydrogenation rates of small toluene molecules and bulkier 1,3,5-trimethylbenzene. High relative hydrogenation rate values (8.1–40.9), reflecting the ratio of total metal surface area to that of externally accessible metal sites, were observed in both parent and transformed zeolites, indicating that the majority of metal species remained confined within the MFI framework. In contrast, attempts to encapsulate metals directly during MFI hydrothermal synthesis were unsuccessful due to early precipitation of metal precursors under the high-pH and high-temperature conditions. Although strategies such as delayed metal addition and replacing OH^- with F^- as the mineralizing agent improved encapsulation selectivity, their effectiveness still fell short of that achieved via interzeolite transformation.

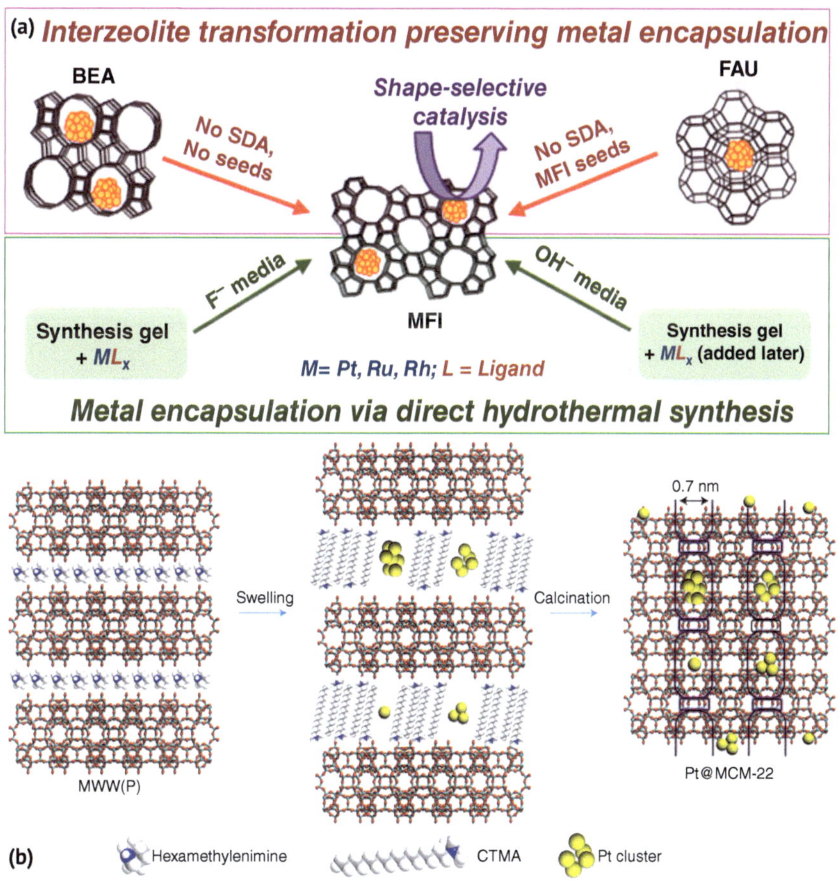

Figure 2.4 (a) Schematic illustration of different synthetic processes of MFI zeolite-encapsulated metal catalysts. *Source*: Reproduced from [41] / with permission of American Chemical Society. (b) Schematic of the conversion process from MWW to Pt@MCM-22. *Source*: Reproduced from [42] / with permission of Springer Nature.

Recently, Corma and co-workers proposed a novel approach for confining single Pt atoms and Pt clusters with excellent thermal stability within siliceous MCM-22, achieved during the structural conversion from two-dimensional MWW layers to the three-dimensional zeolite framework (see Figure 2.4b) [42]. First, the MWW zeolite with a two-dimensional layered structure was swelled and impregnated with Pt clusters, and then the layered zeolite precursor was condensed by calcination. The semi-cage (cup) and super-cage structures of MCM-22, measuring approximately 0.7×0.7 and 0.7×1.8 nm, respectively, provided suitable confinement spaces for incorporating single Pt atoms and Pt clusters. Fast Fourier transform and powder X-ray diffraction analyses confirmed that the hexagonal framework of MCM-22 remained intact following Pt incorporation. Additionally, HAADF-STEM imaging revealed the presence of uniformly dispersed Pt clusters and isolated atoms located within the internal pores of the zeolite. According to the fitting results of EXAFS spectra, the CN of the Pt—Pt bond in Pt@MCM-22 was about 4.7, corresponding to the Pt clusters with a size of less than 0.7 nm (<13 atoms). Moreover, the MCM-22 framework effectively stabilized sub-nanometric Pt clusters even after air treatment at temperatures as high as 540 °C. These small, well-confined Pt species also demonstrated outstanding size-selective catalytic performance in olefin hydrogenation reactions.

2.2.1.4 Other Methods

Compared to conventional hydrothermal synthesis, the dry gel conversion (DGC) technique offers improved efficiency, yielding higher zeolite output with reduced waste generation. Ding and co-workers designed a two-step DGC approach to encapsulate Pt NPs within the MFI zeolite crystals [46]. Hierarchical mesoporous ZSM-5 was first prepared through alkaline treatment (see Figure 2.5a). Pt species were then introduced into the mesopores via impregnation. The resulting Pt/alk-ZSM-5 composite was subsequently dispersed into the precursor gel for Silicalite-1 (S-1, MFI zeotype), which not only occupied the larger voids of the alkali-treated ZSM-5, but also formed a silica shell around the composite, facilitating encapsulation. Finally, the DGC method was used to repair and fill mesopores in the crystal structure so that the metal Pt was encapsulated inside the MFI zeolite crystals. HRTEM analysis revealed that the resulting Pt@MFI samples featured uniformly distributed Pt NPs of approximately 3 nm in size, which were fully confined within the zeolite crystals after calcination in air at 550 °C. Owing to the zeolite's shape-selective properties and effective encapsulation of metal species, the Pt@MFI catalyst exhibited high activity toward nitrobenzene hydrogenation, while showing negligible reactivity for 2,3-dimethylnitrobenzene, highlighting its excellent molecular sieving effect.

In addition to being confined within the internal pores and channels of zeolites, metal species can also be encapsulated within hollow zeolite shells. The hollow zeolite shells with mesoporous/macroporous structures not only play an important role in metal stabilization and inhibiting metal aggregation, but also give unique shape-selective catalytic performance and improved mass transfer during the catalytic process. Farrusseng and co-workers developed an impregnation–dissolution–recrystallization strategy to prepare hollow zeolite-encapsulated

Au NPs [43]. The S-1 zeolite was first impregnated by $HAuCl_4$ solution, and then the dissolution–recrystallization process of S-1 zeolite was carried out in a tetrapropylammonium hydroxide (TPAOH) solution. After calcination and reduction, the Au NPs were encapsulated in the hollow S-1 crystals. The size of Au NPs could be precisely tuned between 1 and 10 nm with a narrow distribution by adjusting the metal loading. In comparison, such precise size control was not achievable using traditional post-impregnation techniques. The porous structure of hollow zeolite shells could also be affected by etching agents during the dissolution–recrystallization process. For example, the single-hollow zeolite crystal-encapsulated Ni NPs could be obtained by recrystallization in TPAOH solution (see Figure 2.5b), while the multi-hollow zeolite crystal-encapsulated Ni NPs were obtained by recrystallization in tetrabutylammonium hydroxide (TBAOH) solution (see Figure 2.5c) [73]. Through the impregnation–dissolution–recrystallization strategy, a variety of monometallic species (e.g., Pd, Pt, Co, Ni, and Cu) as well as bimetallic NPs (e.g., AuAg, PdAg, PtAg, and PdPt) were successfully encapsulated within various zeolite frameworks, including MFI and BEA [44, 74, 75].

In addition to confining the metal species within hollow single-crystal zeolites, metals can also be immobilized in zeolitic nanovesicle structures. Song and co-workers developed a hollow, alveolus-like nanovesicle assembly composed of metal-encapsulated hollow ZSM-5 nanocrystals via a one-pot hydrothermal synthesis [45]. The Fe-containing solid spheres formed during the early crystallization stage consisted of amorphous subunits. The amorphous subunits gradually transformed into crystalline states during the hydrothermal synthesis, and some zeolite crystal nuclei were formed in the shell of the gel particle. As crystallization progressed, the amorphous subunits in the core were gradually dissolved and recrystallized around the initial nucleus. With extended crystallization time, this hollowing process continued, ultimately leading to the formation of well-defined Fe-containing zeolite nanovesicles (hier-Fe–S-1) once the amorphous regions were fully converted

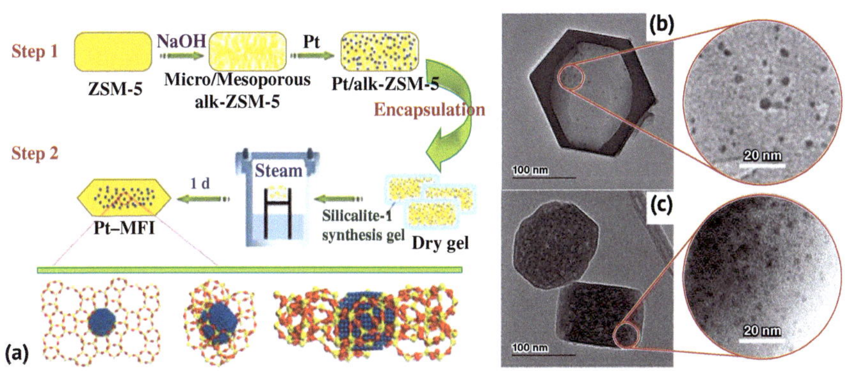

Figure 2.5 (a) Schematic of the synthesis of Pt@MFI by using a two-step DGC method. *Source*: Reproduced from [46] / with permission of American Chemical Society. TEM images of (b) single-hollow Ni@sil-1 sample etched by TPAOH and (c) multi-hollow Ni@sil-1 etched by TBAOH. *Source*: Reproduced from [73] / with permission of John Wiley & Sons.

into crystalline structures. The hollow S-1 zeolite-encapsulated Fe_2O_3 sample was finally obtained by treating the hier-Fe–S-1 sample with TPAOH at 170 °C for three days followed by calcination in air. By using this method, the bimetallic NPs (such as Pt–Fe_2O_3) could be also encapsulated into hollow MFI zeolites. TEM analysis showed that the Fe_2O_3@hier-ZSM-5 featured a hollow zeolitic nanovesicle approximately 450 nm in diameter, with an internal macroporous cavity of around 350 nm. The vesicle shell, composed of smaller hollow zeolite crystals (~50 nm), contained mesopores roughly 17 nm in size. The confined Fe_2O_3 NPs within the hollow zeolites measured about 3.2–4.4 nm, and their size increased only slightly after calcination in air at 800 °C for two hours, highlighting the strong protective capability of the hollow zeolite architecture.

2.2.2 Bottom-up Synthetic Methods

Apart from using top-down synthetic approaches, the metal species can also be in situ encapsulated within zeolite matrixes via the bottom-up synthetic method, in which the encapsulation of metal species is synchronized with the zeolite crystallization process. Generally, the synthetic system of zeolites is basic, so the metal precursors can easily precipitate during the hydrothermal crystallization and fail to be encaged inside zeolite crystals. To avoid the separation of metal species and zeolites as well as to encapsulate metal species into zeolite crystals during the bottom-up synthesis process, additional protective measures to limit the quick precipitation or/and aggregation of metal precursors are required. In recent years, some novel bottom-up synthetic methods, such as the ligand-protected method, organosilane-protected method, and precursor-stabilization method, were developed to prepare the zeolite-encapsulated metal catalysts.

2.2.2.1 Ligand-protected Methods

Iglesia and co-workers reported the encapsulation of noble metals (e.g., Pt, Pd, Ru, and Rh) and metal oxides within different kinds of aluminosilicate small-pore zeolites, such as SOD, GIS, and ANA, via the one-pot hydrothermal synthesis by using metal precursors stabilized by ammonia or organic amine (e.g., ethylenediamine) ligands [47]. During the synthetic process, the ammonia or organic amine molecules could coordinate with metal cations to form stable metal complexes that prevented the formation of metal hydroxides under high temperatures and alkali conditions. The metal species were encapsulated within zeolites along with zeolite crystallization, and the ultrasmall metal clusters or/and NPs were finally encapsulated within the zeolites subjected to calcination and reduction at high temperatures. HAADF-STEM analysis showed that the average particle sizes of Pt@GIS, Pd@GIS, Ru@GIS, and Rh@GIS ranged from 1.4 to 1.7 nm, with the majority of metal species successfully confined within the zeolite crystals. Using a similar approach, a wide range of monometallic species (e.g., Pt, Pd, Rh, Ir, Re, Ag, Co, Ni, and Fe) as well as bimetallic combinations (e.g., AuPd, AuPt, and PdPt) were encapsulated in aluminosilicate zeolites such as LTA, MFI, and FAU, all demonstrating outstanding thermal stability at elevated temperatures [48–50, 76].

Beyond the encapsulation of metal species in aluminosilicate zeolites, the ligand-protected strategy also enables the synthesis of ultrasmall metal clusters within siliceous zeolites. Yu and co-workers successfully prepared uniformly dispersed sub-nanometric Pd clusters inside S-1 nanocrystals under direct hydrothermal conditions, using $[Pd(en)_2]Cl_2$ (en = $NH_2CH_2CH_2NH_2$) as the metal precursor and TPAOH as the structure-directing agent [22]. The theoretical calculation revealed that the TPAOH molecules were located in the straight channel along the *b*-axis, while the metal complexes were more inclined to occupy the intersectional channels of S-1 zeolites. HAADF-STEM analysis showed that, following calcination in air and subsequent H_2 reduction to remove organic templates and ligands, the sub-nanometric Pd clusters were successfully confined within the intersecting voids of the straight and sinusoidal 10-MR channels of the S-1 zeolite, with no detectable Pd species on the external surfaces of the crystals. EXAFS measurements revealed an average Pd—Pd CN of just 2.3, confirming the presence of highly dispersed ultrasmall Pd clusters or single atoms within the zeolite micropores. Furthermore, the incorporation of Na^+ and K^+ cations into the synthesis gel effectively suppressed Pd aggregation during high-temperature treatments, enhancing metal dispersion even further. Subsequently, Yu and co-workers further extended the ligand-protected direct hydrothermal synthesis approach to fabricate S-1 zeolite-encapsulated sub-nanometric hybrid bimetallic clusters, including Pd—$Ni(OH)_2$, Pd—$Co(OH)_2$, and Pd—MnO_x catalysts, with tunable metal loadings and compositions. This was achieved using metal–ligand complexes such as $[Pd(en)_2]Cl_2$, $[Ni(en)_3](NO_3)_2$, $[Co(en)_3](NO_3)_2$, and $(Mn\text{–}TEPA)Cl_2$ (TEPA = tetraethylenepentamine) as precursors (see Figure 2.6a) [51, 55]. HAADF-STEM analysis demonstrated that, akin to monometallic Pd@S-1, the bimetallic clusters were fully encapsulated within the zeolite framework, and the incorporation of secondary metal species (e.g., $Ni(OH)_2$) did not increase the size of the Pd clusters (see Figure 2.6b, c). EXAFS data showed that the average CN of Pd—Pd bond in $0.8Pd0.2Ni(OH)_2$@S-1 was 3.1, closely resembling that of Pd@S-1. Owing to the strong confinement provided by the zeolite matrix, these Pd-based bimetallic clusters exhibited exceptional thermal stability, with no significant aggregation observed after calcination or exposure to various redox environments at 600–700 °C, including steam treatment at 600 °C. By using the similar ethylenediamine ligand-protected method, various zeolite-encapsulated metal catalysts, such as *M*@Y (Co, Ni, Cu, Pt, Pd, Ru, and Rh) [53], Pd@SOD [54], PtSn@MFI [23], Ir@MWW [57], Co@MFI [58], and PdCu@S-1 [59], were also prepared.

Building on their previous strategy, Yu and co-workers developed a ligand-stabilized direct hydrogen reduction method to encapsulate isolated Rh single atoms within S-1 and ZSM-5 zeolites. This was achieved via a one-pot hydrothermal synthesis using $[Rh(en)_3]Cl_3$ as the metal precursor [52]. Compared to the traditional calcination followed by reduction, the ligand-protected direct hydrogen reduction approach enables atomic-scale dispersion of metal species. In this method, organic ligands and structure-directing agents act as stabilizers, effectively preventing metal aggregation during the reduction step. C_s-corrected STEM and EXAFS analyses confirmed that atomically dispersed Rh atoms were confined within 5-MRs of the

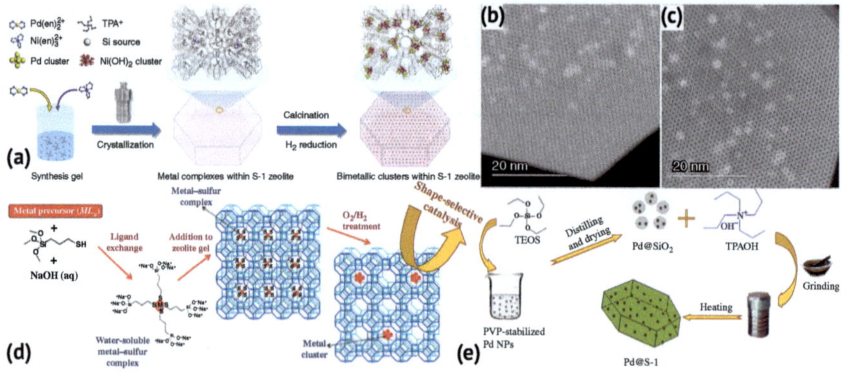

Figure 2.6 (a) Schematic depiction of the formation process for Pd–$Ni(OH)_2$ clusters confined within the S-1 zeolite. C_s-corrected STEM images of Reproduced from [51] / with permission of Elsevier (b) 0.8Pd0.2$Ni(OH)_2$@S-1 sample and (c) Pd@S-1 sample. *Source*: Reproduced from [51] / with permission of Elsevier. (d) Schematic representation of mercaptosilane-mediated incorporation of metal species into zeolites. *Source*: Reproduced from [63] / with permission of American Chemical Society. (e) Preparation of Pd@S-1 sample by the precursor-stabilization method. *Source*: Reproduced from [68] / with permission of American Chemical Society.

zeolite and stabilized by coordination with framework oxygen atoms. The strong confinement provided by the zeolite ensured that the Rh single atoms remained stable even after treatment in H_2 and N_2 atmospheres at 700 °C. Following this approach, sub-nanometric Pt—Zn clusters enriched with isolated Zn sites were also synthesized within S-1 zeolites using $[Pt(en)_2]Cl_2$ and $[Zn(en)_3](OAc)_2$ as precursors through the same ligand-assisted direct hydrogen reduction strategy [19].

Except for ethylenediamine, other organic ligands were also used to stabilize metal precursors during the one-pot hydrothermal synthesis of zeolite-encapsulated metal catalysts. For example, Xiao and co-workers incorporated isolated $Fe^{\delta+}$ species into the S-1 zeolite framework through a direct hydrothermal synthesis route, using ethylenediaminetetraacetic sodium (EDTA) as a chelating agent to facilitate the incorporation [62]. The characteristic bands at 515, 1007, and 1155 cm^{-1} in ultraviolet Raman spectra of FeS-1–EDTA were observed at different crystallization periods, suggesting the existence of framework tetrahedral Fe species. The bands at 515 and 1155 cm^{-1} attributed to the skeletal $Fe^{\delta+}$—O—Si species with symmetric and asymmetric stretching vibrations could also be detected, further indicating the presence of stable $Fe^{\delta+}$—O—Si linkages. The single-site Fe structure was also determined by EXAFS of Fe K-edge. Apart from ethylenediamine and EDTA, other organic ligands such as diethylenetriamine [60], tetraethylenepentamine [77], and pentaethylenehexamine [61] were also used in the synthesis of zeolite-encapsulated metal catalysts, such as Ni@CHA, Cu@SPO-34, and Rh@ZSM-5.

2.2.2.2 Organosilane-protected Methods

Beyond organic amines, organosilanes containing coordinating functional groups (such as amino or thiol groups) can also serve as effective stabilizing agents for the synthesis of zeolite-encapsulated metal catalysts. The alkoxysilane moieties in these compounds participate in the zeolite formation process by covalently integrating into the framework, while the coordinating groups bind to metal precursors, preventing aggregation during synthesis. For instance, Iglesia and co-workers successfully encapsulated a range of noble metals (Ir, Rh, Pt, Pd, and Ag) within NaA (LTA zeotype) zeolites in situ, using (3-mercaptopropyl)trimethoxysilane (MPTMS) as a protective ligand (see Figure 2.6d) [63]. On the one hand, the sulfhydryl (—SH) group in the MPTMS molecules interacted with $PtCl_6^{2-}$ to form Pt(II)–S_4 complexes and protected metal species from hydrolytic agglomeration; on the other hand, the alkoxysilane moiety in MPTMS reacts with the zeolite precursors, forming covalent Si—O—Si or Si—O—Al bonds that facilitate zeolite nucleation and enable the incorporation of metal complexes into the growing framework. As a result, ultrasmall metal species become effectively confined within the NaA zeolite matrix and can subsequently undergo thermal treatments in O_2 and H_2 without significant aggregation. The EXAFS measurement revealed that the CN of the Pt—Pt bond was 6.6 ± 1.2, corresponding to an average metal size of about 1 nm, which was also determined by the chemisorption and TEM measurements. In addition, the encapsulated metals in NaA zeolites were well resistant to sulfur poisoning and thermal sintering.

Corma and co-workers synthesized high-silica CHA zeolite-encapsulated Pt NPs (~1 nm) through an in situ method using *N,N,N*-trimethyl-1-adamantammonium cations as the structure-directing agent and MPTMS as the metal-stabilizing ligand [64]. The encapsulated Pt species displayed outstanding resistance to sintering under various environments (H_2, O_2, and H_2O), maintaining their particle size even after exposure to H_2 and steam at 650 °C. X-ray absorption spectroscopy and C_s-corrected STEM revealed that the Pt structures could reversibly transition between ~1 nm NPs and isolated single atoms under alternating reducing and oxidizing conditions. Similarly, Li and co-workers employed MPTMS to embed small Pd NPs within MFI zeolites [65]. HAADF-STEM imaging confirmed that Pd@MFI contained uniformly distributed Pd NPs with an average size of 1.7–1.8 nm confined within the zeolite. In contrast, Pd/MFI prepared via conventional impregnation showed much larger Pd particles (2–10 nm) located on the external surfaces of the zeolite crystals. The Pd@MFI exhibited high activity for the nitrobenzene hydrogenation, but quite low activity for 1-nitronaphthalene hydrogenation, further confirming that most of the Pd NPs were encapsulated in zeolite matrixes, rather than supported onto the outer surface of zeolite crystals. Li and co-workers later utilized 3-[2-(2-aminoethylamino)ethylamino]propyltrimethoxysilane to synthesize zeolite-encapsulated single-atom metal catalysts, including Ni@FAU and Pt@Y, achieving atomic-level dispersion of metal species within the zeolites [66].

2.2.2.3 Precursor-stabilization Methods

In addition to organic ligands, amorphous silica can act as a protective medium to stabilize metal species during the formation of zeolite-encapsulated metal catalysts. Xiao and co-workers introduced a precursor-stabilization approach to synthesize Pd@S-1 by solvent-free crystallization of Pd@SiO_2 precursors using TPAOH as the structure-directing agent [68]. The amorphous SiO_2 was formed via controlled hydrolysis of tetraethyl orthosilicate (TEOS) in a polyvinylpyrrolidone-stabilized

Pd NP solution, followed by TPAOH addition (see Figure 2.6e). The resulting solid mixture was ground at room temperature in air and then crystallized at 180 °C for 72 hours. After calcination and H_2 reduction, the final Pd@S-1 material was obtained. Tomographic TEM imaging confirmed that Pd NPs, with an average diameter of 6.9 nm (3.7–11.7 nm), were encapsulated within the mesoporous cavities of S-1 crystals, forming a clear core–shell structure. Catalytic testing revealed that Pd@S-1 was active in benzaldehyde hydrogenation but showed no reactivity toward 3,5-diisopropylbenzaldehyde, whereas Pd/S-1 prepared via impregnation catalyzed both reactions. These results demonstrate the complete encapsulation of Pd NPs within the S-1 framework in the Pd@S-1 sample.

In addition to amorphous silica, zeolite seed crystals can also act as precursors to stabilize metal species during the synthesis of zeolite-supported metal catalysts. Xiao and co-workers developed a synthetic strategy for fabricating Pd@zeolite materials by employing Pd-loaded seed crystals as the starting components [67]. The seed crystals acted as support to anchor metal NPs and transferred them in the newly formed zeolites. In this process, the zeolite shell grew around the surface of the seed crystal and isolated individual Pd NPs to prevent the aggregation of metal species during the crystallization. Importantly, this synthesis method can control the size of metal NPs. The HRTEM measurements further showed that the Pd NPs were highly dispersed inside zeolite crystals, with a diameter range from 0.9 to 3.1 nm, which was very comparable to the sizes of the Pd species supported on the seed crystals, suggesting that the Pd NPs were stable during the zeolite crystallization. The Pd@Beta catalyst exhibited outstanding thermal stability, with the size of Pd NPs remaining unchanged after calcination at 600 °C for four hours. In contrast, Pd NPs located on the external surfaces of conventionally prepared Beta zeolite grew significantly larger (2–13 nm) under the same conditions. In a subsequent work, the same precursor-stabilization strategy was applied to encapsulate Pt NPs (0.8–3.2 nm in size) within Beta zeolite. TEM images of samples at various crystallization stages supported a core–shell growth mechanism driven by Pt-containing zeolite seeds. The encapsulated Pt NPs retained their original size after calcination in air at 600 °C, highlighting the ability of the rigid zeolite framework to effectively anchor metal species and inhibit sintering. Using a similar method, other zeolite-encapsulated metal catalysts (Pt@Beta, Ag@Beta, Rh@Beta, Pt@Beta–Fe, Pt@MOR, Rh@MOR, Ag@MOR, and Pd@S-1) [26], Ni@HZSM-5 [27], and AuPd@ZSM-5 [25] were also prepared.

2.3 Catalytic Conversion of Low-carbon Molecules

2.3.1 CO/CO_2 Hydrogenation

The catalytic conversion of syngas through CO hydrogenation offers a promising route for producing valuable chemicals from nonpetroleum feedstocks. Among these approaches, Fischer–Tropsch (FT) synthesis is the most widely used; however, achieving precise control over product selectivity remains a significant challenge due to the uncontrolled C—C coupling of hydrocarbon intermediates on conventional metal catalysts. To address this limitation, bifunctional catalysts combining metal components with zeolites have gained considerable attention in recent years. These systems have enabled notable advances in tailoring syngas conversion toward specific hydrocarbon products such as light olefins, aromatics, and gasoline-range compounds. For instance, zeolite-encapsulated Ru NPs (Ru@NaY), synthesized via a sequential process involving polyvinylpyrrolidone coordination, stabilization, and $NaBH_4$ reduction, demonstrated a remarkable gasoline-range hydrocarbon selectivity of 64.3% in the FT reaction [69]. This performance significantly surpassed that of the conventional Ru/NaY catalyst (32.8%) and exceeded the expectations based on the Anderson–Schulz–Flory (ASF) distribution. The catalytic performance remained unchanged even after 76 hours on stream, giving a C_5–C_{11} selectivity of about 64% and a CO conversion rate of 16.2 $mol_{CO} \cdot mol_{Ru}^{-1} \cdot h^{-1}$. The enhanced catalytic performance of Ru@NaY was primarily attributed to the shape-selective nature of the zeolite micropores, which effectively restricted excessive C—C coupling and thereby regulated the length of the hydrocarbon chains. Additionally, the rigid zeolite framework provided structural confinement that suppressed the migration and sintering of Ru NPs, contributing to the catalyst's high stability and selectivity. Similarly, the catalytic performance of Fe@NaY in FT synthesis was also significantly improved by encapsulating the Fe NPs within zeolite Y crystals, affording a high selectivity of light olefins of up to 36.2% at a CO conversion of 91.2%, which was about threefold higher than that of impregnated Fe/NaY catalyst (12.5%) [70]. In addition, the metal oxide@zeolite double-shelled hollow spheres (MO@ZEO DSHSs) have also been used as the high-efficiency catalysts for the hydrogenation of CO reaction, in which the Fe_2O_3@H–ZSM-5 DSHSs composed of a hollow Fe_2O_3 sphere as core and hollow ZSM-5 sphere as shell displayed a superior C_5–C_{11} hydrocarbon selectivity of 64% at a CO conversion of 79% due to their unique structure, breaking the limitations of the ASF distribution [78].

Apart from the production of hydrocarbons, the oxygen-containing compounds can also be prepared by CO hydrogenation catalyzed by zeolite-encapsulated metal catalysts. Recently, Xiao and co-workers developed a bimetallic RhMn@S-1 catalyst, in which the active species were confined within the S-1 zeolite framework [72]. This catalyst achieved highly efficient syngas conversion to C_2-oxygenates, with an average productivity of 105 $mol \cdot mol_{Rh}^{-1} \cdot h^{-1}$—substantially exceeding those of RhMn/S-1 $(42.3\ mol \cdot mol_{Rh}^{-1} \cdot h^{-1})$ and RhMn/$Al_2O_3$$(11.7\ mol \cdot mol_{Rh}^{-1} \cdot h^{-1})$, where the metal species were located on the external surfaces of the supports (see Figure 2.7a). Notably, RhMn@S-1 also exhibited excellent stability, maintaining a CO conversion of ~42% and a C_2-oxygenate productivity of $103 \pm 3\ mol \cdot mol_{Rh}^{-1} \cdot h^{-1}$ over 220 hours of continuous operation at 320°C. This superior performance was attributed to the synergistic effect between the zeolite confinement and the Mn—O—$Rh^{\delta+}$ interfacial structure, which together promoted C—C bond formation and stabilized the active sites.

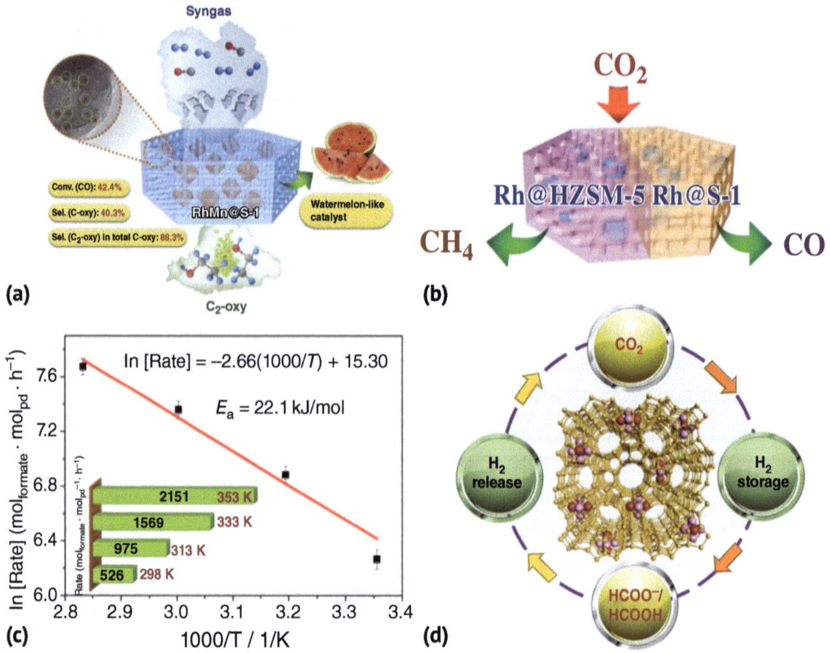

Figure 2.7 (a) Schematic illustration of catalytic conversion of syngas to C_2-oxygenates over RhMn@S-1 catalyst. *Source*: Reproduced from [72] / with permission of Elsevier. (b) Illustration of catalytic conversion of CO_2 to CO or CH_4 via controlling nanoporous environments of Rh@MFI. *Source*: [71] / with permission of American Chemical Society. (c) Catalytic activity of $PdMn_{0.6}$@S-1 in CO_2 hydrogenation to formic acid at various temperatures. Reaction conditions: 5-mg catalyst, 2-mL NaOH solution (1.5 M), and 20-bar CO_2/20-bar H_2. *Source*: Reproduced from [55] / with permission of John Wiley & Sons. (d) Illustration of the hydrogen storage and release process mediated by CO_2 over the PdMn@S-1 catalyst. *Source*: Reproduced from [55] / with permission of John Wiley & Sons.

Selective hydrogenation of CO_2 is a "two birds, one stone" strategy that not only transforms CO_2 into valuable chemicals but also decreases the CO_2 concentration in the atmosphere. For the past few years, the applications of zeolite-encapsulated metal catalysts in selective hydrogenation of CO_2 have aroused extensive attention due to their superior catalytic performance and outstanding thermal stability. For instance, Xiao and co-workers synthesized a series of Rh NPs encapsulated within MFI zeolites (Rh@MFI), featuring tunable nanoporous environments [71]. This was achieved through a seed-directed crystallization approach using metal-containing zeolite precursors. Interestingly, the neutral Rh@S-1 catalyst exhibited a high CO selectivity of 79.8% with a CO_2 conversion of 51.6% at 500 °C. In contrast, the acidic Rh@HZSM-5 catalyst demonstrated an exceptionally high CH_4 selectivity of 98.2% at a CO_2 conversion of 68.2% under identical conditions. This stark contrast in product selectivity is strongly influenced by the distinct nanoporous environments of the zeolites. The S-1 framework, being neutral, suppressed hydrogen spillover and facilitated the rapid desorption of CO, thereby preventing its further hydrogenation to hydrocarbons. In comparison, Rh@HZSM-5 promoted deeper hydrogenation due to its strong CO adsorption capacity and the presence of abundant Brønsted acid sites, favoring the formation of methane (see Figure 2.7b). Not unexpectedly, the KZSM-5 zeolite-encapsulated Rh NPs exhibited much higher CO selectivity, because the introduction of K^+ within HZSM-5 could reduce the acidic sites of zeolites and minimize the hydrogen spillover.

Beyond the production of CO and CH_4, significant efforts have been directed toward the selective hydrogenation of CO_2 into value-added chemicals such as formic acid, methanol, and ethanol. Yan and co-workers developed a series of S-1 zeolite-encapsulated sub-nanometric bimetallic Pd–Mn clusters (PdMn@S-1) for CO_2 hydrogenation to formic acid/formates [55]. The incorporation of Mn promoted electron enrichment on the Pd surface, facilitating formate formation and enhancing the generation of bicarbonate intermediates—key species in the reaction pathway. Owing to the ultrasmall size of the metal clusters and the synergistic interaction between Pd and Mn, the optimized $PdMn_{0.6}$@S-1 catalyst delivered an exceptional formate production rate of 2151 $mol_{formate} \cdot mol_{Pd}^{-1} \cdot h^{-1}$ at 80 °C (see Figure 2.7c). Moreover, it demonstrated outstanding catalytic performance for hydrogen release from formic acid dehydrogenation, achieving an initial turnover frequency (TOF) of 6860 $mol_{H2} \cdot mol_{Pd}^{-1} \cdot h^{-1}$ with 100% H_2 selectivity at 60 °C in the absence of additives. These results highlight the dual functionality of zeolite-encapsulated metal catalysts in CO_2-based hydrogen storage and release (see Figure 2.7d). Hu and co-workers synthesized Cu/ZnO_x(*Y*)@Na–ZSM-5 catalysts to achieve selective hydrogenation of CO_2 to methanol, in which the Cu/ZnO_x NPs with an average size of about 2 nm were encapsulated within the Na–ZSM-5 zeolite [79]. The representative Cu/ZnO_x(1.38)@Na–ZSM-5 catalyst achieved an impressive methanol space–time yield of 44.88 $g_{MeOH} \cdot g_{Cu}^{-1} \cdot h^{-1}$, significantly outperforming both the conventionally supported Cu/ZnO_x/Na–ZSM-5 catalyst $(13.32\ g_{MeOH} \cdot g_{Cu}^{-1}\ h^{-1})$ and the industrial Cu/ZnO/Al_2O_3 benchmark $(8.46\ g_{MeOH} \cdot g_{Cu}^{-1} \cdot h^{-1})$ under identical conditions. Notably, the zeolite-encapsulated catalyst maintained stable performance over a 100-hour continuous operation, underscoring its excellent durability. The superior activity was primarily attributed to the confinement effect of the zeolite framework, which created a well-defined microenvironment that enhanced the synergy between Cu and ZnO_x species. Remarkably, Ding et al. also developed a Cu@Na–Beta catalyst that selectively converted CO_2 to ethanol as the sole organic product [80]. Owing to the strong interaction between the Cu NPs and the surrounding zeolitic matrix, this catalyst delivered a single-pass ethanol yield of approximately 14% at 300 °C, corresponding to a space–time yield of ~398 $mg \cdot g_{cat}^{-1} \cdot h^{-1}$.

2.3.2 Catalytic Conversion of Methane

Metal species confined within zeolite frameworks have also demonstrated outstanding performance in the catalytic transformation of methane into valuable chemicals, including methanol, acetic acid, and syngas. Tao and co-workers developed a

Pd@ZSM-5 catalyst featuring isolated Pd_1O_4 sites anchored to the internal surfaces of the zeolite micropores [34]. This catalyst enabled the direct conversion of CH_4 to CH_3OH via partial oxidation using H_2O_2 in an aqueous medium at temperatures ranging from 50 to 95 °C. Moreover, introducing CuO into the Pd@ZSM-5 catalyst enhanced the decomposition of excess H_2O_2 following methanol formation, thereby further improving catalytic efficiency. The optimized catalyst, comprising 0.01 wt.% Pd@ZSM-5 and 2 wt.% CuO, achieved a high methanol production rate of approximately 2.78 $mol_{methanol} \cdot mol_{Pd}^{-1} \cdot s^{-1}$ at 95 °C, along with a methanol selectivity of up to 86.4%. Thermodynamic calculations indicated that methanol formation was more favorable than the generation of methyl peroxide, accounting for the high selectivity observed. Recently, Xiao and co-workers introduced a "molecular fence" strategy to fabricate an organosilane-modified AuPd@ZSM-5 catalyst for the selective partial oxidation of methane to methanol [25]. In this design, AuPd NPs were encapsulated within the zeolite crystals, while the external surface of the zeolite was coated with a hydrophobic organosilane layer. This hydrophobically modified AuPd@ZSM-5 catalyst exhibited excellent methanol selectivity of 92% at a methane conversion rate of 17.3%, achieving a high methanol productivity of 91.6 $mmol \cdot g_{AuPd}^{-1} \cdot h^{-1}$. The outstanding catalytic performance was attributed to the dual role of the organosilane modifiers: they not only facilitated the diffusion of hydrogen, oxygen, and methane to the active sites but also served as a molecular barrier that prevented H_2O_2 from diffusing out of the zeolite, thereby enhancing methanol formation.

In addition to the preparation of methanol, acetic acid can also be produced by methane oxidation. Flytzani-Stephanopoulos and co-workers synthesized Rh–ZSM-5_{washed} catalyst to catalyze the conversion of methane to acetic acid [28]. They found that the parent Rh–ZSM-5 and Rh–ZSM-5_{washed} catalysts (the metal species on the outside of zeolites were removed by repeated washing with water) showed comparable activity and similar product selectivity, and no catalytic performance could be observed on silica-supported Rh or Rh_2O_3 NPs during the catalytic conversion of methane. These findings indicate that isolated mononuclear Rh species confined within zeolite micropores serve as the active sites responsible for methane conversion. The Rh–ZSM-5_{washed} catalyst, enriched with dispersed Rh atoms inside the micropores, achieved an acetic acid yield of approximately 22 000 $\mu mol \cdot g_{catalyst}^{-1}$ at 150 °C. Similarly, Tao and co-workers developed an Rh@ZSM-5 catalyst featuring atomically dispersed Rh_1O_5 species anchored within the ZSM-5 framework, which demonstrated exceptional activity for methane conversion to acetic acid and methanol under mild conditions [35]. This catalyst exhibited a production rate of roughly 0.10 $mol_{acetic\ acid} \cdot mol_{Rh}^{-1} \cdot s^{-1}$ at 150 °C, with an acetic acid selectivity of around 70%—over 1000 times higher than that of free Rh cations. DFT calculations showed that the Rh_1O_5 single-site species facilitated the activation of the initial C—H bond in methane, while the resulting CH_3 intermediate reacted with CO and OH to form acetic acid via a low-barrier pathway. More recently, Xiao and co-workers developed H-type ZSM-5-encapsulated Ni NPs (Ni@HZSM-5), showing a high density of reactive hydrogen species generated via hydrogen spillover within the zeolite micropores [27]. This design enhanced methane reducibility compared to conventional dry reforming processes, delivering a CO_2-to-CH_4 conversion ratio of 2.9 $mol_{CO2} \cdot mol_{CH4}^{-1}$ and a CO yield of 3.9 $mol_{CO} \cdot mol_{CH4}^{-1}$. Owing to the strong confinement provided by the zeolite, Ni@HZSM-5 showed excellent catalytic stability, maintaining consistent methane reducibility over 30 hours of continuous operation.

2.3.3 Dehydrogenation of Formic Acid

Formic acid is considered a promising hydrogen storage material due to its non-toxic, relatively high H_2 content and excellent stability, and zeolite-encapsulated Pd-based catalysts showed outstanding catalytic performance for formic acid dehydrogenation. Yu and co-workers reported that the Pd@S-1 catalyst, in which the highly dispersed sub-nanometric Pd clusters were encapsulated within the S-1 zeolite nanocrystals via the one-pot hydrothermal synthetic method, exhibited a superhigh activity in the complete decomposition of formic acid [22]. Notably, the incorporation of Na^+ and K^+ ions into the zeolite micropores promoted the formation of smaller Pd clusters and imparted a degree of basicity to the zeolite, both of which enhanced H_2 generation from formic acid decomposition. Among the catalysts tested, the optimized Pd@S-1-in-K catalyst achieved the TOF of 3027 h^{-1} at 50 °C—approximately six times higher than that of commercial Pd/C. Building on this, Yu and co-workers developed a series of zeolite-encapsulated bimetallic catalysts, Pd–*M*@S-1 (*M* = Ni, Co, and Mn), with tunable metal ratios, as highly efficient systems for formic acid dehydrogenation [51, 55]. The introduction of secondary metal species, such as $Ni(OH)_2$, $Co(OH)_2$, or MnO_x, significantly boosted catalytic activity by generating electron-rich Pd surfaces and enhancing the bimetallic interfacial synergy. Among them, the 0.8Pd0.2$Ni(OH)_2$@S-1 catalyst delivered an outstanding initial TOF of 5803 h^{-1} at 60 °C without any additives—14 times higher than commercial Pd/C and nine times greater than monometallic Pd@S-1. Impressively, this catalyst maintained stable performance over 10 consecutive cycles and showed no activity loss even after 300 days of exposure to ambient conditions.

2.3.4 Dehydrogenation of Light Alkanes

In recent years, the dehydrogenation of light alkanes has garnered increasing interest, driven by the rising demand for light olefins such as propylene and ethylene, alongside the expansion of shale gas resources. However, conventional supported metal catalysts often face challenges such as rapid deactivation and limited catalytic efficiency. Zeolite-encapsulated metal catalysts have emerged as promising alternatives for light alkane dehydrogenation, offering exceptionally high metal utilization and outstanding thermal stability. For example, Corma and co-workers developed a Pt@MCM-22 catalyst in which single Pt atoms and Pt clusters were confined within siliceous MCM-22 zeolites [42]. This catalyst demonstrated markedly enhanced activity and long-term stability in propane dehydrogenation (PDH) at 550 °C. Generally, the introduction of secondary metals into Pt species is considered as an effective strategy to improve the propylene selectivity and catalyst stability during

PDH reactions due to the synergistic effect between bimetallic species. For example, K–PtSn@MFI catalyst with highly stable sub-nanometric Pt–Sn clusters inside siliceous MFI zeolites exhibited an initial propylene selectivity of approximately 90%, and retained around 70% of its initial propane conversion even after 65 hours of continuous operation [56]. In sharp contrast, the initial catalytic activities of monometallic Pt@MFI and K–Pt@MFI catalysts significantly decreased by more than 50% within 20 hours. Recently, Yu and co-workers introduced a ligand-stabilized direct hydrogen reduction approach to encapsulate sub-nanometric bimetallic Pt–Zn clusters within S-1 zeolites. The optimized PtZn4@S-1–H catalyst demonstrated exceptional performance in PDH, achieving an ultrahigh propylene selectivity of 99.3% and a specific activity of 65.5 $mol_{C3H6} \cdot g_{Pt}^{-1} \cdot h^{-1}$ at 550°C [19]. The incorporation of Zn species into the zeolite framework enhanced the electron density around Pt atoms, facilitating propane adsorption and propylene desorption while suppressing the polymerization of unsaturated coke precursors. This electronic modification contributed to the high propylene selectivity observed during the PDH process. The PtZn4@S-1–H catalyst also exhibited remarkable long-term stability, showing no significant deactivation even after 13 000 minutes of continuous operation (see Figure 2.8a). Its deactivation rate constant was as low as 0.001 h^{-1}—42 and 200 times lower than those of Pt@S-1–H and PtZn4/Al_2O_3, respectively, under identical conditions. Additionally, the introduction of Cs^+ ions into the PtZn4@S-1–H catalyst further enhanced regeneration stability, with catalytic activity remaining consistent across four successive oxidation–reduction cycles.

In addition to conventional transition metals, rare-earth elements have also been utilized as effective promoters to enhance the catalytic performance of zeolite-confined Pt-based catalysts in PDH reactions. Ryoo et al. developed a series of Pt–rare-earth intermetallic compounds embedded within mesoporous zeolites enriched with surface silanol nests [36]. The presence of silanol nests was considered a key factor in facilitating the formation of Pt–rare-earth intermetallic compounds. These surface defects in mesoporous zeolites enabled rare-earth elements to exist as highly reactive single-atom species with elevated chemical potential compared to their bulk oxide counterparts. This unique environment promoted the diffusion of rare-earth atoms onto Pt surfaces during hydrogen reduction, leading to the formation of Pt–rare-earth intermetallic phases. The representative PtLa/mz-deGa catalyst displayed a high initial propane conversion of 40% at 580°C and experienced extremely slow deactivation, giving 8% propane conversion even after 30 days of continuous reaction, whereas the initial propane conversion of PtLa/mz that failed to alloy La into Pt NPs was only 22% and sharply dropped below 5% within 1.5 hours.

In addition to Pt-based catalysts, zeolite-encapsulated non-noble metal catalysts have also demonstrated outstanding activity and stability in PDH reactions. For instance, Liu and co-workers developed a Co–MFI catalyst featuring atomically dispersed cobalt species integrated into the MFI zeolite framework [58]. The $Co_{1.12}$–MFI showed a much higher initial C_3H_6 formation rate of 8.2 $mmol \cdot g_{cat}^{-1} \cdot h^{-1}$ than those of other controlled Co-containing catalysts (such as Co/S-1-IM synthesized

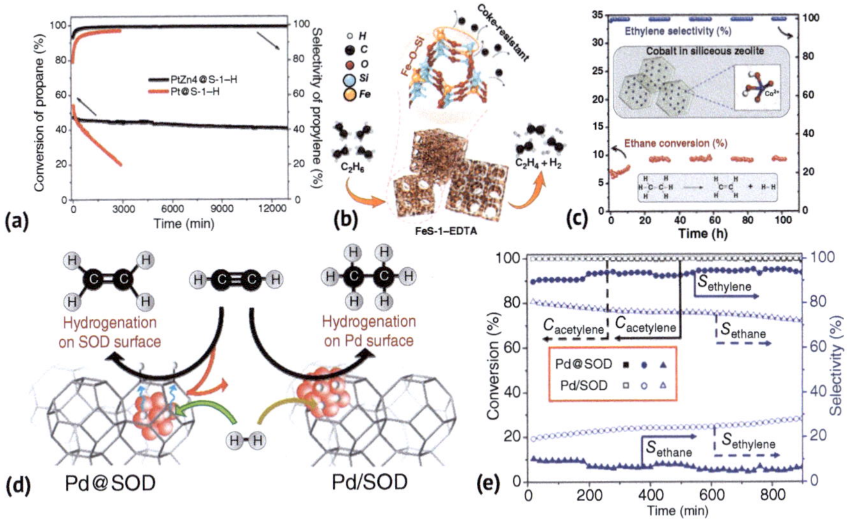

Figure 2.8 (a) Catalytic stability evaluation of PtZn4@S-1-H and Pt@S-1-H catalysts during PDH at 550 °C under a WHSV of 3.6 h^{-1}. *Source*: Reproduced from [19] / with permission of John Wiley & Sons. (b) Schematic of EDH over FeS-1-EDTA catalyst. *Source*: Reproduced from [62] / with permission of American Chemical Society. (c) The catalytic properties of EDH over Co@S-1 catalyst. *Source*: Reproduced from [40] / with permission of Elsevier. (d) Illustration of the synergy between metal active sites and the spatial confinement provided by small-pore zeolites in acetylene hydrogenation *Source*: Reproduced from [54] / with permission of John Wiley & Sons. (e) Performance comparison between Pd@SOD and Pd/SOD catalysts in selective hydrogenation of acetylene. Reaction conditions: gas hourly space velocity (GHSV) of 30 000 h^{-1}, 150 °C, 0.6% C_2H_2, and 6% H_2 (N_2 balance) *Source*: Reproduced from [54] / with permission of John Wiley & Sons.

by the impregnation method, Co/Al_2O_3, and Co/SiO_2). DFT calculation indicated that the isolated $(\equiv Si{-}O)_2Co(HO{-}Si\equiv)_2$ species with tetrahedrally coordinated Co species in Co–MFI were the active site for PDH and possessed higher efficiency and selectivity in the PDH reaction.

Beyond PDH, zeolite-encapsulated metal catalysts have also achieved notable progress in ethane dehydrogenation (EDH). For example, isolated $Fe^{\delta+}$ species were successfully incorporated into the siliceous MFI zeolite framework (FeS-1–EDTA) via a one-pot hydrothermal synthesis using EDTA as a chelating agent [62]. In this system, Fe existed exclusively as the single-site $Fe^{\delta+}$ species, with no evidence of FeO_x aggregates (see Figure 2.8b). The FeS-1–EDTA catalyst delivered an ethylene selectivity of 97.5% at an ethane conversion of 26.3%, achieving an ethene productivity of approximately 0.86 $mol \cdot g_{Fe}^{-1} \cdot h^{-1}$—substantially outperforming FeS-1 $(0.18\ mol \cdot g_{Fe}^{-1} \cdot h^{-1})$ and Fe/S-1 $(0.16\ mol \cdot g_{Fe}^{-1} \cdot h^{-1})$. The catalyst also exhibited excellent stability, maintaining consistent activity over 200 hours of continuous operation. This high performance was attributed to the uniformly dispersed and structurally stable Fe sites, which facilitated ethane activation and enabled rapid desorption of H_2 and C_2H_4 within the micropores, effectively suppressing coke formation. In a related development, Xiao and co-workers prepared a Co@S-1 catalyst through a mechanically assisted spontaneous dispersion method, in which isolated Co atoms were securely anchored within the siliceous zeolite framework [40]. This catalyst achieved 13.2% ethane conversion and 85.9% ethylene selectivity in the EDH reaction. Notably, after 10 hours of continuous use, the catalyst maintained excellent performance, reaching 92.4% ethylene selectivity at 13% ethane conversion (see Figure 2.8c). This performance surpassed that of previously reported non-noble metal catalysts and even the benchmark PtSn/Al_2O_3 system.

2.3.5 Selective Hydrogenation of Unsaturated Hydrocarbons

The selective hydrogenation of unsaturated hydrocarbons into value-added chemicals is a topic of significant interest in the chemical industry but remains challenging due to the propensity for over-hydrogenation and formation of undesired by-products. Recently, zeolite-encapsulated metal catalysts have emerged as promising candidates for this transformation, offering enhanced selectivity toward the desired products. For example, Gong and co-workers demonstrated that the Pd@SOD catalyst with sub-nanometric Pd clusters encapsulated into SOD zeolite exhibited an ultrahigh ethylene selectivity of over 94.5% in acetylene hydrogenation as a consequence of the unique spatial restriction effect of SOD zeolite (see Figure 2.8d) [54]. In stark contrast, the Pd/SOD catalyst prepared via impregnation exhibited a much lower ethylene selectivity of only around 21.5% (see Figure 2.8e). Hydrogen molecules could diffuse through the channels of SOD zeolites and become activated on the encapsulated Pd sites. The resulting hydrogen atoms then migrated to the external surface of the sodalite via hydrogen spillover, forming OH species that subsequently reacted with acetylene to produce ethylene. Meanwhile, the small pore size of the SOD zeolite effectively restricted acetylene diffusion into the internal channels, thereby suppressing over-hydrogenation. In contrast, for Pd/SOD catalyst, acetylene was directly adsorbed onto the exposed Pd surface, where it readily reacted with hydrogen. The produced ethylene, being strongly bound to the Pd sites, had difficulty desorbing, leading to further hydrogenation and the formation of ethane. Similarly, Li and co-workers developed zeolite-encapsulated non-noble metal catalysts for the selective hydrogenation of acetylene to ethylene [60]. The catalyst featured four-coordinated cationic Ni(II) species confined within a CHA zeolite framework, exhibiting exceptional efficiency and ultrahigh selectivity toward ethylene formation. The representative Na–Ni@CHA catalyst achieved outstanding performance, delivering 97% ethylene selectivity at 100% acetylene conversion at 180 °C—significantly surpassing that of the impregnated NiO/Na–CHA catalyst. Additionally, they also encapsulated isolated Pt cations within zeolite Y (Pt@Y), where the atomically dispersed $Pt^{\delta+}$ species were stabilized by oxygen atoms located in the 6-MRs connecting sodalite cages and supercages [66]. The Pt@Y catalyst exhibited excellent performance in the selective hydrogenation of α,β-unsaturated aldehydes to unsaturated alcohols, achieving 90%

selectivity for allyl alcohol at full acrolein conversion. This outstanding catalytic behavior was attributed to the unique structure of Pt@Y, which favored heterolytic hydrogen dissociation at $Pt^{\delta+}\cdots O^{2-}$ interfacial sites. The resulting H^+ and $H^{\delta-}$ species served as the active intermediates, enabling high selectivity in the hydrogenation process.

2.4 Conclusion and Perspectives

Over the past decade, zeolite-encapsulated metal catalysts have garnered significant interest from both academia and industry, establishing themselves as essential heterogeneous catalysts across a wide range of reactions due to their outstanding activity, remarkable product selectivity, exceptional durability, and reusability. In this chapter, we summarized the state-of-the-art strategies for the synthesis of zeolite-encapsulated metal catalysts and their important applications in the catalytic conversion of low-carbon molecules. Continuous innovation and in-depth research have brought about a boom in related fields. Nevertheless, several challenges remain to be overcome, and there is substantial potential for further enhancement and exploration in future studies.

The development of innovative and straightforward synthetic approaches remains a key driver for progress in the field of zeolite-encapsulated metal catalysts. As outlined in earlier sections, two primary strategies are employed: top-down and bottom-up syntheses. In top-down methods, achieving complete encapsulation of metal species within the zeolites is often challenging. As a result, metal species remaining on the external surfaces are vulnerable to sintering, agglomeration, and leaching when exposed to harsh reaction environments. In contrast, it is easy for uniformly distributed metal species to be completely encapsulated within zeolite crystals by the bottom-up synthetic method under optimized conditions, showing better stability and shape selectivity in catalysis. However, until now, the bottom-up synthetic method has only been successfully applied to the encapsulation of metal species within limited types of zeolites (such as MFI, FAU, and MWW), and the metal loadings are usually less than 1 wt.% to ensure high dispersion of metal species. Developing advanced bottom-up synthetic strategies capable of encapsulating a wide range of metal species at high loadings within zeolites of diverse topologies remains a significant challenge. Achieving this goal relies heavily on the precise optimization of synthesis parameters—such as crystallization temperature, duration, and gel composition—as well as on the compatibility between metal precursors and zeolite frameworks. For industrial-scale applications, cost-efficiency is another critical consideration. There is a pressing need for synthetic approaches that enhance catalyst yield, maximize metal precursor utilization, and minimize both energy consumption and environmental impact. Although noble metals typically outperform non-noble counterparts in many catalytic processes, their high cost and limited availability hinder broader application. Thus, future research is expected to focus on strategies to enhance the efficiency and catalytic performance of noble metals or to develop high-performance alternatives based on earth-abundant, low-cost non-noble metals.

It is important to recognize that the encapsulation of metal species within zeolites presents a double-edged sword in catalytic applications. On the one hand, this strategy significantly enhances the dispersion, utilization, and stability of metal species; on the other hand, it may introduce mass transfer limitations by restricting the diffusion of reactants into, and products out of, the zeolite micropores. In some cases, the accumulation of reactants or products can block access to active sites or clog the pore structure, leading to rapid catalyst deactivation. Introducing mesopores or macropores into microporous zeolites, or reducing the crystal size of the zeolite, has proven effective in mitigating these diffusion constraints. However, such modifications may weaken the confinement effect, potentially resulting in increased metal particle size and reduced thermal stability. To overcome these trade-offs, more sophisticated design strategies are needed. For example, achieving full encapsulation of ultrasmall metal species within the micropores of hierarchical zeolites—rather than within mesopores—may offer a promising route to simultaneously maintain strong confinement and improve mass transport. In addition, further efforts should focus on tailoring and optimizing the synergistic interactions between metal species and the zeolite framework, which could not only enhance catalytic activity but also enable novel reaction pathways.

We hope this chapter provides valuable insights into the synthesis of zeolite-encapsulated metal catalysts and their applications in the catalytic transformation of low-carbon molecules, inspiring further interest and exploration among researchers in related fields.

Acknowledgments

This work was supported by the National Key R&D Program of China (Grant No. 2022YFA1506000).

References

1 Liang, J., Liang, Z., Zou, R. et al. (2017). Heterogeneous catalysis in zeolites, mesoporous silica, and metal-organic frameworks. *Adv. Mater.* 29: 1701139.
2 Li, Y., Li, L., and Yu, J. (2017). Applications of zeolites in sustainable chemistry. *Chem* 3: 928–949.
3 Shamzhy, M., Opanasenko, M., Concepcion, P. et al. (2019). New trends in tailoring active sites in zeolite-based catalysts. *Chem. Soc. Rev.* 48: 1095–1149.
4 Meng, X. and Xiao, F.-S. (2014). Green routes for synthesis of zeolites. *Chem. Rev.* 114: 1521–1543.
5 Shi, J., Wang, Y., Yang, W. et al. (2015). Recent advances of pore system construction in zeolite-catalyzed chemical industry processes. *Chem. Soc. Rev.* 44: 8877–8903.

6 Yang, X.-Y., Chen, L.-H., Li, Y. et al. (2017). Hierarchically porous materials: synthesis strategies and structure design. *Chem. Soc. Rev.* 46: 481–558.

7 Li, Y. and Yu, J. (2021). Emerging applications of zeolites in catalysis, separation and host-guest assembly. *Nat. Rev. Mater.* 6: 1156–1174.

8 Sun, Q., Xie, Z., and Yu, J. (2018). The state-of-the-art synthetic strategies for SAPO-34 zeolite catalysts in methanol-to-olefin conversion. *Natl. Sci. Rev.* 5: 542–558.

9 Wang, Y., Wang, C., Wang, L. et al. (2021). Zeolite fixed metal nanoparticles: new perspective in catalysis. *Acc. Chem. Res.* 54: 2579–2590.

10 Wang, H., Wang, L., and Xiao, F.-S. (2020). Metal@zeolite hybrid materials for catalysis. *ACS Cent. Sci.* 6: 1685–1697.

11 Wang, N., Sun, Q., and Yu, J. (2019). Ultrasmall metal nanoparticles confined within crystalline nanoporous materials: a fascinating class of nanocatalysts. *Adv. Mater.* 31: 1803966.

12 Sun, Q., Wang, N., Xu, Q. et al. (2020). Nanopore-supported metal nanocatalysts for efficient hydrogen generation from liquid-phase chemical hydrogen storage materials. *Adv. Mater.* 32: 2001818.

13 Sun, Q., Wang, N., and Yu, J. (2021). Advances in catalytic applications of zeolite-supported metal catalysts. *Adv. Mater.* 33: 2104442.

14 Chai, Y., Shang, W., Li, W. et al. (2019). Noble metal particles confined in zeolites: synthesis, characterization, and applications. *Adv. Sci.* 6: 1900299.

15 Wu, S.-M., Yang, X.-Y., and Janiak, C. (2019). Confinement effects in zeolite-confined noble metals. *Angew. Chem. Int. Ed.* 58: 12340–12354.

16 Kosinov, N., Liu, C., Hensen, E.J.M. et al. (2018). Engineering of transition metal catalysts confined in zeolites. *Chem. Mater.* 30: 3177–3198.

17 Wang, L., Xu, S., He, S. et al. (2018). Rational construction of metal nanoparticles fixed in zeolite crystals as highly efficient heterogeneous catalysts. *Nano Today* 20: 74–83.

18 Liu, L. and Corma, A. (2021). Confining isolated atoms and clusters in crystalline porous materials for catalysis. *Nat. Rev. Mater.* 6: 244–263.

19 Sun, Q., Wang, N., Fan, Q. et al. (2020). Subnanometer bimetallic platinum-zinc clusters in zeolites for propane dehydrogenation. *Angew. Chem. Int. Ed.* 59: 19450–19459.

20 Qi, L., Babucci, M., Zhang, Y. et al. (2021). Propane dehydrogenation catalyzed by isolated Pt atoms in ≡SiOZn-OH nests in dealuminated zeolite beta. *J. Am. Chem. Soc.* 143: 21364–21378.

21 Wang, N., Sun, Q., Zhang, T. et al. (2021). Impregnating subnanometer metallic nanocatalysts into self-pillared zeolite nanosheets. *J. Am. Chem. Soc.* 143: 6905–6914.

22 Wang, N., Sun, Q., Bai, R. et al. (2016). In situ confinement of ultrasmall Pd clusters within nanosized silicalite-1 zeolite for highly efficient catalysis of hydrogen generation. *J. Am. Chem. Soc.* 138: 7484–7487.

23 Liu, L., Lopez-Haro, M., Lopes, C.W. et al. (2020). Structural modulation and direct measurement of subnanometric bimetallic PtSn clusters confined in zeolites. *Nat. Catal.* 3: 628–638.

24 Chai, Y., Han, X., Li, W. et al. (2020). Control of zeolite pore interior for chemoselective alkyne/olefin separations. *Science* 368: 1002–1006.

25 Jin, Z., Wang, L., Zuidema, E. et al. (2020). Hydrophobic zeolite modification for in situ peroxide formation in methane oxidation to methanol. *Science* 367: 193–197.

26 Zhang, J., Wang, L., Zhang, B. et al. (2018). Sinter-resistant metal nanoparticle catalysts achieved by immobilization within zeolite crystals via seed-directed growth. *Nat. Catal.* 1: 540–546.

27 Zhu, Q., Zhou, H., Wang, L. et al. (2022). Enhanced CO_2 utilization in dry reforming of methane achieved through nickel-mediated hydrogen spillover in zeolite crystals. *Nat. Catal.* 5: 1030–1037.

28 Shan, J., Li, M., Allard, L.F. et al. (2017). Mild oxidation of methane to methanol or acetic acid on supported isolated rhodium catalysts. *Nature* 551: 605–608.

29 Tian, Y., Duan, H., Zhang, B. et al. (2021). Template guiding for the encapsulation of uniformly subnanometric platinum clusters in beta-zeolites enabling high catalytic activity and stability. *Angew. Chem. Int. Ed.* 60: 21713–21717.

30 Wang, N., Bai, R., Hui, Y. et al. (2019). Synergetic effect of ultrasmall metal clusters and zeolites promoting hydrogen generation. *Adv. Sci.* 6: 1802350.

31 Lu, J., Aydin, C., Browning, N.D. et al. (2012). Imaging isolated gold atom catalytic sites in zeolite NaY. *Angew. Chem. Int. Ed.* 51: 5842–5846.

32 Snyder, B.E., Vanelderen, P., Bols, M.L. et al. (2016). The active site of low-temperature methane hydroxylation in iron-containing zeolites. *Nature* 536: 317–321.

33 He, J., Wu, Z., Gu, Q. et al. (2021). Zeolite-tailored active site proximity for the efficient production of pentanoic biofuels. *Angew. Chem. Int. Ed.* 60: 23713–23721.

34 Huang, W., Zhang, S., Tang, Y. et al. (2016). Low-temperature transformation of methane to methanol on Pd_1O_4 single sites anchored on the internal surface of microporous silicate. *Angew. Chem. Int. Ed.* 55: 13441–13445.

35 Tang, Y., Li, Y., Fung, V. et al. (2018). Single rhodium atoms anchored in micropores for efficient transformation of methane under mild conditions. *Nat. Commun.* 9: 1231.

36 Ryoo, R., Kim, J., Jo, C. et al. (2020). Rare-earth-platinum alloy nanoparticles in mesoporous zeolite for catalysis. *Nature* 585: 221–224.

37 Iida, T., Zanchet, D., Ohara, K. et al. (2018). Concerted bimetallic nanocluster synthesis and encapsulation via induced zeolite framework demetallation for shape and substrate selective heterogeneous catalysis. *Angew. Chem. Int. Ed.* 57: 6454–6458.

38 Petrov, A.W., Ferri, D., Krumeich, F. et al. (2018). Stable complete methane oxidation over palladium based zeolite catalysts. *Nat. Commun.* 9: 2545.

39 Yao, J., He, Y., Zeng, Y. et al. (2022). Ammonia pools in zeolites for direct fabrication of catalytic centers. *Nat. Commun.* 13: 935.

40 Liu, L., Li, H., Zhou, H. et al. (2023). Rivet of cobalt in siliceous zeolite for catalytic ethane dehydrogenation. *Chem* 9: 637–649.

41 Goel, S., Zones, S.I., and Iglesia, E. (2014). Encapsulation of metal clusters within MFI via interzeolite transformations and direct hydrothermal syntheses and catalytic consequences of their confinement. *J. Am. Chem. Soc.* 136: 15280–15290.

42 Liu, L., Díaz, U., Arenal, R. et al. (2017). Generation of subnanometric platinum with high stability during transformation of a 2D zeolite into 3D. *Nat. Mater.* 16: 132–138.

43 Li, S., Burel, L., Aquino, C. et al. (2013). Ultimate size control of encapsulated gold nanoparticles. *Chem. Commun.* 49: 8507–8509.

44 Li, S., Tuel, A., Laprune, D. et al. (2015). Transition-metal nanoparticles in hollow zeolite single crystals as bifunctional and size-selective hydrogenation catalysts. *Chem. Mater.* 27: 276–282.

45 Dai, C., Zhang, A., Liu, M. et al. (2016). Hollow alveolus-like nanovesicle assembly with metal-encapsulated hollow zeolite nanocrystals. *ACS Nano* 10: 7401–7408.

46 Gu, J., Zhang, Z., Hu, P. et al. (2015). Platinum nanoparticles encapsulated in MFI zeolite crystals by a two-step dry gel conversion method as a highly selective hydrogenation catalyst. *ACS Catal.* 5: 6893–6901.

47 Goel, S., Wu, Z., Zones, S.I. et al. (2012). Synthesis and catalytic properties of metal clusters encapsulated within small-pore (SOD, GIS, ANA) zeolites. *J. Am. Chem. Soc.* 134: 17688–17695.

48 Wu, Z., Goel, S., Choi, M. et al. (2014). Hydrothermal synthesis of LTA-encapsulated metal clusters and consequences for catalyst stability, reactivity, and selectivity. *J. Catal.* 311: 458–468.

49 Otto, T., Zones, S.I., Hong, Y. et al. (2017). Synthesis of highly dispersed cobalt oxide clusters encapsulated within LTA zeolites. *J. Catal.* 356: 173–185.

50 Otto, T., Zones, S.I., and Iglesia, E. (2018). Synthetic strategies for the encapsulation of nanoparticles of Ni, Co, and Fe oxides within crystalline microporous aluminosilicates. *Microporous Mesoporous Mater.* 270: 10–23.

51 Sun, Q., Wang, N., Bing, Q. et al. (2017). Subnanometric hybrid Pd-$M(OH)_2$, M = Ni, Co, clusters in zeolites as highly efficient nanocatalysts for hydrogen generation. *Chem* 3: 477–493.

52 Sun, Q., Wang, N., Zhang, T. et al. (2019). Zeolite-encaged single-atom rhodium catalysts: highly-efficient hydrogen generation and shape-selective tandem hydrogenation of nitroarenes. *Angew. Chem. Int. Ed.* 58: 18570–18576.

53 Liu, Y., Li, Z., Yu, Q. et al. (2019). A general strategy for fabricating isolated single metal atomic site catalysts in Y zeolite. *J. Am. Chem. Soc.* 141: 9305–9311.

54 Wang, S., Zhao, Z.-J., Chang, X. et al. (2019). Activation and spillover of hydrogen on sub-1nm palladium nanoclusters confined within sodalite zeolite for the semi-hydrogenation of alkynes. *Angew. Chem. Int. Ed.* 58: 7668–7672.

55 Sun, Q., Chen, B.W.J., Wang, N. et al. (2020). Zeolite-encaged Pd-Mn nanocatalysts for CO_2 hydrogenation and formic acid dehydrogenation. *Angew. Chem. Int. Ed.* 59: 20183–20191.

56 Liu, L., Lopez-Haro, M., Lopes, C.W. et al. (2019). Regioselective generation and reactivity control of subnanometric platinum clusters in zeolites for high-temperature catalysis. *Nat. Mater.* 18: 866–873.

57 Liu, L., Lopez-Haro, M., Meira, D.M. et al. (2020). Regioselective generation of single-site iridium atoms and their evolution into stabilized subnanometric iridium clusters in MWW zeolite. *Angew. Chem. Int. Ed.* 59: 15695–15702.

58 Hu, Z.-P., Qin, G., Han, J. et al. (2022). Atomic insight into the local structure and microenvironment of isolated Co-motifs in MFI zeolite frameworks for propane dehydrogenation. *J. Am. Chem. Soc.* 144: 12127–12137.

59 Luo, Q., Wang, H., Wang, L. et al. (2022). Alloyed PdCu nanoparticles within siliceous zeolite crystals for catalytic semihydrogenation. *ACS Mater. Au* 2: 313–320.

60 Chai, Y., Wu, G., Liu, X. et al. (2019). Acetylene-selective hydrogenation catalyzed by cationic nickel confined in zeolite. *J. Am. Chem. Soc.* 141: 9920–9927.

61 Shang, W., Gao, M., Chai, Y. et al. (2021). Stabilizing isolated rhodium cations by MFI zeolite for heterogeneous methanol carbonylation. *ACS Catal.* 11: 7249–7256.

62 Yang, Z., Li, H., Zhou, H. et al. (2020). Coking-resistant iron catalyst in ethane dehydrogenation achieved through siliceous zeolite modulation. *J. Am. Chem. Soc.* 142: 16429–16436.

63 Choi, M., Wu, Z., and Iglesia, E. (2010). Mercaptosilane-assisted synthesis of metal clusters within zeolites and catalytic consequences of encapsulation. *J. Am. Chem. Soc.* 132: 9129–9137.

64 Moliner, M., Gabay, J.E., Kliewer, C.E. et al. (2016). Reversible transformation of Pt nanoparticles into single atoms inside high-silica chabazite zeolite. *J. Am. Chem. Soc.* 138: 15743–15750.

65 Chai, Y., Liu, S., Zhao, Z.-J. et al. (2018). Selectivity modulation of encapsulated palladium nanoparticles by zeolite microenvironment for biomass catalytic upgrading. *ACS Catal.* 8: 8578–8589.

66 Deng, X., Qin, B., Liu, R. et al. (2021). Zeolite-encaged isolated platinum ions enable heterolytic dihydrogen activation and selective hydrogenations. *J. Am. Chem. Soc.* 143: 20898–20906.

67 Zhang, J., Wang, L., Shao, Y. et al. (2017). A Pd@zeolite catalyst for nitroarene hydrogenation with high product selectivity by sterically controlled adsorption in the zeolite micropores. *Angew. Chem. Int. Ed.* 56: 9747–9751.

68 Wang, C., Wang, L., Zhang, J. et al. (2016). Product selectivity controlled by zeolite crystals in biomass hydrogenation over a palladium catalyst. *J. Am. Chem. Soc.* 138: 7880–7883.

69 Wang, C., Fang, W., Wang, L. et al. (2021). Fischer-Tropsch reaction within zeolite crystals for selective formation of gasoline-ranged hydrocarbons. *J. Energy Chem.* 54: 429–433.

70 Amoo, C.C., Li, M., Noreen, A. et al. (2020). Fabricating Fe nanoparticles embedded in zeolite Y microcrystals as active catalysts for Fischer-Tropsch synthesis. *ACS Appl. Nano Mater.* 3: 8096–8103.

71 Wang, C., Guan, E., Wang, L. et al. (2019). Product selectivity controlled by nanoporous environments in zeolite crystals enveloping rhodium nanoparticle catalysts for CO_2 hydrogenation. *J. Am. Chem. Soc.* 141: 8482–8488.

72 Wang, C., Zhang, J., Qin, G. et al. (2020). Direct conversion of syngas to ethanol within zeolite crystals. *Chem* 6: 646–657.

73 Laprune, D., Tuel, A., Farrusseng, D. et al. (2017). Highly dispersed nickel particles encapsulated in multi-hollow silicalite-1 single crystal nanoboxes: effects of siliceous deposits and phosphorous species on the catalytic performances. *ChemCatChem* 9: 2297–2307.

74 Li, S., Tuel, A., Rousset, J.-L. et al. (2016). Hollow zeolite single-crystals encapsulated alloy nanoparticles with controlled size and composition. *ChemNanoMat* 2: 534–539.

75 Prates, A.R.M., Meunier, F., Dodin, M. et al. (2019). Hydrogenation size-selective Pt/hollow beta catalysts. *Chem. Eur. J.* 25: 2972–2977.

76 Otto, T., Ramallo-López, J.M., Giovanetti, L.J. et al. (2016). Synthesis of stable monodisperse AuPd, AuPt, and PdPt bimetallic clusters encapsulated within LTA-zeolites. *J. Catal.* 342: 125–137.

77 Martínez-Franco, R., Moliner, M., and Concepción, P. (2014). Synthesis, characterization and reactivity of high hydrothermally stable Cu-SAPO-34 materials prepared by "one-pot" processes. *J. Catal.* 314: 73–82.

78 Xiao, J., Cheng, K., Xie, X. et al. (2022). Tandem catalysis with double-shelled hollow spheres. *Nat. Mater.* 21: 572–579.

79 Cui, W.-G., Li, Y.-T., Yu, L. et al. (2021). Zeolite-encapsulated ultrasmall Cu/ZnO_x nanoparticles for the hydrogenation of CO_2 to methanol. *ACS Appl. Mater. Interfaces* 13: 18693–18703.

80 Ding, L., Shi, T., Gu, J. et al. (2020). CO_2 hydrogenation to ethanol over Cu@Na-Beta. *Chem* 6: 2673–2689.

3

Synthesis of Hierarchical Zeolites for Catalytic Conversion of Low-carbon Molecules

Chengyi Dai[1], Yetong Li[1], Chunshan Song[2,]*

[1]*School of Chemical Engineering, Northwest University, Chang'an Campus, Xi'an, Shaanxi, China*
[2]*Department of Chemistry, Faculty of Science, The Chinese University of Hong Kong, Shatin, New Territories, Hong Kong SAR, China*

*Corresponding author: chunshansong@cuhk.edu.hk

3.1 Introduction

Zeolitic materials constitute a significant class of crystalline microporous aluminosilicate minerals, characterized by three-dimensional frameworks formed from angle-sharing TO_4 tetrahedra (where T denotes tetrahedral coordination framework atoms such as Si, Al, or other heteroatoms). The diversity of zeolite microporous structures is dependent on the manner in which the TO_4 tetrahedra are interconnected [1]. According to the definition established by the International Union of Pure and Applied Chemistry (IUPAC), porous materials can be categorized into three types based on pore size: microporous (<2 nm), mesoporous (2–50 nm), and macroporous (>50 nm). Zeolite molecular sieves can be conceptualized as aggregates comprising a limited number of structural units, including monocyclic, bicyclic, and small cage-like configurations. These configurations contain channels or pores and/or supercages, which range from 0.4 to 1.2 nm in size [2]. The widespread utilization of zeolites in catalysis (as active components and supports), ion exchange (e.g., detergents), and adsorption stems from their defining properties: uniform micropores, abundant acid sites, high surface area, cation exchange capacity, and remarkable stability under hydrothermal conditions. This extensive use is attributed to their uniform and ordered microporous structure, high density of acidic sites, large specific surface area, exchangeable cations, and exceptional hydrothermal stability. Presently, these catalysts account for over 40% of the solid catalysts utilized in refining, petrochemical processing, and the production of fine chemicals. While the small pore size of zeolites ensures excellent conformality in numerous reactions, the abundance of Brønsted and Lewis acids provides active sites for catalytic reactions, and the large specific surface area enhances the adsorption of guest molecules; conversely, the limited pore size poses significant challenges in the diffusion and mass transfer of reactants, potentially restricting applicability in certain catalytic reactions involving macromolecules [3]. Diffusion serves as the primary mechanism for mass transfer and is a critical determinant of catalytic properties (activity, selectivity, and lifetime) and the efficiency of zeolite utilization across a variety of reactions and separations. Moreover, the slow transport of substances through the microporous channels of zeolites can lead to unwanted side reactions, severe coking, rapid catalyst deactivation, and, most importantly, the wastage of valuable zeolite material and energy resources. A primary approach to overcoming diffusion constraints centers on shortening the effective diffusion path length within the crystal. This is achieved by engineering hierarchical zeolites—materials integrating multiple, interconnected pore systems (micro-, meso-, and macro-) within a single architecture. Such hierarchical structuring is pivotal for enhancing performance in demanding applications like low-carbon molecule conversion, positioning these materials as highly promising for industrial catalysis [1].

Numerous types of hierarchical systems, often referred to as networks, exist across various domains, including social, technological, and biological systems. Notable examples of biological systems include the organization of human societies, the distribution of data in informatics (as demonstrated in Figure 3.1), and the widely recognized human transportation system [5]. Clauset et al. [6] stated, "Recent studies have shown that networks typically exhibit a hierarchical organization in which vertices are divided into groups, which are further subdivided into groups, and so on at multiple scales. In many cases, these groups correspond to known functional units, such as ecological niches in food webs, modules in biochemical networks (protein interaction networks, metabolic networks, or genetic regulatory networks), or communities in social networks." Additionally, many flow systems in nature display hierarchical patterns, sometimes featuring different but distinctly defined structures. Figure 3.2 summarizes several typical examples to highlight the diversity of grading characteristics.

Figure 3.2a, b depicts the morphologies of tree roots and leaves, respectively. A similar grading system is evident in the diagrams, where the root system gathers mineral-rich water through extensive, finely distributed flow channels, subsequently transporting the collected water and nutrients to a higher grading level

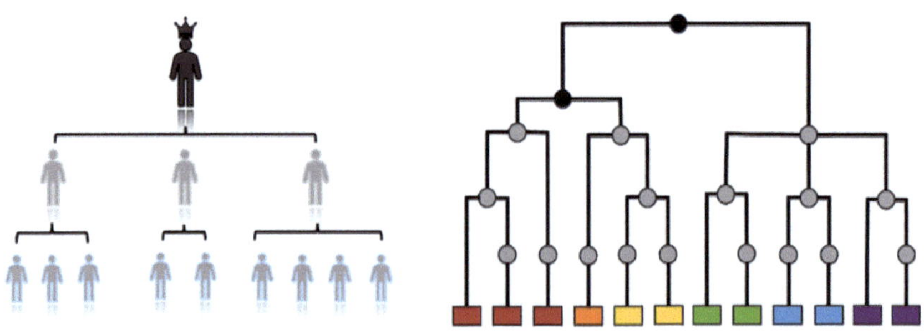

Figure 3.1 Social organization hierarchy (left) and information hierarchy (right). *Source*: Adapted from [4].

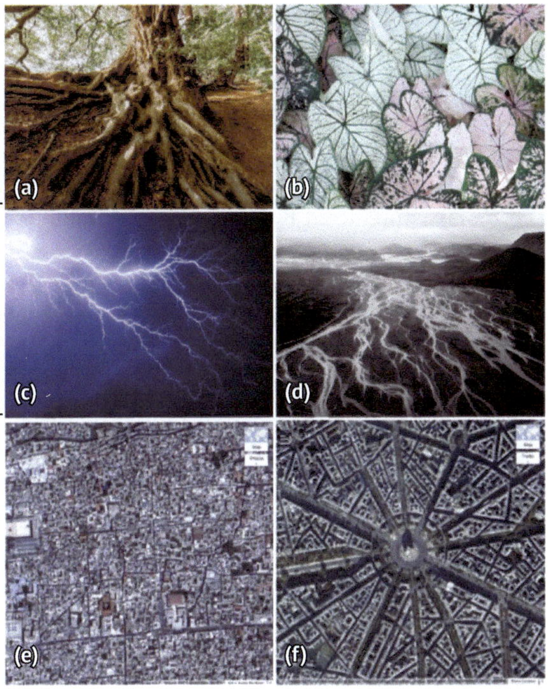

Figure 3.2 Examples of hierarchical systems in nature. (a) Tree roots. *Source*: Felix Mittermeier / pexels / CC0. (b) Tree leaves. *Source*: Madison Inouye / pexels / CC0. (c) Lightning. *Source*: Anton Kudryashov / pexels / CC0 (d) River delta. *Source*: ArtHouse Studio / pexels / CC0. (e) Satellite image of the streets of Damascus. *Source:* [7] / Springer Nature. (f) Satellite image of a Parisian street. *Source:* [7] / with permission of Springer Nature.

(wider trunk transport channels). Ultimately, water is released through the leaves, with the distribution of water flow primarily accomplished through a two-stage hierarchical system of "main channels" (in this case, leaf veins), which facilitate the flow of water into a vast network of smaller capillaries. As shown in Figure 3.2c, lightning displays a strikingly graded, branching structure that enables rapid energy transfer. River deltas exhibit a remarkably similar hierarchical pattern—so much so that their morphology closely resembles that of lightning (Figure 3.2d). In this pattern, an increase in the horizontal distribution of water leads to a corresponding decrease in river flow rate, which in turn reduces the amount of sediment discharged into the ocean.

The road network of a city necessitates an equally meticulous design of hierarchical structures. A large city must facilitate access to all locations while simultaneously improving the transportation efficiency. The same principle applies to zeolites. As illustrated in Figure 3.2e, the congested streets of Damascus are analogous to microporous (<2 nm) zeolites, where large-volume molecules encounter significant entry barriers, hindering access to the interior and reducing the effective utilization of active centers within the pores. This limitation substantially impacts the lifetime of the zeolite. Conversely, the hierarchical structure of the zeolite resembles the road network of Paris, as depicted in Figure 3.2f. This structure includes smaller micropores (narrow streets) and larger meso- and macropores (wide streets), forming a network of interconnected micro-, meso-, and macropores. Compared with purely microporous zeolites, this configuration significantly enhances the rate of mass transfer within the zeolite, improves the accessibility of micropores, and prolongs the zeolite's operational lifespan [8].

Hierarchical flow distribution features directly impact the average reaction rates in structured catalytic microreactors, as predicted by theoretical calculations [9]. Consequently, constructing interconnected pore hierarchical networks in zeolite crystals is ideal. It is crucial that the synthesized hierarchical zeolite retains its high-quality properties, meaning it should incorporate additional porosity without compromising the intrinsic microporosity of the zeolite.

3.2 Synthesis Methods

In essence, crystalline zeolite materials, despite displaying well-defined pore systems, present pore sizes within the microporous range and, as a consequence, are not characterized by graded porosity. The defining characteristics of hierarchical zeolites are the intrinsic micropores present within the zeolite skeleton and the incorporated inter-/intracrystalline meso-/macropores. Hierarchical zeolites are usually synthesized via two methods: (1) the "bottom-up" method, which uses a template or synthesis; and (2) the "top-down" method, which typically involves dealumination and desilication. To introduce additional porosity to the zeolite material to obtain hierarchical zeolites, two preparation strategies, i.e., the template methods and the post-processing methods, are described below.

3.2.1 Template Methods

The synthesis of hierarchical zeolites depends on structure-directing agents (SDAs) that can create microporous zeolite structures, as well as templating agents that generate meso- and macroporous pores, contributing additional porosity. A common approach to synthesizing hierarchical zeolites entails combining a template with the zeolite precursor; removal of the template then generates the desired enhanced porosity. The chosen synthesis strategy is primarily influenced by the rigidity of the template, which may be classified as either solid or liquid. This classification results in the differentiation between the hard template method [10, 11] and the soft template method [12].

3.2.1.1 Hard Template Method

The hard template method, as indicated by its name, necessitates the use of a relatively rigid, structurally strong nonporous or porous solid material as a sacrificial template, thereby creating additional porosity during zeolite crystallization. Since these solid template materials are chemically inert, they do not influence the intrinsic structure of the zeolite [1]. The synthesis step in the hard template method involves mixing the zeolite precursor with a hard template agent, followed by in situ crystallization, and subsequently removing the hard template agent through calcination or leaching to obtain hierarchical zeolites containing mesopores and macropores. Dry gel conversion or conventional crystallization enables the integration of the templating agent within the zeolite crystals, provided the following criteria are fulfilled: (1) the surface properties must align with the chemical properties of the reactants; (2) the templating agent must remain stable at the synthesis temperature; and (3) the zeolite component obtained after the removal of the templating agent must exhibit structural stability. The production of zeolites through the hard template route typically yields materials characterized by high crystallinity, uniform additional porosity, and a porous structure that can be customized to meet specific requirements. Presently, the preferred materials for hard templates include carbon-containing materials [10, 11, 13–20], polymers [21–35] (polystyrene [PS] microspheres/resins), inorganic solid materials [5, 36–40], and biomaterials [4, 41–56]. These materials incorporate carbon nanoparticles [10, 13–15, 17–20] and nanotubes [16]. Carbon-containing materials are utilized as hard templates due to their advantageous properties, which include chemical inertness, structural and morphological diversities, cost-effectiveness, and ease of removal.

Jacobsen and co-workers investigated carbon nanoparticles as hard templates for the synthesis of intracrystalline mesoporous MFI zeolites [10]. The study revealed that confined-space crystallization could be achieved by introducing an equal or lesser quantity of zeolite precursor gels than the overall pore volume of the carbon matrix when carbon material is created using an initial impregnation method. However, the incompatibility between the hydrophobic carbon template and the aqueous zeolite precursor solution hindered the nucleation and crystallization of zeolite within the template. This issue was addressed by the dry gel transformation method, which relies on reducing the zeolite precursor to a dry powder before transforming it in a water vapor-rich environment. This approach effectively suppressed the mobility of the precursor gel in the reaction mixture, leading to the crystallization of the dry gel composite. The results indicated that utilizing an excess of mesoporous carbon nanoparticles enhanced the precise nucleation of zeolite crystals within the mesopores of the carbon nanoparticles [13]. The size of the final crystallites in zeolites was determined more by the inert mesoporous structure than by the gel's composition. A large amount of synthetic gel was added, leading to the spread of carbon nanoparticles throughout the zeolite precursor solution. As the zeolite crystals grew during the crystallization process, these nanoparticles were captured within them (Figure 3.3a, b) [14]. The broad applicability of this method resulted in the successful synthesis of various graded zeolites with different frameworks and compositions, including MFI (TS-1 [15], ZSM-5 [13]), MTW (ZSM-12 [19]), MEL (ZSM-11, TS-2, and Silicalite-2) [17], and BEA (Beta) [10], as well as topologies such as hierarchical aluminum phosphates AFI (AlPO-5 [20]) and CHA (AlPO-34 [20]). The resulting mesoporous zeolites exhibited enhanced diffusion and catalytic properties compared to traditional microporous materials. In a pioneering study, Schmidt et al. utilized carbon nanotubes for the first time to synthesize Silicalite-1 zeolite. This was facilitated by the presence of straight, homogeneous mesoporous channels that penetrate the entire zeolite

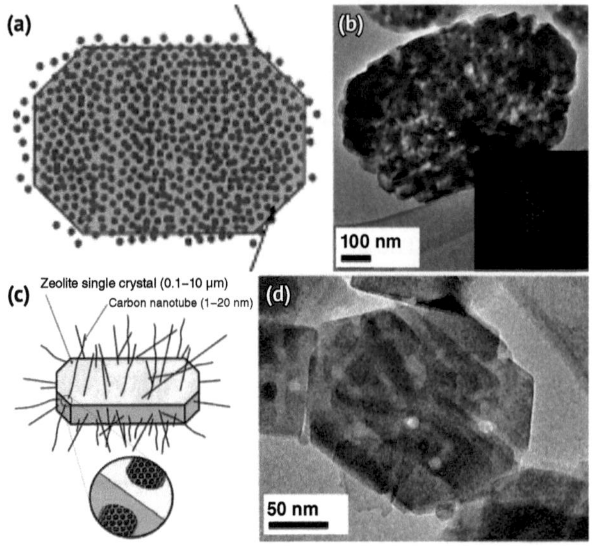

Figure 3.3 Schematic diagrams of using carbon material as a hard template. (a) Schematic of zeolite material using carbon nanoparticles as templates. (b) TEM image of ZSM-5 molecular sieves prepared using carbon nanoparticles as templates. *Source*: Reproduced from [14] / with permission of American Chemical Society. (c) Schematic representation of Silicalite-1 molecular sieves prepared using carbon nanotubes (1–12 nm). (d) TEM image of Silicalite-1 molecular sieve prepared using carbon nanotubes. *Source*: Reproduced from [16] / with permission of American Chemical Society.

single crystal, with widths of about 12–30 nm, corresponding to the diameter of carbon nanotubes (Figure 3.3c, d) [16].

Schmidt et al. synthesized SAPO-34 zeolites with hierarchical structures using carbon nanoparticles and carbon nanotubes as secondary templates. When carbon nanoparticles were employed, mesopores were formed inside the zeolite but did not create mesoporous transport channels on the surface (Figure 3.4). In contrast, the use of carbon nanotubes as secondary templates successfully generated transport pores on the zeolite surface (Figure 3.4), resulting in improved diffusion rates and enhanced catalytic stability [57]. More recently, Sun et al. developed micrometer-scale mesoporous Beta zeolite through the dry gel transformation method, utilizing ordered mesoporous carbon (OMMC) with a large pore size of 220 nm as a template [58]. The interconnected mesopores and macropores of this zeolite significantly improve the accessibility of active sites and resistance to coking. Typically, the synthesis of ordered mes-macroporous Beta zeolites involved impregnating OMMC templates with an excess of zeolite Beta precursor gels, followed by aging, drying, and steam-assisted crystallization processes (Figure 3.5a). Figure 3.5b presents a representative OMMS–Beta (3) sample, crystallized over three days to produce 1.5–2-μm crystals. These crystals are composed of tightly aligned, highly ordered 200-nm zeolite spheres. The micropores within each zeolite sphere are interconnected, forming two distinct macropores on top of the zeolite Beta's intrinsic micropores. This structure

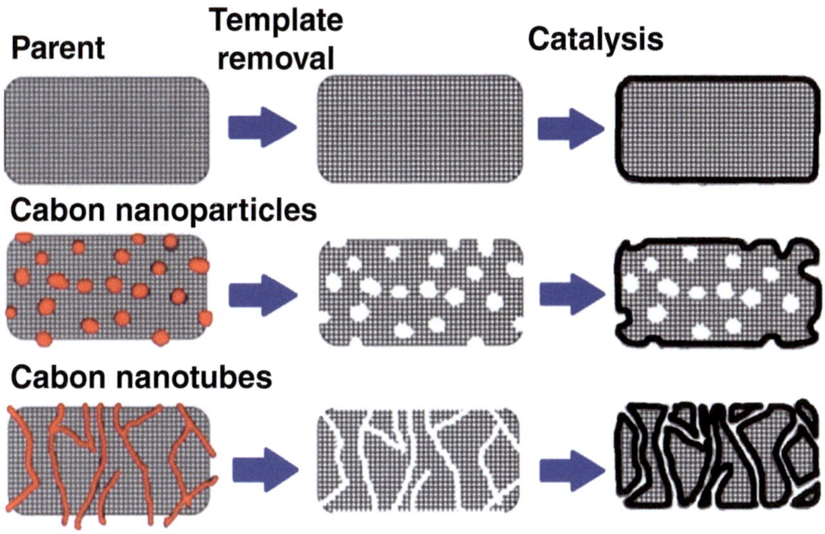

Figure 3.4 Synthesis route of carbon nanoparticles and carbon nanotubes and schematic diagram representing catalytic efficiency. *Source*: [57] / with permission of Elsevier.

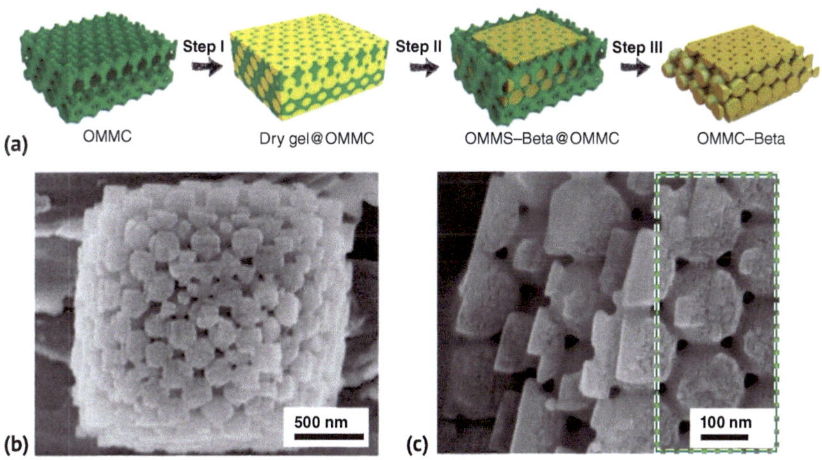

Figure 3.5 Steam-assisted crystallization of OMMS–Beta and its structure configuration. (a) Schematic diagram of the synthesis route of OMMS–Beta. (b) SEM images of OMMS–Beta. (c) The green dashed box highlights the densely packed and intergrown spheres that are imprinted in a highly ordered arrangement. *Source*: [58] / Springer Nature.

creates an ordered, fully interconnected meso- and macroporous network that precisely replicates the OMMC template (Figure 3.5c).

The synthesis of hierarchical zeolites has also been accomplished using polymer microspheres [22–25, 27–29, 31–34] and resins [21, 26, 30, 35]. Holland et al. reported the preparation of three-dimensional macroporous zeolites utilizing PS microspheres as hard templates [22]. Macroporous zeolite monomers with varied compositions and frameworks were synthesized via two routes: the first utilized evaporation-induced in situ co-assembly of PS microspheres and zeolite nanocrystals [32], and the second employed a casting technique where different zeolite nanocrystals were introduced into ordered PS microsphere templates [23, 24, 29]. The fundamental principle involves the preparation of hollow zeolite spheres through the alternate deposition of oppositely charged zeolite nanocrystals and polyelectrolytes onto positively charged PS spheres, utilizing a layer-by-layer deposition strategy [27]. This approach facilitates the synthesis of zeolite monomers with diverse topologies, including hollow [28, 34] and macroporous [25] structures, by manipulating the zeolite nanocrystals used and the thickness of the deposited layer. Zheng et al. synthesized a ZSM-5 zeolite microsphere with macropores using PS nanomicrospheres as templates assisted by water/oil emulsion [59]. These zeolite microspheres are assembled from uniformly hollow zeolite nanospheres with freely adjustable diameters from 210 to 740 nm. The synthesis method is divided into four main steps. The solution mixing this PDDA-treated PS pellets with the zeolite precursor was first added to the vigorously stirred oil phase to generate high shear, which caused the PDDA-treated PS pellets to be uniformly distributed in the ZSM-5 precursor solution and encapsulated by the oil phase (the PDDA-modified PS pellets ensured that the precursor solution could easily wet the surface of the template). The PS opal crystal structure was then formed by evaporative self-assembly within each confined droplet. Then by steam-assisted crystallization and finally calcination of the templating agent, ZSM-5 hollow zeolite spheres with graded microporous–macroporous pores were obtained. This structure provides smaller diffusion channels and higher active site accessibility.

The synthesis of mesoporous zeolite materials has also been documented using ion-exchange resins as hard templates. The process initiates with the adsorption of a zeolite synthesis gel onto anion exchange resin beads. This is subsequently followed by the crystallization and calcination of the resulting mixture, leading to the formation of mesoporous Silicalite-1 zeolite microspheres [26]. Hierarchical palladium-containing zeolites have been reported to date [30]. However, the positively charged precursors in typical AlPO synthesis gels preclude the use of ion-exchange resins as templates [21]. Addressing this limitation, Naydenov et al. developed an enhanced approach enabling the incorporation of diverse metal ions into AlPO frameworks, demonstrated by their synthesis of vanadium-modified mesoporous AlPO-5 spheres [35].

In addition to carbonaceous materials, polymers, and resins, inorganic solid materials such as mesoporous silica spheres [36–38] and $CaCO_3$ nanoparticles [5, 39, 40] have been utilized as templates for synthesizing graded zeolites. Among these, the use of mesoporous silica spheres as hard templates is particularly beneficial for reducing environmental pollution. This advantage arises from the dual functionality of silica spheres, which serve as both mesoporous/macroporous templates and a source of zeolite silica. This characteristic allows for their continuous absorption during zeolite crystallization, thereby eliminating the need for the removal of the templating agent in the template method. Current synthetic approaches encapsulate zeolite crystals onto mesoporous silica spheres via layer-by-layer deposition, yielding either hollow capsules with dense, uniform zeolitic shells [36] or robust zeolite membranes characterized by three-dimensional, sealed [37], and interconnected [38] macroporous architectures.

Li et al. recently synthesized hollow ZSM-5 zeolites using Al-containing Stöber-type silica spheres as hard templates [60]. The synthesis of hollow ZSM-5 can be categorized into two primary methods. The first method, template-mediated hollowing, involves crystallization at the template–liquid interface, and subsequent template removal generates hollow ZSM-5. This approach typically results in zeolite single crystals with large grain sizes (> 1 μm) and thicker shells. To achieve ZSM-5 with smaller particle sizes, Al-containing silica spheres serve as solid templates. Additionally, selective dissolution of aluminum-deficient regions facilitates cavity formation, ultimately yielding ZSM-5 zeolites with hollow structures. Furthermore, Xie and co-workers described the preparation of mesoporous MFI zeolites using cost-effective and readily available spherical $CaCO_3$ nanoparticles as hard templates [39]. It was shown that nanosized $CaCO_3$ with an average particle size of 100 nm could be incorporated as a solid template within a synthesis system with no solvent for producing hierarchical zeolites via physical mixing, mechanical milling, and heating of the solid feedstock [40]. An advantage of using templates, which is $CaCO_3$, is their ability to impart additional favorable properties to the zeolite active site. For example, when the composite undergoes calcination at elevated temperatures (>600 °C), the $CaCO_3$ present within the zeolite crystals is converted into alkaline CaO. The resulting CaO/zeolite product combines the strong acidity of zeolite with the strong alkalinity of CaO, making it suitable for catalytic processes that require both acidic and basic sites [5].

In addition to the previously mentioned hard framework, bio-framework has also been developed as a hard template material due to its advantages of being resourceful, readily available, inexpensive, renewable, and environmentally friendly. The synthesis of zeolites featuring novel shapes, complex functional modes, and graded porosities using natural biological tissues and template materials constitutes a significant advancement in the development of zeolites with improved performance. A variety of biological materials, including bacteria [41], eggshells [4], spider silk [46], insects [47], bagasse [55], diatoms [53], leaves [48, 49], loofah sponges [51], wood [42–44], cotton threads [52], rice hulls [50, 56], starch breads [54], and starch gels [45], have been identified as potential sacrificial templates for creating additional pores in zeolites.

3.2.1.2 Soft Template Method

Soft templates mainly include silanized polymers [61], hydrophilic cationic polymers [62–64], cationic organosilanes [65], and surfactants [66, 67]. Soft templates exhibit greater versatility than hard templates owing to their tunable dimensions and adaptable functionalities. During crystallization, the soft template not only provides physical support but also engages in direct interactions with silica precursors [68–70]. The effective implementation of soft templates necessitates the following conditions: (1) stability in alkaline environments, since zeolite synthesis is typically performed under alkaline conditions; (2) possession of a positive charge to foster strong interactions with negatively charged silica species; (3) an appropriate mesoscale size in aqueous systems; and (4) cost-efficient production. Additionally, the configuration of soft templates can be tailored by modifying functional groups or adjusting surfactant packing parameters. Thus, the soft template approach constitutes a versatile platform for designing customized mesopores.

Pinnavaia and co-worker reported that $-SiO_3$ in silanized polymers can be bonded to zeolite surfaces through the formation of covalent Si—O—Si linkages (Figure 3.6a) [61]. As the zeolite crystals develop, the doped polymer experiences phase separation from the zeolite matrix, leading to the establishment of an intracrystalline polymer network that is covalently linked to the zeolite backbone. Mesoporous MFI zeolites formed after the polymer template removal, with each particle containing small, randomly oriented mesopores (Figure 3.6b). Moreover, lattice fringes extending continuously throughout individual particles confirm their single-crystalline nature (Figure 3.6c). In addition to silane polymers, hydrophilic cationic polymers have been utilized as mesoporous templates because of their high positive charge, considerable alkali resistance, and robust interactions with silicone-based species in high-temperature processes (up to 200 °C). The initial study regarding the application of hydrophilic cationic polymers as mesoporous templates was conducted by Xiao et al. [62–64]. A mesoscale hydrophilic cationic polymer (polydiallyldimethylammonium chloride [PDADMAC]) was used to assist in the development of intracrystalline mesopores within the Beta zeolite single crystals [62]. The significant charge density of this soft template

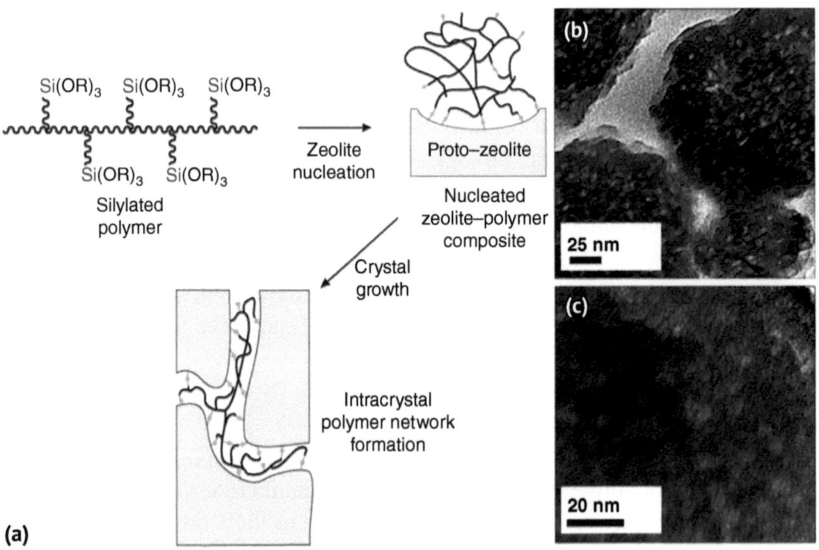

Figure 3.6 Schematic of the soft template route using a non-surfactant polymer. (a) Preparation of intracrystalline mesoporous zeolites using silanized polymers as mesoporous agents. *Source*: [61] / with permission of John Wiley & Sons. (b) TEM image of the particulate sample. (c) TEM image of the flake sample. *Source*: Reproduced from [61] / with permission of John Wiley & Sons.

allows for effective prevention of boundary incompatibility between the zeolite gel and the templating agent. The pore size of the zeolite can be adjusted by varying the amount of the soft template incorporated. The synthesis of cationic organosilanes was first detailed by Ryoo and co-workers [65]. Cationic organic molecules are capable of neutralizing the negative charge of the zeolite skeleton or occupying the voids within the zeolite structure, thereby facilitating the stabilization of the zeolite phase under the desired conditions. Furthermore, the use of specifically formulated amphiphilic organosilane surfactants and non-surfactant polymers allows for precise regulation of intermediate looseness. However, synthesizing these materials presents challenges and incurs significant costs. Thus, the continued introduction of mesopores into zeolites using cost-effective and easily scalable porogenic agents, such as conventional cationic surfactants, is preferable. Surfactants are considered essential templates for the synthesis of hierarchical zeolites, primarily due to their effectiveness in creating mesopores during the synthesis of ordered mesoporous materials. Employing a combination of TPOAC ([3-(trimethoxysilyl)propyl]-octadecyldimethylammonium chloride; amphiphilic organosilane) and CTABr (standard surfactant) as co-templates, Wang et al. fabricated hierarchical mesoporous ZSM-5 [67]. They found that adjusting the TPOAC/CTABr ratio modulated the porosity, with CTABr displaying greater inhibition of MFI nucleation/crystallization than TPOAC. This differential effect underscores the importance of surfactant–zeolite precursor interactions under hydrothermal conditions, which dictate structural outcomes. Relevant to this, Chen et al. reported that in a silicate–CTABr system, elevating the temperature from ambient to 150 °C and then to 165 °C induced phase transitions: initial MCM-41 (hexagonal) transformed to lamellar and ultimately yielded microporous MFI [66].

3.2.2 Post-processing Methods

In addition to the aforementioned stencil method, zeolite crystals can be synthesized to form secondary mesopores within the zeolite structure, thereby creating additional porosity [71–93]. Generally, post-synthetic treatments of previously formed zeolites include demetallization [75, 84, 87, 90, 92] (extraction of backbone atoms) or hierarchical methods, which are achieved via steam treatment [81, 83, 85, 86, 89], acid treatment [71, 76, 93], alkali treatment [72–74, 77–80, 82], fluoride treatment [88, 91], and dissolution–recrystallization [94–97]. Post-synthetic modification represents an industrially prevalent technique pivotal to the commercial success of zeolites. This approach enables the production of high-performance catalysts and adsorbents with tailored compositions, optimized acid site distributions, and enhanced stability [98]. While its principal advantages lie in operational simplicity and low implementation costs, significant drawbacks include: (1) reliance on harsh processing conditions that pose environmental risks; (2) irreversible structural defects including partial framework collapse; and (3) limited control over selective extraction and precise pore architecture engineering.

3.2.2.1 Demetallization

The demetallization method is a process that focuses on conventional pure microporous zeolites, whereby a component (metallic or semimetallic) is selectively extracted from the zeolite skeleton through a chemical extraction process. The process of demetallization typically necessitates the utilization of potent acids or bases [71–74, 76–80, 82, 93], or alternatively, steam [81, 83, 85, 86, 89]. In such instances, the application of steam or strong acids serves to extract metals, such as aluminum, whereas the use of strong bases or hydrofluoric acid (HF) is employed to remove silicon.

The steam treatment process was established in the mid-1960s for the purpose of dealuminizing zeolites. This process primarily involves the treatment of zeolite crystals in high-temperature water vapor environments (greater than 500 °C). Under these conditions, aluminum and silicon become more prone to migration. The process ultimately leads to the removal of aluminum from the zeolite framework, resulting in the formation of voids (hydroxyl clusters) or partial amorphization of the skeleton. Experimental evidence reveals vacancies partially filled by mobile silicon from amorphous domains, whereas unfilled sites evolve into mesoporous voids (Figure 3.7) [83, 85, 86]. The extraction of aluminum and the formation of mesopores depend on both the temperature and duration of the steam exposure [81]. On the one hand, the aluminum removed from the skeleton neutralizes the zeolite skeleton

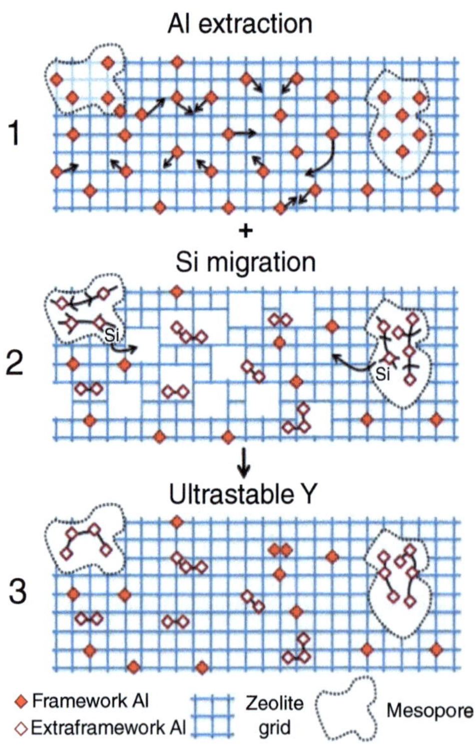

Figure 3.7 Schematic representation of intermediate pore formation during steam dealumination [83, 85, 86]. Copyright 2014 John Wiley & Sons, Inc.

charge, stabilizing the remaining aluminum within the framework, which can result in a limited amount of aluminum being extracted over a specific period. On the other hand, dealumination and silicon migration increase the Si/Al ratio in the framework while facilitating the formation of mesopores. The overall Si/Al ratio of the dealuminized zeolite remains relatively constant, as the micro- and mesoporous regions within the zeolite are often obscured by detritus and exo-aluminum species in the amorphous phase of the zeolite. Mild acids (e.g., dilute HNO_3 or weak organic acids such as oxalic acid) selectively dissolve extra-framework species without generating secondary mesopores. Lohse et al. achieved 10-nm mesopores in Y zeolites through steam treatment, though residual amorphous aluminosilicate fragments within these pores caused significant blockage. Subsequent leaching with HCl effectively removed these obstructions, expanding the mesopore diameter to 20 nm [89].

Barrer and Makki pioneered acid leaching by using mineral acids to extract aluminum from plagioclase-structured zeolites, establishing this method's foundation [71]. While effectively creating intracrystalline mesopores through selective aluminum removal in microporous zeolites, the technique inevitably generates extra-framework alumina residues. Sun and co-workers applied hydrothermal synthesis coupled with acid post-treatment to fabricate hierarchical SAPO-34 with a butterfly-bridged pore architecture [93]. Their comparative acid study revealed distinct impacts on pore–acid site synergy, demonstrating that nitric or oxalic acid treatment produces trimodal porosity: micropores, 40–50-nm mesopores, and macropores (62–500 nm). In contrast, treatment of SAPO-34 with butanedioic acid did not yield hierarchical SAPO-34. Furthermore, it was established that acid treatment effectively enhanced the number of acidic sites in SAPO-34. However, the aforementioned acid leaching process has been shown to lead to a decrease in the quantity of acidic sites. This notable disadvantage has been found to increase the framework Si/Al ratio and acid strength [76]. Although acid leaching results in the formation of mesopores, it is also associated with modifications to the number and strength of acid sites, complicating the separate study of the effects of these changes on catalytic properties.

Alkaline leaching is frequently utilized for the selective extraction of silica from zeolite structures in alkaline environments [72–74, 77–80, 82]. The removal of aluminum atoms from zeolites with elevated Si/Al ratios presents significant challenges, especially when acid leaching techniques are applied. In contrast, alkaline desilication has been shown to be a successful method for preparing hierarchical zeolites that are suitable for applications involving zeolites with high Si/Al ratios [82]. The initial publication highlighting alkaline leaching as a technique to improve adsorption capacity and catalytic activity was authored by Young. Following this, Mao et al. performed an extensive study on the desilication process [72]. Sodium carbonate was utilized as the leaching agent to extract silicon atoms from the framework, while sodium hydroxide was used to adjust the ambient pH, allowing for precise control of the desilication procedure [73, 74]. Ogura et al. revealed that treatment with NaOH resulted in the development of mesopores within the zeolite, thereby enhancing diffusion characteristics and catalytic performance [77, 78]. Subsequent studies by Groen et al. [79, 80] revealed monodisperse mesopores with a broad size distribution centered at ~10 nm [80].

HF serves as a key etchant for silica removal in zeolites. Valtchev et al. pioneered a fluoride leaching technique combining physical and chemical steps to generate well-defined secondary porosity [88]. Their method first employs high-energy U-ion bombardment to induce localized amorphization within the zeolite structure. Subsequent HF treatment selectively dissolves these less-ordered amorphous domains at a higher rate than the crystalline framework, thereby creating a network of macropores throughout the crystal [88]. Application of this modified material in *m*-xylene isomerization demonstrated improved catalytic activity and enhanced selectivity for para- and meta-isomers. An alternative approach developed by Qin et al. utilizes ammonium fluoride (NH_4F) as the etchant [91]. In this method, the bifluoride ion (HF_2^-), generated in situ via NH_4F hydrolysis, facilitates the extraction of both silicon and aluminum atoms from the zeolite framework. This NH_4F treatment enables uniform etching agent distribution within the crystal, with dissolution kinetics finely controllable through etchant concentration and temperature [91]. Compared to conventional dealumination (steam/acid) or desilication (alkali) methods, this non-selective NH_4F etching offers distinct advantages for fabricating hierarchical zeolites with tailored porosity, particularly effective for materials with an Si/Al ratio near unity. However, it's noteworthy that preexisting defects or non-zeolitic impurities within the parent material are preferentially etched due to their higher reactivity. This selective dissolution can potentially increase the population of Brønsted acid sites, as the etching process may expose or generate new acidic hydroxyl groups associated with the newly formed surfaces or defect sites.

3.2.2.2 Delamination

In contrast to metallization, which results in the formation of intracrystalline mesopores or macropores, delamination techniques facilitate the creation of intercrystalline mesopores or macropores in zeolites. The generation of hierarchical zeolites, which consist of aggregates of nanolayers, can be accomplished by reducing the dimensions of individual crystals from the micrometer scale to the nanometer scale in three dimensions [90]. The interactions among the layers of these hierarchical zeolites, including van der Waals interactions, hydrogen bonding, and electrostatic interactions, are relatively weak. Upon template removal, the precursors in hierarchical zeolites with a hierarchical structure typically undergo coagulation. At this stage, the structure or composition of the precursor must experience a post-treatment process, which may involve swelling, stripping [75, 99], or interlayer stacking with amorphous silica [92]. This is when the lamellar precursor generally receives further modifications, specifically the incorporation of surfactant molecules. For instance, during a conventional delamination process, the 2D zeolite precursor is first treated in an alkaline solution aimed at enlarging the interlayer voids, a process aided by the application of surfactants. The resultant material usually displays an increased degree of intergranular (interlayer) mesoporosity (Figure 3.8) [75]. The initial swelling process occurs at elevated concentrations of hydroxide (OH^-) and surfactant, in addition to increased temperatures. This method has been shown to reduce desilication and structural degradation [75]. Tsapatsis and co-workers [84] pioneered gentler protocols for lamellar precursor swelling at ambient temperature to minimize

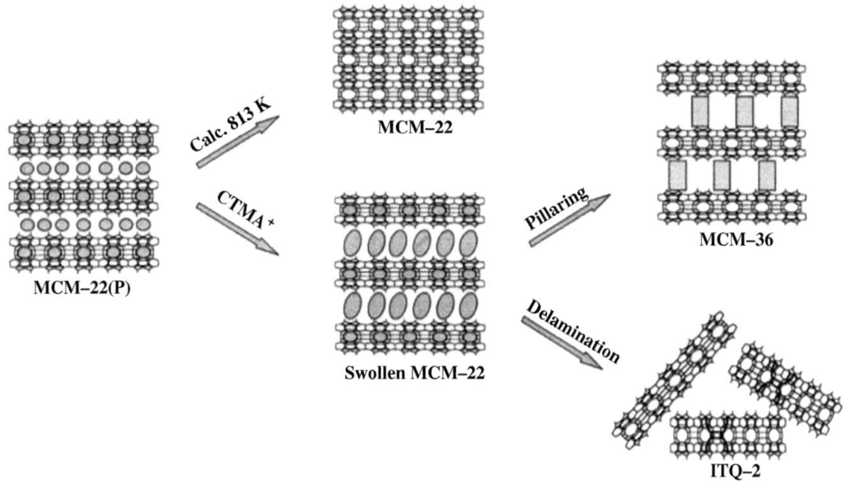

Figure 3.8 Schematic representation of the distribution of MCM-22(P) through an intermediate solvation step for delamination and column bracing to introduce intergranular (interlayer) mesopores in the layered zeolite precursor, facilitating the formation of ITQ-2 and MCM-36. *Source*: [75] / with permission of Elsevier.

structural degradation. Complementing this approach, Zones and co-workers formulated an innovative halide-based solubilizing system (F^-/Cl^-) effective under near-neutral conditions. This method significantly enhanced catalytic site accessibility while preventing framework damage typically associated with highly alkaline leaching treatments [87]. The formation of hierarchical zeolites with intergranular porosity and a hierarchical structure was accomplished through the ultrasonication of swollen precursors, specifically MCM-22(P) (MCM-22 precursor), PREFER, and Nu-6(1), leading to the production of two-dimensional zeolites, namely ITQ-2, ITQ-6, and ITQ-18 [75].

The use of delamination methods allows for the application of zeolite surfaces with high acidity for adsorption and catalysis. However, it should be noted that delamination methods are applicable only to a limited range of zeolites. Key challenges in zeolite delamination center on excessive surfactant requirements and the risk of structural degradation from alkaline dissolution during swelling.

3.2.2.3 Recrystallization

Goto et al. pioneered the recrystallization approach involving zeolite dissolution–recombination in 2002 [94], with Ivanova et al. subsequently reporting analogous methodology in 2004 [95]. The initial phase requires partial framework dissolution and elimination of structural debris, while subsequent recombination of dissolved species generates mesopores, macropores, or void architectures.

In a pioneering study, Dai et al. employed selective dissolution and surface recrystallization techniques to report the first instance of a hollow ZSM-5 with a bilayer shell structure [96]. A single layer of hollow ZSM-5 was selected as the seed and prepared by the addition of Silicalite-1 and ZSM-5 zeolites using the layer-by-layer method, as illustrated in Figure 3.9. Following TPAOH treatment, the Silicalite-1 layer underwent dissolution and recrystallization at the surface of the sample, resulting in the formation of a cavity situated centrally between the two ZSM-5 layers (hollow ZSM-5@void@ZSM-5). Additionally, the method can be employed for the preparation of metal-encapsulated hollow zeolite nanocrystals. The synthesis comprised two distinct stages. The synthesis commenced with the preparation of metal-containing zeolite hollow agglomerate nanocrystals via a template-free hydrothermal method. The metal-containing zeolites were subsequently converted into metal-encapsulated hollow zeolite crystals through a two-step process involving selective desilication–recrystallization and leaching–hydrolysis. The resulting materials possess microporous (0.4–0.6 nm) walls of hollow zeolite crystals and mesoporous (5–17 nm) shells of nanovesicles, with macropores of ~350 nm at the core (Figure 3.10) [97].

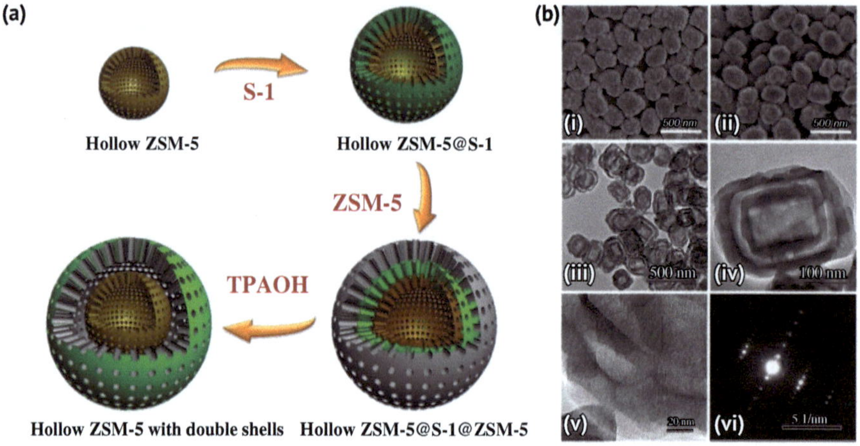

Figure 3.9 Synthetic process of hollow ZSM-5 with double shells. (a) Schematic diagram of the synthesis route of hollow ZSM-5@S-1. *Source*: [96] / with permission of John Wiley & Sons. (b) SEM and TEM of the hollow ZSM-5@S-1. *Source*: Reproduced from [96] / with permission of John Wiley & Sons.

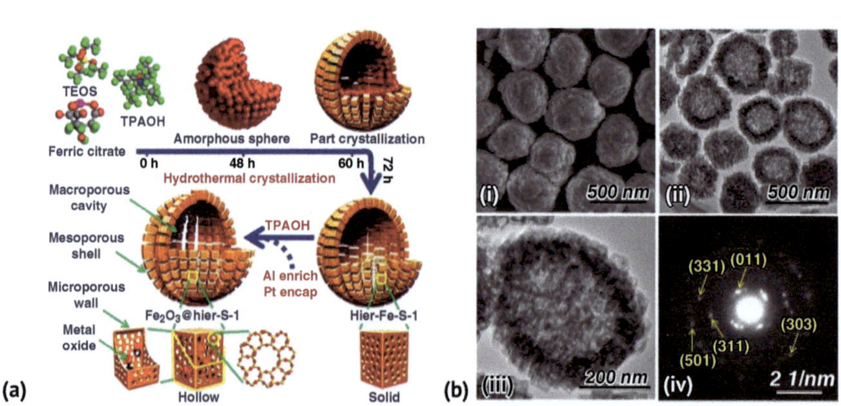

Figure 3.10 (a) Synthetic strategy for the hollow nanovesicle assembly with hollow metal-oxide-encapsulated MFI-type nanocrystal. (b) SEM and TEM of hollow nanovesicle assembly. *Source*: Reproduced from [97] / with permission of American Chemical Society.

3.3 Application of Hierarchical Zeolites in Low-carbon Catalytic Conversion

3.3.1 Fischer–Tropsch Synthesis

The Fischer–Tropsch synthesis (FTS) process is a widely utilized multiphase catalytic process that employs syngas ($CO + H_2$) derived from nonpetroleum feedstocks, including natural gas, coal, and biomass, to generate clean hydrocarbon fuels or chemicals [100]. The primary challenge currently facing FTS is the restriction of product distribution, which typically adheres to the Anderson–Schultz–Flory (ASF) distribution diagram [101]. The most extensively researched of these is the FTS catalyst, which is capable of synthesizing compounds that exhibit a high degree of C_{5+} selectivity [102]. Initially, cobalt nanoparticle (Co NP)-loaded zeolite catalysts were the preferred option because of their high selectivity for *n*-paraffins [102–104]. However, conventional microporous zeolites have been observed to result in higher methane and C_2–C_4 selectivity due to limitations in the diffusion process [105]. It is, therefore, thought that hierarchical zeolites, which increase the selectivity of long-chain hydrocarbons, represent a promising class of carriers for bifunctional FTS catalysts. In addition, the sintering of metal nanoparticles is a major cause of deactivation of cobalt-based FTS catalysts [106, 107]. Maximizing the dispersion of active components across the carrier surface inhibits metal sintering [108]. Sartipi et al. reported that hierarchical mesoporous MFI zeolites have a large external surface area, which enables them to load a substantial number of Co NPs and enhances the participation efficiency of the reactive centers, thereby resulting in high catalytic activity in the FTS process [109, 110]. Additionally, graded porous ultra-stable Y (USY) zeolites have been employed as catalytic carriers for FTS [111, 112]. The utilization of mesoporous H–USY zeolite as a carrier material for Co NPs has the potential to increase their catalytic activity [111]. Compared with microporous zeolites, the larger specific surface area allows for the loading of a greater number of Co NPs, which in turn adds more active sites. Furthermore, the hierarchical structure has the effect of reducing the selectivity of the products below C_4, thus improving both the activity and selectivity properties of the reaction [113]. The advantage of mesopores in FTS was further reported by Peng et al. Co-loaded mesoporous Y zeolite showed better reaction conversion than Co-loaded microporous Y zeolite. The MCM-22 zeolite is a layered zeolite with an MWW structure, consisting of layers connected by oxygen bridges along the *c*-axis, with two independent multidimensional pore systems [112]. In addition, its special sheet morphology yields a high external surface area [114, 115]. This micro-mesoporous material, possessing adjustable acid–base properties, represents a well-suited support for Co NPs [116]. Compared with the Co/SiO_2 samples, the Co/MCM-22 samples presented smaller and more uniform Co nanoclusters, which utilized amorphous SiO_2 as a carrier. The Co/MCM-22 system exhibited the highest selectivity for long-chain hydrocarbons (C_{5+}) when utilizing MCM-22 carriers with an optimal Si/Al ratio. The enhanced C_{5+} selectivity is attributed to the distinctive characteristics of Co NPs, indicating that the structural configuration of MCM-22 exerts a pivotal influence on the determination and regulation of selectivity in FTS [116].

Han et al. achieved substantial Co NP loading on hierarchically porous ZSM-5 nanosponges (Co/MFI-NS) and Beta zeolite nanosponges (Co/Beta-NS), where confined growth produced extended Co nanowires/networks within mesochannels (Figure 3.11a–c). FTS performance evaluation revealed reaction conversion dependency on cobalt loading (Figure 3.11d). Crucially, both Co/zeolite-NS catalysts exhibited enhanced CO conversion relative to amorphous Co/MCM-41 controls at equivalent metal loadings, demonstrating the superior efficacy of crystalline zeolite frameworks as cobalt supports [117].

The FTS reaction kinetics are governed by catalyst architecture. Optimal catalytic activity is achieved when Co NPs maintain diameters between 4 and 6 nm. Hierarchical zeolites with ordered porosity, high surface areas, and monodisperse pore dimensions serve as exceptional supports—their mesoconfined structures enhance cobalt dispersion and regulate product distributions through spatial confinement effects. The introduced meso-/macropores further facilitate reactant diffusion and active site accessibility. Consequently, these hierarchically structured catalysts exhibit significantly improved activity and selectivity compared to purely microporous analogs. This establishes zeolites integrating high surface areas, uniform mesopores, and interconnected macroporosity as strategic catalytic architectures for advanced FTS processes.

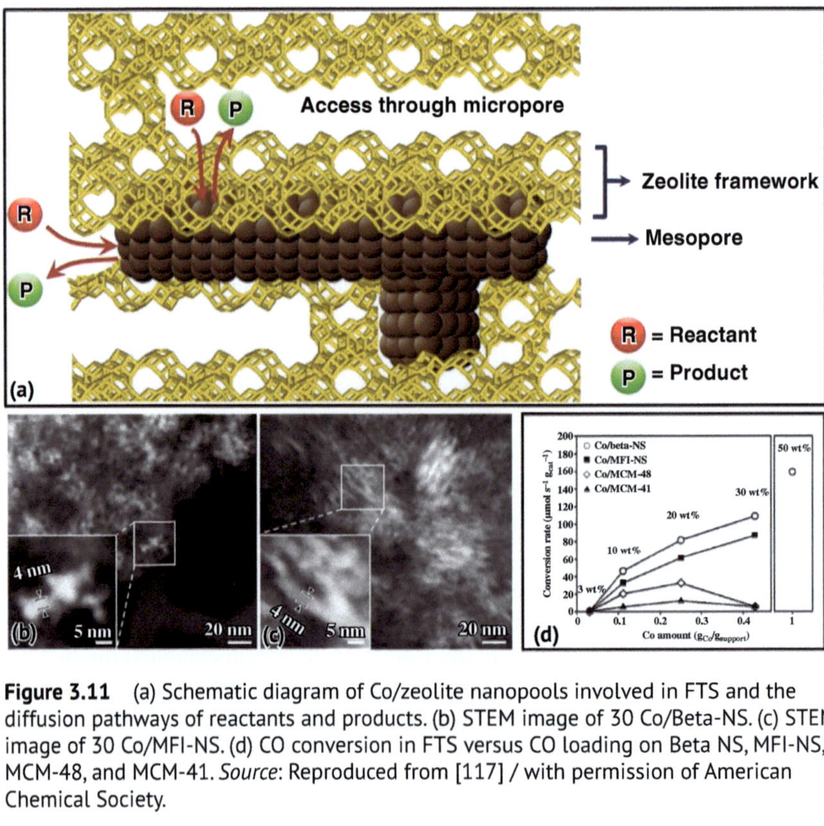

Figure 3.11 (a) Schematic diagram of Co/zeolite nanopools involved in FTS and the diffusion pathways of reactants and products. (b) STEM image of 30 Co/Beta-NS. (c) STEM image of 30 Co/MFI-NS. (d) CO conversion in FTS versus CO loading on Beta NS, MFI-NS, MCM-48, and MCM-41. *Source*: Reproduced from [117] / with permission of American Chemical Society.

3.3.2 Methanol Conversion

Against the backdrop of escalating global energy demand and depleting crude oil reserves, the pursuit of alternative energy sources has intensified. Syngas conversion—derived from coal [118, 119], natural gas [120], or biomass [121, 122] via gasification/reforming—has emerged as a strategically vital pathway for sustainable fuel and petrochemical feedstock production [123]. Complementing FTS, methanol-to-hydrocarbons (MTH) technology represents a commercially mature approach for valorizing nonpetroleum C_1 resources [124]. This process enables synthesis of critical petrochemical intermediates, particularly olefin monomers and polymers, directly from methanol [125].

Zeolite-catalyzed MTH systems demonstrate superior product selectivity over FTS [126], with pioneering work by Chang and Silvestri establishing early zeolite applications [127]. Reaction dynamics are governed by zeolite pore architecture:

- Mesoporous/macroporous frameworks selectively yield C_5–C_7 aromatics/paraffins.
- Microporous topologies (ERI, LTT, CHA, and KFI) favor C_2–C_4 olefin production [128, 129].

Catalyst-mediated pathway modulation allows targeted synthesis of:

- gasoline-range hydrocarbons (MTG [methanol to gasoline]) [130–146];
- light olefins (MTO [methanol to olefins]) [5, 10, 15, 16, 26, 28, 31–33, 48, 49, 53, 57, 58, 81, 85, 88, 98, 100, 103, 106, 109–111, 121, 123, 126, 128, 130, 138, 147–159];
- propylene (C_3H_6) (MTP [methanol to propylene]) [150, 156, 160–174];
- aromatics (MTA [methanol to aromatics]) [147, 148, 151, 157, 175–196];
- dimethyl ether (DME) via dehydration (MTD [methanol dehydration to dimethyl ether]) [197–200].

3.3.2.1 Methanol-to-gasoline Hydrocarbons

Initially developed by ExxonMobil, the zeolite-catalyzed MTG process converts methanol into high-octane gasoline [136]. While various zeolites have been screened for gasoline synthesis, ZSM-5 demonstrates superior catalytic performance for premium-grade fuel production [137, 138, 146]. Its three-dimensional micropore network facilitates efficient diffusion of gasoline-range hydrocarbons [130]. However, this process suffers from two critical limitations: (1) significant co-production of environmentally persistent aromatic compounds, particularly 1,2,4,5-tetramethylbenzene (durene) [133, 145]; and (2) severe coking-induced deactivation necessitating frequent regeneration cycles [134]. To mitigate the generation of aromatics, new MTG zeolite catalysts have been developed. Di et al. [141] devised a two-step crystallization process for synthesizing ZSM-5/MCM-48 aluminum silicate composite zeolites. This method involved the initial incorporation of pre-crystallized colloidal zeolites into the MCM-48 synthesis system. Post-secondary crystallization established interconnected ZSM-5/MCM-48 heterostructures with phase-integrated dielectric properties. The resulting catalysts demonstrated high catalytic activity and enhanced selectivity for C_5–C_{10} hydrocarbons. Furthermore, the improved transfer performance of the microporous ZSM-5/MCM-48 composite zeolite with a hierarchical structure also contributed to inhibiting the formation of aromatic hydrocarbons and reducing coke deposition.

Severe diffusion limitations of reactants and products within zeolite micropores are the primary cause of short catalyst life [139]. Enhanced diffusion facilitated by hierarchical porosity is pivotal for extending catalyst longevity [140]. Hydrothermal synthesis offers a route to hierarchical ZSM-5 zeolites without requiring extra templates or seed crystals [143]. Crucially, for optimal mass transfer efficiency, the introduced mesopores must form an interconnected network accessible directly from the crystal exterior. Fathi et al. [142] demonstrated this principle by modifying HZSM-5 via post-treatment with NaOH, Na_2CO_3, and $CaCO_3$ solutions. These treatments successfully generated intracrystalline mesopores, significantly boosting the catalytic performance in MTH conversion, particularly for gasoline-range hydrocarbon (C_{5+}) production. Notably, Na_2CO_3 treatment yielded the most favorable acidity profile and superior performance, increasing C_{5+} yield by 43% after four hours and extending catalyst lifespan by approximately 11.5% compared to the parent zeolite. The beneficial impact of mesoporosity extends beyond ZSM-5. Liu et al. [144] comprehensively studied mesoporous Beta zeolites in MTH reactions. Their findings revealed that mesoporous Beta exhibited significantly higher selectivity toward C_4–C_7 aliphatic hydrocarbons than conventional Beta across a wide conversion range (20–100%). This improvement stems from the hierarchical pore structure, ensuring full reactant access to active sites throughout the crystal. Unlike conventional Beta, where rapid coke deposition on the external surface blocks access to internal sites, the hierarchical Beta structure minimizes methylbenzene residence time within micropores, effectively suppressing the aromatic-based catalytic cycle responsible for deactivation.

Choi et al. [135] fabricated hierarchical MFI zeolites comprising 2-nm-thick nanosheets, matching the *b*-axis dimension of a single MFI unit cell (Figure 3.12a). During MTG conversion, these hierarchical nanosheet assemblies delivered substantially prolonged catalyst lifetime and dramatically reduced deactivation kinetics relative to conventional MFI counterparts. Quantitative analysis revealed markedly suppressed coking rates in hierarchical structures (45-mg·g^{-1} zeolite versus 170-mg·g^{-1} zeolite over 120 hours). Coke formation was observed to occur primarily on the surface of the mesopores in the graded MFI zeolite nanosheets, while in the conventional MFI zeolite, coke formation was largely confined to the micropores (Figure 3.12b, c) [131, 132]. The improved longevity of the catalyst in the MFI nanosheets was attributed to the optimized mass transfer of coke precursors from the zeolite micropores outward, which led to slower coke deposition on the outer surface and prevented the clogging of the micropores.

3.3.2.2 Methanol to Olefins

Low-carbon olefins, such as ethylene (C_2H_4) and C_3H_6, serve as significant chemical feedstocks across various industrial sectors. Conventional methods for producing light olefins include thermal cracking of crude oil, catalytic cracking, and alkane

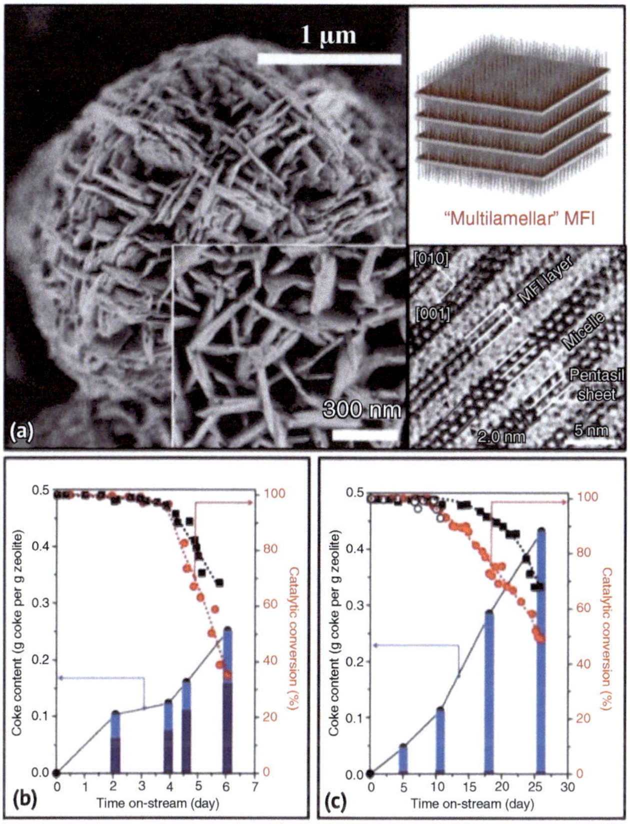

Figure 3.12 (a) SEM images of monolayer MFI molecular sieves with a flake morphology. (b) Coke deposition and catalytic conversion of conventional MFI zeolite. (c) Monolayer MFI zeolite for methanol–gasoline conversion. *Source*: [135] / with permission of Springer Nature.

dehydrogenation. However, the limited supply of crude oil necessitates the urgent development of alternative technologies [201]. The MTO process represents a successful non-petrochemical route for producing light olefins from abundant nonpetroleum resources, bridging the gap between the non-petrochemical and modern petrochemical industries [149]. The MTO process demonstrates superior economic advantages over the conventional naphtha cracking process. In recent years, extensive research has focused on various types of zeolites, including MFI, CHA, MOR, and other topologies. The SAPO-34 in CHA zeolite features larger pore cavity (9.4 Å in diameter) and smaller eight-membered ring micropores (3.8 × 3.8 Å) compared to the ZSM-5 in MFI zeolite. Moreover, the SAPO-34 in CHA zeolite exhibits enhanced catalytic performance in the MTO reaction due to its unique framework, achieving up to 100-wt.% conversion of methanol and over 90% selectivity of C_2–C_4 light olefins [202–206]. However, the zeolite-catalyzed reaction is hindered by slow mass transfer in the catalytic center, increasing the likelihood of secondary reactions that ultimately result in coke formation. Additionally, coke deposition obscures the active sites of the catalyst [207], leading to reduced catalytic lifetime and decreased selectivity for light olefins in the MTO reaction [208, 209].

Significant efforts have been made to enhance the diffusion performance of SAPO-34 catalysts. Previous research has shown that SAPO-34 zeolites with a smaller crystallite size provide a notable advantage in reducing the rate of coke formation and improving mass transfer performance [155, 210–212]. The reduced crystal size of nanosized SAPO-34 significantly enhances methanol conversion efficiency and extends catalyst lifetime in the MTO process [213–215]. This performance improvement stems from shortened diffusion paths within the nanocrystals, which mitigate coke formation and consequently prolong catalyst operation [216]. Further enhancement is achieved through hierarchical porosity. For instance, Sun et al. [154] synthesized hierarchical porous SAPO-34 nanocrystals via a one-step hydrothermal method using the organosilane surfactant TPOAC as a mesopore template (Figure 3.13a–c). This catalyst demonstrated a fourfold increase in operational lifetime and over 10% higher selectivity toward light olefins (C_2H_4 + C_3H_6) compared to conventional microporous SAPO-34 (Figure 3.13d) [154]. Similarly, Wang et al. [217] prepared a hierarchically ordered mesoporous SAPO-34 (3DOm-i SAPO-34) by hydrothermal synthesis within a confined OMMC template. The resulting material, featuring an ordered mesoporous structure imprinted by spherical zeolite

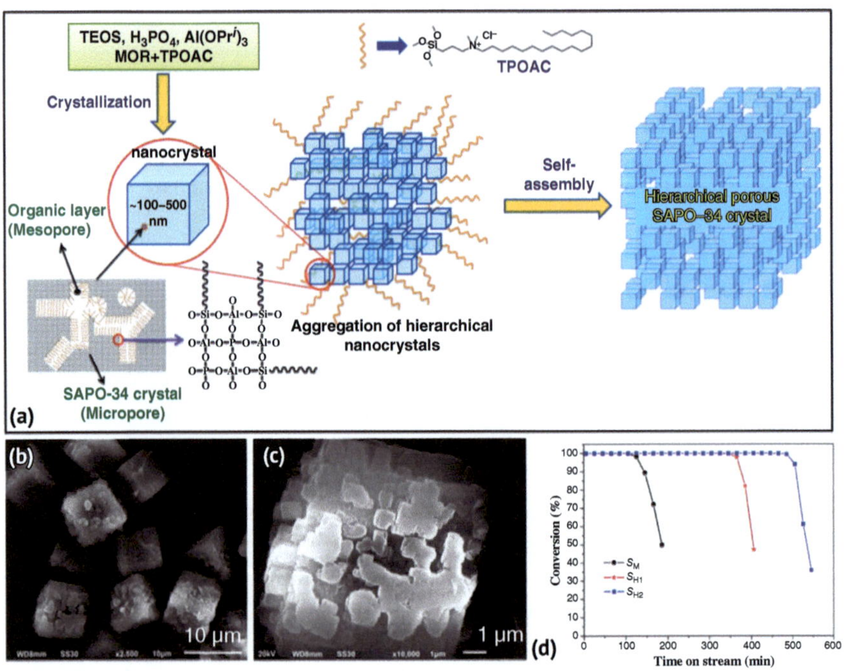

Figure 3.13 (a) Schematic synthesis of cubic aggregates of SAPO-34 nanocrystals. (b, c) SEM images of hierarchical porous SAPO-34. (d) Variation in methanol conversion over conventional microporous SAPO-34 (SM) and hierarchical porous SAPO-34 catalysts with run time. *Source*: (a, d) [154] / with permission of Royal Society of Chemistry. (b, c) Reproduced from [154] / with permission of Royal Society of Chemistry.

nanocrystals, also exhibited superior catalyst longevity and light olefin selectivity relative to its microporous counterpart [217].

Introducing meso- and/or macropores into microporous SAPO-34 significantly enhances molecular diffusion kinetics, suppresses coke deposition, and improves catalyst operational stability [218]. Various strategies under the templating approach have successfully yielded hierarchical SAPO-34 materials encompassing micro- to macroporosity [154, 219]. For instance, Sun et al. [152] employed a one-step nanogel-assisted hydrothermal method using solely tetraethylammonium hydroxide (TEAOH) to fabricate hollow-structured hierarchical SAPO-34. Catalytic evaluation in the MTO reaction revealed substantially improved performance compared to conventional microporous SAPO-34, manifested by a quadrupling of catalyst lifetime and a greater than 10% increase in C_2H_4 and C_3H_6 selectivity. Alternatively, Liu et al. [220] developed a mild post-synthetic etching technique employing TEAOH to generate graded porous SAPO-34. The resulting material contained inherent micropores alongside introduced mesopores (~3.5 nm) and macropores (~100 nm). This hierarchical architecture extended the single-run catalyst life in MTO while preserving the characteristic high methanol conversion and light olefin selectivity. Furthermore, Chen et al. [221] applied fluoride-based post-treatment to create hierarchical SAPO-34. This method preferentially etches highly defective, silicon-rich regions at crystalline domain interfaces, generating mesopores aligned with the crystallographic planes. These interconnected mesopores enhance overall micropore accessibility, leading to superior light olefin selectivity in MTO. The fluoride etching approach also proves versatile for further optimizing hierarchical SAPO-34 catalysts with enhanced MTO performance [222, 223].

Interconnected meso-/macropores in hierarchical SAPO-34 substantially modulate catalytic functionality. Schmidt et al. [57] introduced mass transfer pores in SAPO-34 using carbon nanoparticles and carbon nanotubes as auxiliary templates. The carbon nanotube template material (M/T-3) exhibited mesopores accessible from the external surface of the zeolite. In contrast, the mesopores created by the carbon nanoparticles (M/T-2) were located inside the zeolite and did not form a three-dimensional interconnected mesopore network. After 120 minutes, the methanol conversion of the parent and M/T-2 samples decreased dramatically, indicating a tendency toward deactivation. Conversely, M/T-3 achieved complete conversion of methanol and DME after about 300 minutes, suggesting that the incorporation of mesopores on the external surface of SAPO-34 can significantly enhance catalyst lifetime. Reports have also indicated that the catalytic performance of hierarchical porous SAPO-34 catalysts with open mesopores is improved [158, 159, 224]. Milina et al. [225, 226] established critical relationships between mesopore architecture (dimension/distribution/interconnectivity) and MTO catalyst longevity. Their analysis confirmed that fully interconnected hierarchical networks—not isolated mesopores—are essential for prolonged operational stability. These findings position intracrystalline pore connectivity as a vital longevity determinant, underscoring the need for precision engineering of hierarchical structures. In a parallel work, Sun et al. [153] pioneered the synthesis of trimodal (micro-, meso-, and macro-) SAPO-34 zeolites. This novel architecture delivered 6× extended catalyst life and near-5% higher C_2H_4/C_3H_6 selectivity during MTO conversion versus conventional analogs.

In addition to the SAPO-34 catalyst, other zeolites, such as ZSM-5 [227], SSZ-13 [228], ZSM-22 [229], ZSM-58 [230], SAPO-5 [7], SAPO-11 [231, 232], SAPO-18 [232], and SAPO-35 [231], have also been documented for use in the MTO process.

ZSM-5 has garnered significant interest in methanol conversion due to its unique intersecting two-dimensional sinusoidal (5.1 × 5.5Å) and straight (5.3 × 5.6Å) apertures. However, the diffusion limitations associated with single micropores have driven the development of strategies to introduce additional meso- and macropores into ZSM-5, offering a promising solution for the MTO reaction. Recently, Zhang et al. [227] introduced an innovative preparation strategy for hierarchical porous ZSM-5 microspheres, which extended the MTO reaction lifetime to 47 hours (methanol conversion >99%) compared to conventional nano-ZSM-5 and microporous ZSM-5. This method utilized spray freeze-drying (SFD) to treat the precursor gel, forming amorphous aluminosilicate microspheres. The incorporation of TPAOH facilitated the creation of an interconnected macroporous network shaped by ice. These microspheres were subsequently transformed into SFD-ZSM-5 zeolite microspheres with a hierarchical structure through steam-assisted crystallization (SAC). Experimental findings demonstrated that the hierarchical SFD-ZSM-5 exhibited superior anti-coking properties, with a catalyst lifetime extended by 290% (16–47 hours) compared to HT-ZSM-5, while achieving a C_2–C_4 olefin selectivity of 44.4%.

3.3.2.3 Methanol to Propylene

C_3H_6 serves as a vital feedstock for producing polymers and chemical intermediates like polypropylene, acrylonitrile, propylene oxide, and oxo alcohols [171]. The two primary commercial methods for C_3H_6 production are thermal naphtha cracking and petroleum refining through fluid catalytic cracking units. Both methods rely on crude oil resources [161]. Given the increasing demand for C_3H_6 and the rapid depletion of oil reserves, it is crucial to develop new processes that can achieve high C_3H_6 yields [169]. The MTP process serves as a promising alternative for C_3H_6 production. Therefore, the development of catalysts that can efficiently generate C_3H_6 is of significant importance.

The development of an optimal catalyst for the MTP reaction is dependent on achieving stability, high conversion of methanol, and selectivity of C_3H_6 [170]. The application of ZSM-5 zeolite as a catalyst generally leads to low C_3H_6 selectivity, accompanied by rapid catalyst deactivation due to the production of substantial amounts of coke [165]. To improve selectivity of C_3H_6 and catalyst longevity compared to a ZSM-5 catalyst, extensive research has been undertaken on the MTP reaction [160, 163, 164, 173]. Rownaghi et al. [164] indicated that nano-zeolite crystals demonstrated enhanced catalytic activity, stability, and selectivity for C_3H_6 relative to microporous powdered zeolites. This improvement was attributed to the smaller size of the zeolite crystals, which increased mass transfer efficiency and thereby accelerated the overall reaction rate. It was suggested that diffusion limitations among zeolites significantly influence catalytic activity [160]. The short-chain organosilane 3-aminopropyltrimethoxysilane (APTES) has been shown to undergo pronounced self-condensation on the surface of nanocrystals, allowing for the synthesis of hierarchical mesoporous zeolite nanocrystals [173]. APTES was utilized to direct the formation of zeolite nanocrystals and ensure their assembly in situ into the micro- to meso-macroporous structure of the hierarchical layers. The resulting catalyst lifetimes were significantly improved, exhibiting nearly an eightfold increase compared to pure microporous ZSM-5 catalysts. Additionally, the catalysts demonstrated a higher C_3H_6 selectivity of 41.89%. Interconnected macro-mesoporous networks accelerate mass transfer kinetics and enhance active site availability.

High-mesoporosity ZSM-5 demonstrated exceptional C_3H_6 selectivity (>42.2%), prolonged operational stability, and enhanced coking resistance in MTP conversions [150, 166–168, 172]. These findings corroborate that zeolite mesopores significantly enhance molecular diffusion while suppressing coke accumulation [156]. To probe the pore connectivity effects, Mei et al. [162] engineered graded ZSM-5 with architecturally distinct mesopores via desilication/starch templating:

- Open mesopores delivered 42.2% C_3H_6 selectivity and 10:1 P/E ratio.
- Closed mesopores showed negligible performance gains due to random intracrystalline distribution.

This stark contrast establishes interconnected mesochannels as an indispensable design requirement for optimized MTP catalysis. Jin et al. recently synthesized HZSM-5 zeolite with a mesoporous-rich structure using a dissolution–recrystallization method [174]. Unlike the conventional alkali treatment process, which detaches silica from the zeolite surface and leaves numerous acid centers exposed, inhibiting the shape selectivity of micropores, this method produced T-HZSM-5 zeolites with abundant mesoporous structures, enhanced crystal accessibility, and a silica-rich external surface. In the MTP reaction, the parent HZSM-5 exhibited deactivation after 20 hours of reaction, whereas the methanol conversion of T-HZSM-5 remained above 90% after 52 hours. Furthermore, T-HZSM-5 demonstrated superior C_3H_6 selectivity compared to the parent HZSM-5. This improvement is attributed to its higher weak acid content, which enhances olefin cycling, and the silica-rich external surface, which suppresses side reactions involving C_3H_6. Additionally, the extensive mesopore network of T-HZSM-5 reduces the rate of coke deposition, thereby increasing the catalyst's stability and prolonging its operational lifespan.

3.3.2.4 Methanol to Aromatics

Aromatics serve as a crucial raw material for the polymer and organic industries [175]. Benzene, toluene, and xylene (collectively referred to as BTX) are the most commonly utilized aromatics [176]. Growing global aromatics demand is primarily met through petrochemical routes like catalytic reforming and cracked gasoline hydrogenation, sourced from crude oil [189]. The MTA process, employing microporous solid acid catalysts, offers a viable nonpetroleum alternative for producing valuable chemicals. In 1993, Mo–ZSM-5 was introduced as a highly efficient catalyst for the MTA reaction [177]. Following this, the reaction has undergone extensive investigation. The interaction and hydrogenation of surface metal oxides [182] (e.g., ZnO [183], Ga_2O_3, and SnO_2 [184]) with zeolites have been shown to enhance the aromatization of methanol over ZSM-5.

The performance of catalysts in the MTA process is considerably affected by several factors, including acidic strength [147], pore structure [179, 180], crystal size [192, 194], and morphology [148]. Consequently, synthesizing catalysts combining optimized acidity with enhanced diffusivity is paramount for MTA reaction efficiency. Integrating secondary porosity into zeolite frameworks represents a viable strategy to alleviate diffusion constraints. This structural modification substantially mitigates mass transfer barriers for polyaromatics [151, 181, 185–187, 191, 196], while concurrently suppressing coke accumulation within micropores during MTA conversions [188]. Wang et al. [193] successfully synthesized a hierarchical ZSM-5 with a hollow structure, composed of an aggregation of small coffin-like ZSM-5 crystals (Figure 3.14a, b). The synthesized hierarchical ZSM-5 with a hollow structure (ZSM-5-h [38]) exhibited a catalyst lifetime of up to 186 hours, which is 1.69 times longer than that of the conventional ZSM-5 zeolite (ZSM-5-c1) (Figure 3.14c). This improvement is attributed to a markedly lower coking rate compared to ZSM-5-c1 (0.0061 versus 0.0096 $mg{\cdot}h^{-1}$). Additionally, it demonstrated higher aromatic selectivity (40.1%) than ZSM-5-c1 (33.9%), achieved by inhibiting the generation of C_{5+} aliphatic hydrocarbons. Wang et al. [190] demonstrated the successful synthesis of core–shell MFI zeolite microspheres, featuring a micrometer-scale outer layer and densely packed nanocrystalline domains with vertically aligned channels. This catalyst exhibited enhanced methanol conversion and an extended operational lifespan, underscoring the advantages

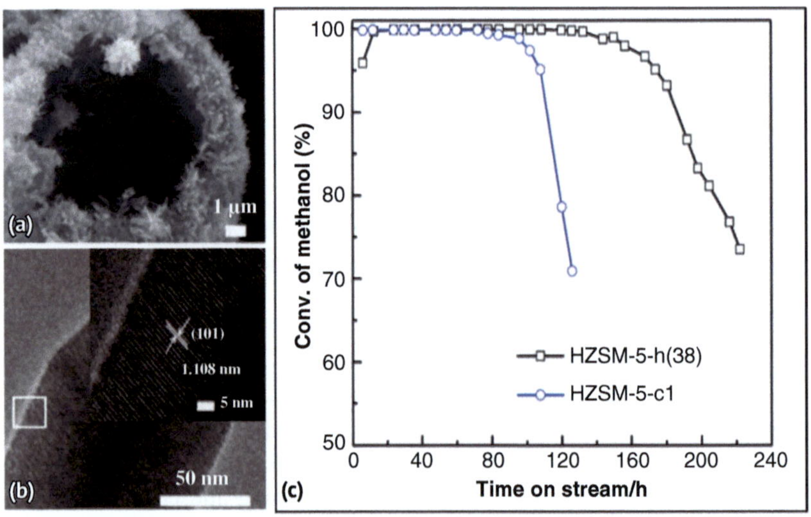

Figure 3.14 (a) SEM image of HZSM-5-h(38). (b) TEM image of a single nanocrystal cross-section. *Source*: Reproduced from [193] / with permission of The Royal Society of Chemistry. (c) Methanol conversion and running times of different ZSM-5 zeolites. *Source*: [193] / with the permission of Royal Society of Chemistry.

of the hierarchical porous structure over pure micro- and nanocrystals in heterogeneous catalytic processes. Zhou et al. [195] engineered MFI zeolites with tunable mesopore architectures and acid properties, methodically correlating these parameters with MTA catalytic performance. The researchers found that desilication of the zeolites led to a significant increase in mesopore volume and a decrease in the density of strong acids. The resulting mesoporous catalysts have been shown to effectively inhibit coke formation and enhance catalytic stability during the MTA reaction. Tian et al. [157] synthesized hierarchical ZSM-5 zeolites by employing block copolymers as scaffolds to transform zeolite and mesoporous copolymer precursors into homogeneous gels. The persistence of the dry gel during the hydrothermal synthesis of Z5/F127, prepared with F127 block copolymer as a soft template, was found to be critical in pore formation during skeleton crystallization. This process significantly influenced the distribution of active sites and the anti-deactivation properties of the hierarchical ZSM-5 molecular sieves. Additionally, Z5/F127 exhibited a large specific surface area, enhancing the interaction between the catalyst and reactants in the MTA reaction by increasing the contact area with the active centers. The increase in L-acid centers in the zeolite further improved the selectivity for BTX. Compared with conventional microporous zeolites, mesoporous Z5/F127 demonstrated superior catalytic activity and substantially enhanced anti-deactivation performance in MTA reactions.

3.3.2.5 Methanol Dehydration to Dimethyl Ether

In general, the MTD process requires a significantly lower dehydration temperature for methanol compared to the MTG, MTO, and MTP processes. DME serves as an intermediate product in the MTH reaction, which takes place at elevated temperatures. Under these conditions, DME reacts with methanol and water to produce olefins, paraffins, aromatics, and other chemicals [199, 200]. Beyond reaction temperature, catalyst structural parameters—including hierarchical porosity, crystal dimensions, and morphology—significantly govern MTD reaction efficiency. ZSM-5 remains the predominant MTD catalyst system. Rownaghi et al. [197] systematically evaluated the crystallite size and mesopore effects on MTD performance across mesoporous ZSM-5 variants. Their results demonstrated maximized catalytic activity and DME yield in homogeneous nanocrystalline catalysts due to diminished mass transfer resistance, establishing reduced zeolite crystallite size as an effective MTD activity enhancement strategy. In the MTD process, hierarchical porous ZSM-5 with intracrystalline mesopores demonstrated extended catalyst lifespan and 100% selectivity for DME. Furthermore, the coke tolerance of ZSM-5 was significantly improved, effectively prolonging the catalyst lifespan [198].

Optimal zeolite catalysts for methanol reforming exhibit sustained high conversion and product selectivity. Crucially, conventional microporous frameworks suffer from limited reactant/product transport—the primary constraint on catalyst longevity. Employing hierarchical architectures mitigates intracrystalline diffusion barriers, accelerating mass transfer to active sites. This structural optimization enhances catalytic efficiency, suppresses side reactions, minimizes coking, and prolongs operational stability.

3.4 Conclusion and Future Prospects

The field of hierarchical zeolites has undergone substantial growth in recent decades. To address the limitations of molecular diffusion within the micropores of conventional molecular sieves, researchers have developed numerous innovative methodologies. This paper categorized these methods into two distinct groups: template methods and post-processing methods. In the template method, the use of a template enhances interaction between the template and the silicate or aluminosilicate species present in the zeolite synthesis gel. The synthesis method employed determines the formation of various mesopores and macropores. It is evident that achieving improved catalytic performance in zeolites requires identifying an optimal balance between the number, location, size distribution, and interconnectivity of the introduced mesopores, intrinsic micropores, and additional pores. Among the post-treatment methods, demetallization technology has been utilized for about half a century. Dealumination was initially developed and later adopted widely in the industrial sector. However, the removal of aluminum from the framework led to a decrease in the number of acid sites. Desilication represents another post-processing method that creates hierarchical pores through the dissolution of silicon within the zeolite framework. Additional methods, such as delamination techniques, expose zeolite surfaces with higher acidity, thereby enhancing the potential for adsorption and catalysis. Despite existing synthetic routes for hierarchical zeolites, fabricating hydrothermally stable variants and rationally designing inter-pore synergies remain fundamental challenges.

Hierarchical pore architectures in zeolites substantially boost catalytic efficacy by accelerating molecular diffusion (>40%), shortening reactant transport paths, and maximizing active site accessibility—collectively enhancing reaction performance. For low-carbon catalytic conversions (e.g., FTS and methanol upgrading), hierarchical zeolites outperform microporous counterparts with 30–80% higher activity, 20–60% enhanced selectivity, and 2–5× prolonged longevity due to accelerated molecular transport and truncated diffusion pathways.

References

1 Chen, L.H., Sun, M.H., Wang, Z. et al. (2020). Hierarchically structured zeolites: from design to application. *Chem. Rev.* 120 (20): 11194–11294.
2 Dai, C., Du, K., Song, C. et al. (2020). Recent progress in synthesis and application of zeolite-encapsulated metal catalysts. *Adv. Catal.* 67: 91–133.
3 Chen, J.-L., Zhou, X.-X., and Shi, J.-L. (2018). Research progress on hierarchically porous zeolites: structural control, synthesis and catalytic applications. *J. Inorg. Mater.* 33 (2): 113–128.
4 Anderson, M.W., Holmes, S.M., Hanif, N. et al. (2000). Hierarchical pore structures through diatom zeolitization. *Angew. Chem. Int. Ed.* 39 (15): 2707–2710.

5 Schwieger, W., Machoke, A.G., Weissenberger, T. et al. (2016). Hierarchy concepts: classification and preparation strategies for zeolite containing materials with hierarchical porosity. *Chem. Soc. Rev.* 45 (12): 3353–3376.
6 Clauset, A., Moore, C., and Newman, M.E.J. (2008). Hierarchical structure and the prediction of missing links in networks. *Nature* 453 (7191): 98–101.
7 Bai, Y., Zeng, Q., Sun, J. et al. (2021). Synthesis of composite zeolites composed of SAPO-5 and SAPO-34 and its application in methanol dehydration to light olefins. *J. Porous Mater.* 28 (4): 1281–1289.
8 Pérez-Ramírez, J. (2012). Imagination has no limits. *Nat. Chem.* 4 (4): 250–251.
9 Okkels, F. and Bruus, H. (2007). Scaling behavior of optimally structured catalytic microfluidic reactors. *Phys. Rev. E: Stat. Nonlin. Soft Matter Phys.* 75: 016301.
10 Schmidt, I., Madsen, C., and Jacobsen, C.J.H. (2000). Confined space synthesis. A novel route to nanosized zeolites. *Inorg. Chem.* 39 (11): 2279–2283.
11 Petkovich, N.D. and Stein, A. (2013). Controlling macro- and mesostructures with hierarchical porosity through combined hard and soft templating. *Chem. Soc. Rev.* 42 (9): 3721–3739.
12 Kresge, C.T., Leonowicz, M.E., Roth, W.J. et al. (1992). Ordered mesoporous molecular sieves synthesized by a liquid-crystal template mechanism. *Nature* 359 (6397): 710–712.
13 Madsen, C., Madsen, C., and Jacobsen, C.J.H. (1999). Nanosized zeolite crystals—convenient control of crystal size distribution by confined space synthesis. *Chem. Commun.* 1999 (8): 673–674.
14 Jacobsen, C.J.H., Madsen, C., Houzvicka, J. et al. (2000). Mesoporous zeolite single crystals. *J. Am. Chem. Soc.* 122 (29): 7116–7117.
15 Schmidt, I., Krogh, A., Wienberg, K. et al. (2000). Catalytic epoxidation of alkenes with hydrogen peroxide over first mesoporous titanium-containing zeolite. *Chem. Commun.* 2000 (21): 2157–2158.
16 Schmidt, I., Boisen, A., Gustavsson, E. et al. (2001). Carbon nanotube templated growth of mesoporous zeolite single crystals. *Chem. Mater.* 13 (12): 4416–4418.
17 Kustova, M.Y., Hasselriis, P., and Christensen, C.H. (2004). Mesoporous MEL-type zeolite single crystal catalysts. *Catal. Lett.* 96 (3): 205–211.
18 Chou, Y.H., Cundy, C.S., Garforth, A.A. et al. (2006). Mesoporous ZSM-5 catalysts: preparation, characterisation and catalytic properties. Part I: Comparison of different synthesis routes. *Microporous Mesoporous Mater.* 89 (1): 78–87.
19 Wei, X. and Smirniotis, P.G. (2006). Synthesis and characterization of mesoporous ZSM-12 by using carbon particles. *Microporous Mesoporous Mater.* 89 (1): 170–178.
20 Egeblad, K., Kustova, M., Klitgaard, S.K. et al. (2007). Mesoporous zeolite and zeotype single crystals synthesized in fluoride media. *Microporous Mesoporous Mater.* 101 (1): 214–223.
21 Prasad, S. and Liu, S.-B. (1995). Generation of the precursor species in the synthesis of $AlPO_4$ molecular sieves. *Microporous Mater.* 4 (5): 391–394.
22 Holland, B.T., Abrams, L., and Stein, A. (1999). Dual templating of macroporous silicates with zeolitic microporous frameworks. *J. Am. Chem. Soc.* 121 (17): 4308–4309.

23 Huang, L., Wang, Z., Sun, J. et al. (2000). Fabrication of ordered porous structures by self-assembly of zeolite nanocrystals. *J. Am. Chem. Soc.* 122 (14): 3530–3531.

24 Wang, Y.J., Tang, Y., Ni, Z. et al. (2000). Synthesis of macroporous materials with zeolitic microporous frameworks by self-assembly of colloidal zeolites. *Chem. Lett.* 29 (5): 510–511.

25 Rhodes, K.H., Davis, S.A., Caruso, F. et al. (2000). Hierarchical assembly of zeolite nanoparticles into ordered macroporous monoliths using core–shell building blocks. *Chem. Mater.* 12 (10): 2832–2834.

26 Tosheva, L., Valtchev, V., and Sterte, J. (2000). Silicalite-1 containing microspheres prepared using shape-directing macro-templates. *Microporous Mesoporous Mater.* 35–36: 621–629.

27 Wang, X.D., Yang, W.L., Tang, Y. et al. (2000). Fabrication of hollow zeolite spheres. *Chem. Commun.* 2000 (21): 2161–2162.

28 Valtchev, V. and Mintova, S. (2001). Layer-by-layer preparation of zeolite coatings of nanosized crystals. *Microporous Mesoporous Mater.* 43 (1): 41–49.

29 Zhu, G., Qiu, S., Gao, F. et al. (2001). Template-assisted self-assembly of macro–micro bifunctional porous materials. *J. Mater. Chem.* 11 (6): 1687–1693.

30 Naydenov, V., Tosheva, L., and Sterte, J. (2002). Palladium-containing zeolite beta macrostructures prepared by resin macrotemplating. *Chem. Mater.* 14 (12): 4881–4885.

31 Valtchev, V. (2002). Silicalite-1 hollow spheres and bodies with a regular system of macrocavities. *Chem. Mater.* 14 (10): 4371–4377.

32 Valtchev, V. (2002). Preparation of regular macroporous structures built of intergrown silicalite-1 nanocrystals. *J. Mater. Chem.* 12 (6): 1914–1918.

33 Valtchev, V. (2002). Core–shell polystyrene/zeolite A microbeads. *Chem. Mater.* 14 (3): 956–958.

34 Yang, W., Wang, X., Tang, Y. et al. (2002). Layer-by-layer assembly of nanozeolite based on polymeric microsphere: zeolite coated sphere and hollow zeolite sphere. *J. Macromol. Sci., Part A* 39 (6): 509–526.

35 Naydenov, V., Tosheva, L., and Sterte, J. (2003). Vanadium modified AlPO-5 spheres through resin macrotemplating. *Microporous Mesoporous Mater.* 66 (2): 321–329.

36 Dong, A., Wang, Y., Tang, Y. et al. (2002). Hollow zeolite capsules: a novel approach for fabrication and guest encapsulation. *Chem. Mater.* 14 (8): 3217–3219.

37 Dong, A., Wang, Y., Tang, Y. et al. (2002). Mechanically stable zeolite monoliths with three-dimensional ordered macropores by the transformation of mesoporous silica spheres. *Adv. Mater.* 14 (20): 1506–1510.

38 Wang, Y. and Caruso, F. (2004). Macroporous zeolitic membrane bioreactors. *Adv. Funct. Mater.* 14 (10): 1012–1018.

39 Zhu, H., Liu, Z., Wang, Y. et al. (2008). Nanosized $CaCO_3$ as hard template for creation of intracrystal pores within silicalite-1 crystal. *Chem. Mater.* 20 (3): 1134–1139.

40 Ren, L., Wu, Q., Yang, C. et al. (2012). Solvent-free synthesis of zeolites from solid raw materials. *J. Am. Chem. Soc.* 134 (37): 15173–15176.

41 Zhang, B., Davis, S.A., Mendelson, N.H. et al. (2000). Bacterial templating of zeolite fibres with hierarchical structure. *Chem. Commun.* 2000 (9): 781–782.

42 Daw, R. (2002). Zeolites branch out. *Nature* 418 (6897): 491.

43 Dong, A., Wang, Y., Tang, Y. et al. (2002). Zeolitic tissue through wood cell templating. *Adv. Mater.* 14 (12): 926–929.

44 Wang, Y., Tang, Y., Dong, A. et al. (2002). Self-supporting porous zeolite membranes with sponge-like architecture and zeolitic microtubes. *Adv. Mater.* 14 (13–14): 994–997.

45 Zhang, B., Davis, S.A., and Mann, S. (2002). Starch gel templating of spongelike macroporous silicalite monoliths and mesoporous films. *Chem. Mater.* 14 (3): 1369–1375.

46 Cook, G., Timms, P.L., and Göltner-Spickermann, C. (2003). Exact replication of biological structures by chemical vapor deposition of silica. *Angew. Chem. Int. Ed.* 42 (5): 557–559.

47 Huang, L., Wang, H., Hayashi, C.Y. et al. (2003). Single-strand spider silk templating for the formation of hierarchically ordered hollow mesoporous silica fibers. *J. Mater. Chem.* 13 (4): 666–668.

48 Valtchev, V., Smaihi, M., Faust, A.-C. et al. (2003). Biomineral-silica-induced zeolitization of *Equisetum arvense*. *Angew. Chem. Int. Ed.* 42 (24): 2782–2785.

49 Valtchev, V.P., Smaihi, M., Faust, A.-C. et al. (2004). *Equisetum arvense* templating of zeolite beta macrostructures with hierarchical porosity. *Chem. Mater.* 16 (7): 1350–1355.

50 Katsuki, H., Furuta, S., Watari, T. et al. (2005). ZSM-5 zeolite/porous carbon composite: conventional- and microwave-hydrothermal synthesis from carbonized rice husk. *Microporous Mesoporous Mater.* 86 (1): 145–151.

51 Zampieri, A., Mabande, G.T.P., Selvam, T. et al. (2006). Biotemplating of *Luffa cylindrica* sponges to self-supporting hierarchical zeolite macrostructures for bio-inspired structured catalytic reactors. *Mater. Sci. Eng. C* 26 (1): 130–135.

52 Liu, W., Zeng, C., Zhang, L. et al. (2007). Facile and versatile preparation of silicalite-1 hollow structures using cotton threads as templates. *Mater. Chem. Phys.* 103 (2): 508–514.

53 Valtchev, V., Gao, F., and Tosheva, L. (2008). Porous materials *via* egg-constituents templating. *New J. Chem.* 32 (8): 1331–1337.

54 Wang, L., Yin, C., Shan, Z. et al. (2009). Bread-template synthesis of hierarchical mesoporous ZSM-5 zeolite with hydrothermally stable mesoporosity. *Colloids Surf. A: Physicochem. Eng. Asp.* 340 (1): 126–130.

55 Ocampo, F., Cunha, J.A., de Lima Santos, M.R. et al. (2010). Synthesis of zeolite crystals with unusual morphology: application in acid catalysis. *Appl. Catal. A: Gen.* 390 (1): 102–109.

56 de la Iglesia, Ó., Sánchez, J.L., and Coronas, J. (2011). Hierarchical silicalite-1 structures based on pyrolized materials. *Mater. Lett.* 65 (19): 3124–3127.

57 Schmidt, F., Paasch, S., Brunner, E. et al. (2012). Carbon templated SAPO-34 with improved adsorption kinetics and catalytic performance in the MTO-reaction. *Microporous Mesoporous Mater.* 164: 214–221.

58 Sun, M.-H., Chen, L.-H., Yu, S. et al. (2020). Micron-sized zeolite beta single crystals featuring intracrystal interconnected ordered macro-meso-microporosity displaying superior catalytic performance. *Angew. Chem. Int. Ed.* 59 (44): 19582–19591.

59 Zheng, M., Chen, Y., Liu, Z. et al. (2024). Hierarchically macroporous zeolite ZSM-5 microspheres for efficient catalysis. *Chem. Res. Chin. Univ.* 40 (4): 704–711.

60 Li, B., Kwok, K.M., and Zeng, H.C. (2021). Versatile hollow ZSM-5 nanoreactors loaded with tailorable metal catalysts for selective hydrogenation reactions. *ACS Appl. Mater. Interfaces* 13 (17): 20524–20538.

61 Wang, H. and Pinnavaia, T.J. (2006). MFI zeolite with small and uniform intracrystal mesopores. *Angew. Chem. Int. Ed.* 45 (45): 7603–7606.

62 Xiao, F.-S., Wang, L., Yin, C. et al. (2006). Catalytic properties of hierarchical mesoporous zeolites templated with a mixture of small organic ammonium salts and mesoscale cationic polymers. *Angew. Chem.* 118 (19): 3162–3165.

63 Liu, S., Cao, X., Li, L. et al. (2008). Preformed zeolite precursor route for synthesis of mesoporous X zeolite. *Colloids Surf. A: Physicochem. Eng. Asp.* 318 (1): 269–274.

64 Wang, L., Zhang, Z., Yin, C. et al. (2010). Hierarchical mesoporous zeolites with controllable mesoporosity templated from cationic polymers. *Microporous Mesoporous Mater.* 131 (1): 58–67.

65 Choi, M., Cho, H.S., Srivastava, R. et al. (2006). Amphiphilic organosilane-directed synthesis of crystalline zeolite with tunable mesoporosity. *Nat. Mater.* 5 (9): 718–723.

66 Chen, X., Huang, L., and Li, Q. (1997). Hydrothermal transformation and characterization of porous silica templated by surfactants. *J. Phys. Chem. B* 101 (42): 8460–8467.

67 Wang, Y., Fang, Y., He, T. et al. (2011). Hydrodeoxygenation of dibenzofuran over noble metal supported on mesoporous zeolite. *Catal. Commun.* 12 (13): 1201–1205.

68 Bolshakov, A., van Diepen, M., van Hoof, A.J.F. et al. (2019). Hierarchically porous (alumino)silicates prepared by an imidazole-based surfactant and their application in acid-catalyzed reactions. *ACS Appl. Mater. Interfaces* 11 (43): 40151–40162.

69 Bolshakov, A., van de Poll, R., van Bergen-Brenkman, T. et al. (2020). Hierarchically porous FER zeolite obtained via FAU transformation for fatty acid isomerization. *Appl. Catal. B: Environ.* 263: 118356.

70 Zhang, P., Li, S., and Zhang, C. (2020). Sustainable synthesis of hierarchically porous ZSM-5 zeolite from iron ore tailings without secondary templates. *Arab. J. Sci. Eng.* 45 (1): 187–196.

71 Barrer, R.M. and Makki, M.B. (1964). Molecular sieve sorbents from clinoptilolite. *Can. J. Chem.* 42 (6): 1481–1487.

72 Le Van Mao, R., Xiao, S., Ramsaran, A. et al. (1994). Selective removal of silicon from zeolite frameworks using sodium carbonate. *J. Mater. Chem.* 4 (4): 605–610.

73 Le Van Mao, R., Ramsaran, A., Xiao, S. et al. (1995). pH of the sodium carbonate solution used for the desilication of zeolite materials. *J. Mater. Chem.* 5 (3): 533–535.

74 Le Van Mao, R., Le, S.T., Ohayon, D. et al. (1997). Modification of the micropore characteristics of the desilicated ZSM-5 zeolite by thermal treatment. *Zeolites* 19 (4): 270–278.

75 Corma, A., Fornes, V., Martínez-Triguero, J. et al. (1999). Delaminated zeolite: combining the benefits of zeolites and mesoporous materials for catalytic uses. *J. Catal.* 186 (1): 57–63.

76 Morin, S., Ayrault, P., Gnep, N.S. et al. (1998). Influence of the framework composition of commercial HFAU zeolites on their activity and selectivity in *m*-xylene transformation. *Appl. Catal. A: Gen.* 166 (2): 281–292.

77 Ogura, M., Shinomiya, S.-y., Tateno, J. et al. (2000). Formation of uniform mesopores in ZSM-5 zeolite through treatment in alkaline solution. *Chem. Lett.* 29 (8): 882–883.

78 Ogura, M., Shinomiya, S.-y., Tateno, J. et al. (2001). Alkali-treatment technique—new method for modification of structural and acid-catalytic properties of ZSM-5 zeolites. *Appl. Catal. A: Gen.* 219 (1): 33–43.

79 Groen, J.C., Jansen, J.C., Moulijn, J.A. et al. (2004). Optimal aluminum-assisted mesoporosity development in MFI zeolites by desilication. *J. Phys. Chem. B* 108 (35): 13062–13065.

80 Groen, J.C., Peffer, L.A.A., Moulijn, J.A. et al. (2005). Mechanism of hierarchical porosity development in MFI zeolites by desilication: the role of aluminium as a pore-directing agent. *Chem. Eur. J.* 11 (17): 4983–4994.

81 Salman, N., Rüscher, C.H., Buhl, J.C. et al. (2006). Effect of temperature and time in the hydrothermal treatment of HY zeolite. *Microporous Mesoporous Mater.* 90 (1): 339–346.

82 Groen, J.C., Moulijn, J.A., and Pérez-Ramírez, J. (2007). Alkaline posttreatment of MFI zeolites. From accelerated screening to scale-up. *Ind. Eng. Chem. Res.* 46 (12): 4193–4201.

83 Wang, J., Groen, J.C., Yue, W. et al. (2007). Single-template synthesis of zeolite ZSM-5 composites with tunable mesoporosity. *Chem. Commun.* 2007 (44): 4653–4655. https://doi.org/10.1002/cctc.201300345.

84 Maheshwari, S., Jordan, E., Kumar, S. et al. (2008). Layer structure preservation during swelling, pillaring, and exfoliation of a zeolite precursor. *J. Am. Chem. Soc.* 130 (4): 1507–1516.

85 Taralkar, U.S., Niphadkar, P.S., and Joshi, P.N. (2009). Synthesis of Si-MCM-41 from ternary SiO_2–CTAOH–H_2O system via dry gel conversion route. *J. Sol-Gel Sci. Technol.* 51 (2): 244–250. https://doi.org/10.1002/cctc.201300345.

86 Zhou, J., Hua, Z., Shi, J. et al. (2009). Synthesis of a hierarchical micro/mesoporous structure by steam-assisted post-crystallization. *Chem. Eur. J.* 15 (47): 12949–12954. https://doi.org/10.1002/cctc.201300345.

87 Eilertsen, E.A., Ogino, I., Hwang, S.-J. et al. (2011). Nonaqueous fluoride/chloride anion-promoted delamination of layered zeolite precursors: synthesis and characterization of UCB-2. *Chem. Mater.* 23 (24): 5404–5408.

88 Valtchev, V., Balanzat, E., Mavrodinova, V. et al. (2011). High energy ion irradiation-induced ordered macropores in zeolite crystals. *J. Am. Chem. Soc.* 133 (46): 18950–18956.

89 Lutz, W. (2014). Zeolite Y: synthesis, modification, and properties—a case revisited. *Adv. Mater. Sci. Eng.* 2014: 724248.

90 Roth, W.J., Nachtigall, P., Morris, R.E. et al. (2014). Two-dimensional zeolites: current status and perspectives. *Chem. Rev.* 114 (9): 4807–4837.

91 Qin, Z., Gilson, J.-P., and Valtchev, V. (2015). Mesoporous zeolites by fluoride etching. *Curr. Opin. Chem. Eng.* 8: 1–6.

92 Roth, W.J., Gil, B., Makowski, W. et al. (2016). Layer like porous materials with hierarchical structure. *Chem. Soc. Rev.* 45 (12): 3400–3438.

93 Ren, S., Liu, G., Wu, X. et al. (2017). Enhanced MTO performance over acid treated hierarchical SAPO-34. *Chin. J. Catal.* 38 (1): 123–130.

94 Goto, Y., Fukushima, Y., Ratu, P. et al. (2002). Mesoporous material from zeolite. *J. Porous Mater.* 9 (1): 43–48.

95 Ivanova, I.I., Kuznetsov, A.S., Yuschenko, V.V. et al. (2004). Design of composite micro/mesoporous molecular sieve catalysts. *Pure Appl. Chem.* 76 (9): 1647–1657.

96 Dai, C., Zhang, A., Liu, M. et al. (2015). Hollow ZSM-5 with silicon-rich surface, double shells, and functionalized interior with metallic nanoparticles and carbon nanotubes. *Adv. Funct. Mater.* 25 (48): 7479–7487.

97 Dai, C., Zhang, A., Liu, M. et al. (2016). Hollow alveolus-like nanovesicle assembly with metal-encapsulated hollow zeolite nanocrystals. *ACS Nano* 10 (8): 7401–7408.

98 Vermeiren, W. and Gilson, J.P. (2009). Impact of zeolites on the petroleum and petrochemical industry. *Top. Catal.* 52 (9): 1131–1161.

99 Wang, L., Wang, Y., Liu, Y. et al. (2008). Post-transformation of MWW-type lamellar precursors into MCM-56 analogues. *Microporous Mesoporous Mater.* 113 (1): 435–444.

100 Torres Galvis, H.M. and de Jong, K.P. (2013). Catalysts for production of lower olefins from synthesis gas: a review. *ACS Catal.* 3 (9): 2130–2149.

101 Khodakov, A.Y., Chu, W., and Fongarland, P. (2007). Advances in the development of novel cobalt Fischer–Tropsch catalysts for synthesis of long-chain hydrocarbons and clean fuels. *Chem. Rev.* 107 (5): 1692–1744.

102 Zhang, Q., Kang, J., and Wang, Y. (2010). Development of novel catalysts for Fischer–Tropsch synthesis: tuning the product selectivity. *ChemCatChem* 2 (9): 1030–1058.

103 Sun, B., Qiao, M., Fan, K. et al. (2011). Fischer–Tropsch synthesis over molecular sieve supported catalysts. *ChemCatChem* 3 (3): 542–550.

104 Zhang, Q., Cheng, K., Kang, J. et al. (2014). Fischer–Tropsch catalysts for the production of hydrocarbon fuels with high selectivity. *ChemSusChem* 7 (5): 1251–1264.

105 Bessell, S. (1993). Support effects in cobalt-based Fischer-Tropsch catalysis. *Appl. Catal. A: Gen.* 96 (2): 253–268.

106 Tsakoumis, N.E., Rønning, M., Borg, Ø. et al. (2010). Deactivation of cobalt based Fischer–Tropsch catalysts: a review. *Catal. Today* 154 (3): 162–182.

107 Eggenhuisen, T.M., Munnik, P., Talsma, H. et al. (2013). Freeze-drying for controlled nanoparticle distribution in Co/SiO_2 Fischer–Tropsch catalysts. *J. Catal.* 297: 306–313.

108 Prieto, G., Zečević, J., Friedrich, H. et al. (2013). Towards stable catalysts by controlling collective properties of supported metal nanoparticles. *Nat. Mater.* 12 (1): 34–39.

109 Sartipi, S., Parashar, K., Makkee, M. et al. (2013). Breaking the Fischer–Tropsch synthesis selectivity: direct conversion of syngas to gasoline over hierarchical Co/H-ZSM-5 catalysts. *Catal. Sci. Technol.* 3 (3): 572–575.

110 Sartipi, S., van Dijk, J.E., Gascon, J. et al. (2013). Toward bifunctional catalysts for the direct conversion of syngas to gasoline range hydrocarbons: H-ZSM-5 coated Co versus H-ZSM-5 supported Co. *Appl. Catal. A: Gen.* 456: 11–22.

111 Sartipi, S., Alberts, M., Meijerink, M.J. et al. (2013). Towards liquid fuels from biosyngas: effect of zeolite structure in hierarchical-zeolite-supported cobalt catalysts. *ChemSusChem* 6 (9): 1646–1650.

112 Peng, X., Cheng, K., Kang, J. et al. (2015). Impact of hydrogenolysis on the selectivity of the Fischer–Tropsch synthesis: diesel fuel production over mesoporous zeolite-Y-supported cobalt nanoparticles. *Angew. Chem. Int. Ed.* 54 (15): 4553–4556.

113 Martens, J.A., Verboekend, D., Thomas, K. et al. (2013). Hydroisomerization of emerging renewable hydrocarbons using hierarchical Pt/H-ZSM-22 catalyst. *ChemSusChem* 6 (3): 421–425.

114 Leonowicz, M.E., Lawton, J.A., Lawton, S.L. et al. (1994). MCM-22: a molecular sieve with two independent multidimensional channel systems. *Science* 264 (5167): 1910–1913.

115 Lawton, S.L., Leonowicz, M.E., Partridge, R.D. et al. (1998). Twelve-ring pockets on the external surface of MCM-22 crystals. *Microporous Mesoporous Mater.* 23 (1): 109–117.

116 Ravishankar, R., Li, M.M., and Borgna, A. (2005). Novel utilization of MCM-22 molecular sieves as supports of cobalt catalysts in the Fischer–Tropsch synthesis. *Catal. Today* 106 (1): 149–153.

117 Han, J., Cho, J., Kim, J.-C. et al. (2018). Confinement of supported metal catalysts at high loading in the mesopore network of hierarchical zeolites, with access via the microporous windows. *ACS Catal.* 8 (2): 876–879.

118 Laurendeau, N.M. (1978). Heterogeneous kinetics of coal char gasification and combustion. *Prog. Energy Combust. Sci.* 4 (4): 221–270.

119 Wen, W.-Y. (1980). Mechanisms of alkali metal catalysis in the gasification of coal, char, or graphite. *Catal. Rev.* 22 (1): 1–28.

120 Hickman, D.A. and Schmidt, L.D. (1993). Production of syngas by direct catalytic oxidation of methane. *Science* 259 (5093): 343–346.

121 Sutton, D., Kelleher, B., and Ross, J.R.H. (2001). Review of literature on catalysts for biomass gasification. *Fuel Process. Technol.* 73 (3): 155–173.

122 Asadullah, M., Ito, S.-I., Kunimori, K. et al. (2002). Biomass gasification to hydrogen and syngas at low temperature: novel catalytic system using fluidized-bed reactor. *J. Catal.* 208 (2): 255–259.

123 Stöcker, M. (2008). Biofuels and biomass-to-liquid fuels in the biorefinery: catalytic conversion of lignocellulosic biomass using porous materials. *Angew. Chem. Int. Ed.* 47 (48): 9200–9211.

124 Keil, F.J. (1999). Methanol-to-hydrocarbons: process technology. *Microporous Mesoporous Mater.* 29 (1): 49–66.

125 Chang, C.D. (1983). Hydrocarbons from methanol. *Catal. Rev.* 25 (1): 1–118.

126 Haag, W.O., Lago, R.M., and Rodewald, P.G. (1982). Aromatics, light olefins and gasoline from methanol: mechanistic pathways with ZSM-5 zeolite catalyst. *J. Mol. Catal.* 17 (2): 161–169.

127 Chang, C.D. and Silvestri, A.J. (1977). The conversion of methanol and other O-compounds to hydrocarbons over zeolite catalysts. *J. Catal.* 47 (2): 249–259.

128 Vora, B., Chen, J.Q., Bozzano, A. et al. (2009). Various routes to methane utilization—SAPO-34 catalysis offers the best option. *Catal. Today* 141 (1): 77–83.

129 Kim, W. and Ryoo, R. (2014). Probing the catalytic function of external acid sites located on the MFI nanosheet for conversion of methanol to hydrocarbons. *Catal. Lett.* 144 (7): 1164–1169.

130 Védrine, J.C., Auroux, A., Bolis, V. et al. (1979). Infrared, microcalorimetric, and electron spin resonance investigations of the acidic properties of the H-ZSM-5 zeolite. *J. Catal.* 59 (2): 248–262.

131 Bibby, D.M., Milestone, N.B., Patterson, J.E. et al. (1986). Coke formation in zeolite ZSM-5. *J. Catal.* 97 (2): 493–502.

132 Bibby, D.M., Howe, R.F., and McLellan, G.D. (1992). Coke formation in high-silica zeolites. *Appl. Catal. A: Gen.* 93 (1): 1–34.

133 Kumar, N., Nieminen, V., Demirkan, K. et al. (2002). Effect of synthesis time and mode of stirring on physico-chemical and catalytic properties of ZSM-5 zeolite catalysts. *Appl. Catal. A: Gen.* 235 (1): 113–123.

134 Bjørgen, M., Joensen, F., Spangsberg Holm, M. et al. (2008). Methanol to gasoline over zeolite H-ZSM-5: improved catalyst performance by treatment with NaOH. *Appl. Catal. A: Gen.* 345 (1): 43–50.

135 Choi, M., Na, K., Kim, J. et al. (2009). Stable single-unit-cell nanosheets of zeolite MFI as active and long-lived catalysts. *Nature* 461 (7261): 246–249.

136 Degnan, T.F., Chen, N.Y., and Somorjai, G.A. (2009). Heinz Heinemann's legacy at ExxonMobil: an illustrious career in industrial catalysis. *Catal. Lett.* 133 (1): 227–230.

137 Mokrani, T. and Scurrell, M. (2009). Gas conversion to liquid fuels and chemicals: the methanol route-catalysis and processes development. *Catal. Rev.* 51 (1): 1–145.

138 Schulz, H. (2010). "Coking" of zeolites during methanol conversion: basic reactions of the MTO-, MTP- and MTG processes. *Catal. Today* 154 (3): 183–194.

139 Rownaghi, A.A. and Hedlund, J. (2011). Methanol to gasoline-range hydrocarbons: influence of nanocrystal size and mesoporosity on catalytic performance and product distribution of ZSM-5. *Ind. Eng. Chem. Res.* 50 (21): 11872–11878.

140 Liu, Z., Fan, W., Ma, J. et al. (2012). Adsorption, diffusion and catalysis of mesostructured zeolite HZSM-5. *Adsorption* 18 (5): 493–501.

141 Di, Z., Yang, C., Jiao, X. et al. (2013). A ZSM-5/MCM-48 based catalyst for methanol to gasoline conversion. *Fuel* 104: 878–881.

142 Fathi, S., Sohrabi, M., and Falamaki, C. (2014). Improvement of HZSM-5 performance by alkaline treatments: comparative catalytic study in the MTG reactions. *Fuel* 116: 529–537.

143 Wan, Z., Wu, W., Chen, W. et al. (2014). Direct synthesis of hierarchical ZSM-5 zeolite and its performance in catalyzing methanol to gasoline conversion. *Ind. Eng. Chem. Res.* 53 (50): 19471–19478.

144 Liu, Z., Dong, X., Zhu, Y. et al. (2015). Investigating the influence of mesoporosity in zeolite beta on its catalytic performance for the conversion of methanol to hydrocarbons. *ACS Catal.* 5 (10): 5837–5845.

145 Olsbye, U., Svelle, S., Lillerud, K.P. et al. (2015). The formation and degradation of active species during methanol conversion over protonated zeotype catalysts. *Chem. Soc. Rev.* 44 (20): 7155–7176.

146 Noor, P., Khanmohammadi, M., Roozbehani, B. et al. (2018). Introduction of table sugar as a soft second template in ZSM-5 nanocatalyst and its effect on product distribution and catalyst lifetime in methanol to gasoline conversion. *J. Energy Chem.* 27 (2): 582–590.

147 Sexton, B.A., Hughes, A.E., and Bibby, D.M. (1988). An XPS study of coke distribution on ZSM-5. *J. Catal.* 109 (1): 126–131.

148 Shen, K., Wang, N., Chen, X. et al. (2017). Seed-induced and additive-free synthesis of oriented nanorod-assembled meso/macroporous zeolites: toward efficient and cost-effective catalysts for the MTA reaction. *Catal. Sci. Technol.* 7 (21): 5143–5153.

149 Stöcker, M. (1999). Methanol-to-hydrocarbons: catalytic materials and their behavior. *Microporous Mesoporous Mater.* 29 (1): 3–48.

150 Sun, C., Du, J., Liu, J. et al. (2010). A facile route to synthesize endurable mesopore containing ZSM-5 catalyst for methanol to propylene reaction. *Chem. Commun.* 46 (15): 2671–2673.

151 Sun, L., Wang, Y., Chen, H. et al. (2018). Direct synthesis of hierarchical ZnZSM-5 with addition of CTAB in a seeding method and improved catalytic performance in methanol to aromatics reaction. *Catal. Today* 316: 91–98.

152 Sun, Q., Wang, N., Bai, R. et al. (2016). Seeding induced nano-sized hierarchical SAPO-34 zeolites: cost-effective synthesis and superior MTO performance. *J. Mater. Chem. A* 4 (39): 14978–14982.

153 Sun, Q., Wang, N., Guo, G. et al. (2015). Synthesis of tri-level hierarchical SAPO-34 zeolite with intracrystalline micro–meso–macroporosity showing superior MTO performance. *J. Mater. Chem. A* 3 (39): 19783–19789.

154 Sun, Q., Wang, N., Xi, D. et al. (2014). Organosilane surfactant-directed synthesis of hierarchical porous SAPO-34 catalysts with excellent MTO performance. *Chem. Commun.* 50 (49): 6502–6505.

155 Tago, T., Konno, H., Nakasaka, Y. et al. (2012). Size-controlled synthesis of nano-zeolites and their application to light olefin synthesis. *Catal. Surv. Asia* 16 (3): 148–163.

156 Tao, Y., Kanoh, H., Abrams, L. et al. (2006). Mesopore-modified zeolites: preparation, characterization, and applications. *Chem. Rev.* 106 (3): 896–910.

157 Tian, H., Chen, Z., Yang, X. et al. (2023). Catalytic performance of methanol to aromatics over hierarchical ZSM-5 zeolite synthesized by a soft template method. *Energy Fuels* 37 (18): 14180–14191.

158 Varzaneh, A.Z., Towfighi, J., and Sahebdelfar, S. (2016). Carbon nanotube templated synthesis of metal containing hierarchical SAPO-34 catalysts: impact of the preparation method and metal avidities in the MTO reaction. *Microporous Mesoporous Mater.* 236: 1–12.

159 Varzaneh, A.Z., Towfighi, J., Sahebdelfar, S. et al. (2016). Carbon nanotube templated synthesis of hierarchical SAPO-34 catalysts with different structure directing agents for catalytic conversion of methanol to light olefins. *J. Anal. Appl. Pyrol.* 121: 11–23.

160 Prinz, D. and Riekert, L. (1988). Formation of ethene and propene from methanol on zeolite ZSM-5: I. Investigation of rate and selectivity in a batch reactor. *Appl. Catal.* 37: 139–154.

161 Corma, A., Melo, F., Sauvanaud, L. et al. (2004). Different process schemes for converting light straight run and fluid catalytic cracking naphthas in a FCC unit for maximum propylene production. *Appl. Catal. A: Gen.* 265 (2): 195–206.

162 Mei, C., Wen, P., Liu, Z. et al. (2008). Selective production of propylene from methanol: mesoporosity development in high silica HZSM-5. *J. Catal.* 258 (1): 243–249.

163 Liu, J., Zhang, C., Shen, Z. et al. (2009). Methanol to propylene: effect of phosphorus on a high silica HZSM-5 catalyst. *Catal. Commun.* 10 (11): 1506–1509.

164 Rownaghi, A.A., Rezaei, F., and Hedlund, J. (2011). Yield of gasoline-range hydrocarbons as a function of uniform ZSM-5 crystal size. *Catal. Commun.* 14 (1): 37–41.

165 Bleken, F.L., Chavan, S., Olsbye, U. et al. (2012). Conversion of methanol into light olefins over ZSM-5 zeolite: strategy to enhance propene selectivity. *Appl. Catal. A: Gen.* 447–448: 178–185.

166 Hu, Z., Zhang, H., Wang, L. et al. (2014). Highly stable boron-modified hierarchical nanocrystalline ZSM-5 zeolite for the methanol to propylene reaction. *Catal. Sci. Technol.* 4 (9): 2891–2895.

167 Zhang, Q., Hu, S., Zhang, L. et al. (2014). Facile fabrication of mesopore-containing ZSM-5 zeolite from spent zeolite catalyst for methanol to propylene reaction. *Green Chem.* 16 (1): 77–81.

168 Ahmadpour, J. and Taghizadeh, M. (2015). Catalytic conversion of methanol to propylene over high-silica mesoporous ZSM-5 zeolites prepared by different combinations of mesogenous templates. *J. Nat. Gas Sci. Eng.* 23: 184–194.

169 Galadima, A. and Muraza, O. (2015). Recent developments on silicoaluminates and silicoaluminophosphates in the methanol-to-propylene reaction: a mini review. *Ind. Eng. Chem. Res.* 54 (18): 4891–4905.

170 Rostamizadeh, M. and Taeb, A. (2015). Highly selective Me-ZSM-5 catalyst for methanol to propylene (MTP). *J. Ind. Eng. Chem.* 27: 297–306.

171 Khanmohammadi, M., Amani, S., Garmarudi, A.B. et al. (2016). Methanol-to-propylene process: perspective of the most important catalysts and their behavior. *Chin. J. Catal.* 37 (3): 325–339.

172 Chen, H., Wang, Y., Sun, C. et al. (2018). Synthesis of hierarchical ZSM-5 zeolites with CTAB-containing seed silicalite-1 and its catalytic performance in methanol to propylene. *Catal. Commun.* 112: 10–14.

173 Li, H., Wang, Y., Fan, C. et al. (2018). Facile synthesis of a superior MTP catalyst: hierarchical micro-meso-macroporous ZSM-5 zeolites. *Appl. Catal. A: Gen.* 551: 34–48.

174 Jin, R., Ma, K., Xu, S. et al. (2022). Effect of acid distribution and pore structure of ZSM-5 on catalytic performance. *React. Chem. Eng.* 7 (10): 2152–2162.

175 Inui, T., Makino, Y., Okazumi, F. et al. (1987). Selective aromatization of light paraffins on platinum-ion-exchanged gallium-silicate bifunctional catalysts. *Ind. Eng. Chem. Res.* 26 (4): 647–652.

176 Gnep, N.S., Doyemet, J.Y., Seco, A.M. et al. (1988). Conversion of light alkanes to aromatic hydrocarbons: II. Role of gallium species in propane transformation on GaZSM5 catalysts. *Appl. Catal.* 43 (1): 155–166.

177 Wang, L., Tao, L., Xie, M. et al. (1993). Dehydrogenation and aromatization of methane under non-oxidizing conditions. *Catal. Lett.* 21 (1): 35–41.

178 Olah, G.A. (2005). Beyond oil and gas: the methanol economy. *Angew. Chem. Int. Ed.* 44 (18): 2636–2639.

179 Li, B., Li, S., Li, N. et al. (2006). Structure and acidity of Mo/ZSM-5 synthesized by solid state reaction for methane dehydrogenation and aromatization. *Microporous Mesoporous Mater.* 88 (1): 244–253.

180 Park, J.W. and Seo, G. (2009). IR study on methanol-to-olefin reaction over zeolites with different pore structures and acidities. *Appl. Catal. A: Gen.* 356 (2): 180–188.

181 Xu, C., Liu, H., Jia, M. et al. (2011). Methane non-oxidative aromatization on Mo/ZSM-5: effect of adding triethoxyphenylsilanes into the synthesis system of ZSM-5. *Appl. Surf. Sci.* 257 (7): 2448–2454.

182 Conte, M., Lopez-Sanchez, J.A., He, Q. et al. (2012). Modified zeolite ZSM-5 for the methanol to aromatics reaction. *Catal. Sci. Technol.* 2 (1): 105–112.

183 Lopez-Sanchez, J.A., Conte, M., Landon, P. et al. (2012). Reactivity of Ga_2O_3 clusters on zeolite ZSM-5 for the conversion of methanol to aromatics. *Catal. Lett.* 142 (9): 1049–1056.

184 Xin, Y., Qi, P., Duan, X. et al. (2013). Enhanced performance of Zn–Sn/HZSM-5 catalyst for the conversion of methanol to aromatics. *Catal. Lett.* 143 (8): 798–806.

185 Hu, J., Wu, S., Liu, H. et al. (2014). Effect of mesopore structure of TNU-9 on methane dehydroaromatization. *RSC Adv.* 4 (51): 26577–26584.

186 An, J.-G., Gao, X., Jin, J.-J. et al. (2015). Mesoporous zeolite ZSM-5 synthesized via gel conversion with polyethyleneglycol as template and its catalytic performance. *J. Inorg. Mater.* 30 (11): 1148–1154.

187 Hu, H., Lyu, J., Wang, Q. et al. (2015). Alkylation of benzene with methanol over hierarchical porous ZSM-5: synergy effects of hydrogen atmosphere and zinc modification. *RSC Adv.* 5 (41): 32679–32684.

188 Zhang, L., Jiang, Z.-X., Yu, Y. et al. (2015). Synthesis of core–shell ZSM-5@meso-SAPO-34 composite and its application in methanol to aromatics. *RSC Adv.* 5 (69): 55825–55831.

189 Niziolek, A.M., Onel, O., and Floudas, C.A. (2016). Production of benzene, toluene, and xylenes from natural gas via methanol: process synthesis and global optimization. *AIChE J.* 62 (5): 1531–1556.

190 Wang, N., Qian, W., Shen, K. et al. (2016). Bayberry-like ZnO/MFI zeolite as high performance methanol-to-aromatics catalyst. *Chem. Commun.* 52 (10): 2011–2014.

191 Zhou, M., Wang, F., Xiao, W. et al. (2016). The comparison of mesoporous HZSM-5 zeolite catalysts prepared by different mesoporous templates and their catalytic performance in the methanol to aromatics reaction. *React. Kinet. Mech. Catal.* 119 (2): 699–713.

192 Niu, X., Gao, J., Wang, K. et al. (2017). Influence of crystal size on the catalytic performance of H-ZSM-5 and Zn/H-ZSM-5 in the conversion of methanol to aromatics. *Fuel Process. Technol.* 157: 99–107.

193 Wang, K., Dong, M., Li, J. et al. (2017). Facile fabrication of ZSM-5 zeolite hollow spheres for catalytic conversion of methanol to aromatics. *Catal. Sci. Technol.* 7 (3): 560–564.

194 Yang, L., Liu, Z., Liu, Z. et al. (2017). Correlation between H-ZSM-5 crystal size and catalytic performance in the methanol-to-aromatics reaction. *Chin. J. Catal.* 38 (4): 683–690.

195 Zhou, F., Gao, Y., Ma, H. et al. (2017). Catalytic aromatization of methanol over post-treated ZSM-5 zeolites in the terms of pore structure and acid sites properties. *Mol. Catal.* 438: 37–46.

196 Jia, Y., Wang, J., Zhang, K. et al. (2018). Hierarchical ZSM-5 zeolite synthesized via dry gel conversion-steam assisted crystallization process and its application in aromatization of methanol. *Powder Technol.* 328: 415–429.

197 Rownaghi, A.A., Rezaei, F., Stante, M. et al. (2012). Selective dehydration of methanol to dimethyl ether on ZSM-5 nanocrystals. *Appl. Catal. B: Environ.* 119–120: 56–61.

198 Yang, Q., Zhang, H., Kong, M. et al. (2013). Hierarchical mesoporous ZSM-5 for the dehydration of methanol to dimethyl ether. *Chin. J. Catal.* 34 (8): 1576–1582.

199 Masih, D., Rohani, S., Kondo, J.N. et al. (2017). Low-temperature methanol dehydration to dimethyl ether over various small-pore zeolites. *Appl. Catal. B: Environ.* 217: 247–255.

200 Ortega, C., Rezaei, M., Hessel, V. et al. (2018). Methanol to dimethyl ether conversion over a ZSM-5 catalyst: intrinsic kinetic study on an external recycle reactor. *Chem. Eng. J.* 347: 741–753.

201 Chang, C.D. (1984). Methanol conversion to light olefins. *Catal. Rev.* 26 (3–4): 323–345.

202 Lok, B.M., Messina, C.A., Patton, R.L. et al. (1984). Silicoaluminophosphate molecular sieves: another new class of microporous crystalline inorganic solids. *J. Am. Chem. Soc.* 106 (20): 6092–6093.

203 Inui, T., Phatanasri, S., and Matsuda, H. (1990). Highly selective synthesis of ethene from methanol on a novel nickel–silicoaluminophosphate catalyst. *J. Chem. Soc., Chem. Commun.* 1990 (3): 205–206.

204 Chen, D., Moljord, K., and Holmen, A. (2012). A methanol to olefins review: diffusion, coke formation and deactivation on SAPO type catalysts. *Microporous Mesoporous Mater.* 164: 239–250.

205 Chen, J., Li, J., Wei, Y. et al. (2014). Spatial confinement effects of cage-type SAPO molecular sieves on product distribution and coke formation in methanol-to-olefin reaction. *Catal. Commun.* 46: 36–40.

206 Pinilla-Herrero, I., Olsbye, U., Márquez-Álvarez, C. et al. (2017). Effect of framework topology of SAPO catalysts on selectivity and deactivation profile in the methanol-to-olefins reaction. *J. Catal.* 352: 191–207.

207 Qi, G., Xie, Z., Yang, W. et al. (2007). Behaviors of coke deposition on SAPO-34 catalyst during methanol conversion to light olefins. *Fuel Process. Technol.* 88 (5): 437–441.

208 Chen, D., Moljord, K., Fuglerud, T. et al. (1999). The effect of crystal size of SAPO-34 on the selectivity and deactivation of the MTO reaction. *Microporous Mesoporous Mater.* 29 (1): 191–203.

209 Chen, D., Rebo, H.P., Grønvold, A. et al. (2000). Methanol conversion to light olefins over SAPO-34: kinetic modeling of coke formation. *Microporous Mesoporous Mater.* 35–36: 121–135.

210 Yang, S.-T., Kim, J.-Y., Chae, H.-J. et al. (2012). Microwave synthesis of mesoporous SAPO-34 with a hierarchical pore structure. *Mater. Res. Bull.* 47 (11): 3888–3892.

211 Wang, C., Yang, M., Tian, P. et al. (2015). Dual template-directed synthesis of SAPO-34 nanosheet assemblies with improved stability in the methanol to olefins reaction. *J. Mater. Chem. A* 3 (10): 5608–5616.

212 Yang, B., Zhao, P., Ma, J. et al. (2016). Synthesis of hierarchical SAPO-34 nanocrystals with improved catalytic performance for methanol to olefins. *Chem. Phys. Lett.* 665: 59–63.

213 Lee, K.Y., Chae, H.-J., Jeong, S.-Y. et al. (2009). Effect of crystallite size of SAPO-34 catalysts on their induction period and deactivation in methanol-to-olefin reactions. *Appl. Catal. A: Gen.* 369 (1): 60–66.

214 Hirota, Y., Murata, K., Miyamoto, M. et al. (2010). Light olefins synthesis from methanol and dimethylether over SAPO-34 nanocrystals. *Catal. Lett.* 140 (1): 22–26.

215 Yang, G., Wei, Y., Xu, S. et al. (2013). Nanosize-enhanced lifetime of SAPO-34 catalysts in methanol-to-olefin reactions. *J. Phys. Chem. C* 117 (16): 8214–8222.

216 Nishiyama, N., Kawaguchi, M., Hirota, Y. et al. (2009). Size control of SAPO-34 crystals and their catalyst lifetime in the methanol-to-olefin reaction. *Appl. Catal. A: Gen.* 362 (1): 193–199.

217 Wang, J., Yang, M., Shang, W. et al. (2017). Synthesis, characterization, and catalytic application of hierarchical SAPO-34 zeolite with three-dimensionally ordered mesoporous-imprinted structure. *Microporous Mesoporous Mater.* 252: 10–16.

218 Li, Y., Huang, Y., Guo, J. et al. (2014). Hierarchical SAPO-34/18 zeolite with low acid site density for converting methanol to olefins. *Catal. Today* 233: 2–7.

219 Yang, H., Liu, Z., Gao, H. et al. (2010). Synthesis and catalytic performances of hierarchical SAPO-34 monolith. *J. Mater. Chem.* 20 (16): 3227–3231.

220 Liu, X., Ren, S., Zeng, G. et al. (2016). Coke suppression in MTO over hierarchical SAPO-34 zeolites. *RSC Adv.* 6 (34): 28787–28791.

221 Chen, X., Vicente, A., Qin, Z. et al. (2016). The preparation of hierarchical SAPO-34 crystals via post-synthesis fluoride etching. *Chem. Commun.* 52 (17): 3512–3515.

222 Xi, D., Sun, Q., Xu, J. et al. (2014). In situ growth-etching approach to the preparation of hierarchically macroporous zeolites with high MTO catalytic activity and selectivity. *J. Mater. Chem. A* 2 (42): 17994–18004.

223 Chen, X., Xi, D., Sun, Q. et al. (2016). A top-down approach to hierarchical SAPO-34 zeolites with improved selectivity of olefin. *Microporous Mesoporous Mater.* 234: 401–408.

224 Rimaz, S., Halladj, R., and Askari, S. (2016). Synthesis of hierarchal SAPO-34 nano catalyst with dry gel conversion method in the presence of carbon nanotubes as a hard template. *J. Colloid Interface Sci.* 464: 137–146.

225 Milina, M., Mitchell, S., Crivelli, P. et al. (2014). Mesopore quality determines the lifetime of hierarchically structured zeolite catalysts. *Nat. Commun.* 5 (1): 3922.

226 Milina, M., Mitchell, S., Cooke, D. et al. (2015). Impact of pore connectivity on the design of long-lived zeolite catalysts. *Angew. Chem. Int. Ed.* 54 (5): 1591–1594.

227 Zhang, Y., Zhang, K., Shang, C. et al. (2023). Synthesis of hierarchical ZSM-5 microspheres with superior performance for catalytic methanol-to-olefin conversion. *AIChE J.* 69 (2): e17913.

228 Zhu, X., Kosinov, N., Kubarev, A.V. et al. (2017). Probing the influence of SSZ-13 zeolite pore hierarchy in methanol-to-olefins catalysis by using nanometer accuracy by stochastic chemical reactions fluorescence microscopy and positron emission profiling. *ChemCatChem* 9 (18): 3470–3477.

229 Cui, Z.-M., Liu, Q., Song, W.-G. et al. (2006). Insights into the mechanism of methanol-to-olefin conversion at zeolites with systematically selected framework structures. *Angew. Chem.* 118 (39): 6662–6665.

230 Kumita, Y., Gascon, J., Stavitski, E. et al. (2011). Shape selective methanol to olefins over highly thermostable DDR catalysts. *Appl. Catal. A: Gen.* 391 (1): 234–243.

231 Zhu, Z., Hartmann, M., and Kevan, L. (2000). Catalytic conversion of methanol to olefins on SAPO-*n* ($n = 11$, 34, and 35), CrAPSO-*n*, and Cr–SAPO-*n* molecular sieves. *Chem. Mater.* 12 (9): 2781–2787.

232 Aguayo, A.T., Gayubo, A.G., Vivanco, R. et al. (2005). Role of acidity and microporous structure in alternative catalysts for the transformation of methanol into olefins. *Appl. Catal. A: Gen.* 283 (1): 197–207.

4

Deactivation and Regeneration of Zeolites in Catalytic Conversion of Low-carbon Molecules

Zaizhe Cheng[1,2], Ying Li[1,3], Shouying Huang[1], Meiyan Wang[1], Xinbin Ma[1]*

[1]*Key Laboratory for Green Chemical Technology of Ministry of Education, School of Chemical Engineering and Technology, Tianjin University, Tianjin, China*
[2]*State Key Laboratory of Green Chemical Synthesis and Conversion, Institute of Industrial Catalysis, Zhejiang University of Technology, Hangzhou, Zhejiang, China*
[3]*Inner Mongolia Key Laboratory of Green Chemical Engineering, College of Chemical Engineering, Inner Mongolia University of Technology, Hohhot, Inner Mongolia, China*

*Corresponding author: huangsy@tju.edu.cn

4.1 Introduction

Zeolites, functioning as solid acid catalysts, are indispensable in modern chemical industry due to their tunable acidity, shape-selective properties, and exceptional thermal stability. These characteristics drive their widespread applications in critical chemical processes such as fluid catalytic cracking (FCC), methanol-to-hydrocarbons (MTH) conversion, toluene disproportionation, alkylation, and dimethyl ether (DME) carbonylation. Furthermore, the intrinsic microporous architecture and highly ordered frameworks of zeolites endow them with substantial surface areas, spatially confined channels, and enzyme-like microenvironments, enabling effective dispersion and stabilization of metal species. Therefore, zeolites have also been widely used as supports in metal-catalyzed conversions of low-carbon feedstocks, including alkane dehydrogenation, methane dry reforming, methyl nitrite (MN) carbonylation, etc.

Despite extensive applications of zeolite-based catalysts, their deactivation poses persistent challenges in chemical industries. In acid-catalyzed reactions, coke deposition emerges as the predominant deactivation mechanism through pore blockage and active site coverage. While metal-catalyzed systems exhibit more complex degradation pathways involving metal sintering, migration, leaching, valence state change, as well as coke formation. The economic and environmental implications of frequent catalyst regeneration or replacement necessitate urgent resolution of these deactivation phenomena.

Recent advancements in zeolite deactivation research have focused on elucidating the fundamental mechanisms and establishing qualitative/quantitative relationships among catalyst structures, reaction parameters, and deactivation behaviors. Driven by the rapid development of advanced characterization techniques and computational modeling, significant progress has been achieved in multiscale analysis of coke properties, spatial deposition patterns, and real-time species evolution under practical reaction conditions. These breakthroughs provide critical insights and guidance for rational catalyst design and optimization of reaction–regeneration protocols, offering substantial potential to enhance the economic viability and sustainability of industrial catalytic processes.

4.2 Deactivation of Zeolite Catalysts in Acid-catalyzed Reactions

As mentioned above, coke formation serves as the primary cause of catalyst deactivation when zeolites are employed as solid acid catalysts for the conversion of low-carbon molecules (e.g., methanol and DME). In this section, we first systematically summarize the state-of-the-art characterization methodologies for deactivated catalysts and coke characteristics. Subsequently, coke formation mechanisms are critically analyzed, including the initiation of C—C bond formation, the debated roles of formaldehyde and ketene intermediates in monocyclic aromatic hydrocarbon (MAH) growth, and the progressive evolution toward polycyclic aromatic hydrocarbons (PAHs). Furthermore, mitigation strategies such as acid site passivation, hierarchical pore engineering, and controlled pre-coking are rigorously evaluated for their universality and operational durability. Emerging regeneration techniques, particularly ozone-assisted low-temperature oxidation and hydrogenative cracking, are comparatively assessed regarding their potential to minimize CO_2 emissions while preserving framework integrity. By integrating mechanistic understanding of coke formation, structure–activity relationships, and regeneration, we establish critical connections between fundamental research and industrial implementation challenges. We hope multidisciplinary roadmap can be provided for developing coke-resistant zeolite catalysts through acidity modulation and scalable hierarchical design, coupled with energy-efficient regeneration protocols.

4.2.1 Characterization of Coke on Deactivated Zeolites

4.2.1.1 Characterization Techniques of Coke Structure

Elementary analysis is usually the first step in characterizing coke composition on spent zeolite catalysts. As coke is primarily constituted of nonmetallic elements, such as carbon, hydrogen, oxygen, nitrogen, sulfur, etc., the analysis is often conducted via combustion in an oxidizing atmosphere under programmed temperature conditions. Commonly used qualitative and quantitative methods include gravimetry and off-gas analysis, which typically detect gases such as CO_x, NO_x, SO_x, and H_2O. By using techniques such as thermogravimetric analysis (TGA), chromatography, mass spectrometry (MS), infrared (IR) spectroscopy, or their combinations (e.g., TG–MS, TG–FTIR, and GC [gas chromatography]–MS), the content and elemental composition of carbonaceous residues on zeolites can be determined [1–3]. Although these characterization techniques cannot precisely identify the exact molecular structure of coke, the combustion temperature always provides qualitative information for coke species (e.g., aliphatic or aromatic hydrocarbon) and location (e.g., internal versus external) [4, 5]. Typically, aromatic species with a low H/C ratio or coke deposits located within the internal channels of zeolites combust at higher temperatures. Moreover, the combustion behaviors of coke and the associated combustion kinetics will provide valuable guidance for the regeneration of coked zeolites in industry [6].

Further insight into coke relies on several spectroscopic techniques, including IR spectroscopy, ultraviolet–visible (UV–vis) spectroscopy, Nuclear Magnetic Resonance (NMR) spectroscopy, etc. Among them, IR spectroscopy is particularly powerful for elucidating the molecular structure of coke on deactivated zeolites, as it is sensitive to C=C stretching modes of unsaturated aromatics (1300–1700 cm^{-1}) and the C—H stretching modes of paraffinic species (2800–3200 cm^{-1}) [7, 8]. For instance, alkyl-substituted aromatic species have been identified as the primary coke species on zeolite-catalyzed reactions, such as DME carbonylation [8] as well as methanol or ethanol transformation reactions [9, 10], as demonstrated by the growing intensities of bands around 2934 and 1603 cm^{-1}.

More importantly, IR spectroscopy can also be employed to monitor the evolution of carbonaceous species on working zeolite by designing an in situ or operando experiment in a cell, which provides detailed evidence for deducing the coke mechanism. For instance, to elucidate the different deactivation behaviors on H–MOR and Cu/H–MOR catalysts, Huang et al. [8] conducted kinetic experiments using diffuse reflectance infrared Fourier transform spectroscopy (DRIFTS). Their findings revealed that the presence of Cu accelerates the formation of non-condensed single aromatic species, while inhibiting their further condensation into more complex structures.

UV–vis spectroscopy is highly sensitive to π–π^* transitions of unsaturated organic compounds, particularly carbocations, making it a complementary technique to IR spectroscopy for characterizing olefinic or aromatic coke species on zeolites. Specifically, UV–vis spectroscopy is capable of distinguishing aromatic species of different ring numbers [8]. The evolution of coke species, from unsaturated hydrocarbons to aromatic species, has been identified during reactions such as DME carbonylation, methanol to hydrocarbons, and alkane conversion [8, 11–13]. For example, Jentoft and co-workers [13] found that cationic species derived from butene are responsible for catalyst deactivation, as evidenced by in situ diffuse reflectance UV–vis spectrum. By combining with in situ IR spectroscopy, the authors further identified alkyl-substituted cyclopentenyl cations as key coke precursors, characterized by an intense absorption band at 292–296 nm [12].

NMR spectroscopy is another widely used technique for investigating zeolite deactivation. Benefiting from the development of sophisticated techniques such as magic angle spinning (MAS) and cross-polarization (CP), high-revolution solid-state NMR (ssNMR) enables the detection of ^{1}H and ^{13}C nuclei in coke. Theoretically, both ^{1}H and ^{13}C MAS NMRs can be employed to characterize the coke structure. However, given that adsorbed water on zeolite strongly perturbates the signals of ^{1}H MAS NMR, ^{13}C MAS NMR is more widely applied to investigate coke by measuring the chemical shifts of carbon nuclei, which generally range from 0 to 300 ppm. Typically, aliphatic compounds ($\delta = 10–40$ ppm), olefinic/aromatic compounds ($\delta = 125–145$ ppm), polyalkylated aromatics ($\delta = 140–150$ ppm), oxygen-containing hydrocarbon species ($\delta < 50$ ppm), and carbocations ($\delta = 150–180$, > 200 ppm) can be easily distinguished. For example, during the DME carbonylation reaction, more intense signals at 10–40 ppm (aliphatic compounds) and 130 ppm (aromatics) were observed as the reaction progressed on MOR [8]. These observations can be interpreted as the formation of alkyl-substituted aromatics [14], consistent with intermediates in the "hydrocarbon pool mechanism," which has been proved in MTH reaction on zeolites. Moreover, the broadening of the aliphatic signal after 10 hours indicated the growing heterogeneity of the alkyl-substituted aromatic species in the coke. Notably, a similar chemical shift of olefins and aromatics in ^{13}C MAS NMR makes it challenging to distinguish between them, requiring a combination of other complementary techniques such as UV–vis for accurate identification and more comprehensive understanding of coke.

Although the aforementioned spectroscopic techniques enable nondestructive analysis of zeolite matrix and coke deposits, it remains challenging for accurate determination of complete coke structure on the basis of response to characteristic functional groups, owing to the complexity of coke compounds. To address this limitation, Guisnet and Magnoux [15] developed a method to extract coke from deactivated zeolite. The procedure involves complete dissolution of the zeolite matrix in HF solution, followed by sequential separation steps. The "soluble coke" (including olefinic, naphthenic, single-ring, and double-ring aromatic compounds) can be first extracted by dichloromethane, while the insoluble coke species are subsequently collected via centrifugation. The mixture of "soluble coke" can be further separated using GC or high-performance liquid chromatography (HPLC), and then analyzed by an MS detector. On the other hand, insoluble coke typically appearing as black particles composed of polyaromatic species, can be further characterized by Transmission Electron Microscopy (TEM) coupled with Electron Energy Loss spectroscopy (EELS). The morphological feature of insoluble coke can be visualized in the TEM image, and the characteristic peaks (in the range of 280–300 eV) in the EELS spectra are associated with various carbon forms including graphite, amorphous carbon, pentacene, coronene, etc. (see Figure 4.1) [16]. Specifically, the peak at ~286 eV is assigned to $1s \rightarrow \pi^*$ transition and the peak at ~293 eV is due to $1s \rightarrow \sigma^*$ transition [17]. Additionally, matrix-assisted laser desorption/ionization time-of-flight mass spectrometry (MALDI-TOF MS) provides more detailed information of both "soluble coke" and "insoluble coke," including molar mass

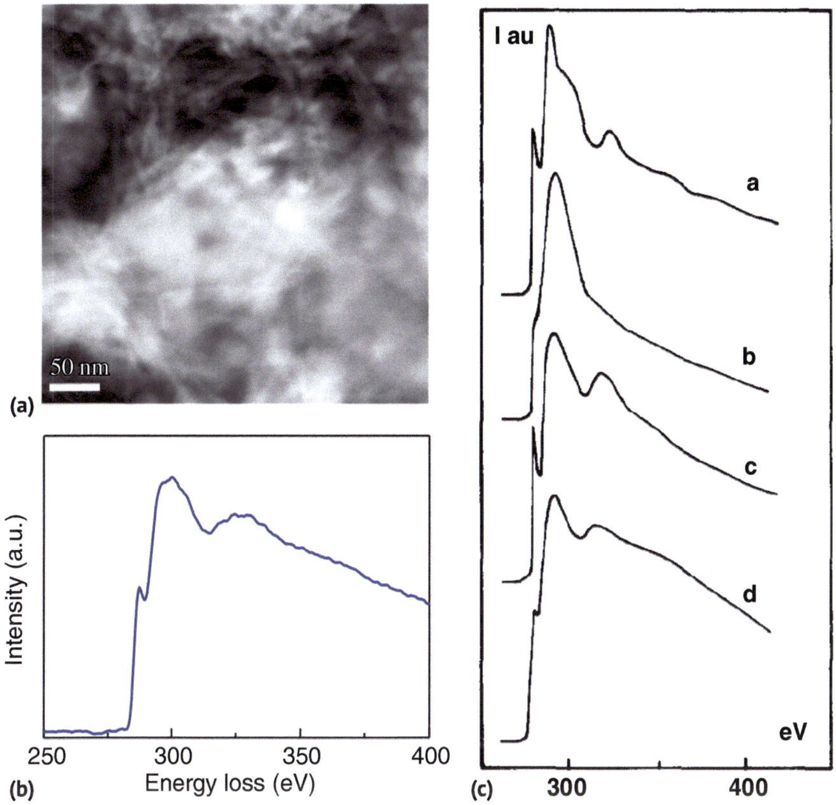

Figure 4.1 (a) TEM image. (b) EELS spectrum of insoluble coke in the spent H–MOR. *Source*: [8] / with permission of Elsevier. (c) Carbon K-edge EELS spectra taken on reference compounds: a: graphite, b: amorphous carbon, c: coronene, and d: pentacene. *Source*: [16] / with permission of Elsevier.

distribution of individual coke species, particularly suitable for polyaromatic coke species with *m*/*z* ranging from 300 to 1200 [18, 19]. Furthermore, matrix-assisted laser desorption/ionization Fourier-transform ion cyclotron resonance mass spectrometry (MALDI FT-ICR MS) is also applied for coke analysis due to its superior resolution and accuracy [20]. Liu and co-workers [21] integrated MALDI FT-ICR MS with isotope labeling to identify the coke species on SAPO-34 catalysts after MTO reaction, unveiling a complete pathway for aromatic hydrocarbon evolution.

4.2.1.2 The Characterization of Coke Location

Different from supported catalysts, coke location in zeolites is considerably more complex owing to the diverse topological structures and spatial distribution of acid sites. Furthermore, the formation of coke is also strongly influenced by the nature of reactant molecules. Influenced by the aforementioned factors, the location of coke can vary significantly within different channels and/or on the external surface of the zeolite matrix. As coke position is an important clue to elucidate deactivation mechanism and largely governs the deactivation behavior, studying its distribution provides guidance in prolonging zeolite lifetime. To date, a variety of characterization techniques, including TEM, X-ray photoelectron spectroscopy (XPS), X-ray powder diffraction (XRD), IR spectroscopy, and adsorption techniques (physical or chemical), are widely utilized to clarify the coke location.

TEM provides high-resolution visualization of external coke deposits surrounding zeolite crystals, while the combined application of energy-dispersive X-ray spectroscopy (EDS) enables elemental mapping of carbon distribution. Furthermore, EELS offers detailed characterization of the chemical structure of external coke. A notable demonstration by Wang et al. [17] used EELS to analyze external coke on the (0 1 0) and (1 0 0) planes of anisotropic MFI zeolites, revealing nonuniform coke distribution on different planes. As a typical surface analysis technology, XPS is extensively employed to distinguish internal and external coke depositions. The relative positioning of coke can be determined through C/Si ratio analysis: an increasing C/Si ratio from XPS with total coke content indicates preferential accumulation on external surfaces, whereas a stable ratio suggests predominant deposition within zeolitic channels [8].

The presence of bulky coke molecules within the zeolite matrix may give rise to zeolite framework distortion, manifesting as shifts in the relative positions and changes in the intensities of XRD peaks. This sensitivity makes XRD a powerful tool for characterizing the internal coke. The extent of zeolite framework deformation usually correlates with the amount of coke and its C/H ratio, where the coke species with high C/H ratios induce greater structural stress on the zeolite framework. Such relationships enable quantitative assessment of coke accumulation through unit cell parameter variations. Building on this principle, Svelle and co-workers [22] developed a straightforward descriptor using the difference in lengths of the a- and b-unit cells to assess the degree of catalyst deactivation simply through XRD analysis of spent catalysts. This descriptor showed strong correlations with critical deactivation indicators, including residual acidity, BET surface area, and total coke content. Weckhuysen and co-workers [23] also reported lattice expansion of small-pore zeolites during MTO process, with H–SSZ-13 exhibiting a 0.9% expansion along c-axis—the most significant distortion observed among tested zeolites. The integration of operando XRD and UV–vis enabled simultaneous spatiotemporal tracking of framework deformation dynamics (via XRD) and chemical speciation analysis of coke (through UV–vis) in the reactor, providing mechanistic insights into deactivation processes.

The textural properties of zeolites including specific surface area and pore volume, also serve as typical indicators for identifying the location and quantity of coke deposits. These parameters can be determined by adsorption techniques such as N_2 adsorption–desorption isotherms analysis. The bulky coke species deposited within zeolite channels leads to measurable reductions in both specific surface area and pore volume. However, it remains a great challenge to distinguish the coke species in different channels. A promising solution involves probing the accessibility of residual acid sites in different channels by using basic probe molecules with varied kinetic diameters, which is opposite to the trend of coke amounts. Note that carbon deposits may physically induce blockage at channel intersections and chemically interact with basic molecules, introducing non-negligible systematic errors. As an alternative strategy, IR spectroscopy analysis of hydroxyl region provides direct insight into coke on Brønsted sites. Band deconvolution analysis provides detailed information of coke-induced changes in the bridging hydroxyl, including shifts in vibrational frequency and reductions in band intensity.

With the development of state-of-the-art time- and space-resolved techniques, the spatial distribution characteristics of coke within catalyst bed and individual zeolite crystal have become increasingly discernible. For instance, Weckhuysen and co-worker [24] employed spatiotemporal UV–vis spectroscopy equipped with high-temperature probes to investigate the coke distribution in Mg–ZSM-5 catalyst bed during MTO reaction (see Figure 4.2a). Compared with conventional ZSM-5 zeolite, the extended lifetime of Mg–ZSM-5 catalyst is associated with a slower progression of coke zone through the catalyst bed and reduced aromatic species, attributable to the decrease in Brønsted acid site density. The operando-computed XRD tomography was employed to monitor the progressive deactivation of industrial-scale zeolite extrudates [25]. It is found that the coke deposition initiates on the external surface of extrudates and propagates radially inward through the catalyst particles,

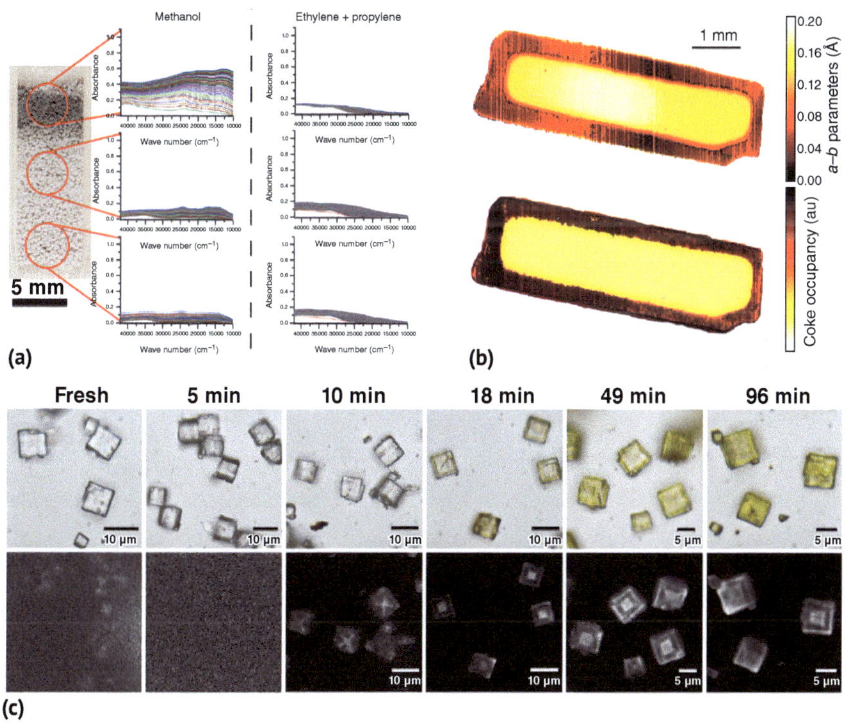

Figure 4.2 (a) The operando UV–vis spectra and corresponding reactor bed photo after two hours of conversion of methanol or ethylene + propylene on Mg–ZSM-5. *Source*: [24] / The Royal Society of Chemistry/CC BY-NC 3.0. (b) The XRD–CT slices through a horizontally mounted extrudate. *Source*: [25] / Elsevier / CC BY 4.0. (c) Bright field images and CFM images of SAPO-34 crystals after MTO conversion at 673 K for various reaction times. Fluorescence from the interior of the crystal. *Source*: Reproduced from [26] / with permission of Elsevier.

accompanied by a corresponding deactivation pattern (see Figure 4.2b). At a more microscopic scale, confocal fluorescence microscopy (CFM) technique is capable of visualizing heterogeneous spatial distribution of coke species within individual SAPO-34 crystals (see Figure 4.2c) [26]. Experimental observations revealed preferential accumulation of bulky coking compounds at the shell layer of crystals, leading to both increased intracrystalline diffusion resistance and reduced accessibility to active acidic sites. Combined with the tip-enhanced fluorescence microscopy (TEFL), the dynamic coke evolution on working catalyst was monitored. Significantly, a novel technique, named single-crystal electron diffraction (ED), has been developed for spatially mapping coke distribution within zeolitic micropores. In comparison to conventional XRD, ED enables ab-initio structural determination at sub-micrometer dimensions. In addition, the sensitivity for light atoms, particularly located near heavy atoms, is enhanced through improved electron–matter interactions. van Bokhoven and co-workers successfully implemented this technique to analyze the coke patterns in the micropores of ZSM-5 catalysts used in MTH conversion (see Figure 4.3) [27]. The results revealed polycyclic aromatic structure preferentially aligned within the straight channels, with the high-quality data permit enabling precise quantification of its occupancy to approximately 40% through advanced refinement technique. Herein, coke-associated signals correlated with catalytic activity decline demonstrated that initial coke nucleation predominantly occurs at the intersections of sinusoidal and straight channels, followed by gradual pore blockage within straight channels during advanced deactivation stages. In particular, atom probe tomography (APT) has been used to perform three-dimensional quantitative analysis of elemental distribution in catalyst materials with sub-nanometer-scale spatial resolution. Ramasamy and co-workers [28] developed APT to image the spatial distribution of hydrocarbon molecules within the pore network of spent HZSM-5 catalyst from surface to bulk at a single-particle level (see Figure 4.4), providing evidences for heterogeneity in Al distribution and

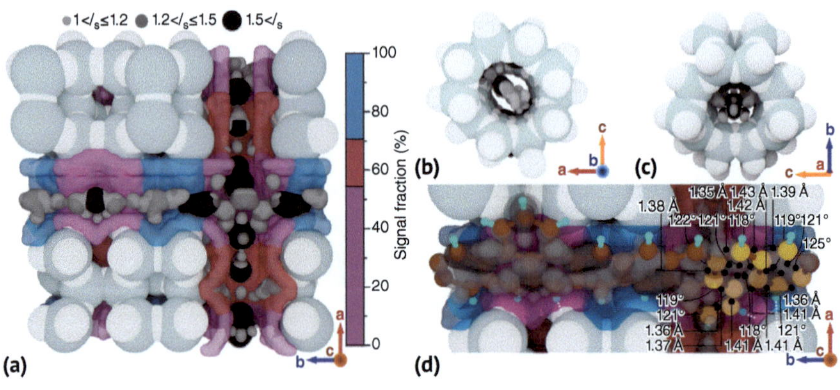

Figure 4.3 (a) Distribution of coke-associated signal in the zeolite unit cell as extracted from the direct ED maps of severely coked Zc catalyst. Views along the (b) straight and (c) sinusoidal channels. (d) The most intense coke-associated signals that protrude through the straight channel can be modeled as a molecule comprising three fused aromatic rings, with indicated bond lengths and angles. *Source*: [27] / John Wiley & Sons / CC BY-NC-ND 4.0.

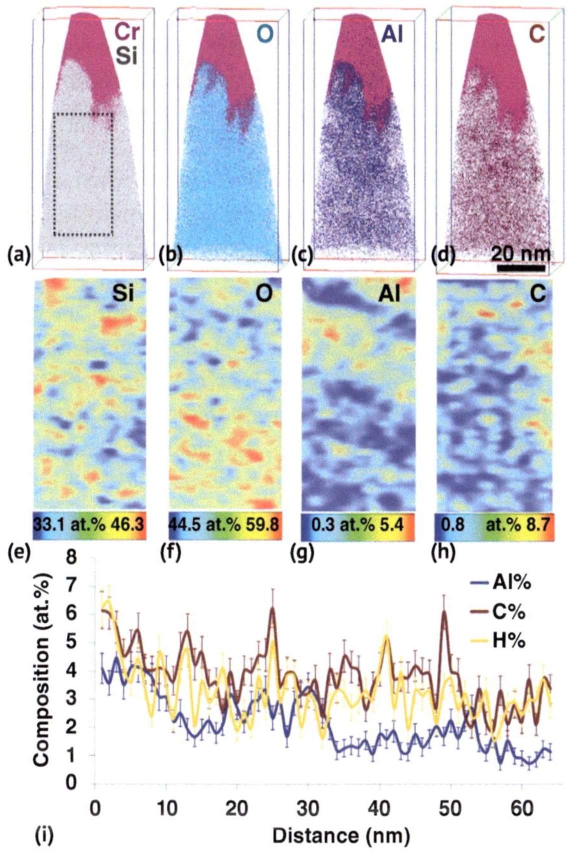

Figure 4.4 APT results of spent HZSM-5. Ion maps showing the Cr cap in pink along with (a) Si, (b) O, (c) Al, and (d) C. The 2D ion maps plotted using a slice of size 1 nm × 25 nm × 50 nm of APT reconstruction showing the (e) Si, (f) O, (g) Al, and (h) C distributions. (i) The 1D concentration profile from the surface of the HZSM-5 to the bulk. *Source*: (a, b, c, d, i) [28] / Springer Nature / CC BY 4.0.

pronounced carbon clustering phenomena. This technical advancement provides a novel methodology to investigate catalytic deactivation mechanisms with unprecedented spatial resolution, particularly those caused by hydrocarbon coking or active-site poisoning in zeolite systems.

4.2.2 Structure–Deactivation Relationship

Upon characterization and differentiation of diverse coke species, it is important to establish a relationship between zeolite architecture and deactivation behavior, which is a prerequisite for elucidating the coke formation mechanism. Zeolites with different topological frameworks and varied elemental compositions exhibit diverse chemical properties and textural characteristics, which in turn influence the nature of coke species and deactivation behaviors.

4.2.2.1 Acidic Properties

Although the mechanism remains under debate, coke formation is generally accepted as the predominant cause in acid-catalyzed reactions. Consequently, the acidic properties of catalysts play a crucial role in deactivation processes. Generally, a positive correlation exists between acid site density and coke content. For example, MCM-36 demonstrated threefold lower acid density compared to MCM-22, which significantly inhibits secondary reactions such as aromatization and hydrogen transfer. Consequently, the formation of aromatics and coke precursors was suppressed during the MTH process [29]. Comparative studies by Liu and co-workers [30] demonstrated that despite the conversion of methanol over HZSM-22 and SAPO-11, both proceeded via an alkene methylation-cracking pathway, the more acidic HZSM-22 catalyst exhibited higher reactivity than SAPO-11, while the moderate acidic strength of SAPO (silicoaluminophosphate) benefited the stability of the MTH reaction. Given that the SiO_2/Al_2O_3 ratio is intrinsically associated with acid density in zeolites, numerous researches have focused on its influence on deactivation. In HZSM-5 with high SiO_2/Al_2O_3 ratio, the deposited carbonaceous species are mainly mono- or bi-aromatics, which exhibit limited impact on catalytic activity. In contrast, low SiO_2/Al_2O_3 promotes the formation of polycyclic aromatics containing three to four fused rings, leading to substantial deactivation within 24 hours [31]. Fan and co-workers [32] achieved enhanced catalytic stability by selective removal of the acid sites in side pockets of H–MCM-22, suggesting the critical importance of both acid density and spatial distribution in catalyst design.

Beyond the well-documented Brønsted acid sites, Lewis acid sites (LASs) and silanol defects also function as active contributors to coke formation. Migliori and co-workers [33] found that strong LASs promote the formation of by-products (mainly methane) and coke deposition during methanol dehydration. Beato and co-workers [34] found that apart from the commonly recognized deactivation of ZSM-5 due to coking at external surface and irreversible dealumination, coking at internal defects of MFI also plays an important role, which is demonstrated to be in positive correlation with the silanol density. Gascon and co-workers [35] also reported that the silanol defects in Sigma-1 are responsible for faster deactivation rate by analyzing the functional surface groups. Selective removal of silanol defects under a mild treatment in the presence of NaOH/CTAB has been proved to be an efficient way to enhance zeolite's lifetime.

4.2.2.2 Textural Structure

Textural structure of zeolites is another important factor, which governs the deactivation behavior through spatial confinement and mass transfer modulation. For example, the hierarchical pore system of MCM-36 facilitates the diffusion of bulky species in MTH process, resulting in a lower deactivation rate compared to the purely microporous MCM-22 catalyst [29]. The presence of mesopores shortens the diffusion path, benefiting the escape of coke precursors. Theoretically, decreasing the crystallite size of zeolites has an analogous effect. Liu and co-workers [36] systematically investigated crystallite size effects using SAPO-34 catalysts. The ultrathin sheetlike SAPO-34 (20 nm) exhibited the slowest coking rate and longest catalyst lifetime in MTO reaction. The improved performance stemmed from enhanced mass transfer as well as increased accessibility of active sites. Weckhuysen and co-workers [37] manipulated the crystal size of ZSM-11 from 750 to 150 nm, resulting in an eightfold increase in catalyst lifespan. The reduced crystal size and porous architecture greatly affected the evolution of hydrocarbon pool (HCP) and coke speciation during the initial reaction stages. Analogously, both coke formation and deactivation behaviors were compared on nano- and micro-ZSM-5 catalysts [38]. The nanosized catalyst showed a sevenfold extension in service duration. Further characterizations revealed contrasting deactivation mechanisms: microcrystalline catalysts underwent fast deactivation due to rapid internal pore blockage and active sites coverage, whereas coke formed preferentially on the external surface of nanocrystalline catalyst. Notably, the external carbonaceous deposits on nanosized catalysts evolved into porous graphitic structure through non-catalytic reaction with methylated benzenes, showing non-detrimental effect on the catalytic performance [2].

In addition to the aforementioned factors, channel or cage size also governs the variety of coke species and deactivation behavior. A representative case is the ITQ-33 framework, which features 18- and 10-ring channel systems. After deactivation in MTH reaction, the quantity of carbonaceous deposits (soluble in CH_2Cl_2) retained within the catalyst pores resembled the level typically observed in conventional 10-ring zeolites. However, complementary characterization by FTIR and DRUV–vis spectroscopy revealed the concurrent formation of heavier, insoluble polyaromatic coke species. Thus, it can be interpreted by the spatial confinement effects that methylbenzenes are formed predominantly in the 10-ring channels, while bulkier, insoluble species predominantly accumulate in the larger 18-ring channel [39]. These observations align with the study by Svelle and co-workers [40], who reported channel size as the dominant parameter of catalyst lifetime and coke characteristics. Notably, in structurally constrained frameworks like ZSM-22 and SAPO-34, where internal void accessibility is limited, activity loss correlates strongly with the accumulation of the insoluble coke species. In large-pore zeolites such as mordenite and Beta catalysts, preferential accumulation of insoluble coke species is observed. Additionally, two distinct deactivation behaviors were observed over zeolites with different topological structures during methanol conversion [41]: HZSM-5 exhibits two-staged deactivation kinetics, whereas HSAPO-34 displays an exponential-type deactivation profile. Mechanistic studies suggested that the deactivation of HZSM-5 stemmed from the "overloading effect" of methylbenzenes, whereas HSAPO-34 deactivation arose from progressive accumulation of inactive methyladamantanes. Spatial analysis further revealed the deactivation proceeded from "external to internal" in HSAPO-34. Notably, enhancement of catalyst diffusivity significantly extended the lifetime of HZSM-5, but showed comparatively limited efficacy in HSAPO-34.

The formation of coke species exhibits analogous dependence on the cage geometry of zeolite frameworks. Even subtle variations in cage dimensions (size, shape, and pore structure) result in the generation of different HCP species, thereby deeply affecting the deactivation behaviors. Three SAPOs with eight-membered ring but different cage structures, SAPO-34, SAPO-18, and SAPO-35, were investigated in MTO reaction. Both reaction temperature and cage size have a significant effect on coke species. At 300 °C, monocyclic aromatic species dominated across all zeolites, irrespective of cage geometry. However, when the temperature was raised to 400 °C, polycyclic and bicyclic aromatics appeared on SAPO-34 and SAPO-18, whereas SAPO-35 exclusively retained monocyclic aromatics. This marked difference indicated pronounced spatial confinement imposed by the restrictive topology of SAPO-35 zeolite [42]. Additionally, Pinilla-Herrero et al. [43] systematically compared four eight-ring SAPOs with cavity windows of various dimensions (SAPO-35 [LEV], SAPO-56 [AFX], STA-7 [SAV], and SAPO-42 [LTA]) in MTO reaction. Their findings also confirmed framework topology-dependent deactivation behaviors and coke mechanisms. The narrow channels of SAPO-35 effectively prevented the formation of bulky aromatics, while the large cages of SAPO-56 allowed the formation of heavy molecules inside the catalyst. Weckhuysen and co-workers [44] systematically investigated the influence of framework topology on MTO reaction using three small-pore zeolites (CHA, DDR, and LEV), all featuring different cages interconnected by eight-ring windows. The large cages of CHA promoted the formation of alkylated benzenes and naphthalenes. In the DDR cage, 1-methylnaphthalene preferentially formed, while LEV with more constrained cages exhibited limited HCP species. The zeolite channel geometry–reactive intermediate relationships were studied in detail using anisotropic zeolite ZSM-5 crystals for the MTH process [45]. Slightly more constrained sinusoidal channels favored the olefin cycle by promoting the homologation of alkanes, whereas the straight channels facilitated the aromatic cycle, achieving a higher degree of alkylation of aromatics. Notably, dynamic nuclear polarization revealed that the preferential formation of heavy aromatics occurs in the sinusoidal channels, providing further insight into catalyst deactivation.

4.2.3 The Coke Formation Mechanism

Coke deposition on zeolites primarily proceeds through dehydrogenation and hydrogen transfer reactions, where hydrogen is continuously abstracted from retained carbonaceous species trapped within the zeolite framework. The olefin methylation-cracking mechanism has been established as the main route for the coke formation from methanol [46], evolving through the following stages: (1) introduction period: C—C bond formation via direct coupling of methanol derivatives initiates light olefin production [47]; (2) transition stage: sequential cyclization and hydrogen transfer reactions generate MAHs; (3) autocatalytic stage: a steady reaction period sustained by the HCP mechanism including both alkene- and aromatic-based cycles [48, 49]; and (4) deactivation stage: progressive accumulation and growth of HCP intermediates to form bulky PAHs, which poison the acidic sites and block pore channels, leading to catalyst deactivation.

4.2.3.1 Formation of First C—C Bond

The formation of the initial C—C bond is crucial for both main reaction pathway and coke formation mechanism. Limited by the experimental techniques available in the past decades, the presence of trace impurities (such as toluene, ethanol, or acetone) was assumed to be responsible for the direct C—C bond formation [50]. More than 20 direct mechanisms have been proposed, including carbene-mediated pathway [51], free radical chain reactions [52], oxonium ylides mechanism [53], and methane–formaldehyde mechanisms [54] (see Figure 4.5). However, most of them were deemed kinetically unfavorable due to high reaction energy barriers or insufficient experimental validation [55, 56]. In 2006, Hunger and co-workers systematically demonstrated that traces of organic impurities have no significant influence on HCP species or product distribution during the MTH reaction [57]. The high reactivity of surface methoxy species (SMS) on solid acid catalysts was identified [58, 59], prompting renewed scientific interest in direct C—C coupling mechanisms. Subsequent studies have provided both experimental and theoretical evidences supporting the direct mechanism during the initial stages of MTH reaction.

Among the various proposed direct mechanisms, the SMS formed on Brønsted acid sites is the most experimentally verified intermediate, playing a critical role in C—C bond formation during zeolite-catalyzed hydrocarbon conversion. Based on NMR spectroscopy and carbene-trapping experiments, Hunger and co-workers first proposed carbene-insertion mechanism involving SMS [59, 60]. Their work suggested that the carbene-like character of SMS originates from the polarization of the C—H bond induced by adjacent oxygen atoms within the zeolite framework (see Figure 4.6a–c) [59]. Subsequently, Kondo and co-workers [61, 62] conducted IR experiments and found that C—C bond-containing hydrocarbon species formed via coupling between the carbene-like SMS intermediate and either methanol or DME, providing complementary evidence for the carbene-insertion mechanism (see Figure 4.6b). While the methane–formaldehyde mechanism, first proposed by Hutchings et al. [53] in 1987, originally suggested the direct C—C bond coupling between methane and formaldehyde derived from methanol disproportionation (see Figure 4.6d), but subsequent theoretical studies revealed its thermodynamic improbability due to an unrealistically high energy barrier [55]. Fan and co-workers [63, 64] developed the methoxymethyl cation mechanism based on experimental and theoretical studies. This mechanism involves the formation of a methoxymethyl cation ($CH_3OCH_2^+$) from SMS and DME, which then undergoes direct C—C coupling with

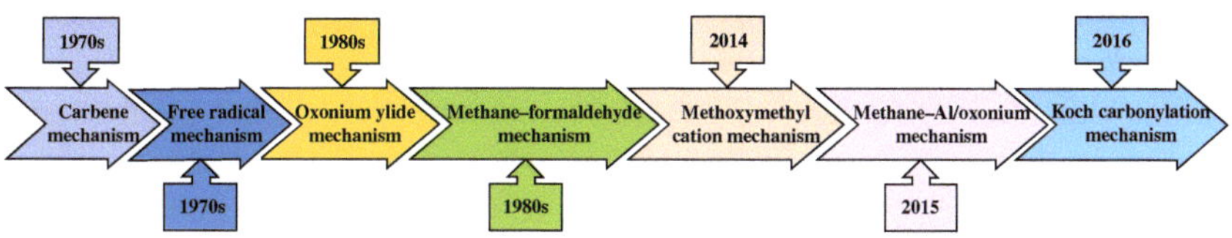

Figure 4.5 The development of the direct mechanism in MTH conversion in recent decades. *Source*: [49] / with permission of Elsevier.

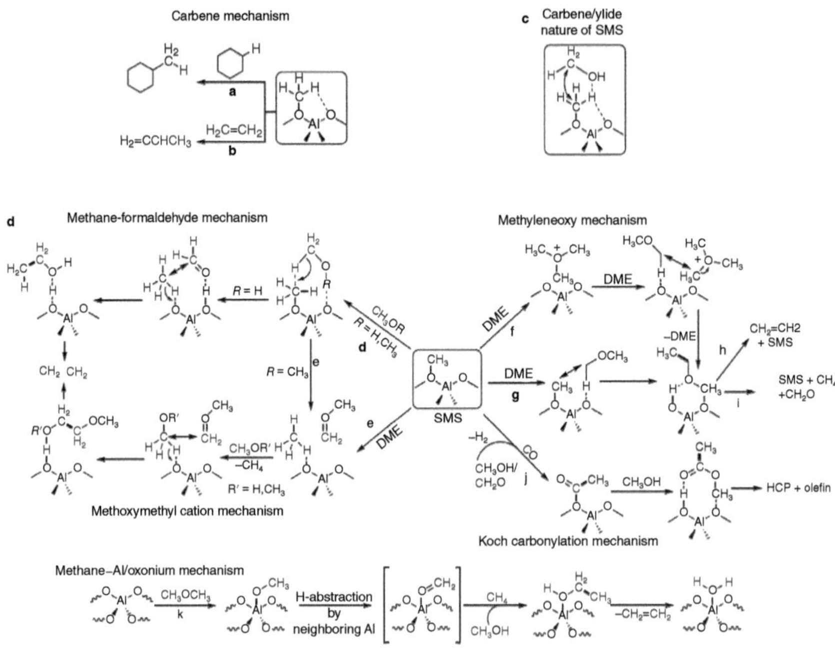

Figure 4.6 Several proposed direct mechanistic routes during the early stages of the zeolite-catalyzed MTH process. The proposed mechanisms of methanol conversion reaction between SMS and (a) cyclohexane and (b) ethylene via the carbene-type intermediate. (c) A possibility for C—H insertion into methanol by carbene-like SMS species. Schematic illustration of (d) the conventional methane–formaldehyde mechanism and (e) its modified analog namely the methoxymethyl cation mechanism. The methyleneoxy mechanism of the MTH reaction, involving either the (f) TMO or (g) SMS intermediate. The SMS-involving route initiates the formation of both (h) ethylene and (i) methane/formaldehyde simultaneously. (j) Plausible routes for the direct formation of the C—C bond during the early stages of MTH reaction through Koch-type carbonylation of SMS. (k) The methane–aluminum/oxonium mechanism for the conversion of methanol over alumina. *Source*: [70] / Springer Nature.

another DME or methanol molecule to form $CH_3OCH_2CH_2OR$ ($R = H, CH_3$) (see Figure 4.6e). Furthermore, Liu and co-workers [65] proposed the methyleneoxy mechanism (see Figure 4.6f, g) based on in situ ssNMR spectroscopy and GC–MS data. In this mechanism, the C—C bond formation initiates through the interaction of the reactants (methanol or DME) with surface intermediates such as SMS and trimethyloxonium (TMO) ions. The crucial C—H bond activation step is assisted by framework oxygen, generating active methyleneoxy species (*R*—O—CH_2—H...zeolite) that ultimately drive the simultaneous formation of olefin (see Figure 4.6h) and either methane or formaldehyde (see Figure 4.6i). Another interesting work by Copéret and co-workers [66] reported the direct C—C bond formation between DME and methane, catalyzed by extra-framework aluminum species in acidic zeolites (see Figure 4.6k). This mechanism involves the formation of a transient aluminum-oxonium species through hydrogen abstraction from an aluminum-bound methoxy species. This aluminum–oxonium species then undergoes C—C coupling with methane to yield a surface-bound ethanolic species.

Lercher and co-workers [67] and Weckhuysen and co-workers [68] nearly simultaneously proposed a novel Koch carbonylation mechanism in the MTO reaction, involving SMS carbonylation to generate acetate species containing the first C—C bond (see Figure 4.6j). These acetyl groups react with water or methanol to generate acetate acid or MA species, respectively. Operando UV–vis diffuse reflectance spectroscopy (DRS) and ssNMR studies on H-SAPO-34 catalysts provided direct evidence for acetyl intermediates during MTO reaction, confirming the involvement of SMS in initiating the first C—C bond. Further investigations by Dai and co-workers [69] on H-ZSM-5 catalysts corroborated this carbonylation mechanism in the early stages of MTO reaction. The presence of acetaldehyde intermediates containing C—C bonds was confirmed using online MS and ^{13}C MAS NMR spectroscopy, while the successive participation of SMS, CO, and H_2 in acetaldehyde formation was experimentally verified for the first time. These findings suggest that both methanol and formaldehyde can act as carbonylation raw materials under MTH reaction conditions.

4.2.3.2 Formation of MAHs

Following a short induction period associated with the formation of the direct C—C bond, it is essential to discuss the subsequent formation of MAHs, in which olefinic species and aldehydes serve as significant coke precursors.

Among the various proposed mechanisms for olefin generation, the formation of the initial C—C bond through carbonylation of SMS to produce ketene stands out as particularly plausible. This pathway appears more favorable compared to alternative routes such as the methane–formaldehyde mechanism or its derivatives [71]. Notably, during DME carbonylation, the presence of large amount of CO significantly promotes ketene formation, which accelerates catalyst deactivation rates even at lower temperature compared to MTO process [72]. Although the high active ketenes ($R_2C{=}C{=}O$) cannot be directly observed experimentally, their formation can be inferred indirectly from the detection of surface acetate species [47]. Theoretical studies suggest that ketenes undergo successive methylation and decarbonylation steps to form monocyclic olefins during MTH reaction over zeolites [71]. Notably, these "short-lived" ketenes are also considered as primary coke intermediates on H-MOR during DME carbonylation reaction, accounting for pronounced catalyst deactivation [72]. Together, experimental and computational evidence underscores the critical role of ketenes as intermediates for subsequent coke formation. However, their short lifetime and low concentration make it exceedingly challenging to detect them in MTO reaction. Hemberger and co-workers [73] provided experimental evidence for the formation of methylketene ($CH_3{-}CH{=}C{=}O$) via ketene methylation over HZSM-5 zeolite by using operando synchrotron photoelectron photoion coincidence (PEPICO) spectroscopy. This work offers direct spectroscopic confirmation of the ketene-to-ethylene pathway, which involves methylation followed by decarbonylation to produce ethylene as the primary product. Furthermore, olefins derived from ketene intermediates can subsequently undergo various reactions, including oligomerization, cyclization, hydrogen transfer, and alkylation, leading to the formation MAHs.

Aldehydes, particularly formaldehyde, play a critical role in the formation of MAHs and contribute to catalyst deactivation in the MTH process [74, 75]. The generation of formaldehyde under MTH conditions occurs through three primary pathways, including hydride transfer between two methanol molecules [76], thermally induced or reactor-wall-catalyzed decomposition of methanol [67], and hydrogen transfer from methanol to alkenes on LASs [77]. Once formed, formaldehyde reacts with surface acetyl groups to form acrylic acid and molecular intermediates. Subsequent decarboxylation of these carboxylic acids yields alkenes, which will transform into aromatics and expedite the formation of coke [78]. Formaldehyde also participates in the polymerization of aromatic compounds (e.g., diphenylmethane) [79]. Lercher and co-workers demonstrated that formaldehyde promoted the formation of H-deficient by-products, particularly monoenes and dienes [80]. Further probing kinetics revealed that dienes exhibit higher reactivity with HCHO, leading to faster deactivation rate in comparison with monoenes [81].

4.2.3.3 Formation of PAHs

The progressive accumulation of MAHs leads to the formation of bulky PAHs, contributing to the rapid deactivation of zeolite catalysts. To gain comprehensive mechanistic insights into zeolite deactivation, elucidating the PAH formation pathway becomes imperative. Previous studies have proposed that PAHs are formed through the oligomerization of methylbenzene with alkenes, followed by subsequent cyclization and hydrogen transfer reactions [82, 83]. However, the specific formation pathways of PAHs remain a subject of ongoing debate.

Li and co-workers [84] employed operando UV–Raman spectroscopy to investigate the formation of PAHs in H–ZSM-5 during methanol dehydration, suggesting methylbenzenium carbenium ions (MB^+) as critical precursors. Then, Xu and co-workers [85] documented a concurrent decline in cyclopentenyl cation and methylbenzene concentrations with the formation of dicyclic naphthalene species during MTO reaction. Through advanced 2D $^{13}C-^{13}C$ proton-driven spin diffusion (PDSD) NMR analyses, the authors proposed π-interactions mechanism for the formation of PAHs. The π-interactions induced spatial proximities between cyclopentenyl cations and aromatics within confined channels and cages of zeolites, thereby facilitating the formation of naphthalene. Recently, Liu and co-workers [86] proposed two possible pathways for the transformation of methylbenzenes to methylnaphthalenes through pioneering characterization of key intermediate species, including tetrahydro-1,8-dimethylnaphthalene, dihydro-1,5,6-trimethyl-1*H*-indene, and 1,2-dimethyl-3-(2-butenyl)benzene (along with their possible isomers), which were conclusively identified as the precursors of naphthalene derivatives (see Figure 4.7). Theoretical calculation confirmed the feasibility of these routes. The formation of alkylbenzenes with bulky side chains markedly boosted methylnaphthalene generation, while the predominant reaction pathway depended on the

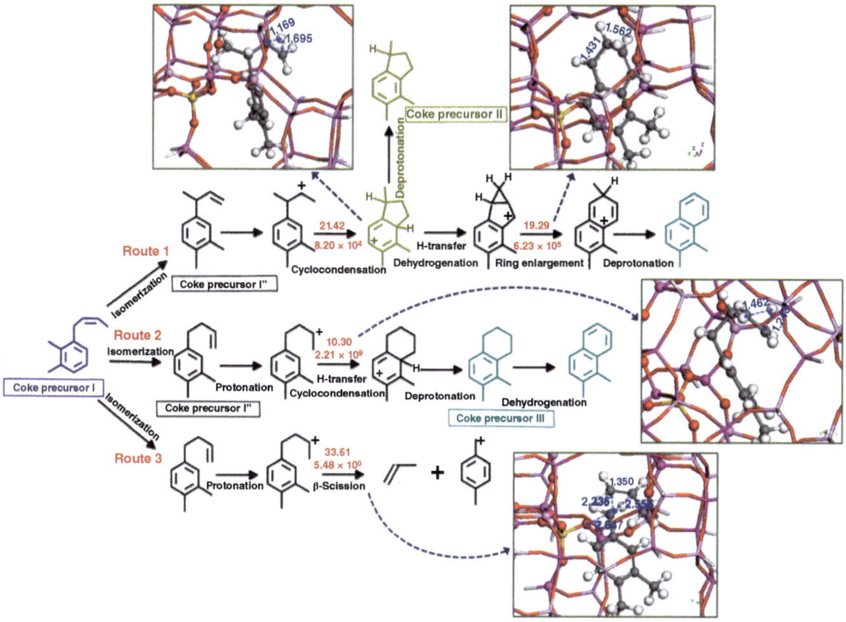

Figure 4.7 The possible routes of coke precursor evolution over H–SAPO-34. Calculated free energy barriers and reaction rate constants at 350 °C are given in kcal·mol^{-1} and s^{-1}, respectively. Optimized transition state structures of these critical steps are presented near the corresponding reactions. *Source*: [86] / The Royal Society of Chemistry.

different methanol conversion stages. The deactivation inevitability stemmed from kinetically preferential methylnaphthalene formation (lower energy barrier) competing against propene elimination (higher energy barrier) within the confined organic transformation networks.

4.2.4 Mitigation of Deactivation

Based on the structure–deactivation relationship described above, zeolite property and reaction conditions are two primary factors that greatly affect coke formation and deactivation behavior. We have summarized the reported strategies to mitigate zeolite deactivation from these two aspects in Figure 4.8.

4.2.4.1 The Zeolite Catalyst Design and Modification

In terms of zeolite design, weakening acidity as well as shortening the diffusion pathway of reactants, intermediates, and products is an efficient approach for mitigating zeolite deactivation.

Post-treatment of zeolite in inorganic or organic acid solution (such as HNO_3, H_2SO_4, and citric acid) results in framework dealumination, resulting in a reduction of acid sites. Furthermore, basic molecules (e.g., pyridine and ammonia) can be used to cover the strong acid sites, which are typically responsible for coke formation. For instance, pre-adsorption of pyridine selectively shields the Brønsted acid sites in

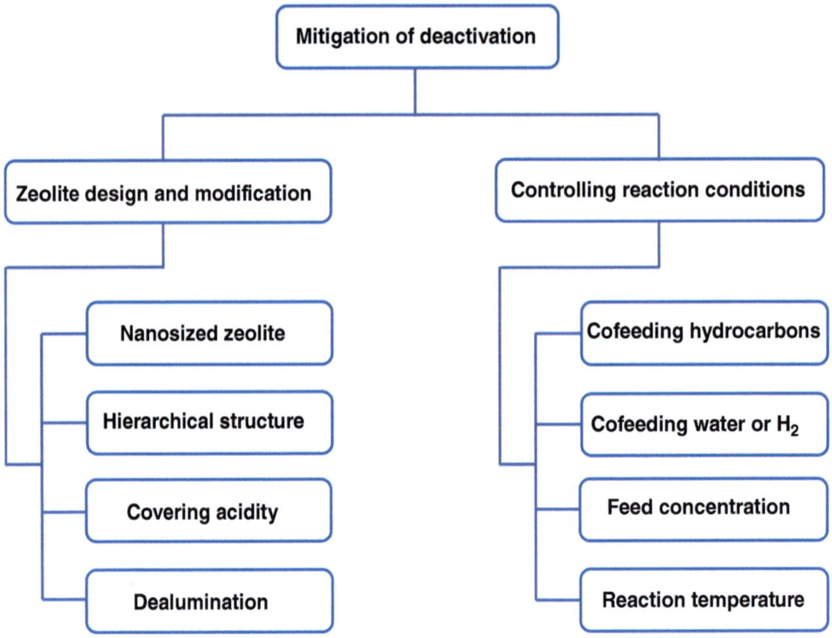

Figure 4.8 The reported strategies to mitigate deactivation on zeolites.

the 12-membered ring of MOR, thereby inhibiting the coke formation during DME carbonylation, which significantly enhances catalyst stability [87, 88]. Besides, inorganic salts, e.g., $ZnCl_2$, are capable of transforming Brønsted acid sites into LASs, slowing down the coke formation (−42%) and expanding catalytic lifetime (+80%) in MTO reaction [89].

During the MTO reaction, weakening external acidity is theoretically beneficial to zeolite lifetime by decreasing the external coke. Weckhuysen and co-workers [90] constructed core–shell-structured H-ZSM-5/Silicalite-1 composites, in which the Silicalite-1 shell prevented the formation of extended graphite-like coke. Analogously, Kwak and co-workers [91] deposited a thin film of gadolinium oxide on the surface of ZSM-5 crystals to enhance its surface basicity. Uniformly distributed Gd atoms on the surface help to modify the surface acid–base properties of the parent ZSM-5 catalyst, reducing the adsorption strength of coke precursors (i.e., carbenium and carboxylate ions). This effect also hinders the growth of coke precursors into polyaromatic structured coke on the external surface, which thereby contributes to extending the lifetime of the catalyst without altering the pool mechanism. A similar result was achieved on CeO_2-coated FER zeolite [92]. The selective coating of CeO_2 on external surface not only prolongs the life of zeolite, but also regulates the catalytic selectivity. Coincidentally, the ZrO_2-coated SAPO-34 also showed a longer stability than unmodified SAPO-34. The removal of acid sites on external surface contributed to suppressing coke formation on the pore mouths of SAPO-34 [93].

The abundant intracrystalline mesoporosity is advantageous for improving the diffusion of coke precursors in zeolite crystal, resulting in a longer catalyst lifetime. Dealumination or/and desilication are the commonly used top-down methods to introduce mesoporous structure by widening the original micropores. Upon acid or alkali treatment, the ZSM-5 catalyst lifetime can be efficiently prolonged by several times due to the increased mesoporosity and decreased acid density [94]. In contrast, engineering hierarchical zeolites by direct synthesis is a bottom-up strategy. Han and co-workers [95] systematically compared the catalytic properties of hierarchically porous Beta-MS and conventional microporous Beta-C for the MTH process. On the one hand, the enhanced diffusion of coke intermediates in Beta-MS results in a remarkably longer lifetime. On the other hand, the mesoporosity also enables Beta-MS to exhibit a higher coke tolerance. This strategy is also applicable to monolithic zeolite. Xie and co-workers [96] introduced intracrystalline mesopores into the full-zeolitic monoliths by recrystallizing the shaped binder/zeolite body with the assistance of controlled organic template steam. A remarkable mass transfer advantage led to higher propylene yield and slower deactivation rate in C_4 olefin cracking reaction. Moreover, the prolonged cycle time was more than 2000hours and three times as long as that of commercial catalyst in MTH reaction under industrial condition.

In addition, well-interconnected macropores are also beneficial for improving catalyst lifetime (e.g., in MTO reaction), owing to the enhanced mass transfer properties [97]. Weissenberger et al. [98] prepared mesoporous ZSM-5, macroporous ZSM-5, and mesoporous/macroporous composite, respectively. Both mesoporous and microporous ZSM-5 zeolites exhibited three times longer catalyst lifetime compared to a conventional catalyst. It must be noted that macropores and mesopores play different roles in improving the catalyst stability. The macropores reduced coke formation rate while additional mesopores predominantly enhanced the resistance against coke deposits. Therefore, the combination of intracrystalline meso- and macropores leads to a tenfold increase in catalyst lifetime.

Synthesis of nanosized zeolite is another strategy to decrease coke formation by shortening diffusion pathway. Corma and co-workers [99] synthesized a nanosized SSZ-13 zeolite via a one-pot procedure by adding a surfactant. The obtained material exhibits a much longer lifetime in MTO reaction, mainly due to the presence of smaller crystallites and intercrystalline mesopores. The ease of diffusion results in more efficient use of the catalyst as well as slower deactivation behavior.

4.2.4.2 Controlling Reaction Conditions

Perez-Uriarte et al. [100] studied the effects of operating conditions (i.e., co-feeding methanol and water, temperature, space–time, and feed concentration) on reaction performance of the transformation of DME to light olefins. The catalyst deactivation caused by coke is accelerated by increasing the DME concentration in feed and is attenuated by the formation of water in reaction medium (as a consequence of the increased temperature and/or space–time). On the one hand, water may trigger the dealumination of zeolite, resulting in irreversible deactivation. But, on the other hand, coke formation is usually inhibited in the presence of water. Therefore, the primary deactivation factor, namely coke formation or dealumination, mainly depends on the concentration of water [101]. Jun and co-workers reported that the stability of HZSM-5 is expected to be improved by adding a proper amount of water into the feed, despite a 20% decrease in methanol conversion due to catalyst deactivation caused by dealumination. This positive effect of water on extending the lifetime of HZSM-5 in the conversion of DME to olefins was also confirmed by Cordero-Lanzac et al. [102], suggesting that co-feeding an appropriate amount of water is an effective strategy to enhance catalyst stability. Liu and co-workers [103] reported a high-pressure MTO process with co-feeding H_2 and H_2O, which prolonged the catalytic lifetime of SAPO-34. The methanol handling capacity under these conditions was about 200 times greater than that under normal pressure condition. This ultralong lifetime originates from the hydrogenation ability of the acid sites on SAPO-34 for aromatic species, which slows down the condensation of coke species. A synergistic effect between H_2 and H_2O on prolonging the catalyst lifetime was observed at higher H_2O partial pressure, probably due to the lower barriers of hydrogenation in the presence of H_2O. Valecillos et al. investigated the intrinsic effect of co-feeding water on the formation of active/deactivating species in the MTH reaction over ZSM-5 zeolite. They found that a moderate amount of water selectively suppressed the deactivation rather than the main reaction by sweeping the retained species from the silanol sites and favoring the formation of olefins. The in situ UV–vis spectra revealed that the rate of formation and growth of discrete retained species decreases with the addition of water, with a more pronounced reduction observed for coke compared to coke precursors or active species [104].

When comparing DME and MeOH as feedstocks in MTO reaction, MeOH exhibits a profound negative impact on catalyst lifetime. This is attributed to the slower methylation rate of MeOH, coupled with the exclusive formation of formaldehyde from MeOH, which promotes hydrogen transfer reactions. Consequently, the increased formation of alkanes, arenes, and coke accounts for the substantial differences in deactivation behavior observed in the MTH process when using different oxygenate feedstocks [105].

Reaction temperature affects the nature of coke intermediates and generally, a low reaction temperature is indeed beneficial to the stability. At low temperature, deactivation occurs mainly due to the formation of inert adamantane species, while poly-condensed aromatic hydrocarbons play a major role at high temperature [35], accompanied by the blockage of pores in zeolites.

Under specific conditions, a controlled amount of coke deposition in zeolite can enhance catalytic activity or stability. Zhang et al. [106] demonstrated that the residual carbon in the catalyst modified its textural and acidic properties, which significantly improved the catalytic stability of Zn/HZ in methanol-to-aromatics (MTA) reaction. Kirchmann and co-workers [107] found that maximum olefin yield could be achieved directly by bringing the used catalyst back to the reaction stream while retaining an optimal amount of coke during regeneration. Furthermore, Li and co-workers [108] conducted a pre-coking process to mitigate the breakage of T—O—T bonds in the SAPO-34 catalyst during long-term continuous regeneration in the MTO conversion. The integration of catalyst pre-coking with water co-feeding has been established as an efficient strategy to enhance the catalytic efficiency and long-term stability of zeolite catalysts, offering valuable insights for the development of other stable zeolite catalysts in industry.

4.2.4.3 Multicomponent Coupling Reactions

Synergistic multicomponent coupling reactions offer a promising strategy to mitigate catalyst deactivation caused by coke deposition in zeolite-catalyzed systems. A representative example is the co-conversion of methanol and *n*-butane for olefin production, where the thermal integration was achieved by coupling exothermic methanol conversion ($\Delta H < 0$) with endothermic *n*-butane cracking ($\Delta H > 0$). This energy compensation strategy allows for a thermally balanced operation at an optimal methanol to *n*-butane molar ratio of 3 [109]. The olefins formed during *n*-butane cracking play a significant role for both activating the methanol transformation mechanism and suppressing coke formation. Similarly, water produced from methanol dehydration effectively inhibits hydrogen transfer reactions during *n*-butane cracking. This synergistic effect not only enhances the overall efficiency of butane cracking and olefin production, but also significantly reduces the catalyst deactivation caused by coking [110]. Furthermore, the coke-induced deactivation was kinetically modeled as a function of reactant and intermediate concentrations in the reaction system. These results demonstrate that the deactivation rate in this coupled system is notably slower compared to conventional MTO process [111].

For the MTP reaction, co-feeding *n*-butanol with methanol presents a positive effect on catalyst lifetime [112]. In particular, increasing the inlet methanol fraction dramatically accelerates catalyst deactivation, and co-feeding butanol significantly mitigates activity loss by enhancing methanol conversion. The deactivation kinetics study ascribes deactivation to the reactions between methanol and olefins/aromatics, and acknowledges the mitigating impact of water content in the reaction medium.

4.2.5 Regeneration of Deactivated Zeolites

Given that coke deposition is the primary cause of acidic zeolite deactivation in the transformation of light carbon molecules, coke removal is a prerequisite for catalyst regeneration. Consequently, developing reliable, efficient, and economic regeneration methods is highly desirable for industrial applications. In this context, various regeneration techniques, including oxidation (using air, ozone, and oxynitride), gasification (with carbon dioxide and water steam), and hydrogenation (with hydrogen), as well as studies on regeneration kinetics are summarized. These insights contribute to the design and optimization of catalysts and their associated processes.

4.2.5.1 Oxidation

The most widely used regeneration method to remove coke in industry is combustion in air or oxygen [113, 114]. Due to the highly exothermic nature of the regeneration process, it is crucial to remove the excess heat in the reactor–regenerator system, especially for exothermic reactions, such as MTO, MTH, and DME carbonylation reactions.

In addition, the water generated during combustion may cause dealumination of zeolite at high temperature, leading to irreversible damages of the catalysts [115].

The rate of coke removal during air combustion is dependent on multiple factors, such as the type of zeolite, the location of coke, and its nature. Generally, the rate is proportional to oxygen concentration. On the other hand, the pore structure affects the intracrystalline diffusion of oxygen, as well as the accessibility of coke to oxygen. Therefore, coke combustion typically occurs more rapidly over zeolites with larger pore size. Magnoux and Guisnet [116] compared the coke combustion rates in different catalysts and found that the oxidation rates of coke in HY and H–MOR were faster than those in HZSM-5. In addition to topological structure, the porous characteristics of zeolites also affect their regeneration behavior. Kaskel and co-workers [117] compared the regeneration behaviors of microporous and hierarchical ZSM-5 in MTH reaction. The microporous sample showed an almost constant reaction rate after each reactivation cycle, whereas the hierarchical sample showed a decline in reaction rate over three deactivation/reactivation cycles. This divergence is attributed to differences in coke location, which lead to varying degrees of damage to zeolitic structure during the reactivation process. Coke deposits on the external surface of zeolite crystals are more easily combusted due to the abundance of oxygen [118]. In addition, the H/C of the coke has a significant effect on combustion, with coke having a lower H/C tending to burn at a higher temperature.

Combustion of coke in air or oxygen atmosphere is usually carried out at high temperature. However, the highly exothermic nature of this process can easily lead to overheating, potentially causing dealumination of zeolite. In contrast, considering the strong oxidizing properties of ozone, coke can be removed at much lower temperature (50–200 °C) in the presence of ozone [119]. Querini [120] found that a combination of ozone treatment followed by hydrogen treatment is an effective method for regenerating coked zeolite. Ozone not only eliminates a large portion of coke but also alters its characteristics, making it easier to be removed. In addition to ozone, NO_x species are also potential and cost-effective oxidization agents capable of removing coke at low temperatures. Janssens and co-workers [121] found that coke removal can take place at temperatures approximately 200 °C lower in the presence of NO_2 compared to air atmosphere. Recently, Batiot-Dupeyrat and co-workers [115] developed a nonthermal plasma technology for regenerating the coked zeolite. This strategy utilizes short-lived oxygenated species as active species, allowing regeneration to be carried out at room temperature.

Kinetic tools are applicable for the simulation of reaction–regeneration systems, enabling the optimization of treatment time, temperature, atmosphere composition, and other parameters. For instance, in the joint transformation of methanol and *n*-butane, a methodology for determining reactivation kinetics integrates kinetic models for the main reaction, deactivation, and regeneration by coke combustion with air [122]. By simulating the reaction–regeneration cycles, optimal conditions, such as reaction temperature, space–time, time on stream, and regeneration parameters, as well as the average olefin production rate can be established. Moreover, coke oxidation kinetics can be also employed to investigate how the characteristics of the deposited coke vary throughout the catalyst bed, which is a complex outcome influenced by the interplay between operation conditions and zeolite [123]. Based on the established kinetic model, variations in the oxidation characteristics of coke along the axial position of the catalyst bed can be determined, providing valuable insights for optimizing regeneration parameters.

In conventional regeneration process, the lower section of the catalyst bed exhibits slower regeneration kinetics compared to the upper region, primarily due to lower temperature and reduced coke deposition in the lower zone. This incomplete regeneration leaves approximately 13% of residual coke unremoved, failing to meet industrial regeneration requirement [124]. To address this issue, moving-bed regenerator concept has been widely adopted for the regeneration of coked zeolites. Liao and co-workers developed a multi-zone regeneration process featuring optimized temperature uniformity across the catalyst bed. By increasing the inlet temperature and oxygen content in the lower zone, the coke conversion efficiency was increased from 87 to 95.7% (see Figure 4.9) [124].

4.2.5.2 Steam Gasification

The oxidization of coke is widely used in industrial applications for regenerating deactivated zeolites. However, this method does not convert coke into valuable hydrocarbon species, and releases large amounts of CO_2 into the atmosphere, contributing to significant environmental issues such as global warming. In this context, an alternative way to regenerate deactivated zeolite with low CO_2 emission is steam gasification, in which steam acts as a gasifying agent to produce syngas (H_2 and CO). Since steam gasification is typically conducted at high temperature, the hydrothermal conditions may cause permanent damage to zeolite structure. Hence, the zeolite used in steam gasification must have excellent hydrothermal stability to withstand the harsh conditions.

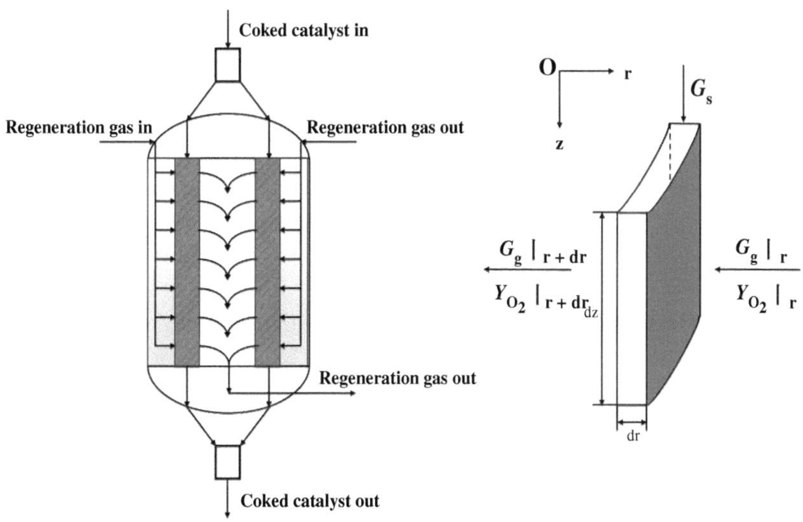

Figure 4.9 Scheme of the regenerator and differential element. *Source:* [124] / with permission of Elsevier.

Liu and co-workers [125] compared the regeneration methods using air combustion and steam gasification for spent SAPO-34. Both methods successfully reactivated the spent SAPO-34 catalyst, but the regenerated zeolite via steam gasification exhibited superior ethylene selectivity. The main species of residual coke after steam gasification are methylated benzenes, which were active HCP species that enhanced ethylene selectivity. In contrast, more oxygenated compounds were observed in partially regenerated SAPO-34 catalyst by air combustion, which accelerated catalyst deactivation. Furthermore, a combination of density functional theory calculations and structured illumination microscopy analysis revealed that naphthalenic cations, as active HCP species in ethylene production, exhibit high stability within SAPO-34 zeolites at high temperature. Consequently, directly converting coke to naphthalenic species within SAPO-34 zeolites via steam cracking represents a promising strategy to boost light olefins selectivity. Fluidized bed reactor–regenerator pilot experiments demonstrated that an unexpectedly high light olefins selectivity of 85% could be achieved in MTO reaction with 88% of the products being valuable CO and H_2 and negligible CO_2 emissions under industry-like continuous operations [126]. Kinetic study on steam regeneration of the coked SAPO-34 suggested that the composition of residual coke on the zeolite shifted from pyrene and phenanthrene to naphthalene as the regeneration temperature increased [127]. Meanwhile, the reduced acidity, BET surface area, and pore volume of the regenerated samples can be recovered, thereby prolonging the catalyst's lifetime. The activation energy of steam regeneration is approximately 177.8 kJ·mol^{-1}, which is higher than that of air regeneration, indicating that the steam regeneration process is more difficult to initiate.

4.2.5.3 Hydrogen Treatment

Hydrogen treatment is also capable of removing coke deposits on zeolites. It has been reported that the relative coke removal rate at 800 °C follows the order: $O_2 > H_2O > CO_2 > H_2$ [128]. Therefore, regeneration with hydrogen requires elevated temperature and high pressure. Weitkamp and co-workers [129, 130] studied the effects of temperature and hydrogen pressure on the regeneration of a spent Pt/zeolite catalyst used for isobutane/butane alkylation. They found that the activity can be fully restored under a hydrogen pressure of 15 bar and temperatures up to 300 °C.

Further studies on the regeneration mechanism revealed that hydrogen reactivation procedure is particularly effective for the coke with a H/C atomic ratio greater than 1 [131]. However, it is less suitable for coke with condensed structure, unless pretreated with steam under a high temperature. Liu and co-workers [103] found that phenanthrene and pyrene efficiently converted to light aromatics under higher hydrogen and H_2O partial pressure, likely due to the reduced energy barriers of hydrogenation reactions in the presence of H_2O. Additionally, in the presence of hydrogen, external coke deposits were only partially cracked from bulky alkyl polyaromatic compounds to less bulky polyaromatic compounds, while the intracrystalline coke near the Brønsted acid sites was preferentially removed [132]. This indicates the Brønsted acid sites play an important role in hydrogen regeneration treatments.

Kim and co-workers [133] systematically evaluated oxidative versus reductive methods for recovering deactivated Mo/H–ZSM-5 catalysts following methane dehydroaromatization. Under oxidative conditions, the bifunctional metal–acid sites were reversibly regenerated at 450 °C, effectively restoring the catalytic activity for benzene formation. In contrast, the reductive regeneration required a temperature exceeding 700 °C, applicable to the catalysts with exceptional thermal stability. The latter process facilitates coke conversion to methane through hydrogenation [134, 135]. Periodic switching between methane and hydrogen flows through catalyst bed can achieve a remarkable long-term stability of 1000 hours (see Figure 4.10) [136]. Based on the deactivation–regeneration cycle analysis, a triple-bed circulating fluidized reactor system was developed to achieve both effective continuous regeneration of Mo/HZSM-5 catalysts and efficient sustained conversion of CH_4 to benzene [135]. Furthermore, a pilot-scale fluidized bed system integrating methane dehydroaromatization with hydrogen regeneration had been successfully implemented (see Figure 4.11) [137]. A stable eight-hour continuous operation verified that enhanced catalyst activity and stability can be attained under conditions of elevated regeneration temperature, reduced reaction and regeneration space velocities, and increased catalyst circulation rates.

In summary, the reactivation with hydrogen is substantially enhanced at elevated temperature and high pressure, restoring major activities. Decoking is particularly efficient for the aluminum-rich catalyst, attaining performance comparable to conventional coke combustion. Hydrogen eliminates internal and external coke from the catalyst with low Si/Al ratio, yielding light hydrocarbons (e.g., methane and ethane) and aromatics (e.g., methylated benzenes).

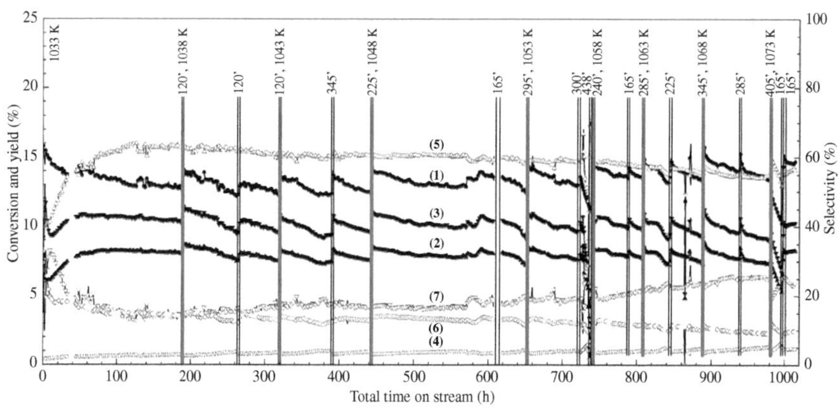

Figure 4.10 Long-term stability test of Mo/HZSM-5 catalysts in methane dehydroaromatization reaction at 1033–1073 K with a periodic switch of CH_4/N_2 to H_2: (1) methane conversion, (2) benzene yield, (3) aromatic yield, (4) ethylene selectivity, (5) benzene selectivity, (6) naphthalene selectivity, and (7) coke selectivity. *Source*: [136] / with permission of Elsevier.

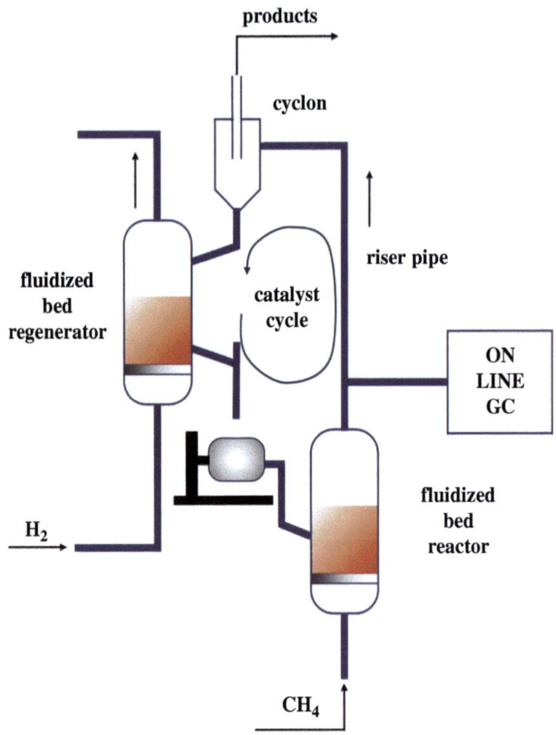

Figure 4.11 A schematic of CH_4 dehydroaromatization–H_2 regeneration fluidized bed system. *Source*: [137] / with permission of Elsevier.

4.2.6 Conclusion

The deactivation of acidic zeolites in low-carbon molecule conversion processes, including MTH, DME carbonylation, and methane dehydroaromatization, is predominantly governed by coke deposition. This section synthesized critical advancements in coke formation mechanisms, advanced characterization, deactivation mitigation strategies, and regeneration protocols, bridging molecular-level insights with industrial applications. State-of-the-art characterization techniques such as MALDI FT-ICR MS and APT have unraveled the chemical complexity and spatial nonuniform distribution of coke deposits. Mechanistic investigations have highlighted the evolution from SMS to MAHs via direct C—C bond formation, progressing to PAHs through oligomerization cascades.

In catalyst design, hierarchical pore engineering (e.g., MCM-36, nanosized SAPO-34) significantly extends catalyst lifetimes by enhancing intracrystalline diffusion and disrupting coke precursor accumulation. Targeted passivation of external acid sites via using rare-earth oxide coatings (e.g., CeO_2 and Gd_2O_3) suppresses external graphitic coke formation, whereas controlled pre-coking stabilizes zeolite frameworks against dealumination. During the reaction, co-feeding strategies

(H_2 or steam) mitigate deactivation by hydrogenating aromatic intermediates or modulating acid site accessibility.

Conventional oxidative regeneration (i.e., air combustion) achieves efficient coke removal but accelerates irreversible structural degradation, particularly in low-silica zeolites with inherent hydrothermal instability. Ozone-assisted low-temperature oxidation emerges as a framework-preserving alternative, selectively degrading aliphatic coke. Cutting-edge nonthermal plasma technology offers room-temperature regeneration via reactive oxygen species, while steam gasification converts coke into syngas (H_2/CO), aligning with circular economy principles.

4.3 Deactivation of Metal–Zeolite Catalysts: Metal-catalyzed and Bifunctional Systems

4.3.1 Propane Dehydrogenation to Propene

The conversion of propane to propylene through dehydrogenation is a crucial industrial process for producing propylene, a valuable basic feedstock in chemical industry. This process has garnered growing interest, owing to advantages such as high propylene yield, short process flow, and low investment cost. The direct dehydrogenation of propane is a thermodynamically exothermic reaction ($\Delta H = 124.3$ kJ · mol^{-1}) with an increase in volume. Thus, the process is typically conducted at high temperature (500–700 °C) and low pressure, thereby facilitating a thermodynamically unfavorable reaction. The reaction proceeds through dissociation of alkane to produce an adsorbed H atom; dissociation of β-H in the alkane to form H_2 and the subsequent generation of C═C bonds. Additionally, side reaction involving C—C cleavage would occur at high temperature, further leading to coke deposition on the catalyst surface [138]. Over the past few decades, noble metal-based catalysts such as Pt, as well as non-noble metal oxide catalysts (e.g., ZnO, CoO, and VO_x), have shown remarkable activity in this reaction, attributable to their great ability for activating C—H dissociation while relative inertness toward C—C cleavage [139]. Nonetheless, these catalysts suffer from significant deactivation, primarily due to sintering and aggregation of metal species at high reaction temperature, as well as coke deposition.

Nowadays, microporous zeolites with three-dimensional and uniform sub-nanopore channels are used as supports to disperse active metals in propane dehydrogenation [140]. Zeolite-supported metal and metal oxide catalysts not only show outstanding activity, but can also effectively overcome the problems of carbon deposition and metal sintering during the reaction, thereby significantly enhancing catalyst stability. In this section, we discuss the deactivation mechanism of metal/zeolite-catalyzed propane dehydrogenation involving metal sintering and coke formation. Subsequently, we outline the corresponding strategies to mitigate deactivation and summarize the catalyst regeneration methods.

4.3.1.1 Deactivation Mechanisms

i. Coke Formation

The primary cause of deactivation during propane dehydrogenation is coke deposition. At high reaction temperature, side reactions such as propane cracking and hydrogenolysis generate various adsorbed carbon cation intermediates, which exhibit significantly higher reactivity than paraffinic compounds. These intermediates subsequently react with alkenes to form hydrocarbon species and coke deposits, ultimately covering active sites and changing the electronic structure of active site [141]. Additionally, coke accumulation physically obstructs zeolite channels, thereby limiting reactant access to active sites.

In Pt-based catalysts, smaller Pt particles show enhanced activity but simultaneously facilitate coke formation, owing to the abundance of uncoordinated sites. These sites are highly active for C—H and C—C bonds activation, ultimately leading to covering of Pt site by carbonaceous deposits [142] (see Figure 4.12). Additionally, the interaction between Pt and coke also induces a downward shift of the Pt d-band center, which suppresses C—H activation capability and consequently decreases catalytic activity.

Coke formation mechanisms on metal oxide catalysts differ substantially from those on metallic systems due to structural complexity. Ethylene generated from cracking reactions strongly interacts with active sites, subsequently converting to coke deposits on LASs (*M*—O). This site blocking fundamentally alters the catalytic properties of metal oxides.

Using DFT-based kinetic Monte Carlo simulation, Su and co-workers discriminated biphasic coking mechanism: an initial rapid deactivation, dominated by interactions between C_1/C_2 species and metal sites, resulting in active site blockage; a subsequent steady-state period, during which the remaining active sites efficiently catalyze propylene production with sustained activity [138].

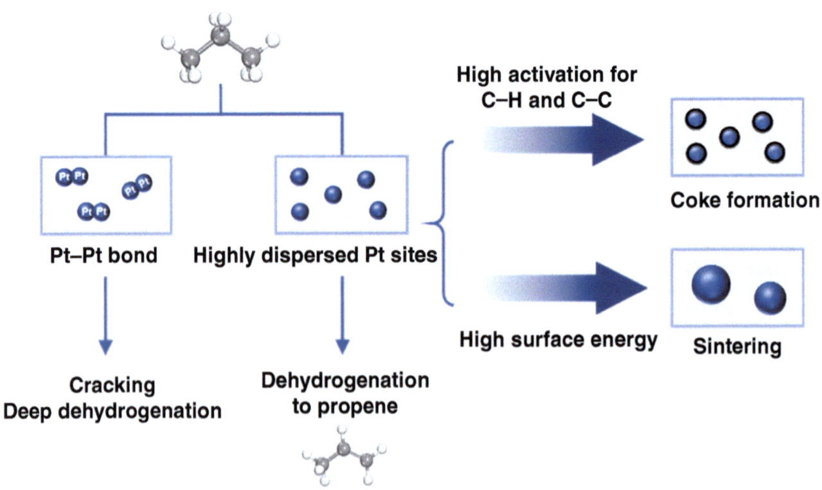

Figure 4.12 The deactivation mechanism for Pt-based catalyst in propane dehydrogenation to propene.

ii. Catalytic Sintering

Active species sintering predominantly occurs in metal catalysts. Sintering of Pd species at elevated temperature during reaction and regeneration is another factor contributing to deactivation. It is generally accepted that side reactions, such as cracking and deep dehydrogenation, are catalyzed by Pt—Pt sites, leading to a reduction in selectivity and activity [143]. To achieve optimal performance, highly dispersed Pt sites are required. However, these Pt species are highly unstable due to their high surface energy when subjected to high reaction temperature.

There are primarily two mechanisms involved in metal sintering. One is Ostwald ripening, which takes place among nanoparticles with discernible diameters. Driven by differences in free energy, smaller particles in close proximity are consumed to facilitate the movement of mobile species, resulting in the formation of larger particles [144]. The other mechanism is Brownian motion of metal particles on the support, which leads to particle interactions and ultimately the formation of larger aggregates through coalescence. Both mechanisms may coexist, with Ostwald ripening likely being the dominant one [145].

iii. Reduction of Metal Oxides

Metal oxides such as ZnO often experience rapid deactivation during propane dehydrogenation, owing to oxides reduction accompanied by the elimination of water [146, 147]. Anchoring isolated Zn^{2+} ions in zeolites (e.g., ZSM-5) results in a great resistance to reduction, attributable to a strong interaction between Zn^{2+} ions and zeolite framework O^{-}. Consequently, enhanced stability can be achieved.

4.3.1.2 Deactivation-resistant Strategies

i. Encapsulation of the Sub-nanometric Metals in Zeolite Pores or Channels

Trapping metals in zeolite crystals is an efficient strategy for preparing stable metal-nanoparticle catalysts and preventing metal sintering at high temperature. For Pt-based catalysts, Corma and co-workers encapsulated single Pt atoms and Pt clusters by using hexadecyltrimethylammonium surfactant as a protecting reagent and transforming a two-dimensional pure siliceous layered precursor (MCM-22) into three-dimensional zeolite. The Pt species were confined to the cages of zeolites with sizes of 0.2–0.7 nm and showed excellent stability in propane dehydrogenation to propene reaction [148]. Zhang and co-workers successfully confined Pt within HZSM-5 crystals using Pt/SiO_2 as the sole silicon source. The high electron density of Pt, due to local environment created by zeolite framework, reduced the strength of the Pt—(C=C) bond. This configuration facilitated olefin desorption while suppressing ethane formation from deep dehydrogenation, thereby effectively inhibiting coke formation [149]. Contrary to the above researches, Xiao and coworkers proposed that the encapsulated Pt with an average size of 1.05 nm showed lower propane conversion and propylene selectivity than Pt nanoparticles (~1.45 nm) loaded on the external surface of zeolite. DFT calculations showed that the enclosed Pt clusters had a lower d-band caused by the zeolite framework, which resulted in decreased activity, deep dehydrogenation, and coke formation [150].

To further improve the activity and stability of Pt-based catalysts, introducing a promoter to Pt–*M* alloy is an effective strategy. Encapsulation within zeolite is still applicable for Pt–*M* alloy nanoparticles/clusters/single sites. Corma and co-workers embedded bimetallic PtSn cluster within the sinusoidal channels of K–MFI zeolite, using a one-pot method with the assistance of ligand protection. This configuration demonstrated excellent durability in 70 hours, owing to the electronic effect between metals and the confinement of zeolite channels [151]. By using the ligand-protected direct H_2 reduction, the atomically dispersed sub-nanometric Pt–Zn clusters were encapsulated in the intersectional void spaces of the MFI zeolite (see Figure 4.13a–e) [152]. The introduction of Zn species provided an electron-enrich environment of Pt, thereby

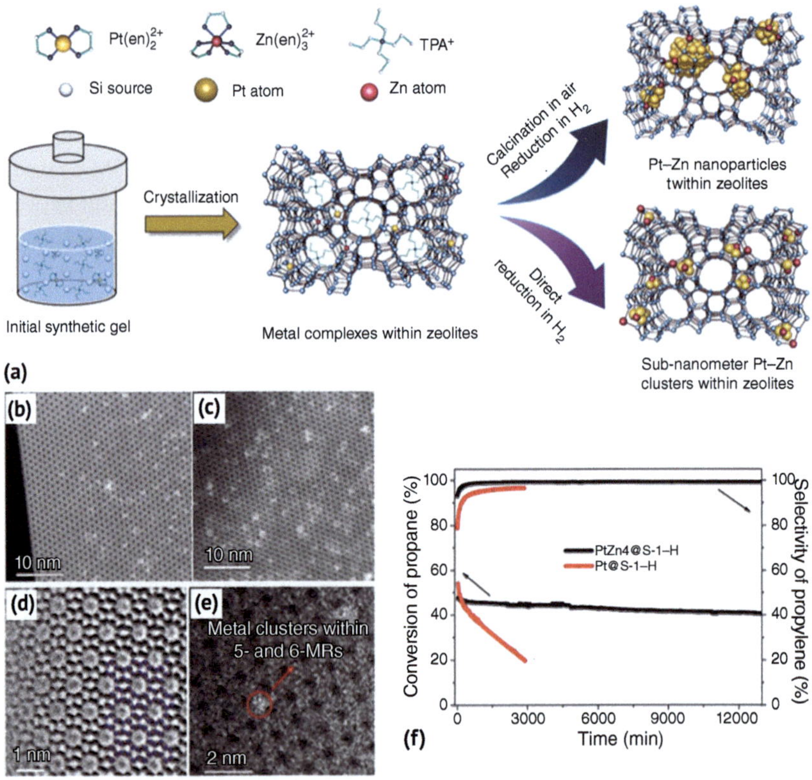

Figure 4.13 (a) Schematic representation of the synthesis of zeolite-encaged sub-nanometer Pt–Zn cluster catalysts; C_s-corrected HAADF-STEM images of (b) Pt@S-1-H and (c) PtZn4@S-1-H samples; (d) C_s-corrected STEM-ABF image along the PtZn4@S-1-H [010] projection with the model superimposed; (e) high-resolution C_s-corrected HAADF-STEM image of PtZn4@S-1-H sample; and (f) long-term stability test for PDH conversion over PtZn4@S-1-H and Pt@S-1-H catalysts (WHSV=3.6 h^{-1}, 550 °C). *Source*: [152] / John Wiley & Sons.

facilitating olefin desorption and suppressing the formation of coke precursors. The obtained PtZn4@S-1–H sample remained stable even after reaction for 216 hours (see Figure 4.13f). A similar method has been applied to prepare different bimetallic Pt–*M* clusters in different zeolites, such as the PtSn cluster in Silicalite-1 [153].

For non-noble metal oxide systems, confinement using zeolitic topology also shows a promising potential for improving catalyst stability. For example, Silicalite-1-confined CoO exhibited a remarkable durability for 120 hours with five regeneration cycles. The great performance was attributed to the π-binding of propylene on zeolite-confined CoO clusters, which inhibited excessive dehydrogenation compared to the di-σ-binding of propylene on metallic Co [154]. Yuan and co-workers confined VO_x in dealuminated Beta zeolite steadily by forming V—O—Si bonds. The monomeric VO_x species were responsible for great activity and stability [155].

ii. **Incorporation of Metal Atoms into Zeolite Framework**

In zeolites, the tetrahedral Si atoms in the framework can be partially substituted by heteroatoms (e.g., Al, Sn, Ti, etc.), imparting a specific electronic and geometric configuration to active or promoter metals. For propane dehydrogenation, enhanced dispersion of Pt species can be achieved through heteroatom-mediated anchoring effects. Zhu and co-workers synthesized Pt–Sn/Sn–Beta catalysts by incorporating Sn into zeolite framework and then grafting Pt precursor onto the Sn-modified support. The strong electronic Pt–Sn interaction resulted in highly dispersed and stabilized Pt–Sn bimetallic clusters. Notably, the Pt–$Sn_{2.00}$/Sn–Beta catalyst exhibited superior performance (deactivation rate of 0.006 h^{-1}, TOF of 114 s^{-1}), while maintaining excellent regeneration ability at 570 °C [156]. In TS-1-supported PtZn catalyst, the heteroatom Ti in zeolite framework could promote the dispersity of Pt and modify the electronic density of Pt and Zn, which result in an increased durability of the catalyst [157]. Li and co-workers synthesized highly dispersed Zn species stabilized by silanol groups within dealuminated zeolite. These Zn species served as effective anchoring sites for Pt species [158]. Furthermore, the synergistic geometrical and electronic interactions between Pt–Zn species enhanced the catalytic robustness of the materials during propane dehydrogenation reaction [159]. A recent study demonstrated that anchoring La and ZnPd ensembles into the silanol nests of Beta zeolite, combined with Zn-mediated isolation of Pd atoms, yielded catalysts exhibiting exceptional stability with an extremely low deactivation constant of 0.0149 h^{-1} [160]. Bell and co-workers embedded isolated Pt atoms within the nests of ≡ SiOZn—OH groups, which were formed by introducing Zn into dealuminated BEA zeolite (see Figure 4.14a) [161]. The catalyst retained its initial activity and nearly 100% propene selectivity after 160 hours of continuous operation. Separately, Ji and co-workers constructed isolated Pt atoms tetrahedrally coordinated to framework oxygen atoms in the hierarchical heteroatomic ferrosilicate zeolite (H–Fe–S-1-3) (see Figure 4.14b, d–g). The incorporation of Fe enhanced the stability of Pt—O bonds through

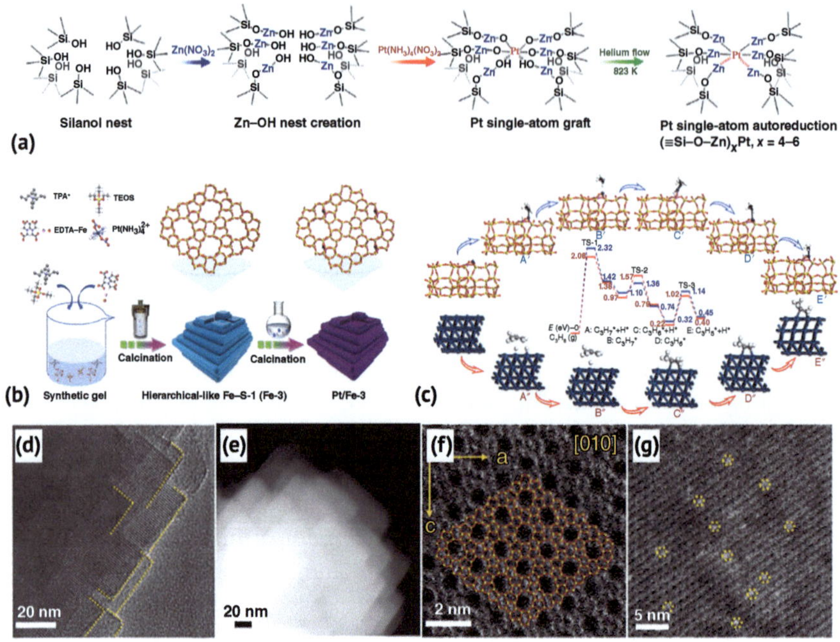

Figure 4.14 (a) Simplified illustration of the formation of a Pt single-atom catalyst. *Source*: [161] / with permission of American Chemical Society. (b) Schematic representation of the synthesis of a hierarchical-like Pt/Fe-3. (c) Calculated energy profiles upon PDH reaction with reduced Pt/Fe-3 (blue) and Pt/Fe-3-C1 (red, Pt[111]). (d) High-resolution TEM image. (e, f) HAADF-STEM and AC-HAADF-STEM images on the [010] plane. (g) AC-HAADF-STEM image of Pt distribution in Pt/Fe-3 and the corresponding enlarged images. *Source*: [162] / with permission of American Chemical Society.

Fe—O—Si linkages (≡ Si—O—Pt—O—Fe ≡), outperforming the conventional Si—O—Si configuration. These isolated Pt-based catalysts exhibited remarkable stability, maintaining performance for over 72 hours in pure propane at 580 °C. Moreover, the ≡ Si—O—Pt—O—Fe ≡ structure suppressed deep dehydrogenation of propylene by elevating the activation energy barriers of secondary dehydrogenation pathways (see Figure 4.14c). Concurrently, it promoted propylene desorption due to weakened interactions between C_3H_5 intermediate and the isolated Pt sites via a single bond, thereby improving propylene selectivity [162].

In addition to optimizing Pt dispersion through promoter metal doping in zeolite framework, the incorporation of rare-earth metals into the framework facilitates the formation of intermetallic rare-earth–Pt alloy phases via the specific chemical environment provided by their tetrahedral coordination. Ryoo et al. demonstrated that rare-earth elements anchored in silanol nest exhibited notably higher chemical potentials compared to their bulk oxide counterparts. This increased potential drove the migration of rare-earth metals onto Pt surfaces, resulting in the formation of Pt–rare-earth intermetallic compounds that displayed high stability and selectivity in propane dehydrogenation [163].

When the zeolite was supported with 0.50-wt.% Pt and 0.50-wt.% Y, an exceptionally extended catalytic lifetime of 20 days was achieved with high propane conversion and propylene selectivity at 580 °C [164].

Beyond Pt-based catalysts, other metals such as Fe [165], Co [166, 167], and Sn [168] have also been successfully incorporated into zeolite framework as active sites in propane dehydrogenation reaction. For example, Xiao and co-workers engineered isolated Fe atoms within an MFI siliceous zeolite by using ligand protection. This catalyst showed superior coke resistance during a 200-hour test, maintaining extremely high activity (26.3%) and selectivity (97.5%) at 600 °C without deactivation. The excellent stability originated from effective modulation and stabilization of the Fe^{3+} sites, avoiding coke formation and deactivation caused by Fe^{3+} reduction. Additionally, the isolated Fe species balanced deep dehydrogenation and product desorption, sustaining catalytic activity while suppressing coke deposition [169]. Dai and co-workers anchored isolated Co^{2+} ions into hydroxyl nests of BEA zeolite via vacuum impregnation. The Co^{2+} sites selectively activated the C—H bonds of propane without cleaving the C—C bonds, enabling stable performance across eight reaction–regeneration cycles [167]. Moreover, the tetrahedrally coordinated Co^{2+} by Si—O unit in the zeolite framework inhibited its reduction to metallic Co [170]. Yue et al. introduced Sn into Beta zeolite framework. The resulting Sn–Beta catalysts exhibited extraordinary stability under harsh reaction conditions (72 hours), with no detectable deactivation due to strong Sn–zeolite framework interaction [168].

iii. Modifying the Acidity and Physical Structure of Zeolite

In metal/zeolite-catalyzed propane dehydrogenation to propene, LASs serve as the active centers for propene production, whereas Brønsted acid sites promote undesired cracking reactions. Since most zeolites inherently contain Brønsted acid sites, modulating their acidity has emerged as a critical strategy for improving catalyst stability. A widely adopted approach involves generating hydroxyl nests via dealumination, which reduces Brønsted acidity and thereby prolongs catalyst lifetime. Furthermore, neutralizing residual Brønsted acid sites can be achieved through ion exchange of protons in the bridging hydroxyl groups by K^+ ions [151, 171, 172]. This substitution not only suppresses cracking pathways but also enhances catalytic performance, as K^+ ions act as promoters in propane dehydrogenation process.

Shortening diffusion pathways and reducing diffusion resistance to remove the coke intermediates are effective strategies for enhancing catalytic stability. Shin and co-workers fabricated PtZn bimetallic nanoclusters supported on two-dimensional zeolite. The enhanced molecular diffusion in the 2D architecture significantly improved resistance to coke-induced deactivation [173]. Bell and co-workers achieved highly stable catalytic performance by strategically dispersing Pt atoms on self-pillared pentasil zeolite nanosheets via ≡ Si—O—Zn—OH cluster grafting. As the desorption of coke precursor was facilitated on the external surface of Pt species, almost no deactivation was observed over 288 hours of continuous operation at 823 K with a WHSV of 1350 h^{-1} [174]. Liu and co-workers demonstrated that PtSn supported on small-sized TS-1 showed better activity and stability due to enhanced mass transportation [175]. Moreover, other hierarchical porous zeolites or mesoporous zeolites such as nanosheet Silicalite-1 [176], SBA-15 [177], MCM-41 [178], and hollow porous Silicalite-1 [179] were also employed as support in propane dehydrogenation to improve the diffusion properties.

4.3.1.3 Catalyst Regeneration

Despite significant advancements in preventing coke formation, coke deposition and metal sintering remain as persistent challenges in propane dehydrogenation for propene production. In industrial settings, periodic catalyst regeneration becomes imperative.

A widely employed regeneration protocol involves oxidative treatment or oxychlorination followed by hydrogen reduction to remove carbonaceous deposits and restore catalytic activity [180–182]. However, the sintering-induced activity loss during regeneration exhibits irreversible characteristics. Consequently, successive regeneration cycles progressively lead to active site depletion and structural changes in zeolite framework, ultimately diminishing activity and selectivity [183]. Notably, when metal species are spatially confined within zeolite channels, where restricted pore dimensions prevent metal particle sintering, coke deposition emerges as the predominant deactivation mechanism. In such configurations, simple oxidative regeneration through air combustion (without subsequent H_2 reduction) enables complete removal of carbon deposits and full restoration of catalytic performance [156]. Recently, Datye and co-workers found that the activity of Pt-based alloy catalysts could be regenerated in pure H_2 without prior oxidative treatment [184]. A similar phenomenon was also observed in zeolite-confined isolated Pd systems, where the activity was fully reclaimed after reductive regeneration (10% H_2/He, 450 °C, and six hours) [161]. Mechanistic studies revealed that carbonaceous deposits migrate from metallic nanoparticles to the silica support during H_2 treatment, with the silica matrix functioning as a carbon reservoir. These findings suggest that intermittent hydrogen exposure during catalytic operation may constitute a viable strategy for prolonging the operational lifetime of Pd/zeolite catalysts. Nevertheless, this approach presents a critical technical constraint: the coke deposits accumulated on support become more difficult to remove compared to those adsorbed on metallic sites. There is a positive correlation between coke content on support and the requisite oxidation temperatures [184]. This elevated temperature inevitably induces sintering of active components during decarbonization processes. Consequently, precisely managing the coke burning temperature is crucial for simultaneously achieving effective coke elimination and optimal performance [184, 185].

4.3.2 Dry Reforming of Methane

Dry reforming of methane (DRM) is a highly promising technology for syngas (CO and H_2) production by simultaneously converting two greenhouse gases, CH_4 and CO_2, thereby providing a sustainable pathway for low-carbon industrial development. The resulting syngas exhibits an ideal H_2/CO ratio (~1), making it directly applicable as feedstock for Fischer–Tropsch synthesis to produce liquid fuels and value-added chemicals [186]. As an endothermic process, DRM typically proceeds at elevated temperatures (650–1000 °C), necessitating external energy input to overcome the inherent thermodynamic stability of the reactants. However, the product yield was limited by the reactant activation. Transition metal-based catalysts are commonly used in DRM, due to their great ability to dissociate CH_4 and CO_2. Notably, the harsh operating conditions (high temperatures) and reliance on metallic active sites induce deactivation mechanisms—such as coking and metal sintering—that parallel those observed in propane dehydrogenation systems. Consequently, strategies to mitigate catalyst degradation in these two processes share common approaches, including structural confinement of active sites and enhanced carbon resistance [187].

Noble metals (e.g., Ru, Rh, Pd, Pt, and Ir) are often selected as active components in DRM catalysts due to their excellent catalytic performance. However, their industrial application is constrained by prohibitive costs and limited natural abundance [188–190]. Non-noble transition metals such as Ni, Co, and Fe also exhibit high catalytic activities while offering greater economic viability for DRM catalyst design [191, 192]. Nevertheless, rapid catalyst deactivation remains a critical barrier to their practical implementation [193].

4.3.2.1 Deactivation Mechanism

The predominant challenge in DRM lies in developing catalysts that sustain long-term stability while exhibiting robust resistance to deactivation. Extensive experimental studies have systematically investigated active metal species, supports, and promoters. Nevertheless, designing an ideal catalyst that is capable of effectively mitigating deactivation remains an unresolved scientific frontier.

As we mentioned above, the rigorous operational conditions inevitably accelerate two deactivation pathways: (1) metal agglomeration and (2) carbon deposition [194]. Carbon deposition is widely recognized as one of the predominant causes of DRM catalyst deactivation. Carbon deposition exhibits a structure-sensitive interdependence with metal sizes, where coking preferentially occurs on larger metallic particles because the exposed edge sites facilitate graphitic carbon nucleation. Conversely, geometric constraints in smaller particles suppress edge-mediated carbon growth, thereby inhibiting the graphitization process [190, 195]. Consequently, reducing the initial size of Ni nanoparticles and preventing their sintering at reaction temperature represent one of the critical challenges in developing highly active and stable Ni-based catalyst.

Therefore, researchers have explored effective strategies to mitigate catalyst deactivation. Their efforts include developing advanced synthesis techniques, engineering catalyst architectures, and elucidating fundamental deactivation mechanisms, particularly those involving carbon deposition and metal agglomeration. To address metal sintering issues, zeolite-supported catalysts have emerged as promising systems where the unique pore structures effectively confine metal nanoparticles. In addition, the exceptional thermal stability, molecular shape selectivity, and uniform microporosity of zeolites collectively contribute to their superior performance in high-temperature DRM processes [196–198]. Furthermore, recent studies demonstrate that tailoring the acid–base properties of zeolite surfaces can significantly modify metal–support interactions, thereby improving catalyst stability and activity.

4.3.2.2 Deactivation-resistant Strategies

Similar to challenges encountered in propane dehydrogenation, suppressing carbon deposition and inhibiting active metal sintering during high-temperature catalytic processes remain critical challenges requiring innovative solutions in the rational design of high-performance catalysts with balanced activity and durability.

Various methods have been developed to address these challenges [188, 199]. The first approach involves modulating the acid–base properties of zeolites. Mechanistic studies reveal that the activation energies of CO_2 and CH_4 are critically dependent on the surface acidity of the catalyst. Excessive acidity accelerates CH_4 cracking and promotes various side reactions, leading to coke-induced deactivation. Hambali et al. demonstrated that Ni/MFI catalysts prepared by impregnation achieved optimal stability when the acidity of the zeolite was tuned to moderate strength [200]. Kweon et al. engineered basic sites in N–ZSM-5 and N–Beta zeolites via substitution of framework oxygen atoms with nitrogen, employing a high-temperature nitridation process with ammonia as the nitrogen source. Owing to the increased basicity of the catalyst and high dispersion of active metals, nitrided Ni/N–ZSM-5 exhibited superior catalytic DRM stability compared to the non-nitrided catalyst [201].

The second solution involves confining active metals within the zeolite channels to prevent metal sintering. The spatial restriction mitigates deactivation through three effects: reducing particle size of metals, inhibiting metal agglomeration, and modulating the electronic structure of the metals. Kweon et al. synthesized different Ni–BEA zeolites via hydrothermal interzeolite conversion of MWW-type precursors, demonstrating that Ni species stabilized by framework silanol defects exhibited superior stability and activity in DRM reaction. DFT calculations revealed that the improved stability originated from strong interactions between Ni clusters and these structural defects [202]. Similarly, Zhang et al. [203] suggested that Ni active species could be embedded in de-alumina Beta to improve the metal–support interactions, resulting in excellent sintering resistance and catalytic stability. Kobayashi et al. successfully encapsulated Pt metal in birdcage-structured zeolites and demonstrated their high catalytic activities at low temperatures. Moreover, birdcage-structured zeolites effectively avoided the sintering of Pt nanoparticles and exhibited excellent catalytic stability at 620 °C [204]. Kong et al. developed a Ni nanoparticle-embedded Silicalite-1 zeolite catalyst. The formed Ni—O ··· Si coordination sites preferentially facilitated CH_x oxidation to CO over direct CH_x dissociation into surface carbon, resulting in remarkable stability during the 100-hour operation at 650 °C with undetectable carbon deposition [205]. Wang et al. successfully prepared microporous Silicalite-2-encapsulated metal Ni catalysts with a peasecod-like structure, achieving precise control over Ni nanoparticle size (2.6 nm average) by adjusting the degree of Ni—O—Si copolymerization (see Figure 4.15a). The dual strategy of physical coating (layered walls of micropores of S-2) and chemical anchoring through strong

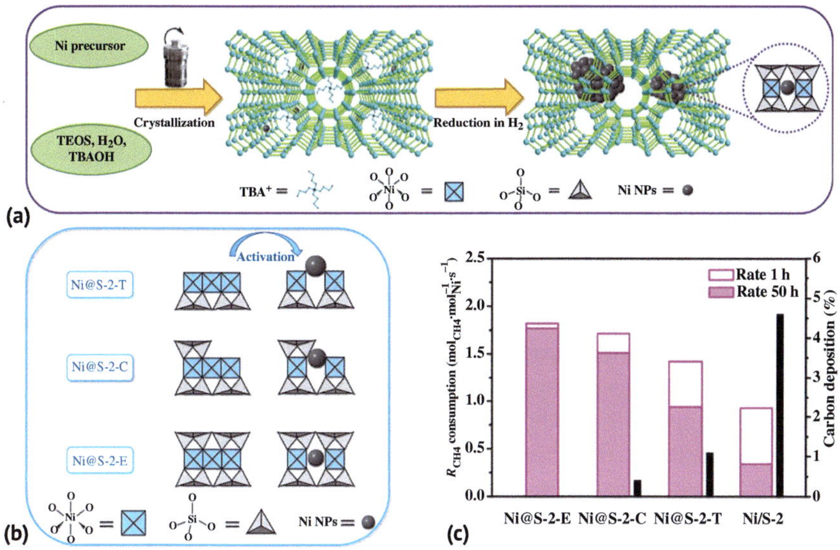

Figure 4.15 (a) Schematic preparation of Ni@S-2 catalyst by a facile one-pot approach; (b) the illustration of migration of Ni NPs; and (c) rates of CH_4 consumption for 1 and 50 hours of TOS and weight losses (%) of spent catalysts. *Source*: [206] / with permission of Elsevier.

interaction between Ni and Ni—O—Si avoids Ni sintering and carbon deposition, which exhibited superior stability at 650 °C for 50 hours (see Figure 4.15b, c) [206]. Another popular strategy for encapsulating metal nanoparticles within zeolites is to design a core–shell structure, where the metal nanoparticles are positioned at the core of the structure and encapsulated by a protective layer of zeolitic material that serves as the shell [207–210]. Han et al. encapsulated Ni within TPAOH-treated Silicalite-1, which possesses abundant mesoporous structures and silanol defects [211]. This structure improved the accessibility of active Ni species as well as the products of intra-diffusion. Moreover, the strong interaction between Ni and Si protected Ni from sintering at high temperature. Li et al. designed a Ni@S-1@SiO_2 dual-shell catalyst, in which nickel particles exposed on the S-1 surface were confined within a mesoporous silica shell [212]. As a result, the reduced mobility and stable valence state of Ni species enhanced the resistance to sintering and carbon deposition.

4.3.2.3 Regeneration Strategy

The deactivated catalyst is commonly regenerated by oxidization at a temperature of ~600 °C, followed by reduction using H_2. The stable zeolite framework plays a crucial role in stabilizing the metal nanoparticles during the continuous reaction and regeneration cycles at high temperatures. For example, Xiao and co-workers regenerated the CoNi@ZSM-5 catalyst by calcination in air at 600 °C for two hours. After three reaction–regeneration cycles, the performance maintained for a total time period of 980 hours [213]. Similarly, Liu et al. performed cyclic regeneration of a spent Ni catalyst supported on fibrous silica zeolite-Y using compressed air at 500°C for 2.5 hours. The catalyst exhibited stable activity after two regeneration cycles [214]. However, regeneration efficacy exhibits significant dependence on the selection of coke-removal gas media. Chong et al. compared the regeneration of spent Ni/SBA-15 catalyst using air and CO_2/N_2, and concluded that the catalytic performance of air regeneration was better than CO_2/N_2 regeneration. This disparity arises from distinct regeneration mechanisms: CO_2-mediated regeneration initiates the Boudouard reaction ($CO_2 + \text{coke} \rightarrow CO + \text{coke} - O$), where carbon redeposition limits the carbon removal efficiency. In contrast, O_2 regeneration drives complete carbon oxidation ($C + O_2 \rightarrow CO_2$) through an irreversible exothermic pathway, achieving near-total carbon deposit elimination [215].

While oxidative regeneration effectively removes carbon deposits from spent catalysts, this approach entails unavoidable costs related to oxidation–reduction cycles and persistent activity loss. In response, recent advancements have focused on developing non-oxidative regeneration strategies to streamline catalyst recovery processes. Hsu and co-workers demonstrated self-regeneration of deactivated Ni/SiO_2 DRM catalysts under inert N_2 atmosphere, bypassing the conventional reduction process. The deposited carbonaceous materials were oxidized to CO_x through mobile O species in SiO_2 support, with oxygen replenishment via the O in NiO generated during the self-regeneration process [216].

4.3.3 Carbonylation of MN to Dimethyl Carbonate

In recent decades, dimethyl carbonate (DMC) has been widely used in industrial applications as an eco-friendly chemical intermediate [217–219]. The vapor-phase carbonylation of MN to produce DMC (see Figure 4.16), developed by UBE Company, has gained significant interest due to its notable advantages including low environmental impact and high atomic efficiency [220, 221]. Pd-based catalysts have been extensively utilized in this route. Compared to chloride-containing catalytic systems, the chloride-free zeolite-supported Pd-based catalysts fundamentally avoid catalyst deactivation and equipment corrosion caused by chloride loss, while simultaneously improving the purity of the DMC product, thereby garnering significant research attention.

The utilization of zeolite as the support enhances the stability of chloride-free Pd(II)-based catalysts, which can be attributable to the following reasons: the oxygen anions connected to the zeolite framework play a critical role in stabilizing

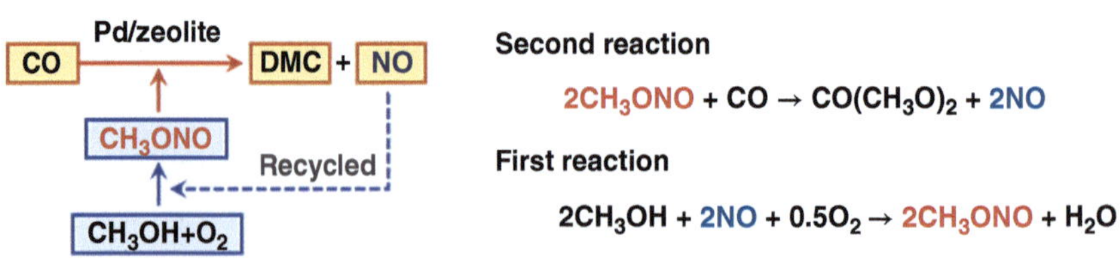

Figure 4.16 The schematic diagram of vapor-phase MN carbonylation to DMC route.

the valence state of Pd during the redox processes [222]. Meanwhile, the confinement effect of the unique channels in zeolites not only significantly improves the dispersion of Pd during the preparation process, but also effectively suppresses the agglomeration of Pd following its reduction under the reaction conditions [223, 224].

4.3.3.1 Deactivation Mechanism

Numerous studies demonstrated that the valence states of Pd species (+2 or 0) determine the catalytic activity of the catalysts. On reduced Pd-based catalyst, dimethyl oxalate (DMO) emerges as the predominant species, while DMC becomes the primary product after oxidation of the catalyst by using MN and HCl [225, 226]. Therefore, the Pd(II) species is considered as the active species for MN carbonylation to DMC, while the Pd(0) species is more favorable for DMO production (see Figure 4.17). However, during DMC synthesis, the Pd(II) species could be gradually reduced to Pd(0). Therefore, the Pd–zeolite catalysts become inactive due to the reduction and migration of Pd(II) species, followed by the aggregation of Pd(0) metallic particles [227].

4.3.3.2 Deactivation-resistant Strategies and Regeneration

To promote industrialization of the vapor-phase MN carbonylation to DMC process, enhancing the stability of isolated Pd(II) sites is crucial for designing long-lived chloride-free Pd-based catalysts. By strengthening the interaction between Pd(II) and zeolite supports, the aggregation of Pd species and catalyst deactivation can be prevented effectively.

The stabilization of Pb(II) species can be maintained through the addition of additives. Guo et al. [229] demonstrated that Cu^{2+} addition improved Pd dispersion

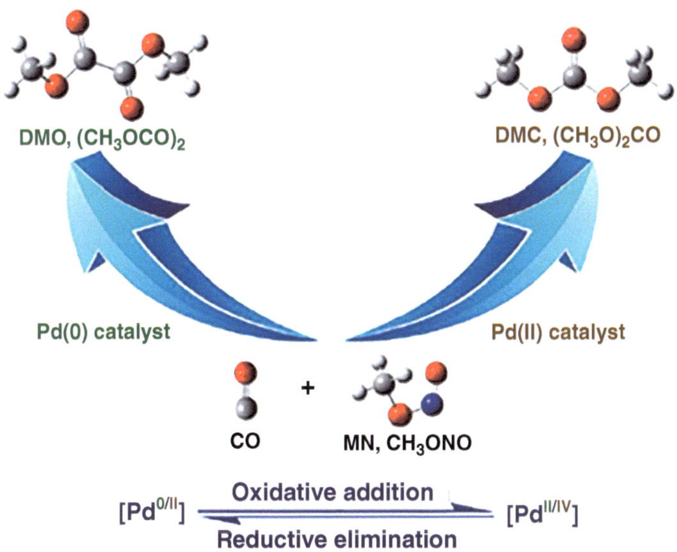

Figure 4.17 The selectivity toward DMC or DMO over Pd-based catalysts. *Source:* [228] / with permission of American Chemical Society.

and stabilized the valence state of Pd^{2+}. Meanwhile, the appropriate Pd–CO affinity enabled successful completion of the catalytic cycle. These factors contributed to an enhanced lifetime of the catalyst. Wang and co-workers enhanced the stability of Pd/NaY by Mo modification. The catalyst demonstrated remarkable performance over 30 hours, maintaining a high CO conversion rate of 97% and a DMC selectivity of 80%. The incorporated Mo not only inhibited Pd sintering by partially healing silanols, but also increased and stabilized the Pd^{2+} active sites caused by the Pt–Mo interaction [230].

Moreover, the nature of Pd^{2+} species and the topology of zeolites have been identified as crucial factors for the lifetime of the catalyst. Guided by DFT calculations, Tan et al. engineered mononuclear-isolated Pd^{2+} centers on Y zeolite through ultrasonic-assisted ammonia evaporation method. The stabilized Pd^{2+} species showed remarkable stability over a period of 100 hours [228]. Mintova and co-workers replaced the traditional NaY zeolite with EMT zeolite as the catalyst support. The resulting catalyst displayed good stability during the continuous test for 130 hours. The good stability was due to the strengthened interaction between Pd and the EMT, which inhibited the sintering of Pd clusters and maintained the Pd^{2+} active sites [231].

Given that the reduction of Pd(II) to Pd(0) is the primary deactivation mechanism, re-oxidization is employed to restore the catalysts' activity. For example, Mintova and co-workers regenerated a deactivated Pd/FAU catalyst by calcining it in air at 200 °C for two hours. After regeneration, the CO conversion improved from 21 to 76% (see Figure 4.18). However, as the reaction proceeded, the CO reactant gradually reduced the Pd(II) species back to Pd(0), leading to a decrease in activity similar to that observed with the fresh catalyst [231]. Note that the aggregation of Pd species could occur simultaneously under high temperature during the re-oxidization process, potentially affecting the catalyst's long-term stability.

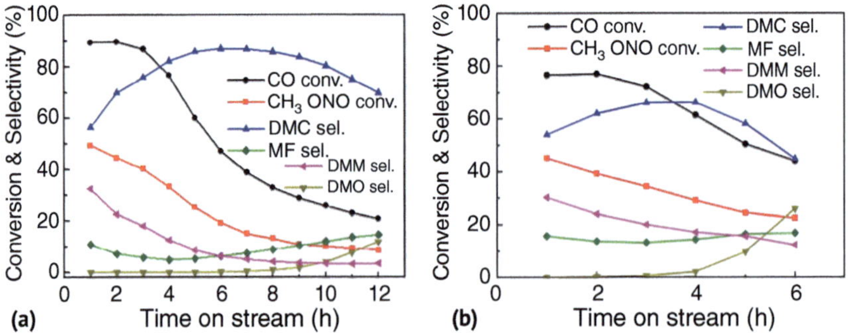

Figure 4.18 Conversion of CO and MN (CH_3ONO), and selectivities of DMC, methyl formate (MF), dimethoxymethane (DMM), and DMO as a function of time in the indirect oxidative carbonylation of methanol to DMC on catalysts: (a) Pd/FAU and (b) recalcined Pd/FAU. *Source*: [231] / with permission of Elsevier.

4.3.4 Conclusion

In this section, we summarized the current research progress on metal–zeolite catalysts in three typical low-carbon molecule conversion reactions (propane dehydrogenation to propene, DRM, and carbonylation of MN to DMC), mainly focusing on deactivation mechanism, deactivation migration strategy, and catalyst regeneration. Generally, coke deposition and metal sintering are the primary causes of metal/zeolite catalyst deactivation in propane dehydrogenation and DRM. Significant efforts have been devoted to understanding and alleviating these detrimental effects. Effective strategies including the incorporation of metal promoters, immobilization of metals within zeolite crystals, and structural modification of zeolites were proposed to prevent deactivation, showing remarkable potential in enhancing catalyst stability. On the other hand, the change in the oxidation state of the active metal species and further agglomeration represent another significant factor contributing to the deactivation in metal–zeolite catalysts. A typical example is the Pd–zeolite-catalyzed MN carbonylation to DMC. Utilizing the strong interaction between Pd and zeolite to stabilize, fixing the position of Pd within the zeolite, and regulating the electron density of Pd(II) are key strategies for maintaining the stability of the catalyst.

4.4 Perspectives

Advances in deactivation research have significantly enhanced deep understanding of the deactivation behaviors and mechanisms of zeolite catalysts. As a widely applied industrial catalyst system, there remain numerous challenges and issues to be addressed in the future. For instance: (1) designing specific zeolitic architectures to constrain coke evolution or facilitate intermediates emigration, inhibiting the deposition of coke to the maximum extent; (2) evaluating the effects of harsh regeneration conditions on active sites' stability or developing milder and greener regeneration strategies; (3) utilizing advanced spatiotemporally resolved characterization techniques to uncover deactivation kinetics and mechanisms under realistic reaction conditions; and (4) quantitatively analyzing the dynamic evolution and mass transfer of active species, coke, and coke precursors. Ultimately, all these efforts will provide theoretical guidance for the development of highly efficient, stable, and recyclable catalysts.

References

1 Karge, H.G. (1991). Coke formation on zeolites. *Stud. Surf. Sci. Catal.* 58: 531–570.
2 Lee, S. and Choi, M. (2019). Unveiling coke formation mechanism in MFI zeolites during methanol-to-hydrocarbons conversion. *J. Catal.* 375: 183–192.
3 Wang, B. and Manos, G. (2007). A novel thermogravimetric method for coke precursor characterisation. *J. Catal.* 250: 121–127.
4 Castaño, P., Elordi, G., Olazar, M. et al. (2011). Insights into the coke deposited on HZSM-5, Hβ and HY zeolites during the cracking of polyethylene. *Appl. Catal. B: Environ.* 104: 91–100.

5 Díaz, M., Epelde, E., Valecillos, J. et al. (2021). Coke deactivation and regeneration of HZSM-5 zeolite catalysts in the oligomerization of 1-butene. *Appl. Catal. B: Environ.* 291: 120076.

6 Zhou, J.B., Zhao, J.P., Zhang, J.L. et al. (2020). Regeneration of catalysts deactivated by coke deposition: a review. *Chin. J. Catal.* 41: 1048–1061.

7 Bauer, F. and Karge, H.G. (2007). Characterization of coke on zeolites. In: *Characterization II* (eds. H.G. Karge and J. Weitkamp), Molecular Sieves (Vol. 5), 249–364. Berlin: Springer.

8 Cheng, Z.Z., Huang, S.Y., Li, Y. et al. (2019). Carbonylation of dimethyl ether over MOR and Cu/H-MOR catalysts: comparative investigation of deactivation behavior. *Appl. Catal. A: Gen.* 576: 1–10.

9 Ferreira Madeira, F., Ben Tayeb, K., Pinard, L. et al. (2012). Ethanol transformation into hydrocarbons on ZSM-5 zeolites: influence of Si/Al ratio on catalytic performances and deactivation rate. Study of the radical species role. *Appl. Catal. A: Gen.* 443: 171–180.

10 Wu, Z.Y., Wang, L.M., Li, Z.X. et al. (2023). Unveiling the promotion of Brønsted acid sites in Cs dispersion and consequential Si-O-Cs species formation for methyl acrylate synthesis from methyl acetate and formaldehyde over Cs/Beta zeolite catalyst. *Chem. Eng. J.* 474: 145655.

11 van Vreeswijk, S.H., Monai, M., Oord, R. et al. (2022). Nano-scale insights regarding coke formation in zeolite SSZ-13 subject to the methanol-to-hydrocarbons reaction. *Catal. Sci. Technol.* 12: 1220–1228.

12 Wulfers, M.J. and Jentoft, F.C. (2013). Identification of carbonaceous deposits formed on H-mordenite during alkane isomerization. *J. Catal.* 307: 204–213.

13 Wulfers, M.J., Tzolova-Müller, G., Villegas, J.I. et al. (2012). Evolution of carbonaceous deposits on H-mordenite and Pt-doped H-mordenite during *n*-butane conversion. *J. Catal.* 296: 132–142.

14 France, L.J., Apperley, D.C., Ditzel, E.J. et al. (2011). An investigation of the nature and reactivity of the carbonaceous species deposited on mordenite by reaction with methanol. *Catal. Sci. Technol.* 1: 932–939.

15 Guisnet, M. and Magnoux, P. (1989). Coking and deactivation of zeolites influence of the pore structure. *Appl. Catal.* 54: 1–27.

16 Gallezot, P., Leclercq, C., Guisnet, M. et al. (1988). Coking, aging, and regeneration of zeolites: VII. Electron microscopy and EELS studies of external coke deposits on USHY, H-OFF, and H-ZSM-5 zeolites. *J. Catal.* 114: 100–111.

17 Wang, N., Sun, W.J., Hou, Y.L. et al. (2018). Crystal-plane effects of MFI zeolite in catalytic conversion of methanol to hydrocarbons. *J. Catal.* 360: 89–96.

18 Ben Tayeb, K., Hamieh, S., Canaff, C. et al. (2019). The radical internal coke structure as a fingerprint of the zeolite framework. *Microporous Mesoporous Mater.* 289: 109617.

19 Rodriguez, A.C., Sad, M.E., Cruchade, H. et al. (2021). Study of catalyst deactivation during 1,3-butanediol dehydration to produce butadiene. *Microporous Mesoporous Mater.* 320: 111066.

20 Yang, H.J., Park, K., Shin, S. et al. (2013). Characterization of heme ions using MALDI-TOF MS and MALDI FT-ICR MS. *Int. J. Mass Spectrom.* 343: 37–44.

21 Wang, N., Zhi, Y.C., Wei, Y.X. et al. (2020). Molecular elucidating of an unusual growth mechanism for polycyclic aromatic hydrocarbons in confined space. *Nat. Commun.* 11: 1079.

22 Rojo-Gama, D., Nielsen, M., Wragg, D.S. et al. (2017). A straightforward descriptor for the deactivation of zeolite catalyst H-ZSM-5. *ACS Catal.* 7: 8235–8246.

23 Goetze, J., Yarulina, I., Gascon, J. et al. (2018). Revealing lattice expansion of small-pore zeolite catalysts during the methanol-to-olefins process using combined operando X-ray diffraction and UV-vis spectroscopy. *ACS Catal.* 8: 2060–2070.

24 Goetze, J. and Weckhuysen, B.M. (2018). Spatiotemporal coke formation over zeolite ZSM-5 during the methanol-to-olefins process as studied with operando UV-vis spectroscopy: a comparison between H-ZSM-5 and Mg-ZSM-5. *Catal. Sci. Technol.* 8: 1632–1644.

25 Wragg, D.S., Kalantzopoulos, G.N., Pappas, D.K. et al. (2021). Mapping the coke formation within a zeolite catalyst extrudate in space and time by operando computed X-ray diffraction tomography. *J. Catal.* 401: 1–6.

26 Gao, S.S., Xu, S.T., Wei, Y.X. et al. (2018). Insight into the deactivation mode of methanol-to-olefins conversion over SAPO-34: coke, diffusion, and acidic site accessibility. *J. Catal.* 367: 306–314.

27 Wennmacher, J.T.C., Mahmoudi, S., Rzepka, P. et al. (2022). Electron diffraction enables the mapping of coke in ZSM-5 micropores formed during methanol-to-hydrocarbons conversion. *Angew. Chem. Int. Ed.* 61: e202205413.

28 Devaraj, A., Vijayakumar, M., Bao, J. et al. (2016). Discerning the location and nature of coke deposition from surface to bulk of spent zeolite catalysts. *Sci. Rep.* 6: 37586.

29 Lacarriere, A., Luck, F., Swierczynski, D. et al. (2011). Methanol to hydrocarbons over zeolites with MWW topology: effect of zeolite texture and acidity. *Appl. Catal. A: Gen.* 402: 208–217.

30 Wang, J.B., Li, J.Z., Xu, S.T. et al. (2015). Methanol to hydrocarbons reaction over HZSM-22 and SAPO-11: effect of catalyst acid strength on reaction and deactivation mechanism. *Chin. J. Catal.* 36: 1392–1402.

31 Lee, K.Y., Kang, M.Y., and Ihm, S.K. (2012). Deactivation by coke deposition on the HZSM-5 catalysts in the methanol-to-hydrocarbon conversion. *J. Phys. Chem. Solids* 73: 1542–1545.

32 Wang, P.F., Huang, L.Z., Li, J.F. et al. (2015). Catalytic properties and deactivation behavior of H-MCM-22 in the conversion of methanol to hydrocarbons. *RSC Adv.* 5: 28794–28802.

33 Catizzone, E., Aloise, A., Migliori, M. et al. (2017). The effect of FER zeolite acid sites in methanol-to-dimethyl-ether catalytic dehydration. *J. Energy Chem.* 26: 406–415.

34 Barbera, K., Bonino, F., Bordiga, S. et al. (2011). Structure-deactivation relationship for ZSM-5 catalysts governed by framework defects. *J. Catal.* 280: 196–205.

35 Yarulina, I., Goetze, J., Gucuyener, C. et al. (2016). Methanol-to-olefins process over zeolite catalysts with DDR topology: effect of composition and structural defects on catalytic performance. *Catal. Sci. Technol.* 6: 2663–2678.
36 Yang, G.J., Wei, Y.X., Xu, S.T. et al. (2013). Nanosize-enhanced lifetime of SAPO-34 catalysts in methanol-to-olefin reactions. *J. Phys. Chem. C* 117: 8214–8222.
37 Shen, Y.F., Le, T.T., Fu, D.L. et al. (2018). Deconvoluting the competing effects of zeolite framework topology and diffusion path length on methanol to hydrocarbons reaction. *ACS Catal.* 8: 11042–11053.
38 Wan, Z.J., Li, G.K., Wang, C.F. et al. (2018). Relating coke formation and characteristics to deactivation of ZSM-5 zeolite in methanol to gasoline conversion. *Appl. Catal. A: Gen.* 549: 141–151.
39 Bjorgen, M., Grave, A.H., Saepurahman et al. (2012). Spectroscopic and catalytic characterization of extra large pore zeotype H-ITQ-33. *Microporous Mesoporous Mater.* 151: 424–433.
40 Rojo-Gama, D., Signorile, M., Bonino, F. et al. (2017). Structure-deactivation relationships in zeolites during the methanol-to-hydrocarbons reaction: complementary assessments of the coke content. *J. Catal.* 351: 33–48.
41 Qi, L., Li, J.Z., Wang, L.Y. et al. (2017). Comparative investigation of the deactivation behaviors over HZSM-5 and HSAPO-34 catalysts during low-temperature methanol conversion. *Catal. Sci. Technol.* 7: 2022–2031.
42 Chen, J.R., Li, J.Z., Wei, Y.X. et al. (2014). Spatial confinement effects of cage-type SAPO molecular sieves on product distribution and coke formation in methanol-to-olefin reaction. *Catal. Commun.* 46: 36–40.
43 Pinilla-Herrero, I., Olsbye, U., Márquez-Álvarez, C. et al. (2017). Effect of framework topology of SAPO catalysts on selectivity and deactivation profile in the methanol-to-olefins reaction. *J. Catal.* 352: 191–207.
44 Goetze, J., Meirer, F., Yarulina, I. et al. (2017). Insights into the activity and deactivation of the methanol-to-olefins process over different small-pore zeolites as studied with operando UV-vis spectroscopy. *ACS Catal.* 7: 4033–4046.
45 Fu, D.L., Paioni, A.L., Lian, C. et al. (2020). Elucidating zeolite channel geometry–reaction intermediate relationships for the methanol-to-hydrocarbon process. *Angew. Chem. Int. Ed.* 59: 20024–20030.
46 Li, J.Z., Wei, Y.X., Qi, Y. et al. (2011). Conversion of methanol over H-ZSM-22: the reaction mechanism and deactivation. *Catal. Today* 164: 288–292.
47 Chowdhury, A.D. and Gascon, J. (2018). The curious case of ketene in zeolite chemistry and catalysis. *Angew. Chem. Int. Ed.* 57: 14982–14985.
48 Gao, Y.L., Chen, S.L., Wei, Y.Q. et al. (2017). Kinetics of coke formation in the dimethyl ether-to-olefins process over SAPO-34 catalyst. *Chem. Eng. J.* 326: 528–539.
49 Liu, R.Z., Shao, X., Wang, C. et al. (2023). Reaction mechanism of methanol-to-hydrocarbons conversion: fundamental and application. *Chin. J. Catal.* 47: 67–92.
50 Song, W.G., Marcus, D.M., Fu, H. et al. (2002). An oft-studied reaction that may never have been: direct catalytic conversion of methanol or dimethyl ether to hydrocarbons on the solid acids HZSM-5 or HSAPO-34. *J. Am. Chem. Soc.* 124: 3844–3845.

51 Chang, C.D. and Silvestri, A.J. (1977). The conversion of methanol and other O-compounds to hydrocarbons over zeolite catalysts. *J. Catal.* 47: 249–259.
52 Clarke, J.K.A., Darcy, R., Hegarty, B.F. et al. (1986). Free radicals in dimethyl ether on H-ZSM-5 zeolite. A novel dimension of heterogeneous catalysis. *J. Chem. Soc., Chem. Commun.* 1986: 425–426.
53 Hutchings, G.J., Gottschalk, F., Hall, M.V.M. et al. (1987). Hydrocarbon formation from methylating agents over the zeolite catalyst ZSM-5. Comments on the mechanism of carbon–carbon bond and methane formation. *J. Chem. Soc., Faraday Trans.* 83: 571–583.
54 Tajima, N., Tsuneda, T., Toyama, F. et al. (1998). A new mechanism for the first carbon–carbon bond formation in the MTG process: a theoretical study. *J. Am. Chem. Soc.* 120: 8222–8229.
55 Lesthaeghe, D., Van Speybroeck, V., Marin, G.B. et al. (2007). The rise and fall of direct mechanisms in methanol-to-olefin catalysis: an overview of theoretical contributions. *Ind. Eng. Chem. Res.* 46: 8832–8838.
56 Lesthaeghe, D., Van Speybroeck, V., Marin, G.B. et al. (2006). What role do oxonium ions and oxonium ylides play in the ZSM-5 catalysed methanol-to-olefin process? *Chem. Phys. Lett.* 417: 309–315.
57 Jiang, Y., Wang, W., Reddy Marthala, V.R. et al. (2006). Effect of organic impurities on the hydrocarbon formation via the decomposition of surface methoxy groups on acidic zeolite catalysts. *J. Catal.* 238: 21–27.
58 Jiang, Y.J., Hunger, M., and Wang, W. (2006). On the reactivity of surface methoxy species in acidic zeolites. *J. Am. Chem. Soc.* 128: 11679–11692.
59 Wang, W. and Hunger, M. (2008). Reactivity of surface alkoxy species on acidic zeolite catalysts. *Acc. Chem. Res.* 41: 895–904.
60 Wang, W., Buchholz, A., Seiler, M. et al. (2003). Evidence for an initiation of the methanol-to-olefin process by reactive surface methoxy groups on acidic zeolite catalysts. *J. Am. Chem. Soc.* 125: 15260–15267.
61 Yamazaki, H., Shima, H., Imai, H. et al. (2012). Direct production of propene from methoxy species and dimethyl ether over H-ZSM-5. *J. Phys. Chem. C* 116: 24091–24097.
62 Yamazaki, H., Shima, H., Imai, H. et al. (2011). Evidence for a "carbene-like" intermediate during the reaction of methoxy species with light alkenes on H-ZSM-5. *Angew. Chem. Int. Ed.* 50: 1853–1856.
63 Li, J.F., Wei, Z.H., Chen, Y.Y. et al. (2014). A route to form initial hydrocarbon pool species in methanol conversion to olefins over zeolites. *J. Catal.* 317: 277–283.
64 Wei, Z.H., Chen, Y.Y., Li, J.F. et al. (2016). Methane formation mechanism in the initial methanol-to-olefins process catalyzed by SAPO-34. *Catal. Sci. Technol.* 6: 5526–5533.
65 Wu, X.Q., Xu, S.T., Zhang, W.N. et al. (2017). Direct mechanism of the first carbon–carbon bond formation in the methanol-to-hydrocarbons process. *Angew. Chem. Int. Ed.* 56: 9039–9043.
66 Comas-Vives, A., Valla, M., Copéret, C. et al. (2015). Cooperativity between Al sites promotes hydrogen transfer and carbon–carbon bond formation upon dimethyl ether activation on alumina. *ACS Cent. Sci.* 1: 313–319.

67 Liu, Y., Müller, S., Berger, D. et al. (2016). Formation mechanism of the first carbon–carbon bond and the first olefin in the methanol conversion into hydrocarbons. *Angew. Chem. Int. Ed.* 55: 5723–5726.

68 Chowdhury, A.D., Houben, K., Whiting, G.T. et al. (2016). Initial carbon–carbon bond formation during the early stages of the methanol-to-olefin process proven by zeolite-trapped acetate and methyl acetate. *Angew. Chem. Int. Ed.* 55: 15840–15845.

69 Yang, L., Yan, T.T., Wang, C.M. et al. (2019). Role of acetaldehyde in the roadmap from initial carbon–carbon bonds to hydrocarbons during methanol conversion. *ACS Catal.* 9: 6491–6501.

70 Yarulina, I., Chowdhury, A.D., Meirer, F. et al. (2018). Recent trends and fundamental insights in the methanol-to-hydrocarbons process. *Nat. Catal.* 1: 398–411.

71 Plessow, P.N. and Studt, F. (2017). Unraveling the mechanism of the initiation reaction of the methanol to olefins process using ab initio and DFT calculations. *ACS Catal.* 7: 7987–7994.

72 Cheng, Z.Z., Huang, S.Y., Li, Y. et al. (2021). Role of Brønsted acid sites within 8-MR of mordenite in the deactivation roadmap for dimethyl ether carbonylation. *ACS Catal.* 11: 5647–5657.

73 Wu, X.K., Zhang, Z.H., Pan, Z.Y. et al. (2022). Ketenes in the induction of the methanol-to-olefins process. *Angew. Chem. Int. Ed.* 61: e202207777.

74 Hwang, A., Kumar, M., Rimer, J.D. et al. (2017). Implications of methanol disproportionation on catalyst lifetime for methanol-to-olefins conversion by HSSZ-13. *J. Catal.* 346: 154–160.

75 Zhang, N.N., Chen, S.L., Zhu, R.Y. et al. (2022). Evolution pathway and role of intermediates acetaldehyde and acetone formed during the methanol-to-olefins reaction. *Energy Fuels* 36: 12708–12718.

76 Sun, X.Y., Mueller, S., Liu, Y. et al. (2014). On reaction pathways in the conversion of methanol to hydrocarbons on HZSM-5. *J. Catal.* 317: 185–197.

77 Müller, S., Liu, Y., Kirchberger, F.M. et al. (2016). Hydrogen transfer pathways during zeolite catalyzed methanol conversion to hydrocarbons. *J. Am. Chem. Soc.* 138: 15994–16003.

78 Liu, Z.H., Dong, X.L., Liu, X. et al. (2016). Oxygen-containing coke species in zeolite-catalyzed conversion of methanol to hydrocarbons. *Catal. Sci. Technol.* 6: 8157–8165.

79 Martinez-Espin, J.S., De Wispelaere, K., Westgård Erichsen, M. et al. (2017). Benzene co-reaction with methanol and dimethyl ether over zeolite and zeotype catalysts: evidence of parallel reaction paths to toluene and diphenylmethane. *J. Catal.* 349: 136–148.

80 Liu, Y., Kirchberger, F.M., Müller, S. et al. (2019). Critical role of formaldehyde during methanol conversion to hydrocarbons. *Nat. Commun.* 10: 1462.

81 Shi, Z.C., Arora, S.S., Trahan, D.W. et al. (2022). Methanol to hydrocarbons conversion: why dienes and monoenes contribute differently to catalyst deactivation? *Chem. Eng. J.* 437: 134229.

82 Wang, S., Chen, Y.Y., Qin, Z.F. et al. (2019). Origin and evolution of the initial hydrocarbon pool intermediates in the transition period for the conversion of methanol to olefins over H-ZSM-5 zeolite. *J. Catal.* 369: 382–395.
83 Wang, S., Chen, Y.Y., Wei, Z.H. et al. (2016). Evolution of aromatic species in supercages and its effect on the conversion of methanol to olefins over H-MCM-22 zeolite: a density functional theory study. *J. Phys. Chem. C* 120: 27964–27979.
84 An, H.Y., Zhang, F., Guan, Z.H. et al. (2018). Investigating the coke formation mechanism of H-ZSM-5 during methanol dehydration using operando UV-Raman spectroscopy. *ACS Catal.* 8: 9207–9215.
85 Wang, C., Hu, M., Chu, Y.Y. et al. (2020). π-Interactions between cyclic carbocations and aromatics cause zeolite deactivation in methanol-to-hydrocarbon conversion. *Angew. Chem. Int. Ed.* 59: 7198–7202.
86 Yu, B.W., Zhang, W.N., Wei, Y.X. et al. (2020). Capture and identification of coke precursors to elucidate the deactivation route of the methanol-to-olefin process over H-SAPO-34. *Chem. Commun.* 56: 8063–8066.
87 Li, Y., Sun, Q., Huang, S.Y. et al. (2018). Dimethyl ether carbonylation over pyridine-modified MOR: enhanced stability influenced by acidity. *Catal. Today* 311: 81–88.
88 Liu, J.L., Xue, H.F., Huang, X.M. et al. (2010). Stability enhancement of H-mordenite in dimethyl ether carbonylation to methyl acetate by pre-adsorption of pyridine. *Chin. J. Catal.* 31: 729–738.
89 Valecillos, J., Epelde, E., Albo, J. et al. (2020). Slowing down the deactivation of H-ZSM-5 zeolite catalyst in the methanol-to-olefin (MTO) reaction by P or Zn modifications. *Catal. Today* 348: 243–256.
90 Mores, D., Stavitski, E., Verkleij, S.P. et al. (2011). Core-shell H-ZSM-5/silicalite-1 composites: Brønsted acidity and catalyst deactivation at the individual particle level. *Phys. Chem. Chem. Phys.* 13: 15985–15994.
91 Kim, S., Kim, Y.T., Hwang, A. et al. (2017). Coke-tolerant gadolinium-promoted HZSM-5 catalyst for methanol conversion into hydrocarbons. *ChemCatChem* 9: 1569–1573.
92 Park, S.J., Jang, H.G., Lee, K.Y. et al. (2018). Improved methanol-to-olefin reaction selectivity and catalyst life by CeO_2 coating of ferrierite zeolite. *Microporous Mesoporous Mater.* 256: 155–164.
93 Fujimoto, Y., Shu, Y., Miyake, K. et al. (2021). SAPO-34 zeolite nanocrystals coated with ZrO_2 as catalysts for methanol-to-olefin conversion. *ACS Appl. Nano Mater.* 4: 8321–8327.
94 Zhou, F., Gao, Y., Wu, G. et al. (2017). Improved catalytic performance and decreased coke formation in post-treated ZSM-5 zeolites for methanol aromatization. *Microporous Mesoporous Mater.* 240: 96–107.
95 Liu, Z.H., Dong, X.L., Zhu, Y.H. et al. (2015). Investigating the influence of mesoporosity in zeolite beta on its catalytic performance for the conversion of methanol to hydrocarbons. *ACS Catal.* 5: 5837–5845.
96 Zhou, J., Teng, J.W., Ren, L.P. et al. (2016). Full-crystalline hierarchical monolithic ZSM-5 zeolites as superiorly active and long-lived practical catalysts in methanol-to-hydrocarbons reaction. *J. Catal.* 340: 166–176.

97 Weissenberger, T., Reiprich, B., Machoke, A.G.F. et al. (2019). Hierarchical MFI type zeolites with intracrystalline macropores: the effect of the macropore size on the deactivation behaviour in the MTO reaction. *Catal. Sci. Technol.* 9: 3259–3269.

98 Weissenberger, T., Machoke, A.G.F., Bauer, J. et al. (2020). Hierarchical ZSM-5 catalysts: the effect of different intracrystalline pore dimensions on catalyst deactivation behaviour in the MTO reaction. *ChemCatChem* 12: 2461–2468.

99 Li, Z., Navarro, M.T., Martinez-Triguero, J. et al. (2016). Synthesis of nano-SSZ-13 and its application in the reaction of methanol to olefins. *Catal. Sci. Technol.* 6: 5856–5863.

100 Perez-Uriarte, P., Ateka, A., Gamero, M. et al. (2016). Effect of the operating conditions in the transformation of DME to olefins over a HZSM-5 zeolite catalyst. *Ind. Eng. Chem. Res.* 55: 6569–6578.

101 Kim, S., Kim, Y.T., Zhang, C. et al. (2017). Effect of reaction conditions on the catalytic dehydration of methanol to dimethyl ether over a K-modified HZSM-5 catalyst. *Catal. Lett.* 147: 792–801.

102 Cordero-Lanzac, T., Ateka, A., Perez-Uriarte, P. et al. (2018). Insight into the deactivation and regeneration of HZSM-5 zeolite catalysts in the conversion of dimethyl ether to olefins. *Ind. Eng. Chem. Res.* 57: 13689–13702.

103 Zhao, X.B., Li, J.Z., Tian, P. et al. (2019). Achieving a superlong lifetime in the zeolite-catalyzed MTO reaction under high pressure: synergistic effect of hydrogen and water. *ACS Catal.* 9: 3017–3025.

104 Valecillos, J., Elordi, G., Aguayo, A.T. et al. (2021). The intrinsic effect of co-feeding water on the formation of active/deactivating species in the methanol-to-hydrocarbons reaction on ZSM-5 zeolite. *Catal. Sci. Technol.* 11: 1269–1281.

105 Martinez-Espin, J.S., Morten, M., Janssens, T.V.W. et al. (2017). New insights into catalyst deactivation and product distribution of zeolites in the methanol-to-hydrocarbons (MTH) reaction with methanol and dimethyl ether feeds. *Catal. Sci. Technol.* 7: 2700–2716.

106 Zhang, G.Q., Zhang, X., Bai, T. et al. (2015). Coking kinetics and influence of reaction-regeneration on acidity, activity and deactivation of Zn/HZSM-5 catalyst during methanol aromatization. *J. Energy Chem.* 24: 108–118.

107 Haas, A., Hauber, C., and Kirchmann, M. (2019). Time-resolved product analysis of dimethyl ether-to-olefins conversion on SAPO-34. *ACS Catal.* 9: 5679–5691.

108 Yang, L., Wang, C., Zhang, L.N. et al. (2021). Stabilizing the framework of SAPO-34 zeolite toward long-term methanol-to-olefins conversion. *Nat. Commun.* 12: 4661.

109 Aguayo, A.T., Castano, P., Mier, D. et al. (2011). Effect of cofeeding butane with methanol on the deactivation by coke of a HZSM-5 zeolite catalyst. *Ind. Eng. Chem. Res.* 50: 9980–9988.

110 Mier, D., Aguayo, A.T., Gayubo, A.G. et al. (2010). Synergies in the production of olefins by combined cracking of n-butane and methanol on a HZSM-5 zeolite catalyst. *Chem. Eng. J.* 160: 760–769.

111 Mier, D., Gayubo, A.G., Aguayo, A.T. et al. (2011). Olefin production by cofeeding methanol and *n*-butane: kinetic modeling considering the deactivation of HZSM-5 zeolite. *AIChE J.* 57: 2841–2853.

112 Li, X.G., Huang, X., Zhang, Y.L. et al. (2020). Effect of *n*-butanol cofeeding on the deactivation of methanol to olefin conversion over high-silica HZSM-5: a mechanism and kinetic study. *Chem. Eng. Sci.* 226: 115859.

113 Guisnet, M. and Magnoux, P. (1997). Deactivation by coking of zeolite catalysts. Prevention of deactivation. Optimal conditions for regeneration. *Catal. Today* 36: 477–483.

114 Trimm, D.L. (2001). The regeneration or disposal of deactivated heterogeneous catalysts. *Appl. Catal. A: Gen.* 212: 153–160.

115 Jia, L.Y., Farouha, A., Pinard, L. et al. (2017). New routes for complete regeneration of coked zeolite. *Appl. Catal. B: Environ.* 219: 82–91.

116 Magnoux, P. and Guisnet, M. (1988). Coking, ageing and regeneration of zeolites: VI. Comparison of the rates of coke oxidation of HY, H-mordenite and HZSM-5. *Appl. Catal.* 38: 341–352.

117 Schmidt, F., Hoffmann, C., Giordanino, F. et al. (2013). Coke location in microporous and hierarchical ZSM-5 and the impact on the MTH reaction. *J. Catal.* 307: 238–245.

118 Doka Nassionou, G.A., Magnoux, P., and Guisnet, M. (1998). Comparative study of coking and regeneration of HEMT and HFAU zeolites. *Microporous Mesoporous Mater.* 22: 389–398.

119 Richard, R., Julcour, C., and Manero, M.H. (2017). Towards a new oxidation process using ozone to regenerate coked catalysts. *Ozone: Sci. Eng.* 39: 366–373.

120 Querini, C.A. (2000). Isobutane/butene alkylation: regeneration of solid acid catalysts. *Catal. Today* 62: 135–143.

121 Barbera, K., Sørensen, S., Bordiga, S. et al. (2012). Role of internal coke for deactivation of ZSM-5 catalysts after low temperature removal of coke with NO_2. *Catal. Sci. Technol.* 2: 1196–1206.

122 Aguayo, A.T., Gayubo, A.G., Ateka, A. et al. (2012). Joint transformation of methanol and *n*-butane into olefins on an HZSM-5 zeolite catalyst in reaction-regeneration cycles. *Ind. Eng. Chem. Res.* 51: 13073–13084.

123 Seierensen, M.D.P. (2017). The establishment of a coke-burn kinetic model for zeolite catalysts. *Chem. Eng. Sci.* 168: 465–479.

124 Jiang, B.B., Zhou, B.J., Jiang, Y.T. et al. (2017). Kinetic and regenerator modeling of the coke combustion in the moving bed MTP process. *Chem. Eng. Res. Des.* 122: 52–62.

125 Zhou, J.B., Zhang, J.L., Zhi, Y.C. et al. (2018). Partial regeneration of the spent SAPO-34 catalyst in the methanol-to-olefins process via steam gasification. *Ind. Eng. Chem. Res.* 57: 17338–17347.

126 Zhou, J.B., Gao, M.B., Zhang, J.L. et al. (2021). Directed transforming of coke to active intermediates in methanol-to-olefins catalyst to boost light olefins selectivity. *Nat. Commun.* 12: 17.

127 An, H.Q., Li, H., Zhou, J.B. et al. (2021). Kinetics of steam regeneration of SAPO-34 zeolite catalyst in methanol-to-olefins (MTO) process. *Chin. J. Chem. Eng.* 35: 231–238.

128 Walker, P.L., Matsumoto, S., Hanzawa, T. et al. (1983). Catalysis of gasification of coal-derived cokes and chars. *Fuel* 62: 140–149.

129 Josl, R., Klingmann, R., Traa, Y. et al. (2004). Regeneration of zeolite catalysts deactivated in isobutane/butene alkylation: an in situ FTIR investigation at elevated H_2 pressure. *Catal. Commun.* 5: 239–241.

130 Klingmann, R., Josl, R., Traa, Y. et al. (2005). Hydrogenative regeneration of a Pt/La-Y zeolite catalyst deactivated in the isobutane/*n*-butene alkylation. *Appl. Catal. A: Gen.* 281: 215–223.

131 Tsai, T.C. (2006). Reactivation of acidic sites in mordenite used in toluene disproportionation. *Appl. Catal. A: Gen.* 301: 292–298.

132 Jong, S.J., Pradhan, A.R., Wu, J.F. et al. (1998). On the regeneration of coked H-ZSM-5 catalysts. *J. Catal.* 174: 210–218.

133 Han, S.J., Kim, S.K., Hwang, A. et al. (2019). Non-oxidative dehydroaromatization of methane over Mo/HZSM-5 catalysts: a detailed analysis of the reaction-regeneration cycle. *Appl. Catal. B: Environ.* 241: 305–318.

134 Honda, K., Yoshida, T., and Zhang, Z.G. (2003). Methane dehydroaromatization over Mo/HZSM-5 in periodic CH_4–H_2 switching operation mode. *Catal. Commun.* 4: 21–26.

135 Xu, Y.B., Lu, J.Y., Wang, J.D. et al. (2011). The catalytic stability of Mo/HZSM-5 in methane dehydroaromatization at severe and periodic CH_4–H_2 switch operating conditions. *Chem. Eng. J.* 168: 390–402.

136 Sun, C.Y., Fang, G.Z., Guo, X.G. et al. (2015). Methane dehydroaromatization with periodic CH_4–H_2 switch: a promising process for aromatics and hydrogen. *J. Energy Chem.* 24: 257–263.

137 Yan, P., Zhang, Z.G., Li, D.P. et al. (2018). Development of a CH_4 dehydroaromatization–catalyst regeneration fluidized bed system. *Chin. J. Chem. Eng.* 26: 1928–1936.

138 Lian, Z., Ali, S., Liu, T.F. et al. (2018). Revealing the Janus character of the coke precursor in the propane direct dehydrogenation on Pt catalysts from a kMC simulation. *ACS Catal.* 8: 4694–4704.

139 Chen, S., Chang, X., Sun, G.D. et al. (2021). Propane dehydrogenation: catalyst development, new chemistry, and emerging technologies. *Chem. Soc. Rev.* 50: 3315–3354.

140 Qu, Z.Q. and Sun, Q.M. (2022). Advances in zeolite-supported metal catalysts for propane dehydrogenation. *Inorg. Chem. Front.* 9: 3095–3115.

141 Lian, Z., Si, C.W., Jan, F. et al. (2021). Coke deposition on Pt-based catalysts in propane direct dehydrogenation: kinetics, suppression, and elimination. *ACS Catal.* 11: 9279–9292.

142 Zhu, J., Yang, M.L., Yu, Y.D. et al. (2015). Size-dependent reaction mechanism and kinetics for propane dehydrogenation over Pt catalysts. *ACS Catal.* 5: 6310–6319.

143 Zhang, W., Wang, H.Z., Jiang, J.W. et al. (2020). Size dependence of Pt catalysts for propane dehydrogenation: from atomically dispersed to nanoparticles. *ACS Catal.* 10: 12932–12942.

144 Hansen, T.W., Delariva, A.T., Challa, S.R. et al. (2013). Sintering of catalytic nanoparticles: particle migration or Ostwald ripening? *Acc. Chem. Res.* 46: 1720–1730.

145 He, Z., Yang, J.N., and Liu, L.C. (2024). Design of supported metal catalysts and systems for propane dehydrogenation. *JACS Au* 4: 4084–4109.

146 Chen, C., Sun, M., Hu, Z. et al. (2019). New insight into the enhanced catalytic performance of ZnPt/HZSM-5 catalysts for direct dehydrogenation of propane to propylene. *Catal. Sci. Technol.* 9: 1979–1988.

147 Zhao, D., Guo, K., Han, S.L. et al. (2022). Controlling reaction-induced loss of active sites in ZnO_x/silicalite-1 for durable nonoxidative propane dehydrogenation. *ACS Catal.* 12: 4608–4617.

148 Liu, L.C., Díaz, U., Arenal, R. et al. (2016). Generation of subnanometric platinum with high stability during transformation of a 2D zeolite into 3D. *Nat. Mater.* 16: 132–138.

149 Qiu, B.C., Zhang, Y.K., Liu, Y. et al. (2021). The state of Pt active phase and its surrounding environment during dehydrogenation of ethane to ethylene. *Appl. Surf. Sci.* 554: 149611.

150 Zhang, Z.Y., Xu, W.L., Ye, X.M. et al. (2021). Enormous passivation effects of a surrounding zeolitic framework on Pt clusters for the catalytic dehydrogenation of propane. *Catal. Sci. Technol.* 11: 5250–5259.

151 Liu, L.C., Lopez-Haro, M., Lopes, C.W. et al. (2020). Structural modulation and direct measurement of subnanometric bimetallic PtSn clusters confined in zeolites. *Nat. Catal.* 3: 628–638.

152 Sun, Q., Wang, N., Fan, Q. et al. (2020). Subnanometer bimetallic platinum-zinc clusters in zeolites for propane dehydrogenation. *Angew. Chem. Int. Ed. Engl.* 59: 19450–19459.

153 Wu, B., Shi, Y.C., Zhang, J.M. et al. (2024). Stabilizing ultra-small bimetallic PtSn clusters within S-1 crystals for effective propane dehydrogenation with low Pt loading. *Chem. Eng. J.* 498: 155205.

154 Song, S.J., Li, J., Wu, Z.J. et al. (2021). In situ encapsulated subnanometric CoO clusters within silicalite-1 zeolite for efficient propane dehydrogenation. *AIChE J.* 68: e17451.

155 Chen, C., Sun, M.L., Hu, Z.P. et al. (2020). Nature of active phase of VO catalysts supported on SiBeta for direct dehydrogenation of propane to propylene. *Chin. J. Catal.* 41: 276–285.

156 Xu, Z.K., Yue, Y.Y., Bao, X.J. et al. (2019). Propane dehydrogenation over Pt clusters localized at the Sn single-site in zeolite framework. *ACS Catal.* 10: 818–828.

157 Zhu, X.X., Wang, X.P., and Su, Y.P. (2021). Propane dehydrogenation over PtZn localized at Ti sites on TS-1 zeolite. *Catal. Sci. Technol.* 11: 4482–4490.

158 Xie, L.J., Chai, Y.C., Sun, L.L. et al. (2021). Optimizing zeolite stabilized PtZn catalysts for propane dehydrogenation. *J. Energy Chem.* 57: 92–98.

159 Zhang, B.F., Li, G.Z., Zhai, Z.W. et al. (2021). PtZn intermetallic nanoalloy encapsulated in silicalite-1 for propane dehydrogenation. *AIChE J.* 67: e17295.

160 Zhu, X.X., Chen, B.B., and Wang, X.P. (2024). High stable PtZnLa/Beta catalyst with low Pt loading for propane dehydrogenation. *Chem. Eng. J.* 498: 155623.

161 Qi, L., Babucci, M., Zhang, Y. et al. (2021). Propane dehydrogenation catalyzed by isolated Pt atoms in identical with ≡SiOZn—OH nests in dealuminated zeolite beta. *J. Am. Chem. Soc.* 143: 21364–21378.

162 Liu, H., Zhou, J., Chen, T.X. et al. (2023). Isolated Pt species anchored by hierarchical-like heteroatomic Fe-silicalite-1 catalyze propane dehydrogenation near the thermodynamic limit. *ACS Catal.* 13: 2928–2936.

163 Ryoo, R., Kim, J., Jo, C. et al. (2020). Rare-earth-platinum alloy nanoparticles in mesoporous zeolite for catalysis. *Nature* 585: 221–224.

164 Park, H., Park, H., Kim, J.C. et al. (2021). Sodium-free synthesis of mesoporous zeolite to support Pt-Y alloy nanoparticles exhibiting high catalytic performance in propane dehydrogenation. *J. Catal.* 404: 760–770.

165 Xu, G., Zhang, X., Dong, Z. et al. (2023). Ferric single-site catalyst confined in a zeolite framework for propane dehydrogenation. *Angew. Chem. Int. Ed. Engl.* 62: e202305915.

166 Long, J.P., Tian, S.Y., Wei, S. et al. (2023). Direct dehydrogenation of propane over Co@silicalite-1 zeolite: steaming-induced restructuring of Co^{2+} active sites. *Appl. Surf. Sci.* 614: 156238.

167 Wei, S., Dai, H., Long, J.P. et al. (2023). Nonoxidative propane dehydrogenation by isolated Co^{2+} in BEA zeolite: dealumination-determined key steps of propane C–H activation and propylene desorption. *Chem. Eng. J.* 455: 140726.

168 Yue, Y.Y., Fu, J., Wang, C.M. et al. (2021). Propane dehydrogenation catalyzed by single Lewis acid site in Sn-Beta zeolite. *J. Catal.* 395: 155–167.

169 Yang, Z., Li, H., Zhou, H. et al. (2020). Coking-resistant iron catalyst in ethane dehydrogenation achieved through siliceous zeolite modulation. *J. Am. Chem. Soc.* 142: 16429–16436.

170 Wu, L.Z., Ren, Z.Z., He, Y.S. et al. (2021). Atomically dispersed Co^{2+} sites incorporated into a silicalite-1 zeolite framework as a high-performance and coking-resistant catalyst for propane nonoxidative dehydrogenation to propylene. *ACS Appl. Mater. Interfaces* 13: 48934–48948.

171 Sattler, A., Paccagnini, M., Liu, L. et al. (2021). Assessment of metal-metal interactions and catalytic behavior in platinum-tin bimetallic subnanometric clusters by using reactive characterizations. *J. Catal.* 404: 393–399.

172 Yan, M.X., Wu, B.S., Yang, Y. et al. (2023). Highly efficient Sn-modified Pt/KY catalyst for *n*-octane reforming: the synergistic effect of Pt in different electronic states. *Catal. Sci. Technol.* 13: 1677–1685.

173 Min, H.K., Oh, S., Kim, Y.W. et al. (2023). Progressive formation of active and stable PtZn bimetallic nanoclusters by exsolution during propane dehydrogenation. *Chem. Eng. J.* 466: 143241.

174 Qi, L., Zhang, Y.F., Babucci, M. et al. (2022). Dehydrogenation of propane and *n*-butane catalyzed by isolated $PtZn_4$ sites supported on self-pillared zeolite pentasil nanosheets. *ACS Catal.* 12: 11177–11189.

175 Li, J.C., Li, J.M., Zhao, Z. et al. (2017). Size effect of TS-1 supports on the catalytic performance of PtSn/TS-1 catalysts for propane dehydrogenation. *J. Catal.* 352: 361–370.

176 Wannapakdee, W., Yutthalekha, T., Dugkhuntod, P. et al. (2019). Dehydrogenation of propane to propylene using promoter-free hierarchical Pt/silicalite-1 nanosheets. *Catalysts* 9: 174.
177 Fan, X.Q., Yang, Y., Song, J.X. et al. (2024). Stable Pt catalysts anchored by Mn-SBA-15 support for propane dehydrogenation. *Appl. Catal. A: Gen.* 670: 119559.
178 Gao, J.X., Zong, X.P., Yue, X. et al. (2024). Exploring the properties of silica nanomaterials on supporting bimetallic PtZn sites for direct propane dehydrogenation. *ACS Appl. Nano Mater.* 7: 7018–7027.
179 Yang, F.W., Zhang, J., Chen, J.W. et al. (2024). Promoting propane dehydrogenation over Zn/hollow porous silicalite-1 catalysts via modulating the electronic structures of Pt. *Fuel* 364: 131163.
180 Deng, L.D., Zhou, Z.J., and Shishido, T. (2020). Behavior of active species on Pt-Sn/SiO_2 catalyst during the dehydrogenation of propane and regeneration. *Appl. Catal. A: Gen.* 606: 117826.
181 Li, S.Y., Li, Q., Li, B.C. et al. (2024). Efficient dehydrogenation of propane to propene over PtIn nanoclusters encapsulated in hollow-structured silicalite-1. *ACS Catal.* 14 (23): 17825–17836.
182 Lu, J.Y., Xu, Z.K., Yue, Y.Y. et al. (2024). Pt-Zn clusters encapsulated in hierarchical MFI zeolite for efficient propane dehydrogenation. *Chem. Eng. J.* 484: 149369.
183 Peng, M., Jia, Z.M., Gao, Z.R. et al. (2022). Antisintering Pd_1 catalyst for propane direct dehydrogenation with in situ active sites regeneration ability. *ACS Catal.* 12: 2244–2252.
184 Alcala, R., Dean, D.P., Chavan, I. et al. (2023). Strategies for regeneration of Pt-alloy catalysts supported on silica for propane dehydrogenation. *Appl. Catal. A: Gen.* 658: 119157.
185 Ren, G.M., Xiong, S.Y., Li, X.Q. et al. (2024). Regeneration of sintered platinum at mild temperature for propane dehydrogenation. *J. Catal.* 429: 115276.
186 Baharudin, L., Rahmat, N., Othman, N.H. et al. (2022). Formation, control, and elimination of carbon on Ni-based catalyst during CO_2 and CH_4 conversion via dry reforming process: a review. *J. CO_2 Util.* 61: 102050.
187 Alhassan, A.M., Hussain, I., Taialla, O.A. et al. (2023). Advances in catalytic dry reforming of methane (DRM): emerging trends, current challenges, and future perspectives. *J. Clean. Prod.* 423: 138638.
188 Abdulrasheed, A., Jalil, A.A., Gambo, Y. et al. (2019). A review on catalyst development for dry reforming of methane to syngas: recent advances. *Renew. Sustain. Energy Rev.* 108: 175–193.
189 Ballesteros-Plata, D., Infantes-Molina, A., Rodríguez-Castellón, E. et al. (2022). Improving noble metal catalytic activity in the dry reforming of methane by adding niobium. *Fuel* 308: 121996.
190 Usman, M., Wan Daud, W.M.A., and Abbas, H.F. (2015). Dry reforming of methane: influence of process parameters—a review. *Renew. Sustain. Energy Rev.* 45: 710–744.

191 Huang, J.J., Yan, Y., Saqline, S. et al. (2020). High performance Ni catalysts prepared by freeze drying for efficient dry reforming of methane. *Appl. Catal. B: Environ.* 275: 119109.

192 Zhang, Y.M., Zu, Y., He, D.D. et al. (2022). The tailored role of "defect" sites on γ-alumina: a key to yield an efficient methane dry reforming catalyst with superior nickel utilization. *Appl. Catal. B: Environ.* 315: 121539.

193 Li, L.Z., Chen, J., Zhang, Q. et al. (2020). Methane dry reforming over activated carbon supported Ni-catalysts prepared by solid phase synthesis. *J. Clean. Prod.* 274: 122256.

194 Muraza, O. and Galadima, A. (2015). A review on coke management during dry reforming of methane. *Int. J. Energy Res.* 39: 1196–1216.

195 Abdel, K.N., Zeaiter, J., Kwapinski, W. et al. (2017). Thermodynamic analysis of methane dry reforming: effect of the catalyst particle size on carbon formation. *Energy Convers. Manage.* 150: 614–622.

196 Frontera, P., Macario, A., Aloise, A. et al. (2013). Effect of support surface on methane dry-reforming catalyst preparation. *Catal. Today* 218–219: 18–29.

197 Frontera, P., Macario, A., Aloise, A. et al. (2012). Catalytic dry-reforming on Ni–zeolite supported catalyst. *Catal. Today* 179: 52–60.

198 Singh, R., Dhir, A., Mohapatra, S.K. et al. (2019). Dry reforming of methane using various catalysts in the process: review. *Biomass Convers. Biorefin.* 10: 567–587.

199 Dai, Y.Q., Lu, P., Cao, Z.M. et al. (2018). The physical chemistry and materials science behind sinter-resistant catalysts. *Chem. Soc. Rev.* 47: 4314–4331.

200 Hambali, H.U., Jalil, A.A., Abdulrasheed, A.A. et al. (2020). Enhanced dry reforming of methane over mesostructured fibrous Ni/MFI zeolite: influence of preparation methods. *J. Energy Inst.* 93: 1535–1543.

201 Kweon, S., An, H., Shin, C.H. et al. (2021). Nitrided Ni/N-zeolites as efficient catalysts for the dry reforming of methane. *J. CO_2 Util.* 46: 101478.

202 Kweon, S., Bae, J., Cho, Y. et al. (2022). Defect-stabilized nickel on beta zeolite as a promising catalyst for dry reforming of methane. *Catal. Sci. Technol.* 12: 3106–3115.

203 Zhang, J.Y., Li, Y.Y., Song, H.H. et al. (2024). Tuning metal-support interactions in nickel–zeolite catalysts leads to enhanced stability during dry reforming of methane. *Nat. Commun.* 15: 8566.

204 Kobayashi, T., Furuya, T., Fujitsuka, H. et al. (2019). Synthesis of birdcage-type zeolite encapsulating ultrafine Pt nanoparticles and its application in dry reforming of methane. *Chem. Eng. J.* 377: 120203.

205 Kong, W.B., Fu, Y., Shi, L. et al. (2021). Nickel nanoparticles with interfacial confinement mimic noble metal catalyst in methane dry reforming. *Appl. Catal. B: Environ.* 285: 119837.

206 Wang, J.Y., Fu, Y., Kong, W.B. et al. (2021). Design of a carbon-resistant Ni@S-2 reforming catalyst: controllable Ni nanoparticles sandwiched in a peasecod-like structure. *Appl. Catal. B: Environ.* 282: 119546.

207 Das, S., Pérez-Ramírez, J., Gong, J. et al. (2020). Core–shell structured catalysts for thermocatalytic, photocatalytic, and electrocatalytic conversion of CO_2. *Chem. Soc. Rev.* 49: 2937–3004.

208 Han, J.W., Kim, C., Park, J.S. et al. (2014). Highly coke-resistant Ni nanoparticle catalysts with minimal sintering in dry reforming of methane. *ChemSusChem* 7: 451–456.

209 Li, Z.W., Mo, L.Y., Kathiraser, Y. et al. (2014). Yolk–satellite–shell structured Ni–Yolk@Ni@SiO_2 nanocomposite: superb catalyst toward methane CO_2 reforming reaction. *ACS Catal.* 4: 1526–1536.

210 Richard, A.R. and Fan, M. (2017). Low-pressure hydrogenation of CO_2 to CH_3OH using Ni-In-Al/SiO_2 catalyst synthesized via a phyllosilicate precursor. *ACS Catal.* 7: 5679–5692.

211 Han, C.L., Zhu, X.X., Chen, B.B. et al. (2024). A strategy of constructing the Ni@ silicalite-1 catalyst structure with high activity and resistance to sintering and carbon deposition for dry reforming of methane. *Fuel* 355: 129548.

212 Li, M., Liu, W.M., Mao, Y.R. et al. (2024). Design dual confinement Ni@S-1@ SiO_2 catalyst with enhanced carbon resistance for methane dry reforming. *Int. J. Hydrog. Energy* 83: 79–88.

213 Zhu, Q.Y., Liu, Y.F., Qin, X.D. et al. (2023). Zeolite fixed cobalt-nickel nanoparticles for coking and sintering resistance in dry reforming of methane. *Chem. Eng. Sci.* 280: 119030.

214 Abdulkadir, B.A., Miskan, S.N., Abd Wahab, A.T. et al. (2024). Dry reforming of methane over Ni/fibrous zeolite-Y: optimization and regeneration studies. *Chem. Eng. Commun.* 211: 937–951.

215 Chong, C.C., Cheng, Y.W., Setiabudi, H.D. et al. (2020). Dry reforming of methane over Ni/dendritic fibrous SBA-15 (Ni/DFSBA-15): optimization, mechanism, and regeneration studies. *Int. J. Hydrog. Energy* 45: 8507–8525.

216 Chen, L.W., Ng, K.H., and Hsu, L.C. (2024). Self-regenerative Ni/SiO_2 for dry reforming of methane (DRM): 1000 h-longevity assessment and operando insights to coke removal under N_2 atmosphere. *Chem. Eng. J.* 499: 155907.

217 Aricò, F. and Tundo, P. (2010). Dimethyl carbonate as a modern green reagent and solvent. *Russ. Chem. Rev.* 79: 479–489.

218 Schaffner, B., Schaffner, F., Verevkin, S.P. et al. (2010). Organic carbonates as solvents in synthesis and catalysis. *Chem. Rev.* 110: 4554–4581.

219 Tan, H.Z., Wang, Z.Q., Xu, Z.N. et al. (2018). Review on the synthesis of dimethyl carbonate. *Catal. Today* 316: 2–12.

220 Nishihira, K., Tanaka, S., Kawashita, T. et al. (1993). Carbonic diester preparation useful as solvents, etc. by vapor phase catalytic reaction of alkyl nitrite and carbon monoxide in presence of solid catalyst. US Patent Number 0559212.

221 Nishihira, K., Yoshida, S., Tanaka, S. et al. (1993). Continuous production of dimethyl carbonate by reaction of carbon monoxide and methyl nitrite, with recovery of product using dimethyl oxalate. US Patent Number 0655433.

222 Yasushi, Y., Tokuo, M., Shuji, T. et al. (1997). Catalysis and characterization of Pd/NaY for dimethyl carbonate synthesis from methyl nitrite and CO. *J. Chem. Soc., Faraday Trans.* 93: 3721–3727.

223 Bergeret, G., Tri, T.M., and Gallezot, P. (1983). X-ray study of palladium location in Y zeolite during in situ hydrogen reduction, benzene adsorption, and benzene hydrogenation. *J. Phys. Chem.* 87: 1160–1165.

224 Zhang, Z.C., Chent, H., and Sachtler, W.M.H. (1991). Migration and coalescence of Pd carbonyl clusters in zeolite Y. *J. Chem. Soc., Faraday Trans.* 87: 1413–1418.

225 Shin-Ichiro, U., Kikuo, A., and Tokuo, M. (1999). Oxidative reactions by a palladium–alkyl nitrite system. *J. Organomet. Chem.* 576: 279–289.

226 Tokuo, M. and Asumaru, N. (1997). Dimethyl carbonate synthesis and other oxidative reactions using alkyl nitrites. *Catal. Surv. Jpn.* 1: 77–88.

227 Yamamoto, Y. (2010). Vapor phase carbonylation reactions using methyl nitrite over Pd catalysts. *Catal. Surv. Asia* 14: 103–110.

228 Tan, H.Z., Chen, Z.N., Xu, Z.N. et al. (2019). Synthesis of high-performance and high-stability Pd(II)/NaY catalyst for CO direct selective conversion to dimethyl carbonate by rational design. *ACS Catal.* 9: 3595–3603.

229 Guo, R., Qin, Y.Y., Qiao, L.Y. et al. (2017). Enhancement of the catalytic performance in Pd-Cu/NaY catalyst for carbonylation of methyl nitrite to dimethyl carbonate: effects of copper doping. *Catal. Commun.* 88: 94–98.

230 Huang, K., Yuan, S.C., Mei, R.Y. et al. (2024). Mo-promoted Pd/NaY catalyst for indirect oxidative carbonylation of methanol to dimethyl carbonate. *Chin. J. Catal.* 60: 327–336.

231 Wang, C.Z., Xu, W.S., Qin, Z.X. et al. (2021). Highly active Pd containing EMT zeolite catalyst for indirect oxidative carbonylation of methanol to dimethyl carbonate. *J. Energy Chem.* 52: 191–201.

5

Conversion of Low-carbon Molecules over Zeolites

Jiale Xiao, Wei Fang, Chengtao Wang, Liang Wang, Fengshou Xiao*

College of Chemical and Biological Engineering, Zhejiang University, Hangzhou, Zhejiang, China

*Corresponding author: ctwang@zju.edu.cn

5.1 Introduction

The dual imperatives of dwindling petroleum reserves and escalating environmental impacts from fossil resource utilization have intensified research into sustainable strategies of carbon utilization. C_1 chemistry—centered on the catalytic transformation of abundant single-carbon molecules (CO, CO_2, CH_4, and CH_3OH) derived from natural resources or industrial by-products—offers a promising pathway to value-added chemicals and fuels [1–15]. Particularly, the direct conversion of these feedstocks into high-demand products (olefins, oxygenates, and liquid hydrocarbons) through cascade catalysis has emerged as a frontier in heterogeneous catalysis.

Zeolites, with their molecular sieving capabilities and acid–base tunability, serve as ideal matrices for constructing multifunctional catalysts [16–26]. The integration of active metal species (e.g., Fe, Co, Ru, and Rh [rhodium]) within zeolitic architectures generates confined nanoreactors that simultaneously regulate reactant diffusion, intermediate stabilization, and product selectivity [27–30]. This spatial confinement strategy has proven critical in overcoming thermodynamic limitations and kinetic barriers inherent to C_1 activation processes. At the same time, the advantageous combination of both zeolite and metal nanoparticles significantly enhances performances of the catalysts.

5.2 Low Temperature Oxidation of Methane to Methanol

Amidst volatile crude oil prices and dwindling reserves, methane—an abundant carbon resource found in natural gas, shale gas, and biogas—has garnered growing interest. Unlike direct combustion for heat generation, the selective conversion of methane into high-value chemicals has a more sustainable and economically viable pathway [31]. The conversion of methane to methanol presents a promising pathway for methane utilization, as methanol serves as a versatile platform chemical that can be further processed into olefins, fuels, and aromatics via well-established methods [32]. This process is often regarded as a "holy grail" reaction that scientists have been pursuing for decades, due to the easy transport and usage [33–36]. Yet, there are two unresolved problems constituting challenges for modern chemistry: the carbon–hydrogen bond exhibits a high bond dissociation energy, reaching up to 105 kcal · mol^{-1}. Additionally, the bond displays significantly low polarity, owing to the comparable electronegativity values of carbon ($\chi_C = 2.55$) and hydrogen ($\chi_H = 2.20$) [37–39]. When methane undergoes a neat functionalization reaction as a reactant, it is likely to form products that are more active than the initial methane, which will greatly affect the conversion rate and selectivity of the products [40]. Therefore, the chemical inertness of methane has made its efficient conversion a worldwide challenge, leading to its low yield (selectivity) and stability issues in the industrial application of methane conversion.

The primary obstacles in this process involve activating the inert C—H bonds of methane and preventing the overoxidation of the methanol product to CO_2. Selecting an appropriate oxidant has a significant challenge: while low-cost O_2 is desirable, its use often leads to unsatisfactory selectivity due to the thermodynamic propensity for CO_2 formation. Conversely, alternative oxidants that are expensive or toxic hinder commercial viability. The utilization of metal nanoparticles confined within zeolite frameworks (metal@zeolite), operating at low temperatures with oxidants such as O_2, H_2O, or hydrogen peroxide (H_2O_2), emerges as a highly promising strategy for advancing the methanol economy [41]. Many studies have reported that methane can be directly oxidized into methanol in the presence of metal-based catalysts and zeolite-based catalysts [32, 42–55].

Copper-exchanged ZSM-5 and mordenite (MOR) zeolites have attracted considerable research interest owing to their capacity to catalyze methane oxidation to methanol using molecular oxygen. In 2005, Schoonheydt and co-workers demonstrated that these Cu–zeolite systems could selectively convert methane to methanol at 125 °C. This capability strongly suggested the presence of highly activated oxygen species within the materials. Furthermore, they hypothesized that the distinct microporous architecture and specific Si/Al ratios characteristic of pentasil zeolites like ZSM-5 and MOR might be particularly effective in stabilizing bis(μ-oxo)dicopper cores [56]. Román-Leshkov and co-workers proposed a tandem reaction sequence which uses Cu–MOR to achieve the conversion of methane into acetic acid by coupling oxidation with carbonylation reactions (see Figure 5.1a, b). Integrated spectroscopic and reactivity analyses revealed that while the Cu^{II}—O—Cu^{II} sites, known for methane oxidation activity in Cu–MOR and Cu–ZSM-5, exhibit oxidative capability, they lack carbonylation functionality. Concurrently, a distinct Cu^{2+} active site emerges in MOR frameworks at the Cu/Al ratios below 0.2. These newly identified sites not only activate methane but also mediate product migration to Brønsted acid sites within the eight-membered ring (MR) channels, facilitating subsequent carbonylation (see Figure 5.1c–e) [57]. Separately, Flytzani-Stephanopoulos and co-workers demonstrated that atomically dispersed Rh species anchored on

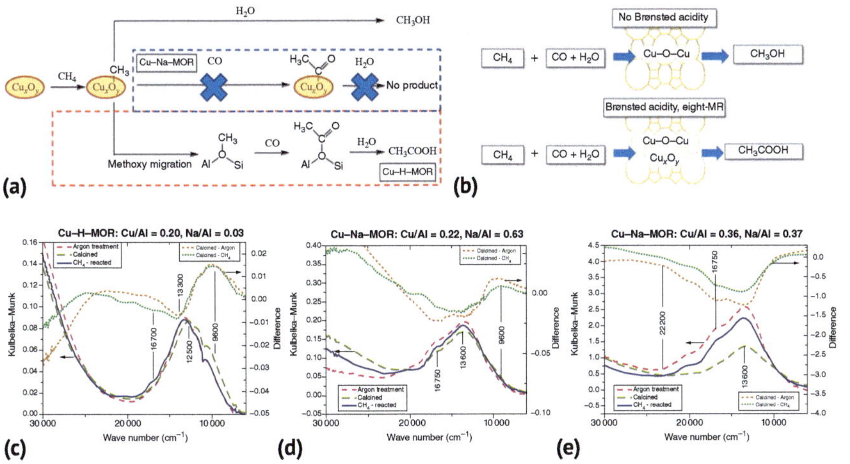

Figure 5.1 (a, b) Reaction pathways for methane oxidation and carbonylation over Cu–MOR catalysts prepared from acid or sodium precursors. Diffuse reflectance UV-visible spectra of (c) Cu–H–MOR (Cu/Al = 0.20, Na/Al = 0.03), (d) Cu–Na–MOR (Cu/Al = 0.22, Na/Al = 0.63), and (e) Cu–Na–MOR (Cu/Al = 0.36, Na/Al = 0.37). Samples were pretreated under argon (823 K, 3 h; pink), calcined in O_2 (823 K, 5 h; dashed gold), and reacted with CH_4 (473 K, 2 h; navy). Difference spectra: calcined minus argon (dash-dot orange), calcined minus CH_4-treated (dotted green). *Source*: Reproduced from [57] with permission of American Chemical Society.

H-ZSM-5 or TiO_2 supports in aqueous phase catalyze the direct conversion of methane to methanol and acetic acid [50]. Using O_2 and CO under mild conditions (150 °C), three-hour batch reactor tests with these catalysts achieved productivities of approximately 22 000 μmol · g_{cat}^{-1} for acetic acid and 230 μmol · g_{cat}^{-1} for methanol, with selectivities ranging from 60% to 100%.

van Bokhoven and co-workers developed a stepwise protocol for the selective partial oxidation of methane to methanol (~97% selectivity) over copper-exchanged zeolites, utilizing water as the oxidant [49]. This method involves catalyst activation in helium at 400 °C, followed by sequential exposure to 7 bar methane and water at 200 °C, achieving a consistent methanol yield of 0.204 mol per mole of copper. In situ X-ray absorption spectroscopy (XAS) and Fourier transform infrared spectroscopy (FTIR) of adsorbed probe molecules revealed a reversible copper redox cycle. Methane exposure induced a characteristic Cu^{I} pre-edge feature at 8983.6 eV, confirming the reduction of Cu^{II} sites with concomitant methanol formation. Subsequent water introduction diminished this Cu^{I} signal while generating an 8997 eV peak attributed to hydrated Cu^{II} species, demonstrating re-oxidation. Furthermore, such a change in the copper oxidation state was also detected in the FTIR spectra of adsorbed carbon monoxide and nitric oxide. In the FTIR spectra of adsorbed CO, the peak intensities at 2179 and 2151 cm^{-1} were both increased or decreased during the reaction with CH_4 and oxidation with H_2O, respectively. Complementary FTIR studies of adsorbed NO showed bands at 1804, 1731, and 1827 cm^{-1} (assigned to Cu^{I}–NO adducts) and at 1909, 1948, and 1995 cm^{-1} (assigned to Cu^{II}–NO species). After methane interaction with activated Cu-MOR at 200 °C, the Cu^{I}–NO band intensity increased while the Cu^{II}–NO signals decreased. This redistribution of copper oxidation states, corroborated by CO adsorption FTIR and XAS data, confirmed the catalytic redox mechanism.

Departing from conventional O_2 re-oxidation, the authors propose that water mediates a two-electron transfer process, regenerating reactive Cu^{II} sites while co-producing H_2. Density functional theory (DFT) calculations elucidate water's oxidative role, revealing that H_2O addition proceeds through a low-energy transition state ($\Delta E = 85\ \text{kJ} \cdot \text{mol}^{-1}$). This facilitates Brønsted acid site-assisted methanol formation from methoxy species, yielding Cu^I–OH_2–Cu^I complexes stabilized by three hydrogen bonds. Water exhibits dual functionality: (1) catalyst regeneration via oxygen atom transfer for $Cu^I \rightarrow Cu^{II}$ re-oxidation; and (2) thermodynamically favorable methanol desorption. Consequently, a catalytic cycle is established: initial methane oxidation at Cu^{II}-oxide centers, followed by water-driven re-oxidation of Cu^I with concomitant H_2 evolution.

Cyclic gas-phase methane oxidation over Cu/Fe-containing zeolite catalysts using O_2 or H_2O may suppress deep oxidation [49, 56, 58–61] but necessitates batch-type operations for oxidant activation and methanol desorption at elevated temperatures (200–500 °C). H_2O_2, a more environmentally friendly oxidant, can be employed for the selective oxidation of methane, though it is more expensive than oxygen. It has also been reported that H_2O_2 can be synthesized in situ from hydrogen and oxygen using metal nanoparticles, but its efficiency is lower than the direct usage of H_2O_2 [33, 43, 49, 62, 63]. This is due to that the H_2O_2 (or other peroxide intermediates) produced in situ tends to dissolve in the reaction solution, forming a less reactive dilute solution [64]. Compared with the conventional supported metal nanoparticle catalysts, the fixed structural metal@zeolite catalysts offered an opportunity to concentrate crucial intermediates within the zeolite crystals and optimize the catalytic process.

In a previous work, AuPd alloy nanoparticles were fixed within ZSM-5 zeolites to form a core–shell structure [53]. A hydrophobic molecular fence was engineered on zeolite surfaces via organosilane functionalization to regulate H_2O_2 diffusion kinetics during methane partial oxidation (see Figure 5.2a). This catalyst system, denoted by AuPd@zeolite–*R* (where "@" signifies nanoparticle encapsulation within zeolite micropores, and *R* represents the silane's organic moiety), leverages hydrophobicity gradients to manipulate reactant distribution. During catalysis, in situ-generated hydrophilic H_2O_2 from H_2/O_2 is retained within zeolite crystals due to diffusion barriers imposed by the hydrophobic sheath, thereby elevating intraparticle oxidant concentration. At the same time, hydrophobic methane molecules selectively permeate the modified surface, accessing encapsulated AuPd active sites (see Figure 5.2b). The results showed that the H_2O_2 predominantly accumulated in the reaction solution over the supported AuPd catalyst or AuPd@ZSM-5 without a hydrophobic sheath. In contrast, more than 78% of H_2O_2 was retained within the AuPd@ZSM-5–*R* zeolite crystals, with less than 22% remaining in the aqueous solution (see Figure 5.2c).

The selective H_2O_2 confinement arises from the hydrophobic molecular fence grafted on zeolite surfaces, which restricts oxidant diffusion through zeolitic micropores, thereby enriching intraparticle H_2O_2 concentrations. Crucially, gaseous reactants (CH_4, H_2, and O_2) readily permeate the modified interface, enabling efficient methane-to-methanol conversion. This engineered catalyst significantly

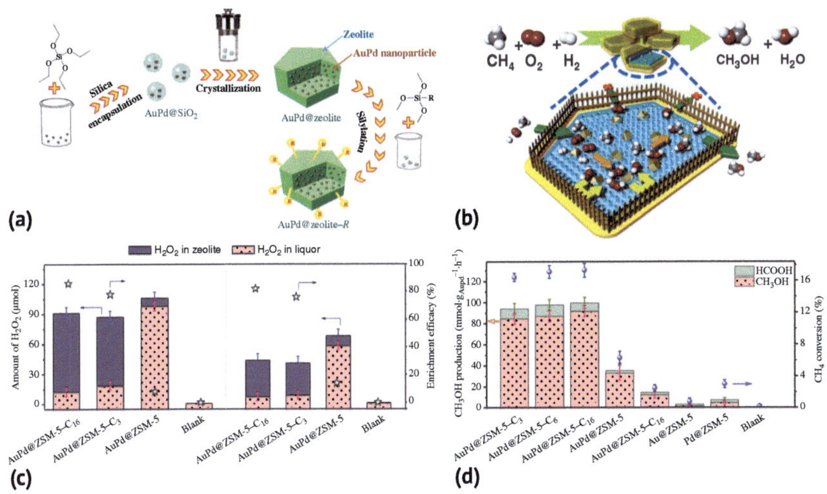

Figure 5.2 (a) AuPd@zeolite-*R* synthesis protocol. (b) Hydrophobic molecular fence mechanism depicting reactant diffusion pathways. (c) H_2O_2 distribution during: H_2O_2 synthesis (left) and CH_4 oxidation (right). Enrichment efficiency = (intracrystalline H_2O_2/total H_2O_2) × 100%. (d) Catalytic performance comparison in CH_4 oxidation with H_2/O_2. *Source*: Reproduced from [53] with permission of American Association for the Advancement of Science-AAAS.

outperforms conventional supported metal nanoparticles and unmodified AuPd@ zeolite counterparts. The competitive diffusion dynamics further promote rapid methanol egress prior to overoxidation, effectively suppressing undesired CO_x formation. In this case, the methane conversion rate of the AuPd@ZSM-5 catalyst was 6.3%. After hydrophobic treatment, the conversion rate increased by at least 2.5-fold, reaching 17.3% with the AuPd@ZSM-5-*R* catalyst. In addition, the methanol selectivity exceeded 90% with a yield as high as 91.6 mmol $\cdot g_{AuPd}^{-1} \cdot h^{-1}$ (see Figure 5.2d).

5.3 Syngas Conversion

The fuels and basic chemicals needed for modern life are increasingly in demand, and their current production is dependent on petroleum. A promising nonpetroleum route is to produce them from syngas, a mixture of CO and H_2, that can be easily produced from coal, natural gas, and biomass [65–67]. As a bridge, highly valuable chemicals, such as olefins, aromatics, methanol, ethanol, and other oxygenates, could be produced through syngas conversion processes, each of which depends on different types of catalysts.

5.3.1 Syngas to Ethanol

As a versatile chemical feedstock and renewable carbon resource, ethanol has garnered considerable research interest [68–71]. Its potential to mitigate energy crises by substituting fossil fuels, while concurrently reducing greenhouse gas and pollutant emissions, positions it as a sustainable alternative. Production routes include sugar fermentation, petroleum refining, and syngas (CO/H_2) conversion, with the latter representing the most environmentally benign and cost-effective pathway. The fundamental reaction for alcohol synthesis from syngas is given by Eq. (5.1) [72]. This exothermic, contraction-phase equilibrium reaction favors low-carbon alcohol formation under elevated pressures or reduced temperatures according to Le Chatelier's principle. However, the synthesis of alcohol compounds from syngas is accompanied by numerous competing side reactions, including: (1) formation of alkanes and alkenes of varying chain lengths; (2) water–gas shift reactions; and (3) subsequent isomerization of alcohol products to form aldehydes and esters. This complex network of parallel and sequential reactions underscores the remarkable intricacy of syngas-to-alcohols conversion processes,

$$nCO + 2nH_2 \rightarrow C_nH_{2n}OH + (n-1)\ H_2O + H(n > 1) \tag{5.1}$$

$$\Delta G = -38.386n + 11.098 + (5.982n - 0.144) \times 10^{-2}$$

The most effective way to avoid side reactions is to choose suitable catalysts to improve selectivity. Therefore, the development of catalysts has become the key to producing alcohols from synthetic gas. The appropriate catalyst is closely related to the adsorption and activation characteristics of CO and H_2 on the catalyst, as well as the selectivity of the products. From the perspective of reactants, the effective synthesis of alcohols requires sufficient activation of CO and H_2. From a product perspective, low-carbon alcohols contain not only alkyl groups but also hydroxyl groups. The adsorption activation process of the former requires the breaking of C—O bonds in CO molecules and hydrogenation to form CH_x intermediates, while the latter requires maintaining CO bonds and inserting CO into the already formed CH_x intermediates. Therefore, the type and quantity of adsorption states of CO and H_2 on the catalyst surface will directly affect the distribution of reaction products. This process cannot be effectively accomplished by a single active site alone but rather requires synergistic cooperation between multiple active centers with distinct functionalities. The complexity of the reaction mechanism necessitates a composite catalytic system where different active sites work in concert to facilitate the alcohol formation.

Since the early twentieth century, extensive research has been conducted on the synthesis of alcohols from synthetic gas, resulting in the development of four typical catalytic systems: modified methanol synthesis catalysts [73], modified Fischer–Tropsch catalysts [74, 75], Mo-based catalysts [76, 77], and Rh-based catalysts [65, 71]. Among them, modified methanol catalysts mainly composed of low-temperature and low-pressure Cu-based catalysts and high-temperature and high-pressure Cr-based catalysts, were first discovered due to the introduction of alkali metal impurities, which led to the formation of mixed alcohols in the methanol synthesis process. Subsequent studies have extensively explored modified methanol synthesis catalysts; however, increasing the C_{2+} alcohol selectivity in syngas conversion remains a persistent challenge for these catalytic systems. Despite considerable research efforts, achieving preferential formation of C_{2+} alcohols over methanol continues to pose significant difficulties in catalyst design. Mo-based catalysts have strong anti-poisoning effects on S, and the catalyst is not prone to carbon deposition. However, this catalyst requires high reaction pressure, and the products contain a large amount of hydrocarbons and carbon dioxide, resulting in low selectivity for total alcohols. Modified Fischer–Tropsch synthesis (FTS) catalysts, primarily comprising Co-based and Fe-based catalysts, offer several advantages including cost-effectiveness, moderate operating conditions (220–260 °C, 3–6 MPa), and relatively high selectivity for C_{2+} oxygenates. However, the modified FTS catalysts still suffer from low alcohol selectivity with predominant hydrocarbon formation. Rh-based catalysts have high activity and selectivity for alcohol synthesis, with ethanol as the main reaction product. Since Bhasin et al. [78] reported Rh/SiO_2 as a catalyst for CO hydrogenation to produce ethanol in the 1980s, Rh-based catalysts have been widely studied in direct alcohol production from syngas and are considered one of the most potential catalysts for synthesizing C_{2+} oxygenates (such as ethanol and acetic acid) from syngas. However, the high cost of Rh compounds and the insufficient catalytic performance (both in activity and ethanol selectivity) of conventional Rh-based catalysts have hindered their industrial application. Consequently, rational design of catalyst architectures—including precise control of nanoparticle size, strategic incorporation of metal promoters, and optimization of support materials—becomes essential for enhancing alcohol production efficiency and metal utilization in syngas-to-ethanol conversion systems.

In the preparation process of Rh-based catalysts, the Rh content, supports, promoter, and preparation method all have significant impacts on their structure. Therefore, to synthesize Rh-based catalysts with excellent performance, it is often necessary to add alkali metal (e.g., Li) or transition metal (e.g., Mn, Fe, La, and Ce) additives to promote the formation of oxygen-containing compounds and the generation of hydrocarbon products (e.g., methane) during the reaction process [69, 79, 80]. For example, Mn promoter is one of the most typical transition metals for modifying Rh-based catalysts, which has received widespread attention from researchers for a long time. The roles of Mn promoter are as follows: (1) the interaction between Rh and Mn can enhance the adsorption of CO, greatly weaken the C—O bond, promote the dissociation of Rh from CO, improve catalytic activity, and also stabilize the C_2 oxygen-containing intermediate on the Rh surface; (2) Mn plays a role in electron pulling, and partially oxidizes the Rh atom at the Rh–MnO interface to Rh^+, to form new sites for CO insertion; and (3) Rh and Mn can form alloys, thereby improving the selectivity for C_{2+} oxygenates such as ethanol.

Rh-based catalysts with well-designed structures (such as the size of Rh nanoparticles; and supports) are beneficial for the formation of oxygenates. To minimize the content of Rh in the catalyst, different types of supports are often used in Rh-based catalysts, and research has found that the properties of the supports are crucial for increasing the yield of C_{2+} oxygenate. Generally, in the selection of Rh-based catalyst supports, when strong alkaline oxides (such as ZnO and MgO) are used as supports, methanol becomes the main oxygenate for CO hydrogenation. When acidic oxides (e.g., SiO_2, γ-Al_2O_3, and V_2O_5) are used as supports, large amounts of methane and long-chain alkanes are produced [81, 82]. Unlike these conventional supports, Bao and co-workers reported that Rh particles confined within carbon nanotubes (CNTs) significantly enhance the conversion of syngas to ethanol [65]. CNTs have entered the field of view as a nanoscale metal catalyst carrier. CNTs have a clear hollow internal structure, which provides the possibility for other substances to enter their internal cavity, and they exhibit excellent mechanical properties, thermal stability, and electronic conductivity. This demonstrates an order-of-magnitude enhancement in ethanol production within the CNTs ($30\ mol \cdot mol_{Rh}^{-1} \cdot h^{-1}$) versus external sites. The confinement effect induces dual catalytic advantages: (1) strengthened Rh nanoparticle–nanotube interactions promote CO dissociation kinetics; and (2) elevated intratubular hydrogen concentration accelerates hydrogenation rates. Consequently, synergistic enhancement in CO activation and hydrogenation drives significant C_2 oxygenate yield improvement. Recently, Lu and co-workers reported a dual interface synergistic catalyst in which FeO_x and Rh were loaded on the ZrO_2 support [71]. The ZrO_2 surface was pretreated with ethylene glycol (EG) to precisely control the selective deposition of FeO_x on the Rh nanoparticles. Through STEM-EDS and HADDF-STEM observations, Rh particles are atomically dispersed and form a highly dispersed dual interface structure. This atomically precise assembly structure promotes the synergistic effect of Rh–ZrO_2 and Rh–FeO_x interfaces at the molecular level. Operando spectroscopy and computational modeling provide mechanistic insights: the Rh–ZrO_2 interface facilitates dissociative CO activation to CH_x species via formate intermediates, while the adjacent Rh–FeO_x interface mediates C—C coupling through non-dissociative CO insertion. In the end, the catalyst achieved 90% ethanol selectivity in oxygenates and 668.2 $mg \cdot g^{-1} \cdot h^{-1}$ ethanol space–time yield with a CO conversion rate of 51%, significantly better than the existing reports.

Zeolite materials are commonly used but highly effective catalytic materials in hydrogenation catalysis processes in the chemical industry. Growing evidence indicates that multifunctional catalysts integrating zeolitic architectures with active metal components offer a promising strategy to address selectivity challenges in syngas conversion. Wang and co-workers demonstrated a tandem catalytic system achieving 90% ethanol selectivity, comprising three functional units operating synergistically within a single reactor: K^+-modified ZnO–ZrO_2, modified MOR zeolite, and Pt–Sn/SiC (see Figure 5.3a–c) [83]. This system executes consecutive transformations: K^+–ZnO–ZrO_2 catalyzes syngas-to-methanol conversion, MOR with eight-MR channels mediates methanol carbonylation to acetic acid, and Pt–Sn/SiC selectively hydrogenates acetic acid to ethanol.

Furthermore, Tsubaki and co-workers proposed completing the carbonylation reaction on zeolite catalysts as an intermediate reaction to achieve a highly selective synthesis of ethanol from nonpetroleum pathway synthesis gas [84]. Syngas derived from municipal waste, coal, or natural gas undergoes a cascade conversion: (1) hydrogenation/dehydration to dimethyl ether (DME) via conventional methanol synthesis catalysts (e.g., Cu/ZnO–Al_2O_3) coupled with solid acid catalysts; (2) selective carbonylation of DME to methyl acetate (MA) with 100% selectivity over zeolite catalysts; and (3) final hydrodeoxygenation to ethanol on Cu-based

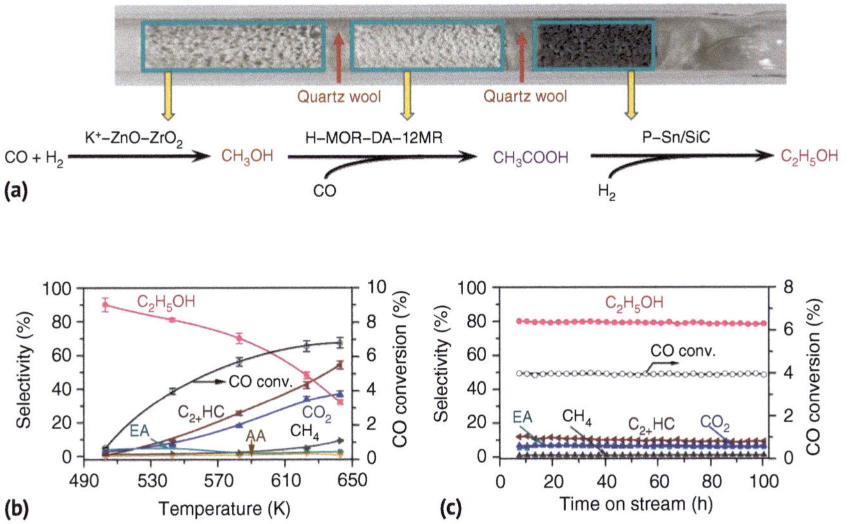

Figure 5.3 (a) Reaction pathways for the direct synthesis of ethanol from syngas. Catalytic performances of the triple tandem system, with the weights of K^+–ZnO–ZrO_2, H–MOR–DA–12MR, and Pt–Sn/SiC catalysts being 0.66, 0.66, and 0.66 g, respectively. (b) Effect of reaction temperature. Reaction conditions: H_2/CO = 1:1; P = 5 MPa; F = 25 mL · min^{-1}; and time on stream = 20h. (c) Catalyst stability. Reaction conditions: H_2/CO = 1:1; P = 5 MPa; F = 25 mL · min^{-1}; and T = 543 K. C_{2+} HC, C_{2+} hydrocarbons; EA, ethyl acetate; AA, acetic acid; and r(CO), the rate of CO conversion. *Source*: [83] / with permission of Springer Nature / CC BY 4.0.

catalysts [85, 86]. The efficiency and stability of zeolite catalysts during DME carbonylation critically determine overall process viability. To enhance these properties, Tsubaki and co-workers synthesized nano-filamentous mordenite (NF-MOR; 70nm crystal diameter) by crystal-orientation-controlled growth using selective inhibitors [87]. Although self-assembled nano-MOR zeolites were developed by a facile inhibitor-assisted strategy, the catalytic performance of self-assembled nano-MOR catalysts was remarkably outstanding in DME carbonylation. The NF-MOR exhibited significantly enhanced activity (3780 versus 1368mmol · $[kg \cdot h]^{-1}$) compared to conventional ellipsoidal MOR (ES-MOR). In a dual-bed configuration with CZ/SiO_2, the NF-MOR–CZ/SiO_2 system achieved efficient one-step ethanol synthesis, demonstrating 85% DME conversion and ~50% ethanol selectivity while maintaining exceptional catalytic stability over 30 hours.

Catalysts for syngas conversion—whether single-atom, cluster, or oxide-based systems (carbides, oxides, and alloys)—invariably suffer from sintering and agglomeration during prolonged operation, leading to irreversible deactivation. Zeolites remain pivotal in catalysis due to their shape-selective confinement, uniform porosity, tunable acidity, and exceptional thermal/hydrothermal stability. Crucially, they anchor metallic species, simultaneously enhancing product selectivity and imparting anti-sintering capabilities.

To address suboptimal oxygenate selectivity and rapid deactivation in syngas conversion, Xiao and co-workers engineered a zeolite-encapsulated catalyst via solvent-free crystallization [7]. RhMn nanoparticles were confined within siliceous

MFI crystals (RhMn@S-1), forming a core–shell architecture. STEM/TEM (transmission electron microscopy) tomography (Figure 5.4a–c) confirmed intracrystalline nanoparticle localization with narrow size distribution (1.2–5.9nm; mean: 2.7nm). In contrast, supported RhMn/S-1 exhibited similar metal dispersion but surface-bound positioning (see Figure 5.4a–c). The supported RhMn catalyst (RhMn/S-1) gave a similar metal size distribution as RhMn@S-1, except for an external surface metal location (see Figure 5.4d–f).

Experimental studies showed that RhMn@S-1 has abundant and stable Rh—O—Mn interface sites, providing moderate CO adsorption (see Figure 5.4g, h). This is due to that the Rh—O—Mn interface sites showed a lower CO adsorption free energy (−1.65eV), compared to −2.19eV on bare Rh atoms and −1.93eV on conventionally supported RhMn species without zeolite encapsulation (see Figure 5.4i). Furthermore, DFT calculations demonstrated that the rigid framework of zeolite crystals efficiently hindered the migration of metal nanoparticles and the ripening of metal species. Such a feature enhanced the catalytic activity and oxygenates selectivity in the

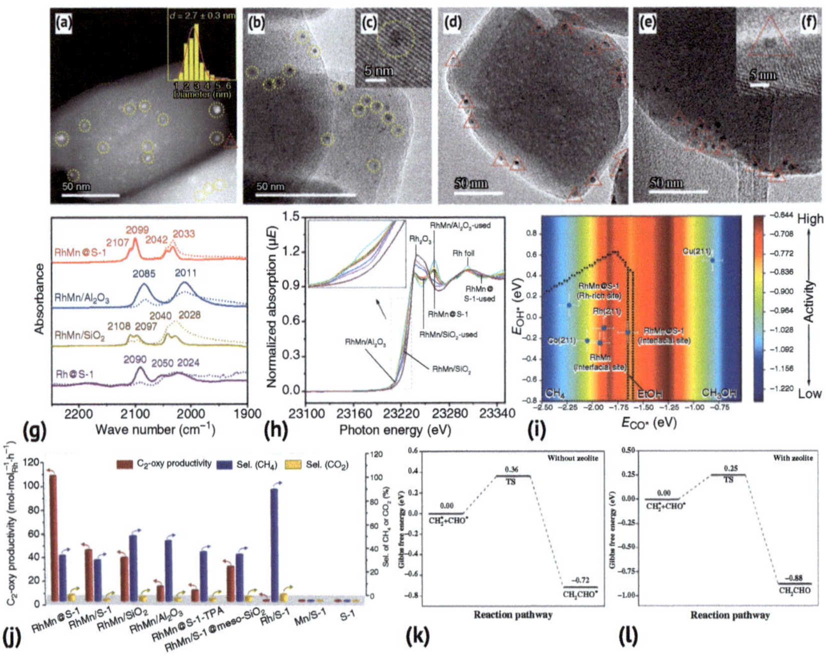

Figure 5.4 (a–c) Tomographic STEM/TEM images of RhMn@S-1. Inset in (b): magnified intracrystalline nanoparticle. (d–f) Tomographic STEM/TEM images of RhMn/S-1. Inset in (e): surface nanoparticle detail. Yellow circles: intracrystalline NPs; red triangles: surface NPs. (g) CO–FTIR spectra (solid: pre-H_2; dashed: post-H_2). (h) Rh K-edge XANES plot of catalysts pre-/post-reaction. (i) Theoretical selectivity mapping (OH/CO adsorption descriptors). Black arrow: high ethanol selectivity zone (DFT uncertainty: ±0.1 eV). (j) Average productivities of C_2 oxygenates and selectivities of methane and CO_2 over various RhMn-based catalysts. (k, l) The reaction energies and energy barriers of the $CH_2^* + CHO^* \rightarrow CH_2CHO^*$ step on RhMn@S-1 and RhMn without zeolite. *Source*: Reproduced from [7] with permission of Elsevier.

CO hydrogenation reaction, achieving an oxygenates selectivity of 40.3% (C_2 oxygenates accounting for 88.3% of the total oxygenates) at a CO conversion of 42.4%, which is higher than the supported RhMn catalyst (see Figure 5.4j).

Furthermore, DFT calculations demonstrated that RhMn nanoparticles without zeolite fixed structure exhibited a C—C coupling energy barrier of 0.36eV, whereas this barrier is reduced to 0.25eV in RhMn@S-1 (see Figure 5.4k, l). This result suggested the important role of zeolite fixed structure in the optimization of C_2 oxygenate yield in syngas conversion. More importantly, in this fixation structure, the metal nanoparticles have a larger size than zeolite micropores, and have to break the zeolite framework structure (crossing several micropores) to immobilize themselves. Therefore, the rigid framework of zeolite crystals efficiently hindered the migration of metal nanoparticles and the ripening of metal species. Such a stabilization also preserved the Rh–MnO_x interfacial structure during the reaction, as evidenced by the IR and X-ray absorption near-edge structure characterization (see Figure 5.4g, h).

5.3.2 Syngas to Olefin

Olefins are considered as critical raw materials in the chemical industry, primarily consisting of light olefins (ethylene, propylene, and butene) and long-chain olefins ($C_{5+}^{=}$). Light olefins serve as the main monomers for polymers, while long-chain terminal olefins (alpha olefins) are essential feedstocks for synthesizing advanced lubricants, high-carbon alcohols, plasticizers, and surfactants [13, 88]. The production of olefins from syngas ($CO + H_2$) can be achieved via two main routes: Fischer–Tropsch to olefins (FTO) route and the OX–ZEO route, in both of which zeolites are typically employed as acid catalysts to facilitate isomerization, aromatization, or cracking reactions, thereby tuning product selectivity [27–29, 89].

The FTO route was pioneered by de Jong's group at Utrecht University, Netherlands, who designed iron nanoparticles modified with trace sodium and sulfur supported on α-alumina (α-Al_2O_3) achieving 60% light olefin selectivity in syngas conversion [88]. Ma and co-workers developed an Na–Zn co-modified Fe-based catalyst via coprecipitation [90], where zinc oxide acted as a structural promoter to enhance iron dispersion, while sodium ions functioned as electronic promoters to optimize olefin desorption kinetics, synergistically improving both CO conversion and olefin selectivity. Sun and co-workers reported an Mn-modified cobalt-based catalyst that delivered 30% CO conversion with 60% light olefin selectivity under mild reaction conditions, accompanied by an exceptional olefin-to-paraffin ratio of 30 [91]. Advanced TEM characterization revealed that prismatic cobalt carbide (Co_2C) nanocrystals serve as a novel active phase for selective olefin formation in this system.

The OX–ZEO strategy was proposed by Bao and co-workers, which has achieved a series of groundbreaking advances in syngas-to-olefins conversion. In 2016, they proposed an innovative OX–ZEO bifunctional catalytic strategy [27], where partially reduced metal oxides efficiently activate CO while zeolite molecular sieves precisely regulate C—C coupling through shape-selective effects, overcoming the product distribution limitations of conventional FTS. Subsequent mechanistic studies revealed a cascade catalytic pathway mediated by ketene intermediates, achieving ethylene selectivity of 73% [92]. Recent investigations on MOR zeolite systems demonstrated that 12-MR channels function as critical reactant transport pathways when optimized to 60nm length—near the mass-transfer limitation threshold—significantly enhancing reactant accessibility to active sites [93]. Through isomorphous substitution of germanium ions, the density and strength of Brønsted acid sites were precisely modulated, simultaneously promoting C—C coupling of ketene intermediates while suppressing over-hydrogenation and polymerization side reactions. This strategy ultimately realized dual breakthroughs with CO conversion and light olefin selectivity both exceeding 80%, successfully resolving the long-standing "trade-off effect" between conversion and selectivity in traditional catalytic systems [94].

Wang and co-workers designed a bifunctional catalytic system by strategically coupling methanol synthesis with C—C coupling (via methanol-to-olefins reactions), engineering an Zr–Zn oxide/SAPO-34 composite that significantly enhanced syngas-to-light-olefins conversion efficiency [29]. Subsequent systematic investigations elucidated the cascade reaction pathway involving CO hydrogenation and C—C coupling through optimizing the synergistic interplay between metal oxides and zeolites [95, 96]. Notably, Xie and co-workers demonstrated that the migration of methanol intermediates from oxide to zeolite active sites is a critical kinetic factor for efficient syngas conversion in tandem processes, as evidenced by their $ZnAlO_x$/AlPO-18 system [97]. Simultaneously, Liu and co-workers deciphered the modulation mechanism of metal oxide crystal phase structures on bifunctional catalyst performance [98].

However, the design of tandem catalysts often results in uncontrolled and spatial distribution of metal species within/on zeolites, thereby limiting their catalytic performance. To address this challenge, Wang and co-workers developed an innovative MO–ZEO (metal oxide–zeolite) architecture [99]. They demonstrated a facile strategy to synthesize double-shelled hollow spheres (DSHs) of Fe_2O_3@H–ZSM-5 through zeolite crystal self-assembly on metal-ion-incorporated carbon spheres. This hierarchical configuration exhibited superior performance in Fischer–Tropsch to gasoline production compared to conventional Fe_2O_3–H–ZSM-5 catalysts with random architectures, achieving enhanced mass transfer and optimized metal–zeolite interfacial interactions.

Recently, Xiao and co-workers developed an FTO catalytic system by physically mixing an NaFe catalyst with MFI zeolite nanosheets and demonstrated that the zeolites played more of a role as diffusion promoters than as acid catalysts [10]. The MFI zeolite nanosheets (*s*-MFI) with a short *b*-axis of ~100nm were successfully synthesized by a urea-assisted method (see Figure 5.5a–c). The ferric carbide catalyst (Na–FeC_x) physically mixed with *s*-ZSM-5 zeolite promoters exhibited a CO conversion of 82.5% with an olefin selectivity of 72% at the lower temperature of 260 °C (see Figure 5.5d, e). Further studies showed that the topological structure, silicon–aluminum ratio, and the microporous environment of the zeolites, as well as the proximity between zeolite and Na–FeC_x component, could significantly affect the FTO performance. Subsequently, the effect of zeolite was studied and

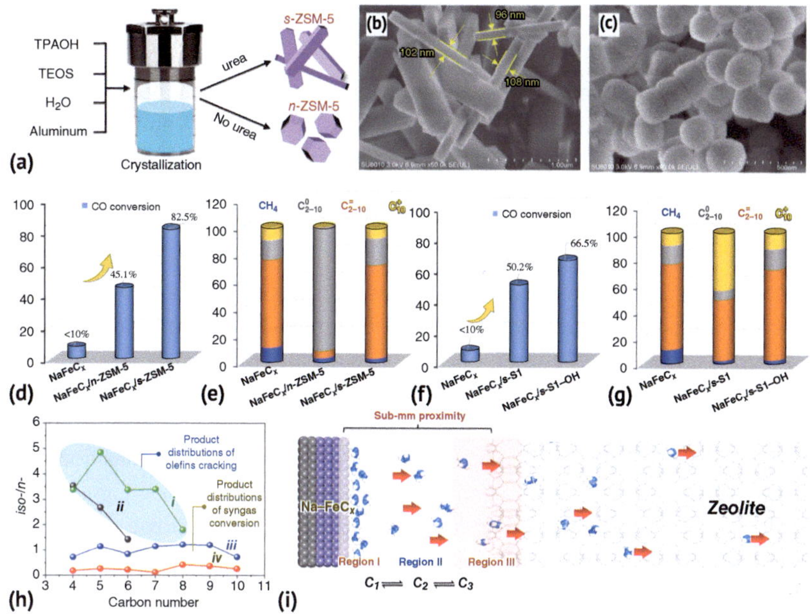

Figure 5.5 (a) The scheme showing the synthesis of *s*(*n*)-ZSM-5. SEM images of (b) *s*-ZSM-5 and (c) *n*-ZSM-5. (d–g) Catalytic data over various catalysts. (h) The ratios of isomer to normal hydrocarbons in the products from different reactions: i: 1-octene isomerization/cracking over *s*-ZSM-5; ii: 1-hexene isomerization/cracking over *s*-ZSM-5; iii: LT–FTO over Na–FeC_x/*s*-ZSM-5; and iv: LT–FTO over Na–FeC_x/*s*-S1–OH. (i) Scheme showing the diffusion of olefins from the Na–FeC_x surface to the zeolite via different regions. *Source*: Reproduced from [10] with permission of Springer Nature.

could be attributed to accelerating the escape of olefin products from the catalyst surface rather than the role of zeolites in previous metal oxide/carbide and zeolite composite catalysts, where those zeolites acted as acid catalysts, such as in the OX–ZEO system.

To validate this concept, the Na–FeC_x/*s*-S1–OH catalyst (siliceous MFI devoid of acid sites) was evaluated, achieving 66.5% CO conversion with 68.7% olefin selectivity (see Figure 5.5f, g). Model cracking experiments of C_4–C_8 α-olefins (1-butene, 1-octene, and 1-hexene) over H–ZSM-5 at 260 °C revealed distinct isomerization product distributions. These results contrast sharply with the high α-olefin selectivity observed in the Na–FeC_x/*s*-ZSM-5-catalyzed FTO process (see Figure 5.5h). Moreover, the used Na–FeC_x was recovered from the reactor and mixed with zeolite nanosheets again for performance evaluation, and catalytic data showed that the CO conversion was still several times higher than the fresh Na–FeC_x. All the results supported the viewpoint that the zeolite should to be a promoter for shipping the olefin molecules from Na–FeC_x (see Figure 5.5i). Further characterization experiments and theoretical calculations also confirmed this conclusion. In the future, more advanced in situ characterizations are needed to further understand the mechanism of this catalytic system.

5.4 CO_2 Conversion

With the development of economy and society, the global energy consumption is gradually increasing, and the demand for primary energies such as coal, oil, and natural gas continue to rise every year, leading to a significant increase in overall carbon emissions [100–102]. For example, in 2018, natural gas demand increased by 4.6% compared to the previous year, while global carbon emissions reached a historic high of 33 billion tons. From the United Nations Framework Convention on Climate Change in 1992 to the Paris Agreement in 2015, requirements have been put forward for carbon emission limits. China's formal pledge on 30 September 2020 commits to carbon dioxide emission peaking by 2030 and carbon neutrality by 2060. Achieving these goals necessitates integrated strategies: (1) enhancing carbon-based energy utilization efficiency; and (2) catalytic conversion of CO_2 into high-value platform molecules. These approaches constitute critical pathways for carbon reduction and sustainable energy transition.

CO_2, transformation into value-added chemicals has garnered significant attention, leveraging its attributes as an abundant, inexpensive, nontoxic, nonflammable, and renewable C_1 building block [103]. Therefore, the energy-efficient introduction of CO_2 in chemical building blocks represents an elegant strategy with a dual benefit: mitigating environmental impact, while introducing oxygen functionalities into base compounds to produce valuable chemicals [104]. However, the inert nature of CO_2, specifically, thermodynamically the most stable form and kinetically inert, requires highly reactive, rich-energy substrates, harder reaction conditions, or highly stable products for conversion.

5.4.1 Reverse Water–Gas Shift Reaction

More and more people are paying attention to the further utilization of carbon dioxide to manufacture chemicals such as fuels. Specifically, CO_2 hydrogenation to liquid hydrocarbons proceeds via a dual-stage process: reverse water–gas shift (RWGS) reaction converting CO_2 to CO (Eq. 5.2), followed by FTS transforming CO into liquid fuels (diesel, sustainable aviation fuel). Concurrently, methanol and methane form as by-products during the RWGS stage [105–108]. Considering the influence of thermodynamics on this reaction, maintaining a high $H_2{:}CO_2$ ratio at high temperatures (400–600 °C) is usually more conducive to achieving better CO_2 conversion while also achieving high CO selectivity [109, 110],

$$\text{RWGS reaction:}\quad CO_2 + H_2 \rightarrow CO + H_2O,\quad \Delta H = 42.1\ \text{kJ}\cdot\text{mol}^{-1} \tag{5.2}$$

$$\text{Side reaction:}\quad CO + 3H_2 \rightarrow CH_4 + H_2O,\quad \Delta H = -206.5\ \text{kJ}\cdot\text{mol}^{-1} \tag{5.3}$$

$$CO_2 + 4H_2 \rightarrow CH_4 + 2H_2O,\quad \Delta H = -164.7\ \text{kJ}\cdot\text{mol}^{-1} \tag{5.4}$$

$$CO_2 + 3H_2 \rightarrow CH_3OH + H_2O,\quad \Delta H = 131\ \text{kJ}\cdot\text{mol}^{-1} \tag{5.5}$$

Metal-oxide-supported catalysts (e.g., Pd, Pt, Rh, Cu, Fe on SiO_2, CeO_2, Al_2O_3, TiO_2, ZrO_2, and ZnO) represent primary systems for RWGS reactions. However, the endothermic nature necessitates high operating temperatures and elevated H_2/CO_2 ratios, often inducing catalyst sintering—particularly in multicomponent Cu-based and noble metal systems [111–113]. Farha and co-workers demonstrated a phase-pure cubic α-Mo_2C catalyst synthesized via a facile, scalable route that overcomes critical limitations: (1) 100% CO selectivity from CO_2 reduction; (2) maintained equilibrium conversion at high space velocity after >500 hours under harsh 600 °C conditions; and (3) exceptional stability [108]. Operando and post-reaction characterization revealed three key attributes: crystallographic phase purity ensures stability, weak CO–Mo_2C interactions enable high selectivity, and interstitial oxygen atoms enhance activity. Furthermore, the classical strong metal–support interaction (SMSI) effect stabilizes nanocatalysts at high temperatures while modulating reaction selectivity. Bao et al. reported an Ru–Mo oxide catalyst that forms an encapsulation structure (Ru@MoO_{3-x}) during the reaction process as the defective MoO_{3-x} overlayers from the MoO_3 support can migrate onto the Ru NPs, and the classical SMSI effect has been observed in the reaction process (see Figure 5.6a). The formed SMSI state changes 100% CH_4 selectivity on fresh Ru particle surfaces to above 99% CO selectivity with excellent activity and long-term catalytic stability. More generally, the changes in the metal particle size [114, 115] and their chemical state [116] cause the regulation of CO_2 reduction selectivity.

Catalysis research in the past decade has prioritized atomically dispersed (noble) metal catalysts to maximize metal utilization efficiency [117–121] and enhance selectivity [122]. Notably, metal particle size critically governs CO_2 hydrogenation outcomes by modulating catalyst stability and active site distribution—key determinants of reactivity and selectivity [115, 123–127]. Christopher and co-workers demonstrated that Rh/TiO_2 systems exhibit divergent catalytic behaviors: isolated atoms and nanoparticles favor distinct parallel reaction pathways, while nanoparticle disintegration dynamically regulates stability (see Figure 5.6b) [115]. Complementary work by Guo et al. [127] (see Figure 5.6c) revealed a dual-regime size dependency: (1) below 1 nm (single-atom to sub-nanometer clusters), RWGS reaction dominates with turnover frequency (TOF) inversely scaling with cluster size; and (2) above 1 nm, methanation prevails, displaying volcano-shaped activity dependence.

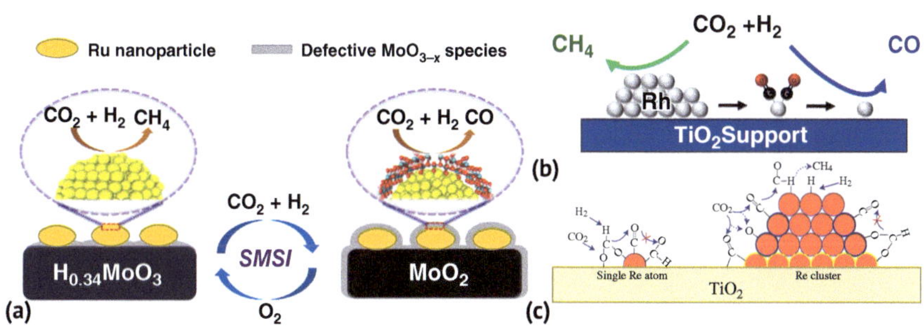

Figure 5.6 (a) Schematic of reaction pathway and structure evolution process between Ru–MoO_3 catalyst and Ru@MoO_{3-x} catalyst during the RWGS reaction. *Source*: Reproduced from [106] with permission of American Chemical Society. (b) Schematic of reaction pathway of isolated Rh atoms and Rh nanoparticles on the TiO_2 support. *Source*: Reproduced from [114] with permission of American Chemical Society. (c) Schematic of reaction pathway of single Re atom and Re cluster on the TiO_2 support. *Source*: Reproduced from [115] with permission of American Chemical Society.

For the supported metal catalysts, much attention has been paid to controlling the hydrogenation selectivity, such as the regulation of metal structural parameters or the construction of metal–support interactions [106, 115, 124, 125, 128–136]. Previously, we found that the metal@zeolite catalyst with a core–shell structure could efficiently stabilize the metal nanoparticles and endow them with the characteristic of "shape-selective catalysis." Based on this, we realized that the hydrogenation selectivity could be further optimized by adjusting the microporous environment to improve the well-known hydrogen spillover [137–139]. Here, we synthesized zeolite-fixed Rh nanoparticles with a core–shell structure using a seed-guided method, denoted as Rh@S-1 catalyst [8]. Catalytic data showed that the Rh@S-1 zeolite with a siliceous framework exhibited high CO selectivity. Even at 500 °C, the CO selectivity maintained at 79.8% with a CO_2 conversion rate of 51.6% (see Figure 5.7a). When silanol groups (defects) were artificially introduced into the zeolite framework (Rh@S-1–OH), methane selectivity increased significantly, with

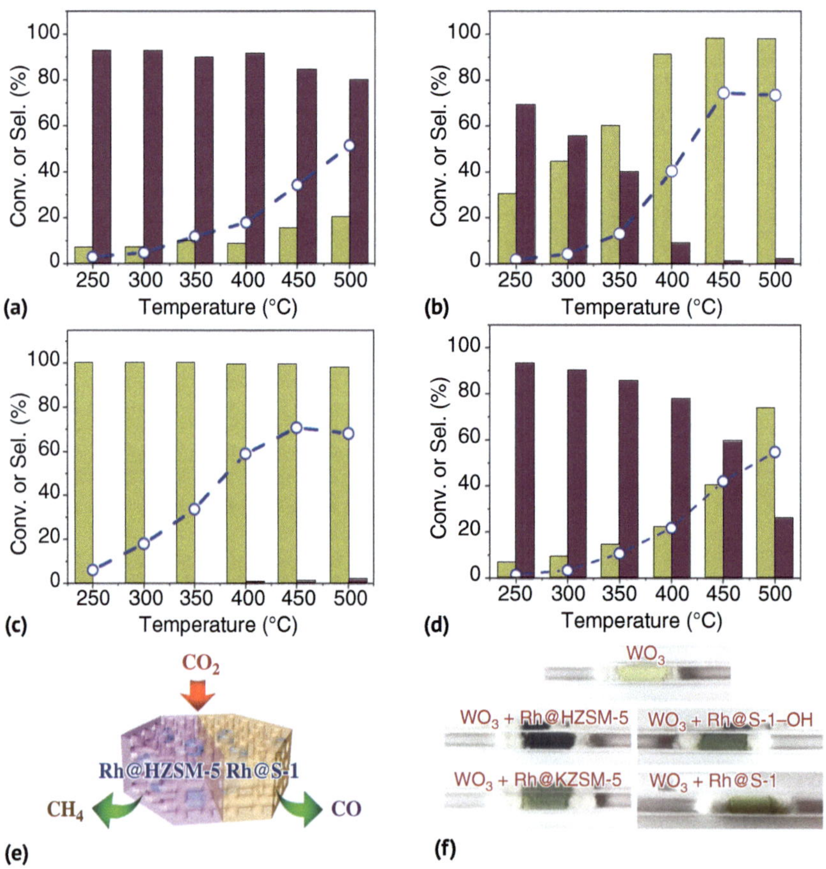

Figure 5.7 Catalytic performances of (a) Rh@S-1, (b) Rh–S-1–OH, (c) Rh@HZSM-5, and (d) Rh@KZSM-5 in CO_2 hydrogenation. (e) Scheme showing the reaction process in Rh@HZSM-5 and Rh@S-1. (f) Photographs of samples made with 1 g of WO_3 mixed with 0.02 g of various catalysts after treatment with H_2 at 30 °C for 10 minutes. *Source*: Reproduced from [8] with permission of American Chemical Society.

CO selectivity dropping below 3% at 500 °C (see Figure 5.7b). For Rh catalysts fixed in aluminosilicate zeolites, performance could also be tuned, with the proton-form (Rh@HZSM-5) and K-form (Rh@KZSM-5) micropores being preferred to produce methane and CO, respectively (see Figure 5.7c, d).

The differences in selectivity among these catalysts, despite having similar Rh nanoparticles, are strongly related to the distinct microenvironments of the zeolite pores (see Figure 5.7e). As reaction selectivity is highly sensitive to active hydrogen species, we assessed the hydrogen spillover ability of these catalysts by identifying color changes in WO_3 (see Figure 5.7f). Zeolites featuring abundant defects (e.g., silanol nests) and Brønsted acid sites facilitated robust hydrogen spillover within micropores. Conversely, defect-free micropores and K^+-exchanged aluminosilicate frameworks exhibited significantly attenuated spillover. This phenomenon correlated directly with CO_2 hydrogenation selectivity: silanol groups and protons enhanced hydrogen spillover, driving consecutive hydrogenation steps that favored methane formation.

Furthermore, a tandem CO_2 hydrogenation carbonylation reaction by shape-selective zeolites was reported [140]. De Vos and co-workers achieved concurrent reaction integration in a single reactor: Pt nanoparticles encapsulated within small-pore LTA zeolite (pore size: ~4.1 Å) selectively produced CO under mild conditions, while insitu carbonylation consumed CO to drive equilibrium shift. Crucially, zeolite confinement suppressed undesired side reactions (e.g., olefin hydrogenation) via molecular sieving effects.

5.4.2 Dry Reforming of Methane and CO_2

Valorizing CO_2 as a chemical/feedstock precursor offers a dual-path strategy for emission mitigation and carbon resource utilization. A pivotal approach involves reducing CO_2 to CO—a versatile platform molecule enabling: (1) direct synthesis of value-added oxygenates via alkyne carbonylation; (2) production of high-performance engineering plastics (e.g., polycarbonates) through polymerization; and (3) conversion to fuels/chemicals via syngas processes (Fischer–Tropsch, methanol synthesis). Notably, industrial CO_2 streams often contain reducing components (e.g., methane, hydrocarbons), exemplified by emissions from combustion engines, waste incineration, biogas, and natural gas sources. Taking the South China Sea, one of the world's four major centers for marine oil and gas accumulation, as an example, natural gas in the South China Sea contains not only low-carbon hydrocarbons such as methane, ethane, and propane, but also a large amount of CO_2. Some natural gases have a CO_2 content of over 50%, and some even reach up to 75% [141].

Dry reforming of methane and CO_2 (DRM) has proved to be an ideal route for CO_2 reduction with methane. Taking DRM as an example, the classical reaction process follows Eq. (5.1), where one CH_4 molecule can reduce one CO_2 molecule, typically defined as a methane reduction capacity (MRC) of 1. By adjusting the reaction conditions, this process can fluctuate up and down within a certain range. However, it is difficult to fully utilize CO_2 and low-carbon alkanes in the reaction process following this ratio, as the content of CO_2 in natural gas is too high. Therefore, in order to achieve maximum CO_2 utilization, it is necessary to significantly increase the reducibility of methane by changing the reaction ratio in the DRM process. The maximum MRC is 3, which means that one methane molecule can reduce three CO_2 molecules, following the idealized reaction process of Eq. (5.3). Compared to the classical DRM process, this idealized process has the advantages in the following: (1) the reduction ability of low-carbon alkanes and the utilization efficiency of CO_2 reach their limits, which can fully utilize the carbon resources in CO_2-rich natural gas, maximize carbon resource utilization efficiency, and reduce carbon emissions; (2) reforming is a high-temperature and strongly endothermic reaction. The energy required for the classical DRM process to reduce CO_2 is 258.6 $kJ \cdot mol_{CO_2}^{-1}$. If the ideal reaction process is followed, the energy required for CO_2 reduction is 111 $kJ \cdot mol_{CO_2}^{-1}$, which will significantly reduce the energy consumption for CO_2 emission reduction:

$$\text{Classical DRM process: } CH_4(g) + CO_2(g) \rightarrow 2CO(g) + 2H_2(g)$$
$$\Delta H^{\circ}_{773\,K} = 258.6\ kJ \cdot mol_{CO_2}^{-1} \quad (5.6)$$
$$\text{RWGS process: } CO_2(g) + H_2(g) \rightarrow CO(g) + 2H_2O(g)$$
$$\Delta H^{\circ}_{773\,K} = 37.2\ kJ \cdot mol_{CO_2}^{-1} \quad (5.7)$$
$$\text{Idealized DRM process: } CH_4(g) + 3CO_2(g) \rightarrow 4CO(g) + 2H_2O(g)$$
$$\Delta H^{\circ}_{773\,K} = 111\ kJ \cdot mol_{CO_2}^{-1} \quad (5.8)$$

However, due to thermodynamic limitations, the idealized reaction of $3CO_2 + CH_4 \rightarrow 4CO + 2H_2O$ cannot occur directly, meaning that the reduction performance of methane is thermodynamically limited to a specific value. Figure 5.8a shows the gas composition distribution of thermodynamic equilibrium states at different temperatures, which can be used to calculate the MRC. Figure 5.8b it can be seen that at the thermodynamic limit, a methane molecule can reduce up to 1.9 CO_2 molecules, which means the MRC is 1.9, and reaching this performance requires a high temperature of 1000 °C.

Currently, at conventional DRM reaction temperatures (650–750 °C), it is difficult for the equilibrium reduction performance of methane to exceed 1.5. In order to break through this limitation and achieve a significant improvement in MRC, scientists have

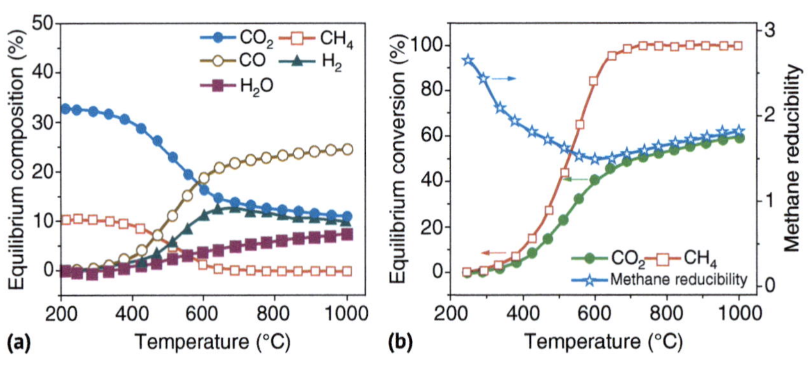

Figure 5.8 (a) The thermodynamic equilibrium-state gas composition at different temperatures. (b) CH_4 and CO_2 equilibrium conversion, and MRCs at different temperatures. The CO_2/CH_4 ratio of 3.3 in the feed gas.

made some attempts. The most successful example is the Super Dry Reforming (Super DRM) route designed by Marin and co-workers at Ghent University in Belgium [142], which cleverly combines multiple cascade reaction pathways to achieve efficient utilization of methane and CO_2. The conversion process they designed is as follows: (1) using $MgAl_2O_4$ spinel-supported Ni nanoparticle catalyst to catalyze methane and CO_2 reforming to obtain CO and H_2 products and their unreacted CO_2; (2) fixing CaO and $Fe_3O_4/MgAl_2O_4$ particles in the second bed of the reactor. The H_2 generated after the first bed will react with iron oxide species to form elemental Fe, while unreacted CO_2 will be adsorbed by CaO to obtain calcium carbonate; and (3) after switching the CO_2–CH_4 mixed gas raw material with an inert gas, CO_2 desorbs from calcium carbonate and further oxidizes elemental Fe to obtain CO products and iron oxide. In summary, after a two-step CO_2 reduction reaction, some hydrogen species are oxidized to water, resulting in a significant increase in CO production and an improvement in MRC. In a feed gas with a CO_2 content of 75%, the maximum MRC can reach 1.9, achieving an average reaction of 1.9 CO_2 molecules per methane molecule and a CO yield of 2.9 $mol_{CO} \cdot mol_{CH_4}^{-1}$, and it is the highest level currently reported. However, in order to promote the redox reaction between CO_2 and elemental iron, this process needs to be carried out at high temperatures (>800 °C) and repeatedly switched between CO_2–CH_4 mixture and inert carrier gas to achieve the coupling of DRM and secondary redox process. This leads to a complex reaction process, increased separation costs, and an overall average CO generation rate of 2.5 $mol_{CO} \cdot mol_{Ni}^{-1} \cdot h^{-1}$, which is still difficult to meet the actual reaction requirements.

In addition to reduction efficiency, the reforming process of CO_2 and low-carbon alkanes also faces challenges such as catalyst coking and deactivation [7, 143–147] as these reactions are mainly catalyzed by Ni-based catalysts, which can easily generate nickel carbide species, leading to catalyst deactivation. In practical operation, it is necessary to stop the reaction and introduce oxidizing gas to remove coke and regenerate the catalyst. In addition, during high-temperature catalytic reactions and catalyst oxidation roasting regeneration processes, Ni nanoparticles are prone to sintering and agglomeration, leading to irreversible deactivation of the catalyst. Therefore, stabilizing Ni nanoparticles is also an important research topic in this field. In previous studies, scientists have focused on modulating the composition of metal particles and enhancing the interaction between metal and support to enhance the anti-carbon deposition and anti-sintering properties of Ni nanoparticles [7, 143, 146]. However, strong coordination with the carrier can cause nickel particles to carry more positive charges, resulting in the presence of many ionic nickel species, thereby reducing DRM reaction activity. In order to achieve a balance between activity and stability, it is necessary to finely regulate the microenvironment of nickel particles and their interactions with carriers or additives. For example, scientists at home and abroad have generally found that adding rare-earth additives such as La can effectively enhance the thermal stability of Ni particles [148–150]. In addition, the team led by Xu et al. from Tsinghua University has developed a nano-magnesium-oxide-supported Ni catalyst, which exhibits superior catalytic lifetime in long-period reactions [151]. McFarland and co-workers at the University of California, Santa Barbara found that NiIn alloy exhibited higher thermal stability and sintering resistance than single-metal Ni catalysts [152]. Yavuz and co-workers from the Korean Academy of Science and Technology found that the corner positions of MgO crystals can effectively stabilize NiMo particles [153], maintaining stable catalytic activity in continuous DRM reactions for over 850 hours. Furthermore, through advanced in situ characterization techniques, researchers have found that sub-nano-Ni clusters can inhibit the deep cracking of CH_4 and are more likely to generate reactive oxygen species to effectively avoid carbon deposition. However, designing and preparing sub-nano-Ni clusters with high activity and stability under reaction conditions is still very difficult.

Zeolite molecular sieves have good thermal stability, and their uniform microporous structure provides an ideal channel for controlling the microenvironment of active sites. In addition, zeolite pores have also been found to be an ideal environment for hydrogen overflow in some reactions. Although there have been many studies on the composite reaction system of nanometal/oxide catalysts and zeolites, zeolites are mostly used as solid acid catalysts for series reactions. In the preliminary research, our project team encapsulated Rh nanoparticles inside silica–alumina zeolite crystals and regulated the catalytic activity and selectivity in the CO_2 hydrogenation reaction by enhancing hydrogen overflow [8]. Therefore, we envision that if Ni nanoparticles can be embedded inside zeolite crystals with a hydrogen overflow environment, the reactivity of H in CH_4 can be significantly improved, and the reduction ability of methane to CO_2 can be increased to nearly 3.

Consequently, Xiao and co-workers engineered Ni nanoparticles confined within aluminosilicate zeolite (denoted by Ni@HZSM-5; see Figure 5.9a–c) [11]. During dry reforming, Ni@HZSM-5 exhibited substantially enhanced activity versus conventional supported Ni/HZSM-5. Quantitative analysis revealed a CO_2 conversion efficiency of 2.9 $mol_{CO_2} \cdot mol_{CH_4}^{-1}$ for the confined system versus 1.9 $mol_{CO_2} \cdot mol_{CH4}^{-1}$ for the supported analog (see Figure 5.9d). This metric signifies that each CH_4 molecule reduces 2.9 CO_2 molecules, generating 3.9 CO molecules (carbon balance >95%), demonstrating exceptional CO_2 utilization—crucial for maximizing carbon efficiency in CO_2-rich natural gas.

To investigate the role of protonic hydrogen in Ni@HZSM-5 during the DRM reaction, we further explored hydrogen diffusion and transformation through a tungsten oxide discoloration experiment. The results revealed that Ni@HZSM-5 exhibited a significantly stronger hydrogen spillover effect compared to Ni/HZSM-5, which enhanced the CO_2 reduction reaction. Furthermore, the zeolitic pore microenvironment was engineered to amplify hydrogen spillover within the microporous framework. This strategic modulation maintained elevated concentrations of active hydrogen species proximal to nickel nanoparticles, thereby maximizing CO_2 reduction to CO while suppressing the competitive water–gas shift reaction.

Based on these results, we proposed that the Ni nanoparticles encapsulated within the zeolite framework form a "zeolite nanoreactor," promoting CO_2 reduction (see Figure 5.9e). The protonic microporous structure of the zeolite enhances hydrogen spillover around the nickel nanoparticles. This study successfully applied the hydrogen spillover effect to the dry reforming system under CO_2-rich conditions, maximizing CO_2 utilization and presenting a new approach for reducing CO_2 emissions and converting carbon resources.

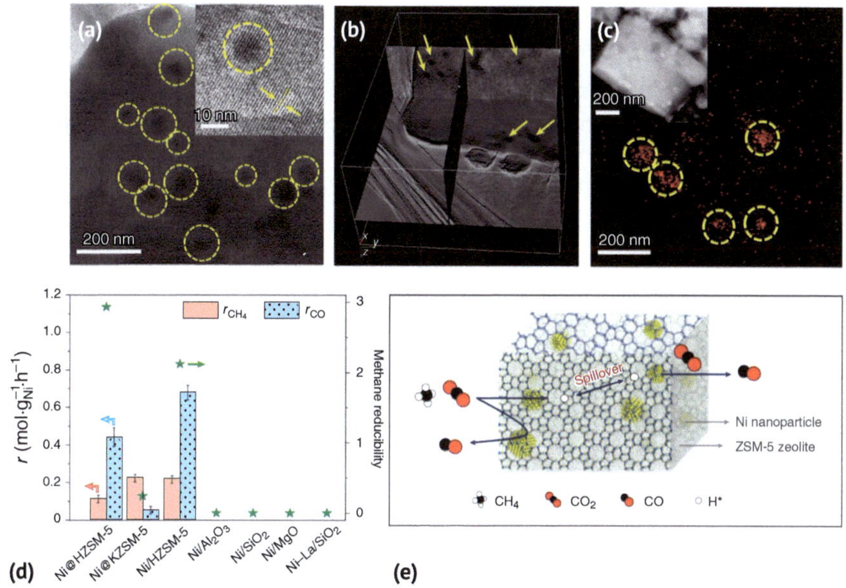

Figure 5.9 (a) TEM image of the Ni@HZSM-5 sample; the yellow arrows highlight the zeolite micropores. (b) Three-dimensional tomographic TEM image and (c) Ni EDS elemental map of Ni@HZSM-5; yellow circles highlight the Ni nanoparticles localized within the zeolite. *Source*: Reproduced from [11] with permission of Springer Nature. (d) Data showing the methane reaction rate r_{CH4}, CO formation rate r_{CO}, and MRC (green stars) on various catalysts under equivalent reaction conditions. (e) DRM process on the model of Ni@ZSM-5. *Source*: Reproduced from [11] / with permission of Springer Nature.

5.5 Conclusion and Outlook

Over the past decades, advancements in zeolite catalysts have significantly accelerated conversion of C_1 molecules (CO, CO_2, and CH_4), driving industrialization to reduce petroleum dependences while meeting escalating demands for energy and chemicals. This chapter has highlighted catalytic pathways transforming C_1 feedstocks into value-added products (methanol, ethanol, olefins, etc.), wherein zeolites serve pivotal functions as both active catalysts and metal–support matrices. Crucially, core–shell-structured metal@zeolite systems demonstrate exceptional capabilities over conventional supported catalysts: enhancing activity, regulating product selectivity, and optimizing stability. Such effects are reflected in the enrichment of key reaction intermediates to improve the oxidation activity of methane, the construction of Rh—O—Mn interface sites in a confined environment to improve the oxygenate yields, the regulation of micropore environment to change the selectivity of CO_2 hydrogenation, and the enhancement of hydrogen spillover to maximize the reduction of CO_2 in DRM. More importantly, the metal@zeolite catalysts showed excellent stability than the supported catalysts due to the rigid framework of zeolite crystals. In addition, we have found the role of zeolite as a diffusion promoter in FTO process. We believe that these findings offer new perspectives for designing efficient zeolite-based catalysts for conversion of low-carbon molecules.

References

1 Zhu, X.C., Hofmann, J.P., Mezari, B. et al. (2016). Trimodal porous hierarchical SSZ-13 zeolite with improved catalytic performance in the methanol-to-olefins reaction. *ACS Catal.* 6: 2163–2177.

2 Liu, Z.D., Wakihara, T., Oshima, K. et al. (2015). Widening synthesis bottlenecks: realization of ultrafast and continuous-flow synthesis of high-silica zeolite SSZ-13 for NO_x removal. *Angew. Chem. Int. Ed.* 54: 5683–5687.

3 Fickel, D.W., D'Addio, E., Lauterbach, J.A. et al. (2011). The ammonia selective catalytic reduction activity of copper-exchanged small-pore zeolites. *Appl. Catal. B: Environ.* 102: 441–448.

4 Liu, F.D., Yu, Y.B., and He, H. (2014). Environmentally-benign catalysts for the selective catalytic reduction of NO_x from diesel engines: structure-activity relationship and reaction mechanism aspects. *Chem. Commun.* 50: 8445–8463.

5 Xie, D., McCusker, L.B., Baerlocher, C. et al. (2013). SSZ-52, a zeolite with an 18-layer aluminosilicate framework structure related to that of the $DeNO_x$ catalyst Cu-SSZ-13. *J. Am. Chem. Soc.* 135: 10519–10524.

6 Choi, M., Na, K., Kim, J. et al. (2009). Stable single-unit-cell nanosheets of zeolite MFI as active and long-lived catalysts. *Nature* 461: 246–249.

7 Wang, C.T., Zhang, J., Qin, G.Q. et al. (2020). Direct conversion of syngas to ethanol within zeolite crystals. *Chem.* 6: 646–657.

8 Wang, C.T., Guan, E.J., Wang, L. et al. (2019). Product selectivity controlled by nanoporous environments in zeolite crystals enveloping rhodium nanoparticle catalysts for CO_2 hydrogenation. *J. Am. Chem. Soc.* 141: 8482–8488.

9 Wang, C.T., Fang, W., Wang, L. et al. (2021). Fischer-Tropsch reaction within zeolite crystals for selective formation of gasoline-ranged hydrocarbons. *J. Energy Chem.* 54: 429–433.

10 Wang, C.T., Fang, W., Liu, Z.Q. et al. (2022). Fischer-Tropsch synthesis to olefins boosted by MFI zeolite nanosheets. *Nat. Nanotechnol.* 17: 714–720.

11 Zhu, Q.Y., Zhou, H., Wang, L. et al. (2022). Enhanced CO_2 utilization in dry reforming of methane achieved through nickel-mediated hydrogen spillover in zeolite crystals. *Nat. Catal.* 5: 1030–1037.

12 Fujisaki, H., Ishizuka, T., Kotani, H. et al. (2023). Selective methane oxidation by molecular iron catalysts in aqueous medium. *Nature* 616: 476–481.

13 Zhou, W., Cheng, K., Kang, J.C. et al. (2019). New horizon in C1 chemistry: breaking the selectivity limitation in transformation of syngas and hydrogenation of CO_2 into hydrocarbon chemicals and fuels. *Chem. Soc. Rev.* 48: 3193–3228.

14 Bao, J., Yang, G.H., Yoneyama, Y. et al. (2019). Significant advances in C1 catalysis: highly efficient catalysts and catalytic reactions. *ACS Catal.* 9: 3026–3053.

15 Mesters, C. (2016). A selection of recent advances in C1 chemistry. *Annu. Rev. Chem. Biomol. Eng.* 7: 223–238.

16 Wang, Y.Q., Wang, C.T., Wang, L.X. et al. (2021). Zeolite fixed metal nanoparticles: new perspective in catalysis. *Acc. Chem. Res.* 54: 2579–2590.

17 Zhang, Q., Yu, J.H., and Corma, A. (2020). Applications of zeolites to C1 chemistry: recent advances, challenges, and opportunities. *Adv. Mater.* 32: e2002927.

18 Wang, Z., Wang, H., Mitra, A. et al. (2001). Pure-silica zeolite low-*k* dielectric thin films. *Adv. Mater.* 13: 746–749.

19 Lei, C., Dong, Z.Y., Martínez, C. et al. (2020). A cationic oligomer as an organic template for direct synthesis of aluminosilicate ITH zeolite. *Angew. Chem. Int. Ed.* 59: 15649–15655.

20 Vattipalli, V., Paracha, A.M., Hu, W.G. et al. (2018). Broadening the scope for fluoride-free synthesis of siliceous zeolites. *Angew. Chem. Int. Ed.* 57: 3607–3611.

21 Dusselier, M. and Davis, M.E. (2018). Small-pore zeolites: synthesis and catalysis. *Chem. Soc. Rev.* 118: 5265–5329.

22 Rojas, A., Martínez-Morales, E., Zicovich-Wilson, C.M. et al. (2012). Zeolite synthesis in fluoride media: structure direction toward ITW by small methylimidazolium cations. *J. Am. Chem. Soc.* 134: 2255–2263.

23 Jones, C.W., Tsuji, K., and Davis, M.E. (1998). Organic-functionalized molecular sieves as shape-selective catalysts. *Nature* 393: 52–54.

24 Dusselier, M., Wouwe, P.V., Dewaele, A. et al. (2015). Shape-selective zeolite catalysis for bioplastics production. *Science* 349: 78–80.

25 Li, Y., Li, L., and Yu, J.H. (2017). Applications of zeolites in sustainable chemistry. *Chem.* 3: 928–949.

26 Corma, A. (1995). Inorganic solid acids and their use in acid-catalyzed hydrocarbon reactions. *Chem. Soc. Rev.* 95: 559–614.

27 Jiao, F., Li, J.J., Pan, X.L. et al. (2016). Selective conversion of syngas to light olefins. *Science* 351: 1065–1068.

28 Pan, X.L., Jiao, F., Miao, D.Y. et al. (2021). Oxide-zeolite-based composite catalyst concept that enables syngas chemistry beyond Fischer-Tropsch synthesis. *Chem. Soc. Rev.* 121: 6588–6609.

29 Cheng, K., Gu, B., Liu, X.L. et al. (2016). Direct and highly selective conversion of synthesis gas into lower olefins: design of a bifunctional catalyst combining methanol synthesis and carbon-carbon coupling. *Angew. Chem. Int. Ed.* 55: 4725–4728.

30 Wang, Z.J., Wei, Y.Z., Qi, J. et al. (2024). Mass transfer modulation by hollow multi-shelled structures for high space-time yield synthesis of light olefins from syngas. *Adv. Funct. Mater.* 34: 2316547.

31 Schwach, P., Pan, X.L., and Bao, X.H. (2017). Direct conversion of methane to value-added chemicals over heterogeneous catalysts: challenges and prospects. *Chem. Soc. Rev.* 117: 8497–8520.

32 Wang, L. and Xiao, F.-S. (2023). Methane oxidation by catalyst reduction. *Nat. Catal.* 6: 866–867.

33 Coperet, C. (2010). C-H bond activation and organometallic intermediates on isolated metal centers on oxide surfaces. *Chem. Soc. Rev.* 110: 656–680.

34 Labinger, J.A. and Bercaw, J.E. (2002). Understanding and exploiting C-H bond activation. *Nature* 417: 507–514.

35 Crabtree, R.H. (2001). Alkane C–H activation and functionalization with homogeneous transition metal catalysts: a century of progress—a new millennium in prospect. *J. Chem. Soc., Dalton Trans.* 17: 2437–2450.

36 Arndtsen, B.A., Bergman, R.G., Mobley, T.A. et al. (1995). Selective intermolecular carbon-hydrogen bond activation by synthetic metal complexes in homogeneous solution. *Acc. Chem. Res.* 28: 154–162.

37 Balasubramanian, R., Smith, S.M., Rawat, S. et al. (2010). Oxidation of methane by a biological dicopper centre. *Nature* 465: 115–119.

38 Ahlquist, M., Nielsen, R.J., Periana, R.A. et al. (2009). Product protection, the key to developing high performance methane selective oxidation catalysts. *J. Am. Chem. Soc.* 131: 17110–17115.

39 Lunsford, J.H. (2000). Catalytic conversion of methane to more useful chemicals and fuels: a challenge for the 21st century. *Catal. Today* 63: 165–174.

40 Caballero, A. and Pérez, P.J. (2013). Methane as raw material in synthetic chemistry: the final frontier. *Chem. Rev.* 42: 8809–8820.

41 Olah, G.A. (2005). Beyond oil and gas: the methanol economy. *Angew. Chem. Int. Ed.* 44: 2636–2639.

42 Sobolev, V.I., Dubkov, K.A., Panna, O.V. et al. (1995). Selective oxidation of methane to methanol on a FeZSM-5 surface. *Catal. Today* 24: 251–252.

43 Periana, R.A., Taube, D.J., Gamble, S. et al. (1998). Platinum catalysts for the high-yield oxidation of methane to a methanol derivative. *Science* 280: 560–564.

44 Hammond, C., Forde, M.M., Ab Rahim, M.H. et al. (2012). Direct catalytic conversion of methane to methanol in an aqueous medium by using copper-promoted Fe-ZSM-5. *Angew. Chem. Int. Ed.* 51: 5129–5133.

45 Tomkins, P., Mansouri, A., Bozbag, S.E. et al. (2016). Isothermal cyclic conversion of methane into methanol over copper-exchanged zeolite at low temperature. *Angew. Chem. Int. Ed.* 55: 5467–5471.

46 Huang, W.X., Zhang, S.R., Tang, Y. et al. (2016). Low-temperature transformation of methane to methanol on Pd_1O_4 single sites anchored on the internal surface of microporous silicate. *Angew. Chem. Int. Ed.* 55: 13441–13445.

47 Snyder, B.E.R., Vanelderen, P., Bols, M.L. et al. (2006). The active site of low-temperature methane hydroxylation in iron-containing zeolites. *Nature* 536: 317–321.

48 Agarwal, N., Freakley, S.J., McVicker, R.U. et al. (2017). Aqueous Au-Pd colloids catalyze selective CH_4 oxidation to CH_3OH with O_2 under mild conditions. *Science* 358: 223–227.

49 Sushkevich, V.L., Palagin, D., Ranocchiari, M. et al. (2017). Selective anaerobic oxidation of methane enables direct synthesis of methanol. *Science* 356: 523–527.

50 Shan, J.J., Li, M.W., Allard, L.F. et al. (2017). Mild oxidation of methane to methanol or acetic acid on supported isolated rhodium catalysts. *Nature* 551: 605–608.

51 Dinh, K.T., Sullivan, M.M., Narsimhan, K. et al. (2019). Continuous partial oxidation of methane to methanol catalyzed by diffusion-paired copper dimers in copper-exchanged zeolites. *J. Am. Chem. Soc.* 141: 11641–11650.

52 Koishybay, A. and Shantz, D.F. (2020). Water is the oxygen source for methanol produced in partial oxidation of methane in a flow reactor over Cu-SSZ-13. *J. Am. Chem. Soc.* 142: 11962–11966.

53 Jin, Z., Wang, L., Zuidema, E. et al. (2020). Hydrophobic zeolite modification for in situ peroxide formation in methane oxidation to methanol. *Science* 367: 193–197.

54 Qi, G.D., Davies, T.E., Nasrallah, A. et al. (2022). Au-ZSM-5 catalyses the selective oxidation of CH_4 to CH_3OH and CH_3COOH using O_2. *Nat. Catal.* 5: 45–54.

55 Wang, S.K., Fung, V., Hülsey, M.J. et al. (2023). H_2-reduced phosphomolybdate promotes room-temperature aerobic oxidation of methane to methanol. *Nat. Catal.* 6: 895–905.

56 Groothaert, M.H., Smeets, P.J., Sels, B.F. et al. (2005). Selective oxidation of methane by the bis(μ-oxo)dicopper core stabilized on ZSM-5 and mordenite zeolites. *J. Am. Chem. Soc.* 127: 1394–1395.

57 Narsimhan, K., Michaelis, V.K., Mathies, G. et al. (2015). Methane to acetic acid over Cu-exchanged zeolites: mechanistic insights from a site-specific carbonylation reaction. *J. Am. Chem. Soc.* 137: 1825–1832.

58 Schroeder, D., Fiedler, A., Hrusak, J. et al. (1992). Experimental and theoretical studies toward a characterization of conceivable intermediates involved in the gas-phase oxidation of methane by bare FeO^+. Generation of four distinguishable $[Fe,C,H_4,O]^+$ isomers. *J. Am. Chem. Soc.* 114: 1215–1222.

59 Smeets, P.J., Groothaert, M.H., and Schoonheydt, R.J. (2005). Cu based zeolites: a UV–vis study of the active site in the selective methane oxidation at low temperatures. *Catal. Today* 110: 303–309.

60 Woertink, J.S., Smeets, P.J., Groothaert, M.H. et al. (2009). A $[Cu_2O]^{2+}$ core in Cu-ZSM-5, the active site in the oxidation of methane to methanol. *Proc. Natl. Acad. Sci. USA* 106: 18908–18913.

61 Alayon, E.M., Nachtegaal, M., Ranocchiari, M. et al. (2012). Catalytic conversion of methane to methanol over Cu–mordenite. *Chem. Commun.* 48: 404–406.

62 Ab Rahim, M.H., Forde, M.M., Jenkins, R.L. et al. (2012). Oxidation of methane to methanol with hydrogen peroxide using supported gold–palladium alloy nanoparticles. *Angew. Chem. Int. Ed.* 52: 1280–1284.

63 Palkovits, R., Antonietti, M., Kuhn, P. et al. (2009). Solid catalysts for the selective low-temperature oxidation of methane to methanol. *Angew. Chem. Int. Ed.* 48: 6909–6912.

64 Xie, J.J., Jin, R.X., Li, A. et al. (2018). Highly selective oxidation of methane to methanol at ambient conditions by titanium dioxide-supported iron species. *Nat. Catal.* 1: 889–896.

65 Pan, X.L., Fan, Z.L., Chen, W. et al. (2007). Enhanced ethanol production inside carbon-nanotube reactors containing catalytic particles. *Nat. Mater.* 6: 507–511.

66 Zhou, W., Kang, J.C., Cheng, K. et al. (2018). Direct conversion of syngas into methyl acetate, ethanol, and ethylene by relay catalysis via the intermediate dimethyl ether. *Angew. Chem. Int. Ed.* 57: 12012–12016.

67 Lin, T.J., Qi, X.Z., Wang, X.X. et al. (2019). Direct production of higher oxygenates by syngas conversion over a multifunctional catalyst. *Angew. Chem. Int. Ed.* 58: 4627–4631.

68 Farrell, A.E., Plevin, R.J., Turner, B.T. et al. (2006). Ethanol can contribute to energy and environmental goals. *Science* 311: 506–508.

69 Luk, H.T., Mondelli, C., Ferre, D.C. et al. (2017). Status and prospects in higher alcohols synthesis from syngas. *Chem. Soc. Rev.* 46: 1358–1426.

70 Spivey, J.J. and Egbebi, A. (2007). Heterogeneous catalytic synthesis of ethanol from biomass-derived syngas. *Chem. Soc. Rev.* 36: 1514–1528.

71 Li, S., Feng, L., Wang, H.W. et al. (2024). Atomically intimate assembly of dual metal–oxide interfaces for tandem conversion of syngas to ethanol. *Nat. Nanotechnol.* 20: 255–264.

72 He, Y.D. (2019). *Handbook of Modern Coal Chemical Technology*. Beijing: Chemical Industry Press.

73 Xu, R., Yang, C., Wei, W. et al. (2004). Fe-modified $CuMnZrO_2$ catalysts for higher alcohols synthesis from syngas. *J. Mol. Catal. A: Chem.* 221: 51–58.

74 Xu, J., Wei, J., Zhang, J.X. et al. (2023). Precisely synergistic synthesis of higher alcohols from syngas over iron carbides. *Chem. Catal.* 3: 100584.

75 Li, Y.W., Gao, W., Peng, M. et al. (2020). Interfacial Fe_5C_2-Cu catalysts toward low-pressure syngas conversion to long-chain alcohols. *Nat. Commun.* 11: 61.

76 Hu, J.T., Wei, Z.Y., Zhang, Y.L. et al. (2023). Edge-rich molybdenum disulfide tailors carbon-chain growth for selective hydrogenation of carbon monoxide to higher alcohols. *Nat. Commun.* 14: 6808.

77 Fang, K.G., Li, D.B., Lin, M.G. et al. (2009). A short review of heterogeneous catalytic process for mixed alcohols synthesis via syngas. *Catal. Today* 147: 133–138.

78 Bhasin, M.M., Bartley, W.J., Ellgen, P.C. et al. (1978). Synthesis gas conversion over supported Rhodium-Iron catalysts. *J. Catal.* 54: 120-128.

79 Liu, G.B., Yang, G.H., Peng, X.B. et al. (2022). Recent advances in the routes and catalysts for ethanol synthesis from syngas. *Chem. Soc. Rev.* 51: 5606–5659.

80 Ao, M., Pham, G.H., Sunarso, J. et al. (2018). Active centers of catalysts for higher alcohol synthesis from syngas: a review. *ACS Catal.* 8: 7025–7050.

81 Yang, N., Yoo, J.S., Schumann, J. et al. (2017). Rh-MnO interface sites formed by atomic layer deposition promote syngas conversion to higher oxygenates. *ACS Catal.* 7: 5746–5757.

82 Mo, X.H., Gao, J., Umnajkaseam, N. et al. (2009). La, V, and Fe promotion of Rh/SiO_2 for CO hydrogenation: effect on adsorption and reaction. *J. Catal.* 267: 167–176.

83 Kang, J.C., He, S., Zhou, W. et al. (2020). Single-pass transformation of syngas into ethanol with high selectivity by triple tandem catalysis. *Nat. Commun.* 11: 827.

84 San, X.G., Zhang, Y., Shen, W.J. et al. (2009). New synthesis method of ethanol from dimethyl ether with a synergic effect between the zeolite catalyst and metallic catalyst. *Energy Fuels* 23: 2843–2844.

85 Cheung, P., Bhan, A., Sunley, G.J. et al. (2006). Selective carbonylation of dimethyl ether to methyl acetate catalyzed by acidic zeolites. *Angew. Chem. Int. Ed.* 45: 1617–1620.

86 Bhan, A., Allian, A.D., Sunley, G.J. et al. (2007). Specificity of sites within eight-membered ring zeolite channels for carbonylation of methyls to acetyls. *J. Am. Chem. Soc.* 129: 4919–4924.

87 Yao, J., Feng, X.B., Fan, J.Q. et al. (2022). Self-assembled nano-filamentous zeolite catalyst to realize efficient one-step ethanol synthesis. *Chem. Eur. J.* 28: e202201783.

88 Torres Galvis, H.M., Bitter, J.H., Khare, C.B. et al. (2012). Supported iron nanoparticles as catalysts for sustainable production of lower olefins. *Science* 335: 835–838.

89 Fang, W., Wang, C.T., Liu, Z.Q. et al. (2022). Physical mixing of a catalyst and a hydrophobic polymer promotes CO hydrogenation through dehydration. *Science* 377: 406–410.

90 Zhai, P., Xu, C., Gao, R. et al. (2016). Highly tunable selectivity for syngas-derived alkenes over zinc and sodium-modulated Fe_5C_2 catalyst. *Angew. Chem. Int. Ed.* 55: 9902–9907.

91 Zhong, L.S., Yu, F., An, Y.L. et al. (2016). Cobalt carbide nanoprisms for direct production of lower olefins from syngas. *Nature* 538: 84–87.

92 Jiao, F., Pan, X.L., Gong, K. et al. (2018). Shape-selective zeolites promote ethylene formation from syngas via a ketene intermediate. *Angew. Chem. Int. Ed.* 57: 4692–4696.

93 Wang, H.D., Jiao, F., Feng, J.Y. et al. (2025). Maximizing the accessibility of acid sites within zeolite catalysts for syngas conversion. *Angew. Chem. Int. Ed.* 64: e202424946.

94 Jiao, F., Bai, B., Li, G. et al. (2023). Disentangling the activity-selectivity trade-off in catalytic conversion of syngas to light olefins. *Science* 380: 727–730.

95 Liu, X.L., Zhou, W., Yang, Y.D. et al. (2018). Design of efficient bifunctional catalysts for direct conversion of syngas into lower olefins *via* methanol/dimethyl ether intermediates. *Chem. Sci.* 9: 4708–4718.

96 Liu, X.L., Wang, M.H., Yin, H.R. et al. (2020). Tandem catalysis for hydrogenation of CO and CO_2 to lower olefins with bifunctional catalysts composed of spinel oxide and SAPO-34. *ACS Catal.* 10: 8303–8314.

97 Su, J.J., Zhang, L., Zhou, H.B. et al. (2023). Unveiling the anti-trap effect for bridging intermediates on $ZnAlO_x$/AlPO-18 bifunctional catalysts to boost syngas to olefin conversion. *ACS Catal.* 13: 2472–2481.

98 Liu, Z.P., Ni, Y.M., Hu, Z.P. et al. (2022). Insights into effects of ZrO_2 crystal phase on syngas-to-olefin conversion over ZnO/ZrO_2 and SAPO-34 composite catalysts. *Chin. J. Catal.* 43: 877–884.

99 Xiao, J.D., Cheng, K., Xie, X.B. et al. (2022). Tandem catalysis with double-shelled hollow spheres. *Nat. Mater.* 21: 572–579.

100 Tollefson, J. (2017). World's carbon emissions set to spike by 2% in 2017. *Nature* 551: 283.

101 Patel, H.A., Byun, J., and Yavuz, C.Y. (2017). Carbon dioxide capture adsorbents: chemistry and methods. *ChemSusChem* 10: 1303–1317.

102 BP p.l.c. (2019). BP Statistical review of world energy 2019. London: BP plc. https://www.bp.com/en/global/corporate/energy-economics.html.

103 Velty, A. and Corma, A. (2023). Advanced zeolite and ordered mesoporous silica-based catalysts for the conversion of CO_2 to chemicals and fuels. *Chem. Soc. Rev.* 52: 1773–1946.

104 Artz, J., Müller, T.E., Thenert, K. et al. (2018). Sustainable conversion of carbon dioxide: an integrated review of catalysis and life cycle assessment. *Chem. Soc. Rev.* 118: 434–504.

105 Wang, J.Y., Li, R.T., Zhang, G.H. et al. (2024). Confinement-induced indium oxide nanolayers formed on oxide support for enhanced CO_2 hydrogenation reaction. *J. Am. Chem. Soc.* 146: 5523–5531.

106 Xin, H., Lin, L., Li, R.T. et al. (2022). Overturning CO_2 hydrogenation selectivity with high activity via reaction-induced strong metal-support interactions. *J. Am. Chem. Soc.* 144: 4874–4882.

107 Zhang, X.B., Han, S.B., Zhu, B.E. et al. (2020). Reversible loss of core–shell structure for Ni–Au bimetallic nanoparticles during CO_2 hydrogenation. *Nat. Catal.* 3: 411–417.

108 Ahmadi Khoshooei, M., Wang, X.J., Vitale, G. et al. (2024). An active, stable cubic molybdenum carbide catalyst for the high-temperature reverse water-gas shift reaction. *Science* 384: 540–546.

109 Choi, Y.J., Sim, G.D., Jung, U. et al. (2024). Copper catalysts for CO_2 hydrogenation to CO through reverse water–gas shift reaction for e-fuel production: fundamentals, recent advances, and prospects. *Chem. Eng. J.* 492: 152283.

110 Triviño, M.L.T., Arriola, N.C., Kang, Y.S. et al. (2024). Transforming CO_2 to valuable feedstocks: emerging catalytic and technological advances for the reverse water gas shift reaction. *Chem. Eng. J.* 487: 150369.

111 Roy, S., Cherevotan, A., and Peter, S.C. (2018). Thermochemical CO_2 hydrogenation to single carbon products: scientific and technological challenges. *ACS Energy Lett.* 3: 1938–1966.

112 Daza, Y.A. and Kuhn, J.N. (2016). CO_2 conversion by reverse water gas shift catalysis: comparison of catalysts, mechanisms and their consequences for CO_2 conversion to liquid fuels. *RSC Adv.* 6: 49675–49691.

113 Porosoff, M.D., Yan, B.H., and Chen, J.G.G. (2016). Catalytic reduction of CO_2 by H_2 for synthesis of CO, methanol and hydrocarbons: challenges and opportunities. *Energy Environ. Sci.* 9: 62–73.

114 Aitbekova, A., Wu, L.H., Wrasman, C.J. et al. (2018). Low-temperature restructuring of CeO_2-supported Ru nanoparticles determines selectivity in CO_2 catalytic reduction. *J. Am. Chem. Soc.* 140: 13736–13745.

115 Matsubu, J.C., Yang, V.N., and Christopher, P. (2015). Isolated metal active site concentration and stability control catalytic CO_2 reduction selectivity. *J. Am. Chem. Soc.* 137: 3076–3084.

116 Li, S.W., Xu, Y., Chen, Y.F. et al. (2017). Tuning the selectivity of catalytic carbon dioxide hydrogenation over iridium/cerium oxide catalysts with a strong metal-support interaction. *Angew. Chem. Int. Ed.* 56: 10761–10765.

117 Yang, X.-F., Wang, A.Q., Qiao, B.T. et al. (2013). Single-atom catalysts: a new frontier in heterogeneous catalysis. *Acc. Chem. Res.* 46: 1740–1748.

118 Vilé, G., Albani, D., Nachtegaal, M. et al. (2015). A stable single-site palladium catalyst for hydrogenations. *Angew. Chem. Int. Ed.* 54: 11265–11269.

119 Lou, Y., Zheng, Y.P., Li, X. et al. (2019). Pocketlike active site of Rh_1/MoS_2 single-atom catalyst for selective crotonaldehyde hydrogenation. *J. Am. Chem. Soc.* 141: 19289–19295.

120 Serna, P. and Gates, B.C. (2011). Zeolite-supported rhodium complexes and clusters: switching catalytic selectivity by controlling structures of essentially molecular species. *J. Am. Chem. Soc.* 133: 4714–4717.

121 Kuo, C.-T., Lu, Y.B., Kovarik, L.B. et al. (2019). Structure sensitivity of acetylene semi-hydrogenation on Pt single atoms and subnanometer clusters. *ACS Catal.* 9: 11030–11041.

122 Kyriakou, G., Boucher, M.B., Jewell, A.D. et al. (2012). Isolated metal atom geometries as a strategy for selective heterogeneous hydrogenations. *Science* 335: 1209–1212.

123 Lu, B.W., Ju, Y.W., Abe, T. et al. (2015). Dispersion and distribution of bimetallic oxides in SBA-15, and their enhanced activity for reverse water gas shift reaction. *Inorg. Chem. Front.* 2: 741–748.

124 Vovchok, D., Zhang, C., Hwang, S. et al. (2020). Deciphering dynamic structural and mechanistic complexity in Cu/CeO_2/ZSM-5 catalysts for the reverse water-gas shift reaction. *ACS Catal.* 10: 10216–10228.

125 Nelson, N.C., Chen, L.X., Meira, D. et al. (2020). In situ dispersion of palladium on TiO_2 during reverse water–gas shift reaction: formation of atomically dispersed palladium. *Angew. Chem. Int. Ed.* 59: 17657–17663.

126 Zhao, Z.Y., Wang, M.Z., Ma, P.J. et al. (2021). Atomically dispersed Pt/CeO_2 catalyst with superior CO selectivity in reverse water gas shift reaction. *Appl. Catal. B: Environ.* 291: 120101.

127 Yang, B., Wang, Y.F., Gao, B. et al. (2023). Size-dependent active site and its catalytic mechanism for CO_2 hydrogenation reactivity and selectivity over Re/TiO_2. *ACS Catal.* 13: 10364–10374.

128 Baddour, F.G., Roberts, E.J., To, A.T. et al. (2020). An exceptionally mild and scalable solution-phase synthesis of molybdenum carbide nanoparticles for thermocatalytic CO_2 hydrogenation. *J. Am. Chem. Soc.* 142: 1010–1019.

129 Xu, J.J., Gong, X.X., Hu, R.R. et al. (2021). Highly active K-promoted Cu/β-Mo_2C catalysts for reverse water gas shift reaction: effect of potassium. *Mol. Catal.* 516: 111954.

130 Sun, G.J., Mottaghi-Tabar, S., Ricardez-Sandoval, L. et al. (2021). Highly active, selective and stable reverse water gas shift catalyst based on high surface area MoC/γ-Al_2O_3 synthesized by reverse microemulsion. *Top. Catal.* 64: 414–430.

131 Liu, X.Y., Pajares, A., Calinao Matienzo, D.J.D. et al. (2020). Preparation and characterization of bulk Mo_XC catalysts and their use in the reverse water-gas shift reaction. *Catal. Today* 356: 384–389.

132 Juneau, M., Vonglis, M., Hartvigsen, J. et al. (2020). Assessing the viability of K-Mo_2C for reverse water–gas shift scale-up: molecular to laboratory to pilot scale. *Energy Environ. Sci.* 13: 2524–2539.

133 Zhang, Q., Pastor-Pérez, L., Jin, W. et al. (2019). Understanding the promoter effect of Cu and Cs over highly effective β-Mo_2C catalysts for the reverse water-gas shift reaction. *Appl. Catal. B: Environ.* 244: 889–898.

134 Zhang, X., Zhu, X.B., Lin, L.L. et al. (2016). Highly dispersed copper over β-Mo_2C as an efficient and stable catalyst for the reverse water gas shift (RWGS) reaction. *ACS Catal.* 7: 912–918.

135 Marquart, W., Raseale, S., Prieto, S. et al. (2021). CO_2 reduction over Mo_2C-based catalysts. *ACS Catal.* 11: 1624–1639.

136 Kang, H., Zhu, L., Li, S.Y. et al. (2023). Generation of oxide surface patches promoting H-spillover in Ru/(TiO_x)MnO catalysts enables CO_2 reduction to CO. *Nat. Catal.* 6: 1062–1072.

137 Conner, W.C. and Falconer, J.L. (2002). Spillover in heterogeneous catalysis. *Chem. Soc. Rev.* 95: 759–788.

138 Im, J., Shin, H., Jang, H. et al. (2014). Maximizing the catalytic function of hydrogen spillover in platinum-encapsulated aluminosilicates with controlled nanostructures. *Nat. Commun.* 5: 3370.

139 Karim, K., Spreafico, C., Kleibert, A. et al. (2017). Catalyst support effects on hydrogen spillover. *Nature* 541: 68–71.

140 Van Dessel, H., Van Minnebruggen, S., Dedapper, J. et al. (2024). Shape-selective zeolites for tandem CO_2 hydrogenation-carbonylation reactions. *Angew. Chem. Int. Ed.* 64: e202418670.

141 APEC Energy Working Group (2010). Assessment of the Capture and Storage Potential of CO_2 Co-produced with Natural Gas in South-East Asia. Final Report, Prepared by UNSW and CO2CRC Technologies for Asia-Pacific Economic Cooperation.

142 Buelens, L.C., Galvita, V.V., Poelman, H. et al. (2016). Super-dry reforming of methane intensifies CO_2 utilization via Le Chatelier's principle. *Science* 354: 449–452.

143 Wang, L.-X., Guan, E., Wang, Z.Q. et al. (2020). Dispersed nickel boosts catalysis by copper in CO_2 hydrogenation. *ACS Catal.* 10: 9261–9270.

144 Zhang, L., Jiang, Y.W., Chen, B.-B. et al. (2020). Exceptional activity for formaldehyde combustion using siliceous Beta zeolite as a catalyst support. *Catal. Today* 339: 174–180.

145 Zhang, J., Shan, Y.L., Zhang, L. et al. (2020). Importance of controllable Al sites in CHA framework by crystallization pathways for NH-SCR reaction. *Appl. Catal. B: Environ.* 277: 119193.

146 Zhu, Q.Y., Wang, Y.Q., Wang, L.X. et al. (2020). Solvent-free crystallization of ZSM-5 zeolite on SiC foam as a monolith catalyst for biofuel upgrading. *Chin. J. Catal.* 41: 1118–1124.

147 Peng, Y.X., Zhang, L., Jiang, Y.W. et al. (2019). Fe-ZSM-5 supported palladium nanoparticles as an efficient catalyst for toluene abatement. *Catal. Today* 332: 195–200.

148 Zhang, J., Chu, Y.Y., Liu, X.L. et al. (2019). Interzeolite transformation from FAU to CHA and MFI zeolites monitored by UV Raman spectroscopy. *Chin. J. Catal.* 40: 1854–1859.

149 Wang, L.X., He, S.X., Wang, L. et al. (2019). Cobalt–nickel catalysts for selective hydrogenation of carbon dioxide into ethanol. *ACS Catal.* 9: 11335–11340.

150 Dai, Z.F., Tang, Y.Q., Sun, Q. et al. (2018). Porous organic polymers constructed from Tröger's base as efficient carbon dioxide adsorbents and heterogeneous catalysts. *ChemCatChem* 10: 1900–1904.

151 Xu, B.-Q., Wei, J.-M., Wang, H.-Y. et al. (2001). Nano-MgO: novel preparation and application as support of Ni catalyst for CO_2 reforming of methane. *Catal. Today* 68: 217–225.

152 Palmer, C., Upham, D.C., Smart, S. et al. (2020). Dry reforming of methane catalysed by molten metal alloys. *Nat. Catal.* 3: 83–89.

153 Song, Y.D., Ozdemir, E., Ramesh, S. et al. (2020). Dry reforming of methane by stable Ni-Mo nanocatalysts on single-crystalline MgO. *Science* 367: 777–781.

6

Zeolites for Low-carbon Hydrocarbon Conversion: Mechanistic Insights

*Chao Wang, Min Hu, Feng Deng, Jun Xu**

National Center for Magnetic Resonance in Wuhan; State Key Laboratory of Magnetic Resonance Spectroscopy and Imaging; and Innovation Academy for Precision Measurement Science and Technology, Chinese Academy of Sciences, Wuhan, Hubei, China

*Corresponding author: xujun@wipm.ac.cn

6.1 Introduction

Zeolites are crystalline aluminosilicates known for their unique porous structures, characterized by regular channels and cavities. Their high surface area, acidic properties, selective adsorption, and high stability make them valuable catalysts in various industrial processes [1–4]. The active sites in zeolite-catalyzed reactions include both Brønsted acid sites (BASs) and Lewis acid sites (LASs). BASs typically form when aluminum atoms in the zeolite framework are balanced by protons, while LASs often arise from dealumination or the introduction of metal heteroatoms such as gallium, germanium, titanium, zinc, and molybdenum [5–10].

The versatility in the synthesis of zeolites allows for the creation of materials with varying pore sizes and shapes, making them suitable for a broad range of applications in the chemical, petrochemical, and environmental sectors. An emerging area of interest is the transformation of low-carbon molecules—such as alcohols, alkanes, and carbon oxides—into value-added products [2, 3, 11]. This conversion is crucial for mitigating greenhouse gas emissions and for the production of renewable fuels and chemicals. Zeolites have proven to be effective catalysts for these reactions due to their ability to stabilize reactive intermediates and promote desired reaction pathways [12, 13].

This chapter will explore the reaction mechanisms underlying the transformation of low-carbon molecules using zeolite catalysts. The focus will be on the activation and conversion of methanol, ethanol, and C_1–C_4 alkanes—processes that are particularly valuable for generating useful products from abundant resources like coal, natural gas, biomass, and waste gases. A thorough understanding of these fundamental processes will enable the development of more sustainable and efficient strategies for converting low-carbon molecules into valuable hydrocarbons.

6.2 Conversion of Low-carbon Alcohols over Zeolites

Low-carbon alcohols, particularly methanol and ethanol, serve as promising feedstocks for hydrocarbon production, offering an alternative to traditional petroleum sources [3]. The widespread availability of these resources, such as coal, natural gas, biomass, and waste gases, significantly increases the attractiveness of converting low-carbon alcohols [14]. The conversion process typically involves several steps, starting with the dehydration of alcohol, followed by secondary reactions that generate hydrocarbons. This sequence not only produces the desired products but also results in the formation of coke species. Despite extensive research on the conversion of low-carbon alcohols to hydrocarbons using zeolite catalysts, understanding the process remains challenging due to the intricate reaction networks involved.

6.2.1 Methanol-to-hydrocarbons Reaction

Since its discovery in the 1970s, the methanol-to-hydrocarbons (MTH) reaction has evolved into a promising pathway for producing light olefins (MTO), gasoline (MTG), and aromatics (MTA) by selecting appropriate zeolites and reaction conditions [15–20]. This process features autocatalytic behavior and consists of four main stages: methanol dehydration, induction period, steady-state period, and deactivation [21]. A central challenge in MTH chemistry is the formation of C—C bonds from C_1 reactants. After four decades of research into this complex reaction, two mechanisms have been proposed to explain the C—C bond formation: the direct C—C coupling mechanism and the indirect hydrocarbon pool mechanism (see Figure 6.1).

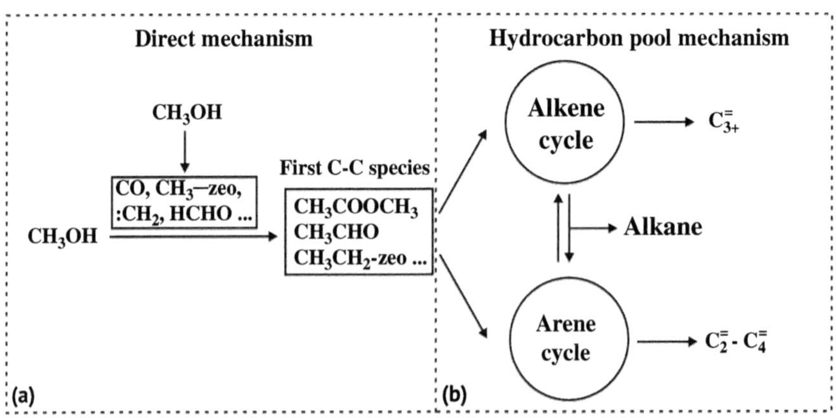

Figure 6.1 Proposed reaction mechanism of MTH: (a) direct and (b) indirect hydrocarbon pool routes.

6.2.1.1 Dehydration of Methanol

The first stage of the MTH reaction is the dehydration of methanol, where methanol molecules interact with the acidic sites of zeolites, establishing an equilibrium that includes methanol, dimethyl ether (DME), and water. Researchers generally agree that BASs are the primary active centers for this dehydration, though LASs may also play a role. Despite extensive studies on methanol dehydration to DME over zeolites, a consensus on the mechanism remains elusive [22–25]. Two main pathways are commonly proposed for the catalytic dehydration of methanol: the associative and dissociative mechanisms (see Figure 6.2) [26, 27].

In the associative mechanism (see Figure 6.2 [left panel]), two methanol molecules undergo a concerted reaction to form DME. Monomers react with gaseous CH_3OH, yielding protonated CH_3OH dimers. These dimers subsequently rotate, facilitating the attainment of a transition state crucial for the kinetically significant steps in DME formation. This pathway enhances molecular interactions, enabling a direct, uninterrupted conversion of methanol into DME. Notably, this process avoids the accumulation of intermediate species on the catalyst surface. This mechanism is preferred under a higher partial pressure and within larger zeolite voids,

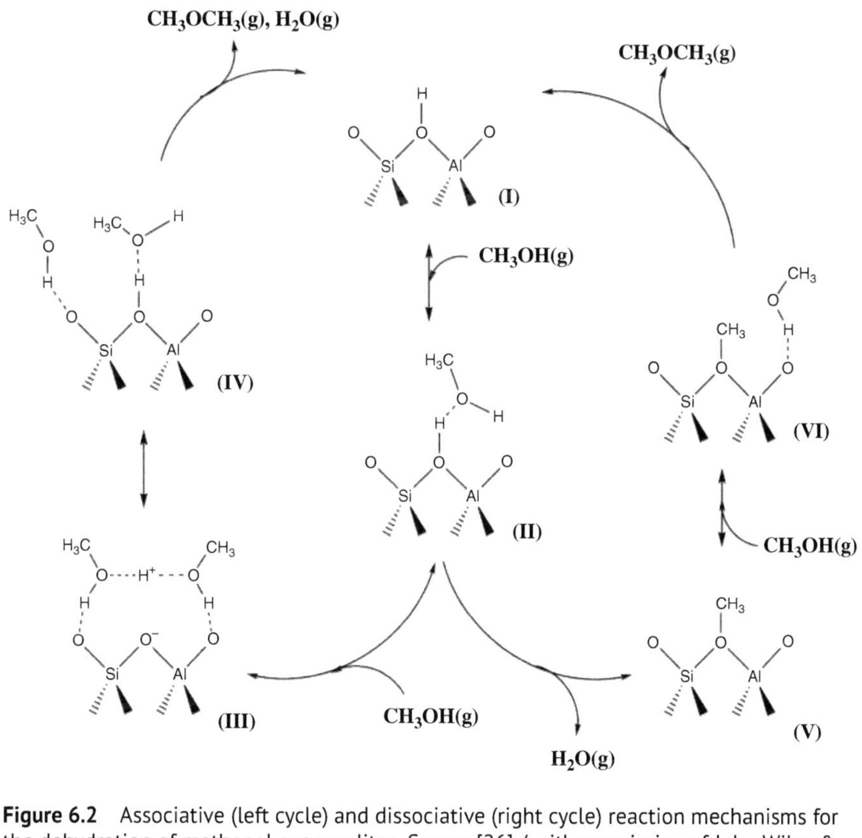

Figure 6.2 Associative (left cycle) and dissociative (right cycle) reaction mechanisms for the dehydration of methanol over zeolites. *Source*: [26] / with permission of John Wiley & Sons.

where the close proximity and favorable orientations of methanol molecules facilitate the efficient elimination of water by-product.

Conversely, the dissociative mechanism (see Figure 6.2 [right panel]) involves the initial dissociation of methanol molecules on the catalyst surface, leading to the formation of surface-bound methoxy species as intermediates. These methoxy intermediates are highly reactive and serve as crucial precursors for the subsequent formation of DME. They undergo further reactions with additional methanol molecules, facilitating the elimination of water and the generation of DME. The dissociative mechanism relies on the formation and reactivity of these intermediate species, which can influence the overall rate and selectivity of the dehydration process. Unlike the associative pathway, the dissociative reaction pathway prefers a lower pressure environment and smaller zeolite pores.

In the dehydration of methanol, the surface methoxy species stands out as a crucial intermediate [28]. This species finds stability within zeolite pores and can be identified using advanced in situ spectroscopic methods, including FTIR [22, 29–33], solid-state NMR [34–44], and inelastic neutron scattering (INS) [45, 46]. Covalently attached to the zeolite framework, the surface methoxy species exhibits carbenium ion-like properties. Among the alkoxide species residing within zeolites, the surface methoxy species is the most stable, maintaining stability even under isolated heating up to 400 °C. However, in the presence of other nucleophilic reagents, it readily engages in reactions with these molecules [38, 39, 47]. For example, it promptly reacts with another methanol molecule to yield DME; in the presence of CO, it generates surface acetyl groups; furthermore, methoxy groups function as methylating agents, involved in methylation reactions with a wide range of molecules, including methanol, olefins, aromatics, and even alkanes [39, 40].

6.2.1.2 Direct C—C Bond Coupling Mechanism

The direct C—C bond coupling mechanism suggests that active C_1 intermediates directly combine to form C—C bonds. Over 20 direct routes have been proposed, each focusing on different active C_1 intermediates such as CO, surface methoxy species, carbene, formaldehyde, trimethoxonium ion, and radicals (see Figure 6.3) [15, 48, 49]. This direct mechanism is considered crucial for the formation of the initial C—C bond in the MTH reaction. However, due to the autocatalytic nature of the process, direct mechanism generally involves high activation energy barriers and slow reaction rates [37, 39, 41, 47, 50–52].

Recent research has highlighted the carbonylation of surface methoxy species as a viable pathway for the initial formation of C—C bonds, supported by both theoretical calculations and experimental evidence [47]. The carbonylation reagent, CO, is generated from the decomposition of methanol during the MTH reaction. As shown in Figure 6.4, surface methoxy species, formed through methanol dehydration, undergo carbonylation with CO, leading to the formation of acetate or methyl acetate. The carbocation-like properties of surface methoxy species enable them to react with nucleophilic species like CO.

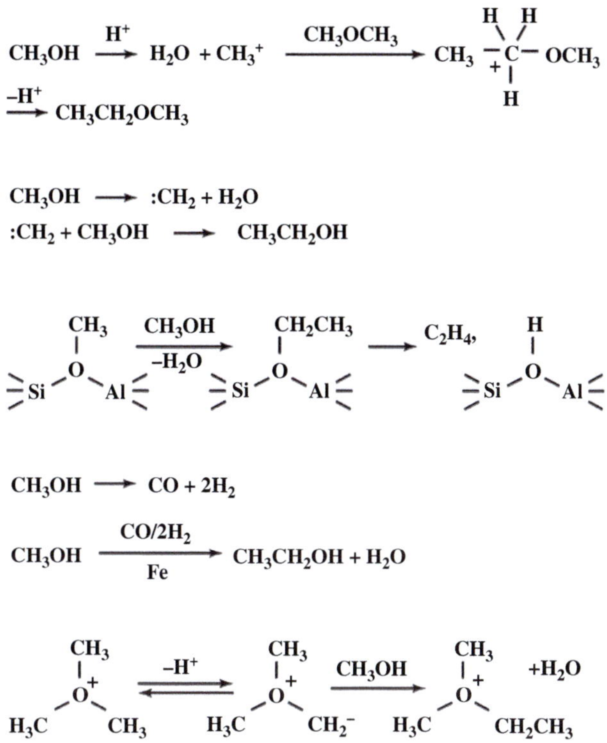

Figure 6.3 Proposed direct C—C bond coupling mechanism responsible for the formation of first C—C bond species in MTH reaction.

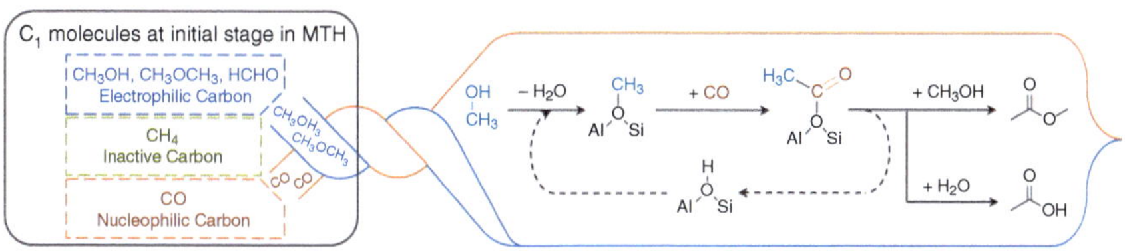

Figure 6.4 Carbonylation of surface methoxy species responsible for the formation of first C—C bond species in MTH reaction. *Source*: [47] / with permission of John Wiley & Sons.

This carbonylation reaction is not unique to the MTH process. It is a critical step in the DME carbonylation over zeolites [53, 54], a significant reaction in coal chemistry. Carbonylation of carbocation/carbocation-like intermediates is also a key organic reaction, known as the Koch reaction, responsible for carbonyl group formation in organic molecules. DFT calculations indicate that the carbonylation of surface methoxy species within ZSM-5 pores occurs readily with a relatively low activation energy of 80 kJ·mol^{-1} [47]. Furthermore, ^{13}C solid-state NMR spectroscopy has confirmed the presence of key intermediates in this reaction process, including surface methoxy species, surface acetate, and methyl acetate [50]. The C—C bond products, methyl acetate or acetic acid, can further transform into light alkenes or aromatics through successive methylation and decarbonylation processes [51], as depicted in Figure 6.5. While the carbonylation reaction provides a plausible explanation for C—C bond formation, its contribution to the MTH reaction is likely marginal due to the limited concentration of CO during the process. The conversion of methanol/DME into CO, while possible, is a significant challenge.

In addition to BASs being crucial for the initial formation of C—C bonds, LASs have also been identified as active centers for C—C bond formation from the C_1 reactants. Copéret and co-workers demonstrated that Lewis acidic Al sites on γ-Al_2O_3 effectively activate DME, leading to the generation of C—C bond species (see Figure 6.6) [55]. This finding suggests that Lewis acidic Al sites within zeolites would also play a significant role in catalyzing the initial C—C bond formation during the MTH process. Xu and co-workers provided experimental evidence for the formation of the first C—C bond catalyzed by Lewis acidic Al sites within zeolites using solid-state NMR spectroscopy and theoretical calculations [42]. Their work confirmed that extra-framework Al centers in ZSM-5 zeolites can facilitate the catalytic transformation of methanol into surface methoxy species, which then react with formaldehyde or methoxymethyl cation to yield C—C bond species (namely acetaldehyde) (see Figure 6.7). The subsequent transfer of hydride from methanol to acetaldehyde results in the formation of ethanol or surface ethoxy species, both of which serve as precursors to ethene. Besides directly forming the first C—C bond species, LASs can enhance the acidity of neighboring BASs, creating a BAS/LAS synergistic effect that potentially accelerates the initial C—C bond formation

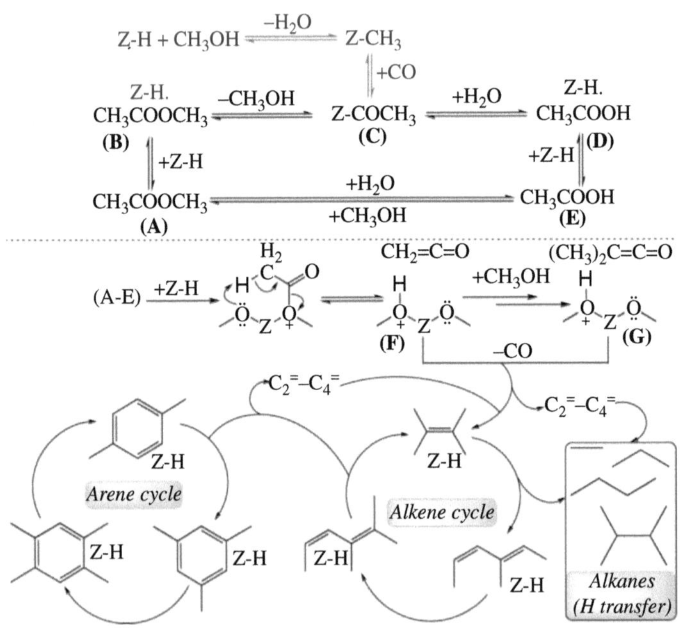

Figure 6.5 Reaction pathways of the methyl-acetate-to-hydrocarbons conversion catalyzed by zeolite. *Z*–H represents acidic zeolites. *Source*: [51] / with permission of John Wiley & Sons.

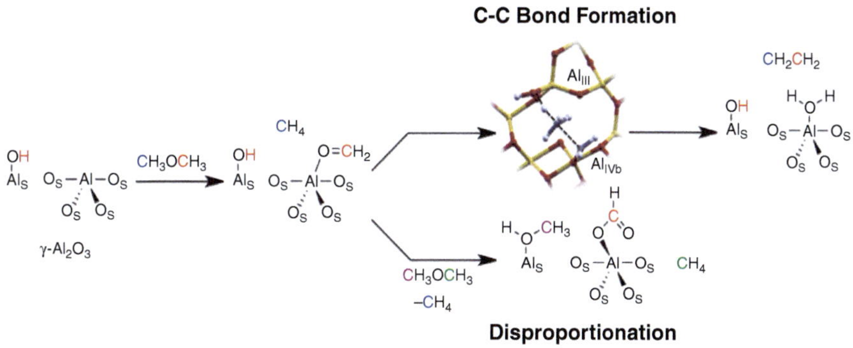

Figure 6.6 Reaction pathway for the formation of C—C bond from DME over Lewis acidic Al sites. *Source*: [55] / with permission of American Chemical Society.

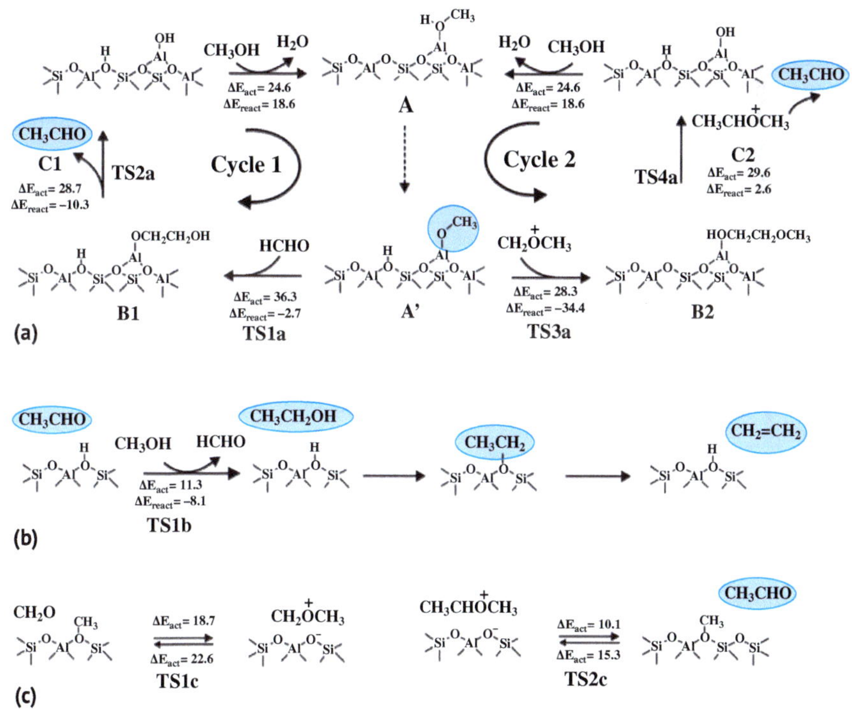

Figure 6.7 Proposed reaction routes for the formation of C—C bond containing species over the ZSM-5 zeolite. The calculated activation energy (ΔE_{act}) and reaction energy (ΔE_{react}) are indicated in kcal·mol^{-1} for the elementary steps. *Source*: [42] / with permission of John Wiley & Sons.

in the MTH reaction [56]. Additionally, LASs within zeolites exhibit a remarkable ability to decompose methanol/DME, generating formaldehyde or CO, both of which can expedite the first C—C bond formation through their respective reaction pathways [57, 58].

6.2.1.3 Hydrocarbon Pool Mechanism

The hydrocarbon pool mechanism, first proposed in the 1980s, is considered the primary pathway for hydrocarbon product formation in the MTH reaction, particularly during the steady-state period [59–61]. This mechanism involves the formation of active intermediates, known as hydrocarbon pool species formed from the initial C—C bond compounds. These species then catalyze the subsequent conversion of methanol into the desired products. The hydrocarbon pool mechanism was initially proposed based on experiments involving the co-feeding of methanol and toluene or higher alcohols [60, 61]. These experiments revealed that C_1 reactants initially combine with hydrocarbon pool species, followed by the release of light olefins as products (see Figure 6.8a). Compared to the direct reaction mechanism, the hydrocarbon pool mechanism involves a lower activation energy. Some researchers believe that trace impurities in the reaction system can even trigger the MTH reaction, bypassing the direct reaction process altogether [62, 63].

Hydrocarbon pools are typically composed of active C—C bond species adsorbed on the BASs of zeolites, including higher alkenes, carbenium ions, and aromatics. The concept of a supramolecular reactive center (SRC) was introduced to explain the hydrocarbon pool mechanism in MTH reactions [65]. This concept proposes that the SRC arises from the interaction between organic hydrocarbon pools and the BASs of

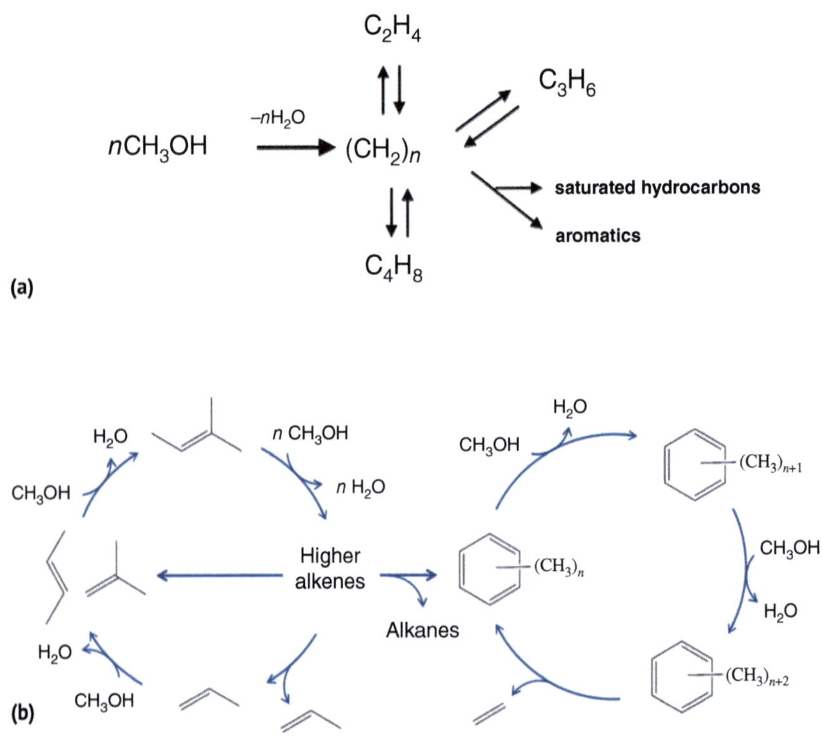

Figure 6.8 (a) Model of hydrocarbon pool mechanism. (b) Dual-cycle concept for the conversion of methanol over zeolites. *Source*: [64] / with permission of The Royal Society of Chemistry.

zeolite, driving the transformation of methanol. Recent studies have provided both experimental and theoretical evidence supporting the existence of SRCs [66, 67]. Hydrocarbon pool species play a dual role: they facilitate the formation of target products while themselves also undergoing transformation into coke species over time. This thermodynamically controlled process leads to the transition of active hydrocarbon pool species into stable but inactive species, ultimately causing the deactivation of the zeolite catalyst.

In the MTH reaction, both higher aliphatic alkenes and aromatics serve as hydrocarbon pools. Alkenes exhibit greater reactivity than methylbenzenes, as observed in the transient $^{12}C/^{13}C$ isotopic switching experiments [68]. This difference in reactivity leads to distinct catalytic behavior for the various hydrocarbon pools within zeolites. A dual-cycle reaction mechanism explains the different catalytic behaviors of alkenes and aromatics in the MTH reaction (see Figure 6.8b) [64]. This mechanism involves two distinct cycles: an alkene-based cycle and an aromatic-based cycle. The alkene-based cycle involves the β-scission of higher alkenes, producing light olefins, particularly propene. The aromatic-based cycle, on the other hand, favors the formation of ethene. Both cycles are interconnected through the aromatization of higher alkenes [64, 68, 69]. The dual-cycle mechanism is now widely accepted as the main route in MTH reactions, applicable to various zeolites beyond ZSM-5. Additionally, this mechanism allows for control over product selectivity by adjusting the distribution of hydrocarbon pool species [70–72]. For instance, selecting zeolites with smaller pores can suppress the aromatic-based cycle, enhancing the selectivity toward C_{3+} alkenes.

Aromatics formation within zeolite pores, particularly in those with larger cages or channels like SSZ-13 and Beta, is a common process in MTH reactions. Aromatics significantly impact the selectivity and efficiency of methanol conversion and influence coke formation within zeolite pores. The pairing and side-chain mechanisms have been proposed to explain the behavior of aromatic-based hydrocarbon pools in MTH reactions [16, 73], as shown in Figure 6.9. In the pairing mechanism, methylbenzenes are methylated multiple times to form methylbenzenium ions, which then undergo ring contraction to produce propyl-substituted cyclopentadienyl cations [74]. The decomposition of these cations generates propene and other by-products, completing the cycle with the formation of tetramethylbenzene. This mechanism involves the exchange of carbon atoms between the aromatic ring and methyl groups. In contrast, the side-chain mechanism involves successive deprotonation and methylation of heptamethylbenzenium ions to form ethyl- or propyl-substituted derivatives, leading to light olefins without altering the carbon atoms in the aromatic ring [60, 61]. The preference for either mechanism depends on the zeolite structure and reaction conditions. Zeolites with larger channels or cages, such as Y and SAPO-34, often favor the side-chain mechanism due to their ability to accommodate critical transition states. In contrast, smaller-pore zeolites like ZSM-5 tend to favor the pairing mechanism because of the smaller transition states involved. Additionally, the shape selectivity of zeolites affects the reactivity of hydrocarbon pool species. For example, hexamethylbenzene reacts more readily over

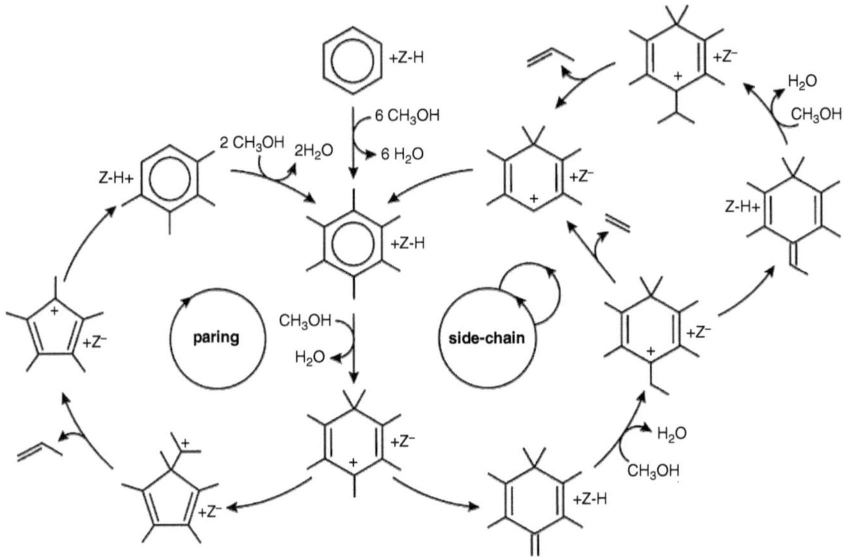

Figure 6.9 Pairing and side-chain mechanism for the formation of light olefins in aromatic-based hydrocarbon pool route on zeolites [16, 76].

SSZ-13, which has CHA cages, compared to ZSM-5, due to easier gem-methylation in SSZ-13 [75].

Carbenium ions play a crucial role in MTH reactions, acting as key intermediates in the hydrocarbon pool formation and operation [77]. Cyclopentenyl cations and methylbenzenium ions are particularly important for transforming methanol into hydrocarbons. These carbenium ions, with their planar geometry and sp^2-hybridized positively charged carbon atoms, are highly reactive intermediates in organic reactions [78]. The confined space of zeolite pores creates a unique environment that stabilizes carbenium ions, facilitating their prevalence in zeolite catalysis [79, 80]. In the MTH reaction, carbenium ions are generated through methylation, protonation, and dehydrogenation processes. They are essential for constructing C—C bonds and driving various sub-reactions, such as hydride transfer, isomerization, cyclization, oligomerization, methylation, and cracking [15, 16, 49, 81]. These reactions contribute to the formation of a diverse range of hydrocarbon pool species, including by-products like coke. The complex interplay between carbenium ions and these reactions determines the final products in MTH reactions.

The study of carbenium ions within zeolite pores is crucial for understanding MTH reactions and optimizing catalytic processes. Techniques such as solid-state NMR, FTIR, and UV–vis spectroscopy are used to analyze these intermediates [77, 82–93]. Solid-state NMR is particularly effective for identifying the structure and behavior of carbenium ions within zeolite pores. As shown in Figure 6.10, solid-state NMR has confirmed the presence of pentamethylbenzenium ions and various substituted cyclopentenyl cations in ZSM-5 pores during MTH reactions [85, 86]. FTIR and UV–vis spectroscopy are typically employed to monitor the time evolution of carbenium ions under reaction conditions [91, 94–96]. By combining

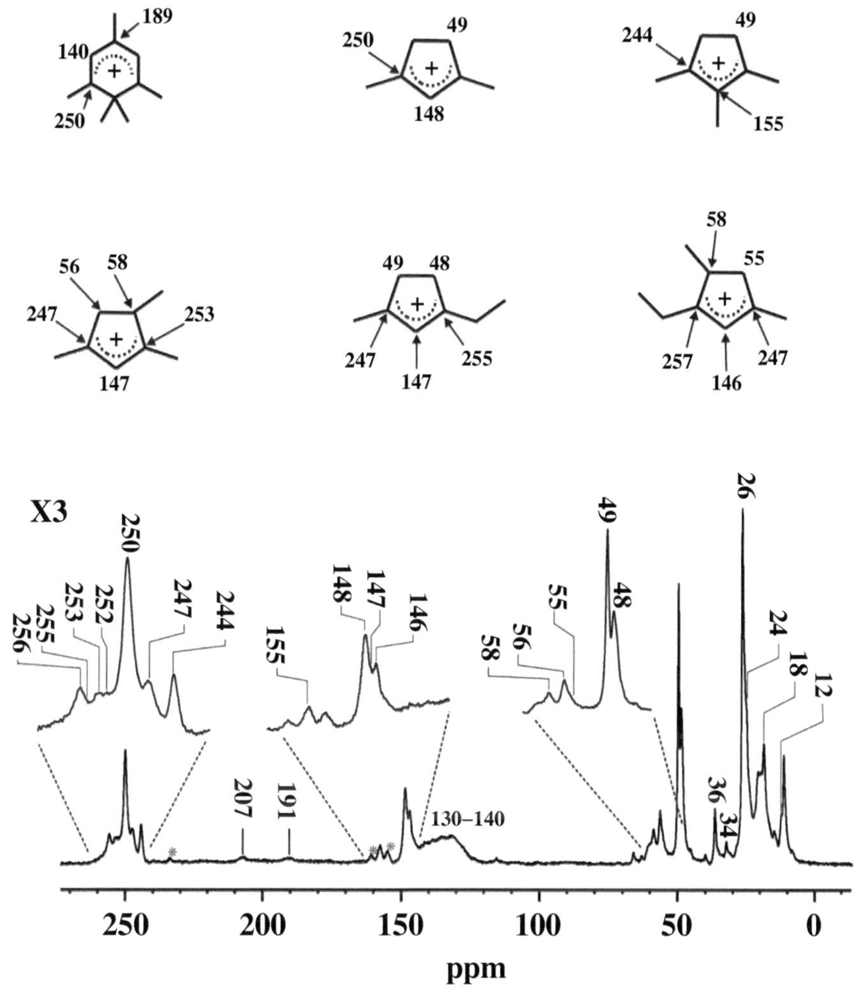

Figure 6.10 Solid-state ^{13}C MAS NMR spectra of the trapped hydrocarbon pools produced from methanol conversion over ZSM-5. The confirmed cyclopentenyl cations are shown in the top [85, 86].

these techniques, researchers have identified cyclic carbenium ions (see Table 6.1), including cyclopentenyl, methylbenzenium, and naphthalenium ions, in MTH reactions over zeolites. Understanding the properties of these carbenium ions can guide the development of high-performance catalysts.

6.2.2 Ethanol Conversion over Zeolites

Bioethanol is a promising renewable feedstock for producing various chemicals. Both BASs and grafted metals on zeolites serve as active sites for transforming ethanol into the desired products. This subsection outlines the mechanisms of acid-catalyzed and metal-catalyzed ethanol transformations over zeolites [103].

Table 6.1 Observed carbenium ions in the MTH reaction over zeolites.

Carbenium ions		**Spectroscopic technique(s)**	**Source**
Five-membered	 R:-CH_3, -CH_2CH_3, -$CH(CH_3)CH_3$	Solid-state NMR; UV–vis; IR	[77, 82–93]
Six-membered	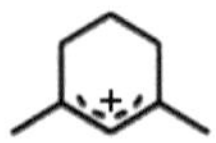	Solid-state NMR; UV–vis; IR	[89, 91, 97]
	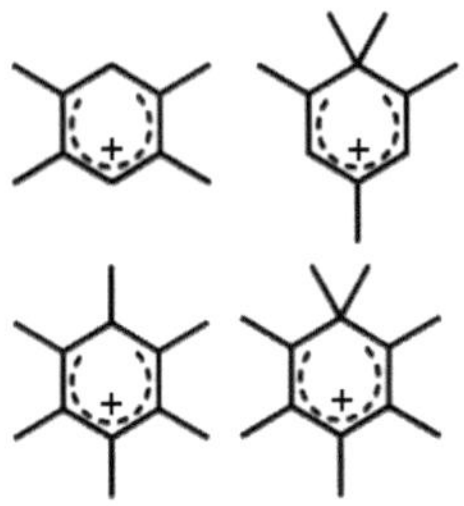	Solid-state NMR; IR	[77, 88, 89, 91, 95, 98–102]
Polycyclic	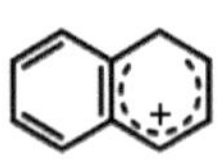	Solid-state NMR	[89, 99, 100]

6.2.2.1 Ethanol to Hydrocarbons over Acidic Zeolite

Since the commercialization of ethanol to hydrocarbons (ETH) in 1913, the dehydration of ethanol to produce ethylene and higher hydrocarbons has attracted significant attention from both academia and industry [104, 105]. Acid-catalyzed ETH over zeolites can generate ethene, butenes, and higher liquid hydrocarbons [106–108]. This process is similar to the MTH reaction discussed earlier. As shown in Figure 6.11, the ETH reaction involves two main stages: the dehydration of ethanol to ethene and the subsequent conversion of ethene to higher hydrocarbons [109].

The initial dehydration of ethanol to ethene can occur via either a stepwise or a concerted route [110]. The stepwise route involves surface ethoxy species as intermediates, while the concerted route involves the simultaneous dehydration of multiple ethanol molecules. In the second stage, ethene undergoes dimerization and polymerization to form butenes and other higher olefins [109]. Additional reactions, such as the cracking of higher olefins into light olefins and the aromatization of alkenes into aromatics, also occur. These olefins and aromatics can be further ethylated by ethene, leading to the formation of light olefins [111], which resembles the hydrocarbon pool mechanism in the MTH reaction. In acid-catalyzed ETH reactions, carbocation or carbocation-like species, including surface ethoxy species, triethoxonium ions, various cyclopentenyl cations, and methylbenzenium ions, are key intermediates. These intermediates have been identified using in situ spectroscopic techniques such as FTIR, UV–vis, and solid-state NMR (see Table 6.2).

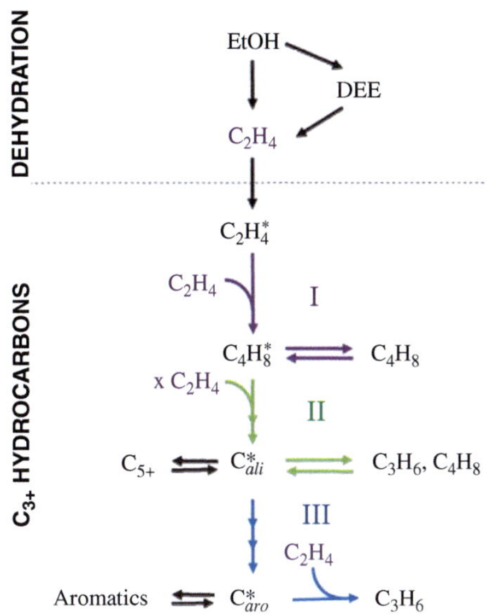

Figure 6.11 Proposed reaction route for the ethanol transformation over zeolites. The asterisk indicates the surface-adsorbed species: C^*_{ali}, aliphatic surface species; C^*_{aro}, aromatic surface species. *Source*: [109] / with permission of John Wiley & Sons.

Table 6.2 Identified intermediates in ETH reaction over zeolites.

Species	Spectroscopic technique(s)	Source
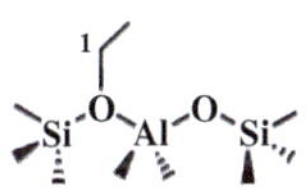	Solid-state NMR; FTIR	[39, 42, 43, 112, 113]
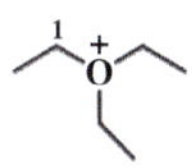	Solid-state NMR	[43, 114]
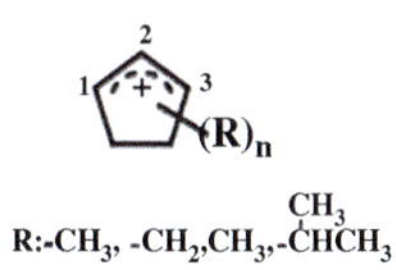	Solid-state NMR; UV–vis	[111]
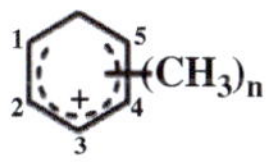	Solid-state NMR; UV–vis	[111]

6.2.2.2 Ethanol Conversion to 1,3-Butadiene over Metal-modified Zeolite

In addition to acidic zeolites, metal-modified zeolites are used to convert ethanol into target products. The production of 1,3-butadiene (1,3-BD) from ethanol is particularly promising due to its application as a monomer in styrene–butadiene rubber [115]. Metal-modified zeolites are effective catalysts for this transformation. By incorporating specific metals into the zeolite framework or as supported species,

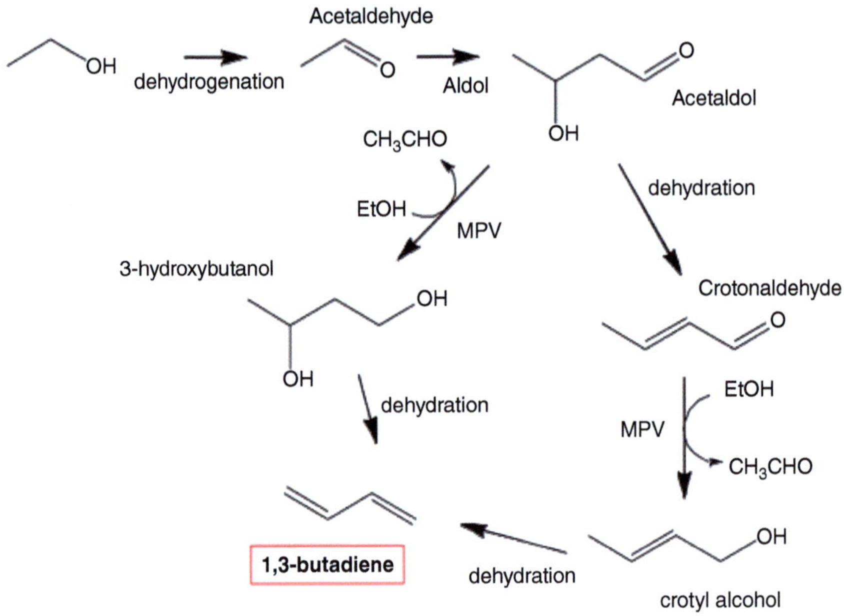

Figure 6.12 Generally accepted mechanism for the production of 1,3-BD from ethanol over metal-modified zeolites. *Source*: [120] / with permission of The Royal Society of Chemistry.

the catalytic activity and selectivity for 1,3-BD production can be enhanced. Metals such as Zn, cerium (Ce), yttrium (Y), and zirconium (Zr) have been studied for their ability to modify zeolites and promote the ethanol-to-1,3-BD reaction [116–119]. These metals influence key reaction steps like dehydrogenation, aldol condensation, dehydration, hydrogenation, and Meerwein–Ponndorf–Verley (MPV) reactions, thereby affecting the overall mechanism and product selectivity.

The reaction mechanism of ethanol-to-1,3-BD over metal-modified zeolites involves several steps (see Figure 6.12) [120]. Initially, ethanol undergoes dehydrogenation, removing hydrogen atoms from the molecule. This critical step converts ethanol into acetaldehyde, a reactive intermediate for further transformations. Subsequently, acetaldehyde participates in an aldol condensation reaction. This process involves the coupling of two acetaldehyde molecules, creating a new carbon–carbon bond and forming 3-hydroxybutanal.

In one pathway, 3-hydroxybutanal undergoes an MPV reaction, where the carbonyl group (C=O) is reduced to an alcohol (C—OH), often using ethanol as the hydrogen source. This yields 3-hydroxybutan-1-ol, which dehydrates to form 1,3-BD. In another pathway, 3-hydroxybutanal can directly dehydrate to form crotonaldehyde. Crotonaldehyde then undergoes an MPV reaction to produce crotyl alcohol, which subsequently dehydrates to yield 1,3-BD. These reactions can occur simultaneously or sequentially, depending on the catalyst and reaction conditions [120]. Optimizing these processes through catalyst design and reaction engineering is essential for maximizing the yield and selectivity of 1,3-BD.

6.3 Transformation of Low-carbon Alkanes over Zeolites

Zeolites are increasingly employed for converting low-carbon alkanes—such as methane, ethane, and propane—into higher-value chemicals and fuels [121]. This transformation is crucial for advancing sustainable energy solutions and improving the efficiency of the chemical industry. However, the process is challenging due to the strong stability of the C—C and C—H bonds in these alkanes. Zeolites offer a promising approach by selectively adsorbing and confining reactants, thus promoting beneficial interactions between the reactants and catalytic sites [122–125].

Despite their potential, traditional zeolites often fall short in achieving the necessary activity and selectivity for low-carbon alkane conversions. This limitation has led to research focused on metal-modified zeolites. Incorporating transition metals, noble metals, or metal oxides into the zeolite framework or pores can significantly improve catalytic performance [58, 126–128]. These metal modifications introduce additional active sites for bond activation and alter the local electronic environment, which enhances the reactivity and selectivity of the catalytic process.

6.3.1 Conversion of Methane over Metal-modified Zeolites

Methane, the primary component of natural gas, is an abundant and cost-effective feedstock for the synthesis of chemicals and fuels [129, 130]. However, its high stability and low reactivity, characterized by the demanding energy required to break the C—H bond, pose significant challenges in transforming methane into more valuable products [131, 132].

The conversion of methane can follow two main pathways: the indirect route and the direct route (see Figure 6.13). The indirect route, a well-established industrial process, involves multiple steps. Methane first reacts with gases like water, carbon dioxide, or oxygen at high temperatures, producing syngas (a mixture of hydrogen

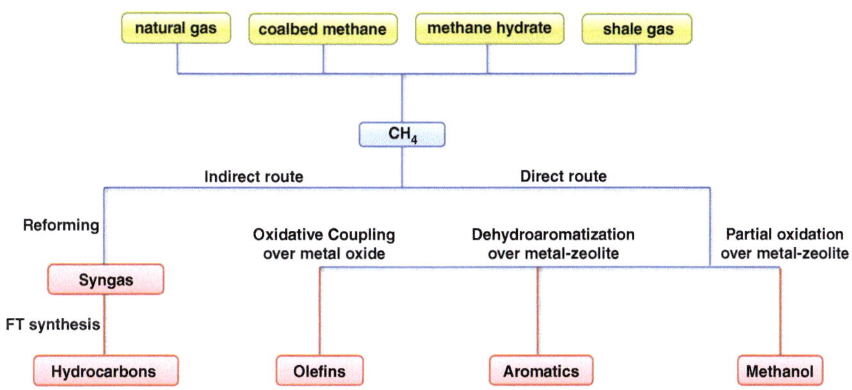

Figure 6.13 Overview of different routes for the direct and indirect conversions of methane to value-added products. *Source*: [3] / with permission of John Wiley & Sons.

and carbon monoxide). Syngas then serves as a versatile feedstock for the Fischer–Tropsch synthesis, enabling the production of liquid fuels and various chemical intermediates [133, 134]. Additionally, syngas can be directly converted to methanol, which can be further transformed into a diverse range of chemicals through processes like MTH conversion [14, 81]. While effective, the indirect route is energy intensive. Consequently, the direct conversion of methane presents a compelling challenge, attracting substantial interest from both industry and academia to enhance efficiency and minimize energy consumption [135].

Metal-containing zeolites, incorporating elements like ruthenium, rhodium, palladium, platinum, gold, gallium, copper, and Zn, are commonly employed as catalysts for the direct transformation of methane into valuable products. Direct methane activation over metal-modified zeolites can occur through various pathways, notably oxidative activation and non-oxidative methane activation. Oxidative activation involves the use of oxidants such as oxygen, nitric oxide, or hydrogen peroxide, leading to the generation of oxygenates, ethane, or higher hydrocarbons [136–138]. In contrast, non-oxidative methane activation involves the direct conversion of methane into hydrocarbon intermediates without the involvement of oxygen [139–143]. The hydrocarbon pool mechanism plays a crucial role in this process, where hydrocarbon intermediates are formed on the catalyst surface, leading to the efficient production of desired products [142, 143].

6.3.1.1 Partial Oxidation of Methane

The partial oxidation of methane (POM) offers a pathway for the direct transformation of methane into oxygenated hydrocarbons under mild conditions. Among various strategies, the POM to methanol, facilitated by metal-containing zeolites and in the presence of oxidizing agents like oxygen, nitrous oxide, or hydrogen peroxide, is considered the most viable approach to enhance methanol production [128, 137, 138].

Nature provides inspiration for this process, as methane-oxidizing bacteria utilize methane monooxygenases (MMOs) within their systems to convert methane into methanol at ambient temperatures [144, 145]. This biological enzyme-catalyzed process is remarkably intricate and inspiring. Drawing inspiration from this natural phenomenon, two active metal centers found in MMOs—Fe and Cu—have emerged as key elements for designing catalysts for methane partial oxidation [138, 139, 146–148]. Research has demonstrated the effectiveness of Fe/ZSM-5 zeolites in catalyzing the oxidation of methane to methanol using nitrous oxide as an oxidant. This oxidation process involves the decomposition of nitrous oxide on the binuclear Fe center of the zeolite, generating α-oxygen species that activate methane [149–152]. Additionally, Cu-modified zeolites have been reported as exceptional catalysts in the POM process, capable of being activated by either oxygen or nitrous oxide to directly and selectively convert methane into methanol [138, 139]. The $\left[Cu_2(\mu\text{-}O)_2\right]^{2+}$ species present on the zeolite surface are identified as the active centers responsible for this transformation [153].

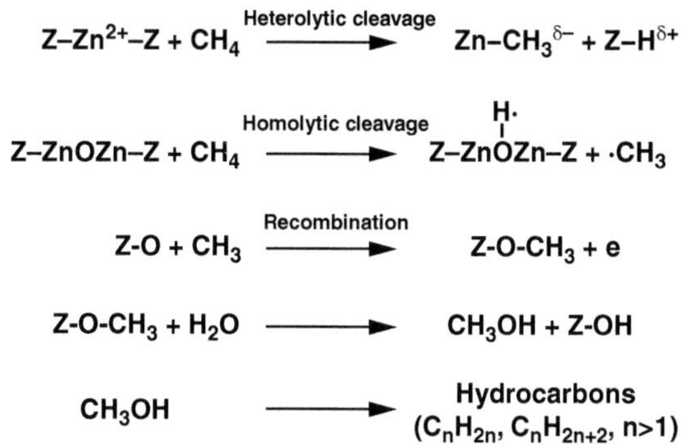

Figure 6.14 Proposed methane activation and consequent conversion pathway on Zn/ZSM-5 zeolites. *Source*: [154] / with permission of The Royal Society of Chemistry.

Besides utilizing strong oxidants, water can also serve as the oxidant to partially oxidize methane into methanol. By using Cu-containing Na–MOR zeolite, a continuous methanol production from methane has been achieved [138]. This production route/cycle consists of three consecutive steps: activation of the zeolite, methane activation over the catalyst, and desorption of methanol by water vapor. After the third reaction cycle, the selectivity of methanol can exceed 96%. In situ XANES spectra show that the activation of methane is accompanied by a decrease in Cu^{2+} and an increase in Cu^{+}, suggesting that Cu^{2+} is the active site for the oxidation of methane. The active Cu^{2+} sites can be regenerated from the oxidation of Cu^{+} by introducing water. This process highlights the potential of water as an oxidant in methane conversion.

During the partial oxidation of methane, various reactive C_1 intermediates form on the surface of metal–zeolite catalysts. For instance, when methane is activated over Zn/ZSM-5, Zn–methyl complexes can arise through a heterolytic cleavage process, whereas methyl radicals can be generated via homolytic cleavage (see Figure 6.14) [154]. The interaction of methyl radicals with zeolite catalysts yields surface methoxy species. Subsequently, the direct hydration of these methoxy species facilitates the synthesis of methanol. Recent studies have revealed that methyl radicals, generated during methane activation over Au-ZSM-5 in an aqueous/solid phase, can engage in additional reactions with carbon monoxide, a by-product of methane overoxidation, leading to the formation of carbon–carbon bonded oxygenates, including acetic acid [128, 155].

The cooperative transformation of methane with CO or CO_2 to produce oxygenates has garnered significant attention [155–157]. In the co-reaction of methane and CO catalyzed by Zn/ZSM-5 (see Figure 6.15, pathway 1), the formation of Zn–methyl intermediates from methane activation, possessing organometallic characteristics, is readily observed. These intermediates subsequently react with CO_2, leading to the formation of acetate species. Proton abstraction by zeolites facilitates the conversion

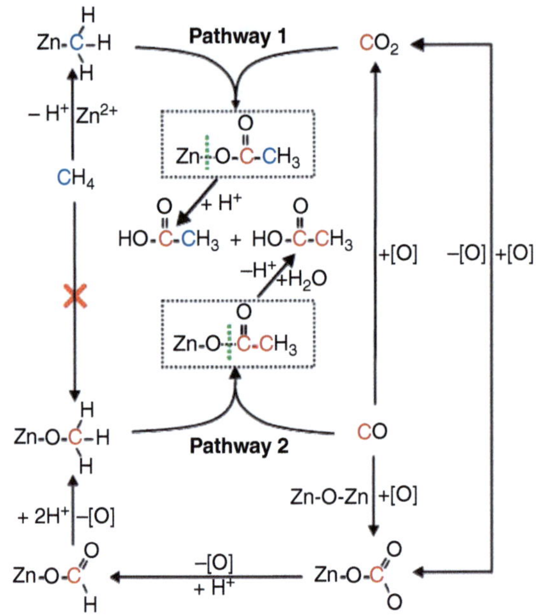

Figure 6.15 Proposed reaction pathways for the formation of acetic acid from methane and carbon monoxide on Zn–ZSM-5 catalyst. Zn^{2+} stands for isolated Zn^{2+} ions in the Zn/ZSM-5 catalyst, Zn—O—Zn clusters are the probable oxidation centers, and H^+ are protons from either the bridging hydroxy groups of the Zn–ZSM-5 catalyst or the heterolytic dissociation of methane. *Source*: [155] / with permission of John Wiley & Sons.

of these acetate species into acetic acid [155, 157]. Notably, the CO_2 involved in this reaction can originate from the oxidation of CO by active oxygen species of metal sites. Alternatively, in the same reaction system, CO can undergo hydrogenation to generate surface methoxy species, facilitated by critical intermediates such as carbonate and formate (see Figure 6.15, pathway 2). These surface methoxy species then interact with CO, resulting in the formation of surface acetate species that ultimately convert into acetic acid.

Despite significant achievements in the selective oxidation of methane to oxygen-containing compounds under mild conditions, the POM process continues to face challenges, including low methane conversion rates, excessive oxidation of desired products, and the complexity of the catalyst systems employed. Overcoming these difficulties remains a key area of focus for future developments in this field.

6.3.1.2 Non-oxidative Methane Dehydroaromatization

The non-oxidative methane dehydroaromatization (MDA) process represents a key direct approach for methane activation over zeolite-based catalysts. Over the years, various reaction mechanisms have been proposed to elucidate the intricacies of the MDA process (see Figure 6.16) [140, 158]. One early mechanism, gaining prominence in the 1990s, was the radical mechanism observed in molybdenum (Mo)-modified zeolites. In this pathway, methane undergoes homolytic cleavage facilitated by the synergy between MoO_x and BASs, generating free methyl radicals.

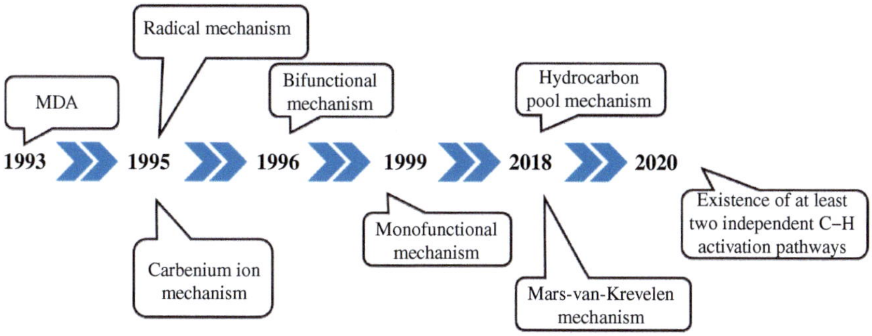

Figure 6.16 Timeline of reaction mechanistic studies of the MDA reaction over metal-modified zeolites. *Source*: [161] / with permission of China National Knowledge Infrastructure.

These radicals then dimerize and dehydrogenate to form ethylene, which subsequently aromatizes to benzene on the BASs [159]. Around the same time, the carbocation mechanism was also proposed. This mechanism suggests that MoO_3 within the zeolite pore channels polarizes the first C—H bond of methane, leading to the formation of a polarized $CH_3^{\delta+}—H^{\delta-}$ [160]. This polarized methane then interacts with the BASs, yielding a carbocation intermediate that reacts with Mo to form a carbene-like intermediate ($CH_2{=}MoO_3$). The coupling and dehydrogenation of these carbene-like intermediates produce ethylene, which ultimately aromatizes to benzene on the BASs of zeolite.

More recently, the bifunctional reaction mechanism has gained significant attention. In this mechanism, the active Mo species facilitate the initial methane C—H bond activation and C—C bond coupling, while the BASs are responsible for the aromatization of C_2 intermediates (see Figure 6.17) [162]. The Mo sites, BASs, and pore structure of Mo/ZSM-5 zeolites play pivotal roles in this process, with ethylene serving as both a primary product and an intermediate, whose selectivity increases as the catalyst deactivates.

Another proposed mechanism is the hydrocarbon pool mechanism, where confined aromatic hydrocarbons and their derivatives act as the hydrocarbon pool, engaging in secondary reactions with primary products $C_x H_y$, derived from methane activation [126, 127, 143], to yield aromatics like benzene (see Figure 6.17). In situ spectroscopic and high-resolution microscopy studies have revealed the presence of polycyclic aromatic hydrocarbons (PAHs) with radical properties within the catalyst pores during the induction period [140], suggesting their potential contribution to the benzene product.

Recent advancements in solid-state NMR spectroscopy have provided valuable insights into the MDA reaction mechanism. Specifically, the application of ^{1}H-$\{^{95}Mo\}$ double-resonance NMR has enabled the identification of the bifunctional sites (BAS–Mo) [127]. However, the Mo species in the bifunctional sites exhibit dynamic behavior as the reaction progresses [163–165], which presents a significant challenge for confirmation. To address this issue, 2D ^{1}H-$\{^{95}Mo\}$ correlation NMR spectroscopy has proven to be a powerful tool, allowing for detailed characterization of

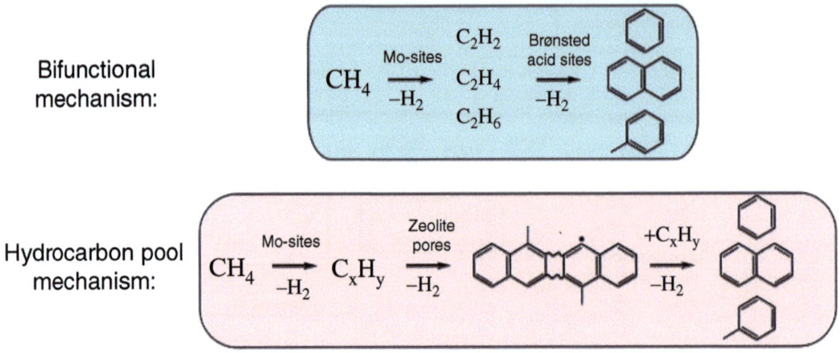

Figure 6.17 Two main proposed reaction mechanisms for the MDA reaction on Mo/ZSM-5. *Source*: [140] / with permission of John Wiley & Sons/CC BY 4.0.

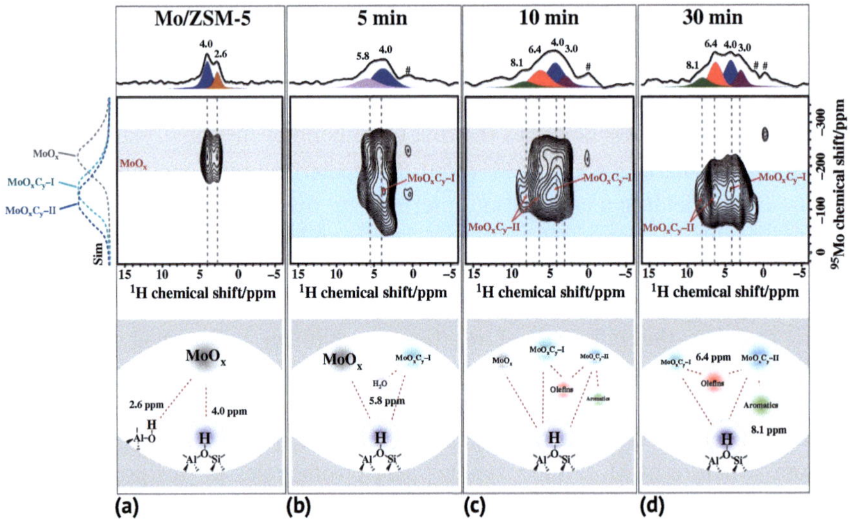

Figure 6.18 The 2D ^{1}H-$\{^{95}$Mo$\}$ PT D-HMQC NMR spectra of fresh ^{95}Mo/ZSM-5 (a) and ^{95}Mo/ZSM-5 samples reacted for (b) 5, (c) 10, and (d) 30min of MDA reaction. "#" denotes T_1 noise and the ^{1}H chemical shifts of specific species are displayed in the diagram (bottom). *Source*: [126] / with permission of John Wiley & Sons.

the local environment surrounding Mo sites and their reactivity within the MDA reaction [126]. As shown in Figure 6.18a, for the fresh Mo/ZSM-5 catalyst, the proximity of MoO_x species to BASs suggests the formation of these bifunctional sites, which serve as precursors for the initiation of the MDA reaction. As the reaction progresses (see Figure 6.18b–d), two types of Mo species, designated as MoO_xC_y-I and MoO_xC_y-II, emerge on catalyst. Both of these species maintain close proximity to BASs, forming distinct bifunctional sites. Notably, the MoO_xC_y-II species arises from the further carburization of MoO_xC_y-I. In contrast to MoO_xC_y-I, the MoO_xC_y-II species exhibits higher activity in the MDA reaction, facilitating the formation of hydrocarbon pool centers and ultimately contributing to the generation of reaction products.

Despite these research efforts and mechanistic insights, the complete reaction pathway of MDA remains elusive. The ambiguity is primarily attributed to an insufficient understanding of the precise structure of the active sites and confined hydrocarbon pool intermediates, coupled with ongoing controversies regarding the nature of the reaction intermediates. Deepening the understanding of the MDA reaction mechanism continues to be a focus for future research in this field.

6.3.2 Transformation of C_2–C_4 Alkanes over Zeolites

The transformation of light alkanes, such as ethane and propane (C_2–C_4), into more valuable products over zeolite catalysts is another area of substantial interest. These alkanes can be converted into olefins, such as ethylene and propylene, which are important building blocks for the production of plastics and chemicals [101, 166–168]

The transformation of C_2–C_4 alkanes over zeolites can occur through several pathways, including dehydrogenation, cracking, and oligomerization [121]. Dehydrogenation involves the removal of hydrogen atoms from the alkane molecule, yielding olefins. Cracking involves the breaking of carbon–carbon bonds, resulting in the formation of smaller olefins and paraffins. Oligomerization involves the coupling of two or more alkane molecules to form higher olefins. The selectivity toward specific products is highly dependent on the reaction conditions and catalyst properties. Zeolites with different acidities and pore sizes can exhibit distinct selectivities, making them valuable catalysts for the transformation of light alkanes.

Both the BASs and the LASs within zeolites are active for the activation of C_{2+} alkanes. Metal alkyls formed on the metal LASs, exhibiting carbonation properties, are commonly proposed as intermediates in the activation of C_{2+} alkanes, which can then undergo β-hydride elimination to form alkenes [169, 170]. On the BASs, carbonium ions and carbenium ions are considered as the transition states or intermediates in the activation of C_{2+} alkanes via monomolecular protolytic and bimolecular carbenium ion pathways (see Figure 6.19). At high temperatures or sufficiently low partial pressures of C_{2+} alkanes, the monomolecular mechanism dominates [125, 171, 172], where the alkane directly reacts with the BASs (see Figure 6.20a). The monomolecular mechanism is characterized by the adsorption of gaseous alkane on active site, the protonation of adsorbed alkane to carbonium ion transient state, and the subsequent dehydrogenation or cracking (see Figure 6.20a) [171, 173]. The host–guest interaction between the alkane and the zeolite framework plays a crucial role in the monomolecular reaction route [122, 123, 172, 174–177], determined by the size of the zeolite pores and the alkane molecules. This host–guest interaction affects the measured/apparent activation energy barrier (E_{meas}), which can be expressed as the sum of the intrinsic activation energy (E_{int}) and the adsorption enthalpy ($-\Delta H_{ads}$) (see Figure 6.20b) based on the following equation:

$$E_{meas} = E_{int} + \Delta H_{ads}$$

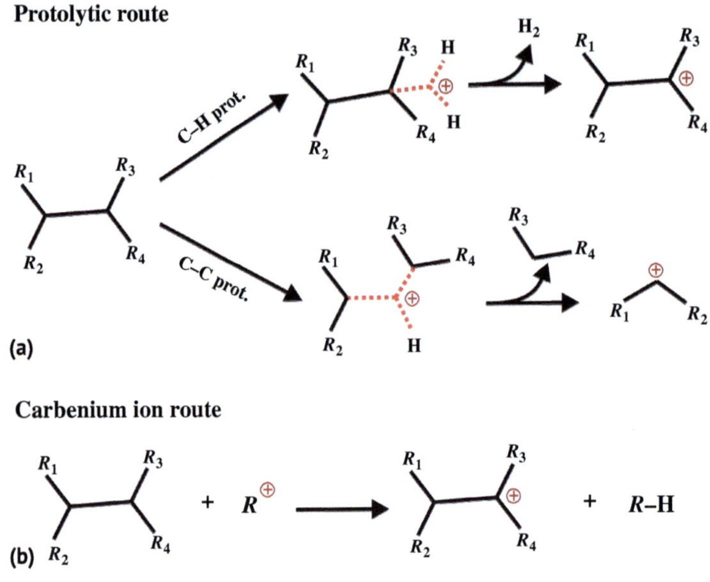

Figure 6.19 Schematic of (a) protolytic route and (b) carbenium ion route for the activation of alkanes over zeolites. *R* denotes alkyl group. The monomolecular alkane activation by protonation (prot) over BASs is responsible for the protolytic reaction of alkane. The activation of alkane by a carbenium ion R⊕ is responsible for the carbenium ion route.

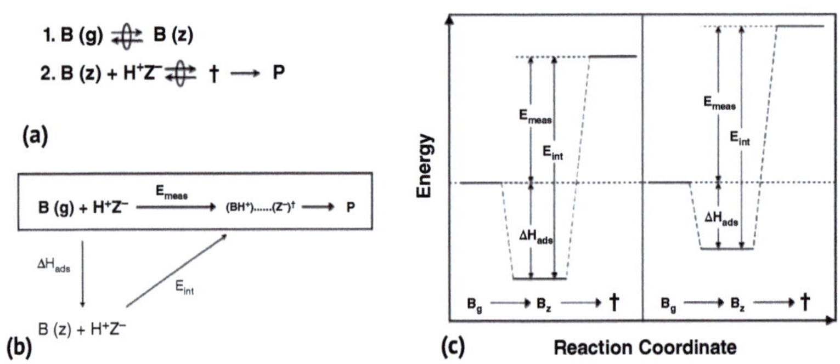

Figure 6.20 (a) Reaction sequence for monomolecular alkane activation on BASs located within zeolite channels (H^+Z^-). *B* denotes a reacted alkane molecule and *P* denotes products. The g in the bracket means gas state and z means adsorption state.
(b) Thermochemical cycle for monomolecular reactions of alkanes in zeolite channels.
(c) Relations between measured activation energies (E_{meas}), adsorption enthalpies (ΔH_{ads}), and intrinsic activation energies (E_{int}) are shown schematically for two different channel environments. *Source*: [123] / with permission of American Chemical Society.

Accordingly, a stronger adsorption of alkanes leads to a lower E_{meas} for the activation of gaseous reactants (see Figure 6.20c). In the monomolecular *n*-hexane cracking over Y, MOR, and ZSM-5, it was found that the zeolite pore size largely affects the measured E_{meas} [122, 178]. In a zeolite void with tighter confinement, the transition state can be stabilized to a greater extent, leading to a lower E_{meas} for the *n*-hexane cracking.

When carbenium ions are generated through the monomolecular mechanism, the bimolecular carbenium ion mechanism can become advantageous due to its lower activation energy compared to the monomolecular pathway (see Figure 6.19b) [179]. This can lead to an autocatalytic process, where the initial formation of carbenium ions significantly accelerates the subsequent transformation of alkanes. Recent in situ ^{13}C MAS NMR studies on propane dehydroaromatization over Ga/ZSM-5 have experimentally confirmed the importance of the bimolecular carbenium ion mechanism [180]. Notably, cyclopentenyl cations (observed at chemical shifts of 250, 155, and 147 ppm) form within the zeolite pores (see Figure 6.21a), and exhibit high reactivity, facilitating propane activation. This behavior implies an induction period

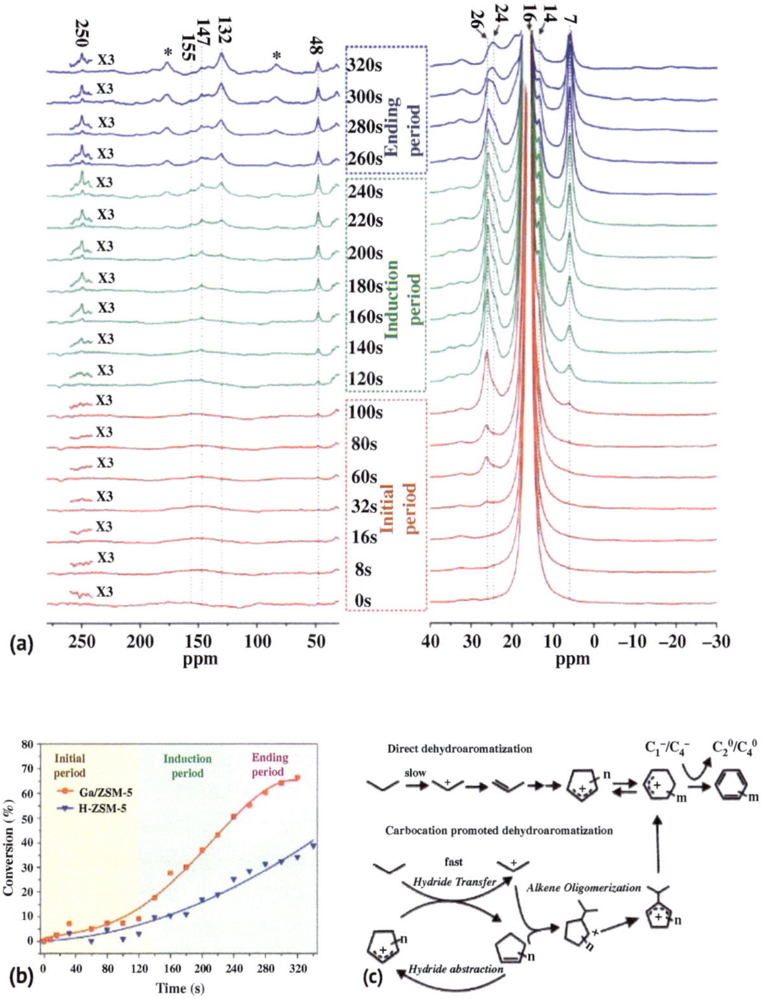

Figure 6.21 (a) The ^{13}C CP/MAS NMR spectra of the adsorbed hydrocarbon species formed in propane aromatization reaction on Ga/ZSM-5 at 350 °C for 0–320 s. (b) Propane conversion with reaction time over H-ZSM-5 and Ga/ZSM-5. (c) Proposed propane direct and cyclopentenyl cations-mediated aromatization routes. *Source*: [180] / with permission of John Wiley & Sons.

similar to that observed in the MTH reaction (see Figure 6.21b, c). Specifically, the cyclopentenyl cations, which form from secondary reactions involving initial products resulting from direct propane dehydrogenation, serve as catalysts for subsequent dehydroaromatization via hydride transfer reactions. In comparison to direct dehydroaromatization, the pathway facilitated by carbenium ions proves to be more viable. These findings underscore the key role of intermediate carbenium ions in driving the overall catalytic cycle and enhancing the efficiency of the propane dehydroaromatization process.

6.4 Conclusion

Zeolites have emerged as highly versatile catalysts for converting low-carbon molecules into more valuable products, including hydrocarbons, oxygenates, and various chemicals. Their exceptional properties, such as high surface area, thermal stability, and tunable acidity, make them ideal candidates for catalyzing a diverse array of reactions. However, the zeolite catalysis process often involves intricate reaction networks within the zeolite voids, posing significant challenges in elucidating the underlying reaction mechanisms.

The conversion of low-carbon alcohols, notably methanol and ethanol, over zeolites has attracted considerable research attention and holds promise for hydrocarbon production through various mechanisms, particularly the MTH and ethanol dehydration pathways. In these reactions, dehydration serves as the initial step, typically occurring via concerted or stepwise mechanisms over the acidic sites within zeolites. Following the formation of these initial products, complex reaction networks, including alkylation, oligomerization, cyclization, hydrogen transfer, and isomerization, occur within the zeolite voids, ultimately regulating the product distribution.

The transformation of low-carbon alkanes, such as methane and C_2–C_4 alkanes, over zeolites has also been extensively explored, with metal-modified zeolites demonstrating potential in activating methane and facilitating the production of higher hydrocarbons. Given the high bond energies of C—H and C—C bonds in alkanes, extreme reaction conditions are typically required. The synergistic interplay between the BASs and metal sites within these zeolites is pivotal for the transformation of low-carbon alkanes, such as in the aromatization of alkanes.

Despite the progress made toward the transformation of low-carbon molecules over zeolites, the intricate reaction mechanisms remain largely elusive. To gain a deeper understanding of these mechanisms, it is imperative to develop and utilize advanced characterization tools, reaction dynamics analysis, and theoretical calculations. These multifaceted approaches will shed light on the intricacies of how zeolites facilitate the conversion of low-carbon molecules, enabling the optimization of processes and the design of more efficient catalysts.

References

1 Li, Y., Li, L., and Yu, J. (2017). Applications of zeolites in sustainable chemistry. *Chem* 3: 928–949.

2 Čejka, J., Morris, R.E., and Serrano, D.P. (2016). Catalysis on zeolites. *Catal. Sci. Technol.* 6: 2465–2466.

3 Zhang, Q., Yu, J., and Corma, A. (2020). Applications of zeolites to C1 chemistry: recent advances, challenges, and opportunities. *Adv. Mater.* 32: 2002927.

4 Yue, B., Liu, S., Chai, Y. et al. (2022). Zeolites for separation: fundamental and application. *J. Energy Chem.* 71: 288–303.

5 Shamzhy, M., Opanasenko, M., Concepción, P. et al. (2019). New trends in tailoring active sites in zeolite-based catalysts. *Chem. Soc. Rev.* 48: 1095–1149.

6 Ravi, M., Sushkevich, V.L., and van Bokhoven, J.A. (2020). Towards a better understanding of Lewis acidic aluminium in zeolites. *Nat. Mater.* 19: 1047–1056.

7 Haag, W.O., Lago, R.M., and Weisz, P.B. (1984). The active site of acidic aluminosilicate catalysts. *Nature* 309: 589–591.

8 Wang, Q.L., Giannetto, G., Torrealba, M. et al. (1991). Dealumination of zeolites II. Kinetic study of the dealumination by hydrothermal treatment of a NH_4NaY zeolite. *J. Catal.* 130: 459–470.

9 Yuan, Y., Lee, J.S., and Lobo, R.F. (2022). Ga^+-chabazite zeolite: a highly selective catalyst for nonoxidative propane dehydrogenation. *J. Am. Chem. Soc.* 144: 15079–15092.

10 Zhang, Q., Gao, S., and Yu, J. (2023). Metal sites in zeolites: synthesis, characterization, and catalysis. *Chem. Rev.* 123: 6039–6106.

11 Song, S., Sun, Y., Yang, K. et al. (2023). Recent progress in metal-molecular sieve catalysts for propane dehydrogenation. *ACS Catal.* 13: 6044–6067.

12 Chai, Y., Dai, W., Wu, G. et al. (2021). Confinement in a zeolite and zeolite catalysis. *Acc. Chem. Res.* 54: 2894–2904.

13 Wang, X., Ma, Y., Wu, Q. et al. (2022). Zeolite nanosheets for catalysis. *Chem. Soc. Rev.* 51: 2431–2443.

14 Tian, P., Wei, Y., Ye, M. et al. (2015). Methanol to olefins (MTO): from fundamentals to commercialization. *ACS Catal.* 5: 1922–1938.

15 Haw, J.F., Song, W.G., Marcus, D.M. et al. (2003). The mechanism of methanol to hydrocarbon catalysis. *Acc. Chem. Res.* 36: 317–326.

16 Ilias, S. and Bhan, A. (2013). Mechanism of the catalytic conversion of methanol to hydrocarbons. *ACS Catal.* 3: 18–31.

17 Hemelsoet, K., Van der Mynsbrugge, J., De Wispelaere, K. et al. (2013). Unraveling the reaction mechanisms governing methanol-to-olefins catalysis by theory and experiment. *ChemPhysChem* 14: 1526–1545.

18 Xu, J., Wang, Q., and Deng, F. (2019). Metal active sites and their catalytic functions in zeolites: insights from solid-state NMR spectroscopy. *Acc. Chem. Res.* 52: 2179–2189.

19 Wang, C., Xu, J., and Deng, F. (2020). Mechanism of methanol-to-hydrocarbon reaction over zeolites: a solid-state NMR perspective. *ChemCatChem* 12: 965–980.

20 Khare, R., Arora, S.S., and Bhan, A. (2016). Implications of cofeeding acetaldehyde on ethene selectivity in methanol-to-hydrocarbons conversion on MFI and its mechanistic interpretation. *ACS Catal.* 6: 2314–2331.

21 Dai, W., Yang, L., Wu, G. et al. (2022). Methanol to hydrocarbons. In: *Heterogeneous Catalysis for Sustainable Energy* (eds. L. Li and J.S.J. Hargreaves), 351–389. Weinheim: Wiley-VCH.

22 Kubelková, L., Nováková, J., and Nedomová, K. (1990). Reactivity of surface species on zeolites in methanol conversion. *J. Catal.* 124: 441–450.

23 Bandiera, J. and Naccache, C. (1991). Kinetics of methanol dehydration on dealuminated H-mordenite: model with acid and basic active centres. *Appl. Catal.* 69: 139–148.

24 Alharbi, W., Kozhevnikova, E.F., and Kozhevnikov, I.V. (2015). Dehydration of methanol to dimethyl ether over heteropoly acid catalysts: the relationship between reaction rate and catalyst acid strength. *ACS Catal.* 5: 7186–7193.

25 Ha, K.-S., Lee, Y.-J., Bae, J.W. et al. (2011). New reaction pathways and kinetic parameter estimation for methanol dehydration over modified ZSM-5 catalysts. *Appl. Catal. A: Gen.* 395: 95–106.

26 Jones, A.J. and Iglesia, E. (2014). Kinetic, spectroscopic, and theoretical assessment of associative and dissociative methanol dehydration routes in zeolites. *Angew. Chem. Int. Ed.* 53: 12177–12181.

27 Di Iorio, J.R., Hoffman, A.J., Nimlos, C.T. et al. (2019). Mechanistic origins of the high-pressure inhibition of methanol dehydration rates in small-pore acidic zeolites. *J. Catal.* 380: 161–177.

28 Van der Mynsbrugge, J., Moors, S.L.C., De Wispelaere, K. et al. (2014). Insight into the formation and reactivity of framework-bound methoxide species in H-ZSM-5 from static and dynamic molecular simulations. *ChemCatChem* 6: 1906–1918.

29 Matam, S.K., Howe, R.F., Thetford, A. et al. (2018). Room temperature methoxylation in zeolite H-ZSM-5: an operando DRIFTS/mass spectrometric study. *Chem. Commun.* 54: 12875–12878.

30 Matam, S.K., Nastase, S.A.F., Logsdail, A.J. et al. (2020). Methanol loading dependent methoxylation in zeolite H-ZSM-5. *Chem. Sci.* 11: 6805–6814.

31 Forester, T.R. and Howe, R.F. (1987). In situ FTIR studies of methanol and dimethyl ether in ZSM-5. *J. Am. Chem. Soc.* 109: 5076–5082.

32 Campbell, S.M., Jiang, X.-Z., and Howe, R.F. (1999). Methanol to hydrocarbons: spectroscopic studies and the significance of extra-framework aluminium. *Microporous Mesoporous Mater.* 29: 91–108.

33 Minova, I.B., Matam, S.K., Greenaway, A. et al. (2019). Elementary steps in the formation of hydrocarbons from surface methoxy groups in HZSM-5 seen by synchrotron infrared microspectroscopy. *ACS Catal.* 9: 6564–6570.

34 Salehirad, F. and Anderson, M.W. (1996). Solid-state ^{13}C MAS NMR study of methanol-to-hydrocarbon chemistry over H-SAPO-34. *J. Catal.* 164: 301–314.

35 Ivanova, I.I., Pomakhina, E.B., Rebrov, A.I. et al. (2001). Surface species formed during aniline methylation on zeolite H–Y investigated by in situ MAS NMR spectroscopy. *J. Catal.* 203: 375–381.

36 Wang, W., Seiler, M., and Hunger, M. (2001). Role of surface methoxy species in the conversion of methanol to dimethyl ether on acidic zeolites investigated by in situ stopped-flow MAS NMR spectroscopy. *J. Phys. Chem. B* 105: 12553–12558.

37 Munson, E.J. and Haw, J.F. (1991). NMR observation of trimethyloxonium formation from dimethyl ether on zeolite HZSM-5. *J. Am. Chem. Soc.* 113: 6303–6305.

38 Wang, W., Buchholz, A., Seiler, M. et al. (2003). Evidence for an initiation of the methanol-to-olefin process by reactive surface methoxy groups on acidic zeolite catalysts. *J. Am. Chem. Soc.* 125: 15260–15267.

39 Jiang, Y., Hunger, M., and Wang, W. (2006). On the reactivity of surface methoxy species in acidic zeolites. *J. Am. Chem. Soc.* 128: 11679–11692.

40 Wang, W. and Hunger, M. (2008). Reactivity of surface alkoxy species on acidic zeolite catalysts. *Acc. Chem. Res.* 41: 895–904.

41 Wu, X., Xu, S., Zhang, W. et al. (2017). Direct mechanism of the first carbon-carbon bond formation in the methanol-to-hydrocarbons process. *Angew. Chem. Int. Ed.* 56: 9039–9043.

42 Wang, C., Chu, Y.Y., Xu, J. et al. (2018). Extra-framework aluminum-assisted initial C-C bond formation in methanol-to-olefins conversion on zeolite H-ZSM-5. *Angew. Chem. Int. Ed.* 57: 10197–10201.

43 Zhou, X., Wang, C., Chu, Y. et al. (2019). Observation of an oxonium ion intermediate in ethanol dehydration to ethene on zeolite. *Nat. Commun.* 10: 1961.

44 Yu, M., Tormene, N., Bolshakov, A. et al. (2021). Selective methanethiol-to-olefins conversion over HSSZ-13 zeolite. *Chem. Commun.* 57: 3323–3326.

45 Howe, R.F., McGregor, J., Parker, S.F. et al. (2016). Application of inelastic neutron scattering to the methanol-to-gasoline reaction over a ZSM-5 catalyst. *Catal. Lett.* 146: 1242–1248.

46 Zachariou, A., Hawkins, A., Lennon, D. et al. (2019). Investigation of ZSM-5 catalysts for dimethylether conversion using inelastic neutron scattering. *Appl. Catal. A: Gen.* 569: 1–7.

47 Liu, Y., Müller, S., Berger, D. et al. (2016). Formation mechanism of the first carbon–carbon bond and the first olefin in the methanol conversion into hydrocarbons. *Angew. Chem. Int. Ed.* 55: 5723–5726.

48 Liu, R., Shao, X., Wang, C. et al. (2023). Reaction mechanism of methanol-to-hydrocarbons conversion: fundamental and application. *Chin. J. Catal.* 47: 67–92.

49 Yarulina, I., Chowdhury, A.D., Meirer, F. et al. (2018). Recent trends and fundamental insights in the methanol-to-hydrocarbons process. *Nat. Catal.* 1: 398–411.

50 Chowdhury, A.D., Houben, K., Whiting, G.T. et al. (2016). Initial carbon–carbon bond formation during the early stages of the methanol-to-olefin process proven by zeolite-trapped acetate and methyl acetate. *Angew. Chem. Int. Ed.* 55: 15840–15845.

51 Chowdhury, A.D., Paioni, A.L., Houben, K. et al. (2018). Bridging the gap between the direct and hydrocarbon pool mechanisms of the methanol-to-hydrocarbons process. *Angew. Chem. Int. Ed.* 57: 8095–8099.

52 Sun, T., Chen, W., Xu, S. et al. (2021). The first carbon-carbon bond formation mechanism in methanol-to-hydrocarbons process over chabazite zeolite. *Chem* 7: 2415–2428.

53 Li, B., Xu, J., Han, B. et al. (2013). Insight into dimethyl ether carbonylation reaction over mordenite zeolite from in-situ solid-state NMR spectroscopy. *J. Phys. Chem. C* 117: 5840–5847.

54 Cheung, P., Bhan, A., Sunley, G.J. et al. (2007). Site requirements and elementary steps in dimethyl ether carbonylation catalyzed by acidic zeolites. *J. Catal.* 245: 110–123.

55 Comas-Vives, A., Valla, M., Copéret, C. et al. (2015). Cooperativity between Al sites promotes hydrogen transfer and carbon–carbon bond formation upon dimethyl ether activation on alumina. *ACS Cent. Sci.* 1: 313–319.

56 Chu, Y., Yi, X., Li, C. et al. (2018). Brønsted/Lewis acid sites synergistically promote the initial C–C bond formation in the MTO reaction. *Chem. Sci.* 9: 6470–6479.

57 Paunović, V., Wu, X., Maggiulli, L. et al. (2024). The formation, reactivity and transformation pathways of formaldehyde in the methanol-to-hydrocarbon conversion. *Catal. Sci. Technol.* 14: 1216–1228.

58 Müller, S., Liu, Y., Kirchberger, F.M. et al. (2016). Hydrogen transfer pathways during zeolite catalyzed methanol conversion to hydrocarbons. *J. Am. Chem. Soc.* 138: 15994–16003.

59 Dessau, R.M. and Lapierre, R.B. (1982). On the mechanism of methanol conversion to hydrocarbons over HZSM-5. *J. Catal.* 78: 136–141.

60 Mole, T., Bett, G., and Seddon, D. (1983). Conversion of methanol to hydrocarbons over ZSM-5 zeolite—an examination of the role of aromatic-hydrocarbons using carbon-13-labeled and deuterium-labeled feeds. *J. Catal.* 84: 435–445.

61 Mole, T., Whiteside, J.A., and Seddon, D. (1983). Aromatic co-catalysis of methanol conversion over zeolite catalysts. *J. Catal.* 82: 261–266.

62 Song, W., Marcus, D.M., Fu, H. et al. (2002). An oft-studied reaction that may never have been: direct catalytic conversion of methanol or dimethyl ether to hydrocarbons on the solid acids HZSM-5 or HSAPO-34. *J. Am. Chem. Soc.* 124: 3844–3845.

63 Vogt, C., Weckhuysen, B.M., and Ruiz-Martínez, J. (2017). Effect of feedstock and catalyst impurities on the methanol-to-olefin reaction over H-SAPO-34. *ChemCatChem* 9: 183–194.

64 Olsbye, U., Svelle, S., Lillerud, K.P. et al. (2015). The formation and degradation of active species during methanol conversion over protonated zeotype catalysts. *Chem. Soc. Rev.* 44: 7155–7176.

65 Haw, J.F. and Marcus, D.M. (2005). Well-defined (supra)molecular structures in zeolite methanol-to-olefin catalysis. *Top. Catal.* 34: 41–48.

66 Wang, C., Wang, Q., Xu, J. et al. (2016). Direct detection of supramolecular reaction centers in the methanol-to-olefins conversion over zeolite H-ZSM-5 by C-13-Al-27 solid-state NMR spectroscopy. *Angew. Chem. Int. Ed.* 55: 2507–2511.

67 Bailleul, S., Yarulina, I., Hoffman, A.E.J. et al. (2019). A supramolecular view on the cooperative role of Brønsted and Lewis acid sites in zeolites for methanol conversion. *J. Am. Chem. Soc.* 141: 14823–14842.

68 Svelle, S., Joensen, F., Nerlov, J. et al. (2006). Conversion of methanol into hydrocarbons over zeolite H-ZSM-5: ethene formation is mechanistically separated from the formation of higher alkenes. *J. Am. Chem. Soc.* 128: 14770–14771.

69 Bjorgen, M., Svelle, S., Joensen, F. et al. (2007). Conversion of methanol to hydrocarbons over zeolite H-ZSM-5: on the origin of the olefinic species. *J. Catal.* 249: 195–207.

70 Svelle, S., Olsbye, U., Joensen, F. et al. (2007). Conversion of methanol to alkenes over medium- and large-pore acidic zeolites: steric manipulation of the reaction intermediates governs the ethene/propene product selectivity. *J. Phys. Chem. C* 111: 17981–17984.

71 Teketel, S., Olsbye, U., Lillerud, K.-P. et al. (2010). Selectivity control through fundamental mechanistic insight in the conversion of methanol to hydrocarbons over zeolites. *Microporous Mesoporous Mater.* 136: 33–41.

72 Westgård Erichsen, M., Svelle, S., and Olsbye, U. (2013). The influence of catalyst acid strength on the methanol to hydrocarbons (MTH) reaction. *Catal. Today* 215: 216–223.

73 Song, W.G., Haw, J.F., Nicholas, J.B. et al. (2000). Methylbenzenes are the organic reaction centers for methanol-to-olefin catalysis on HSAPO-34. *J. Am. Chem. Soc.* 122: 10726–10727.

74 Sullivan, R.F., Sieg, R.P., Langlois, G.E. et al. (1961). A new reaction that occurs in hydrocracking of certain aromatic hydrocarbons. *J. Am. Chem. Soc.* 83: 1156–1160.

75 Lesthaeghe, D., De Sterck, B., Van Speybroeck, V. et al. (2007). Zeolite shape-selectivity in the gem-methylation of aromatic hydrocarbons. *Angew. Chem. Int. Ed.* 46: 1311–1314.

76 Dahl, I.M. and Kolboe, S. (1994). On the reaction-mechanism for hydrocarbon formation from methanol over SAPO-34 .1. Isotopic labeling studies of the co-reaction of ethene and methanol. *J. Catal.* 149: 458–464.

77 Xu, S.T., Zheng, A.M., Wei, Y.X. et al. (2013). Direct observation of cyclic carbenium ions and their role in the catalytic cycle of the methanol-to-olefin reaction over chabazite zeolites. *Angew. Chem. Int. Ed.* 52: 11564–11568.

78 Naredla, R.R. and Klumpp, D.A. (2013). Contemporary carbocation chemistry: applications in organic synthesis. *Chem. Rev.* 113: 6905–6948.

79 Gong, X., Çağlayan, M., Ye, Y. et al. (2022). First-generation organic reaction intermediates in zeolite chemistry and catalysis. *Chem. Rev.* 122: 14275–14345.

80 Chen, W., Yi, X., Liu, Z. et al. (2022). Carbocation chemistry confined in zeolites: spectroscopic and theoretical characterizations. *Chem. Soc. Rev.* 51: 4337–4385.

81 Olsbye, U., Svelle, S., Bjorgen, M. et al. (2012). Conversion of methanol to hydrocarbons: how zeolite cavity and pore size controls product selectivity. *Angew. Chem. Int. Ed.* 51: 5810–5831.

82 Goguen, P.W., Xu, T., Barich, D.H. et al. (1998). Pulse-quench catalytic reactor studies reveal a carbon-pool mechanism in methanol-to-gasoline chemistry on zeolite HZSM-5. *J. Am. Chem. Soc.* 120: 2650–2651.

83 Haw, J.F., Nicholas, J.B., Song, W.G. et al. (2000). Roles for cyclopentenyl cations in the synthesis of hydrocarbons from methanol on zeolite catalyst HZSM-5. *J. Am. Chem. Soc.* 122: 4763–4775.

84 Wulfers, M.J. and Jentoft, F.C. (2014). The role of cyclopentadienium ions in methanol-to-hydrocarbons chemistry. *ACS Catal.* 4: 3521–3532.

85 Wang, C., Yi, X., Xu, J. et al. (2015). Experimental evidence on the formation of ethene through carbocations in methanol conversion over H-ZSM-5 zeolite. *Chem. Eur. J.* 21: 12061–12068.

86 Wang, C., Xu, J., Qi, G. et al. (2015). Methylbenzene hydrocarbon pool in methanol-to-olefins conversion over zeolite H-ZSM-5. *J. Catal.* 332: 127–137.

87 Wang, J., Wei, Y., Li, J. et al. (2016). Direct observation of methylcyclopentenyl cations (MCP^+) and olefin generation in methanol conversion over TON zeolite. *Catal. Sci. Technol.* 6: 89–97.

88 Xiao, D., Xu, S., Han, X. et al. (2017). Direct structural identification of carbenium ions and investigation of host–guest interaction in the methanol to olefins reaction obtained by multinuclear NMR correlations. *Chem. Sci.* 8: 8309–8314.

89 Xiao, D., Xu, S., Brownbill, N.J. et al. (2018). Fast detection and structural identification of carbocations on zeolites by dynamic nuclear polarization enhanced solid-state NMR. *Chem. Sci.* 9: 8184–8193.

90 Xiao, D., Han, X., Bao, X. et al. (2019). Identification of different carbenium ion intermediates in zeolites with identical chabazite topology *via* ^{13}C–^{13}C through-bond NMR correlations. *RSC Adv.* 9: 12415–12418.

91 Hernandez, E.D. and Jentoft, F.C. (2020). Spectroscopic signatures reveal cyclopentenyl cation contributions in methanol-to-olefins catalysis. *ACS Catal.* 10: 5764–5782.

92 Hernandez, E.D., Manookian, B., Auerbach, S.M. et al. (2021). Shape-selective synthesis of alkylcyclopentenyl cations in zeolites and spectroscopic distinction of constitutional isomers. *ACS Catal.* 11: 12893–12914.

93 Zhang, M., Xu, S., Wei, Y. et al. (2016). Methanol conversion on ZSM-22, ZSM-35 and ZSM-5 zeolites: effects of 10-membered ring zeolite structures on methylcyclopentenyl cations and dual cycle mechanism. *RSC Adv.* 6: 95855–95864.

94 Tabor, E., Bernauer, M., Wichterlova, B. et al. (2019). Enhancement of propene oligomerization and aromatization by proximate protons in zeolites; FTIR study of the reaction pathway in ZSM-5. *Catal. Sci. Technol.* 9: 4262–4275.

95 Bjørgen, M., Bonino, F., Kolboe, S. et al. (2003). Spectroscopic evidence for a persistent benzenium cation in zeolite H-beta. *J. Am. Chem. Soc.* 125: 15863–15868.

96 Qian, Q., Vogt, C., Mokhtar, M. et al. (2014). Combined operando UV/Vis/IR spectroscopy reveals the role of methoxy and aromatic species during the methanol-to-olefins reaction over H-SAPO-34. *ChemCatChem* 6: 3396–3408.

97 Dai, W.L., Wang, C.M., Dyballa, M. et al. (2015). Understanding the early stages of the methanol-to-olefin conversion on H-SAPO-34. *ACS Catal.* 5: 317–326.

98 Xu, T., Barich, D.H., Goguen, P.W. et al. (1998). Synthesis of a benzenium ion in a zeolite with use of a catalytic flow reactor. *J. Am. Chem. Soc.* 120: 4025–4026.

99 Borodina, E., Meirer, F., Lezcano-González, I. et al. (2015). Influence of the reaction temperature on the nature of the active and deactivating species during methanol to olefins conversion over H-SSZ-13. *ACS Catal.* 5: 992–1003.

100 Borodina, E., Sharbini Harun Kamaluddin, H., Meirer, F. et al. (2017). Influence of the reaction temperature on the nature of the active and deactivating species during methanol-to-olefins conversion over H-SAPO-34. *ACS Catal.* 7: 5268–5281.

101 Song, W., Nicholas, J.B., Sassi, A. et al. (2002). Synthesis of the heptamethylbenzenium cation in zeolite-β: in situ NMR and theory. *Catal. Lett.* 81: 49–53.

102 Zhang, W., Chen, J., Xu, S. et al. (2018). Methanol to olefins reaction over cavity-type zeolite: cavity controls the critical intermediates and product selectivity. *ACS Catal.* 8: 10950–10963.

103 Koveza, V.A., Potapenko, O.V., and Lavrenov, A.V. (2024). Catalytic conversion of ethanol to aromatic hydrocarbons in the presence of zeolite catalysts. *Catal. Ind.* 16: 39–57.

104 Sun, J. and Wang, Y. (2014). Recent advances in catalytic conversion of ethanol to chemicals. *ACS Catal.* 4: 1078–1090.

105 Zhang, M. and Yu, Y. (2013). Dehydration of ethanol to ethylene. *Ind. Eng. Chem. Res.* 52: 9505–9514.

106 Xin, H., Li, X., Fang, Y. et al. (2014). Catalytic dehydration of ethanol over post-treated ZSM-5 zeolites. *J. Catal.* 312: 204–215.

107 Meng, Q., Xin, H., Zhang, Y. et al. (2015). Catalytic dehydration of ethanol to ethylene over steam-treated ZSM-5 zeolites. *Sci. Adv. Mater.* 7: 2343–2351.

108 Bryant, D.E. and Kranich, W.L. (1967). Dehydration of alcohols over zeolite catalysts. *J. Catal.* 8: 8–13.

109 Van der Borght, K., Batchu, R., Galvita, V.V. et al. (2016). Insights into the reaction mechanism of ethanol conversion into hydrocarbons on H-ZSM-5. *Angew. Chem. Int. Ed.* 55: 12817–12821.

110 Chiang, H. and Bhan, A. (2010). Catalytic consequences of hydroxyl group location on the rate and mechanism of parallel dehydration reactions of ethanol over acidic zeolites. *J. Catal.* 271: 251–261.

111 Chowdhury, A.D., Lucini Paioni, A., Whiting, G.T. et al. (2019). Unraveling the homologation reaction sequence of the zeolite-catalyzed ethanol-to-hydrocarbons process. *Angew. Chem. Int. Ed.* 58: 3908–3912.

112 Wang, W., Jiao, J., Jiang, Y. et al. (2005). Formation and decomposition of surface ethoxy species on acidic zeolite Y. *ChemPhysChem* 6: 1467–1469.

113 Phung, T.K. and Busca, G. (2015). Diethyl ether cracking and ethanol dehydration: acid catalysis and reaction paths. *Chem. Eng. J.* 272: 92–101.

114 Zeng, S., Li, J., Wang, N. et al. (2021). Investigation of ethanol conversion on H-ZSM-5 zeolite by in situ solid-state NMR. *Energy Fuels* 35: 12319–12328.

115 Makshina, E.V., Dusselier, M., Janssens, W. et al. (2014). Review of old chemistry and new catalytic advances in the on-purpose synthesis of butadiene. *Chem. Soc. Rev.* 43: 7917–7953.

116 Mamedov, K. and Davis, R.J. (2023). Cascade reaction of ethanol to butadiene over Ag-promoted, silica- or zeolite-supported Ta, Y, Pr, or La oxide catalysts. *ACS Catal.* 13: 3333–3344.

117 Yan, T., Yang, L., Dai, W. et al. (2018). On the deactivation mechanism of zeolite catalyst in ethanol to butadiene conversion. *J. Catal.* 367: 7–15.

118 Qi, L., Zhang, Y., Conrad, M.A. et al. (2020). Ethanol conversion to butadiene over isolated zinc and yttrium sites grafted onto dealuminated beta zeolite. *J. Am. Chem. Soc.* 142: 14674–14687.

119 Yan, T., Dai, W., Wu, G. et al. (2018). Mechanistic insights into one-step catalytic conversion of ethanol to butadiene over bifunctional Zn–Y/Beta zeolite. *ACS Catal.* 8: 2760–2773.

120 Jones, M.D., Keir, C.G., Iulio, C.D. et al. (2011). Investigations into the conversion of ethanol into 1,3-butadiene. *Catal. Sci. Technol.* 1: 267–272.

121 del Campo, P., Martínez, C., and Corma, A. (2021). Activation and conversion of alkanes in the confined space of zeolite-type materials. *Chem. Soc. Rev.* 50: 8511–8595.

122 van Bokhoven, J.A., Williams, B.A., Ji, W. et al. (2004). Observation of a compensation relation for monomolecular alkane cracking by zeolites: the dominant role of reactant sorption. *J. Catal.* 224: 50–59.

123 Gounder, R. and Iglesia, E. (2009). Catalytic consequences of spatial constraints and acid site location for monomolecular alkane activation on zeolites. *J. Am. Chem. Soc.* 131: 1958–1971.

124 Janda, A., Vlaisavljevich, B., Lin, L.-C. et al. (2015). Adsorption thermodynamics and intrinsic activation parameters for monomolecular cracking of *n*-alkanes on Brønsted acid sites in zeolites. *J. Phys. Chem. C* 119: 10427–10438.

125 Janda, A. and Bell, A.T. (2013). Effects of Si/Al ratio on the distribution of framework Al and on the rates of alkane monomolecular cracking and dehydrogenation in H-MFI. *J. Am. Chem. Soc.* 135: 19193–19207.

126 Gao, W., Wang, Q., Qi, G. et al. (2023). Active ensembles in methane dehydroaromatization over molybdenum/ZSM-5 zeolite identified by 2D ^{1}H–^{95}Mo magic angle spinning nuclear magnetic resonance correlation spectroscopy. *Angew. Chem. Int. Ed.* 62: e202306133.

127 Gao, W., Qi, G., Wang, Q. et al. (2021). Dual active sites on molybdenum/ZSM-5 catalyst for methane dehydroaromatization: insights from solid-state NMR spectroscopy. *Angew. Chem. Int. Ed.* 60: 10709–10715.

128 Qi, G., Davies, T.E., Nasrallah, A. et al. (2022). Au-ZSM-5 catalyses the selective oxidation of CH_4 to CH_3OH and CH_3COOH using O_2. *Nat. Catal.* 5: 45–54.

129 Olivos-Suarez, A.I., Szecsenyi, A., Hensen, E.J.M. et al. (2016). Strategies for the direct catalytic valorization of methane using heterogeneous catalysis: challenges and opportunities. *ACS Catal.* 6: 2965–2981.

130 Cruchade, H., Medeiros-Costa, I.C., Nesterenko, N. et al. (2022). Catalytic routes for direct methane conversion to hydrocarbons and hydrogen: current state and opportunities. *ACS Catal.* 12: 14533–14558.

131 Schwach, P., Pan, X., and Bao, X. (2017). Direct conversion of methane to value-added chemicals over heterogeneous catalysts: challenges and prospects. *Chem. Rev.* 117: 8497–8520.

132 Spivey, J.J. and Hutchings, G. (2014). Catalytic aromatization of methane. *Chem. Soc. Rev.* 43: 792–803.

133 Schulz, H. (1999). Short history and present trends of Fischer–Tropsch synthesis. *Appl. Catal. A: Gen.* 186: 3–12.

134 Dry, M.E. (2002). The Fischer–Tropsch process: 1950–2000. *Catal. Today* 71: 227–241.

135 Bergman, R.G. (2007). C–H activation. *Nature* 446: 391–393.

136 Petrov, A.W., Ferri, D., Krumeich, F. et al. (2018). Stable complete methane oxidation over palladium based zeolite catalysts. *Nat. Commun.* 9: 2545.

137 Jin, Z., Wang, L., Zuidema, E. et al. (2020). Hydrophobic zeolite modification for in situ peroxide formation in methane oxidation to methanol. *Science* 367: 193–197.

138 Sushkevich, V.L., Palagin, D., Ranocchiari, M. et al. (2017). Selective anaerobic oxidation of methane enables direct synthesis of methanol. *Science* 356: 523–527.

139 Tomkins, P., Ranocchiari, M., and van Bokhoven, J.A. (2017). Direct conversion of methane to methanol under mild conditions over Cu-zeolites and beyond. *Acc. Chem. Res.* 50: 418–425.

140 Kosinov, N. and Hensen, E.J.M. (2020). Reactivity, selectivity, and stability of zeolite-based catalysts for methane dehydroaromatization. *Adv. Mater.* 32: 2002565.

141 Choudhary, V.R., Kinage, A.K., and Choudhary, T.V. (1997). Low-temperature nonoxidative activation of methane over H-galloaluminosilicate (MFI) zeolite. *Science* 275: 1286–1288.

142 Kosinov, N., Uslamin, E.A., Coumans, F.J.A.G. et al. (2018). Structure and evolution of confined carbon species during methane dehydroaromatization over Mo/ZSM-5. *ACS Catal.* 8: 8459–8467.

143 Kosinov, N., Wijpkema, A.S.G., Uslamin, E. et al. (2018). Confined carbon mediating dehydroaromatization of methane over Mo/ZSM-5. *Angew. Chem. Int. Ed.* 57: 1016–1020.

144 Baik, M.-H., Newcomb, M., Friesner, R.A. et al. (2003). Mechanistic studies on the hydroxylation of methane by methane monooxygenase. *Chem. Rev.* 103: 2385–2420.

145 Balasubramanian, R. and Rosenzweig, A.C. (2007). Structural and mechanistic insights into methane oxidation by particulate methane monooxygenase. *Acc. Chem. Res.* 40: 573–580.

146 Brezicki, G., Kammert, J.D., Gunnoe, T.B. et al. (2019). Insights into the speciation of Cu in the Cu-H-mordenite catalyst for the oxidation of methane to methanol. *ACS Catal.* 9: 5308–5319.

147 Grundner, S., Markovits, M.A.C., Li, G. et al. (2015). Single-site trinuclear copper oxygen clusters in mordenite for selective conversion of methane to methanol. *Nat. Commun.* 6: 7546.

148 Tao, L., Khramenkova, E., Lee, I. et al. (2023). Speciation and reactivity control of Cu-oxo clusters via extraframework Al in mordenite for methane oxidation. *J. Am. Chem. Soc.* 145: 17710–17719.

149 Sobolev, V.I., Dubkov, K.A., Panna, O.V. et al. (1995). Selective oxidation of methane to methanol on a FeZSM-5 surface. *Catal. Today* 24: 251–252.

150 Panov, G.I., Uriarte, A.K., Rodkin, M.A. et al. (1998). Generation of active oxygen species on solid surfaces. Opportunity for novel oxidation technologies over zeolites. *Catal. Today* 41: 365–385.

151 Snyder, B.E.R., Bols, M.L., Schoonheydt, R.A. et al. (2018). Iron and copper active sites in zeolites and their correlation to metalloenzymes. *Chem. Rev.* 118: 2718–2768.

152 Snyder, B.E.R., Vanelderen, P., Bols, M.L. et al. (2016). The active site of low-temperature methane hydroxylation in iron-containing zeolites. *Nature* 536: 317–321.

153 Groothaert, M.H., Smeets, P.J., Sels, B.F. et al. (2005). Selective oxidation of methane by the bis(μ-oxo)dicopper core stabilized on ZSM-5 and mordenite zeolites. *J. Am. Chem. Soc.* 127: 1394–1395.

154 Xu, J., Zheng, A., Wang, X. et al. (2012). Room temperature activation of methane over Zn modified H-ZSM-5 zeolites: insight from solid-state NMR and theoretical calculations. *Chem. Sci.* 3: 2932–2940.

155 Wang, X., Qi, G., Xu, J. et al. (2012). NMR-spectroscopic evidence of intermediate-dependent pathways for acetic acid formation from methane and carbon monoxide over a ZnZSM-5 zeolite catalyst. *Angew. Chem. Int. Ed.* 51: 3850–3853.

156 Narsimhan, K., Michaelis, V.K., Mathies, G. et al. (2015). Methane to acetic acid over Cu-exchanged zeolites: mechanistic insights from a site-specific carbonylation reaction. *J. Am. Chem. Soc.* 137: 1825–1832.

157 Wu, J.-F., Yu, S.-M., Wang, W.D. et al. (2013). Mechanistic insight into the formation of acetic acid from the direct conversion of methane and carbon dioxide on zinc-modified H–ZSM-5 zeolite. *J. Am. Chem. Soc.* 135: 13567–13573.

158 Kosinov, N., Coumans, F.J.A.G., Uslamin, E.A. et al. (2017). Methane dehydroaromatization by Mo/HZSM-5: mono- or bifunctional catalysis? *ACS Catal.* 7: 520–529.

159 Chen, L.Y., Lin, L.W., Xu, Z.S. et al. (1995). Dehydro-oligomerization of methane to ethylene and aromatics over molybdenum/HZSM-5 catalyst. *J. Catal.* 157: 190–200.

160 Xu, Y., Liu, S., Guo, X. et al. (1994). Methane activation without using oxidants over Mo/HZSM-5 zeolite catalysts. *Catal. Lett.* 30: 135–149.

161 Xin, H., Xi, J., Xiaobo, W. et al. (2022). Research progress in the direct, nonoxidative conversion of methane to olefins/aromatics (II). *J. Fuel Chem. Technol.* 50: 44–53.

162 Liu, S., Wang, L., Ohnishi, R. et al. (1999). Bifunctional catalysis of Mo/HZSM-5 in the dehydroaromatization of methane to benzene and naphthalene XAFS/TG/DTA/MASS/FTIR characterization and supporting effects. *J. Catal.* 181: 175–188.

163 Lezcano-González, I., Oord, R., Rovezzi, M. et al. (2016). Molybdenum speciation and its impact on catalytic activity during methane dehydroaromatization in zeolite ZSM-5 as revealed by operando X-ray methods. *Angew. Chem. Int. Ed.* 55: 5215–5219.

164 Vollmer, I., van der Linden, B., Ould-Chikh, S. et al. (2018). On the dynamic nature of Mo sites for methane dehydroaromatization. *Chem. Sci.* 9: 4801–4807.

165 Li, W., Meitzner, G.D., Borry, R.W. et al. (2000). Raman and X-ray absorption studies of Mo species in Mo/H-ZSM5 catalysts for non-oxidative CH_4 reactions. *J. Catal.* 191: 373–383.

166 Hagen, A. and Roessner, F. (2000). Ethane to aromatic hydrocarbons: past, present, future. *Catal. Rev.* 42: 403–437.

167 Najari, S., Saeidi, S., Concepcion, P. et al. (2021). Oxidative dehydrogenation of ethane: catalytic and mechanistic aspects and future trends. *Chem. Soc. Rev.* 50: 4564–4605.

168 Guisnet, M., Gnep, N.S., and Alario, F. (1992). Aromatization of short chain alkanes on zeolite catalysts. *Appl. Catal. A: Gen.* 89: 1–30.

169 Rane, N., Overweg, A.R., Kazansky, V.B. et al. (2006). Characterization and reactivity of Ga^+ and GaO^+ cations in zeolite ZSM-5. *J. Catal.* 239: 478–485.

170 Phadke, N.M., Mansoor, E., Bondil, M. et al. (2019). Mechanism and kinetics of propane dehydrogenation and cracking over Ga/H-MFI prepared via vapor-phase exchange of H-MFI with $GaCl_3$. *J. Am. Chem. Soc.* 141: 1614–1627.

171 Kotrel, S., Knözinger, H., and Gates, B.C. (2000). The Haag–Dessau mechanism of protolytic cracking of alkanes. *Microporous Mesoporous Mater.* 35–36: 11–20.

172 Li, X., Han, H., Xu, W. et al. (2022). Acid catalysis over low-silica faujasite zeolites. *J. Am. Chem. Soc.* 144: 9324–9329.

173 Zhang, Y., Zhao, R., Sanchez-Sanchez, M. et al. (2019). Promotion of protolytic pentane conversion on H-MFI zeolite by proximity of extra-framework aluminum oxide and Brønsted acid sites. *J. Catal.* 370: 424–433.

174 Narbeshuber, T.F., Vinek, H., and Lercher, J.A. (1995). Monomolecular conversion of light alkanes over H-ZSM-5. *J. Catal.* 157: 388–395.

175 Babitz, S.M., Williams, B.A., Miller, J.T. et al. (1999). Monomolecular cracking of *n*-hexane on Y, MOR, and ZSM-5 zeolites. *Appl. Catal. A: Gen.* 179: 71–86.

176 Rahimi, N. and Karimzadeh, R. (2011). Catalytic cracking of hydrocarbons over modified ZSM-5 zeolites to produce light olefins: a review. *Appl. Catal. A: Gen.* 398: 1–17.

177 Janda, A., Vlaisavljevich, B., Lin, L.-C. et al. (2016). Effects of zeolite structural confinement on adsorption thermodynamics and reaction kinetics for monomolecular cracking and dehydrogenation of *n*-butane. *J. Am. Chem. Soc.* 138: 4739–4756.

178 van Bokhoven, J.A., Tromp, M., Koningsberger, D.C. et al. (2001). An explanation for the enhanced activity for light alkane conversion in mildly steam dealuminated mordenite: the dominant role of adsorption. *J. Catal.* 202: 129–140.

179 Caeiro, G., Carvalho, R.H., Wang, X. et al. (2006). Activation of C_2–C_4 alkanes over acid and bifunctional zeolite catalysts. *J. Mol. Catal. A: Chem.* 255: 131–158.

180 Wang, C., Zhao, X., Hu, M. et al. (2021). Unraveling hydrocarbon pool boosted propane aromatization on gallium/ZSM-5 zeolite by solid-state nuclear magnetic resonance spectroscopy. *Angew. Chem. Int. Ed.* 60: 23630–23634.

Part III

CO Conversion

7

Application of Zeolite in Fischer–Tropsch Synthesis

*Yingluo He[1], Guangbo Liu[2,3], Guohui Yang[1,4], Noritatsu Tsubaki[1]**

[1]*Department of Applied Chemistry, Graduate School of Engineering, University of Toyama, Toyama, Japan*
[2]*Key Laboratory of Biofuels, Qingdao Institute of Bioenergy and Bioprocess Technology, Chinese Academy of Sciences, Qingdao, Shandong, China*
[3]*College of Chemical Engineering, Qingdao University of Science and Technology, Qingdao, Shandong, China*
[4]*State Key Laboratory of Fine Chemicals, Frontiers Science Center for Smart Materials Oriented Chemical Engineering, School of Chemical Engineering, Dalian University of Technology, Dalian, China*

*Corresponding author: tsubaki@eng.u-toyama.ac.jp

7.1 What Is Fischer–Tropsch Synthesis?

7.1.1 Introduction

The Fischer–Tropsch synthesis (FTS) is a thermal catalytic process discovered by Franz Fischer and Hans Tropsch in 1920s. The FTS process uses synthesis gas (syngas, a mixture of carbon monoxide and hydrogen) to synthesize liquid hydrocarbons [1–4]. Syngas is particularly important and used widely in the industry, and it can be made from coal, natural gas, biomass, and other organic compounds such as plastic waste [1, 5–7].

As a key technology within various X-to-liquid (XTL) industrial conversion processes, including coal-to-liquid (CTL), natural gas-to-liquid (GTL), biomass-to-liquid (BTL), and waste-to-liquid (WTL), the development of FTS technology offers a viable alternative to crude oil for producing clean hydrocarbon fuels [8–11]. The hydrocarbon formation in the FTS can be generalized by the following reactions:

$$2nH_2 + nCO \rightarrow C_nH_{2n} + nH_2O \quad (7.1)$$

$$(2n+1)H_2 + nCO \rightarrow C_nH_{(2n+1)} + nH_2O \quad (7.2)$$

From a polymerization perspective, the FTS mechanism can be described as a chain-growth reaction involving the sequential addition of CH_2 groups to a growing carbon chain. The catalysts essential for producing valuable chemicals and fuels typically contain metallic components. However, these metal-based catalysts face

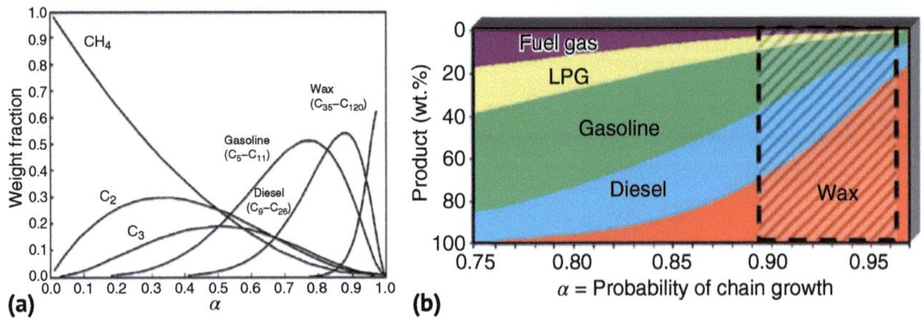

Figure 7.1 (a) The ASF distribution of FTS. *Source*: Reproduced from [15] with permission of Elsevier. (b) The selectivity of product fractions of interest, including fuel gas, LPG, gasoline, diesel, and wax. *Source*: Reproduced from [13] / IFP Energies Nouvelles / CC BY 4.0.

challenges, including poor stability and product distributions that adhere to the Anderson–Schulz–Flory (ASF) model (see Figure 7.1) [8, 12]. This model predicts theoretical selectivity limits of approximately 58, 48, 41, and 40% for light hydrocarbons (C_2–C_4), gasoline (C_5–C_{11}), jet fuel (C_8–C_{16}), and diesel (C_{10}–C_{20}), respectively [13, 14].

Therefore, a breakthrough requires a new catalytic paradigm that can maintain stable performance while overcoming the theoretical product selectivity limits imposed by the ASF distribution. In response, researchers have proposed various mathematical models and theoretical frameworks for composite catalysts, advancing beyond the traditional ASF model and opening new avenues for combining different catalysts with distinct active sites [16–24]. These efforts are aimed at optimizing syngas utilization to establish a more sustainable and versatile route for producing liquid fuels and chemicals from diverse feedstocks.

7.1.2 Reaction Mechanism Hypothesis

In FTS, the reactants CO and H_2 generate monomers in situ (an "initiation" step). These monomers then undergo chain growth, followed by termination reactions that are independent of chain length, yielding products such as olefins, paraffins, and/or oxygenates. Oxygen is removed as either H_2O or CO_2, depending on the catalyst and reaction conditions. Considering these three reaction types (initiation, propagation, and termination) for hydrocarbon chains up to 40 carbons in length, the detailed mechanism of FTS can easily encompass over 10000 elementary reaction steps [25, 26].

Since 1923, numerous reaction pathways of FTS have been reported and proposed by many researchers. In these proposed mechanisms, (1) the carbide mechanism, (2) the CO insertion mechanism, and (3) the enolate, hydroxycarbene ($HCOH^*$), and oxymethylene mechanism are the three most prominent hypotheses (see Figure 7.2) [26–30].

The monomer serves as a chain initiator in the first two mechanisms, the carbide mechanism and the $HCOH^*$ mechanism. Fischer and Tropsch originally proposed that metal or carbide active sites react with H_2 to form methylene (CH_2^*).

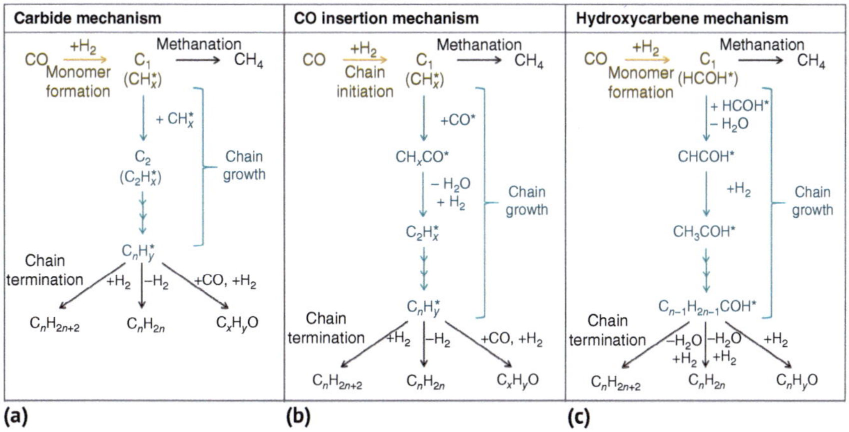

Figure 7.2 Reaction scheme of FTS, summarizing the three most common chain-growth mechanisms: (a) carbide mechanism, (b) CO insertion mechanism, and (c) $HCOH^*$ mechanism. *Source*: Reproduced from [11] with permission of American Chemical Society.

The CH_2^* acts as the chain-growth monomer in further reactions [11]. This "carbon mechanism" that has been substantiated by many experiments, in a modified form, remains widely accepted today [26, 31]. Subsequently, C^* and CH^* also seen as two possible potential monomers, with the growing chain being an alkylene (RC^*) species rather than an alkyl (RCH_2^*) species [30, 32]. CH_x^* monomers are generated via CO activation and hydrogenation in the carbide mechanism, and then transformation to olefins, paraffins, and oxygenates [11].

In the CO insertion mechanism, CO molecule plays the role of monomer, inserting itself into the growing hydrocarbon chain, after the hydrogenation and cleavage of the C—O bond [33]. Consequently, the CO insertion mechanism seems to have a greater advantage in explaining the formation of alcohol oxygenates [34, 35]. The main difference between the carbide mechanism and the CO insertion mechanism is the order of C—C bond formation and C—O bond cleavage. The C—C bond is formed first in the CO insertion mechanism, while the C—O bond is cleaved first in the carbide mechanism [36, 37].

In the abundant co-stages of the catalytic particles, the barrier for CO insertion is lower than that for direct CO dissociation [38]. Unlike the CO insertion mechanism, the $HCOH^*$ mechanism considers that HCO^* can first be formed by CO^* hydrogenation, which is further hydrogenated to form $HCOH^*$. Moreover, $HCOH^*$ undergoes polycondensation with the growth of the $RCOH^*$ chain [39, 40]. However, there are not enough experimental evidences to prove this mechanism yet [11].

7.1.3 Catalysts for FTS

A wide variety of metal-based catalysts, particularly those containing iron (Fe), cobalt (Co), nickel (Ni), or ruthenium (Ru), demonstrated high activity for the FTS reaction [3, 35, 41–43]. The choice of catalyst is a critical factor that determines the final hydrocarbon product distribution. The products of FTS comprise a complex mixture of hydrocarbons and oxygenates; by selecting appropriate catalysts and conditions, linear and branched olefins, alkanes, alcohols, and other products can be synthesized.

Among these, Fe and Co catalysts are the most extensively used in both academic studies and industrial FTS processes [4, 44, 45]. Co-based catalysts are generally considered more selective for producing fuel-range hydrocarbons because they generate minimal oxygenates and aromatics. Co is more active in hydrogenation reactions and can promote hydrocarbon chain-growth reactions involving olefins. Furthermore, Co-based catalysts are often favored over Fe-based catalysts due to their higher activity under milder conditions, especially at lower temperature, and their longer operational lifetime. Typically, active metals are supported on oxides, mainly aluminum, silicon, titanium, or zinc oxide [46–52].

Fe-based catalysts are also widely employed in FTS due to their low cost and applicability across ranges of syngas compositions and reaction conditions, as well as their ability to produce valuable low-carbon olefins and oxygenates [53–55]. The active centers of Fe-based catalysts are different from those of Co-based catalysts, with iron carbide $\left(Fe_xC_y\right)$ being considered the main active site in FTS. The complex nature and dynamic transformation of these active sites make mechanistic studies of Fe-based catalysts considerably more challenging than those of Co-based catalysts. However, Fe-based catalysts perform well over a broad and higher temperature range (300–350 °C), making them suitable candidates for the processes optimized to generate a high proportion of light olefins as intermediates for subsequent upgrading stages. Consequently, Fe-based catalysts have been widely explored for their thermal stability and high-temperature operability, which favor the production of olefins with much lower selectivity toward alkanes compared to other FTS catalysts [30, 56–58].

The product distribution in the FTS typically follows the AFS polymerization model, which inherently limits the selectivity toward any specific hydrocarbon compounds. Therefore, achieving highly selective generation of the desired hydrocarbons range and composition is not feasible with the conventional FTS catalysts alone [3, 22, 59, 60]. Therefore, the concept of combining traditional FTS catalysts with acidic materials, particularly zeolites, was proposed as early as the mid-1970s to overcome the constraints of the products with low ASF distribution [61–63]. In such bifunctional catalyst systems, hydrocarbons or their intermediates, formed at the metallic FTS sites, can undergo further conversion at the acidic sites, thereby deviating from the ASF distribution. The final product composition is determined by the relative reaction rates at the FTS active sites and acidic sites [64].

Herein, combining FTS catalysts and zeolites in a single system creates what are known as bifunctional or hybrid catalysts [48, 65, 66]. These catalysts enable the one-step, selective production of synthetic fuels and/or hydrocarbon chemicals free of heavy waxes directly from CO and H_2. This eliminates the need for a separate hydrotreating stage, making XTL technology more economically promising [64].

7.1.4 Reactors for FTS

FTS is a complex, exothermic catalytic process. A substantial body of studies is aimed at improving hydrocarbon yields while minimizing the production of methane and carbon dioxide. Consequently, key research focuses include the development of novel catalysts and advanced FTS reactors [9, 67–69]. For catalyst development in laboratory, FTS is generally performed in single-tube fixed-bed microreactors or slurry bed reactors. In industrial-scaled CTL and GTL processes (see Figure 7.3), FTS units mainly have the following three types of reactors: fixed-bed reactor or packed bed reactor (PBR), fluidized bed reactor (FBR), and slurry bubble bed reactor (SBCR) [10, 68, 70].

In a PBR, the catalyst is held in a fixed position, which obviates the need for a downstream catalyst separation process. This reactor type is also relatively straightforward to scale up from laboratory-scale tubular reactors to commercial pilot plants. However, a major challenge with traditional PBRs is the efficient and timely removal of large heat by the exothermic FTS reaction [71, 72]. To mitigate this, multi-tubular shell-and-tube reactors have been developed, featuring cooling systems that circulate fluid between the tubes. Feed syngas typically flows downward in a plug-flow manner, catalyst is packed inside the tubes, while the coolant in the shell prevents reactor overheating [73].

In an FBR, the reactant gas is introduced from the bottom of the reactor at a sufficiently high velocity to suspend the catalyst particles, creating a fluid-like state [74]. The catalyst particles used in FBR are typically of the order of hundreds of nanometers, much smaller than the >1-mm particles used in PBRs. This small particle size enhances catalyst dispersion and ensures close contact between the catalyst and the reactant gas, while the lower catalyst density prevents localized heat accumulation. However, FBR imposes stringent requirements on reaction conditions and equipment manufacturing [75, 76].

The SBCR operates differently from the other two reactor types. In an SBCR, the catalyst is dispersed in a liquid solvent (often a heavy wax product), and the three phases (gas, liquid, and solid) are brought into contact within the vessel [2]. Gaseous reactants are bubbled up from the bottom, contacting the liquid phase, which contains the suspended solid catalyst particles. An internal tube bundle serves as a heat

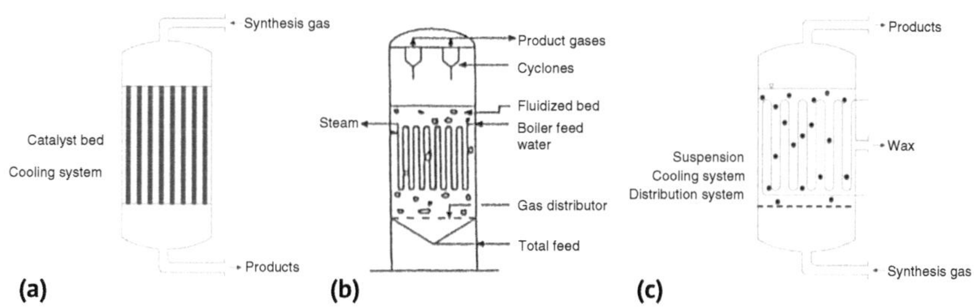

Figure 7.3 (a) Multi-tubular PBR/fixed-bed reactor and (b) FBR. *Source*: Reproduced from [68] / with permission of John Wiley & Sons. (c) SBCR. *Source*: Reproduced from [10] with permission of Elsevier.

exchanger for temperature control (heating or cooling). Gaseous products exit from the top, while the liquid/solid slurry can be withdrawn from side pipes at various heights. SBCR offers excellent heat dissipation and has moderate construction cost, thereby enhancing the overall efficiency of the FTS process [43].

In addition to these three main types, microchannel reactors have emerged as an ideal choice for addressing the challenges of smaller-scale BTL processes, owing to their compact and modular design [77, 78]. Microchannel reactors can efficiently remove a large amount of heat generated by the FTS, enabling higher single-pass conversion rates when used with optimized catalysts. However, conventional reactors have not entirely solved the inherent heat transfer limitations of PBRs or the mass transfer limitations of SBCRs [8, 43].

Currently, new concepts and strategies are being proposed to enhance the mass and heat transfer performances of multiphase reactors [73, 79–82]. Among these, membrane reactors, which leverage the separation function of molecular sieve membranes to improve product yield/selectivity and overcome thermodynamic equilibria, represent a promising direction for the development of next-generation FTS reactors [73, 83, 84]. The industrial challenge often involves a trade-off effect between a high content of active sites and the limitation of diffusion, and membrane reactors offer a potential solution to this issue. We will give a detailed introduction about zeolite-based membrane reactor later.

7.2 Zeolite-based FTS Catalysts

7.2.1 Unique Properties of Zeolites for FTS

Zeolites possess a unique, crystalline microporous structure consisting of an interconnected system of channels and cavities [85–91]. Coupled with their inherent acidic properties, zeolites are used as effective hydrocracking and hydroisomerization catalysts. For instance, H–ZSM-5 zeolite is used commercially for acid-catalyzed cracking reactions and also promotes low-temperature isomerization [64, 92–94]. HY zeolite is widely used in the industrial cracking of crude oil to produce commercial fuel products [95, 96]. These characteristics make zeolites a promising potential component as part of the bifunctional FTS catalysts, enabling a one-step synthesis process of liquid fuels from syngas [64, 97, 98].

Since their discovery, zeolites have been incorporated into FTS catalysts in various forms [64, 73]. These zeolite-containing catalysts introduce secondary reaction pathways that modify the primary FTS product distribution [55, 61, 62, 99]. Among these secondary reactions, hydrocracking and hydroisomerization are the most widely studied [100, 101]. However, different types of zeolites can also catalyze alkylation, oligomerization, cracking, isomerization, and aromatization [102–106]. These specialized reaction capabilities have injected new vitality into FTS research, offering the potential for highly selective and targeted production of valuable products such as gasoline, diesel, jet fuel, olefins, and even aromatics [59, 107–110].

However, the optimal temperature ranges for these secondary reactions can differ significantly. Hydrocarbon isomerization on zeolite typically occurs at lower temperatures (200–250 °C) [111, 112], whereas hydrocracking requires much higher temperatures (400–550 °C) [28, 113]. This temperature mismatch presents a challenge for seamlessly integrating these desirable secondary reactions into the FTS process.

The following subsections will highlight research advances in zeolite-containing FTS catalysts over the past decade, focusing on zeolite-supported catalysts, physically mixed catalysts, and novel core–shell-structured catalysts. The roles of different zeolite types and the effects of various zeolite modification strategies will also be discussed.

7.2.2 Zeolite-supported Catalysts

Supported catalysts are a cornerstone of heterogeneous catalysis, and FTS is no exception [41, 42, 114]. The unique properties of zeolites make them promising and widely studied support materials for FTS catalysts. Since the 1980s, reports on zeolite-supported Co-based, Ru-based, and Fe-based FTS catalysts have shown unexpectedly high selectivity toward liquid fuels compared to traditional oxide-supported catalysts [55, 61, 62]. Building on this foundation, research into zeolite-supported FTS catalysts has expanded significantly, exploring various zeolite types, zeolite modifications, active metals, metal precursors, metal loadings, particle sizes, loading methods, and deactivation mechanisms [63, 115–125].

In 2010, de Jong et al. developed a high-performance, Co-based supported FTS catalyst with high loading and controlled particle size using a melt-infiltration method with a mesoporous silica zeolite (SBA-15) support [101]. They demonstrated that this method could completely fill the pores of the mesoporous support with a Co phase. Also, 60 °C was found to be critical for the success of the melt-infiltration technique. In contrast, using an open crucible led to dehydration of the precursor, which inhibited uniform packing within the support. The dispersion and distribution of the resulting Co_3O_4 nanoparticles after calcination could be controlled using methods similar to those for solution impregnation preparation, such as confinement effects and the choice of calcination atmosphere [114]. Catalysts with Co loadings ranging from 10 to 22 wt.% were prepared using ordered mesoporous SBA-15 supports and commercial silica gel supports. The resulting Co_3O_4 crystallite sizes, ranging from 4 to 10 nm, were proportional to the support pore size. Calcination in pure N_2 produced plug-like nanoparticles that aggregated across multiple pore sizes, whereas calcination in 1% NO/N_2 yielded smaller, individual nanoparticles. After reduction, the Co/SBA-15 catalysts exhibited high activity in FTS, validating melt infiltration as an effective preparation method for supported catalysts.

To further validate the method, in situ tracking of $Co(NO_3)_2{\cdot}6H_2O$ infiltration into the SBA-15 support was performed using differential scanning calorimetry. The results confirmed that any $Co(NO_3)_2{\cdot}6H_2O$ remaining outside the pores could be detected by observing the melting point depression of the confined precursor. Assuming the intrapore phase was crystalline, the maximum precursor loading was equivalent to 70–80% of the pore volume (see Figure 7.4). This led to the conclusion

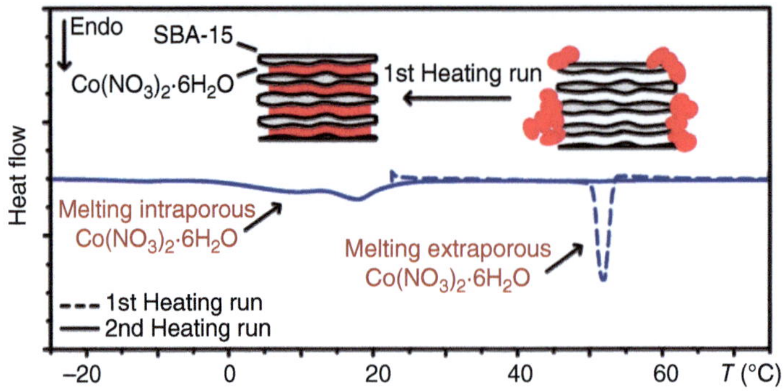

Figure 7.4 Fundamentals of melt infiltration for the preparation of supported metal catalysts. *Source*: Reproduced from [17] with permission of American Chemical Society.

that the support pores were completely filled with an amorphous $Co(NO_3)_2{\cdot}6H_2O$ phase, which has a lower density than its crystalline counterpart [114].

Soon after, Marchetti and co-workers reported an Fe catalyst supported on SBA-15 for FTS. They investigated two different precursor activation methods: pure H_2 and syngas (H_2:CO = 2:1). The activation treatment resulted in different initial Fe species [126]. The H_2-activated catalyst (H_2–Fe/SBA-15) contained α-Fe, Fe_3O_4, and Fe^{2+} species diffused into the SBA-15 wall, while the syngas-activated sample contained χ-Fe_5C_2, Fe_3O_4, and Fe^{2+}. Researchers found that these initial differences disappeared once the catalysts were operating under FTS conditions, the H_2–Fe/SBA-15 catalyst achieved higher CO conversion and higher hydrocarbon yields over a six-day period. Additionally, the H_2–Fe/SBA-15 catalyst showed higher selectivity for light olefins during the first 24 hours on stream, though this selectivity decreased over time. This phenomenon proposed that χ-Fe_5C_2 from α-Fe is more reactive and/or smaller than that formed directly from Fe_2O_3. Researchers speculated that when χ-Fe_5C_2 is formed directly from Fe_2O_3, the crystal planes' higher carbon content becomes dominant. The formation of Fe_xC_y "nodules" on the nanoparticle surface occurs at different rates depending on whether they form on an Fe_3O_4 surface or an α-Fe "shell," with the latter potentially producing numerous small nodules (see Figure 7.5). This would result in more sites for CO adsorption and dissociation, shorter diffusion paths, and a more active catalyst [126]. This "living" active surface constantly changes and regenerates under FTS conditions. This study provided new insights into the preparation, activation, and active site dynamics of Fe-based catalysts supported on porous materials.

Wang and co-workers prepared a series of mesoporous zeolite-supported Ru nanoparticle catalysts for the highly selective gasoline-range hydrocarbons (C_5–C_{11}) production via FTS [60]. They first prepared a Ru-based FTS catalyst with hierarchical porous zeolite, owning both micropores and mesopores by treating a commercial HZSM-5 zeolite with an alkali solution. This material retained the unique acidity and microporous channels of HZSM-5 while gaining enhanced mass transfer properties, making it an excellent catalyst support. They found that using this mesoporous ZSM-5 (meso-ZSM-5) as a support greatly enhanced the gasoline-range

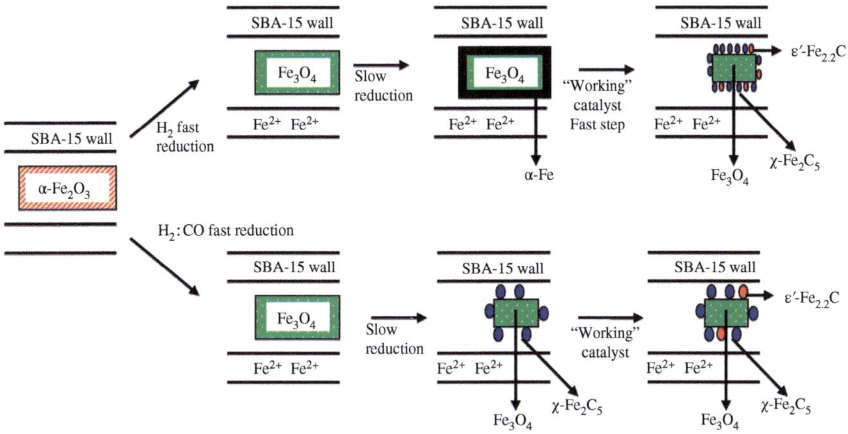

Figure 7.5 Schematic representation of the sequential phase modifications from precursor to "working" catalysts depending on the activation atmospheres. *Source*: Reproduced from [126] with permission of Elsevier.

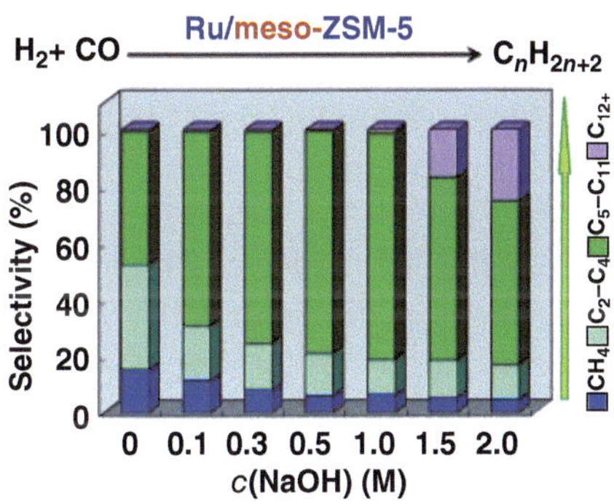

Figure 7.6 Comparison of catalytic behaviors of Ru-based catalysts loaded on ZSM-5 or meso-ZSM-5 zeolites. *Source*: Reproduced from [60] / with permission of John Wiley & Sons.

selectivity of the Ru-based catalyst. Although Ru is more expensive than Co or Fe, its unique properties in FTS make it well suited for fundamental research aimed at rationally designing catalysts with controlled selectivity. Their work demonstrated two key effects of meso-ZSM-5: the mesoporous structure and the unique acidity were crucial for regulating product selectivity toward C_5–C_{11} iso-paraffins. The resulting catalyst yielded a hydrocarbon product with an iso-to-*n*-alkane ratio of approximately 2.7 and a C_5–C_{11} selectivity of around 80%, substantially higher than the theoretical maximum of ~45% predicted by the ASF distribution (see Figure 7.6). In the *n*-hexadecane hydrocracking model reaction, researchers found that the hydrocracking of heavy hydrocarbons proceeded rapidly over their Ru/meso-ZSM-5 catalyst, which possessed mainly Lewis acidity. The larger mesoporous and weaker acidity suppressed excessive cracking, thereby significantly improving the selectivity toward C_5–C_{11} hydrocarbons. Based on this, they subsequently developed an industrially viable Co/meso-ZSM-5 catalyst, which also achieves highly selective productivity of iso-paraffin-rich C_5–C_{11} hydrocarbons.

Gascon and co-workers developed a Co catalyst supported on hierarchical H–ZSM-5 for gasoline fraction hydrocarbons from syngas [127]. Different from Wang and co-workers, they created mesostructure within H–ZSM-5 crystals by a desilication method with an alkali treatment and an acid wash to remove non-framework aluminum [16]. Desiliconization by alkali treatment made the pore size and volume of H–ZSM-5 similar to those of amorphous SiO_2 (traditional support), while acid treatment removed the generated non-framework aluminum and improved its activity. Compared to a conventional Co/SiO_2 catalyst, this bifunctional catalyst exhibited higher selectivity to liquid fractions and reduced wax formation. Experiments confirmed that the addition of zeolite significantly improved the C_5–C_{11} gasoline fraction selectivity due to its strong acid-catalyzed hydrocracking activity. Moreover, the strong interaction of Co species and H–ZSM-5 revealed by H_2-TPR and IR of adsorbed CO revealed a strong Co–zeolite interaction that stabilized low-coordinated Co sites, which in turn increased methane selectivity. This was attributed to the enhanced activity of these coordinatively unsaturated Co sites for hydrogenation and hydrogenolysis reactions.

Additionally, Wang and co-workers developed a mesoporous Y-type zeolite (meso-Y)-supported catalyst, Co/meso-Y, with high performance for syngas conversion to diesel by effectively controlling the hydrogenolysis reaction, in 2015. They identified Co particle size and support porosity as key factors for the outstanding activity and selectivity in FTS. Researchers focused on the effects of H-type zeolite and Na-type zeolite on the distribution of FTS products in this study [108]. H–Y zeolite enables weaker Brønsted acidity than H–ZSM-5, hypothesizing that it would lead to milder hydrocracking and thus higher C_{10}–C_{20} selectivity. However, the H–Y-supported catalyst showed similar diesel selectivity compared to the Co/SiO_2 or Co/Al_2O_3 samples. The Na–Y-supported catalyst, which lacks Brønsted acidity, proved lower C_{21+} selectivity similar to the Co/H–Y sample. The meso-Y-supported catalyst significantly improved the selectivity of diesel as well as decreased the selectivity of undesirable hydrocarbons (see Figure 7.7). Herein, the Co/Na–meso-Y catalyst achieved 60% diesel selectivity, well above the ASF maximum of 39%. In addition, the Co/Na–meso-Y catalyst produced normal alkanes, favoring the improvement of the cetane number. This performance surpassed more expensive catalysts that were known at the time, and was better than the more expensive Ru-based catalyst. Researchers also found that the Co nanoparticles on Co/Na–meso-Y catalyst showed the smallest average particle size (8.4 nm), which may explain its high hydrogenolysis activity. By synthesizing catalysts with various Co particle sizes from 4.9 to 27 nm, they were able to tune the selectivity of *n*-hexadecane hydrogenolysis, demonstrating that this catalytic function could be controlled to favor the production of diesel-range hydrocarbons.

Tsubaki and co-workers developed a metallic bifunctional catalyst prepared by a homemade polygonal barrel sputtering process, which deposits highly dispersed Ru nanoparticles supported on H–Beta zeolite support [128]. They applied this novel preparation process with argon plasma flow, which is effective for the deposition of highly dispersed metal atoms directly onto powder supports

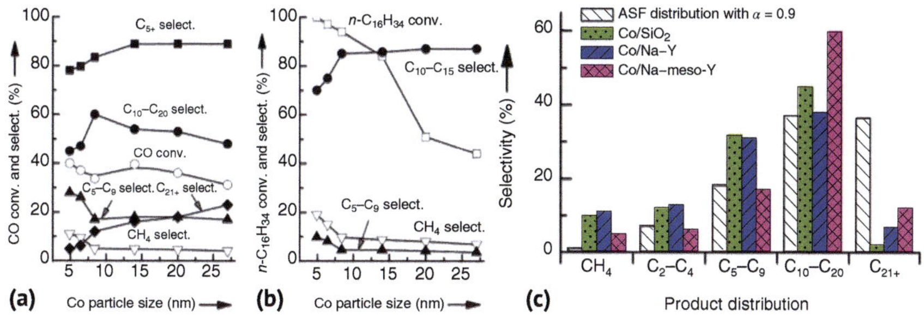

Figure 7.7 Effects of Co particle size on the catalytic behaviors of the Co/Na–meso-Y catalysts. (a) FTS and (b) n-$C_{16}H_{34}$ conversion. (c) Comparison of product selectivity for FTS over Co-based catalysts loaded on SiO_2, Na–Y, and Na–meso-Y. *Source*: Reproduced from [108] / with permission of John Wiley & Sons.

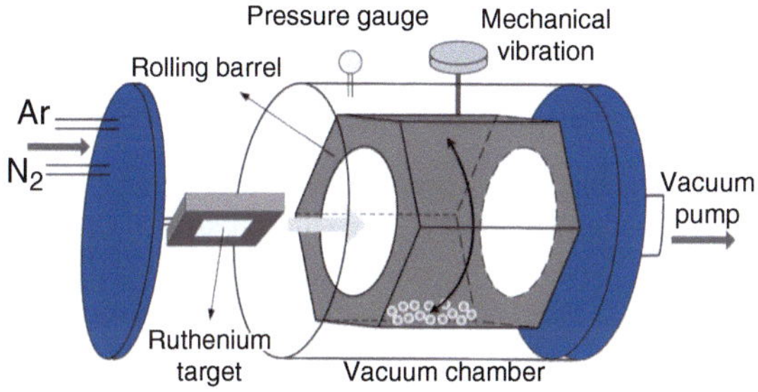

Figure 7.8 Schematic representation of the sputtering apparatus for Ru-based FTS catalyst preparation. *Source*: Reproduced from [128] with permission of American Chemical Society.

(see Figure 7.8). In this method, the sputtered metal atoms are uniformly distributed on the support through continuous hexagonal rotation and mechanical vibration. This "dry" method avoids the production of nitrate-containing wastewater associated with conventional wet impregnation [128]. They prepared a Ru/H–Beta catalyst with highly dispersed Ru nanoparticles (2–4-nm diameter, 31.2% metal dispersion). These sputtered Ru nanoparticles interacted weakly with the acidic zeolite support, a significant difference from impregnated catalysts. The catalyst could be used directly for FTS without any reduction pretreatment, achieving a CO conversion 1.6 times higher than its impregnated counterpart. In addition, the close proximity between the sputtered Ru and the nearby acid sites of zeolite resulted in a gasoline product with a C_{iso}/C_n ratio of 4.6, the highest value reported at the time.

A year later, the same group demonstrated an effective H–USY zeolite-supported Co nanoparticles bifunctional catalyst, also prepared by physical sputtering. The physical sputtering process makes the Co particles nanosized and well dispersed on the acidic USY zeolite [129]. They found that the nanosized cobalt particles were well dispersed and easily activated at a low reduction temperature. This sputtered Co/USY catalyst showed higher CO conversion and iso-paraffin selectivity than a traditional impregnation catalyst. A comparison of different zeolite supports (including H–MOR, H–Beta, and H–USY) revealed that while sputtering improved performance for all, the H–USY support yielded the most significant enhancement.

Researchers concluded that the physical sputtering catalyst synthesis method provides a new way to overcome the problems associated with strong metal–support interaction (SMSI) often seen in heterogeneous catalysis. Unlike the traditional impregnation method, the physical wedge bonding between Co atoms and the zeolite support leads to a weak interaction, which makes the prepared catalyst have high metal dispersion and reducibility [129]. Those results led to the conclusion that the sputtered Co/USY–S catalyst was constituted of Co^0 atoms mostly and only trace amount of Co oxide after reduction under a mild temperature of 260 °C. Compared to the catalyst prepared by the traditional impregnation method, the sputtered catalyst demonstrated higher CO conversion and iso-paraffin selectivity. This synthesis method provides a novel way to overcome the limitations caused by SMSI for the preparation of metal support catalysts with high dispersion and reducibility.

From the above studies, regulating product selectivity by controlling complex reaction pathways is a major challenge in FTS. In 2018, Tsubaki and co-workers reported an integrated catalytic process capable of directly converting syngas into different liquid fuels, including gasoline, jet fuel, and diesel, without subsequent hydro-refining of FTS wax [59]. By combining Co nanoparticles and mesoporous Y-type zeolite (Y_{meso}) and controlling the zeolite's porosity and acidity, they achieved selectivities of 74% for gasoline, 72% for jet fuel, and 58% for diesel (see Figure 7.9). This represented the first one-step, high-selectivity synthesis of jet fuel from syngas. The resulting fuel met the strict standards of the corresponding type of internal combustion engine without further processes. The productivity of liquid fuel types can be controlled by porosity and acidity of Y_{meso} easily. Researchers also established a new product distribution model for these Co-based zeolite catalysts that deviates from the traditional ASF distribution [59]. They identified a cracking contribution parameter that is as important as the chain-growth probability. Rationally controlling these two key parameters could easily enhance the yield of specific products range. This study provided a sample and significant method for the direct synthesis of various types of liquid fuels and offered an important theoretical basis for other complex, multistep reactions.

7.2.3 Zeolite-mixed Catalysts

In addition to using zeolites as a catalyst support, another approach involves physically mixing a conventional FTS active metal with a zeolite. Here, we call this mixed component catalyst as the zeolite-mixed-type catalyst. Unlike supported catalysts, the components in a mixed catalyst are only physically combined, sometimes even in separate packed beds (a dual-bed configuration). This approach prevents direct interaction between the FTS active metal and the zeolite's acidic sites, allowing each component to function independently. This makes it a relatively simple yet powerful bifunctional system, enabling the direct synthesis of high-value, nonfuel chemicals from the FTS reaction.

Corsaro et al. tested a mixed catalyst of commercial Co-based FTS catalyst and HZSM-5 zeolite in a fixed-bed reactor process [130]. The FTS reaction was evaluated at 190 °C, while the HZSM-5 was evaluated at different temperatures

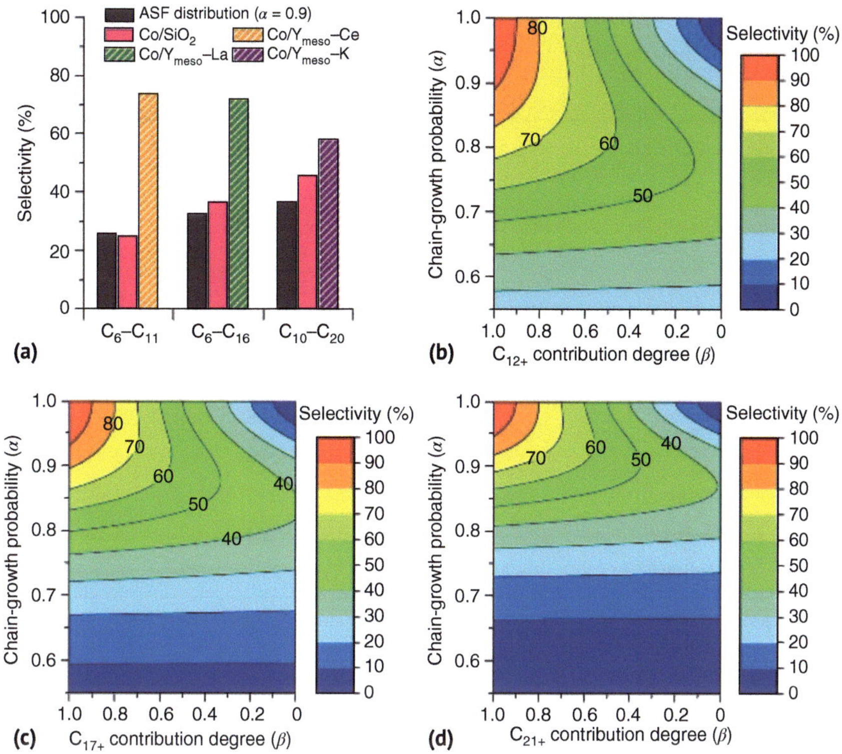

Figure 7.9 Deviation of ideal ASF distribution and bifunctional catalyst distribution models. (a) Comparison of liquid fuel selectivity with ideal ASF distribution. The ideal product selectivity with $\alpha = 0.9$ was obtained from the original ASF prediction. The distribution of Co/Y_{meso} catalysts remarkably deviated from the ASF law. Distribution models for (b) gasoline, (c) jet fuel, and (d) diesel fuel. *Source*: Reproduced from [59] / with permission of Springer Nature.

(250, 300, and 350 °C). The effects of HZSM-5 reaction temperature, flow rate, and aromatics selectivity were studied (see Table 7.1). They found that adding HZSM-5 inhibited the production of CO_2 and CH_4, while significantly promoting the formation of both LPG and aromatics. Improving operating conditions generally increases the amount of LPG produced, while the results for aromatics produced were more inconsistent. Their results indicated that with increasing temperature and loading of the HZSM-5 catalyst, the aromatics yield reached the maximum at 300 °C and 1.25g, respectively. Under optimal conditions, they observed a 4.4% improvement in CO conversion, reduced CO_2 and CH_4 selectivities of 26.3% and 20.6%, respectively, and significant enhancements of C_3, C_4, and aromatics selectivities up to 322.6, 434.5, and 100%, respectively.

In 2017, de Jong and co-workers developed a combination of an Fe-based FTO (FTS-to-olefins) catalyst and a H–ZSM-5 zeolite to produce short-chain olefins and aromatics from syngas (see Figure 7.10) [131]. On the Fe-based catalyst, syngas was converted to produce olefins, a portion of which was subsequently converted to aromatic products on the acidic sites of zeolite. Surprisingly, researchers found that

Table 7.1 Reaction conditions and catalytic performances.

	1	2	3	4	5	6	7
HZSM-5 temperature (K)	523	573	623	623	623	623	623
Flow rate (smL·min^{-1})	70	70	70	100	130	70	70
HZSM-5 loading (g)	2.5	2.5	2.5	2.5	2.5	0.5	1.25
CO conversion (%)	50.5	57.1	49.5	52.3	21.6	78	55.2
CO_2 selectivity [c] (%)	2.1	2.7	1.4	2	1.3	1.3	2.2
C_1 selectivity [c] (%)	6.3	10.4	10.8	10.3	15.7	18.9	13.4
C_2 selectivity [a,c] (%)	0.6	1.1	0.9	0.7	0.9	1.8	0.8
C_3–C_4 selectivity [a,b,c] (%)	12.9	23.7	28.6	37.5	36.8	22.9	27.6
C_{5+} selectivity [a,b,c,d] (%)	43.8	25.8	29.9	24.3	43	43.4	26.4
Aromatics selectivity [d] (%)	34.3	36.2	28.4	25.2	02.3	11.6	29.6
Total gas selectivity (%)	34.9	49.8	52.9	63.4	72	51.7	55.6
Total liquid selectivity (%)	65.1	50.2	47.1	36.6	28	48.3	44.4

Source: Reproduced from [130] / with permission of Taylor and Francis Group.

a. Paraffins and olefins.
b. Includes *n*- and iso-paraffins.
c. Gas.
d. Liquid.

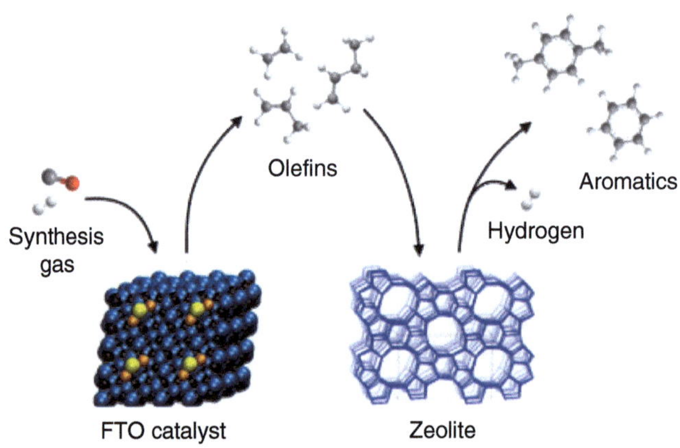

Figure 7.10 Proposed pathway for the conversion of synthesis gas to aromatics: conversion of synthesis gas to olefins on the FTO catalyst and dehydroaromatization of olefins to aromatics on the zeolite. *Source*: Reproduced from [131] / with permission of John Wiley & Sons.

the aromatization reaction proceeded via a dehydrogenation pathway rather than a hydrogen transfer pathway (which would co-produce three alkane molecules for every aromatic molecule). This allowed them to achieve a combined carbon selectivity of 70–80% (the sum of light olefins and aromatics) at a reaction pressure of 0.1MPa. Increasing the hydrogen partial pressure also led to a large amount of hydrogenation of olefins to alkanes. At low pressures, this olefin aromatization pathway enabled the conversion of syngas into chemicals with carbon selectivity up to 73% (55% light olefins and 18% aromatics) and without the formation of unwanted alkanes by adjusting the reaction conditions and reactor configuration (mixed versus stacked bed). In addition, the light olefins selectivity and the all olefins selectivity in the C_{5+} hydrocarbons reduced in the presence of zeolite, indicating that aromatics were formed not only from light olefins but also from heavy olefins.

They also found that adding Na and S as promoters in the Fe-based FTS catalyst improved its activity and selectivity for light olefins [132]. The performance of the Fe-based catalyst without promoter decreased at the beginning, while the activity of the Fe-based catalyst with promoter increased by two times after the addition of zeolite. Furthermore, close mixing of the FTS catalyst with zeolite influenced its activity by affecting the degrees of reduction and carburization of the Fe phase. This study highlighted how the proximity of the two catalytic functions in a bifunctional system can significantly impact performance.

Ma and co-workers achieved high aromatics selectivity and high CO conversion by physically mixing a selective Na–Zn–Fe_5C_2 catalyst with a mesoporous H–ZSM-5 zeolite [110, 133]. The hierarchical pore structure and optimized Brønsted acid site density and strength of the zeolite were considered as key to achieving high aromatics selectivity. Based on their olefin conversion experimental results, olefins are considered to play the role of an irreplaceable central hub. The selectivity of heavy hydrocarbons (C_{12+}) indicates that they were cracked into olefins over the acid center and then converted into aromatics and light alkanes (see Figure 7.11). Notably, the selectivity of CH_4 and CO_2 is less than 35%, ensuring that most of the carbon resources are converted into aromatics (mainly C_6–C_9 light aromatics) and light hydrocarbons [110]. Researchers also precisely adjusted the pore structure and acidity of H–ZSM-5 by making the alkaline treatment concentration to dominate. The appropriate Brønsted acid center density and strength and the hierarchical pore of H–ZSM-5 give the catalyst an unprecedented aromatic productivity. This study revealed its broad development space in the field of FTS.

Ding and co-workers synthesized a hollow ZSM-5 zeolite nanoparticle with tunable acidity, which was combined with Fe_3O_4@MnO_2 to directly convert syngas into aromatics-rich gasoline, showing an excellent selectivity of 73.5% for gasoline-range hydrocarbons [134]. This tandem catalyst achieved an impressive gasoline-range hydrocarbons selectivity (containing 52.8% aromatics) of 70% for CO conversion, and it remained stable for 100 hours. In this study, both the Lewis and Brønsted acid centers of the zeolite support were precisely found to be the active sites to catalyze the polymerization of light hydrocarbons and the cracking of heavy hydrocarbons. In particular, the Brønsted acid centers of hollow ZSM-5 zeolites promoted the aromatics formation reaction. Syngas first converted into a wide range of products, including light hydrocarbons, gasoline, and heavy hydrocarbons via the primary FTS reaction (see Figure 7.12). Weak acidic centers of added ZSM-5 zeolite enabled polymerization and the cracking reaction, so that light and heavy hydrocarbons are enriched to produce gasoline. In particular, the strong acidic centers promoted the complete cracking of heavy hydrocarbons, further improving the selectivity of gasoline-range products. Researchers indicated that weak acids were not the

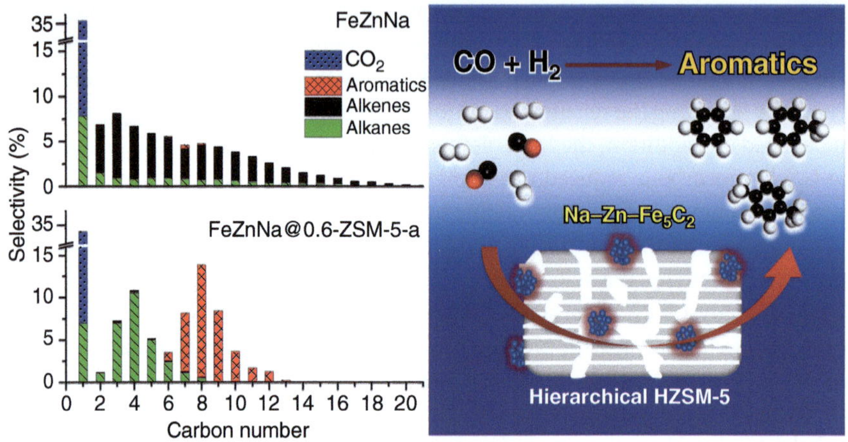

Figure 7.11 Comparison of product distribution over FeZnNa and FeZnNa@0.6-ZSM-5-a with all the carbon-containing products counted. *Source*: Reproduced from [110] with permission of Elsevier.

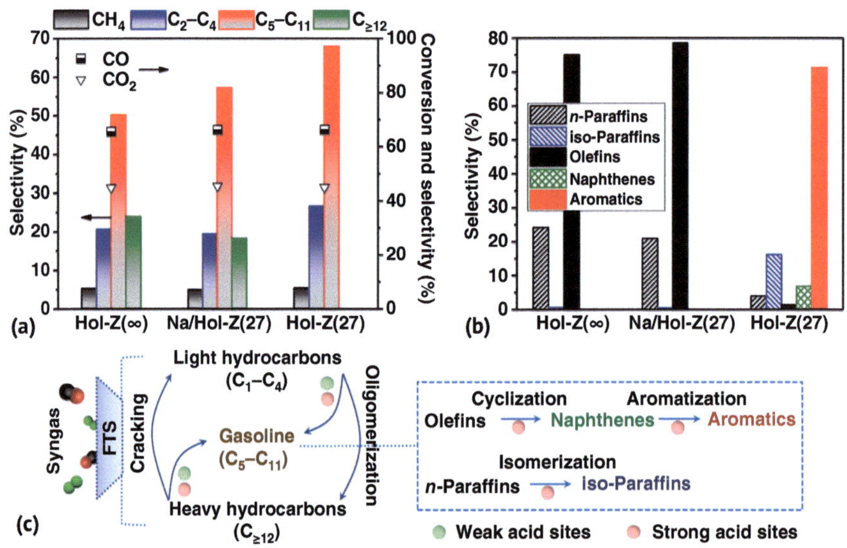

Figure 7.12 Effects of weak and strong acids on the primary FTS reaction. (a) Conversion and product selectivity and (b) hydrocarbon distribution within C_5–C_{11}. (c) Proposed scheme for directly transforming syngas to aromatics-rich gasoline. *Source*: Reproduced from [134] with permission of American Chemical Society.

active centers to catalyze cyclization, isomerization, and aromatization reactions, and only strong acids showed reactivity to produce more high-octane cycloalkanes, iso-alkanes, and aromatics. The introduction of hollow structures further decreased the diffusion length of nano-zeolite particles and improved their stability, showing great activity in producing aromatics-rich gasoline.

Xiao and co-workers reported that an MFI-type zeolite with short *b*-axis and a controlled microporous structure could highly promote the FTO reaction by facilitating the transport of olefin products, in 2022 [109]. In this study, an Fe_xC_y catalyst (Na–FeC_x) and a promoter, MFI-type zeolite, prepared by a sample physical mixture method, enabled a CO conversion of 82.5% and an olefin selectivity of 72% at a low temperature of 260 °C (see Figure 7.13). In contrast, the Na–FeC_x catalyst alone showed poor activation and significantly low olefin yields. Those results indicated that well-designed zeolites, as a promising co-catalyst, could significantly promote FTO by accelerating the product's escape from the catalyst surface. For finding out the role of zeolites in this reaction, researchers tested Na–FeC_x's catalytic performance in combination with various zeolites, and all combinations had higher CO conversions than Na–FeC_x alone. Density functional theory (DFT) demonstrated that acidic sites or silanol groups within the zeolite enhanced the adsorption of ethylene, indicating the important porous environment effect of the zeolite. Furthermore, theoretical studies made in this study demonstrated that a reasonably controlled microporous environment and a short *b*-axis thickness of zeolite can accelerate the escape of olefin product molecules. This feature was conducive to a free catalyst surface generation in CO hydrogenation. In addition to equilibrium changes, future studies should also consider other explanations for the zeolite promotion effect to gain a profound insight into the synergistic effect between

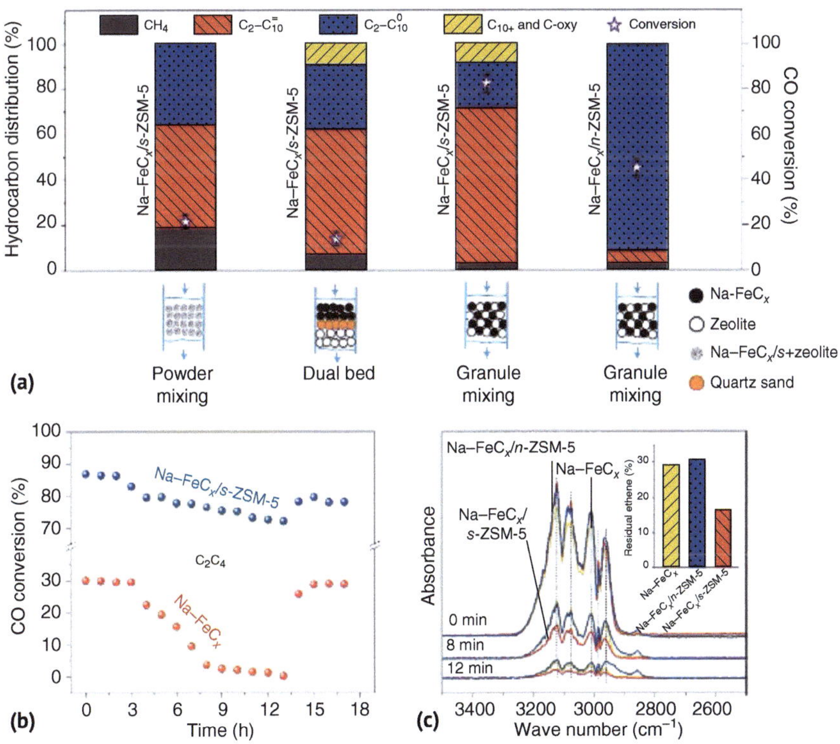

Figure 7.13 Catalytic data for the Na–FeC_x/zeolite catalysts in LT-FTO and ethene desorption DRIFT spectra over Na–FeC_x/zeolite catalysts. *Source*: Reproduced from [109] / with permission of Springer Nature.

Fe_xC_y and zeolites, thereby opening a new path for the development of more efficient catalysts for FTS.

In the same year, Tsubaki and co-workers systematically reported a study on the morphology effect of ZSM-5 zeolite on syngas to aromatics via the FTS route [135]. Researchers synthesized ZSM-5 zeolites with different morphologies, including nano-ellipsoids (Z5-NEs), nano-cubes (Z5-NCs), nanoaggregates (Z5-NAs), nano-beads (Z5-NBs), and nano-pile (Z5-NPs) (see Figure 7.14). These homemade ZSM-5 zeolites all showed better performance in FTS reactions when compared to the commercial ZSM-5 zeolites (Z5-CMs). They found that the catalyst combining FeMnK with ZS-NCs enabled the highest xylene selectivity of 12% and the highest total aromatics selectivity of 42.5% [135]. Studies on the reaction temperature showed that higher temperatures suppressed the selectivity of CO_2, but also limited aromatics selectivity to around 44% due to increased hydrocracking activity of ZSM-5 zeolite. In addition, the prepared ZSM-5 with different morphologies in this study provided a novel view of the design and utilization of zeolites in the FTS.

More recently, Tsubaki and co-workers designed a dynamic "dual-engine-driven" (DED) catalytic system for the direct synthesis of *para*-xylene (PX) from CO_2-containing syngas with ultrahigh yield [136]. The system combined two syngas conversion "engines," an ZnCr component for methanol synthesis and an FeMn component for olefin synthesis via FTS, with an ZSM-5@SiO_2 (Z5@SiO_2) capsule zeolite. The DED catalyst exhibited an extremely high PX STY of 36.1 $g_{px} \cdot kg_{cat}^{-1} \cdot h^{-1}$, approximately eight times that of conventional single-engine catalyst ZnCr–Z5 (see Figure 7.15). DED controls the ZnCr engine to produce methanol and drives the FeMn engine to produce light olefins through the FTS. Both intermediates are then synergistically in situ converted into PX-rich aromatics on the acidic sites of

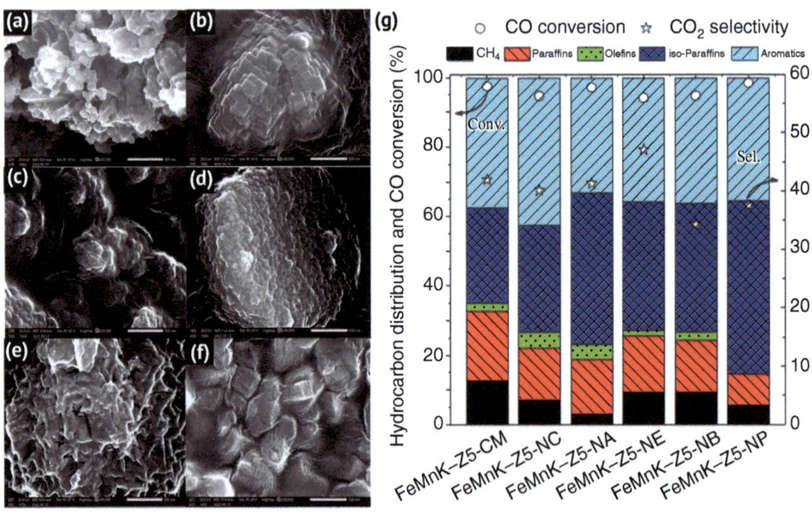

Figure 7.14 SEM images of ZSM-5 zeolites with different morphologies: (a) Z5-NEs, (b) Z5-NCs, (c) Z5-NAs, (d) Z5-NBs, (e) Z5-NPs, and (f) Z5-CMs. *Source*: (a–f) Reproduced with permission from [135] / with permission of John Wiley & Sons. (g) Catalytic performances of various FeMnK–ZSM-5 hybrid catalysts. *Source*: Reproduced from [135] / with permission of John Wiley & Sons.

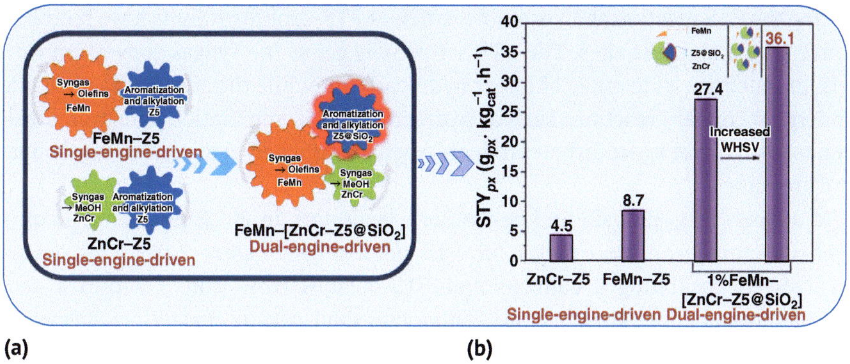

Figure 7.15 (a) Schematic diagram of DED catalytic system and (b) reaction performances of four types of catalysts. *Source*: Reproduced from [136] / with permission of Springer Nature.

Z5@SiO_2. Researchers found that syngas/CO_2 can be quickly and efficiently converted to olefins on the FeMn catalyst, the FTS engine, and then the olefins further undergo aromatization on Z5@SiO_2 sample. Another methanol engine, ZnCr, provides the necessary methanol to drive the benzene alkylation reaction to produce PX. Due to the presence of olefins quickly pre-generated by the olefin engine, side reactions such as methanol to olefins can be greatly suppressed. This customized catalytic system constructed by the DED catalysts solves the conversion–selectivity limitation of the traditional single-engine route to synthesize PX.

Importantly, the researchers found that the activity of the DED catalyst was obviously affected by the binding mode and distance of the active components. The beneficial distance between FeMn and Z5@SiO_2 should be large, while ZnCr and Z5@SiO_2 need to be closer, which will make the synthesis of PX-rich aromatics more efficient. DFT calculations and further characterization results indicated that the contribution of olefin intermediates is significantly higher than that of methanol, which dominates the reaction pathway of aromatic synthesis and reduces the adsorption energy of CH_3O^* intermediates on Z5@SiO_2, enhancing the toluene alkylation reaction with methanol, thereby achieving efficient synthesis of *para*-toluene from syngas. This study realized the industrial production of nonpetroleum PX and promoted the knowledge of designing innovative heterogeneous catalysts for FTS.

7.2.4 Core–Shell-structured Catalysts

While zeolite-supported and mixed catalysts can achieve good FTS activity and enhanced liquid fuel selectivity, they have limitations. For a tandem reaction ($A \rightarrow B \rightarrow C$), two catalyst types are required. In a physical mixture of an FTS catalyst (catalyst A) and a zeolite (catalyst B), syngas is first converted to the linear hydrocarbons on catalyst A. These intermediates must then migrate to catalyst B for next reactions to produce specific hydrocarbons. However, a portion of these primary products may desorb and exit the reactor without ever contacting the zeolite. This leads to wider product distribution and lower selectivity for the target product.

Taking the FTS reaction as an example, efficient FTS-zeolite catalysts have both FTS active sites and acidic sites. The FTS active sites act as the syngas conversion centers, producing a wide range of linear hydrocarbons, while the acidic sites of zeolite undergo secondary reactions such as hydrocracking, isomerization, and aromatization to form target hydrocarbon products, such as branched hydrocarbons, gasoline, and aromatics.

To address this, Tsubaki and co-workers, beginning in 2005, pioneered a core–shell-structured capsule catalyst for FTS process (see Figure 7.16). This design involves encapsulating a conventional FTS catalyst (the "core") within a zeolite membrane (the "shell") [138]. This core–shell capsule catalyst achieves its high-performance preparation of light hydrocarbons, especially gasoline component iso-paraffins, directly synthesized from syngas. In this capsule catalyst, all primary products formed in the core must pass through the zeolite shell before exiting. The hydrocarbon distributions of four types of FTS catalysts have been compared (see Figure 7.17). The hydrocarbon distribution of the mixed catalyst is narrower than that of the Co/SiO_2 catalyst because the waxy products that migrate from the conventional FTS catalyst to the surface of the zeolite catalyst undergo secondary isomerization and hydrocracking to produce lighter hydrocarbons containing iso-paraffins.

Furthermore, researchers found that the sequential isomerization and hydrocracking reactions catalyzed by the zeolite shell decreased selectivity toward long-chain alkanes while significantly increasing it toward short-chain iso-paraffins. Unlike the stochastic product–catalyst interactions in a mechanical mixture, the core–shell architecture of the capsule catalyst mandates that all primary products from the Co/SiO_2 core must diffuse through the zeolite shell. This configuration ensures that all long-chain waxy products undergo secondary conversion. Since the diffusion rate of hydrocarbons through the zeolite is chain-length-dependent,

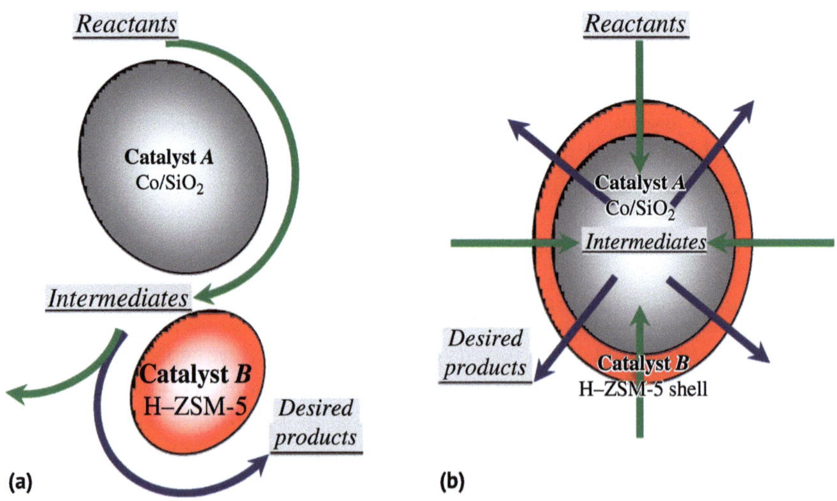

Figure 7.16 (a) Classic catalyst mixture versus new capsule catalyst; and (b) confinement effect of capsule catalyst cuts the intermediates. *Source*: Reproduced from [137] with permission of Elsevier.

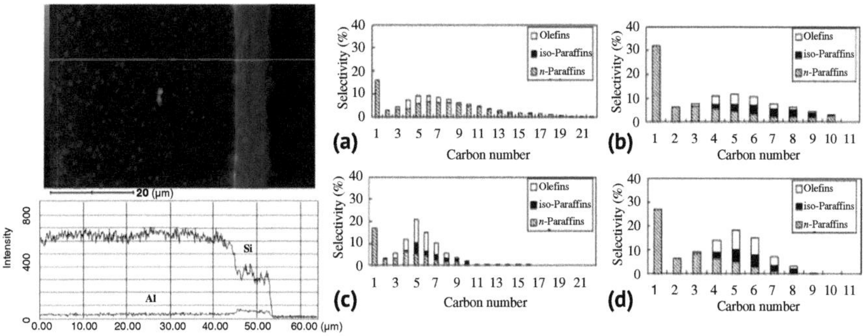

Figure 7.17 A cross-sectional image of one capsule catalyst (2-Co/SiO_2–zeolite) and the intensities of Si-Kα and Al-Kα X-ray signals from the EDX line scan indicated in the image. FTS product distributions on (a) Co/SiO_2, (b) Co/SiO_2 + zeolite–*MX*, (c) 2-Co/SiO_2–zeolite, and (d) 1-Co/SiO_2–zeolite–S catalyst. *Source*: Reproduced from [138] with permission of American Chemical Society.

longer-chain molecules exhibit longer residence times within the shell, leading to their comprehensive cracking and isomerization. Consequently, the capsule-like catalysts yield an extremely narrow distribution of hydrocarbon products, truncating at C_9–C_{10}, whereas the product from the mechanically mixed catalyst still contains trace amounts of C_{13}–C_{20} hydrocarbons. These results demonstrated that the zeolite shell achieves superior short-chain products selectivity, suppressing the production of long-chain hydrocarbons and yielding higher fractions of iso-paraffins and olefins completely. In essence, the architecture of capsule catalyst overcomes the probabilistic nature of secondary reactions seen in mechanical mixtures, leading to a much higher conversion efficiency.

Further research extended this concept to Co/Al_2O_3 cores encapsulated by H–Beta zeolite shells (see Figure 7.18). The encapsulated catalyst completely suppressed the formation of heavy alkanes (C_{12+}) and made intermediate iso-paraffins the main product [139]. Interestingly, the core–shell catalyst also exhibited lower methane selectivity. This was attributed to the hydrophilic nature of the zeolite shell, which can lead to a rich CO and poor H_2 atmosphere near the core, and the possibility that iso-paraffins formed in the shell can diffuse back to the core and reenter the chain-growth process, consuming CH_2 intermediates that would otherwise be hydrogenated to methane. This showed that the core–shell-structured zeolite-coated catalyst extends a complete and confined FTS reaction environment for intermediate iso-paraffins production.

It is also important that the zeolite-coated Co/Al_2O_3 catalyst exhibits lower methane selectivity than the conventional FTS catalyst, a phenomenon linked to the hydrophilicity of the zeolite shell. With the decrease in Si/Al ratio in the zeolite, its hydrophilicity increases. It was observed that the zeolite membrane possesses a nonuniform Si/Al gradient, with a higher concentration of aluminum near the Al_2O_3 substrate, resulting in a strongly hydrophilic inner region. According to the solubility rule of like dissolves like, this localized hydrophilicity promotes water retention, which in turn leads to CO accumulation. The resulting higher local CO/H_2 ratio likely suppresses methane formation. Furthermore, a concentration

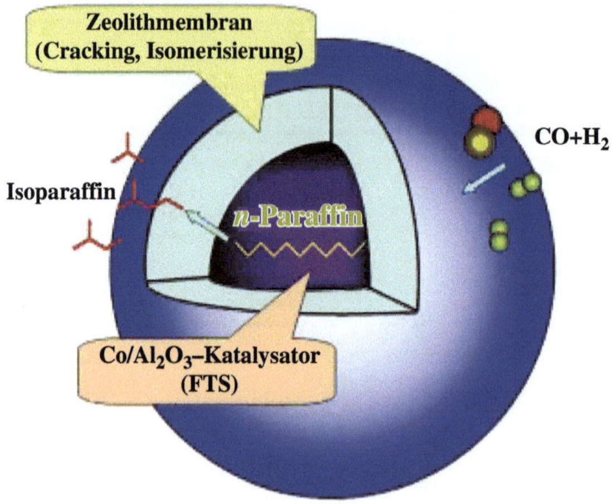

Figure 7.18 Representation of the Co/Al_2O_3@H-Beta core–shell catalyst. *Source*: Reproduced from [139] / with permission of John Wiley & Sons.

gradient can drive the back-diffusion of iso-alkanes and olefins from the zeolite channels to the core catalyst. There, they recombine with CH_2 intermediates, competing with the hydrogenation pathway that forms methane, thus, further reducing its selectivity.

In addition, Ru-based catalysts with different core metals and sizes have also been successfully prepared and studied. Tsubaki and co-workers prepared three H–ZSM-5 zeolite capsule catalysts coated on Ru/SiO_2 cores of varying dimensions [140]. Characterization results showed that defect-free H–ZSM-5 capsules were constructed on the Ru/SiO_2 surface. Researchers found that smaller core particles with larger surface areas facilitate more effective capsule growth. In the FTS reaction, the Ru/SiO_2–*Z* capsule catalysts, especially zeolite capsule catalysts with small particle sizes (Ru/SiO_2–*M*–*Z*), achieved superior iso-alkane selectivity compared to traditional or physically mixed catalysts of Ru/SiO_2 and H–ZSM-5 zeolite. Notably, controlling the size of the core particles, the structure and reaction performance of zeolite capsule catalysts can be regulated, opening a new path for their design and preparation. These findings are directly applicable to Co-based catalyst systems [141]. Given their lower cost and performance comparable to Ru-based catalysts, strategically designed Co-based capsule catalysts hold significant potential for large-scale commercial applications (see Figure 7.19).

Zeolite capsule catalysts with core–shell structures have demonstrated excellent performance in tandem catalytic processes and the one-step synthesis of light iso-alkanes from syngas. However, difficulty and complexity of the hydrothermal synthesis method commonly used to prepare zeolite shells limit the application of various types of zeolite shells with metal/metal oxide core and the industrial scalability of zeolite capsule catalysts. To overcome the limitations of the traditional hydrothermal synthesis method for creating zeolite shells, Tsubaki and co-workers developed alternative preparation methods, including a novel physical adhesion (PA) method [58, 137]. This non-hydrothermal technique (see Figure 7.20) offers a

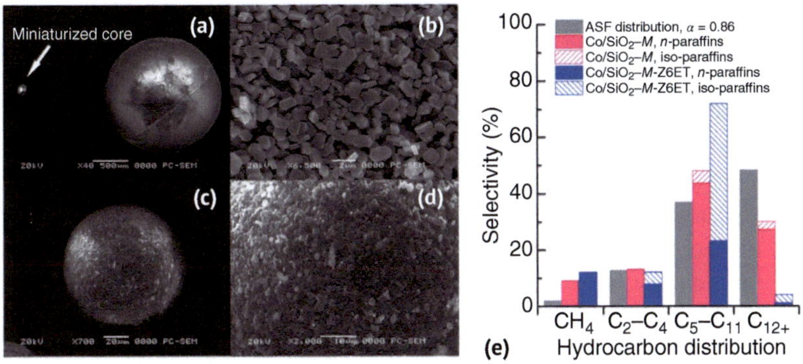

Figure 7.19 SEM images of the Co/SiO_2 cores, pure ZSM-5, and miniaturized capsule catalyst. (a) Comparison of miniaturized core and conventional large core, (b) pure ZSM-5, (c) Co/SiO_2–*M*–Z6ET, and (d) outer surface of Co/SiO_2 –*M*–Z6ET. *Source*: Reproduced with permission from [141] with permission of Royal Society of Chemistry. (e) Comparison of FT product selectivity on Co/SiO_2 –*M* and Co/SiO_2–*M*–Z6ET and ideal ASF distribution. *Source*: Reproduced from [141] / with permission of The Royal Society of Chemistry.

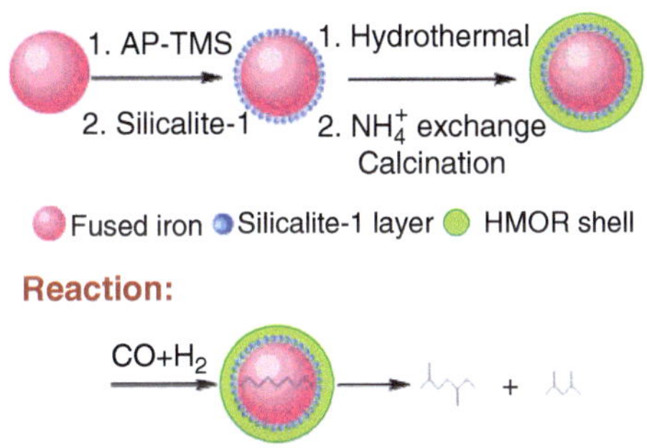

Figure 7.20 A H-mordenite zeolite shell is directly synthesized by PA method, without an organic template, on a Silicalite-1-coated FTS catalyst. *Source*: Reproduced from [58] / with permission of John Wiley & Sons.

low-cost, reliable, and scalable route to producing core–shell catalysts, making their industrial application more feasible. In this research, two kinds of methods for preparing zeolite capsule catalysts using the traditional 10-wt.% Co/SiO_2 FTS catalyst as the core catalyst were proposed. One is a modified hydrothermal synthesis method. The other method is called as the PA method. Both H-ZSM-5 zeolite capsule catalysts prepared by hydrothermal and PA methods achieved high reaction performance for light iso-alkanes synthesis, which were superior to traditional mixed catalysts. The PA method for preparing zeolite capsule catalysts proposed for the first time in this research had low cost, reliable performance, and strong scalability. In addition, the concepts of zeolite capsule catalysts and PA methods could also bring new opportunities for the development of wider types of zeolite-capsulized catalysts for different tandem catalytic processes, making it easier to achieve industrial production of zeolite capsule catalysts.

Nevertheless, Khodakov and co-workers proposed a different synthesis strategy for creating hierarchical metal–zeolite nanocomposites, in 2020 [142]. The synthesis of the nanocomposite in this research was divided into three steps (see Figure 7.21): (1) the parent (core) zeolite is etched with an ammonium fluoride solution. Etching produces small mesopores inside the zeolite crystals; (2) Ru nanoparticles dropped using an oil-in-water microemulsion are deposited in the mesopores of the zeolite; and (3) at last, a zeolite shell of an MFI-type zeolite (Silicalite-1 or ZSM-5) is grown on the parent zeolite crystals, covering the etched surface and the metal nanoparticles. They found that the proximity of metal sites to acid sites within the zeolite shell is the key factor for achieving high gasoline-range fuels with Ru–zeolite composite catalysts. In contrast, the acidity of zeolite showed a limited effect for product distribution. These catalysts aimed to improve the interaction between metal nanoparticles and acidic active sites by encapsulating the metal within a zeolite matrix. The reaction performance of these catalysts was mainly determined by the Ru content, the composition and acidity of the catalyst shell, as well as the existence of secondary mesopores. Specifically, the reaction rate increased with higher Ru content. Furthermore, using meso-ZSM-5 or Silicalite-1 as the core helped to reduce diffusion limitations within the composite catalysts. Interestingly, catalysts with ZSM-5 zeolite in the shell showed a considerable increase in the relative yield of iso-paraffins. This was crucial because the isomerization of hydrocarbons from FTS requires both metal and acidic centers. Consequently, the close proximity of these centers in the zeolite shell is a critical parameter for effective bifunctional catalysts design for the selective production of gasoline fuels via FTS.

Most recently, Weckhuysen and co-workers reported the synthesis of double-shelled hollow spheres (DSHSs) of metal–zeolite composites for FTS. They demonstrated a simple strategy for synthesizing double-shelled, continuous metal oxide@zeolite hollow spheres (denoted as MO@ZEO DSHSs) with controllable structural parameters and chemical composition [143]. The strategy involves generating zeolite crystals by a self-assembly process. After preparation and calcination, the zeolite is grown successfully onto the ion-containing carbon spheres surface. Compared to the traditional mixture of Fe_2O_3/H–ZSM-5 catalysts, the developed Fe_2O_3@H–ZSM-5 DSHSs showed catalytic CO conversion and C_5–C_{11} hydrocarbon selectivity of 79 and 64%, respectively, in the production of FTS gasoline. Thus, researchers

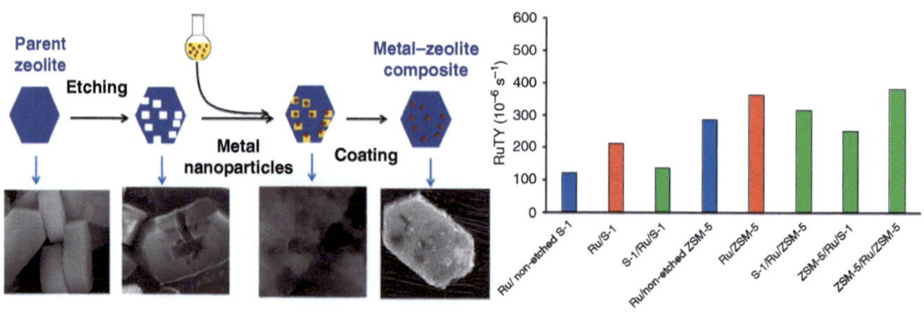

Figure 7.21 Synthesis of the zeolite catalysts with encapsulated Ru nanoparticles and RuTY for Ru–zeolite nanocomposite catalysts. *Source*: Reproduced from [142] with permission of American Chemical Society.

demonstrated the performance of Fe_2O_3@H-ZSM-5 DSHS catalyst, which was significantly synthesized from Fe_2O_3 and H-ZSM-5 composites for high gasoline production in FTS. Each structural engineering element in Fe_2O_3@H-ZSM-5 DSHS involves the designed multi-shell hollow sphere, the core-shell Fe_2O_3@H-ZSM-5 structure, and the "cage" effect of the H-ZSM-5 zeolite shell, which contribute to the improved FTS performance (see Figure 7.22).

On the other hand, after 45 hours of FTS, the Fe_2O_3@H-ZSM-5 DSHS catalyst particles maintained their physical integrity and the hollow iron ball was not destroyed. Researchers thought that it may be the protection effect of the zeolite shell, while the iron spheres are more porous than fresh catalysts. Interestingly, after a two-hour regeneration process (500 °C in air), the catalyst recovered to its fresh structure. Compared to existing bifunctional systems, this DSHSs catalyst proved to be a new preparation method for high-performance metal oxide-zeolite bifunctional catalysts, and indicated the significance of structural engineering at nano-based scales for the catalysts' performance in FTS.

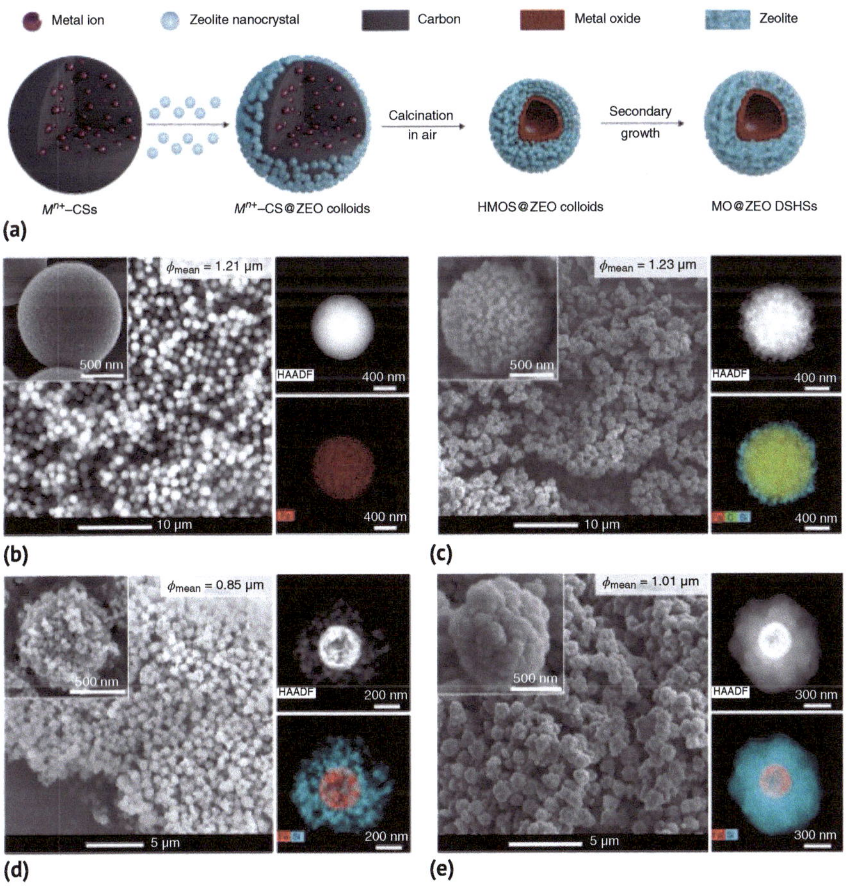

Figure 7.22 Synthesis of MO@ZEO DSHSs of metal–zeolite composites. Panel (a). *Source*: Reproduced from [143] / Springer Nature. Panels (b–e). *Source*: Reproduced from [143] / Springer Nature.

7.3 Zeolite-based Membrane Reactors

Membrane reactors offer another advanced approach for improving FTS performance (see Figure 7.23). By allowing the controlled addition of a reactant along the reactor length (distributed feed) or the selective removal of a product, they can enhance selectivity and improve temperature control. Zeolites, with their well-defined micropores and tunable hydrophilicity/hydrophobicity, are excellent candidates for use as membranes in these reactors [83].

Two main concepts have been explored for FTS [84]. The first is distributed feeding. Since the activity and selectivity of Co-based catalysts are highly dependent on the local H_2/CO ratio, distributing one of the reactants through a membrane can maintain an optimal ratio along the reactor. For example, feeding H_2 through the reactor bed while distributing CO through a tubular membrane can maintain a high H_2/CO ratio, reducing CO-induced inhibition and increasing conversion, although this often favors light hydrocarbons formation. Conversely, distributing H_2 into a primary CO feed stream leads to a lower H_2/CO ratio, which can increase the yield of C_{10+} hydrocarbons and decrease methane selectivity. Using an ZSM-5 membrane can further alter the product distribution via secondary acid-catalyzed reactions (see Table 7.2).

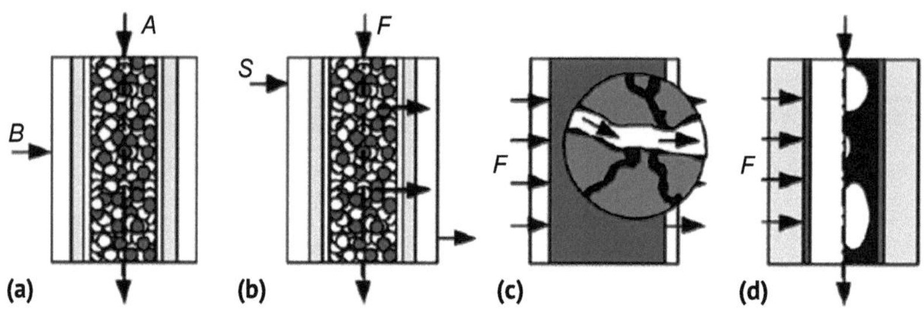

Figure 7.23 Membrane reactors for FTS proposed in the literature. *Source*: Reproduced from [83] with permission of Elsevier.

Table 7.2 Concepts of membrane applications in FTS reactors.

Number	Concept	Reactor	Membrane/support	Catalyst	X_{CO}	S_{C2+}	Y_{C2+}
1(a)	Distributed feeding	PBNMR;	γ-Al_2O_3/α-Al_2O_3	Co/α-Al_2O_3;	+[a]	−[a]	−[a]
		PBNMR		ZSM-5/α-Al_2O_3	−[b]	+[b]	+[b]
1(b)	Selective water removal	PBNMR	Mordenite/ZSM-5/Silicalite/stainless steel	NA			

Number	Concept	Reactor	Membrane/support	Catalyst	X_{co}	S_{C2+}	Y_{C2+}
1(b)	Selective water removal + reactant distribution	PBMR	$Si(OH)_xO_y$/ γ-Al_2O_3/α-Al_2O_3	Fe/Al_2O_3/ 5K/Cu	+[c]		+[c]
1(b)	Selective water removal	PBMR	4A-zeolite/TiO_2/ stainless steel	NA			
1(c1)	Forced-through-flow catalytic membrane	PBMR	Porous catalyst/ copper structure	Co/Al_2O_3	+	+	+
1(c2)	Forced-through-flow catalytic membrane	CNMR	Catalyst/ γ-Al_2O_3/α-Al_2O_3	P/Pt–Co/Al_2O_3	+	−	+

Source: Reproduced from [83] with permission of Elsevier.
Note: PBNMR, Packed bed non-permselective membrane reactor; PBMR, packed bed membrane reactor; CMR, catalytic membrane reactor; and CNMR, catalytic non-permselective membrane reactor.
a. CO-fed across membrane.
b. H_2-fed across membrane.
c. Based on CO_2.

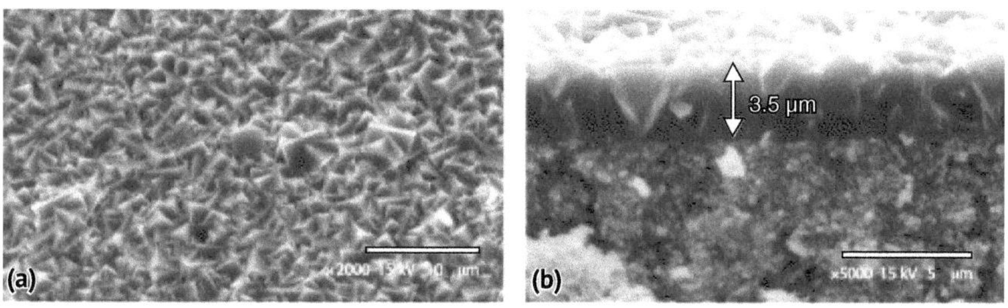

Figure 7.24 SEM images of the zeolite-4A membrane on the UV-pretreated TiO_2-coated stainless-steel support: (a) top view and (b) cross-sectional view. *Source*: Reproduced with permission from [144]/Elsevier.

The second concept is selective water removal. Water is the most abundant by-product of the FTS reaction, and its removal can shift the reaction equilibrium toward hydrocarbons formation, thereby increasing conversion. Moulijn and co-workers synthesized a 4A zeolite membrane (see Figure 7.24) that exhibited high permeability and selectivity for water [144]. Due to the strong adsorption of water within the 4A zeolite cages, the permeation of other gases like CO, H_2, and CH_4 was suppressed. This makes such membranes highly effective for the in situ removal of water from the FTS reaction zone, demonstrating their potential value for process intensification.

7.4 Conclusion and Outlook

Zeolites, characterized by their ordered porous structures and tunable acidity, are increasingly playing a pivotal role in FTS. The key challenges faced by FTS are achieving high selectivity and controlled product distribution. In conventional FTS, Fe-, Co-, or Ru-based catalysts typically yield a broad product spectrum, encompassing methane, liquid hydrocarbons, and even waxes. The incorporation of zeolites, particularly their integration with FTS active components to form composite catalysts, can significantly enhance product selectivity. This part introduced three types of zeolite-containing FTS catalysts: zeolite-supported catalysts, zeolite-mixed catalysts, and zeolite-based core–shell catalysts, and explained the important innovations and new understandings that zeolites have brought to the research on FTS catalysts in recent years. In addition, zeolite membrane reactors have also brought new ideas for the design of FTS reactors.

Overall, the unique pore size exclusion and the acidic site effects of zeolites can restrict the formation of larger molecules, thereby increasing the selectivity toward specific carbon numbers (e.g., gasoline, diesel, and jet fuel). Moreover, these diversified secondary reactions, such as olefin isomerization, oligomerization, or even aromatization, also improve product value. Encapsulating or closely associating FTS active components within the zeolites can improve the catalyst's resistance to sintering and overall stability.

Despite the significant potential demonstrated by zeolites in FTS, several promising research avenues warrant further exploration: (1) development of novel zeolite frameworks: research must focus on discovering new zeolite frameworks with more precise pore sizes, superior structural stability, and more flexible acidity control, aiming for a finer tailoring of product distributions; (2) design of multifunctional integrated catalysts: developing composite catalysts where FTS active sites and zeolite acidic sites are optimally arranged in space is crucial for more efficient tandem reaction; (3) optimization of reaction conditions: leveraging the specific properties of zeolites, further optimization of FTS operating conditions (temperature, pressure, and H_2/CO ratio) is essential to maximize the yield and selectivity of target products; and (4) investigation of deactivation mechanisms: a deeper understanding of the deactivation mechanisms of zeolite composite catalysts under FTS conditions is needed to develop more coke-resistant and sintering-tolerant zeolite modification strategies.

As scientists and engineers around the world continue to deepen their understanding of the structure–performance relationship of zeolites, coupled with advances in material synthesis technology, zeolites will play a more important role in future FTS, providing more efficient and economical solutions for the XTL conversion process.

References

1 Wilhelm, D.J., Simbeck, D.R., Karp, A.D. et al. (2001). Syngas production for gas-to-liquids applications: technologies, issues and outlook. *Fuel Process. Technol.* 71 (1–3): 139–148.
2 Rauch, R., Kiennemann, A., and Sauciuc, A. (2013). Fischer-Tropsch synthesis to biofuels (BtL process). In: *The Role of Catalysis for the Sustainable Production of Bio-fuels and Bio-chemicals* (eds. K.S. Triantafyllidis, A.A. Lappas, and M. Stöcker), 397–443. Amsterdam: Elsevier.
3 Torres Galvis, H.M. and de Jong, K.P. (2013). Catalysts for production of lower olefins from synthesis gas: a review. *ACS Catal.* 3 (9): 2130–2149.
4 Chen, Y., Wei, J., Duyar, M.S. et al. (2021). Carbon-based catalysts for Fischer–Tropsch synthesis. *Chem. Soc. Rev.* 50 (4): 2337–2366.
5 Rostrup-Nielsen, J.R. (2000). New aspects of syngas production and use. *Catal. Today* 63 (2–4): 159–164.
6 Göransson, K., Söderlind, U., He, J. et al. (2011). Review of syngas production via biomass DFBGs. *Renew. Sustain. Energy Rev.* 15 (1): 482–492.
7 Rostrup-Nielsen, J.R. (2002). Syngas in perspective. *Catal. Today* 71 (3–4): 243–247.
8 Martinelli, M., Gnanamani, M.K., LeViness, S. et al. (2020). An overview of Fischer-Tropsch synthesis: XtL processes, catalysts and reactors. *Appl. Catal. A: Gen.* 608: 117740.
9 Glasser, D., Hildebrandt, D., Liu, X. et al. (2012). Recent advances in understanding the Fischer–Tropsch synthesis (FTS) reaction. *Curr. Opin. Chem. Eng.* 1 (3): 296–302.
10 Dry, M.E. (2002). The Fischer–Tropsch process: 1950–2000. *Catal. Today* 71 (3–4): 227–241.
11 Rommens, K.T. and Saeys, M. (2023). Molecular views on Fischer-Tropsch synthesis. *Chem. Rev.* 123 (9): 5798–5858.
12 Tavakoli, A., Sohrabi, M., and Kargari, A. (2008). Application of Anderson–Schulz–Flory (ASF) equation in the product distribution of slurry phase FT synthesis with nanosized iron catalysts. *Chem. Eng. J.* 136 (2–3): 358–363.
13 Boyer, C., Gazarian, J., Lecocq, V. et al. (2016). Development of the Fischer-Tropsch process: from the reaction concept to the process book. *Oil Gas Sci. Technol.: Rev. IFP Energies nouvelles* 71 (3): 44.
14 Zhou, W., Cheng, K., Kang, J. et al. (2019). New horizon in C1 chemistry: breaking the selectivity limitation in transformation of syngas and hydrogenation of CO_2 into hydrocarbon chemicals and fuels. *Chem. Soc. Rev.* 48 (12): 3193–3228.
15 Martín, M.M. (2016). Syngas. In: *Industrial Chemical Process Analysis and Design* (ed. Martín, M.M), 199–297. Boston: Elsevier. https://doi.org/10.1016/B978-0-08-101093-8.00009-4.
16 Sartipi, S., Parashar, K., Makkee, M. et al. (2013). Breaking the Fischer–Tropsch synthesis selectivity: direct conversion of syngas to gasoline over hierarchical Co/H-ZSM-5 catalysts. *Catal. Sci. Technol.* 3 (3): 572–575.

17 Eggenhuisen, T.M., den Breejen, J.P., Verdoes, D. et al. (2010). Fundamentals of melt infiltration for the preparation of supported metal catalysts. The case of Co/SiO_2 for Fischer-Tropsch synthesis. *J. Am. Chem. Soc.* 132 (51): 18318–18325.

18 Bezemer, G.L., Bitter, J.H., Kuipers, H.P.C.E. et al. (2006). Cobalt particle size effects in the Fischer-Tropsch reaction studied with carbon nanofiber supported catalysts. *J. Am. Chem. Soc.* 128 (12): 3956–3964.

19 Torres Galvis, H.M., Bitter, J.H., Davidian, T. et al. (2012). Iron particle size effects for direct production of lower olefins from synthesis gas. *J. Am. Chem. Soc.* 134 (39): 16207–16215.

20 Pedrolo, D.R.S., Ordomsky, V.V., Schwaab, M. et al. (2021). Design of ruthenium-zeolite nanocomposites for enhanced hydrocarbon synthesis from syngas. *J. Mater. Sci.* 56 (32): 18019–18030.

21 Subramanian, V., Zholobenko, V.L., Cheng, K. et al. (2016). The role of steric effects and acidity in the direct synthesis of iso-paraffins from syngas on cobalt zeolite catalysts. *ChemCatChem* 8 (2): 380–389.

22 Cheng, K., Virginie, M., Ordomsky, V.V. et al. (2015). Pore size effects in high-temperature Fischer–Tropsch synthesis over supported iron catalysts. *J. Catal.* 328: 139–150.

23 Xing, C., Yang, G., Wu, M. et al. (2015). Hierarchical zeolite Y supported cobalt bifunctional catalyst for facilely tuning the product distribution of Fischer–Tropsch synthesis. *Fuel* 148: 48–57.

24 Agwara, J.N., Bakas, N.J., Neidig, M.L. et al. (2022). Challenges and opportunities of Fe-based core-shell catalysts for Fischer-Tropsch synthesis. *ChemCatChem* 14 (19): e202200289.

25 Ojeda, M., Nabar, R., Nilekar, A.U. et al. (2010). CO activation pathways and the mechanism of Fischer–Tropsch synthesis. *J. Catal.* 272 (2): 287–297.

26 van Santen, R.A., Markvoort, A.J., Filot, I.A.W. et al. (2013). Mechanism and microkinetics of the Fischer–Tropsch reaction. *Phys. Chem. Chem. Phys.* 15 (40): 17038–17063.

27 Davis, B.H. (2009). Fischer–Tropsch synthesis: reaction mechanisms for iron catalysts. *Catal. Today* 141 (1–2): 25–33.

28 Weitkamp, J. (2012). Catalytic hydrocracking-mechanisms and versatility of the process. *ChemCatChem* 4 (3): 292–306.

29 Rofer-DePoorter, C.K. (1981). A comprehensive mechanism for the Fischer-Tropsch synthesis. *Chem. Rev.* 81 (5): 447–474.

30 Gaube, J. and Klein, H.F. (2008). Studies on the reaction mechanism of the Fischer–Tropsch synthesis on iron and cobalt. *J. Mol. Catal. A: Chem.* 283 (1–2): 60–68.

31 Weststrate, C.J., Sharma, D., Garcia Rodriguez, D. et al. (2020). Mechanistic insight into carbon-carbon bond formation on cobalt under simulated Fischer-Tropsch synthesis conditions. *Nat. Commun.* 11 (1): 750.

32 Weststrate, C.J., van Helden, P., and Niemantsverdriet, J.W. (2016). Reflections on the Fischer-Tropsch synthesis: mechanistic issues from a surface science perspective. *Catal. Today* 275: 100–110.

33 Joyner, R.W. (1977). Mechanism of hydrocarbon synthesis from carbon monoxide and hydrogen. *J. Catal.* 50 (1): 176–180.

34 Inderwildi, O.R., Jenkins, S.J., and King, D.A. (2008). Fischer-Tropsch mechanism revisited: alternative pathways for the production of higher hydrocarbons from synthesis gas. *J. Phys. Chem. C* 112 (5): 1305–1307.

35 Zhuo, M., Tan, K.F., Borgna, A. et al. (2009). Density functional theory study of the CO insertion mechanism for Fischer-Tropsch synthesis over Co catalysts. *J. Phys. Chem. C* 113 (19): 8357–8365.

36 Schweicher, J., Bundhoo, A., and Kruse, N. (2012). Hydrocarbon chain lengthening in catalytic CO hydrogenation: evidence for a CO-insertion mechanism. *J. Am. Chem. Soc.* 134 (39): 16135–16138.

37 Su, H.Y., Zhao, Y., Liu, J.X. et al. (2017). First-principles study of structure sensitivity of chain growth and selectivity in Fischer–Tropsch synthesis using HCP cobalt catalysts. *Catal. Sci. Technol.* 7 (14): 2967–2977.

38 Saif, M., Blay-Roger, R., Nawaz, M.A. et al. (2025). Bio-aromatics: revolutionizing the integrated biomass and plastic waste valorization for high-value aromatic hydrocarbons via bifunctional catalytic pathways of bio-syngas conversion. *Biomass Bioenergy* 196: 107736.

39 Zijlstra, B., Broos, R.J.P., Chen, W. et al. (2020). The vital role of step-edge sites for both CO activation and chain growth on cobalt Fischer-Tropsch catalysts revealed through first-principles-based microkinetic modeling including lateral interactions. *ACS Catal.* 10 (16): 9376–9400.

40 Kummer, J.T. and Emmett, P.H. (1953). Fischer-Tropsch synthesis mechanism studies. The addition of radioactive alcohols to the synthesis gas. *J. Am. Chem. Soc.* 75 (21): 5177–5183.

41 Torres Galvis, H.M., Bitter, J.H., Khare, C.B. et al. (2012). Supported iron nanoparticles as catalysts for sustainable production of lower olefins. *Science* 335 (6070): 835–838.

42 Prieto, G., Zečević, J., Friedrich, H. et al. (2013). Towards stable catalysts by controlling collective properties of supported metal nanoparticles. *Nat. Mater.* 12 (1): 34–39.

43 Nasser, A.H., Guo, L., Elnaggar, H. et al. (2018). Mn–Fe nanoparticles on a reduced graphene oxide catalyst for enhanced olefin production from syngas in a slurry reactor. *RSC Adv.* 8 (27): 14854–14863.

44 Jahangiri, H., Bennett, J., Mahjoubi, P. et al. (2014). A review of advanced catalyst development for Fischer–Tropsch synthesis of hydrocarbons from biomass derived syn-gas. *Catal. Sci. Technol.* 4 (8): 2210–2229.

45 Zhang, Q., Kang, J., and Wang, Y. (2010). Development of novel catalysts for Fischer–Tropsch synthesis: tuning the product selectivity. *ChemCatChem* 2 (9): 1030–1058.

46 Lee, J.H., Bonte, W., Corthals, S. et al. (2019). Zeolite nanoreactor for investigating sintering effects of cobalt-catalyzed Fischer-Tropsch synthesis. *Ind. Eng. Chem. Res.* 58 (13): 5140–5145.

47 Wang, H., Wang, Z., Wang, S. et al. (2021). The effect of the particle size on Fischer–Tropsch synthesis for ZSM-5 zeolite supported cobalt-based catalysts. *Chem. Commun.* 57 (99): 13522–13525.

48 Adeleke, A.A., Liu, X., Lu, X. et al. (2020). Cobalt hybrid catalysts in Fischer-Tropsch synthesis. *Rev. Chem. Eng.* 36 (4): 437–457.

49 Tang, Q., Wang, Y., Zhang, Q. et al. (2003). Preparation of metallic cobalt inside NaY zeolite with high catalytic activity in Fischer–Tropsch synthesis. *Catal. Commun.* 4 (5): 253–258.

50 Davoodian, N., Nakhaei Pour, A., Izadyar, M. et al. (2020). Fischer–Tropsch synthesis using zeolitic imidazolate framework (ZIF-7 and ZIF-8)-supported cobalt catalysts. *Appl. Organomet. Chem.* 34 (9): e5747.

51 Carvalho, A., Marinova, M., Batalha, N. et al. (2017). Design of nanocomposites with cobalt encapsulated in the zeolite micropores for selective synthesis of isoparaffins in Fischer–Tropsch reaction. *Catal. Sci. Technol.* 7 (21): 5019–5027.

52 Kim, J.C., Lee, S., Cho, K. et al. (2014). Mesoporous MFI zeolite nanosponge supporting cobalt nanoparticles as a Fischer-Tropsch catalyst with high yield of branched hydrocarbons in the gasoline range. *ACS Catal.* 4 (11): 3919–3927.

53 Bein, T., Schmiester, G., and Jacobs, P.A. (1986). Characterization of a new iron-on-zeolite Y Fischer-Tropsch catalyst. *J. Phys. Chem.* 90 (20): 4851–4856.

54 Wang, R., Liang, B., Yang, X. et al. (2023). Critical role of sodium migration in iron-based FT-zeolite tandem catalyst system for syngas hydrogenation to gasoline. *Appl. Catal. B: Environ.* 322: 122132.

55 Zwart, J. and Vink, J. (1987). Fischer-Tropsch synthesis using zeolite-supported iron catalysts derived from iron carbonyl complexes. *Appl. Catal.* 33 (2): 383–393.

56 Egiebor, N.O., Cooper, W.C., and Wojciechowski, B.W. (1989). Synthesis of motor fuels from HY-zeolite supported Fischer-Tropsch iron catalysts. *Appl. Catal.* 55 (1): 47–64.

57 Wang, P., Kang, J., Zhang, Q. et al. (2007). Lithium ion-exchanged zeolite faujasite as support of iron catalyst for Fischer-Tropsch synthesis. *Catal. Lett.* 114 (3–4): 178–184.

58 Lin, Q., Yang, G., Li, X. et al. (2013). A catalyst for one-step isoparaffin production via Fischer–Tropsch synthesis: growth of a H-mordenite shell encapsulating a fused iron core. *ChemCatChem* 5 (10): 3101–3106.

59 Li, J., He, Y., Tan, L. et al. (2018). Integrated tuneable synthesis of liquid fuels via Fischer–Tropsch technology. *Nat. Catal.* 1 (10): 787–793.

60 Kang, J., Cheng, K., Zhang, L. et al. (2011). Mesoporous zeolite-supported ruthenium nanoparticles as highly selective Fischer–Tropsch catalysts for the production of C_5–C_{11} isoparaffins. *Angew. Chem. Int. Ed.* 50 (22): 5200–5203.

61 Fraenkel, D. and Gates, B.C. (1980). Shape-selective Fischer-Tropsch synthesis catalyzed by zeolite-entrapped cobalt clusters. *J. Am. Chem. Soc.* 102 (7): 2478–2480.

62 Chen, Y.W., Wang, H.T., and Goodwin, J.G. (1983). Effect of preparation methods on the catalytic properties of zeolite-supported ruthenium in the Fischer-Tropsch synthesis. *J. Catal.* 83 (2): 415–427.

63 Iton, L.E., Beal, R.B., and Hodul, D.T. (1983). A new approach to the generation of metal-bearing, medium-pore, shape-selective zeolites for Fischer-Tropsch catalysis: spectroscopic studies of zeolites. *J. Mol. Catal.* 21 (1–3): 151–171.

64 Sineva, L.V., Asalieva, E.Y., and Mordkovich, V.Z. (2015). The role of zeolite in the Fischer–Tropsch synthesis over cobalt–zeolite catalysts. *Russ. Chem. Rev.* 84 (11): 1176–1189.

65 Tsubaki, N., Yoneyama, Y., Michiki, K. et al. (2003). Three-component hybrid catalyst for direct synthesis of isoparaffin via modified Fischer–Tropsch synthesis. *Catal. Commun.* 4 (3): 108–111.

66 Martínez, A. and López, C. (2005). The influence of ZSM-5 zeolite composition and crystal size on the in situ conversion of Fischer–Tropsch products over hybrid catalysts. *Appl. Catal. A: Gen.* 294 (2): 251–259.

67 Davis, B.H. (2002). Overview of reactors for liquid phase Fischer–Tropsch synthesis. *Catal. Today* 71 (3–4): 249–300.

68 Guettel, R., Kunz, U., and Turek, T. (2008). Reactors for Fischer-Tropsch synthesis. *Chem. Eng. Technol.* 31 (5): 746–754.

69 Wentrup, J., Pesch, G.R., and Thöming, J. (2022). Dynamic operation of Fischer-Tropsch reactors for power-to-liquid concepts: a review. *Renew. Sustain. Energy Rev.* 162: 112454.

70 Eldesouki, M.H., Rashed, A.E., and El-Moneim, A.A. (2023). A comprehensive overview of carbon dioxide, including emission sources, capture technologies, and the conversion into value-added products. *Clean Technol. Environ. Policy* 25 (10): 3131–3148.

71 Davis, B.H. (2005). Fischer-Tropsch synthesis: overview of reactor development and future potentialities. *Top. Catal.* 32 (3–4): 143–168.

72 Rashed, A.E., Nasser, A., Elkady, M.F. et al. (2023). Temperature calibration effect on FTS activity and product selectivity using Fe-MOF catalyst. *Case Stud. Chem. Environ. Eng.* 7: 100300.

73 Liuzzi, D., Fernandez, E., Perez, S. et al. (2022). Advances in membranes and membrane reactors for the Fischer-Tropsch synthesis process for biofuel production. *Rev. Chem. Eng.* 38 (1): 55–76.

74 Yeh, J.T., Resnik, K.P., Rygle, K. et al. (2005). Semi-batch absorption and regeneration studies for CO_2 capture by aqueous ammonia. *Fuel Process. Technol.* 86 (14–15): 1533–1546.

75 Guda, V.K., Steele, P.H., Penmetsa, V.K. et al. (2015). Fast pyrolysis of biomass: recent advances in fast pyrolysis technology. In: *Recent Advances in Thermochemical Conversion of Biomass* (eds. A. Pandey, T. Bhaskar, M. Stöcker et al.), 177–211. Amsterdam: Elsevier.

76 Rajmohan, K.S., Yadav, H., Vaishnavi, S. et al. (2020). Perspectives on bio-oil recovery from plastic waste. In: *Current Developments in Biotechnology and Bioengineering: Resource Recovery from Wastes* (eds. S. Varjani, A. Pandey, E. Gnansounou et al.), 459–480. Amsterdam: Elsevier.

77 Arzamendi, G., Diéguez, P.M., Montes, M. et al. (2010). Computational fluid dynamics study of heat transfer in a microchannel reactor for low-temperature Fischer–Tropsch synthesis. *Chem. Eng. J.* 160 (3): 915–922.

78 Deshmukh, S.R., Tonkovich, A.L.Y., Jarosch, K.T. et al. (2010). Scale-up of microchannel reactors for Fischer-Tropsch synthesis. *Ind. Eng. Chem. Res.* 49 (21): 10883–10888.

79 Pfeifer, P., Schmidt, S., Betzner, F. et al. (2022). Scale-up of microstructured Fischer–Tropsch reactors—status and perspectives. *Curr. Opin. Chem. Eng.* 36: 100776.

80 Barrera, J.L., Hartvigsen, J.J., Hollist, M. et al. (2023). Design optimization of integrated cooling inserts in modular Fischer-Tropsch reactors. *Chem. Eng. Sci.* 268: 118423.

81 Wei, Q., Li, H., Liu, G. et al. (2020). Metal 3D printing technology for functional integration of catalytic system. *Nat. Commun.* 11 (1): 4098.

82 Zhao, J., Peng, X., Jiang, L. et al. (2025). 3D-printing technologies for C1 chemistry. *Chem* 11 (7): 102614.

83 Rohde, M.P., Unruh, D., and Schaub, G. (2005). Membrane application in Fischer–Tropsch synthesis reactors—overview of concepts. *Catal. Today* 106 (1–4): 143–148.

84 Khassin, A.A. (2005). Membrane reactors for the Fischer-Tropsch synthesis. In: *Sustainable Strategies for the Upgrading of Natural Gas: Fundamentals, Challenges, and Opportunities* (eds. E.G. Derouane, V. Parmon, F. Lemos et al.), NATO Science Series II: Mathematics, Physics and Chemistry (Vol. 191), 249–271. Dordrecht: Springer.

85 Ozin, G.A., Kuperman, A., and Stein, A. (1989). Advanced zeolite, materials science. *Angew. Chem. Int. Ed. Engl.* 28 (3): 359–376.

86 Weckhuysen, B.M. and Yu, J. (2015). Recent advances in zeolite chemistry and catalysis. *Chem. Soc. Rev.* 44 (20): 7022–7024.

87 Pérez-Botella, E., Valencia, S., and Rey, F. (2022). Zeolites in adsorption processes: state of the art and future prospects. *Chem. Rev.* 122 (24): 17647–17695.

88 del Campo, P., Martínez, C., and Corma, A. (2021). Activation and conversion of alkanes in the confined space of zeolite-type materials. *Chem. Soc. Rev.* 50 (15): 8511–8595.

89 Wei, Y., Feng, J., Guan, B. et al. (2024). Structural engineering of hierarchical zeolite-based catalysts. *Acc. Mater. Res.* 5 (7): 857–871.

90 Wang, X., Ma, Y., Wu, Q. et al. (2022). Zeolite nanosheets for catalysis. *Chem. Soc. Rev.* 51 (7): 2431–2443.

91 Yu, J. and Zhao, D. (2022). Preface to special topic on new era of zeolite science. *Natl. Sci. Rev.* 9 (9): nwac157.

92 Olson, D.H., Haag, W.O., and Lago, R.M. (1980). Chemical and physical properties of the ZSM-5 substitutional series. *J. Catal.* 61 (2): 390–396.

93 Quann, R.J., Green, L.A., Tabak, S.A. et al. (1988). Chemistry of olefin oligomerization over ZSM-5 catalyst. *Ind. Eng. Chem. Res.* 27 (4): 565–570.

94 Rahimi, N. and Karimzadeh, R. (2011). Catalytic cracking of hydrocarbons over modified ZSM-5 zeolites to produce light olefins: a review. *Appl. Catal. A: Gen.* 398 (1–2): 1–17.

95 Sato, K., Nishimura, Y., Honna, K. et al. (2001). Role of HY zeolite mesopores in hydrocracking of heavy oils. *J. Catal.* 200 (2): 288–297.

96 Castaño, P., Elordi, G., Olazar, M. et al. (2011). Insights into the coke deposited on HZSM-5, Hβ and HY zeolites during the cracking of polyethylene. *Appl. Catal. B: Environ.* 104 (1–2): 91–100.
97 Hammer, H., Joisten, M., Lüngen, S. et al. (1994). New zeolites in Fischer-Tropsch synthesis. *Int. J. Energy Res.* 18 (2): 223–231.
98 Kang, S.H., Bae, J.W., Prasad, P.S.S. et al. (2008). Fischer-Tropsch synthesis using zeolite-supported iron catalysts for the production of light hydrocarbons. *Catal. Lett.* 125 (3–4): 264–270.
99 Chen, Y.W., Wang, H.T., Goodwin, J.G. et al. (1983). Fischer-Tropsch synthesis over zeolite-supported ruthenium catalysts derived from $Ru_3(CO)_{12}$. *Appl. Catal.* 8 (3): 303–314.
100 Oppmann, N. and Jess, A. (2023). Improving the selectivity to liquefied petroleum gas by combining Fischer-Tropsch synthesis with zeolite cracking. *Chem. Eng. Technol.* 46 (5): 908–917.
101 de Jong, K.P., Zečević, J., Friedrich, H. et al. (2010). Zeolite Y crystals with trimodal porosity as ideal hydrocracking catalysts. *Angew. Chem. Int. Ed.* 49 (52): 10074–10078.
102 Liu, Y., Baráth, E., Shi, H. et al. (2018). Solvent-determined mechanistic pathways in zeolite-H-BEA-catalysed phenol alkylation. *Nat. Catal.* 1 (2): 141–147.
103 Muraza, O. (2015). Maximizing diesel production through oligomerization: a landmark opportunity for zeolite research. *Ind. Eng. Chem. Res.* 54 (3): 781–789.
104 Rachwalik, R., Olejniczak, Z., Jiao, J. et al. (2007). Isomerization of α-pinene over dealuminated ferrierite-type zeolites. *J. Catal.* 252 (2): 161–170.
105 Vogt, E.T.C. and Weckhuysen, B.M. (2015). Fluid catalytic cracking: recent developments on the grand old lady of zeolite catalysis. *Chem. Soc. Rev.* 44 (20): 7342–7370.
106 Tan, P. (2016). Active phase, catalytic activity, and induction period of Fe/zeolite material in nonoxidative aromatization of methane. *J. Catal.* 338: 21–29.
107 Wang, C., Fang, W., Wang, L. et al. (2021). Fischer-Tropsch reaction within zeolite crystals for selective formation of gasoline-ranged hydrocarbons. *J. Energy Chem.* 54: 429–433.
108 Peng, X., Cheng, K., Kang, J. et al. (2015). Impact of hydrogenolysis on the selectivity of the Fischer–Tropsch synthesis: diesel fuel production over mesoporous zeolite-Y-supported cobalt nanoparticles. *Angew. Chem. Int. Ed.* 54 (15): 4553–4556.
109 Wang, C., Fang, W., Liu, Z. et al. (2022). Fischer–Tropsch synthesis to olefins boosted by MFI zeolite nanosheets. *Nat. Nanotechnol.* 17 (7): 714–720.
110 Zhao, B., Zhai, P., Wang, P. et al. (2017). Direct transformation of syngas to aromatics over Na-Zn-Fe_5C_2 and hierarchical HZSM-5 tandem catalysts. *Chem* 3 (2): 323–333.
111 Jo, D., Hong, S.B., and Camblor, M.A. (2015). Monomolecular skeletal isomerization of 1-butene over selective zeolite catalysts. *ACS Catal.* 5 (4): 2270–2274.
112 Shakun, A.N. and Fedorova, M.L. (2014). Isomerization of light gasoline fractions: the efficiency of different catalysts and technologies. *Catal. Ind.* 6 (4): 298–306.

113 Costa, C.S., Dao Thi, H., Van Geem, K.M. et al. (2022). Assessment of acidity and the zeolite porous structure on hydrocracking of HDPE. *Sustain. Energy Fuels* 6 (15): 3611–3625.

114 Munnik, P., de Jongh, P.E., and de Jong, K.P. (2015). Recent developments in the synthesis of supported catalysts. *Chem. Rev.* 115 (14): 6687–6718.

115 Nakhaei Pour, A. and Housaindokht, M.R. (2013). Fischer-Tropsch synthesis on iron catalyst promoted with HZSM-5 zeolite: regeneration studies of catalyst. *J. Nat. Gas Sci. Eng.* 14: 49–54.

116 Javed, M., Cheng, S., Zhang, G. et al. (2020). A facile solvent-free synthesis strategy for Co-imbedded zeolite-based Fischer-Tropsch catalysts for direct gasoline production. *Chin. J. Catal.* 41 (4): 604–612.

117 Burgun, U., Zonouz, H.R., Okutan, H. et al. (2022). Effects of rare earth metal promotion over zeolite-supported Fe-Cu-based catalysts on the light olefin production performance in Fischer-Tropsch synthesis. *ACS Omega* 8: 648–662.

118 Martínez, A., Rollán, J., Arribas, M.A. et al. (2007). A detailed study of the activity and deactivation of zeolites in hybrid Co/SiO_2-zeolite Fischer–Tropsch catalysts. *J. Catal.* 249 (2): 162–173.

119 Chalupka, K.A., Maniukiewicz, W., Mierczynski, P. et al. (2015). The catalytic activity of Fe-containing SiBEA zeolites in Fischer–Tropsch synthesis. *Catal. Today* 257 (P1): 117–121.

120 Wang, Y., Jiang, Y., Huang, J. et al. (2016). Effect of preparation methods on hierarchical zeolites for cobalt-based Fischer–Tropsch synthesis. *RSC Adv.* 6 (109): 107498–107506.

121 Martínez, A. and Prieto, G. (2009). The application of zeolites and periodic mesoporous silicas in the catalytic conversion of synthesis gas. *Top. Catal.* 52 (1–2): 75–90.

122 Chalupka, K.A., Casale, S., Zurawicz, E. et al. (2015). The remarkable effect of the preparation procedure on the catalytic activity of CoBEA zeolites in the Fischer–Tropsch synthesis. *Microporous Mesoporous Mater.* 211: 9–18.

123 Bessell, S. (1995). Investigation of bifunctional zeolite supported cobalt Fischer-Tropsch catalysts. *Appl. Catal. A: Gen.* 126 (2): 235–244.

124 Yakovenko, R.E., Bakun, V.G., Agliullin, M.R. et al. (2022). Effects of zeolite type on integrated Fischer–Tropsch synthesis and hydroprocessing. *Pet. Chem.* 62 (8): 950–961.

125 Liu, Z.W., Li, X., Asami, K. et al. (2005). Formation of isoparaffins through Pd/β zeolite application in Fischer-Tropsch synthesis. *Energy Fuels* 19 (5): 1790–1794.

126 Cano, L.A., Cagnoli, M.V., Bengoa, J.F. et al. (2011). Effect of the activation atmosphere on the activity of Fe catalysts supported on SBA-15 in the Fischer–Tropsch synthesis. *J. Catal.* 278 (2): 310–320.

127 Sartipi, S., Parashar, K., Valero-Romero, M.J. et al. (2013). Hierarchical H-ZSM-5-supported cobalt for the direct synthesis of gasoline-range hydrocarbons from syngas: advantages, limitations, and mechanistic insight. *J. Catal.* 305: 179–190.

128 Sun, J., Li, X., Taguchi, A. et al. (2014). Highly-dispersed metallic Ru nanoparticles sputtered on H-beta zeolite for directly converting syngas to middle isoparaffins. *ACS Catal.* 4 (1): 1–8.

129 Lu, P., Sun, J., Zhu, P. et al. (2015). Sputtered nano-cobalt on H-USY zeolite for selectively converting syngas to gasoline. *J. Energy Chem.* 24 (5): 637–641.
130 Corsaro, A., Wiltowski, T., Juchelkova, D. et al. (2014). Conversion of syngas to LPG and aromatics over commercial Fischer-Tropsch catalyst and HZSM-5 in a dual bed reactor. *Pet. Sci. Technol.* 32 (20): 2497–2505.
131 Weber, J.L., Dugulan, I., de Jongh, P.E. et al. (2018). Bifunctional catalysis for the conversion of synthesis gas to olefins and aromatics. *ChemCatChem* 10 (5): 1107–1112.
132 Weber, J.L., Martínez del Monte, D., Beerthuis, R. et al. (2021). Conversion of synthesis gas to aromatics at medium temperature with a Fischer Tropsch and ZSM-5 dual catalyst bed. *Catal. Today* 369: 175–183.
133 Gao, W., Zhu, Q., and Ma, D. (2018). Nanostructured catalyst for Fischer–Tropsch synthesis. *Chin. J. Chem.* 36 (9): 798–808.
134 Xu, Y., Wang, J., Ma, G. et al. (2020). Hollow zeolite nanoparticles combined with Fe_3O_4@MnO_2 tandem catalyst for converting syngas to aromatics-rich gasoline. *ACS Appl. Nano Mater.* 3 (3): 2857–2866.
135 Zhang, B., Yao, J., Wang, Y. et al. (2023). Effects of zeolite morphologies on CO conversion to aromatics via a modified Fischer-Tropsch synthesis pathway. *J. Chem. Technol. Biotechnol.* 98 (1): 98–105.
136 Wu, X., Wang, C., Zhao, S. et al. (2024). Dual-engine-driven realizing high-yield synthesis of *para*-xylene directly from CO_2-containing syngas. *Nat. Commun.* 15 (1): 8064.
137 Yang, G., Xing, C., Hirohama, W. et al. (2013). Tandem catalytic synthesis of light isoparaffin from syngas via Fischer–Tropsch synthesis by newly developed core–shell-like zeolite capsule catalysts. *Catal. Today* 215: 29–35.
138 He, J., Yoneyama, Y., Xu, B. et al. (2005). Designing a capsule catalyst and its application for direct synthesis of middle isoparaffins. *Langmuir* 21 (5): 1699–1702.
139 Bao, J., He, J., Zhang, Y. et al. (2008). A core/shell catalyst produces a spatially confined effect and shape selectivity in a consecutive reaction. *Angew. Chem.* 120 (2): 359–362.
140 Yang, G., Tan, Y., Han, Y. et al. (2008). Increasing the shell thickness by controlling the core size of zeolite capsule catalyst: application in iso-paraffin direct synthesis. *Catal. Commun.* 9 (15): 2520–2524.
141 Yamane, N., Wang, Y., Li, J. et al. (2017). Building premium secondary reaction field with a miniaturized capsule catalyst to realize efficient synthesis of a liquid fuel directly from syngas. *Catal. Sci. Technol.* 7 (10): 1996–2000.
142 Přech, J., Strossi Pedrolo, D.R., Marcilio, N.R. et al. (2020). Core-shell metal zeolite composite catalysts for in situ processing of Fischer-Tropsch hydrocarbons to gasoline type fuels. *ACS Catal.* 10 (4): 2544–2555.
143 Xiao, J., Cheng, K., Xie, X. et al. (2022). Tandem catalysis with double-shelled hollow spheres. *Nat. Mater.* 21 (5): 572–579.
144 Zhu, W., Gora, L., van den Berg, A.W.C. et al. (2005). Water vapour separation from permanent gases by a zeolite-4A membrane. *J. Membr. Sci.* 253 (1–2): 57–66.

8

Application of Zeolites in Syngas to Oxygenated Hydrocarbons Synthesis

Suhan Liu[1], Yubing Li[2], Kang Cheng[1,], Jincan Kang[1], Ye Wang[1]*

[1]*State Key Laboratory of Physical Chemistry of Solid Surfaces, College of Chemistry and Chemical Engineering, Xiamen University, Xiamen, Fujian, China*
[2]*Faculty of Chemical Engineering, Kunming University of Science and Technology, Kunming, Yunnan, China*

*Corresponding author: kangcheng@xmu.edu.cn

8.1 Introduction

Oxygenated hydrocarbons widely serve as liquid fuels, fuel additives, and precursors/intermediates for the production of high-value chemicals, including solvents, plastics, resins, and pharmaceuticals. Direct synthesis of oxygenated hydrocarbons from syngas is a promising way to utilize nonpetroleum carbon resources. The classical process of syngas to methanol synthesis has been commercialized for more than a century and now enormous research has been conducted for the synthesis of C_{2+} oxygenates [1–3]. However, compared to the Fischer–Tropsch reaction that produces mainly linear hydrocarbon paraffins or olefins, the formation of oxygenates with C—C bonds from syngas has to involve both non-dissociative and dissociative adsorptions of CO, and subsequent C—C coupling (see Figure 8.1) [4–6]. This mechanism makes

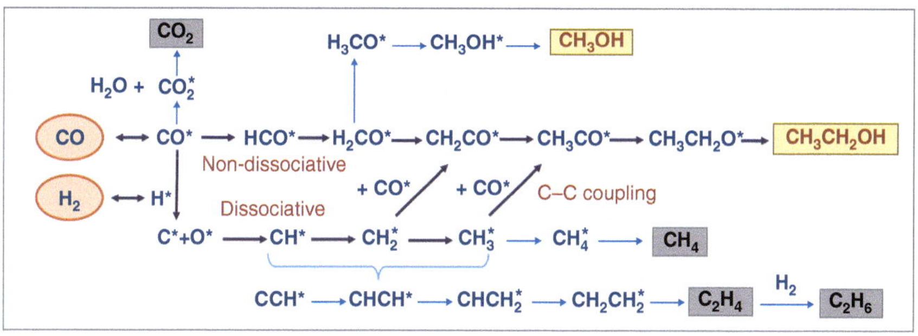

Figure 8.1 Complicated reaction network of syngas conversion over conventional heterogeneous catalysts for ethanol synthesis.

the control of oxygenate selectivity a challenging task in syngas chemistry. Besides, the final product of syngas conversion tends to be saturated alkanes without functional groups, such as methane and ethane, which are thermodynamically favorable.

Bifunctional catalysts composed of metal and zeolites have been widely applied in the direct conversion of syngas to olefins, aromatic hydrocarbons, gasoline, and diesel fuels, which reshaped the research pattern of syngas conversion [7–9]. Moreover, zeolites are increasingly being applied in syngas-related processes; for example, (1) SAPO-34 and ZSM-5 are employed for methanol dehydration to lower olefins; (2) various zeolites are employed for hydroformylation reaction; and (3) MOR is employed for dimethyl ether (DME) carbonylation reaction [10–13]. Some zeolite-participated reactions involve the production of oxygenated hydrocarbons, such as carbonylation and hydroformylation mentioned above. Also, the use of zeolite to regulate the products of syngas conversion to oxygenated hydrocarbons can be traced back to more than 40 years ago. In 1978, Mantovani et al. reported an Rh@NaY catalyst for olefin hydroformylation to aldehydes [14]. In 1984, Fujimoto et al. reported the vapor phase carbonylation of methanol catalyzed by acidic zeolites [15]. However, compared to the application of zeolite in hydrocarbon processing [16, 17], the employment of zeolites to produce oxygenated compounds is much less. Over the past 15 years, we have witnessed significant progress in this area based on the unique functionalities brought by zeolites, such as syngas to acetic acid (AA), methyl acetate (MA), ethanol, DME, and aldehydes [18, 19].

Here, we would like to emphasize the importance of thermodynamics in syngas conversion under typical reaction conditions as the fundamental background of this chapter. Most hydrogenation reactions are exothermic reactions. The methanol synthesis from syngas is strongly restricted at temperatures above 300 °C with CO equilibrium conversion levels lower than 15.5%, while the formation of DME is more thermodynamically favorable than methanol even at 300 °C with CO equilibrium conversion of around 45.6% (see Figure 8.2). Compared to methanol and DME, the thermodynamic restriction is almost negligible for the conversion of syngas to ethanol and ethylene at temperatures lower than 400 °C. The formation of saturated alkanes is even more thermodynamically favorable. The thermodynamic stability of product molecules determines the final product distribution or selectivity to a large extent, and the final products always tend to be thermodynamically stable molecules. This makes the selective synthesis of oxygenated hydrocarbons from syngas quite difficult.

In the context of transportation fuels being gradually replaced by electricity, syngas will be increasingly used for the production of hydrocarbon chemicals, especially oxygenated hydrocarbons, which cannot be replaced by other materials [20]. To date, a wide variety of zeolite-based catalysts have been explored for the direct conversion of syngas to oxygenates with high atomic economy. This chapter will showcase several representative examples: (1) the production of C_2 oxygenates mediated by MOR-catalyzed or Rh-catalyzed carbonylation; (2) the production of DME mediated by zeolite-catalyzed dehydration; and (3) zeolite-confined metal catalysts for aldehyde synthesis.

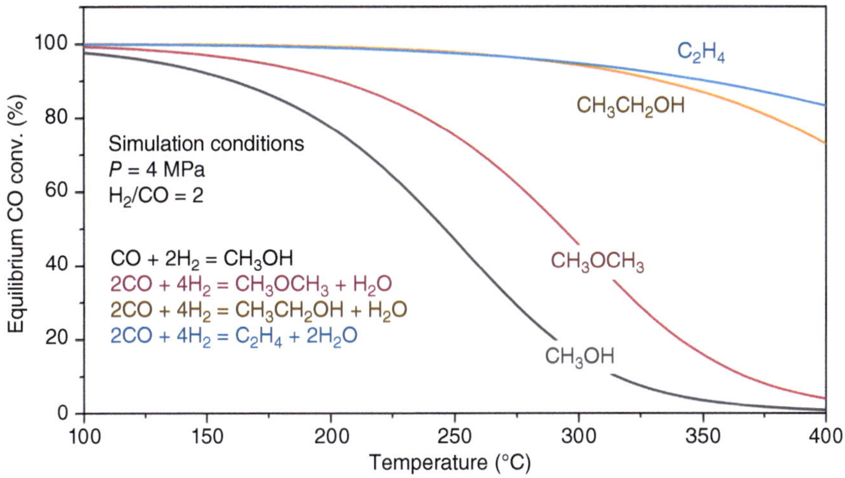

Figure 8.2 Equilibrium CO conversion during the synthesis of methanol, DME, ethanol, and ethane. Simulated conditions: $H_2/CO = 2$ and $P = 4$ MPa. The calculation was based on HSC5 Chemistry software.

8.2 Production of C_2 Oxygenates Mediated by Carbonylation

Currently, AA is primarily produced via homogeneous methanol carbonylation, using rhodium or iridium organometallic complexes and halide-containing promoters [21]. This reaction has been commercialized through the Monsanto and BP Cativa™ processes for more than 50 years [22]. In 1984, Fujimoto et al. reported that gaseous methanol carbonylation to AA in the absence of any halide promoter could be catalyzed by acidic zeolites, such as H–MOR and H–Y [15]. In these systems, methanol or DME adsorbs on acid sites to form methoxy groups, which can react directly with CO to form acetyl carbonium ions (CH_3CO^+). The CH_3CO^+ species could be converted to AA or MA by hydrolysis and methanolysis (see Figure 8.3a). Later, a series of zeolites were evaluated for carbonylation reaction, and MOR exhibited attractive activity and stability [23]. Besides, the fabrication of H–MOR with a transition metal, such as copper and palladium, can improve the activity of methanol carbonylation by CO [24]. Rather than methanol, the carbonylation of DME with CO can also be catalyzed by MOR zeolites.

Zeolite MOR is characterized by its unique pore structure; the framework of MOR consists of 12-membered ring (12-MR; 6.5×7 Å^2) and eight-membered ring (8-MR; 2.6×5.7 Å^2) channels along the *c*-axis, interconnected by 8-MR side pockets along the *b*-axis (see Figure 8.3b–d) [25]. Previous studies have evidenced that the Brønsted acid sites (BASs) in 8-MR side pockets are responsible for the carbonylation reaction, while the presence of BASs in the 12-MR pores usually leads to undesired side reactions to hydrocarbons or coke formation thus causing fast deactivation [26, 27]. To enhance the carbonylation ability of MOR zeolite, several strategies have been developed to selectively remove or poison the acid sites in the 12-MR channels,

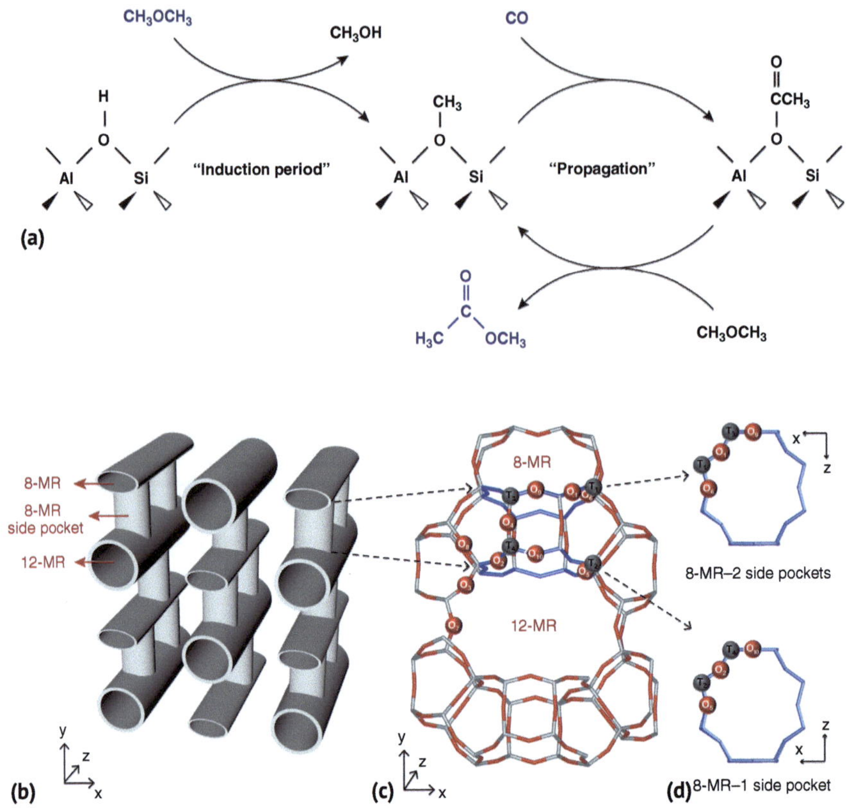

Figure 8.3 DME carbonylation over MOR. (a) Reaction pathways. (b) The 12-MR and 8-MR pores are connected via the 8-MR side pocket. (c) Geometries of the T sites (T_1–T_4) and the O atoms (O_1–O_{10}). (d) Schematic illustrations of the 8-MR side pocket viewed from the *y*/−*y* orientations. *Source*: [25]/with permission of Elsevier.

including steam treatment to remove aluminum of 12-MR, blocking the BASs by pyridine, and ion exchange with bulky organic basic ions, while keeping active sites in the 8-MR side pockets intact [28–30].

Since methanol and DME can be directly synthesized from syngas, researchers tried to directly convert syngas to AA or MA by combining CO hydrogenation and DME/CH_3OH carbonylation reaction in a single reactor. Before realizing this target, Tsubaki and co-workers demonstrated that the combination of DME carbonylation over the Cu–MOR catalyst and MA hydrogenation over the Cu/ZnO catalyst in a dual-bed mode could provide an ethanol selectivity of 44% together with a methanol selectivity of 46% at a CO conversion of 1.9%, operating at a temperature of 200°C, pressure of 1.5MPa, and Ar : DME : CO : H_2 ratio = 1.6 : 1 : 47.4 : 50 (see Figure 8.4a) [31]. In 2018, Wang and co-workers reported a relay catalysis strategy for the selective conversion of syngas into C_{2+} oxygenates by combining different functional catalysts in one reactor [32]. For example, by integrating DME synthesis over $CuZnAlO_x$/H–ZSM-5 and DME carbonylation over steamed H–MOR in a dual-bed system (catalyst denoted as $CuZnAlO_x$/H–ZSM-5|H–MOR), the total

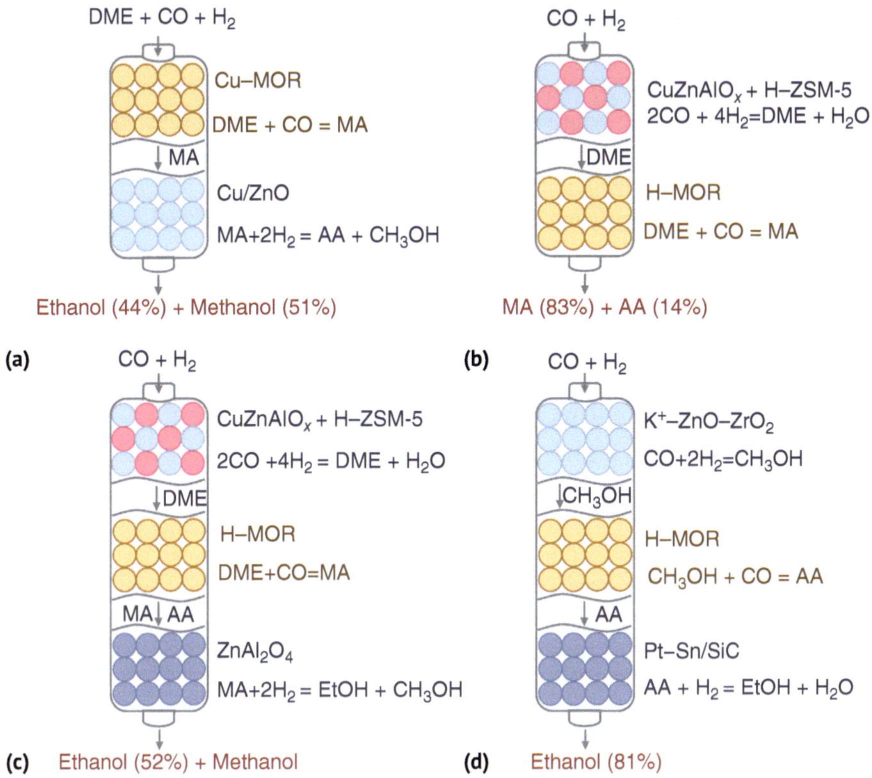

Figure 8.4 Different multi-bed systems for the synthesis of C_2 oxygenates involving syngas. (a) DME and syngas to ethanol and methanol synthesis mediated by MA. (b) Syngas to AA and MA synthesis mediated by DME. (c) Syngas to ethanol and methanol synthesis mediated by DME, MA, and AA. (d) Syngas to ethanol synthesis mediated by AA.

selectivity of MA and AA reached 97% at a CO conversion of 7.5% (see Figure 8.4b). By further integrating $ZnAl_2O_4$ into the third downstream bed, MA and AA could be converted into ethanol with a selectivity of 52%, which is higher than most reported catalysts (see Figure 8.4c).

In tandem catalysis, the choice of each catalyst with a precise function is pivotal to obtaining high selectivity [7]. To increase the ethanol selectivity, Kang et al. designed a triple tandem system, composed of K^+–ZnO–ZrO_2, modified MOR zeolite, and Pt–Sn/SiC working compatibly in syngas stream in one reactor, to improve the ethanol selectivity to 90 and 81% at CO conversions of 0.7 and 4%, respectively (see Figure 8.4d) [33]. The three catalysts selectively and efficiently catalyze syngas conversion to methanol, methanol carbonylation to AA, and hydrogenation of AA to ethanol. To keep the high selectivity of each step, carefully designing the corresponding catalyst is the key to obtaining high ethanol selectivity. Later on, Chen and co-workers also reported a triple tandem system combining $CuZnAlO_x$/H–ZSM-5, Zn–MOR, and $CuZnAlO_x$ catalysts to directly synthesize ethanol from syngas, with ethanol selectivity of 69.5% at a CO conversion of 7.4% [34]. To improve the single-pass CO conversion, Liu and co-workers reported a multi-bed system consisting of $CuZnAlO_x/\gamma\text{-}Al_2O_3$ for DME synthesis, pyridine-modified H–MOR (Py–HMOR) for DME carbonylation, and $CuZnAlO_x$ for MA hydrogenation to ethanol [35]. The $CuZnAlO_x/\gamma\text{-}Al_2O_3$, which is placed in the upper bed in tandem systems, could catalyze the syngas to DME conversion with a DME selectivity of 94% at a CO conversion of 56% at 270 °C. The tandem system eventually provided an ethanol selectivity of 62% with the CO conversion at 52%.

Some researchers also tried to integrate the different functional components closely into one multifunctional catalyst to enhance the synergy. Tsubaki and co-workers reported an ethanol synthesis strategy from syngas and DME by fabricating a capsule catalyst consisting of a Cu/ZnO core and Cu–MOR shell [36]. This kind of space-confined structure arranges reaction steps ideally in tandem order. Conversion of DME and CO into MA was first catalyzed by H–MOR shell with a CO conversion of 0.9%, then MA was hydrogenated by H_2 over Cu/ZnO core into ethanol with a selectivity of 44.2%. However, this core–shell catalyst did not show superior performance in comparison to the corresponding dual-bed systems [31].

Other than the carbonylation-involved reactions above, Rh-based catalysts can also offer high selectivity toward C_2 oxygenates in syngas conversion owing to their unique capabilities of C—O dissociation and C—C coupling [37, 38]. Because of this mechanism, the acid function of zeolites is not necessary for Rh-based catalysts, but the functions of well-defined microporous structure and rigid zeolite framework have been utilized for rhodium catalysts in syngas conversion. Wang et al. demonstrated a watermelon-like RhMn@S-1 catalyst, i.e., RhMn nanoparticles encapsulated within Silicalite-1 zeolite crystals, for the direct conversion of syngas to ethanol (see Figure 8.5) [39]. The RhMn@S-1 catalyst exhibited an oxygenate selectivity of 40.3% at a CO conversion of 42.4% and C_2 oxygenate selectivity was 88.3% in the total oxygenates. Compared to the RhMn/S-1 catalyst prepared by the conventional impregnation method, RhMn@S-1 displayed not only higher C_2 oxygenate productivity and selectivity but also superior stability, maintaining CO conversion at ~42% for 220 hours. This stability is attributed to the rigid framework of zeolite, which stabilizes the active Mn–O–$Rh^{\delta+}$ structure and prevents Rh sintering.

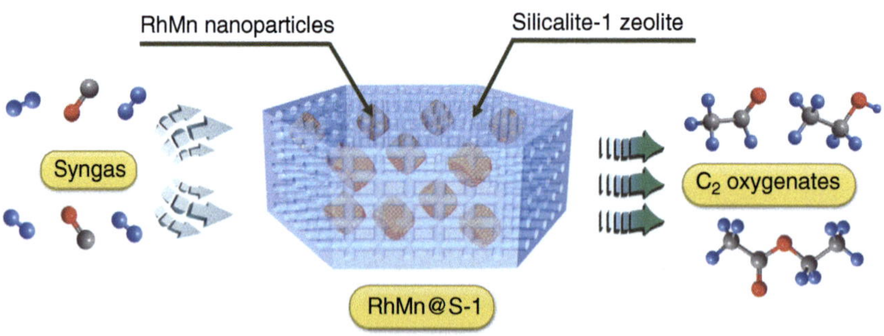

Figure 8.5 The direct conversion of syngas to C_2 oxygenates over RhMn@S-1 catalyst. *Source:* [39]/with permission of Elsevier.

8.3 Production of DME Mediated by Dehydration

DME is one of the most promising environmentally optimized alternatives to conventional fossil fuels and an important platform molecule for the chemical industry [40]. The production of DME is typically catalyzed by thermodynamically feasible methanol dehydration using solid acid catalysts or a two-step process with methanol as the intermediate from syngas [41]. The direct synthesis of DME from syngas without significant thermodynamic restriction ($2CO + 4H_2 \Longleftrightarrow CH_3OCH_3 + H_2O$, $\Delta H_{298\ K} = -204.6\ kJ \cdot mol^{-1}$) can be achieved by combining methanol synthesis and methanol dehydration [42]. Since syngas can be largely produced from biomass, carbon-containing waste, or even CO_2, DME is in the spotlight as a biofuel or sustainable chemical [41, 43]. One should note that the dehydration of methanol to DME cannot be complete due to thermodynamic reasons (see Figure 8.6), also indicating that DME selectivity cannot be 100% in syngas conversion.

Zeolites are featured with their different pore structures and tunable acidity; thus, a variety of zeolites have been employed to combine with methanol synthesis catalysts for DME synthesis [44, 45]. As early as 1991, Sofianos and Scurrell reported that the combination of $CuZnAlO_x$ with H–Y, H–MOR, or H–ZSM-5 could convert syngas directly to DME, but together with the formation of short paraffins as side products [46]. To increase the catalyst efficiency, Tsubaki and co-workers demonstrated a new zeolite capsule catalyst, $CuZnAlO_x$@H–ZSM-5, for the DME synthesis from syngas [47]. Methanol produced in the core must diffuse through the zeolite shell, ensuring sufficient contact with the acid sites for dehydration (see Figure 8.7a). This catalyst exhibited a high DME selectivity of 96.6% without the formation of alkanes, which was much higher than 40.5% over the $CuZnAlO_x$/H–ZSM-5 catalyst prepared by the physical mixing method (see Figure 8.7b).

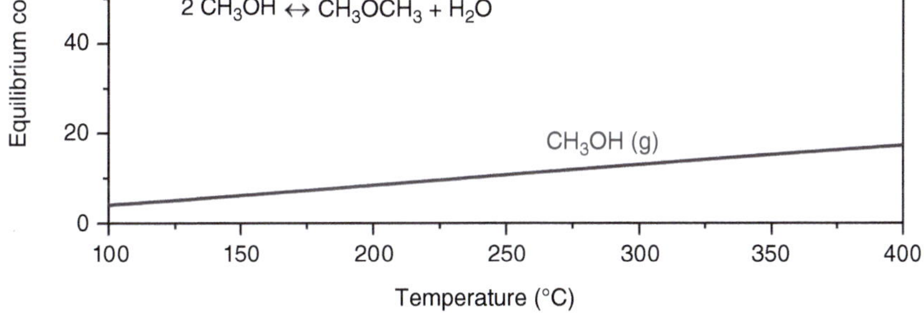

Figure 8.6 Equilibrium composition for the dehydration of methanol to DME. The calculation was based on HSC5 Chemistry software.

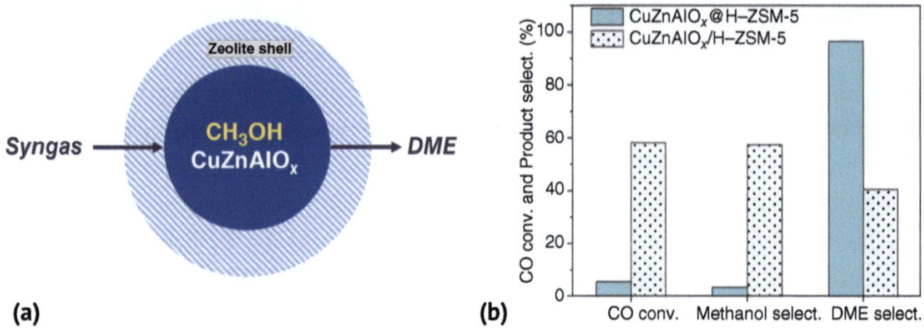

Figure 8.7 (a) Schematic of the direct synthesis of DME from syngas using a zeolite capsule catalyst. (b) Comparison of catalytic performance: core–shell versus physical mixing method.

However, the CO conversion was only 5.6% over the core–shell catalyst, lower than 58% over the physical mixing catalyst, indicating that the strong alkaline environment required for growing the zeolite shell may destroy the active sites on the $CuZnAlO_x$ core, leading to low catalytic activity. Zhao and co-worker successfully constructed a $CuZnAlO_x$ @ H-ZSM-5 capsule catalyst for the efficient synthesis of DME from syngas using an ethanol-assisted physical coating method. This method effectively prevents the corrosion of Cu species by H–ZSM-5 during the hydrothermal synthesis process. The resulting catalyst demonstrates excellent catalytic performance, achieving a CO conversion rate of 76.5% while maintaining a high DME selectivity of 95% [48].

Under typical reaction conditions of methanol to hydrocarbons, coking is the well-known reason for zeolite deactivation. Schulz and Wei reported that the deactivation of H–ZSM-5 at low temperatures (<300 °C) is due to the formation and accumulation of large amounts of monocyclic aromatics within the zeolite pores, caused by irreversible alkylation of benzene rings with lower olefins [49]. At higher temperatures, the alkylation reaction becomes reversible, allowing the deactivated large aromatics to escape from the zeolite after dealkylation into smaller molecules [50]. The activity of H–ZSM-5 for methanol dehydration to DME is generally determined by its BASs, although the contribution of strong Lewis acid sites (LASs) associated with extra-framework aluminum (EFAl) cannot be entirely ignored [51]. Mao et al. modified a series of H–ZSM-5 zeolites with an appropriate amount of MgO, which were combined with $CuZnAlO_x$ for syngas to DME reactions [52]. The MgO reacted with the bridged hydroxyl groups to generate $Mg(OH)^+$ species, resulting in a decrease in BASs and an increase in LASs. As a consequence, the selectivity for by-products such as hydrocarbons and CO_2 was significantly reduced over the bifunctional catalysts, while the DME selectivity and catalyst stability were significantly improved.

While operating at low temperatures <200 °C, catalyst deactivation due to coking can be avoided and the direct conversion of syngas to DME can achieve a higher theoretical CO conversion level. Tsubaki and co-workers introduced a new approach by using alcohols (*R*OH) as solvents and catalysts for syngas to methanol conversion in the presence of Cu-based solid catalysts (see Figure 8.8a, b) [54]. This

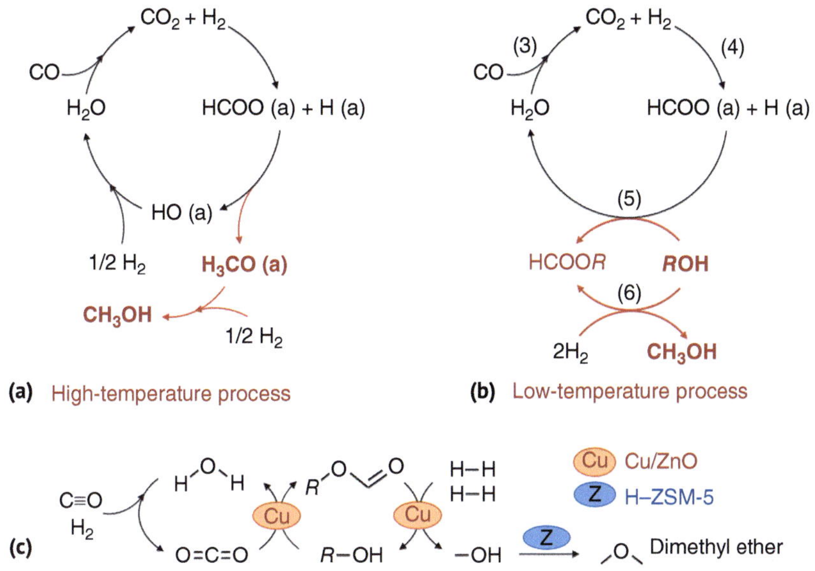

Figure 8.8 (a) Typical high-temperature methanol synthesis. The hydrogenation of formate has to proceed under temperatures ≥200 °C. (b) Decreasing the reaction temperature by the addition of alcohol to form esters as intermediates. *Source*: [41]/American Chemical Society. (c) Proposed mechanism for the novel low-temperature (170 °C) synthesis of DME from syngas over Cu/ZnO–HZSM-5 catalyst with methanol as a solvent catalyst. *Source*: [53]/Springer Nature.

method markedly differs from the conventional methanol synthesis process. Before the formation of methanol, the formate intermediates directly reacted with *R*OH to form esters, which can be hydrogenated to methanol and *R*OH, thus enabling methanol production at a lower temperature of 150°C. Based on this concept, the same group also developed a new low-temperature process for the direct conversion of syngas to DME with auxiliary methanol, catalyzed by a Cu/ZnO–zeolite catalyst (see Figure 8.8c). The CO conversion reached 29% at 170°C and 43% at 180°C, while DME selectivities were as high as 69 and 68%, respectively. This process holds promise for producing high-purity DME with lower energy consumption [53].

One should note that the thermodynamic concentration of DME is limited in syngas conversion because of the formation of H_2O. Besides, H_2O often causes the leaching and oxidation of Cu-based catalysts, thereby accelerating the deactivation of the Cu-based catalysts. Zhou et al. demonstrated a bifunctional catalytic membrane reactor for the efficient dehydration of methanol to DME by removing H_2O in situ [55]. A sandwich FAU–LTA zeolite dual-layer membrane was well designed with acidic H–FAU on the top layer, which catalyzes the dehydration reaction (see Figure 8.9). The formed H_2O can be removed through a hydrophilic Na–LTA layer, which is placed between the alumina support and the H–FAU. A high DME selectivity of 100% at methanol conversion of 91% was successfully achieved. We can expect that this reaction mode could be applied to syngas conversion, which also faces a series of problems brought about by H_2O vapor.

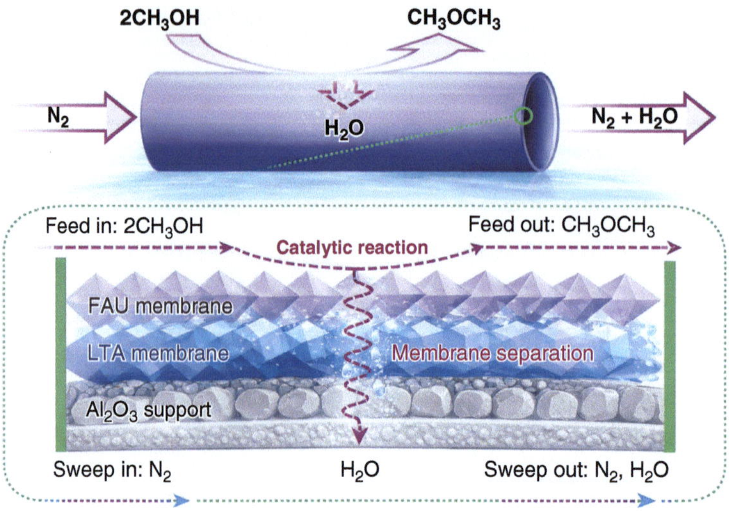

Figure 8.9 Methanol dehydration to DME by in situ removal of water in a reaction/separation bifunctional catalytic membrane reactor based on a zeolite FAU–LTA double-layer membrane. *Source*: [55]/John Wiley & Sons.

8.4 Production of Aldehydes Mediated by Hydroformylation

Hydroformylation, also known as oxo synthesis or oxo process, is an important industrial process for the production of aldehydes from syngas and alkenes [56]. This reaction is one of the most efficient ways for the formation of C—C bonds in organic synthesis and represents an exceptional example of the application of homogeneous catalysis on an industrial scale [57]. The aldehyde-derived products are highly valuable for the chemical industry and are used as plasticizers, detergents, and surfactants on a million-ton scale. Hydroformylation is typically catalyzed by rhodium- or cobalt-based homogeneous catalysts (Figure 8.10), and rhodium catalyst exhibits better performance under mild conditions [58, 59]. A key challenge for hydroformylation is the regioselectivity control during the insertion of CO into different positions of the C=C bond, which can be adjusted to some extent by using phosphine ligands due to their suitable electronic and steric properties [60]. Nevertheless, this homogeneous process unavoidably faces some problems to be solved, such as the use of excessive ligands, inevitable rhodium loss, catalyst separation, recycling, and purification. The development of a heterogeneous process for hydroformylation is always a hot research topic.

Several methods, including anchoring active transition metal or its complexes on inorganic supports, metal–organic frameworks, and polymers, have been demonstrated to modulate the rhodium species for heterogeneous hydroformylation [61–64]. Among the reported catalysts, zeolites exhibit superior performance as they enable high dispersion or even atomically dispersed active sites. Besides, the confinement effect of zeolites can not only stabilize these sites against leaching and aggregation but also modulate the adsorption mode of alkene, thus improving the

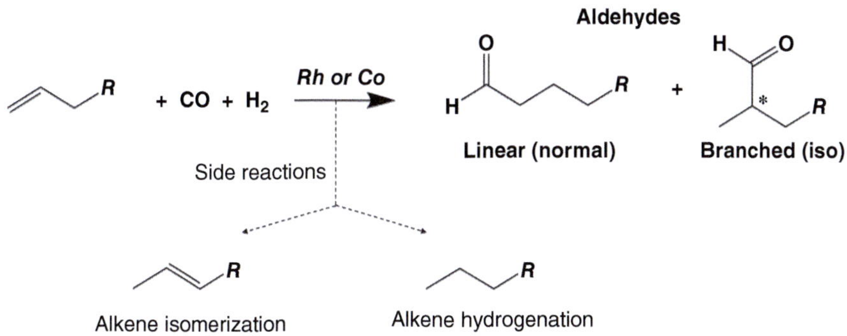

Figure 8.10 Hydroformylation of an alkene (*R* alkyl or aryl group or hydrogen). *Source*: [58]/American Chemical Society.

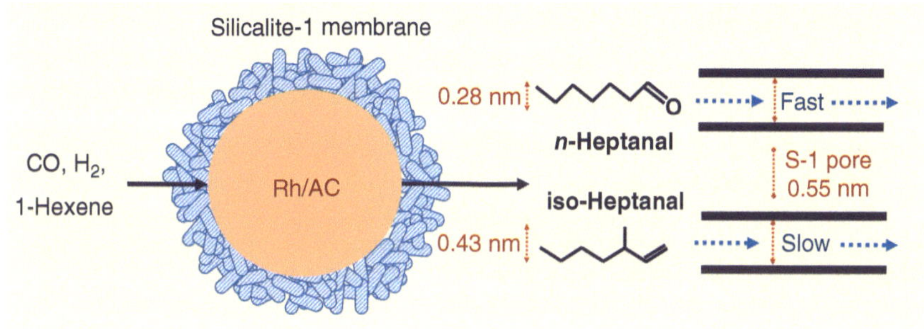

Figure 8.11 Scheme of Rh/AC@S-1 catalyst with Rh/AC encapsulated inside Silicalite-1 membranes and diffusion behavior of *n*-heptanal and iso-heptanal inside the S-1 channels.

regioselectivity [65–67]. Early studies showed that zeolite X-, Y-, and A-supported rhodium complexes as immobilized homogeneous catalysts could well catalyze the hydroformylation reactions [68–70]. However, the dispersion and distribution of rhodium inside zeolite crystals were not well investigated due to limited synthesis methods and characterization techniques at that time.

Recent studies have successfully encapsulated active metal sites inside zeolite membranes for heterogeneous hydroformylation, which also allows for improving the linear-to-branched (l/b) ratio of aldehyde products because linear aldehydes are more diffusive than their isomers in the narrow micropore channels. Li et al. synthesized a Silicalite-1-encapsulated Rh/activated carbon catalyst (Rh/AC@S-1) with millimeter size for the hydroformylation of 1-hexene (see Figure 8.11) [71]. The l/b ratio of heptanals over Rh/AC was only 0.6, while this ratio increased dramatically to 2.3 over Rh/AC@S-1. The suppression of iso-heptanal formation was due to its slow diffusion rate inside the S-1 channels. They found that a thicker S-1 membrane enhanced the normal aldehyde selectivity but caused a reduction in the conversion rate of 1-hexene. Although the modification by tetraethyl orthosilicate (TEOS) could reduce the zeolite crystalline gaps to keep high normal aldehyde selectivity without compromise in 1-hexene conversion, regioselectivity and reactivity still followed a seesaw trend.

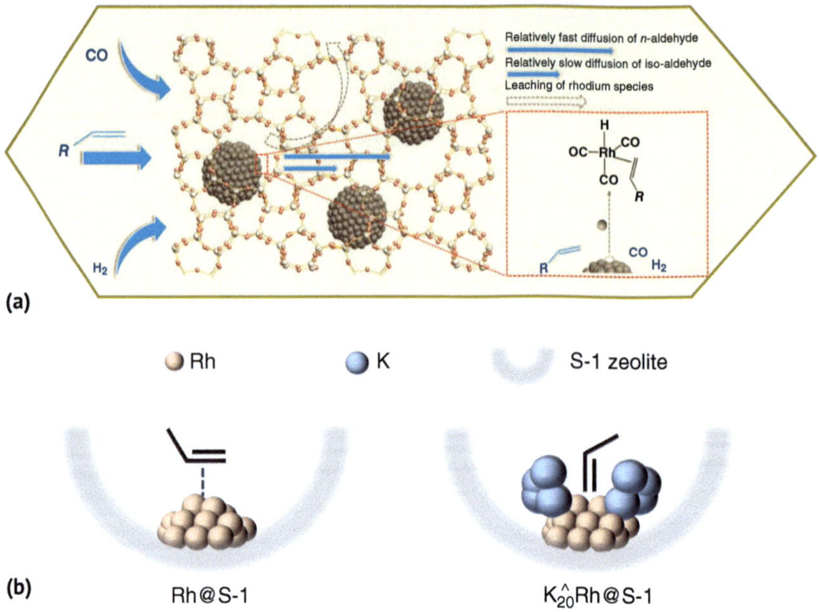

Figure 8.12 (a) Schematic illustration of the proposed hydroformylation course over rhodium oxide catalyst embedded within zeolitic microchannel. *Source*: [72]/ with permission of American Chemical Society. (b) Illustration of different adsorption configurations of propylene on Rh@S-1 and $K^{\wedge}_{20}$Rh@S-1 *Source*: [60]/with permission of Elsevier.

To improve regioselectivity in hydroformylation, Liu et al. encapsulated rhodium oxide within the microporous S-1 microchannels (Rh_2O_3@S-1) with tunable thickness and integrity of the S-1 shell [72]. This core–shell nanostructure was achieved through the epitaxial growth of S-1 seeds pre-anchored with rhodium species; thus, all the rhodium sites were well confined inside the zeolite crystals (see Figure 8.12a). The size of Rh_2O_3@S-1 particles with narrow distribution could be adjusted in the range of 400–700 nm. The Rh_2O_3@S-1 catalyst was applied in the hydroformylation of terminal olefins, including 1-hexene, 1-octene, 1-decene, and 1-dodecene, and all exhibited enhanced regioselectivity compared to the supported Rh_2O_3/S-1 catalyst. In the hydroformylation of 1-decene, Rh_2O_3@S-1 demonstrated an aldehyde yield of 93.6% and a l/b ratio of 6.65. The enhanced regioselectivity was attributed to the distinctive product diffusion rates endowed by the steric hindrance of the S-1 microchannel around Rh_2O_3.

From the perspective of active sites, Zhang et al. reported a strategy of fabricating the surface microenvironment around Rh clusters, which are surrounded by K species and confined within S-1 micropore channels [60]. K species build a sterically hindered microenvironment on the surface of Rh clusters, promoting the linear adsorption of propylene and reducing the activation energy required for the formation of linear aldehydes (see Figure 8.12b). Compared to Rh@S-1, which only provided a 46% regioselectivity toward linear butyraldehyde, the $K^{\wedge}_{20}$Rh@S-1 could increase this regioselectivity to 83%. Besides, K species were found to decrease the acidic sites, thus inhibiting the isomerization of long-chain olefins. From the perspective of zeolites, Liu et al. demonstrated that the rhodium nanoparticles supported on the external surface of siliceous MFI zeolite with abundant silanol nests are very active for hydroformylation [18], which is quite different from the Rh@S-1 catalysts. They claimed that the silanol nests could efficiently enrich olefin molecules to adjacent rhodium nanoparticles on the MFI surface, benefiting the hydroformylation of styrene with syngas in toluene solvent, outperforming the classical Wilkinson's catalyst ($RhCl(PPh_3)_3$).

Dou et al. reported an elaborate synthesis method that utilized the MFI zeolite framework as a rigid inorganic solid ligand that accommodates sub-nanometer Rh clusters [73]. This catalyst enables the efficient hydroformylation of C_6–C_{12} linear α-olefins, which can be supplied via the ethylene oligomerization or Fischer–Tropsch process. In the hydroformylation of 1-decene, the l/b ratio reached 400 at the temperature of 110°C and pressure of 3 MPa. The enhancement in regioselectivity was attributed to the unique coordination environment that surrounds the Rh clusters in the sinusoidal 10-membered ring (10-MR) channels of the MFI zeolite. The authors proposed that the diversity of zeolite framework structures (>200 topologies) can be employed for tuning both the state of Rh clusters and the properties of the olefin molecules in the microporous channels/cavities. Zhang et al. demonstrated that the Rh species with Rh(I)-gem-dicarbonyl centers can be encapsulated within the MEL zeolite framework, exhibiting exceptional hydroformylation of propene. This catalyst offered a regioselectivity of *n*-butanal above 99% and over 99% aldehyde selectivity at a high turnover frequency of 6500 h^{-1}, outperforming most existing heterogeneous and homogeneous catalysts [74]. The zeolite framework was found to guide the reaction pathway of intermediates, confined between the zeolite and Rh centers, leading to the exclusive formation of *n*-butanal (see Figure 8.13). Specifically, the reaction intermediates adsorbed on the Rh sites could be effectively confined into a constrained space between the zeolite framework and the Rh sites. Therefore, even a small difference in adsorption configuration between the linear and branched intermediates inside the appropriately confined microenvironment might lead to distinct shape selectivity. The highly regioselective formation of *n*-butanal from a less bulky linear intermediate is feasible.

Besides the most widely used Rh- and Co-based catalysts, various other transition metals including Ir, Ru, Os, Pt, Pd, Fe, and Ni have also been proven to be efficient for hydroformylation. Au is often considered to be inactive for hydroformylation

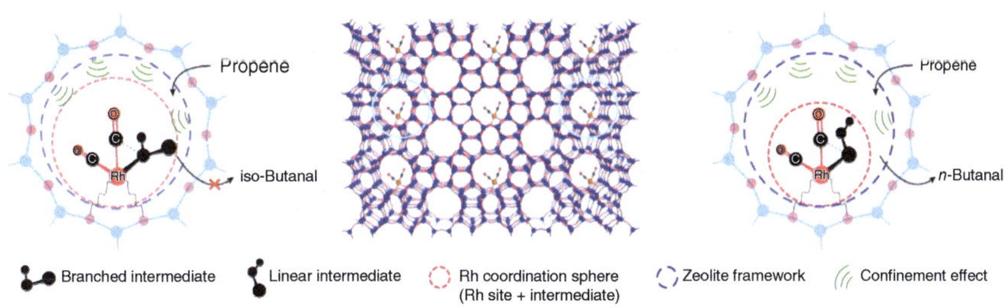

Figure 8.13 Illustration of Rh–zeolite catalysts by extending the classical transition-state shape selectivity in zeolite catalysis. *Source*: [74]/Springer Nature.

reactions due to its inert nature. However, Wei et al. recently reported a zeolite-encaged Au single-atom catalyst Au(0.2%)@S-1, where the formed Au_1–O–SiO_x motifs in zeolite can effectively stabilize isolated Au sites [75]. This catalyst showed a high formation rate of butyraldehyde and noticeable stability after five cycles in the propene hydroformylation reaction, which is even comparable to Rh-based catalysts. Zeolites with a high surface area and well-defined nanostructure can change the coordination environment of metal sites. This offers a promising way to heterogenize homogeneous hydroformylation processes with transition metals, eliminating the need for ligands or organic catalysts and resulting in more economical and efficient methods.

8.5 Conclusion and Outlook

From the perspective of atomic economy and product value, oxygenated hydrocarbons are ideal products for CO and H_2. Syngas conversion to oxygenated hydrocarbons is typically catalyzed by metal (oxide) catalysts and operated under high temperatures of above 200 °C and high pressures. Although the formation of oxygenated hydrocarbons is thermodynamically feasible, keeping the oxygen-containing functional groups in the product is challenging. There are many reaction channels on the surface of metal (oxide) and zeolites, among which the formation of olefin and saturated alkanes is tendentious. For example, dehydroxylation is easily catalyzed by acid sites of zeolites; thus ethanol and propanol will be converted to ethylene and propylene. In the presence of metal (oxide) and high-pressure hydrogen, decarboxylation and olefin hydrogenation are easily catalyzed, thus the preservation of carboxylic acids or olefins as the final product is quite difficult. Compared to the application in hydrocarbon processing, very few successful examples of zeolite catalysis have been implemented in syngas to oxygenates conversion. However, we did notice that researchers have made extensive attempts and significant progress in this area and brought new chemistry to zeolite catalysis.

Vapor phase carbonylation of methanol to AA catalyzed by acid zeolites has been discovered for more than 40 years and is now the essential link for the indirect synthesis of ethanol from syngas. This reaction also provides the possibility of directly converting syngas to C_2 oxygenates by reaction coupling. Several groups have integrated methanol catalysts, such as $CuZnAlO_x$ and $ZnZrO_x$, in the upper bed, and carbonylation catalysts, particularly modified H–MOR, in the lower bed, to realize the direct synthesis of MA or AA from syngas. By placing another metal (oxide) catalyst in the third bed, ethanol can be synthesized with a high selectivity of above 80%. For the hydroformylation of olefins requiring regioselectivity, such as propylene to *n*-butanal, the confinement of active metal sites in a well-defined and constrained space is indispensable. It was demonstrated that the encapsulation of Rh species within an MEL zeolite framework allows a 99% regioselectivity to *n*-butanal and more than 99% selectivity to aldehydes, the best performance that has been reported. This methodology has also been applied to the hydroformylation of long-chain linear α-olefins (C_6–C_{12}) into linear aldehydes with very high l/b aldehyde ratios.

One should note that both carbonylation and hydroformylation reactions are two of the largest homogeneous catalytic processes in the chemical industry. Through zeolite catalysis, these two reactions can be implemented in continuous fixed-bed reactors with advantages in catalyst recycling and product separation.

In syngas conversion involving zeolites, zeolites are often combined with a metal (oxide) catalyst. To keep the oxygen-containing groups, different components are often packed in a multi-bed mode, which weakens the synergy between different catalysts or reactions. For example, if the catalyst packed in the first bed executes methanol synthesis, the CO conversion is restricted by thermodynamics. In syngas conversion over an integrated catalyst or one catalyst particle, oxygenated species can be generated of course, but are easily further converted into olefins or alkanes because of excessive iterative reactions between different sites. Therefore, controlling the diffusion path to avoid side reactions, such as tuning the interface hydrophilicity and building membrane reactors, is one of the future directions. Besides, the compatibility of zeolites and other components should be investigated from the perspectives of system stability, reaction kinetics, and catalyst regeneration. Some groups have already started the relevant research, representing the forefront in this field [76–81]. We also think the potential of zeolite properties, such as high surface area, well-defined structure, and tunable acidity, has not been exploited in the syngas to oxygenated hydrocarbon conversion. We do believe that zeolite catalysis will reshape the research pattern of syngas conversion toward a clean and sustainable way with a high atom economy.

Acknowledgments

The authors appreciate the financial support from the National Key Research and Development Program of Ministry of Science and Technology (No. 2022YFA1504600) and the National Natural Science Foundation of China (Nos. 22121001, 22222206, U25B6005, and 22502077).

References

1 Jeske, K., Rösler, T., Belleflamme, M. et al. (2022). Direct conversion of syngas to higher alcohols via tandem integration of Fischer–Tropsch synthesis and reductive hydroformylation. *Angew. Chem. Int. Ed.* 61: e202201004.
2 Lin, T., Qi, X., Wang, X. et al. (2019). Direct production of higher oxygenates by syngas conversion over a multifunctional catalyst. *Angew. Chem. Int. Ed.* 58: 4627–4631.
3 Luk, H.T., Mondelli, C., Ferre, D.C. et al. (2017). Status and prospects in higher alcohols synthesis from syngas. *Chem. Soc. Rev.* 46: 1358–1426.
4 Rommens, K.T. and Saeys, M. (2023). Molecular views on Fischer–Tropsch synthesis. *Chem. Rev.* 123: 5798–5858.

5 Subramani, V. and Gangwal, S.K. (2008). A review of recent literature to search for an efficient catalytic process for the conversion of syngas to ethanol. *Energy Fuels* 22: 814–839.

6 Xu, D., Wang, Y., Ding, M. et al. (2021). Advances in higher alcohol synthesis from CO_2 hydrogenation. *Chem* 7: 849–881.

7 Cheng, K., Li, Y., Kang, J. et al. (2024). Selectivity control by relay catalysis in CO and CO_2 hydrogenation to multicarbon compounds. *Acc. Chem. Res.* 57: 714–725.

8 Pan, X., Jiao, F., Miao, D. et al. (2021). Oxide–zeolite-based composite catalyst concept that enables syngas chemistry beyond Fischer–Tropsch synthesis. *Chem. Rev.* 121: 6588–6609.

9 Cheng, K., Kang, J., Zhang, Q. et al. (2017). Reaction coupling as a promising methodology for selective conversion of syngas into hydrocarbons beyond Fischer–Tropsch synthesis. *Sci. China Chem.* 60: 1382–1385.

10 Tian, P., Wei, Y., Ye, M. et al. (2015). Methanol to olefins (MTO): from fundamentals to commercialization. *ACS Catal.* 5: 1922–1938.

11 Xie, J. and Olsbye, U. (2023). The oxygenate-mediated conversion of CO_x to hydrocarbons—on the role of zeolites in tandem catalysis. *Chem. Rev.* 123: 11775–11816.

12 Bhan, A. and Iglesia, E. (2008). A link between reactivity and local structure in acid catalysis on zeolites. *Acc. Chem. Res.* 41: 559–567.

13 Shang, W., Qin, B., Gao, M. et al. (2023). Efficient heterogeneous hydroformylation over zeolite-encaged isolated rhodium ions. *CCS Chem.* 5: 1526–1539.

14 Mantovani, E., Palladino, N., and Zanobi, A. (1978). Zeolite entrapped rhodium carbonyl clusters as catalyst for liquid phase hydroformylation of olefins. *J. Mol. Catal.* 3: 285–291.

15 Fujimoto, K., Shikada, T., Omata, K. et al. (1984). Vapor phase carbonylation of methanol with solid acid catalysts. *Chem. Lett.* 13: 2047–2050.

16 Cheng, K., Smulders, L.C.J., van der Wal, L.I. et al. (2022). Maximizing noble metal utilization in solid catalysts by control of nanoparticle location. *Science* 377: 204–208.

17 Přech, J., Pizarro, P., Serrano, D.P. et al. (2018). From 3D to 2D zeolite catalytic materials. *Chem. Soc. Rev.* 47: 8263–8306.

18 Liu, Y., Liu, Z., Hui, Y. et al. (2023). Rhodium nanoparticles supported on silanol-rich zeolites beyond the homogeneous Wilkinson's catalyst for hydroformylation of olefins. *Nat. Commun.* 14: 2531.

19 Zhou, W., Cheng, K., Kang, J. et al. (2019). New horizon in C1 chemistry: breaking the selectivity limitation in transformation of syngas and hydrogenation of CO_2 into hydrocarbon chemicals and fuels. *Chem. Soc. Rev.* 48: 3193–3228.

20 Vogt, E.T.C. and Weckhuysen, B.M. (2024). The refinery of the future. *Nature* 629: 295–306.

21 Haynes, A. (2010). Catalytic methanol carbonylation. *Adv. Catal.* 53: 1–45.

22 Luque-Álvarez, L.A., González-Arias, J., Romero-Sarria, F. et al. (2024). Mechanistic insights into methanol carbonylation to methyl acetate over an efficient organic template-free Cu-exchanged mordenite. *Catal. Sci. Technol.* 14: 128–136.

23 Li, Y., Yu, M., Qi, G. et al. (2024). Designing a mordenite catalyst with enhanced acidity for dimethyl ether carbonylation by engineering open Sn sites. *Engineering* 35: 104–114.

24 Ellis, B., Howard, M.J., Joyner, R.W. et al. (1996). Heterogeneous catalysts for the direct, halide-free carbonylation of methanol. *Stud. Surf. Sci. Catal.* 101: 771–779.

25 Xiong, Z., Qi, G., Zhan, E. et al. (2023). Experimental identification of the active sites over a plate-like mordenite for the carbonylation of dimethyl ether. *Chem* 9: 76–92.

26 Cao, K., Fan, D., Gao, M. et al. (2022). Recognizing the important role of surface barriers in MOR zeolite catalyzed DME carbonylation reaction. *ACS Catal.* 12: 1–7.

27 Zhan, E., Xiong, Z., and Shen, W. (2019). Dimethyl ether carbonylation over zeolites. *J. Energy Chem.* 36: 51–63.

28 Xue, H., Huang, X., Zhan, E. et al. (2013). Selective dealumination of mordenite for enhancing its stability in dimethyl ether carbonylation. *Catal. Commun.* 37: 75–79.

29 Liu, J., Xue, H., Huang, X. et al. (2010). Stability enhancement of H-mordenite in dimethyl ether carbonylation to methyl acetate by pre-adsorption of pyridine. *Chin. J. Catal.* 31: 729–738.

30 Liu, S., Fang, X., Liu, Y. et al. (2020). Dimethyl ether carbonylation over mordenite zeolite modified by alkyimidazolium ions. *Catal. Commun.* 147: 106161.

31 Yang, G., San, X., Jiang, N. et al. (2011). A new method of ethanol synthesis from dimethyl ether and syngas in a sequential dual bed reactor with the modified zeolite and Cu/ZnO catalysts. *Catal. Today* 164: 425–428.

32 Zhou, W., Kang, J., Cheng, K. et al. (2018). Direct conversion of syngas into methyl acetate, ethanol, and ethylene by relay catalysis via the intermediate dimethyl ether. *Angew. Chem. Int. Ed.* 57: 12012–12016.

33 Kang, J., He, S., Zhou, W. et al. (2020). Single-pass transformation of syngas into ethanol with high selectivity by triple tandem catalysis. *Nat. Commun.* 11: 827.

34 Cao, Z., Hu, T., Guo, J. et al. (2019). Stable and facile ethanol synthesis from syngas in one reactor by tandem combination CuZnAl-HZSM-5, modified-H-mordenite with CuZnAl catalyst. *Fuel* 254: 115542.

35 Han, S., Fan, D., Chen, N. et al. (2023). Efficient conversion of syngas into ethanol by tandem catalysis. *ACS Catal.* 13: 10651–10660.

36 Lu, P., Chen, Q., Yang, G. et al. (2019). Space-confined self-regulation mechanism from a capsule catalyst to realize an ethanol direct synthesis strategy. *ACS Catal.* 10: 1366–1374.

37 Hu, J., Zhou, W., Cheng, K. et al. (2022). CO hydrogenation. In: *The Chemical Transformations of C1 Compounds* (eds. X.F. Wu, B. Han, K. Ding et al.), 861–907. Weinheim: Wiley-VCH Verlag GmbH.

38 Cheng, K., Kang, J., King, D.L. et al. (2017). Advances in catalysis for syngas conversion to hydrocarbons. *Adv. Catal.* 60: 125–208.

39 Wang, C., Zhang, J., Qin, G. et al. (2020). Direct conversion of syngas to ethanol within zeolite crystals. *Chem* 6: 646–657.

40 Khodakov, A.Y., Ordomsky, V.V., Palčić, A. et al. (2020). Assessment of metal sintering in the copper-zeolite hybrid catalyst for direct dimethyl ether synthesis using synchrotron-based X-ray absorption and diffraction. *Catal. Today* 343: 199–205.

41 Sun, J., Yang, G., Yoneyama, Y. et al. (2014). Catalysis chemistry of dimethyl ether synthesis. *ACS Catal.* 4: 3346–3356.

42 Cai, M., Subramanian, V., Sushkevich, V.V. et al. (2015). Effect of Sn additives on the CuZnAl-HZSM-5 hybrid catalysts for the direct DME synthesis from syngas. *Appl. Catal. A: Gen.* 502: 370–379.

43 Chen, K., Li, Y., Wang, M. et al. (2021). Functionalized carbon materials in syngas conversion. *Small* 17: 2007527.

44 Saravanan, K., Ham, H., Tsubaki, N. et al. (2017). Recent progress for direct synthesis of dimethyl ether from syngas on the heterogeneous bifunctional hybrid catalysts. *Appl. Catal. B: Environ.* 217: 494–522.

45 Cai, M., Palčić, A., Subramanian, V. et al. (2016). Direct dimethyl ether synthesis from syngas on copper–zeolite hybrid catalysts with a wide range of zeolite particle sizes. *J. Catal.* 338: 227–238.

46 Sofianos, A.C. and Scurrell, M.S. (1991). Conversion of synthesis gas to dimethyl ether over bifunctional catalytic systems. *Ind. Eng. Chem. Res.* 30: 2372–2378.

47 Yang, G., Tsubaki, N., Shamoto, J. et al. (2010). Confinement effect and synergistic function of H-ZSM-5/Cu-ZnO-Al_2O_3 capsule catalyst for one-step controlled synthesis. *J. Am. Chem. Soc.* 132: 8129–8136.

48 Guo, Y. and Zhao, Z. (2020). Ethanol as a binder to fabricate a highly-efficient capsule-structured CuO–ZnO–Al_2O_3@HZSM-5 catalyst for direct production of dimethyl ether from syngas. *ChemCatChem* 12: 999–1006.

49 Schulz, H. and Wei, M. (1999). Deactivation and thermal regeneration of zeolite HZSM-5 for methanol conversion at low temperature (260–290°C). *Microporous Mesoporous Mater.* 29: 205–218.

50 Schulz, H. (2010). "Coking" of zeolites during methanol conversion: basic reactions of the MTO-, MTP- and MTG processes. *Catal. Today* 154: 183–194.

51 Abu-Dahrieh, J., Rooney, D., Goguet, A. et al. (2012). Activity and deactivation studies for direct dimethyl ether synthesis using CuO–ZnO–Al_2O_3 with NH_4ZSM-5, HZSM-5 or γ-Al_2O_3. *Chem. Eng. J.* 203: 201–211.

52 Mao, D., Yang, W., Xia, J. et al. (2005). Highly effective hybrid catalyst for the direct synthesis of dimethyl ether from syngas with magnesium oxide-modified HZSM-5 as a dehydration component. *J. Catal.* 230: 140–149.

53 Reubroycharoen, P., Teppood, S., Vitidsant, T. et al. (2009). A novel, low temperature synthesis method of dimethyl ether over Cu–Zn catalyst based on self-catalysis effect of methanol. *Top. Catal.* 52: 1079–1084.

54 Shi, L., Yang, G., Tao, K. et al. (2013). An introduction of CO_2 conversion by dry reforming with methane and new route of low-temperature methanol synthesis. *Acc. Chem. Res.* 46: 1838–1847.

55 Zhou, C., Wang, N., Qian, Y. et al. (2016). Efficient synthesis of dimethyl ether from methanol in a bifunctional zeolite membrane reactor. *Angew. Chem. Int. Ed.* 55: 12678–12682.

56 Peng, J.-B., Geng, H.-Q., and Wu, X.-F. (2019). The chemistry of CO: carbonylation. *Chem* 5: 526–552.
57 Pospech, J., Fleischer, I., Franke, R. et al. (2013). Alternative metals for homogeneous catalyzed hydroformylation reactions. *Angew. Chem. Int. Ed.* 52: 2852–2872.
58 Franke, R., Selent, D., and Börner, A. (2012). Applied hydroformylation. *Chem. Rev.* 112: 5675–5732.
59 Beller, M., Cornils, B., Frohning, C.D. et al. (1995). Progress in hydroformylation and carbonylation. *J. Mol. Catal. A: Chem.* 104: 17–85.
60 Zhang, J., Sun, P., Gao, G. et al. (2020). Enhancing regioselectivity via tuning the microenvironment in heterogeneous hydroformylation of olefins. *J. Catal.* 387: 196–206.
61 Wang, L., Zhang, W., Wang, S. et al. (2016). Atomic-level insights in optimizing reaction paths for hydroformylation reaction over Rh/CoO single-atom catalyst. *Nat. Commun.* 7: 14036.
62 Amsler, J., Sarma, B.B., Agostini, G. et al. (2020). Prospects of heterogeneous hydroformylation with supported single atom catalysts. *J. Am. Chem. Soc.* 142: 5087–5096.
63 Lang, R., Li, T., Matsumura, D. et al. (2016). Hydroformylation of olefins by a rhodium single-atom catalyst with activity comparable to $RhCl(PPh_3)_3$. *Angew. Chem. Int. Ed.* 55: 16054–16058.
64 Li, T., Chen, F., Lang, R. et al. (2020). Styrene hydroformylation with in situ hydrogen: regioselectivity control by coupling with the low-temperature water–gas shift reaction. *Angew. Chem. Int. Ed.* 59: 7430–7434.
65 Chai, Y., Shang, W., Li, W. et al. (2019). Noble metal particles confined in zeolites: synthesis, characterization, and applications. *Adv. Sci.* 6: 1900299.
66 Wu, S.-M., Yang, X.-Y., and Janiak, C. (2019). Confinement effects in zeolite-confined noble metals. *Angew. Chem. Int. Ed.* 58: 12340–12354.
67 Chai, Y., Dai, W., Wu, G. et al. (2021). Confinement in a zeolite and zeolite catalysis. *Acc. Chem. Res.* 54: 2894–2904.
68 Takahashi, N. and Kobayashi, M. (1984). Comparison of ethylene with propylene hydroformylation Rh-Y zeolite catalyst under atmospheric pressure. *J. Catal.* 85: 89–97.
69 Rode, E.J., Davis, M.E., and Hanson, B.E. (1985). Propylene hydroformylation on rhodium zeolites X and Y: I. Catalytic activity. *J. Catal.* 96: 563–573.
70 Davis, R.J., Rossin, J.A., and Davis, M.E. (1986). Hydroformylation by rhodium zeolite A catalysts. *J. Catal.* 98: 477–486.
71 Li, X., Zhang, Y., Meng, M. et al. (2010). Silicalite-1 membrane encapsulated Rh/activated-carbon catalyst for hydroformylation of 1-hexene with high selectivity to normal aldehyde. *J. Membr. Sci.* 347: 220–227.
72 Liu, C., Zhang, J., Liu, H. et al. (2019). Heterogeneous ligand-free rhodium oxide catalyst embedded within zeolitic microchannel to enhance regioselectivity in hydroformylation. *Ind. Eng. Chem. Res.* 58: 21285–21295.
73 Dou, X., Yan, T., Qian, L. et al. (2024). Regioselective hydroformylation with subnanometre Rh clusters in MFI zeolite. *Nat. Catal.* 7: 666–677.

74 Zhang, X., Yan, T., Hou, H. et al. (2024). Regioselective hydroformylation of propene catalysed by rhodium-zeolite. *Nature* 629: 597–602.

75 Wei, B., Liu, X., Chang, Q. et al. (2022). Single-atom gold species within zeolite for efficient hydroformylation. *Chem Catal.* 2: 2066–2076.

76 Xu, Y., Li, X., Gao, J. et al. (2021). A hydrophobic FeMn@Si catalyst increases olefins from syngas by suppressing C1 by-products. *Science* 371: 610–613.

77 Wang, C., Fang, W., Liu, Z. et al. (2022). Fischer–Tropsch synthesis to olefins boosted by MFI zeolite nanosheets. *Nat. Nanotechnol.* 17: 714–720.

78 Wang, C., Liu, L., Li, H. et al. (2023). Hydrophobic catalysts for syngas conversion. *Matter* 6: 2697–2710.

79 Wang, C., Zheng, M., Hu, M. et al. (2024). Unraveling spatially dependent hydrophilicity and reactivity of confined carbocation intermediates during methanol conversion over ZSM-5 zeolite. *J. Am. Chem. Soc.* 146: 8688–8696.

80 Wang, Y., Wang, G., van der Wal, L.I. et al. (2021). Visualizing element migration over bifunctional metal-zeolite catalysts and its impact on catalysis. *Angew. Chem. Int. Ed.* 133: 17876–17884.

81 Li, Y., Wang, M., Liu, S. et al. (2022). Distance for communication between metal and acid sites for syngas conversion. *ACS Catal.* 12: 8793–8801.

9

Application of Zeolite in Carbonylation

Wei Luo[1,2], *Kaijie Han*[1], *Song Song*[1,*], *Tong Ding*[1], *Ye Tian*[1], *Xingang Li*[1]

[1]*State Key Laboratory of Chemical Engineering, Collaborative Innovation Center of Chemical Science and Engineering (Tianjin), Haihe Laboratory of Sustainable Chemical Transformations, Tianjin Key Laboratory of Applied Catalysis Science and Engineering, School of Chemical Engineering and Technology, Tianjin University, Tianjin, P. R. China*
[2]*PetroChina Guangxi Petrochemical Company, Qinzhou, Guangxi, P. R. China*

*Corresponding author: ssong@tju.edu.cn

9.1 Introduction

The exploitation and utilization of energy have significantly driven the development of the global economy and human society. However, with rapid societal development, the excessive consumption of fossil fuels has exacerbated the energy crisis. Carbon monoxide (CO), as an important feedstock in chemical production, offers distinct advantages, such as wide availability and low cost. It can be obtained through multiple routes, including coal gasification, the reverse water-gas shift reactions, biomass gasification, and electrochemical CO_2 reduction.

The carbonylation reaction, which introduces CO as a carbonyl group into molecules to synthesize carbonyl-containing compounds, represents one of the most effective strategies for synthesizing target carbonyl products. In fact, carbonylation is the second-largest chemical reaction that employs CO as a raw material, following Fischer–Tropsch synthesis. Furthermore, it provides a straightforward approach for the direct incorporation of carbonyl functional groups in organic synthesis.

Over the past decade, significant progress has been achieved in the field of carbonylation. Nevertheless, numerous challenges remain unresolved. In this chapter, we mainly focus on the carbonylation of several low-carbon molecules, including dimethyl ether, methanol, ethanol, and methane. Particular attention is given to the influence of the confinement effect, synergistic effect, acid-base properties of catalysts, and metal-support interaction on catalytic performance, as well as the related reaction mechanisms.

9.2 DME Carbonylation Reaction

As an important basic chemical and alternative energy source, ethanol possesses broad development prospects and considerable market potential. However, traditional ethanol production is hindered by issues, such as low selectivity, high production costs, and significant energy consumption. In recent years, a novel pathway for ethanol synthesis from syngas has attracted increasing attention owing to its mild reaction conditions, high atom economy, and excellent selectivity. This route typically proceeds through three sequential steps: (1) The conversion of syngas to dimethyl ether (DME), (2) The carbonylation of DME to methyl acetate (MA), and (3) The hydrogenation of MA to ethanol.

Among these steps, the carbonylation of DME represents the key reaction, which involves C—O bond cleavage and subsequent C=O bond formation. The bond dissociation energy of the C—O bond in DME (~350 kJ/mol) is comparable to the C—C bond in alkanes (~344 kJ/mol), highlighting the intrinsic challenge of achieving efficient carbonylation of DME. Catalysts employed for this transformation can be broadly divided into homogeneous and heterogeneous systems. Traditional homogeneous catalysts often depend on noble metals and halogen promoters, which not only increase costs but also introduce corrosivity, thereby restricting their large-scale industrial application. In contrast, heterogeneous catalysts—particularly zeolite-based materials—offer advantages such as facile product separation, lower costs, and improved stability, making them attractive alternatives for practical processes.

9.2.1 Zeolite Catalysts

In 2006, Iglesia et al. [1] first reported the application of H-mordenite (H-MOR) and H-ferrierite (H-FER) catalysts for the carbonylation of DME to MA at low temperatures (423–463 K). They observed a stable reaction rate after an initial induction period, with MA selectivity exceeding 99%. Subsequently, Iglesia's group proposed a detailed reaction mechanism based on kinetic studies and infrared spectroscopy [1–3]. During the induction period, DME is activated on Brønsted acid sites (BAS) of H-MOR, producing two surface-adsorbed methyl species and one molecule of H_2O, which accelerates the conversion of DME. However, excessive accumulation of H_2O leads to competitive adsorption with DME, thereby suppressing catalytic activity. The key step of the overall process involves the rapid insertion of adsorbed CO into the methoxy group, forming an acetyl intermediate. The reaction rate increases with the partial pressure of CO but remains independent of DME partial pressure. The acetyl species subsequently reacts with another DME molecule to form MA, while a methyl group is re-adsorbed onto the zeolite surface.

Iglesia's group [1] also investigated the influence of zeolite pore structures on DME carbonylation. Their results revealed that no carbonylation occurs on H-Y and H-Beta with 12-membered ring (12-MR) channels. A small amount of MA is formed on H-ZSM-5 with a 10-MR structure, whereas H-MOR with 8-MR channels exhibits remarkable activity at 423 K. These findings confirm that DME carbonylation on zeolites is a typical channel-shape-selective reaction, with 8-MR zeolites showing superior activity due to their appropriate pore size and geometric configuration.

To date, only a limited number of zeolites—including H-ZSM-35, H-MOR, H-FER, EU-12, and SUZ-4—have demonstrated catalytic activity for DME carbonylation. For example, H-ZSM-35, which contains both 10-MR and 8-MR channels, effectively suppresses the formation of coke precursors such as aromatic compounds, thus achieving high selectivity and stability over 32 hours at 473 K [4]. Similarly, H-SUZ-35 with intersecting 10-MR and 8-MR channels exhibits excellent activity and durability, attributed to the enhanced diffusion of the reactive molecules [5]. Feng et al. [6] reported that DME carbonylation over rod-shaped H-SUZ-4 is strongly influenced by the synergistic effect of acid concentration and the number of 8-MR. Using quantum chemical calculations, they further demonstrated that carbonylation preferentially occurs at the T1–O12 site within the 8-MR.

The structural features of zeolites play a decisive role in their catalytic performance. H-MOR contains 12-MR and 8-MR channels along the c direction, interconnected by 8-MR side pockets along the b direction. The BAS located in these side pockets acts as a unique active center for carbonylation due to spatial constraints [7]. H-FER, on the other hand, is a microporous zeolite with one-dimensional 8-MR channels intersected vertically by 10-MR channels, both of which contribute to efficient DME carbonylation [8]. A novel zeolite, Al-RUB-41, features intersecting 8-MR and distorted 10-MR channels, with vertical 8-MR sites serving as active centers [9]. Furthermore, T-type zeolites with 8-MR and 12-MR channels also show high activity, as strong carbonylation-active acid sites are mainly located in the 8-MR, while coke formation is more likely to occur in the larger 12-MR channels [10]. More recently, Feng et al. [11] identified EU-12 as a promising zeolite for DME carbonylation, where CO preferentially adsorbs with methoxyl groups within the 8-MR to form acetyl intermediates.

Among the reported zeolites, H-MOR and H-FER remain the most extensively studied due to their superior catalytic performance, consistently delivering high yields and excellent selectivity. Kinetic and diffusion studies further reveal that the synergistic effect of favorable diffusion and reaction kinetics contributes significantly to the high carbonylation activity of MOR zeolites, followed by FER zeolites [12]. A brief summary of DME carbonylation over different zeolite catalysts is shown in Table 9.1.

9.2.2 MOR Catalysts

9.2.2.1 H-MOR Catalysts

MOR is among the earliest discovered zeolites, featuring a 12-MR main channel (0.65 × 0.70 nm) aligned along the c-axis and an 8-MR side channel (0.34 × 0.48 nm) along the b-axis (see Figure 9.1). Corma and co-workers [13], using density functional theory (DFT), investigated the role of BAS at different T-sites within the 8-MR and 12-MR channels of H-MOR during DME carbonylation. They identified the T3–O33 site in the 8-MR channel as highly selective for carbonylation. This was attributed to the confined pore size and favorable spatial orientation of the methoxy

Table 9.1 The carbonylation of DME over different types of zeolite catalysts.

Catalyst	Reaction conditions	Conversion (%)	Selectivity (%)	References
HZSM-35 HMOR	$DME/CO/N_2/H$ = 5/50/2.5/42.5, 473 K, 1.0 MPa	10 10	90 90	[4]
HSUZ-4	DME/CO/Ar/He = 5/50/2.5/42.5, 473 K, 1.0 MPa	10	90	[5]
HSUZ-4	DME/CO/Ar = 3/93/3, 493 K, 1.5 MPa	27	99	[6]
HMOR	$DME/CO/N_2$ = 5/35/60, 473 K, 2.0 MPa	85	100	[7]
HFER	$DME/CO/N_2$ = 5/45/50, 493 K, 1.0 MPa	24	95	[8]
Al-RUB-41	$DME/CO/Ar/N_2$ = 1.9/45.6/2.5/50.0, 473 K, 1.5 MPa	35	95	[9]
T-O	DME/CO/Ar = 4.1/92.8/3.1, 473 K, 1.5 MPa	25	75	[10]
HEU-12	Ar/DME/CO = 3.03/4.13/92.84, 493 K, 1.5 MPa	22	90	[11]

group, which facilitates CO insertion. Consequently, the activation energy barrier at T3–O33 was lower than that for competing olefin formation.

Huang et al. [14] further examined the influence of acid sites by modifying the chemical composition of the precursor gel. Their results confirmed that BAS at the T3 positions within the 8-MR serve as the principal catalytic centers for DME carbonylation, where confinement effects reduce the activation barrier. Moreover, both the number and strength of BAS were found to correlate linearly with catalytic activity [15]. In contrast, the 12-MR channels primarily act as diffusion pathways for reactants and products, while the T4 site in the 12-MR was identified as a major deactivation center, favoring the formation of alkanes and aromatic hydrocarbons [16]. BAS within the 12-MR were also shown to play an important role in influencing mass transfer processes [17].

Kinetic analysis by Xiong et al. [18] revealed that the specific activity of T3 sites is approximately four times higher than that of T4 sites at 473 K. Complementary quantum chemical studies by Wang et al. [19] demonstrated that the formation of surface methoxy species in the 8-MR and their subsequent carbonylation are energetically favorable. Notably, the effective rate constant for the 8-MR pathway was estimated to be 13 orders of magnitude greater than that for the 12-MR channel, underscoring the dominant role of 8-MR channels in MOR zeolites for DME carbonylation.

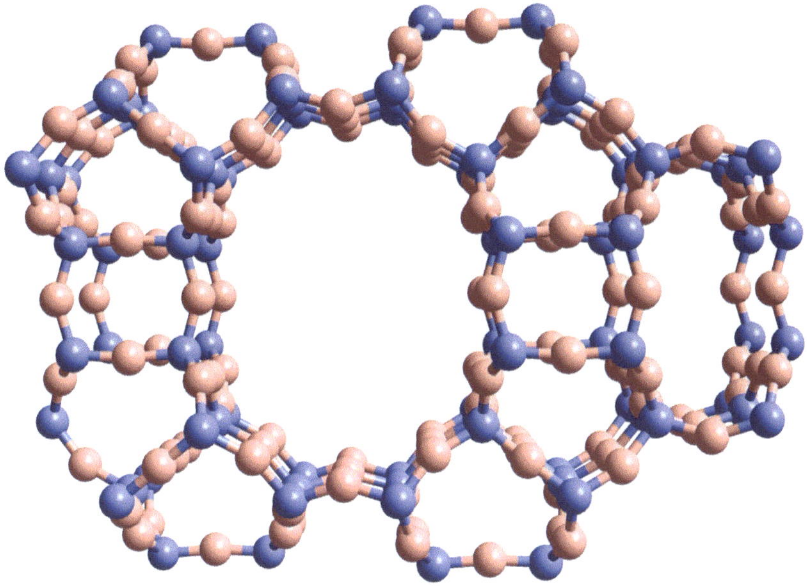

Figure 9.1 The framework image of MOR zeolite viewed along the [001] direction.

The content and distribution of Al species in MOR strongly influence the concentration and spatial arrangement of BAS, making Al regulation a key strategy for enhancing catalytic activity. For example, Wang et al. [20] selectively increased the density of BAS within the 8-MR channels of H-MOR by tailoring the sol–gel composition and employing different structure-directing agents (SDAs). Similarly, Li et al. [21] tuned the Al distribution between the two channels of the MOR by leveraging the competitive effects of Na^+ ions and organic templates, which also led to improved catalytic activity. More recently, Wang et al. [22] introduced aniline and p-phenylenediamine as additives during sol–gel synthesis, with p-phenylenediamine preferentially directing Al incorporation into the 8-MR channels, thereby enhancing activity.

Beyond acid site engineering, various strategies have been developed to improve the catalytic performance of the MOR in DME carbonylation by shortening diffusion paths. Considerable efforts have focused on synthesizing nanoscale or hierarchical MOR to facilitate mass transfer. Li et al. [23] successfully prepared nanosized spherical MOR zeolites using a vapor-assisted method, where organic amines modulated the Al distribution between the 8-MR and 12-MR. This approach resulted in a threefold enrichment of BAS in the 8-MR, significantly improving both activity and stability. Furthermore, the authors demonstrated that catalytic performance correlated with crystal morphology, with the carbonylation rate proportional to the c/b ratio [24]. Shortening channel lengths in nanoscale MOR or nanosheet structures also enhanced MA production rates by reducing diffusion limitations [25].

In another study, Liu et al. [26] synthesized MOR with a wheel-like morphology using K^+ ions as structure-modifying agents to suppress growth along the [001] direction. The resulting H-MOR exhibited superior performance due to improved mass transfer properties. Subsequently, Liu et al. [27] developed a template-free method to assemble H-MOR nanosheets, which provided abundant framework BAS and enhanced molecular diffusion, thereby delivering excellent catalytic performance.

The impact of crystal size has also been highlighted. Wen et al. [28] demonstrated that smaller MOR crystals enhance CO adsorption, thus promoting DME carbonylation activity. He et al. [29] synthesized MOR with diverse morphologies—including nanoflakes, bundles, and flower-like structures—using cetyltrimethylammonium bromide (CTAB) to improve mass transfer, resulting in higher activity. Additionally, Liu et al. [30] introduced secondary porosity into H-MOR via NH_4F-assisted chemical etching, which significantly increased turnover frequency by reducing diffusion resistance.

A brief summary of DME carbonylation over H-MOR catalysts is shown in Table 9.2.

Despite significant progress, several challenges remain for DME carbonylation over H-MOR, including relatively low catalytic efficiency and rapid deactivation. To overcome these limitations, considerable efforts have been devoted to

Table 9.2 The carbonylation of DME over H-MOR catalysts.

Catalyst	Reaction conditions	Conversion (%)	Selectivity (%)	References
H-MOR	DME/CO = 1/49, 473K, 1.5 MPa, 6000 h^{-1}	60	100	[20]
H-MOR	DME/CO = 1/49, 473K, 1.5 MPa, 6000 h^{-1}	69	100	[21]
H-MOR	Ar/DME/CO = 3/5/92, 473K, 1.5 MPa, 1600 h^{-1}	86	98	[22]
H-MOR	DME/CO/H_2 = 5/35/60, 473K, 2.0 MPa, 1500 mL $(g\,h)^{-1}$	90	97	[23]
H-MOR	DME/CO/Ar = 1/47/2, 473K, 1.5 MPa, 3000 h^{-1}	210[a]	99	[24]
H-MOR	DME/CO/N_2/He = 5/50/2.5/42.5, 473K, 1.0 MPa, 1250 mL $(g\,h)^{-1}$	37	>99	[25]
H-MOR	DME/CO/Ar = 3/30/67, 473K, 2.0 MPa, 3600 mL$(g\,h)^{-1}$	18	100	[26]
H-MOR	DME/CO/N_2 = 3/95.5/1.5, 483K, 1.5 MPa, 5280 mL $(g \cdot h)^{-1}$	55	98	[27]

Catalyst	Reaction conditions	Conversion (%)	Selectivity (%)	References
H-MOR	$DME/CO/N_2 = 5/35/60$, 473K, 2.09 MPa, GHSV = 1500 mL $(g\ h)^{-1}$	40	100	[28]
H-MOR	DME/CO = 1/49, 473K, 1.5 MPa, 6000 h^{-1}	67	100	[29]

a. Reaction rate (g g_{cat}^{-1} h^{-1}).

the modification of H-MOR. Among the various strategies, metal incorporation has emerged as one of the most effective approaches. To date, a range of metals—including Cu, Fe, Co, Rh, Ag, and Pt—have been successfully introduced into H-MOR, leading to notable improvements in catalytic performance.

9.2.2.2 Cu-MOR Catalysts

During DME carbonylation over H-MOR, the reaction rate is often restricted by the sluggish insertion of CO into the methyl group to generate the acetyl intermediate, which is regarded as the rate-determining step. Transition metal modification, particularly with copper, has been shown to markedly enhance the yield of MA [31]. Among the various preparation methods, such as impregnation and ion exchange [32], Cu/H-MOR synthesized via ion exchange generally exhibits the highest activity. However, the nature of the active Cu species and the mechanistic role of copper in DME carbonylation remain under debate.

The identity of the active copper species is controversial. Wang et al. [31] proposed that Cu^{2+} species function as Lewis acids to activate CO, thereby increasing MA yield, though their stability is limited under high-temperature and reducing conditions. Other studies suggest that Cu^0 or Cu^+ species serve as the actual active centers [33]. Zhan et al. [34] systematically analyzed the correlation between BAS and Cu species in the 8-MR. Their results revealed that MA yield decreased with increasing Cu^+ content, whereas a positive linear correlation existed between MA yield and Cu^0 content, highlighting the promotional effect of Cu^0. In contrast, Li et al. [35] based on in situ FTIR and DFT studies, proposed a synergistic mechanism in which Cu^0 clusters promote DME dissociation rather than CO activation, with $Cu{-}CH_3$ and CO co-adsorption on Cu^0 clusters facilitating acetyl formation. Meanwhile, Cu^+ not only activates CO but also adsorbs or stabilizes DME; CO adsorption on Cu^+ forms $Cu^+{-}CO$ sites, which favor co-adsorption with DME and subsequent acetyl attack, thereby enhancing MA formation. Over nanosized Cu/H-MOR, both CO activation and DME adsorption occurred predominantly at Cu^+ sites, which shortened the induction period to approximately two hours and improved catalyst stability [36]. Similarly, Corma et al. [37] demonstrated that Cu modification of H-MOR enhanced MA space-time yields, attributing the effect to the preferential adsorption of DME over methanol and the effective CO activation by Cu^+.

In addition to the oxidation state of copper, dispersion also plays a decisive role. Reule et al. [38] reported that highly dispersed Cu, either as single-site oxygen complexes or as nanoparticles, is essential for improved catalytic performance. Noble metals have also been investigated in combination with Cu-Pt-based catalysts, widely used for C—O bond hydrolysis, were shown by DFT calculations to provide active sites for C—C bond formation [39]. Consequently, bimetallic Pt—Cu catalysts exhibited superior activity in DME carbonylation, benefiting from the synergistic effects between small Cu particles and Pt, coupled with the high density of BAS in the 8-MR [40].

A brief summary of DME carbonylation over Cu/H-MOR catalysts is shown in Table 9.3.

9.2.2.3 Other Metal-modified H-MOR Catalysts

H-MOR exhibits high activity and selectivity for DME carbonylation, primarily owing to the steric constraints imposed by its 8-MR channels. An effective strategy for further improving catalytic performance involves regulating the distribution of acid sites between the 8-MR and 12-MR structures. Introducing metal atoms into the framework or pores of MOR not only modifies its physicochemical properties but also enables precise tuning of acid site density and distribution. In particular, controlling the siting of tetrahedral Al atoms is critical, as heteroatom isomorphous substitution allows modulation of BAS associated with framework Al positions.

Iron (Fe), an inexpensive transition metal, can readily substitute Al in the MOR framework. Fe incorporation has been shown to improve DME carbonylation by reducing both acid strength and density within the 12-MR channels, thereby suppressing coke formation [41]. When oxalic acid is used as a complexing agent,

Table 9.3 Carbonylation of DME over Cu/H-MOR catalysts.

Catalyst	Reaction conditions	Conversion (%)	Selectivity (%)	References
Cu-MOR	DME/CO = 1/19, 483 K, 1.8 MPa	100	100	[31]
Cu/H-MOR	Ar/DME/CO = 3.09/2.03/94.88, 493 K, 1.5 MPa, 4800 mL $(g\,h)^{-1}$	98	96	[32]
Cu/H-MOR	DME/CO = 1/47, 473 K, 1.5 MPa, 4500 h^{-1}	91	98	[34]
Cu/H-MOR	CO/DME/He = 76/5/19, 573 K, 1.0 MPa, 5000 mL $(g\,h)^{-1}$	100	75	[36]
Cu-MOR	CO/DME/H_2/He = 50/2.4/2.9/44.7, 483 K, 1.0 MPa	78	94	[38]
Pt/Cu-MOR	DME/CO/N_2/He = 2/50/3/45 483 K, 1.0 MPa, 2760 mL $(g\,h)^{-1}$	85	95	[40]

Fe(OH)Si species preferentially form at T3 sites in the 8-MR, decreasing the number and strength of BAS and leading to a longer induction period. Although this reduces intrinsic catalytic activity, it significantly enhances MA selectivity and catalyst stability [42]. Similar effects have been observed with Ce-substituted MOR. Li et al. [43] reported that Ce incorporation via a one-step hydrothermal method enriched Al—O—Si groups in the 8-MR, leading to enhanced activity. Zirconium (Zr) has also been investigated; its incorporation generates acid sites with strong CO adsorption ability, facilitating the rate-determining insertion step of CO and thereby improving DME conversion [44].

Beyond Cu, other metals such as Co, Mg, Zn, and Ag have also demonstrated beneficial effects. Co-modified MOR shows both higher activity and improved stability compared to unmodified H-MOR. Co^{2+} ions preferentially occupy the A site in the 8-MR—promoting CO and DME adsorption—and the E site in the 12-MR—suppressing coke deposition—thus yielding a dual enhancement in activity and stability [45]. Incorporation of Mg, a basic earth metal widely applied in hydrocarbon conversion, decreases the number of strong acid sites in both 8-MR and 12-MR channels. This suppresses side reactions and results in high MA selectivity and improved catalyst durability [46]. Similarly, Zn doping enhances catalyst stability in DME carbonylation [47, 48]. Bai et al. [49] reported that Zn^{2+} introduced into the 8-MR increases electrostatic interactions and electron transfer by coordinating with O-sites near T3O31/T3O31′ and T3O33, creating an optimal confined environment. Ag has also been identified as a potential active species [50]; however, its role remains controversial. To clarify this, Li et al. prepared Ag@H-MOR and Ag/H-MOR-out catalysts with different Ag locations. Their results demonstrated that encapsulated Ag^0 species within zeolite pores, rather than those on the external surface, are primarily responsible for promoting DME carbonylation [51].

A brief summary of DME carbonylation over metal-modified H-MOR catalysts is shown in Table 9.4.

9.2.2.4 Stability of HMOR

Although HMOR exhibits excellent catalytic performance in DME carbonylation, the BAS located in the 12-MR is particularly prone to inducing carbon deposition, leading to rapid catalyst deactivation. Studies have shown that the Al atom at the T3 site in the 8-MR functions as the true active site for DME carbonylation, whereas the Al atom at the T4 site in the 12-MR is more susceptible to coking [14]. For example, $SiCl_4$ treatment enables the selective substitution of Al in the 12-MR, thereby relocating most Al atoms to the T3 site. This significantly improves catalyst lifetime by enhancing the proportion of active Al in the 8-MR [52].

To improve the catalytic stability of HMOR in DME carbonylation, several strategies have been developed to selectively remove or poison acid sites within the 12-MR channels, including high-temperature steam treatment, acid treatment, alkaline treatment, and pyridine modification.

High-temperature steam treatment is widely used for zeolite dealumination. Under such conditions, Al—O bonds break more readily than Si—O bonds, resulting in the removal of Al and the formation of a more stable Si—O—Si framework.

Table 9.4 Carbonylation of DME over metal-modified H-MOR catalysts.

Catalyst	Reaction conditions	Conversion (%)	Selectivity (%)	References
Fe/HMOR	$CO/DME/H_2 = 35/5/60$, 473 K, 3 MPa	85	100	[41]
Fe/HMOR	$CO/DME/H_2 = 35/5/60$, 473 K, 3 MPa	50	100	[42]
Ce-MOR	DME/CO = 1/49, 473 K, 1.5 MPa, 6000 h^{-1}	60	100	[43]
Zr-MOR	$DME/CO/N_2 = 3/90/7$, 473 K, 1.0 MPa, 2040 mL $(g\ h)^{-1}$	75	95	[44]
Co-HMOR	$DME/CO/H_2 = 5/76/19$, 473 K, 1.0 MPa, 2500 mL $(g\ h)^{-1}$	42	99	[45]
Mg-MOR	CO/DME = 40/1, 473 K, 1.5 MPa, 4000 mL $(g\ h)^{-1}$	30	100	[46]
Zn/MOR	$CO/DME/H_2/He =$ 50/2.4/2.9/44.7, 483 K, 2 MPa	70	100	[47]
Zn/MOR	DME/ CO/ $N_2 = 3/95.5/1.5$, 483 K, 1.5 MPa, 5280 mL $(g\ h)^{-1}$	30	—	[48]
Ag-HMOR	CO/DME = 1/49, 473 K, 1.5 MPa, 6000 h^{-1}	400[a]	100	[50]
Ag@HMOR	CO/DME = 1/49, 473 K, 1.5 MPa, 6000 h^{-1}	30	99	[51]

a. MA STY (g kg_{cat}^{-1} h^{-1}).

For instance, Xue et al. [53] introduced Na into the 8-MR of HMOR followed by steam dealumination, which selectively removed ~86% of the Al in the 12-MR while leaving the 8-MR largely intact. The resultant HMOR exhibited markedly enhanced stability in DME carbonylation. Similarly, acid treatment (e.g., nitric acid) is also effective for dealumination. This method preferentially removes Al from the 4-MR at T3 and T4 sites, while T1 and T2 contribute more directly to catalyst deactivation. Thus, nitric acid treatment improves both MA selectivity and catalyst lifetime [54]. Conversely, alkaline treatment primarily removes Si. Wang et al. [23] reported that the use of templates during alkaline treatment preserved crucial acid sites while simultaneously generating mesopores and macropores. This improved mass transfer efficiency and reduced side reactions, though the method suffers from limited controllability in selectively removing Al from the 12-MR.

Pyridine modification has proven to be an effective approach for selectively poisoning acid sites in the 12-MR. Due to its molecular size, pyridine can access the

12-MR but not the narrower 8-MR side pockets, thus preferentially neutralizing BAS in the 12-MR [55]. NMR studies revealed that pyridine adsorption in the 12-MR decreases self-diffusion efficiency and overall DME conversion but significantly enhances catalytic stability [56]. Importantly, the effect of pyridine is influenced by the Si/Al ratio: high aluminum density in the 12-MR weakens pyridine adsorption, facilitating desorption during reaction conditions [57]. Zhao et al. [58] further demonstrated that the O_2 site between the 8-MR side pocket and the 12-MR is particularly crucial for activity, but it is easily poisoned by pyridine. Fortunately, this site can be regenerated by thermal treatment at 673K, while other pyridine molecules remain in the 12-MR.

Pre-adsorption strategies further optimize pyridine modification. Xue et al. [53] combined pyridine pre-adsorption with steam treatment to preferentially remove up to 86% of the Al in the 12-MR while retaining Al in the 8-MR, thereby greatly enhancing stability. Theoretical simulations confirmed that pyridine pre-adsorption improved CO and DME adsorption equilibrium, increasing CO concentration in the 8-MR [59]. Spectroscopic studies also indicated that pyridine interacts with hydroxyl protons in the 8-MR side pocket, and preferential desorption upon heating explains the high activity and stability of pyridine-modified catalysts [60].

Nevertheless, gradual deactivation of H-MOR-Py still occurs. Xie et al. [61] proposed a new deactivation pathway involving cyclic oxygenates, formed from excessive acetyl groups converting into ketenes in the 8-MR. These intermediates diffuse into the 12-MR, where they cause channel blockage and coke formation. Recent studies also suggest that BAS in the 18-MR contributes to coke deposition, broadening the deactivation picture beyond the 12-MR alone [62].

Other strategies have been developed to remove or passivate BAS specifically in the 12-MR. Liu et al. [63] employed tetramethylammonium chloride for selective ion exchange of protons in the 12-MR, taking advantage of steric hindrance. Similarly, silanization with trimethylchlorosilane (TMCS) selectively covered BAS in the 12-MR while retaining most in the 8-MR, thereby improving catalytic performance in DME carbonylation [64].

Overall, these findings highlight that coke formation is closely linked to BAS located in large-pore channels, especially the 12-MR. Thus, selective removal, passivation, or relocation of these sites—while preserving those in the 8-MR—is critical for achieving long-lived catalysts in DME carbonylation.

9.2.3 FER Catalysts

ZSM-35, FER is a layered zeolite characterized by 5-MR, 6-MR, 8-MR, and 10-MR structures. Its framework consists of 5-MRs aligned perpendicular to the a-axis, connected to 6-MRs through 10-MRs. The pore system is defined by two intersecting channels: the 8-MR (0.35 × 0.48 nm) running parallel to [010] and the 10-MR (0.42 × 0.54 nm) oriented along [001]. This unique topology endows FER with wide catalytic versatility, and it has been employed in reactions such as n-butene skeletal isomerization, alkane hydroisomerization, m-xylene isomerization, isobutene polymerization, styrene epoxidation, propane oxidative dehydrogenation, and DME carbonylation. Compared with MOR, an excellent DME carbonylation catalyst, FER exhibits slightly lower activity but superior stability.

In 2017, Pei and co-workers investigated the adsorption of DME and CO on FER and found that both molecules preferentially occupy the 8-MR channel rather than the 10-MR, with most species concentrated in the 6-MR zone [65]. Subsequent DFT studies revealed that surface methoxy groups (SMGs), generated via the interaction of DME with BAS, preferentially form in the 8-MR and 10-MR. The carbonylation reaction was shown to occur selectively in the 6-MR region of the 8-MR channel [66]. The circular 8-MR of H-FER features low activation barriers for DME carbonylation at the T4–O7 and T2–O5 sites, while the 10-MR channel suppresses coke formation more effectively than the 12-MR in MOR, thereby extending catalyst lifetime.

The modification of FER primarily aims to enhance catalytic performance by fine-tuning acidity and optimizing diffusion characteristics. High-crystallinity FER tends to generate less coke, leading to improved stability and higher selectivity toward MA [67, 68]. Recrystallization has proven effective in improving FER crystallinity. Seed-assisted recrystallization, in particular, enables adjustment of Al distribution, favoring the formation of Al—O—Si—O—Al linkages with strong DME adsorption capacity. For example, FER-S1, with optimized Al siting and acidity in the 8-MR, demonstrates high activity and long-term stability in DME carbonylation [69]. The choice of seeds strongly influences crystallinity and acidity. Comparative studies of FER, MOR, ZSM-5, and USY have shown that highly crystalline FER achieves superior catalytic stability and activity, attributable to the preferential formation of stable tetrahedral Al sites, especially at the T2 position of the Al—O—Si—O—Al framework [70]. Adjusting seed content further allows a balance of BAS in the 8-MR while minimizing defective Lewis acid sites (LAS) [71, 72].

The carbonylation step is governed by CO insertion into SMGs at the T4–O7 and T2–O5 sites to form acetyl intermediates, which represents the rate-limiting step. Thus, precise control over Al distribution during FER synthesis is crucial. Chu et al. [73] prepared FER samples using different SDAs, including piperidine, cyclohexylamine, ethylenediamine, and pyrrolidine, and observed that Al preferentially located at the T1 and T3 sites, resulting in lower MA yields compared with H-MOR. In contrast, Wang et al. [74] reported that FER nanosheets synthesized with CTAB and piperidine promoted Al siting at the T2 and T4 positions, thereby improving catalytic activity and suppressing coke formation due to reduced grain size and optimized acid site distribution. Similarly, hydrothermal synthesis of ZSM-35 with dioxane as the SDA promoted Al enrichment at the T4 site of the 8-MR, while the combined use of CTAB and piperidine yielded string-like, thin FER morphologies with abundant strong acid sites, leading to enhanced activity and stability in DME carbonylation [75].

Diffusion limitations in FER micropores also restrict catalytic performance by hindering access to active sites and promoting coke deposition. Several strategies have been proposed to alleviate these constraints. Mild alkaline treatment improves mass transfer of DME within FER channels and enhances carbonylation activity [76]. Reducing crystal size to the nanoscale shortens diffusion pathways,

Table 9.5 Carbonylation of DME over FER catalysts.

Catalyst	Reaction conditions	Conversion (%)	Selectivity (%)	Stability (h)	References
FER-S1	$DME/CO/N_2$ = 5/45/50, 493K, 1.0 Mpa	30	97	>40	[69]
FER@FER	$DME/CO/N_2$ = 5/45/50, 493K, 1.0 Mpa	3[a]	99	>100	[70]
FER	$DME/CO/N_2$ = 5/45/50, 493K, 1.0 Mpa	50	99.6	>100	[71]
FER	$DME/CO/N_2/Ar$ = 3/45/50/2, 493K, 2.0 Mpa	50	100	>120	[74]
ZSM-35	DME/CO/Ar/He = 5/50/2.5/72.5, 493K, 2.0 Mpa	30	—	>170	[74]
FER	$DME/CO/N_2$ = 5/45/50, 493K, 1.0 Mpa	26	98	>100	[75]
ZSM-35	DME/CO = 1/10, 493K, 1.0 Mpa	45	92	>66	[76]
FER	$N_2/DME/CO$ = 5/5/90, 493K, 1.0 Mpa	36	100	70	[78]

a. MA STY (mmol g_{cat}^{-1} h^{-1}).

thereby improving accessibility to acid sites. For instance, Dai et al. synthesized nanorod-like FER with a large surface area and high site accessibility using pyrrolidine as SDA, aided by ammonium fluoride (NH_4F) [77]. Similarly, Zhang et al. [78] etched FER with NH_4F to introduce mesopores while preserving crystallinity, micropore volume, and BAS density, thereby improving DME conversion and MA selectivity while minimizing coke formation.

A brief summary of DME carbonylation over FER catalysts is shown in Table 9.5.

9.3 Methanol Carbonylation Reaction

Acetic acid (AA) is an essential chemical feedstock widely employed in the synthesis of MA and acetic anhydride. Industrial AA production primarily relies on two routes: the oxidation of ethylene or acetaldehyde and methanol carbonylation. Notably, methanol carbonylation accounts for more than 65% of the global AA output.

The carbonylation of methanol (CH_3OH) with CO to form AA was first reported in a liquid superacid system, where intermediates such as $(OCH_3)^+$ and O(CO) CH_3^+ were detected. The proposed mechanism involves the initial conversion of CH_3OH into $(OCH_3)^+$, which subsequently reacts with CO to generate $O(CO)CH_3^+$. This intermediate then reacts with another methanol molecule to produce AA [79]. Thermodynamic analysis indicates that the reaction is exothermic but associated with a high activation barrier, underscoring the need for efficient catalytic systems.

$$CH_3OH + CO \rightarrow CH_3COOH$$

Current industrial methanol carbonylation processes predominantly utilize rhodium (Rh)- or iridium (Ir)-based organometallic complexes, often in halogenated homogeneous media. However, these systems face several limitations: liquid superacids are highly corrosive and environmentally hazardous; homogeneous processes incur significant energy costs for catalyst/product separation; and halogen-containing systems generate HI, causing severe equipment corrosion. Moreover, precious metal catalysts are both scarce and expensive. While non-noble metal homogeneous catalysts have been investigated, their catalytic activity and selectivity generally remain inferior to noble metal systems. Therefore, the development of efficient, stable, and environmentally benign catalytic alternatives is of paramount importance.

Zeolites, as heterogeneous catalysts, offer several advantages for methanol carbonylation, including controllable acidity, high thermal stability, well-defined pore structures, and facile separation. Among zeolites, MOR demonstrates the highest catalytic activity, followed by FER. The superior performance of zeolites arises mainly from their pore confinement effect, with acidity playing a secondary role. MOR features both 8-MR and 12-MR channels, making it particularly effective for methanol carbonylation. The proposed reaction pathway begins with CH_3OH adsorption on BAS, forming an SMG ($-OCH_3$). Subsequently, CO inserts into the methoxy group to form an acetyl species ($-COCH_3$), which then reacts with a second CH_3OH molecule to yield MA. Quantum chemical studies suggest that hydrocarbon byproducts preferentially form in the 12-MR channels, whereas the 8-MR channel, particularly the T3–O33 site, is the most favorable active center for acetyl formation. In contrast, methoxy groups at other sites tend to undergo side reactions, hindering carbonylation.

Catalyst modification has been explored to enhance MOR performance. For instance, Corma et al. [33] reported that Cu^{2+} incorporation into MOR significantly improved carbonylation activity and MA selectivity, attributed to enhanced CO adsorption as confirmed by infrared spectroscopy. Similarly, Deng et al. [16] employed in situ solid-state NMR to identify carbonylation intermediates within MOR 8-MR channels, further validating the proposed mechanism. Despite these advances, developing stable and selective zeolite catalysts for direct methanol carbonylation under anhydrous conditions remains challenging. Ni et al. [80] addressed this issue by preparing a halogen-free, precious-metal-free pyridine-modified

H-MOR (Py-MOR) catalyst, achieving complete methanol conversion with 95% selectivity toward AA.

Zeolites also serve as excellent hosts for transition-metal species, providing confined microenvironments that stabilize active sites. Rh-based catalysts remain the most active candidates for methanol carbonylation, and incorporating isolated Rh species within zeolite channels enhances both stability and reactivity. Recently, Shang et al. [81] successfully synthesized isolated Rh cations in MFI zeolite via a hydrothermal method, achieving exceptional catalytic performance with a turnover frequency of 3950 mol acetyl $(mol_{Rh} \cdot h)^{-1}$ at 423 K. Beyond stabilizing Rh cations, the MFI framework also provides an electron confinement effect, further facilitating methanol carbonylation.

9.4 Methanol Oxidative Carbonylation Reaction

Dimethyl carbonate (DMC, $CH_3OCOOCH_3$) is a versatile green chemical with wide applications. Owing to the presence of functional groups such as $-OCH_3$ and $-COOCH_3$, DMC can replace hazardous intermediates, including phosgene and dimethyl sulfate, thereby enhancing production safety and minimizing environmental impact. In addition, its good miscibility with organic solvents makes it a promising alternative to toxic solvents in paints, coatings, and cleaning agents.

Various synthetic routes have been developed for DMC, including phosgene synthesis, urea alcoholysis, direct CO_2 utilization, transesterification, and the carbonylation of methyl nitrite and methanol. Among these, methanol oxidation carbonylation has emerged as one of the most attractive approaches due to its simple process, inexpensive raw materials, low environmental burden, favorable thermodynamics, and high atom economy.

$$4CH_3OH + CO + O_2 \rightarrow 2(CH_3O)_2CO + 2H_2O$$

Methanol oxidation carbonylation can be carried out in both liquid- and gas-phase systems. While liquid-phase processes have achieved industrialization, challenges such as complex catalyst recovery, severe equipment corrosion, and short catalyst lifetimes restrict large-scale commercialization. In contrast, gas-phase carbonylation of alcohols into carbonates, pioneered by Dow Chemical in the late 1980s, offers notable advantages. Conducted in fixed-bed reactors, it enables continuous production with improved efficiency over liquid-phase batch processes. Furthermore, the gas-phase method utilizes CO, CH_3OH, and O_2 as feedstocks, offering a simpler, cleaner, and more sustainable route that aligns with green chemistry principles.

Copper-based catalysts are the most widely investigated for methanol oxidation carbonylation, with Cu identified as the key active center. Conventional catalysts such as $CuCl_2$ and Wacker-type systems show high activity but suffer from chlorine-related drawbacks, including equipment corrosion and poor catalyst stability. Consequently, chlorine-free supported catalysts have gained increasing attention. These include Cu-based activated carbon, zeolite, and mesoporous material catalysts. For example, King [82] reported that Cu_2O/Y prepared by impregnation using $Cu(NO_3)_2$ exhibited high activity, confirming that chlorine is not a prerequisite for effective DMC synthesis.

Zeolite-supported Cu catalysts are particularly promising owing to their large pore volume, uniform channels, and high surface area, which provide a stable chemical environment for active copper species. Among them, Y-type zeolite-supported copper catalysts (CuY) has demonstrated superior catalytic activity compared to other zeolites such as ZSM-5, MOR, and X, due to their regular pore structure, favorable adsorption properties, tunable acidity, and high surface area [83]. The catalytic performance of Cu-exchanged zeolites is closely related to the Si/Al ratio. Zhang et al. [84] systematically studied CuY_n (n = 1–10) with different Si/Al ratios, identifying CuY_4 (Si/Al = 6.5) as the most active catalyst. Similarly, Álvarez et al. [85] revealed that the synthesis method strongly affects catalytic behavior: Cu/Y prepared via liquid-ethanol ion exchange using $CuCl_2$ and NaY showed the highest activity $(0.72\ \text{mol DMC} \cdot \text{mol}_{\text{Cu}}^{-1} \cdot \text{h}^{-1})$, attributed to enhanced surface chloride content and Lewis acidity. In addition, ultrasonic treatment improved catalyst surface area, acidity, and Cu^+ dispersion, thereby increasing the fraction of active Cu^+ sites [86].

Mass transfer limitations in zeolite micropores can hinder DMC synthesis. Introducing mesoporosity is an effective strategy to alleviate diffusion constraints. For instance, NaY treated with H_4EDTA–NaOH developed mesopores, enhancing Cu loading and utilization of Cu^+ species, which significantly improved catalytic activity [87]. Activation conditions are equally critical: activation under N_2 alone causes carbon deposition that blocks pores and covers Cu^+ sites, while activation in pure O_2 suppresses Cu^+ formation. Optimal performance was achieved under N_2/O_2 mixed atmospheres, which balance carbon removal with Cu^+ generation [88]. Moreover, H_2O formed as a byproduct can shift the equilibrium unfavorably; its removal promotes the forward reaction. Lee et al. [89] demonstrated that introducing dehydrating agents such as CaO enhanced CH_3OH conversion from 8.7 to 14.6% and improved DMC selectivity from 39.0 to 53.1%.

In summary, methanol oxidation carbonylation represents a highly promising pathway for green and sustainable DMC production. CuY has emerged as the most effective system. Future advances are expected to focus on tuning zeolite micro/mesostructures for improved mass transfer, optimizing Si/Al ratios for balanced acidity, and stabilizing active Cu^+ species to achieve higher catalytic activity, selectivity, and stability under industrial conditions.

9.5 Ethanol Carbonylation Reaction

Propionic acid (PA) and its derivatives are important low-carbon carboxylic acids widely used in food, pharmaceuticals, pesticides, and coatings. Traditionally, PA is produced via the oxidation of propionaldehyde, which offers mild conditions and high yields (>95%). Recently, growing interest has focused on ethanol carbonylation as an alternative route, leveraging the abundance of bioethanol and supporting the development of non-petroleum-based PA synthesis methods. The carbonylation of ethanol to PA proceeds under mild conditions, providing both economic and environmental benefits [90]. The reaction formula is:

$$CH_3CH_2OH + CO \rightarrow C_2H_5COOH$$

Various catalyst systems have been investigated for the gas-phase carbonylation of ethanol, including both precious and non-precious metals. Rh-based catalysts show excellent activity but suffer from high costs [91]. In contrast, Ni-based catalysts are more economical but face challenges such as carbon deposition and metal leaching, limiting their long-term stability [92]. Zeolite-supported Rh catalysts have demonstrated promise: Yacob et al. [93] reported that Rh exchanged into 13X zeolite is effective for PA production when iodide promoters are added. The catalytic performance is strongly influenced by balancing zeolite acidity and Rh activation within the framework. Despite these advances, no catalyst system has yet reached industrial implementation. Therefore, the development of stable, selective, and cost-effective catalysts remains crucial for advancing ethanol carbonylation to PA.

9.6 Ethanol Oxidative Carbonylation Reaction

Diethyl carbonate (DEC) is an environmentally friendly compound used as an intermediate in organic synthesis, a lithium battery electrolyte, and an additive in fuels and lubricants. Its direct synthesis via the oxidative carbonylation of ethanol offers a green and phosgene-free route with excellent atom economy:

$$2CH_3CH_2OH + CO + 0.5O_2 \rightarrow CO(OC_2H_5)_2 + H_2O$$

Research has primarily focused on developing efficient heterogeneous catalysts, with Cu, Pd, and Co being the most studied active metals. Among supports, zeolites have attracted attention due to their tunable acidity and shape selectivity. Molecular modeling and experimental studies reveal that Cu-β zeolite exhibits high selectivity toward DEC, whereas Cu-Y favors the formation of 1,1-diethoxyethane (DEE), highlighting the shape-selective role of zeolite frameworks [94].

The active sites in Cu-zeolite catalysts are Cu^+ cations, typically generated by ion exchange with CuCl. Post-synthetic modifications, such as selective silicon extraction in β zeolite using $NH_3 \cdot H_2O$, enhance the accessibility and dispersion of Cu^+ species, thereby improving DEC yield [95]. With the rapid growth of the polycarbonate industry and lithium battery demand, DEC production via ethanol oxidative carbonylation represents a cost-effective, renewable, and environmentally sustainable route with strong industrial potential.

9.7 Methane Oxidative Carbonylation Reaction

Methane (CH_4), the primary component of natural gas and biogas, is an abundant but highly inert feedstock due to its high C—H bond energy (439 kJ/mol) and low proton affinity. Recently, direct oxidative carbonylation of methane to AA has attracted increasing attention as a promising pathway for natural gas valorization.

A breakthrough was reported by Flytzani-Stephanopoulos et al. [96], who demonstrated that a single-atom Rh–ZSM-5 catalyst could directly convert CH_4 and CO into methanol and AA under mild conditions. Enhancing zeolite porosity has proven effective for improving diffusion and catalyst efficiency: Kolesnichenko et al. [97] showed that introducing mesopores into Rh–ZSM-5 improves AA yield. Moreover, secondary modifiers (Zn, Si, La, and Mg) can fine-tune acidity and stabilize Rh distribution, further enhancing activity [98].

Given the scarcity of precious metals, the development of non-precious metal-based catalysts is essential for achieving the conversion of CH_4 and CO to AA. However, the direct conversion of methane to highly reactive and selective AA using non-precious metal catalysts under mild conditions remains a substantial challenge. Notably, Wang et al. [99] prepared a Fe/ZSM-5(0.25) catalyst via an impregnation method, achieving high yields of AA attributed to mononuclear Fe^{3+} species located primarily in the 5-MRs, where they replace some BAS. Utilizing H_2O_2 as an oxidant and Fe as the active site source offers a new strategy for establishing a methane oxidation carbonylation catalytic system under mild conditions. Similarly, isolated nickel cations anchored within zeolite pores can directly convert CH_4 to CH_3COOH through oxidative carbonylation, where AA formation occurs directly from the coupling of CH_4 and CO, bypassing any intermediate products [100].

The advancement of methane oxidation carbonylation technology not only facilitates the efficient utilization of natural gas resources but also contributes to the green transformation and sustainable development of the chemical industry. By modifying the types of active metals and the topological structure of zeolite, it is possible to develop efficient and stable metal-supported zeolite catalysts. This approach can help mitigate harsh reaction conditions, thereby enhancing the economic viability and sustainability of methane oxidation carbonylation.

9.8 Conclusion and Future Outlook

Introducing carbonyl groups into organic compounds is a vital method for producing oxygen-containing compounds such as aldehydes and alcohols. Developing efficient carbonylation technologies is significant for achieving sustainable energy development, encompassing both carbonylation and oxidative carbonylation reactions, with low-carbon molecules such as DME, methanol, ethanol, and methane being utilized to synthesize high-value products like MA, AA, dimethyl carbonate, PA, and DEC.

In carbonylation reactions, selecting the appropriate catalyst is critical. Thus, creating a catalytic system that offers high activity, selectivity, stability, and low cost has become a primary focus for achieving the industrialization of carbonyl synthesis. Catalysts based on a metal as the main active component include zero-valent metal catalysts, metal oxide catalysts, and metal complex catalysts, which are among the earliest catalysts used for carbonyl synthesis. Nonmetallic catalysts, particularly those using acids as the catalytically active component—such as heteropoly acids and zeolites—also play an important role. Currently, metal-zeolite composite catalysts, which combine the benefits of both zeolites and metal catalysts, are widely employed in carbonylation reactions. The synergistic effect of Brønsted acidity and metal sites can further improve catalyst performance. Despite some progress in carbonyl synthesis, ongoing research will continue to focus on developing catalyst systems with high activity, selectivity, and cost-effectiveness.

References

1 Cheung, P., Bhan, A., Sunley, G. J., and Iglesia, E. (2006). Selective carbonylation of dimethyl ether to methyl acetate catalyzed by acidic zeolites. *Angew. Chem. Int. Ed.* 45: 1617–1620.

2 Bhan, A., Allian A. D., Sunley, G. J., Law, D. J., and Iglesia, E. (2007). Specificity of sites within eight-membered ring zeolite channels for carbonylation of methyls to acetyls. *J. Am. Chem. Soc.* 129: 4919–4924.

3 Bhan, A., and Iglesia, E. (2008). A link between reactivity and local structure in acid catalysis on zeolites. *Acc. Chem. Res.* 41: 559–567.

4 Liu, J., Xue, H., Huang, X., Li, Y., and Shen, W. (2010). Dimethyl ether carbonylation to methyl acetate over HZSM-35. *Catal. Lett.* 139: 33–37.

5 Xiong, Z., Zhan, E., Li, M., and Shen, W. (2020). DME carbonylation over a HSUZ-4 zeolite. *Chem. Commun.* 56: 3401–3404.

6 Feng, X., Cao, J., Su, C., He Z., and Zhao, X. (2022). Tailoring the acid distribution and identifying the active center of rod-shaped HSUZ-4 zeolite for enhancing dimethyl ether carbonylation performance. *Fuel* 315: 123267.

7 Cao, K., Fan, D., Zeng, S., Fan, B., Chen, N., Gao, M., Zhu, D., Wang, L., Tian, P., and Liu, Z. (2021). Organic-free synthesis of MOR nanoassemblies with excellent DME carbonylation performance. *Chinese J. Catal.* 42: 1468–1477.

8 Park, S.Y., Shin, C.H., and Bae, J.W. (2016). Selective carbonylation of dimethyl ether to methyl acetate on ferrierite. *Catal. Commun.* 75: 28–31.

9 Yao, J., Wu, Q., Fan, J., Komiyama, S., Yong, X., Zhang, W., Zhao, T., Guo, Z., Yang, G., and Tsubaki, N. (2021). A carbonylation zeolite with specific nanosheet structure for efficient catalysis. *ACS Nano* 15: 13568–13578.

10 Feng, X., Chen, F., He, Z., Zhao, X., and Cao, J. (2022). Comparison of dimethyl ether carbonylation performance over some zeolites containing 8-member ring pores. *Energy Fuels* 36: 14341–14348.

11 Feng, X., Yao, J., Li, H., Fang, Y., Yoneyama, Y., Yang, G., and Tsubaki, N. (2019). A brand new zeolite catalyst for carbonylation reaction. *Chem. Commun.* 55: 1048–1051.

12 Liu, Z., Yi, X., Wang, G., Tang, X., Li, G., Huang, L., and Zheng, A. (2019). Roles of 8-ring and 12-ring channels in mordenite for carbonylation reaction: from the perspective of molecular adsorption and diffusion. *J. Catal.* 369: 335–344.

13 Boronat, M., Martinez-Sanchez, C., Law, D., and Corma, A. (2008). Enzyme-like specificity in zeolites: a unique site position in mordenite for selective carbonylation of methanol and dimethyl ether with CO. *J. Am. Chem. Soc.* 130: 16316–16323.

14 Huang, X., Ma, M., Li, M., and Shen, W. (2020). Regulating the iocation of framework aluminium in mordenite for the carbonylation of dimethyl ether. *Catal. Sci. Technol.* 10: 7280–7290.

15 Cai, K., Huang, S., Li, Y., Cheng, Z., Lv, J., and Ma, X. (2019). Influence of acid strength on the reactivity of dimethyl ether carbonylation over H-MOR. *ACS Sustain. Chem. Eng.* 7: 2027–2034.

16 Li, B., Xu, J., Han, B., Wang, X., Qi, G., Zhang, Z., Wang, C., and Deng, F. (2013). Insight into dimethyl ether carbonylation reaction over mordenite zeolite from in-situ solid-state NMR spectroscopy. *J. Phys. Chem. C* 117: 5840–5847.

17 Wang, X., Li, R., Yu, C., Liu, Y., Liu, L., Xu, C., Zhou, H., and Lu, C. (2019). Influence of acid site distribution on dimethyl ether carbonylation over mordenite. *Ind. Eng. Chem. Res.* 58: 18065–18072.

18 Xiong, Z., Qi, G., Zhan, E., Chu, Y., Xu, J., Wei, J., Ta, N., Hao, A., Zhou, Y., Deng, F., and Shen, W. (2023). Experimental identification of the active sites over a plate-like mordenite for the carbonylation of dimethyl ether. *Chem* 9: 76–92.

19 Chu, Y., Lo, A-Y., Wang, C., and Deng, F. (2019). Origin of high selectivity of dimethyl ether carbonylation in the 8-membered ring channel of mordenite zeolite. *J. Phys. Chem. C* 123: 15503–15512.

20 Wang, M., Huang, S., Lü, J., Cheng, Z., Li, Y., Wang, S., and Ma, X. (2016). Modifying the acidity of H-MOR and its catalytic carbonylation of dimethyl ether. *Chinese J. Catal.* 37: 1530–1537.

21 Li, Y., Yu, M., Cai, K., Wang, M., Lv, J., Howe, R. F., Huang, S., and Ma X. (2020). Template-induced Al distribution in MOR and enhanced activity in dimethyl ether carbonylation. *Phys. Chem. Chem. Phys.* 22: 11374–11381.

22 Chen, F., Feng, X., Zhang, L., Zhao, J., He, Z., Yi, F., Zhao, X., and Cao, J. (2022). Selective enrichment of Brønsted acid site in 8-membered ring channels of MOR zeolite to enhance the catalytic reactivity of dimethyl ether carbonylation. *Chem. Eng. Sci.* 263: 118110.

23 Li, L., Wang, Q., Liu, H., Sun, T., Fan, D., Yang, M., Tian, P., and Liu, Z. (2018). Preparation of spherical mordenite zeolite assemblies with excellent catalytic performance for dimethyl ether carbonylation. *ACS Appl. Mater. Interfaces* 10: 32239–32246.

24 Li, Y., Li, Z., Huang, S., Cai, K., Qu, Z., Zhang, J., Wang, Y., and Ma, X. (2019). Morphology-dependent catalytic performance of mordenite in carbonylation of dimethyl ether: enhanced activity with high c/b ratio. *ACS Appl. Mater. Interfaces* 11: 24000–24005.

25 Ma, M., Huang, X., Zhan, E., Zhou, Y., Xue, H., and Shen, W. (2017). Synthesis of mordenite nanosheets with shortened channel lengths and enhanced catalytic activity. *J. Mater. Chem. A* 5: 8887–8891.

26 Liu, W., Wang, Y., Bu, L., Zhi, Y., Wang, Z., Yang, M., Chu, K., Huang, Y., Guo, N., Qu, L., and Sang, J. (2023). Wheel-like mordenite nanoassemblies with shortened channel lengths for improved catalytic performance in dimethyl ether carbonylation. *ACS Appl. Nano Mater.* 6: 18005–18015.

27 Liu, Y., Zhao, N., Xian, H., Cheng, Q., Tan, Y., Tsubaki, N., and Li, X. (2015). Facilely synthesized H-mordenite nanosheet assembly for carbonylation of dimethyl ether. *ACS Appl. Mater. Interfaces* 7: 8398–8403.

28 Wen, F., Ding, X., Fang, X., Liu, H., and Zhu, W. (2021). Crystal size sensitivity of HMOR zeolite in dimethyl ether carbonylation. *Catal. Commun.* 154: 106309.

29 He, P., Li, Y., Cai, K., Xiong, X., Lv, J., Wang, Y., Huang, S., and Ma, X. (2020). Nano-assembled mordenite zeolite with tunable morphology for carbonylation of dimethyl ether. *ACS Appl. Nano Mater.* 3: 6460–6468.

30 Liu, S., Cheng, Z., Li, Y., Sun, J., Cai, K., Huang, S., Lv, J., Wang, S., and Ma, X. (2020). Improved catalytic performance in dimethyl ether carbonylation over hierarchical mordenite by enhancing mass transfer. *Ind. Eng. Chem. Res.* 59: 13861–13869.

31 Wang, S., Guo, W., Zhu, L., Wang, H., Qiu, K., and Cen, K. (2015). Methyl acetate synthesis from dimethyl ether carbonylation over mordenite modified by cation exchange. *J. Phys. Chem. C* 119: 524–533.

32 Yang, G., San, X., Jiang, N., Tanaka, Y., Li, X., Jin, Q., Tao, K., Meng, F., and Tsubaki, N. (2011). A new method of ethanol synthesis from dimethyl ether and syngas in a sequential dual bed reactor with the modified zeolite and Cu/ZnO catalysts. *Catal. Today* 164: 425–428.

33 Zhan, E., Xiong, Z., and Shen, W. (2019). Dimethyl ether carbonylation over zeolites. *J. Energy Chem.* 36: 51-63.

34 Zhan, H., Huang, S., Li, Y., Lv, J., Wang, S., and Ma, X. (2015). Elucidating the nature and role of Cu species in enhanced catalytic carbonylation of dimethyl ether over Cu/H-MOR. *Catal. Sci. Technol.* 5: 4378–4389.

35 Li, Y., Huang, S., Cheng, Z., Wang, S., Ge, Q., and Ma, X. (2018). Synergy between Cu and Brønsted acid sites in carbonylation of dimethyl ether over Cu/H-MOR. *J. Catal.* 365: 440–449.

36 Xue, H., Huang, X., Ditzel, E., Zhan, E., Ma, M., and Shen, W. (2013). Dimethyl ether carbonylation to methyl acetate over nanosized mordenites. *Ind. Eng. Chem. Res.* 52: 11510–11515.

37 Blasco, T., Boronat, M., Concepción, P., Corma, A., Law, D., and Vidal-Moya, J. A. (2007). Carbonylation of methanol on metal–acid zeolites: evidence for a mechanism involving a multisite active center. *Angew. Chem. Int. Ed.* 119: 4012–4015.

38 Reule, A.A.C., and Semagina, N. (2016). Zinc hinders deactivation of copper-mordenite: dimethyl ether carbonylation. *ACS Catal.* 6: 4972–4975.

39 Mikhailov, M.N., Mishin, I.V., and Kustov, L.M. (2009). Platinum nanoparticles as active sites for C-C bond activation in high-silica zeolites. *Microporous Mesoporous Mater.* 117: 603–608.

40 Sheng, H., Ma, H., Qian, W., Fei, N., Zhang, H., and Ying, W. (2019). Platinum-copper bimetallic-modified nanoprism mordenite for carbonylation of dimethyl ether. *Energy Fuels* 33: 10159–10166.

41 Zhou, H., Zhu, W., Shi, L., Liu, H., Liu, S., Xu, S., Ni, Y., Liu, Y., Li, L., and Liu, Z. (2015). Promotion effect of Fe in mordenite zeolite on carbonylation of dimethyl ether to methyl acetate. *Catal. Sci. Technol.* 5: 1961–1968.

42 Zhang, L., Feng, X., He, Z., Chen, F., Su, C., Zhao, X., Cao, J., and He, Y. (2022). Enhancing the stability of dimethyl ether carbonylation over Fe-doped MOR zeolites with tunable 8-MR acidity. *Chem. Eng. Sci.* 256: 117671.

43 Li, Y., Huang, S., Cheng, Z., Cai, K., Li, L. D., Milan, E., Lv, J., Wang, Y., Sun, Q., and Ma, X. B. (2019). Promoting the activity of Ce-incorporated MOR in dimethyl ether carbonylation through tailoring the distribution of Brønsted acids. *Appl. Catal. B: Environ.* 256: 117777.

44 Zhao, P., Qian, W., Ma, H., Sheng, H., Zhang, H., and Ying, W. (2021). Effect of Zr incorporation on mordenite catalyzed dimethyl ether carbonylation. *Catal. Lett.* 151: 940–954.

45 Ma, M., Zhan, E., Huang, X., Ta, N., Xiong, Z., Bai, L., and Shen, W. (2018). Carbonylation of dimethyl ether over Co-HMOR. *Catal. Sci. Technol.* 8: 2124–2130.

46 Xu, F., Hong, Z., Lv, J., Chen, C., Zhao, G., Miao, L., Yang, W., and Zhu, Z. (2022). Mg enhances the catalytic selectivity and stability of mordenite for carbonylation of dimethyl ether. *Appl. Catal. A: Gen.* 648: 118928.

47 Reule, A.A.C., Prasad, V., and Semagina, N. (2018). Effect of Cu and Zn ion-exchange locations on mordenite performance in dimethyl ether carbonylation. *Microporous Mesoporous Mater.* 263: 220–230.

48 Zhang, Z., Zhao, N., Ma, K., Cheng, Q., Zhang, J., Zheng, L., Tian, Y., and Li, X. (2019). Isolated zinc in mordenite stabilizing carbonylation of dimethyl ether to methyl acetate. *Chin. Chem. Lett.* 30: 513–516.

49 Bai, H., Ma, M., Bai, B., Bai, L., Bai, Y., Zuo, J., Vinokurov, V., and Huang, W. (2022). Zinc enhances catalytic performance of pyridine-adsorbed HMOR: dimethyl ether carbonylation. *Fuel* 310: 122225.

50 Li, S., Cai, K., Li, Y., Liu, S., Yu, M., Wang, Y., Ma, X., and Huang, S. (2020). Identifying the active silver species in carbonylation of dimethyl ether over Ag-HMOR. *ChemCatChem* 12: 3290–3297.

51 Li, S., Huang, S., Cai, K., Li, Y., Lv, J., and Ma, X. (2022). Importance of metal location in M-H zeolite for synergistically catalyzing dimethyl ether carbonylation. *Chin. J. Chem. Eng.* 41: 350–357.

52 Liu, R., Fan, B., Zhang, W., Wang, L., Qi, L., Wang, Y., Xu, S., Yu, Z., Wei, Y., and Liu, Z. (2022). Increasing the number of aluminum atoms in T_3 sites of a mordenite zeolite by a low-pressure $SiCl_4$ treatment to catalyze dimethyl eher carbonylation. *Angew. Chem. Int. Ed.* 61: e202116990.

53 Xue, H., Huang, X., Zhan, E., Ma, M., and Shen, W. (2013). Selective dealumination of mordenite for enhancing its stability in dimethyl ether carbonylation. *Catal. Commun.* 37: 75–79.

54 Reule, A.A.C., Sawada, J.A., and Semagina, N. (2017). Effect of selective 4-membered ring dealumination on mordenite-catalyzed dimethyl ether carbonylation. *J. Catal.* 349: 98–109.

55 Liu, J., Xue, H., Huang, X., Wu, P., Huang, S., Liu, S., and Shen, W. (2010). Stability enhancement of H-mordenite in dimethyl ether carbonylation to methyl acetate by pre-adsorption of pyridine. *Chinese J. Catal.* 31: 729–738.

56 He, T., Liu, X., Xu, S., Han, X., Pan, X., Hou, G., and Bao, X. (2016). Role of 12-ring channels of mordenite in DME carbonylation investigated by solid-state NMR. *J. Phys. Chem. C* 120: 22526–22531.

57 Li, Y., Sun, Q., Huang, S., Cheng, Z., Cai, K., Lv, J., and Ma, X. (2018). Dimethyl ether carbonylation over pyridine-modified MOR: enhanced stability influenced by acidity. *Catal. Today* 311: 81–88.

58 Zhao, N., Tian, Y., Zhang, L., Cheng, Q., Lyu, S., Ding, T., Hu, Z., Ma, X., and Li, X. (2019). Spacial hindrance induced recovery of over-poisoned active acid sites in pyridine-modified H-mordenite for dimethyl ether carbonylation. *Chinese J. Catal.* 40: 895–904.

59 Zhao, J., Qian, W., Ma, H., Ying, W., Yuan, P., and Zhang, H. (2023). Theoretical study for adsorption–diffusion on H-MOR and pyridine pre-adsorbed H-MOR of dimethyl ether carbonylation. *ACS Omega* 8: 22067–22076.

60 Cao, K., Fan, D., Li, L., Fan, B., Wang, L., Zhu, D., Wang, Q., Tian, P., and Liu, Z. (2020). Insights into the pyridine-modified MOR zeolite catalysts for DME carbonylation. *ACS Catal.* 10: 3372–3380.

61 Xie, M., Fang, X., Liu, H., Chen, Z., Li, B., Yang, L., and Zhu, W. (2023). Cyclic oxygenate-based deactivation mechanism in dimethyl ether carbonylation reaction over a pyridine-modified H-MOR catalyst. *ACS Catal.* 13: 14327–14333.

62 Cheng, Z., Huang, S., Li, Y., Cai, K., Wang, Y., Wang, M., Lv, J., and Ma, X. (2021). Role of Brønsted acid sites within 8-MR of mordenite in the deactivation roadmap for dimethyl ether carbonylation. *ACS Catal.* 11: 5647–5657.

63 Liu, S., Liu, H., Ma, X., Liu, Y., Zhu, W., and Liu, Z. (2020). Identifying and controlling the acid site distributions in mordenite zeolite for dimethyl ether carbonylation reaction by means of selective ion-exchange. *Catal. Sci. Technol.* 10: 4663–4672.

64 Liu, R., Zeng, S., Sun, T., Xu, S., Yu, Z., Wei, Y., and Liu, Z. (2022). Selective removal of acid sites in mordenite zeolite by trimethylchlorosilane silylation to improve dimethyl ether carbonylation stability. *ACS Catal.* 12: 4491–4500.

65 Feng, P., Zhang, G., Zang, K., Li, X., Xu, L., and Chen, X. (2017). A theoretical study on the selective adsorption behavior of dimethyl ether and carbon monoxide on H-FER zeolites. *Chem. Phys. Lett.* 684: 279–284.

66 Feng, P., Zhang, G., Chen, X., Zang, K., Li, X., and Xu, L. (2018). Specific zone within 8-membered ring channel as catalytic center for carbonylation of dimethyl ether and methanol over FER zeolite. *Appl. Catal. A: Gen.* 557: 119–124.

67 Ham, H., Kim, J., Lim, J.H., Sung, W. C., Lee, D. H., and Bae, J. W. (2018). Selective ethanol synthesis via multi-step reactions from syngas: ferrierite-based catalysts and fluidized-bed reactor application. *Catal. Today* 303: 93–99.
68 Jung, H.S., Xuan, N.T., and Bae, J.W. (2021). Carbonylation of dimethyl ether on ferrierite zeolite: effects of crystallinity to coke distribution and deactivation. *Microporous Mesoporous Mater.* 310: 110669.
69 Ham, H., Jung, H.S., Kim, H.S., Kim, J., Cho, S. J., Lee, W. B., Park, M.J., and Bae, J. W. (2020). Gas-phase carbonylation of dimethyl ether on the stable seed-derived ferrierite. *ACS Catal.* 10: 5135–5146.
70 Kim, J., Ham, H., Jung, H.S., Wang, Y., He, Y., Tsubaki, N., Cho, S. J., Han, G. Y., and Bae, J. W. (2018). Dimethyl ether carbonylation to methyl acetate over highly crystalline zeolite seed-derived ferrierite. *Catal. Sci. Technol.* 8: 3060–3072.
71 Jung, H.S., Ham, H., and Bae, J.W. (2020). Highly stable seed-derived ferrierite for carbonylation of dimethyl ether to methyl acetate: effects of seed content to catalytic stability. *Catal. Today* 339: 79–85.
72 Kim, E.J., Gao, X., Tian, J., and Bae, J. W. (2023). Nanosized seed-derived ferrierite zeolite for a gas-phase carbonylation of dimethyl ether to methyl acetate. *Catal. Today* 411: 113822.
73 Chu, W., Liu, X., Yang, Z., Nakata, H., Tan, X., Liu, X., Xu, L., Guo, P., Li, X., and Zhu, X. (2021). Constrained Al sites in FER-type zeolites. *Chinese J. Catal.* 42: 2078–2087.
74 Tuo, J., Wang, J., Gong, X., Zhai, C., Xu, H., Xue, T., Jiang, J., Guan, Y., and Wu, P. (2024). Ferrierite nanosheets with preferential Al locations as catalysts for carbonylation of dimethyl ether. *Fuel* 357: 130001.
75 Jung, H.S., Kim, J., Park, C., Bo Lee, W., and Wook Bae, J. (2024). Unprecedented contributions of thinner ferrierite zeolite morphology for superior gas-phase dimethyl ether carbonylation activity. *Chem. Eng. J.* 481: 148657.
76 Li, X., Liu, X., Liu, S., Xie, S., Zhu, X., Chen, F., and Xu, L. (2013). Activity enhancement of ZSM-35 in dimethyl ether carbonylation reaction through alkaline modifications. *RSC Adv.* 3: 16549–16557.
77 Dai, W., Ruaux, V., Deng, X., Tai, W., Wu, G., Guan, N., Li, L., and Valtchev, V. (2021). Synthesis and catalytic application of nanorod-like FER-type zeolites. *J. Mater. Chem. A* 9: 24922–24931.
78 Zhang, D., Li, K., Chen, J., Sun, C., Li, Z., Lei, J., Ma, Q., Zhang, P., Liu, Y., and Yang, L. (2023). Improved catalytic performance in gas-phase dimethyl ether carbonylation over facile NH4F etched ferrierite. *RSC Adv.* 13: 35379–35390.
79 Smith, G., Dworkin, A., Pagni, R., and Zingg, S. (1989). Broensted superacidity of hydrochloric acid in a liquid chloroaluminate. Aluminum chloride-1-ethyl-3-methyl-1H-imidazolium chloride (55. O m/o $AlCl_3$). *J. Am. Chem. Soc.* 111: 525–530.
80 Ni, Y., Shi, L., Liu, H., Zhang, W., Liu, Y., Zhu, W., and Liu, Z. (2017). A green route for methanol carbonylation. *Catal. Sci. Technol.* 7: 4818–4822.
81 Shang, W., Gao, M., Chai, Y., Wu, G., Guan, N., and Li, L. (2021). Stabilizing isolated rhodium cations by MFI zeolite for heterogeneous methanol carbonylation. *ACS Catal.* 11: 7249–7256.

82 King, S. (1997). Oxidative carbonylation of methanol to dimethyl carbonate by solid-state ion-exchanged CuY catalysts. *Catal. Today* 33: 173–182.

83 Zhou, L., Li, S., Qi, G., Su, Y., Li, J., Zheng, A., Yi, X., Wang, Q., and Deng, F. (2016). Methanol carbonylation over copper-modified mordenite zeolite: a solid-state NMR study. *Solid State Nucl. Magn. Reson.* 80: 1–6.

84 Zhang, R., Li, J., and Wang, B. (2013). The effect of Si/Al ratios on the catalytic activity of CuY zeolites for DMC synthesis by oxidative carbonylation of methanol: a theoretical study. *RSC Adv.* 3: 12287–12298.

85 Álvarez, M., Cueto, J., Serrano, D.P., Marín, P., and Ordóñez, S. (2024). Influence of the synthesis method of Cu/Y zeolite catalysts for the gas phase oxidative carbonylation of methanol to dimethyl carbonate. *Catal. Today* 428: 114464.

86 Woo, J., Seo, J., Kim, H., Lee, D., Park, Y., Yi, C., Park, Y., and Moon, J. (2018). CuY zeolite catalysts prepared by ultrasonication-assisted ion-exchange for oxidative carbonylation of methanol to dimethyl carbonate. *Ultrason. Sonochem.* 44: 146–151.

87 Zhang, G., Liang, J., Yin, J., Yan, L., Narkhede, N., Zheng, H., and Li, Z. (2020). An efficient strategy to improve the catalytic activity of CuY for oxidative carbonylation of methanol: modification of NaY by H4EDTA-NaOH sequential treatment. *Microporous Mesoporous Mater.* 307: 110500.

88 Wang, Y., Liu, Z., Tan, C., Sun, H., and Li, Z. (2020). High catalytic activity of CuY catalysts prepared by high temperature anhydrous interaction for the oxidative carbonylation of methanol. *RSC Adv.* 10: 3293–3300.

89 Lee, D.H., You, J., Woo, J.M., Seo, J., Park, Y. C., Lee, J. S., Kim, H., Moon, J. H., and Park, S. B. (2018). Influence of dehydrating agents on the oxidative carbonylation of methanol for dimethyl carbonate synthesis over a Cu/Y-zeolite catalyst. *Chin. J. Chem. Eng.* 26: 1059–1063.

90 Gao, P., Wang, J., Zhang, X., Ding, C., Wang, S., Li, Y., Wang, M., Meng, Y., Zhang, K., and Liu, P. (2021). Influence of dispersion and stabilization of active metals on Ni-Cu/AC catalyst on gas phase carbonylation of ethanol. *Fuel* 292: 120308.

91 Yacob, S., Park, S., Kilos, B.A., Barton, D. G., and Notestein, J. M. (2015). Vapor-phase ethanol carbonylation with heteropolyacid-supported Rh. *J. Catal.* 325: 1–8.

92 Yang, W., Zhao, H., Song, H., Yang, J., Zhao, J., Yan, L., and Chou, L. (2023). Vapor-phase ethanol carbonylation over Ni supported on activated carbon: influence of the textural properties and oxygen-containing groups. *Chem. Eng. J.* 476: 146595.

93 Yacob, S., Kilos, B.A., Barton, D.G., and Notestein, J. M. (2016). Vapor phase ethanol carbonylation over Rh supported on zeolite 13X. *Appl. Catal. A: Gen.* 520: 122–131.

94 Zhang, P., Huang, S., Yang, Y., Meng, Q., Wang, S., and Ma, X. (2010). Effect of SSIE structure of Cu-exchanged β and Y on the selectivity for synthesis of diethyl carbonate by oxidative carbonylation of ethanol: a comparative investigation. *Catal. Today* 149: 202–206.

95 Zhang, P., Huang, S., Wang, S., and Ma, X. (2011). Effect of extra-framework silicon on the catalytic activity of Cuβ zeolite catalyst for synthesis of diethyl carbonate by oxidative carbonylation of ethanol. *Chem. Eng. J.* 172: 526–530.

96 Shan, J., Li, M., Allard, L.F., Lee, S., and Flytzani-Stephanopoulos, M. (2017). Mild oxidation of methane to methanol or acetic acid on supported isolated rhodium catalysts. *Nature* 551: 605–608.

97 Kolesnichenko, N.V., Snatenkova, Y.M., Batova, T.I., Yashina, O. V., and Golubev, K. B. (2022). Oxidative carbonylation of methane to acetic acid over micro-mesoporous rhodium-modified zeolites. *Microporous Mesoporous Mater.* 330: 111581.

98 Batova, T.I., Stashenko, A.N., Obukhova, T.K., Snatenkova, Y. M., Khramov, E. V., Sadovnikov, A. A., Golubev, K. B., and Kolesnichenko, N. V. (2024). Oxidative carbonylation of methane into acetic acid: effect of metal (Zn, Cu, La, and Mg) doping on Rh/ZSM-5 activity. *Microporous Mesoporous Mater.* 366: 112953.

99 Wang, C., Sun, Y., Wang, L., Feng, W., Miao, Y., Yu, M., Wang, Y., Gao, X., Zhao, Q., Ding, Z., Feng, Z., Yu, S., Yang, J., Hu, Y., and Wu, J. (2023). Oxidative carbonylation of methane to acetic acid on an Fe-modified ZSM-5 zeolite. *Appl. Catal. B: Environ.* 329: 122549.

100 Liu, J., Wei, Y., Li, R., Liu, Y., Yu, H., Zhou, X., Wu, B., Lin, T., and Zhong, L. (2024). Isolated Ni sites anchored on zeolites for direct synthesis of acetic acid from methane oxidative carbonylation. *Appl. Catal. B: Environ.* 350: 123951.

Part IV

CO_2 Conversion

10

Application of Zeolites in CO_2 to Aromatics

*Cederick Cyril Amoo, Jian Sun**

Dalian National Laboratory for Clean Energy, Dalian Institute of Chemical Physics, Chinese Academy of Sciences, Dalian, China

*Corresponding author: sunj@dicp.ac.cn

10.1 Introduction

Aromatics are basically defined as a special class of chemical compounds consisting of conjugated planar ring systems possessing unique pi (π) and sigma (σ) bonds in link. A more recognized and long-lasting definition of aromaticity can be identified by Huckel's "4n+2 rule" for the number of electrons in planar, cyclic molecules [1, 2]. The simplest form of an aromatic is a benzene ring with the molecular formula C_6H_6 [3, 4]. Some popular classes of aromatic compounds are presented (see Figure 10.1). Benzene, Toluene, and Xylene and their derivatives, known as the BTX, are the most recognized aromatic compounds. In the petroleum sector, they may be classified under monocyclic aromatics (MA) and polycyclic aromatics (PA) based on the number of benzene rings [5, 6]. Some benefits derived from aromatics include their use as blends in liquid fuels due to their high-octane ratings as well as functioning as feedstock for the polymer industries [7]. Conventionally, aromatics will be produced as extracts from fossil fuels. However, with the recent increase in their demand as well as concerns about the over-dependence on fossil fuels, the need for alternative sources for aromatic productions has become a necessity. CO_2 conversion is a catalytic process that is currently considered promising for the production of valued products traditionally produced from fossils, which include aromatics [8, 9]. The process can reduce the threat associated with recent climate changes as an alternative to valued hydrocarbon production [10–12]. Products obtained from CO_2 conversion are normally defined by the catalyst system. Conventionally, the CO_2 conversion process will proceed over transitional and other main group elemental composite catalysts. The high stability of CO_2 has challenged the dissociation and subsequent production of valued compounds. In this respect,

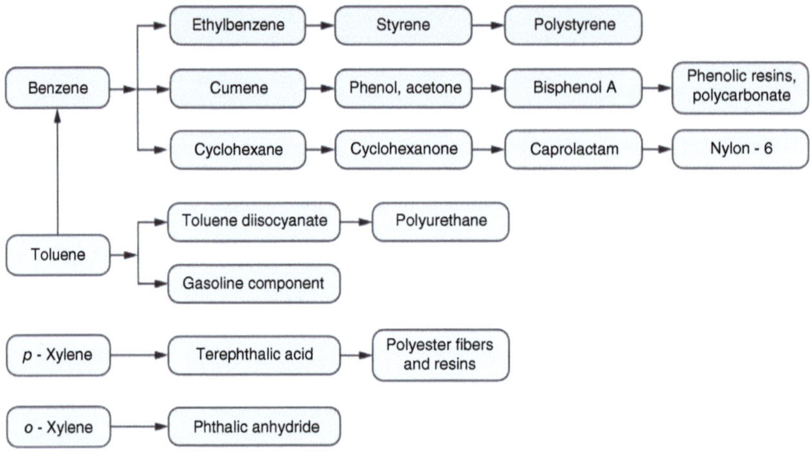

Figure 10.1 BTX aromatics and their common derivatives. *Source*: [14]/with permission of Elsevier.

photo- and electro-catalytic processes are currently limited in the product distribution to lower carbon number molecules ($<C_4$) [13]. Nonetheless, the production of aromatics from CO_2 is favored via a thermocatalytic approach.

Thermocatalytic CO_2 conversion processes have mostly been employed for producing heavy hydrocarbons (C_{5+}) from CO_2. Some major processes recognized for the direct production of aromatics from CO_2 are identified as the hydrothermal CO_2 conversion and the thermocatalytic CO_2 hydrogenation. The hydrothermal CO_2 conversion involves the dissolution of CO_2 in a solvent to form bicarbonates, which are subsequently converted to the various valued products over specific reducing agents and catalysts [15]. The process utilizes H_2O as a proton source and is merited by being fast in addition to eliminating transport and storage systems for H_2O. Depending on the reaction conditions and target products, some reducing agents and catalysts can include Fe, Mg, Mn, Al, Zn, Ni, and Cu as catalysts. Some recent reports have exploited the characteristics of some porous materials, such as zeolites, inclusive as a catalyst to enhance aromatics selectivity. Zeolites have also functioned well as a major interfaces in enhancing aromatics selectivity in CO_2 hydrogenation [16]. The CO_2 hydrogenation is more popular for the production of aromatics, as the production of other valued products via this route already has an industrial recognition. The rates and final product distribution are defined by the catalyst interfaces. Zeolites in this respect can perform oligomerization and aromatization functions.

Generally, zeolite is defined as a porous material consisting of a framework of TO_4 (where T = Si, Al, P, B, and Ge) tetrahedrally linked. They have been applied widely in the adsorption, ion exchange, and catalysis sectors. Currently, there are over 247 recorded zeolite structures, each exploited for different functions [17]. Some popular zeolite structures recognized for aromatics synthesis are shown (see Figure 10.2). In the catalysis sector, zeolites are attractive mainly due to their unique pore geometry that offer shape-selective ability for molecules, their amphiprotic nature that is often exploited for hydride transfer and acid-base reactions, as well as their surface

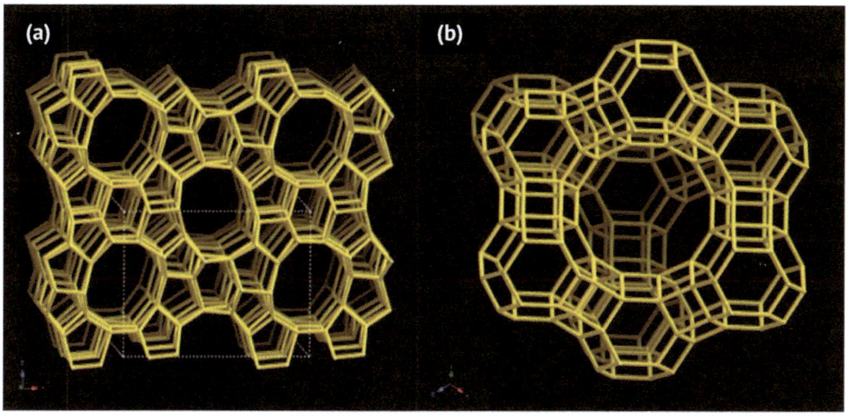

Figure 10.2 Some popular zeolite structures for aromatics synthesis: (a) MFI and (b) FAU. Panels were reproduced with permission from the IZA website.

area and high thermal stability that generally creates a platform for enhancing catalyst performances [18, 19]. Some recognized applications include refining processes such as alkylation, isomerization, and hydrocarbon cracking. Various platforms for aromatic production can recognize zeolites as essential catalysts. As stated, zeolites possess different topologies that orient products towards different products even among the aromatic compounds. Additionally, the zeolite types function optimally for the production of aromatics depending on the reaction process.

In this work, we will detail the conversion of CO_2 to aromatics over zeolite-based catalysts. Different processes and routes employed for CO_2 conversion to aromatics are highlighted. Common zeolites and their basic characteristics that majorly influence the aromatics formation are discussed from recent literature. Additionally, operando and in-situ characteristics that emerge for aromatic formation during the various reactions, as well as modelling that predicts the reaction rates and limitations, are added. The work concluded with a summary of recent findings and the outlook for future research.

10.2 Zeolite Catalysis in Aromatics Production

Zeolites are extremely popular as catalysts in the production of aromatics across various platforms. Among their applications, the pyrolysis of carbon resources such as coal or biomass is recognized as a renewable approach for the production of aromatics. By definition, the pyrolysis processes follow the destructive distillation of the carbon sources in the absence of air. The products, which may be gases, liquids, or solids, normally contain a high carbon content as their basis. While temperature remains the most important factor in this process, the introduction of catalysts can subsequently enhance the selective production of certain molecules. Zeolite catalysts are most recognized in the pyrolysis process for the selective production of aromatics. In a study by Liu et al., Zn-modified ZSM-5 could effectively convert jatropha oil to aromatics with a 50.33% yield [20]. Among the various metallic variations, the Zn-modified ZSM-5 possessed ample acidity, dominated by Lewis and weak acid sites. This characteristic functioned properly for PA formation. Light aromatics selectivity was also optimized using rubber seed oil as feed and alkali-treated ZSM-5 as catalysts. The reaction results indicate that zeolite favored the selective production of BTX with the least selectivity for PA at low temperatures [21]. The catalytic co-pyrolysis of biomass and plastic has also received attention, employing various zeolites for the production of aromatics [22, 23]. The vapors from coal pyrolysis have also been upgraded to light aromatics using zeolite Y [24]. Aside from pyrolysis, other processes have been employed for converting biomass to aromatics. The dealkylation of biomass to various aromatics has been demonstrated by many.

Syngas conversion has become a core procedure for aromatics production. Solid acids such as zeolites are used to enhance aromatics selectivity. A bifunctional composite consisting of Fe_3O_4@MnO_2 and hollow HZSM-5 synthesized aromatics at 57% at a CO conversion of >90%. The aromatization was attributed to the diffusion of the olefin's molecules on the acid sites of the zeolites [25]. Zhou et al. fabricated Mo-doped ZrO_2 particles coupled with ZSM-5 for aromatics production from syngas. The aromatics selectivity reached a high of 76% at a CO conversion of 22% [26].

Some light hydrocarbons, such as CH_4 and olefins, can be converted to aromatics. The direct non-oxidative methane conversion (DNMC) is a process that requires only CH_4 as feed for the production of C_{2+} hydrocarbons. Zeolites loaded with some transitional metals are popular for the direct dehydroaromatization of CH_4 via this process. Likewise, the aromatization of various hydrocarbons has also been reported. Also, aromatics can be produced from oxygenated compounds via the conversion of methanol to aromatics or the cracking of C_{2+} oxygenates. The effects of ZSM-5 pore properties on MeOH conversion to aromatics were investigated by Shen et al. [27]. It was revealed that the catalysts with a larger pore hierarchy demonstrated higher selectivity for aromatics. The reaction set revealed olefins as immediate intermediates for the formation of aromatics from MeOH.

10.3 Heterogeneous CO_2 Hydrogenation to Aromatics

With concerns about recent climate changes, it has become necessary to find suitable utilization for CO_2, changing its status from a liability to a benefit [28]. Among plausible processes recognized to suppress this anthropogenic gas, the heterogeneous CO_2 hydrogenation to valued products is one popular process for the effective utilization of CO_2. While the potential of this process cannot be doubted, considering the boundless selectivity for various hydrocarbons and oxygenates, the stability of CO_2 presents a hindrance for the formation of long-chain products such as aromatics [29, 30]. CO_2 is recognized as a highly stable gas and thus requires a high enthalpy for its dissociation. This has challenged the effective reduction of CO_2 and subsequent transformation to other chemicals [31, 32]. As identified earlier, reports on most photo- and electro-driven CO_2 hydrogenation reactions, even in the presence of catalysts, mostly demonstrate selectivity to low-carbon products. However, the thermocatalytic reaction can break this limitation with selectivity to long-chained carbon products by employing catalytic phases that favor oligomerization of lighter hydrocarbons. Zeolites are predominant among materials recognized for presenting the characteristics for effective oligomerization of lighter hydrocarbons.

Aromatics are prominent long-chained products obtainable from CO_2 hydrogenation. Like other long-chained products, their production is a challenge in the CO_2 hydrogenation reaction. Transitional and main group metals with some alkali promotions are ubiquitous for this reaction. Very few literature studies have efficiently produced substantial long-chained hydrocarbons directly over a metallic composite, and that excludes aromatics [33–35]. A thorough review of the concerned literature reveals two routes that can effectively produce aromatics directly from the thermocatalytic CO_2 hydrogenation: a methanol-mediated route and Fischer–Tropsch synthesis route, both of which recognize zeolites as a base material for aromatization [9]. In this section, we highlight the functions of zeolites in the CO_2 hydrogenation process over the two major routes, with our main focus on zeolite catalysts. The effects of proximity and zeolite characteristics will be discussed to reveal and conclude some basic facts involving the application of zeolites in the CO_2 hydrogenation.

10.3.1 CO_2-MTA Route

The CO_2 hydrogenation to aromatics via the CO_2-MTA route employs methanol and other oxygenate intermediates prior to zeolite catalysis. *Enormous progress has been made* in the conversion of CO_2 to methanol with high selectivity. Cu, Zn, Zr, In, and Co composites have been modelled, designed, and fabricated as catalysts for enhanced selectivity to MeOH [36–41]. The methanol can be subsequently converted to aromatics on zeolites and other zeolite catalysts. Different zeolites present structures that orient the methanol conversion to various hydrocarbon products [42, 43]. A review of the literature suggests the selectivity to hydrocarbons is mainly defined by pore geometry and acid site distribution. The modelled performance of zeolites in methanol conversion to hydrocarbons indicates that light hydrocarbon selectivity is favored over 8-MR ring zeolites such as SAPO-34, SAPO-18, and SSZ-39, which possess channel sizes less than 4Å, whereas heavy hydrocarbons (C_{5+}) are favored over >10-MR zeolites (see Figure 10.3) [44]. From literature, the tandem catalysis following the CO_2 hydrogenation to aromatics via a MeOH intermediate route employs ZSM-5 zeolite the most.

10.3.1.1 Effects of Proximity

Different coupling procedures have been used in developing zeolite-metal catalysts. The final catalyst system will assume a configuration with the metal and zeolite interacting at a certain proximal distance. Independent coupling procedures will define the retention of the intermediates, which is key to identifying species involved in the final reaction. Popular configurations identified for fabricating catalysts in CO_2 hydrogenation to aromatics include powder mixing of the zeolites and metallic

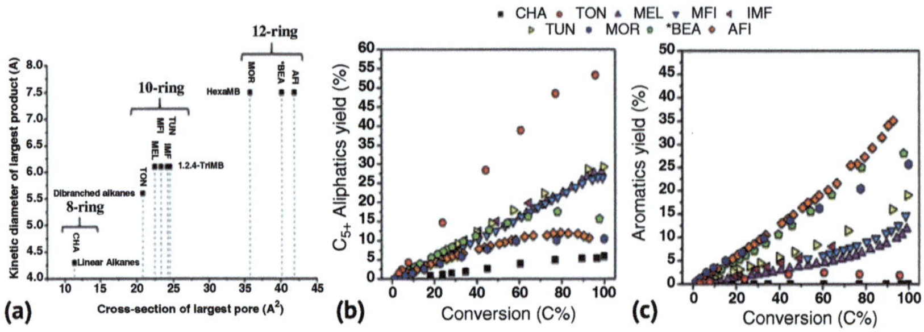

Figure 10.3 (a) Cross-section of the pore of some zeolites versus the kinetic diameter of heavy products. (b) C_{5+} aliphatic yield over various zeolites as a function of conversion, 423 K, P (methanol) = 0.01 MPa. (c) Aromatics yield over various zeolites as a function of conversion, 423 K, P (methanol) = 0.01 MPa. *Source*: [44]/The Royal Society of Chemistry/CC BY-NC 3.0.

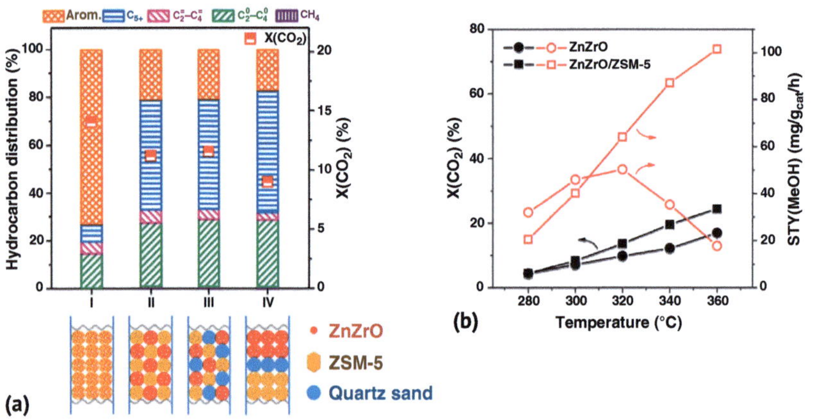

Figure 10.4 (a) Hydrogenation of CO_2 over ZnZrO/ZSM-5, integrated catalyst components ZnZrO and H-ZSM-5 with different proximity (1.0 g, 320 °C, 4.0 MPa, 1200 mL · g_{cat}^{-1} · h^{-1}). (b) Hydrogenation of CO_2 over ZnZrO and ZnZrO/ZSM-5 at reaction temperatures of 280–360 °C (0.5 g, 4 MPa, 2400 mL · g_{cat}^{-1} · h^{-1} for ZnZrO; 1.0 g, 4 MPa, 1200 mL · g_{cat}^{-1} · h^{-1} for ZnZrO/ZSM-5). *Source*: [16]/with permission of Elsevier.

phases, granular mixing, and stack-bed mode methods. A detailed work by Li et al. [16] investigated tandem catalysis of ZnZrO/ZSM-5 with the independent phases in different proximities (see Figure 10.4). When ZnZrO and H-ZSM-5 were powdered and mixed, the aromatics selectivity reached over 65% (see Figure 10.4a, i). However, when mixed in granule form, the aromatics selectivity reached only 21% (see Figure 10.4a, ii). Further reductions in aromatics selectivity were observed when some quartz sand particulates (200–450 nm) were mixed with granules of both catalysts (see Figure 10.4a, iii) and when used to separate the ZnZrO and H-ZSM-5 (see Figure 10.4a, iv). These results prove that the proximity between the MeOH-producing phase and H-ZSM-5 can obviously affect the conversion of CO_2 and the selectivity of aromatics. This work concludes that the formation of aromatics is favored when the two phases are in closest proximity, and these in turn favor rapid transfer of intermediates from the surface of ZnZrO to the H-ZSM-5 with higher efficiency. Upon increasing the spatial distance between ZnZrO and H-ZSM-5 (see Figure 10.4a, ii and iii), the transfer of intermediates becomes difficult. Considering the spatial separation between ZnZrO and H-ZSM-5 by a quartz sand layer (see Figure 10.4a, iv), MeOH is the intermediate that desorbed from the surface of ZnZrO and then diffused into the pore structure. The high selectivity of aromatics over ZnZrO/ZSM-5 tandem catalyst suggests the probable presence of other intermediates. A similar investigation consisting of $ZnAlO_x$ and HZSM-5 concluded that the $ZnAlO_x$ and HZSM-5 prepared by powder grinding of the final forms of the two materials demonstrated higher activity for aromatics selectivity with the granular mix and dual-bed options, following the patterns of the previous results [29].

Further insights by Zhang et al. [30] confirmed enhanced selectivity for aromatics over catalysts with the zeolite crystals and metallic phases in proximity. The immediate intermediates identified included MeOH which was converted to alkenes prior to aromatization, with both processes catalyzed by the zeolite. CO_2 hydrogenation to aromatics via an oxygenated-mediated pathway is favored when the metallic phases of the catalysts are in close contact with the zeolite crystals. From the reports, a larger gap lowers the aromatization process.

10.3.1.2 Effects of Physical Properties

The physical properties of zeolite can influence the aromatization process in various ways. Key physical properties that influence aromatics production include different surface and pore properties, morphology, and framework topologies. HZSM-5 with different morphologies was investigated in tandem with the $ZnZr_x$ catalyst for aromatics production from the hydrogenation of CO_2 [45]. The reaction consisted of hollow, spherical, sheet, and chain morphologies of ZSM-5 (see Figure 10.5). An aromatics selectivity of 51.9% was recorded, with 46.7% being BTX over the sheet type H-ZSM-5. A short b-axis length was identified as the unique physical characteristic for sheet H-ZSM-5 over the other forms, which was conducive to material transport and diffusion, making it easier for the guest molecules to enter the active site and preventing side reactions from occurring. The effects of zeolite crystal sizes on various reactions have been performed thoroughly studied. Studies investigating the influences of zeolite crystal sizes have indicated that the nano-sized crystals perform better for aromatics formation [46, 47]. These results conclude that a larger active surface area and larger pores can be conducive to aromatics selectivity.

10.3.1.3 Effects of Chemical Properties

The major chemical characteristic exploited from zeolites for producing aromatics is the acid site distribution. Other chemical influential factors normally gravitate towards this characteristic. In tuning the acid site distribution, the Si/Al ratio can be varied for the same zeolite, while cation exchange can also be introduced to alter the acid distribution. In some rare instances, deposited materials have been on the zeolite surfaces to suppress or express specific acid sites [48]. The effects of the Si/Al ratio on the direct CO_2 hydrogenation to aromatics have been thoroughly

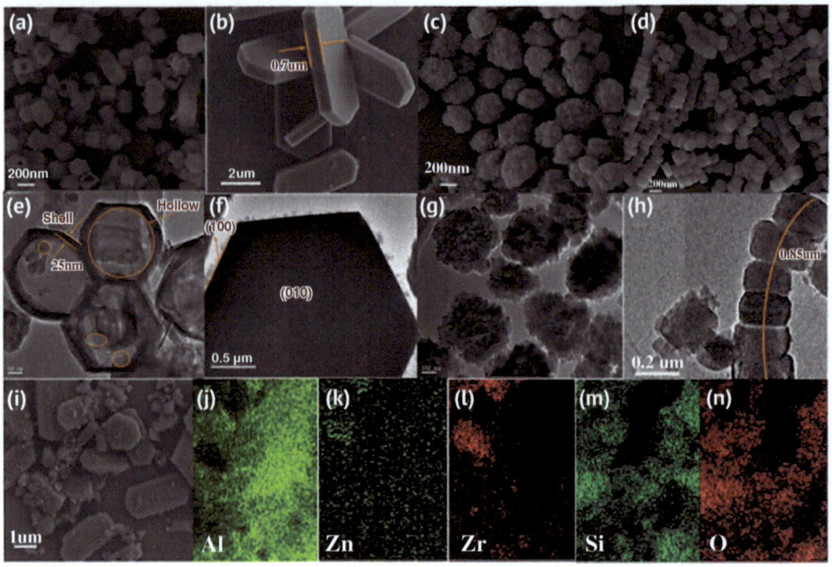

Figure 10.5 SEM images of (a) hollow, (b) sheet, (c) spherical, (d) chain HZSM-5, TEM images of (e) hollow, (f) sheet, (g) spherical, (h) chain HZSM-5, and (i–n) SEM images and SEM-EDS elemental mapping of $ZnZr_7O(500)$-sheet-Z5 tandem catalysts. *Source*: Reproduced with permission from [45]/Elsevier.

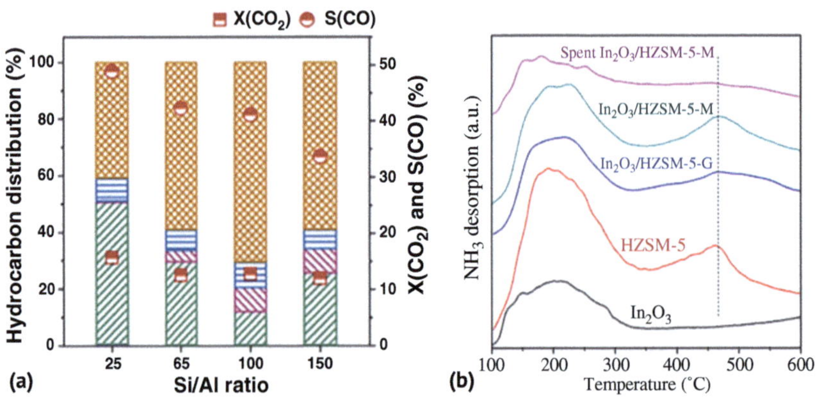

Figure 10.6 (a) CO_2 conversion over ZnZrO/ZSM-5 (H-ZSM-5 with different Si/Al ratios), hydrocarbon distribution, and CO selectivity. *Source*: Reproduced from [16]/with permission of Elsevier. (b) NH_3-TPD profiles of ZSM-5 and In_2O_3-incorporated ZSM-5. *Source*: Reproduced from [49]/Springer Nature.

investigated, with the performances translated from the acid distribution. An investigation by Li et al. showed that the maximum aromatics selectivity was achieved using a zeolite with a Si/Al ratio of 100 among the variations tested (see Figure 10.6a). Generally, under the same preparation conditions, an increasing Si/Al ratio should correspond to decreasing acidity. Also, combining the catalytic phases can alter the acid distribution on the zeolite. This has been confirmed by the distributions of acid characterization (see Figure 10.6b) [49]. Introducing cations can also alter the chemical properties and be beneficial for the selectivity of aromatics. "Zn" was used to improve the strong acid sites while inducing extra Lewis-type acids to the zeolites. A comparative analysis of the results indicates that ZSM-5 generally "Zn" type enhances the selectivity for aromatics over the acid type distribution as the major difference [50]. Assessments from the various reports indicate that no linear correlation exists between the Si/Al ratio and aromatics selectivity [16]. Rather, each specific investigation realizes a threshold for optimum selectivity for aromatics. A general conclusion drawn from this reaction indicates that aromatic formation may not require a large quantity of acid sites and may be prominent on Lewis acid sites. Without a linear relationship, the effects of acid site distribution on aromatics selectivity via CO_2-MTA will be strict to the specific reaction.

10.3.1.4 Reaction Mechanism over CO_2-MTA

This section reveals the cycles and surface molecules that are paramount for aromatics formation during CO_2 hydrogenation. Aromatics produced via this route use various oxygenates as intermediates. Some recognized intermediates prior to the zeolite phase catalysis include MeOH, HCOO*, and CH_3O Generally, these oxygenates go through a series of hydroprocessing steps to form olefins, which are hydrogenated to form aromatics. The density of each species depends on the metallic phases of the catalysts. Operando drifts performed on $ZnAlO_x$ + ZSM-5 suggests initial species formed after the initial reduction of CO_2 are HCOO*. The adsorbed HCOO* are then hydrogenated to CH_3O* [29]. Similar species were identified by Tian et al., with the inclusion of the influence of the temperature gradient. It is observed that the peak intensities for HCOO* and CH_3O* increase with temperature, suggesting the ease of their formation at higher temperature on the metallic phase of the catalyst [50]. Irrespective of the metallic phases, the oxygenate intermediates are seen to be reduced after the inclusion of the zeolites. Catalysis on the zeolite phase shows MeOH, olefins, and aromatics as main products. A conclusion by Gao et al. suggests the formation of MeOH prior to the olefin formation and aromatization process, however, it suggests the CH_xO* as the main intermediate proceeding to the zeolite [49]. During zeolite stage catalysis, a series of dehydration steps form intermediates that react to form olefins. The olefins undergo cyclization to form aromatics (see Figure 10.7).

10.3.2 CO_2-FTS Route

The CO_2-FTS route is equally popular for the direct synthesis of aromatics directly from CO_2. The process follows two sub-reactions in chronology. First, the reverse water-gas shift (RWGS), which reduced CO_2 to CO, followed by the FTS reaction, which hydrogenates CO to aromatics. Fe-based catalysts are popular since different phases of Fe can simultaneously reduce CO_2 and follow up with subsequent FTS reactions. A well-performing Fe-based catalyst is dominated by two major phases. The Fe_3O_4, which reduced CO_2 to CO and Fe-carbides, which facilitate the FTS process [51].

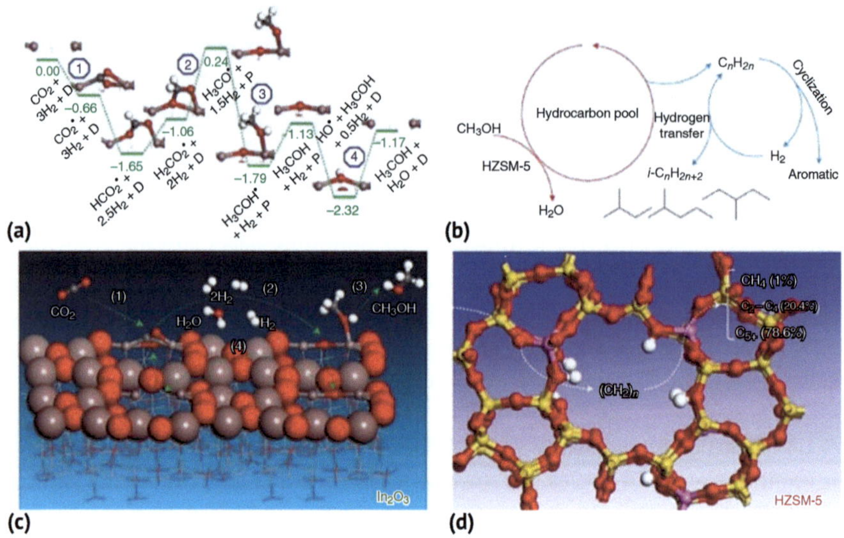

Figure 10.7 Schematics representing the in-situ steps following methanol formation and follow-up hydrogenation to aromatics [49]. (a) Energy profile from DFT calculations for CO_2 hydrogenation to CH_3OH on the In_2O_3(110) surface (D and P stands for defective and perfect surfaces with and without the oxygen vacancy, respectively). (b) Schematic of the hydrocarbon-pool mechanism for CH_3OH conversion into hydrocarbons inside HZSM-5. (c) Schematic for the formation of CH_3OH from CO_2 at the oxygen-vacancy site on the In_2O_3 catalyst surface, which involves four major steps: (1) CO_2 adsorption at the oxygen-vacancy site, (2) sequential hydrogenation of the adsorbed CO_2 species to CH_3OH, (3) CH_3OH desorption, which leads to the surface without the oxygen vacancy, and (4) hydrogenation of the surface to regenerate the oxygen-vacancy site. (d) Schematic for hydrocarbon formation from CH_3OH at the acidic site inside the pores of the HZSM-5 catalyst via the hydrocarbon-pool mechanism. *Source*: Reproduced from [49]/Springer Nature.

The product distribution could be enhanced with metallic promoters to orient the distribution toward olefins and longer-chained hydrocarbons, but not aromatics [52, 53]. Zeolites are essential to produce aromatics directly from this route. Different topologies vary the aromatization activity in terms of quality and quantity of aromatics produced. Therefore, this part highlights and compounds the effects of proximity and basic zeolite characteristics on direct aromatics production from CO_2 aromatics.

10.3.2.1 Effects of Proximity

Proximity effects between the metallic and zeolite phases of the catalysts used via the CO_2-FTS route differ from the CO_2-MTA route. The main intermediates sought for the aromatics production via the CO_2-FTS route are olefins. However, their production, transportation, and aromatization are mainly defined by the availability and stability of the independent phases. A recent work by Song et al. [54] investigated the effects of proximity between the two phases on the general performance and hydrocarbon distribution (see Figure 10.8). They observed that a physical mixture of the powdered forms of the two catalysts led to a reduced CO_2 conversion and hydrocarbon distribution. The reason identified was that the acidity

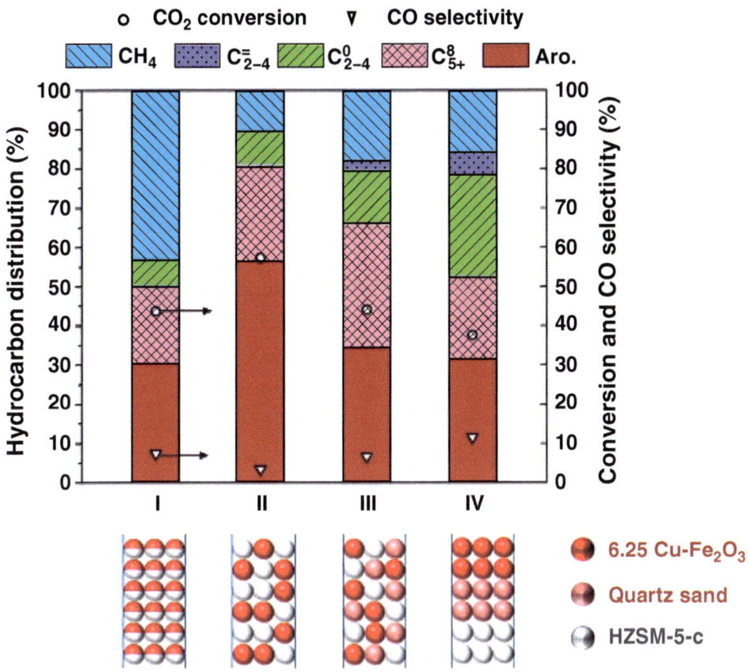

Figure 10.8 Effect of proximity on aromatics selectivity, I: powder mixing, II: granule mixing, III: granule mixing with quartz sand, and IV: dual bed. Reaction conditions: 320 °C, 3.0 MPa, $H_2/CO_2/N_2$ = 72/24/4, 1000 mL · g^{-1} · h^{-1}, and 2 g of composite catalysts. Notes: $C^{=}_{2\text{-}4}$ refers to lower olefins, $C^{O}_{2\text{-}4}$ refers to lower paraffins, C^{a}_{5+} refers to C_{5+} aliphatic, and Aro. refers to aromatics. *Source*: Reproduced from [54]/with permission of American Chemical Society.

associated with zeolite easily poisons the surfaces of the metallic catalyst by reducing its performance. When the catalysts are granularly mixed, the distance between the metallic phases and the zeolite is quite close. However, the independence of each catalytic phase is kept secured, distinct from each other. When quartz sand is either mixed or used to separate the catalysts in dual-bed mode, the general performance is reduced with an equivalent reduction for aromatics. A similar investigation was done by Wang et al. while investigating a NaFe@C catalyst with ZSM-5 [55]. The results realized were consistent with the results from the previous study in terms of performance and specific selectivity for aromatics. The conclusion drawn from their work was that although autonomous activity was ensured by the granular mix of the Fe-based and zeolite phases of the catalysts, the consumption of the alkenes from the reactor shifted the chemical thermodynamic equilibrium forward, which is beneficial for CO_2 conversion. This favored the early removal of H_2 species produced from the dehydrogenation and aromatization process, ultimately enhancing the selectivity for aromatics. The close proximity is essential for the direct aromatics' formation from CO_2; however, keeping the catalysts autonomous is essential to ensure uninterrupted phase activities.

10.3.2.2 Effects of Physical Properties

The physical properties refer to the structural characteristics of the zeolites that mainly influence aromatics formation with the intermediates proceeding from the CO_2-FTS. Noreen et al. [56] investigated the effects of different 10-MR zeolite structures on the direct synthesis of gasoline from CO_2. The SAPO-11 and ZSM-5 zeolites, possessing "1D" and "3D" structures, respectively, demonstrated oligomerization on the hydrocarbons produced from the NaFe catalyst. The aromatic selectivity was facilitated by the 3D structures of the ZSM-5, which enhance the cyclization of the intermediate hydrocarbons. The 1D structure enhances the oligomerization of lighter products in a linear fashion, leading to the formation of straight-chained hydrocarbons with less branched and aromatic hydrocarbons in products. A thorough investigation was done by Ramirez et al. [57] using different zeolite structures to reveal the selective descriptors facilitating their oligomerization processes directly from CO_2 hydrogenation. Their work further confirmed that the structure of ZSM-5 best performed aromatization. The effects of porosity were investigated using a hierarchical ZSM-5 against other variants of zeolite. The introduced mesoporosity can enhance the diffusion of intermediates and maintain a high conversion and aromatization activity. The low-carbon deposits on the spent hierarchical zeolite prove that the larger pores can reduce the deactivation rate of the zeolite. A series of hollow-type ZSM-5 was also designed and investigated in the CO_2-FTS route for the formation of aromatics [55]. The series prepared using various concentrations of NaOH revealed different pore properties that altered the activity. Among the catalysts, the sample treated with 0.2 M NaOH demonstrated the highest surface area, which was crucial for the optimum aromatic selectivity (see Figure 10.9).

10.3.2.3 Effects of Chemical Properties

The effects of the chemical properties of the zeolites are limited to the acidity, which is broadened between strength and quality. Brønsted and Lewis acid sites

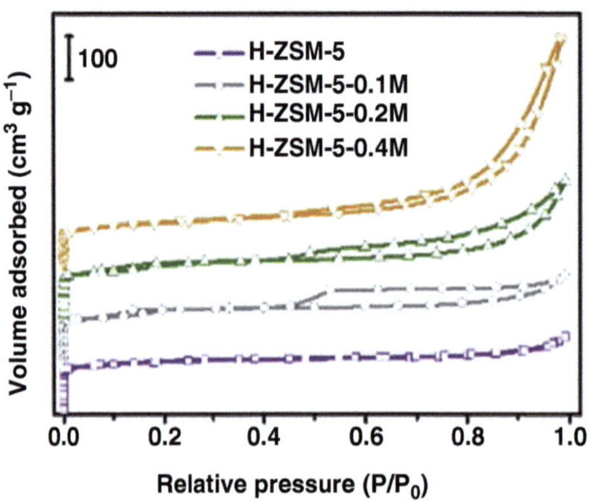

Figure 10.9 N_2 adsorption-desorption isotherms of the parent H-ZSM-5 and NaOH-treated H-ZSM-5. *Source*: Reproduced from [55]/with permission of Elsevier.

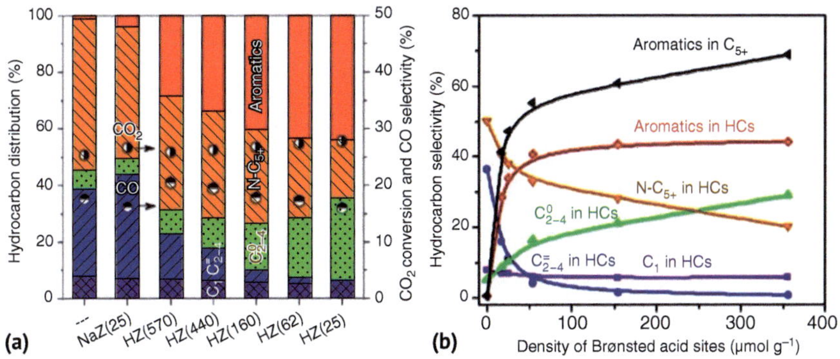

Figure 10.10 (a) Catalytic performance over a series of NaFe/ZSM-5 catalysts with distinct acidities. (b) Relationship between product selectivity and BAS density. N-C_{5+} denotes the C_{5+} hydrocarbons excluding aromatics. $C^{0}_{2\text{-}4}$ and $C^{=}_{2\text{-}4}$ denote the paraffins and olefins of C_{2-4} hydrocarbons, respectively. HCs: hydrocarbons. Reaction conditions: H_2/CO_2 = 2, 320 °C, 3 MPa, 4000 mL $\cdot$ g^{-1}_{cat} $\cdot$ h^{-1}. *Source*: Reproduced from [58]/with permission of Elsevier.

are familiar in the aromatization process. The effects of the acid strength can vary depending on the reaction conditions. From the literature, the acidity of the zeolites used in the CO_2 hydrogenation can be modified by changing the Si/Al, introducing metallic phases by ion exchange, or depositing porous matter on their surfaces. A work by Wei et al. [58] revealed Brønsted acid sites as the major acid sites functioning for aromatics production in CO_2 hydrogenation (see Figure 10.10). The work revealed a linear correlation between aromatic selectivity and decreasing Si/Al. Further conclusions observed were that excess acidity can lead to reduced aromatization activity. The acid distribution from the zeolites is also reported to be altered by various cations. An extra work series of hollow-type ZSM-5 was also designed and investigated in the CO_2-FTS route for the formation of aromatics. The series prepared using various concentrations of NaOH revealed different acid properties [55]. The sample treated with 0.2 M NaOH demonstrated ample acid sites, which were crucial for aromatization (see Figure 10.11). Comparing the conclusion with previous work by Wei et al., the overall activity is more defined by ample surface area rather than by acid sites present.

10.3.2.4 Reaction Mechanism over CO_2-FTS

The activity of the zeolites in oligomerization can be translated by using previous studies on the oligomerization potential of olefins on zeolites. Olefin oligomerization over zeolites has been presented over the previous years for the production of aromatics. With the exception of ethene, the general conclusion has been that lighter hydrocarbons undergo oligomerization via a carbocation intermediate on the zeolites. Basically, there is a hydride transfer to the olefins on the surfaces of the zeolites, forming unstable charged molecules. These cations undergo rearrangements and further hydroprocessing, defined by the acidity and pore geometry of the zeolites, for the formation of hydrocarbons [59]. Over the zeolites in CO_2 hydrogenation, the selectivity for aromatics cannot be limited to oligomerization of lighter

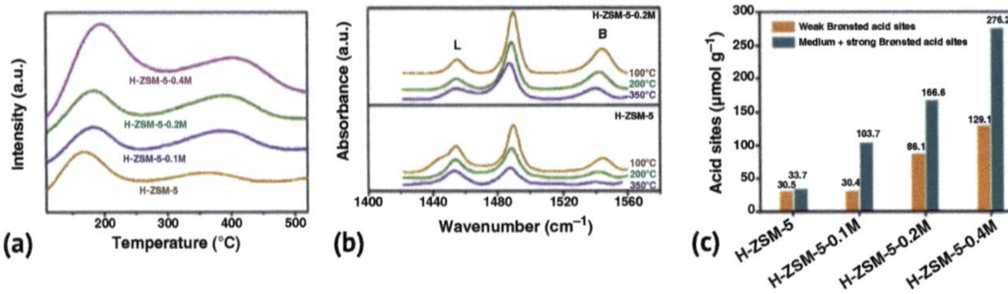

Figure 10.11 (a) NH_3-TPD profiles of H-ZSM-5 and H-ZSM-5-X, (b) FTIR spectra of adsorbed pyridine on H-ZSM-5 and H-ZSM-5-0.2M at different temperatures, and (c) the number of Brønsted acid sites in H-ZSM-5 and H-ZSM-5-X. *Source*: Reproduced from [55]/ with permission of Elsevier.

hydrocarbons, but more so the complete FTS part of the reaction. In the FTS part of the reaction, CO is transformed via a dissociative or insertion pathway, subsequently leading to the formation of olefins and organic-carbonylated species. On the zeolites, the olefins can be transformed into aromatics aided by the acid sites [55, 58]. The detailed studies conducted by Ramirez et al. [60] reveal that the zeolite can further facilitate the formation of more olefins in addition to those produced directly on the metallic surface in a direct FTS step. Projections from the literature reveal that the sinusoidal channels favor the olefins/paraffins cycle, while the straight channels aromatize the intermediates. Olefins are recognized as the main intermediates for aromatics formation, aside other molecules, which can enhance the selectivity for aromatics in the final products. Among the various zeolite studies, the topology of ZSM-5 zeolite with cross-sectional 3D structure is the favorite for aromatics formation (see Figure 10.12).

10.3.3 Hydrothermal Conversion

The hydrothermal CO_2 conversion involves the reduction of CO_2 under hydrothermal conditions to some valuable products. In brief, this process involves the use of H_2O as a solvent in the absence of H_2 for the reduction of CO_2 to valuable products. Metals and catalysts are essential components for the reduction of CO_2. The metallic species act as reduction agents that break down H_2O into metallic irons and H_2 as shown in Eq. (10.1) [61]. Some valued products usually obtained via this process include hydrocarbons, formic acid, carboxylic acids, and MeOH [62, 63]. This is also a recognized route with prospects for the direct synthesis of aromatics and other cyclic products from CO_2. Conventionally, the metallic catalysts can convert CO_2 and H_2O directly to cyclic products. However, recent works have shown improvement in the selective production of longer-chained hydrocarbons and cyclic products by zeolite-based catalysts owing to distinct properties such as acidity. In this section, we highlight and reveal a few studies done in this respect on the influence of zeolites in direct hydrothermal CO_2 conversion to aromatics.

$$Fe + H_2O \rightarrow Fe^{2+} + H_2 \tag{10.1}$$

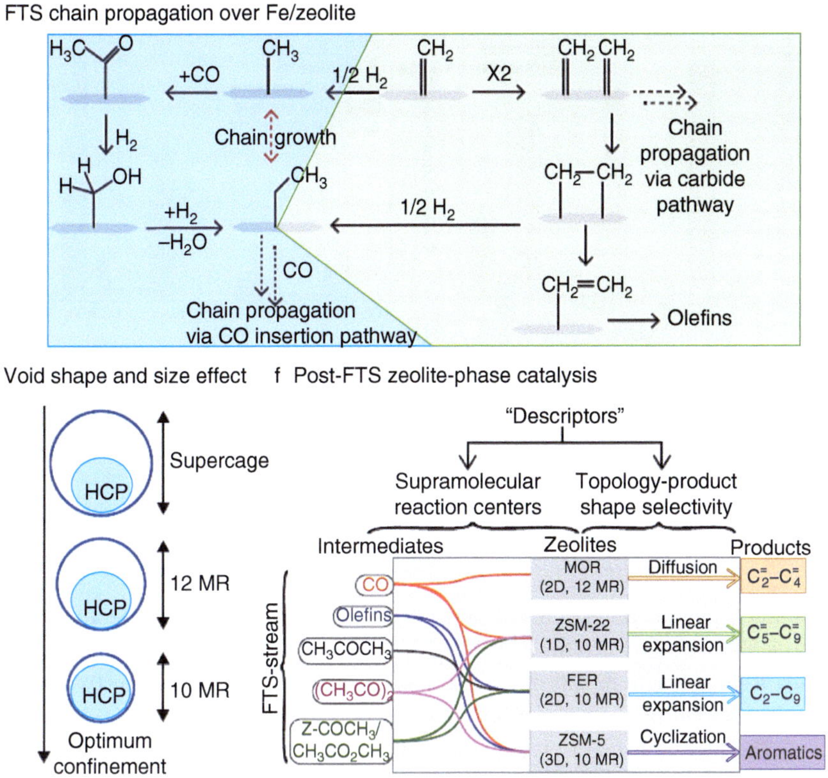

Figure 10.12 The reaction sequence illustrating the formation of olefins over Fe/Zeolite composite and zeolite topology for aromatics formation. *Source*: Reproduced from [57]/ Springer Nature/CC BY 4.0.

10.3.4 Influence of Zeolite Characteristics

The major characteristics, being acidity, framework topology, and pore dimensions, are distinct features of zeolites normally exploited in hydrocarbon processing. The hydrothermal CO_2 conversion is indifferent, as variations in these characteristics can result in different product distribution and overall activity. Generally, Fe-based catalysts can be configured for the direct synthesis of long-chained hydrocarbons and cyclic products [64]. However, subsequent improvements have been realized with the introduction of zeolites as catalysts. A typical investigation done by Gomez et al. revealed the direct influence of zeolite and zeolite properties on aromatics production from hydrothermal CO_2 conversion [63]. A clear comparison between the bulk Fe catalysts and zeolite-supported options reveals higher selectivity for longer-chained products over zeolite-based catalysts. The formation of more cyclic compounds over the zeolites suggests the zeolites are able to catalyze more cyclization reactions as a secondary reaction, improving the selectivity for the target compounds. Among the zeolites, HY had a higher tendency for the formation of large carbon number molecules, and this is ascribed to the larger pore sizes (see Figure 10.13). Similar works exploiting the zeolite phases.

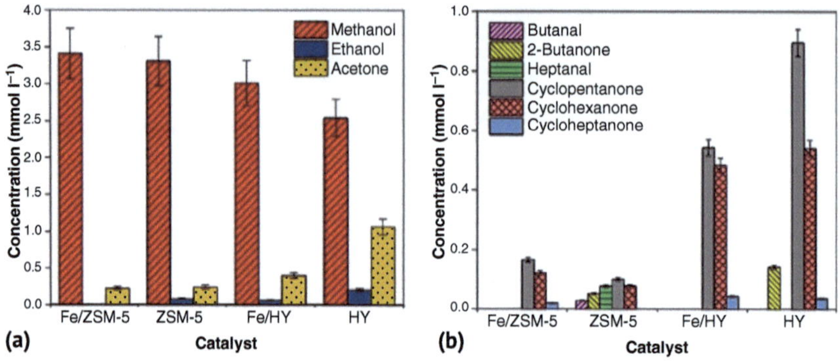

Figure 10.13 Distribution of liquid-phase products over zeolite and zeolite-supported iron catalysts. Left-hand plot shows low-carbon-number products; right-hand plot shows higher-carbon-number products. Reaction conditions: T = 300 °C, CO_2: H_2O mole ratio = 0.26, $t = 4$ h, 0.56 g catalyst. *Source*: Reproduced from [63]/John Wiley & Sons.

10.4 Conclusions and Outlook

In summary, this section compiles CO_2 conversion to aromatics over zeolite-based catalysts. The details reveal the popular technologies for aromatics production using zeolites, followed with details on zeolite applications in CO_2 conversion. Two major technologies identified were the hydrothermal CO_2 conversion and CO_2 hydrogenation. The hydrothermal CO_2 conversion to aromatics normally uses transitional metals as main catalysts; however, recent introduction of zeolites has proved to boost the formation of aromatics and cyclo-compounds selectivity. The acidity, framework topology, and pore geometry of the zeolites collectively govern the enhanced selectivity toward aromatics. In the CO_2 hydrogenation reaction, zeolites are recognized as the main component facilitating aromatics formation. Two main routes were identified: the methanol-mediated path and the FTS-mediated path. Methanol was recognized as the main intermediate for aromatics formation. A review of the zeolite characteristics reveals proximity to be crucial and that a very close proximity is beneficial for aromatics production. Further revelations indicated that aromatic formation may not require a large quantity of acid sites and may be prominent on Lewis acid sites. In the case of the CO_2-FTS route, olefins are recognized as the main intermediates. Proximity is crucial; however, phase distinction is necessary to ensure independent activity. Brønsted acid sites were identified as the main acid type for enhanced aromatics formation in all these reactions.

In all these reactions, ZSM-5 dominated as the main zeolite, facilitating aromatics formation, and the reason could be because it is structured to enhance aromatics formation. This has limited the activities over zeolites. Therefore, more studies will be required in the design and optimization of other zeolite materials in this respect to well rationalize aromatization activity over zeolites. Additionally, the stream of productions proceeding to zeolites normally consists of other compounds. For optimum aromatics production, more studies will be required to develop zeolite-based catalysts capable of catalyzing enough of these compounds into aromatics.

References

1 Fortenberry, R.C., Novak, C.M., Lee, T.J. et al. (2018). Identifying molecular structural aromaticity for hydrocarbon classification. *ACS Omega* 3: 16035–16039.

2 Artigas, A., Hagebaum-Reignier, D., Carissan, Y. et al. (2021). Visualizing electron delocalization in contorted polycyclic aromatic hydrocarbons. *Chem. Sci.* 12: 13092–13100.

3 Varjani, S.J., Joshi, R.R., Senthil Kumar, P. et al. (2018). Polycyclic aromatic hydrocarbons from petroleum oil industry activities: effect on human health and their biodegradation. In: *Waste Bioremediation* (ed. S. Varjani, E. Gnansounou, B. Gurunathan, et al.), 185–199. Singapore: Springer.

4 Busca, G. (2021). Production of gasolines and monocyclic aromatic hydrocarbons: from fossil raw materials to green processes. *Energies* 14: 4061–4093.

5 Sadeghbeigi, R. (2012). Chapter 3: FCC feed characterization. In: *Fluid Catalytic Cracking Handbook*, 3e, 51–86. Oxford, United Kingdom: Butterworth-Heinemann.

6 Yang, C., Brown, C.E., Hollebone, B. et al. (2017). Chapter 4: Chemical fingerprints of crude oils and petroleum products. In: *Oil Spill Science and Technology*, 2e (ed. M. Fingas), 209–304. Massachusetts, USA/Oxford, United Kingdom: Gulf Professional Publishing, Elsevier Inc.

7 Drewniak, A., Tomczyk, M.D., Hanusek, L. et al. (2018). The effect of aromatic diimide side groups on the π-conjugated polymer properties. *Polymers* 10: 487–501.

8 Artz, J., Müller, T.E., Thenert, K. et al. (2018). Sustainable conversion of carbon dioxide: an integrated review of catalysis and life cycle assessment. *Chem. Rev.* 118: 434–504.

9 Sibi, M.G., Verma, D., and Kim, J. (2022). Direct conversion of CO_2 into aromatics over multifunctional heterogeneous catalysts. *Catal. Rev.* 66: 863–922.

10 Zhang, Q., Pastor-Pérez, L., Zhang, X. et al. (2021). CO_2 conversion to value-added gas-phase products: technology overview and catalysts selection. In: *Engineering Solutions for CO_2 Conversion* (ed. T.R. Reina, H. Arellano-Garcia, and J.A. Odriozola), 175–203. New Jersey, USA: Wiley Online Library

11 Bobadilla, L.F., Azancot, L., and Odriozola, J.A. (2021). CO_2 utilization enabled by microchannel reactors. In: *Engineering Solutions for CO_2 Conversion* (ed. T.R. Reina, H. Arellano-Garcia, and J.A. Odriozola), 205–225. New Jersey, USA: Wiley Online Library.

12 Campello, I., Sepúlveda-Escribano, A., and Ramos-Fernández, E.V. (2021). Metal–organic frameworks (MOFs) for CO_2 cycloaddition reactions. In: *Engineering Solutions for CO_2 Conversion* (ed. T.R. Reina, H. Arellano-Garcia, and J.A. Odriozola), 407–427. Wiley Online Library.

13 Xu, S. and Carter, E.A. (2019). Theoretical insights into heterogeneous (photo) electrochemical CO_2 reduction. *Chem. Rev.* 119: 6631–6669.

14 Speight, J.G. (2017). Chapter 3: Industrial organic chemistry. In: *Environmental Organic Chemistry for Engineers* (ed. J.G. Speight). 87–151. John Wiley & Sons.

15 Andérez-Fernández, M., Pérez, E., Martín, Á. et al. (2022). Synergistic hydrothermal conversion of aqueous solutions of CO_2 and biomass waste liquefaction into formate. *ACS Sustain. Chem. Eng.* 10: 16948–16957.

16 Li, Z., Qu, Y., Wang, J. et al. (2019). Highly selective conversion of carbon dioxide to aromatics over tandem catalysts. *Joule* 3: 570–583.

17 Azhari, N.J., Nurdini, N., Mardiana, S. et al. (2022). Zeolite-based catalyst for direct conversion of CO_2 to C_{2+} hydrocarbon: a review. *J. CO2 Util.* 59: 101969–101985.

18 Amoo, C.C., Xing, C., Tsubaki, N. et al. (2022). Tandem reactions over zeolite-based catalysts in syngas conversion. *ACS Cent. Sci.* 8: 1047–1062.

19 Gao, X., Deng, S., and Kawi, S. (2022). Zeolite-based catalytic membrane reactors for thermo-catalytic conversion of CO_2. *iScience* 25: 105343–105368.

20 Liu, X., Wu, Y., Zhang, J. et al. (2022). Catalytic pyrolysis of nonedible oils for the production of renewable aromatics using metal-modified HZSM-5 catalysts. *ACS Omega* 7: 18953–18968.

21 Wang, F., Zheng, Y., Huang, Y. et al. (2017). Optimizing catalytic pyrolysis of rubber seed oil for light aromatics and anti-deactivation of ZSM-5. *J. Anal. Appl. Pyrolysis* 126: 180–187.

22 Ghorbannezhad, P., Park, S., and Onwudili, J.A. (2020). Co-pyrolysis of biomass and plastic waste over zeolite- and sodium-based catalysts for enhanced yields of hydrocarbon products. *Waste Manage.* 102: 909–918.

23 Dyer, A.C., Nahil, M.A., and Williams, P.T. (2022). Biomass: polystyrene co-pyrolysis coupled with metal-modified zeolite catalysis for liquid fuel and chemical production. *J. Mater. Cycles Waste Manage.* 24: 477–490.

24 Lv, P., Yan, L., Liu, Y. et al. (2020). Catalytic conversion of coal pyrolysis vapors to light aromatics over hierarchical Y-type zeolites. *J. Energy Inst.* 93: 1354–1363.

25 Xu, Y., Liu, J., Wang, J. et al. (2019). Selective conversion of syngas to aromatics over Fe_3O_4@MnO_2 and hollow HZSM-5 bifunctional catalysts. *ACS Catal.* 9: 5147–5156.

26 Zhou, W., Shi, S., Wang, Y. et al. (2019). Selective conversion of syngas to aromatics over a Mo-ZrO_2/H-ZSM-5 bifunctional catalyst. *ChemCatChem* 11: 1681–1688.

27 Shen, X., Kang, J., Niu, W. et al. (2017). Impact of hierarchical pore structure on the catalytic performances of MFI zeolites modified by ZnO for the conversion of methanol to aromatics. *Catal. Sci. Technol.* 7: 3598–3612.

28 Li, W., Wang, H., Jiang, X. et al. (2018). A short review of recent advances in CO_2 hydrogenation to hydrocarbons over heterogeneous catalysts. *RSC Adv.* 8: 7651–7669.

29 Ni, Y., Chen, Z., Fu, Y. et al. (2018). Selective conversion of CO_2 and H_2 into aromatics. *Nat. Commun.* 9: 3457–3464.

30 Zhang, J., Zhang, M., Chen, S. et al. (2019). Hydrogenation of CO_2 into aromatics over a $ZnCrO_x$–zeolite composite catalyst. *Chem. Commun.* 55: 973–976.

31 Song, G., Li, M., Xu, L. et al. (2022). Tuning the integration proximity between Na promoter and $FeMnO_x$ coupled with rationally modified HZSM-5 to promote selective CO_2 hydrogenation to aromatics. *Ind. Eng. Chem. Res.* 61: 6820–6830.

32 Zhang, J., Zheng, K., Wang, X. et al. (2022). Promotional effect of dispersant modification to ZnCr on CO_2 hydrogenation into aromatics over hybrid catalysts. *Ind. Eng. Chem. Res.* 61: 10766–10774.

33 Yao, B., Xiao, T., Makgae, O.A. et al. (2020). Transforming carbon dioxide into jet fuel using an organic combustion-synthesized Fe-Mn-K catalyst. *Nat. Commun.* 11: 6395–6947.

34 Zhang, L., Dang, Y., Zhou, X. et al. (2021). Direct conversion of CO_2 to a jet fuel over CoFe alloy catalysts. *Innovation* 2: 100170–100180.

35 Nasriddinov, K., Min, J.-E., Park, H.-G. et al. (2022). Effect of Co, Cu, and Zn on FeAlK catalysts in CO_2 hydrogenation to C_{5+} hydrocarbons. *Catal. Sci. Technol.* 12: 906–915.

36 Li, K. and Chen, J.G. (2019). CO_2 hydrogenation to methanol over ZrO_2-containing catalysts: insights into ZrO_2 induced synergy. *ACS Catal.* 9: 7840–7861.

37 Borisut, P. and Nuchitprasittichai, A. (2019). Methanol production via CO_2 hydrogenation: sensitivity analysis and simulation—based optimization. *Front. Energy Res.* 7: 81

38 Azhari, N.J., Erika, D., Mardiana, S. et al. (2022). Methanol synthesis from CO_2: a mechanistic overview. *Res. Eng.* 16: 100711–100730.

39 Shi, Y.-F., Kang, P.-L., Shang, C. et al. (2022). Methanol synthesis from CO_2/CO mixture on Cu–Zn catalysts from microkinetics-guided machine learning pathway search. *J. Am. Chem. Soc.* 144: 13401–13414.

40 Dieterich, V., Buttler, A., Hanel, A. et al. (2020). Power-to-liquid via synthesis of methanol, DME or fischer–tropsch-fuels: a review. *Energy Environ. Sci.* 13: 3207–3252.

41 Ting, K.W., Toyao, T., Siddiki, S.M.A.H. et al. (2019). Low-temperature hydrogenation of CO_2 to methanol over heterogeneous TiO_2-supported Re catalysts. *ACS Catal.* 9: 3685–3693.

42 Wang, S., Zhang, L., Wang, P. et al. (2022). Highly selective hydrogenation of CO_2 to propane over $GaZrO_x$/H-SSZ-13 composite. *Nat. Catal.* 5: 1038–1050.

43 Li, N., Zhu, Y., Jiao, F. et al. (2022). Steering the reaction pathway of syngas-to-light olefins with coordination unsaturated sites of $ZnGaO_x$ spinel. *Nat. Commun.* 13: 2742–2750.

44 Sharma, P., Sebastian, J., Ghosh, S. et al. (2021). Recent advances in hydrogenation of CO_2 into hydrocarbons via methanol intermediate over heterogeneous catalysts. *Catal. Sci. Technol.* 11: 1665–1697.

45 Tian, H., He, H., Jiao, J. et al. (2022). Tandem catalysts composed of different morphology HZSM-5 and metal oxides for CO_2 hydrogenation to aromatics. *Fuel* 314: 123119–123132.

46 Xin, Q., Guo, H., Wang, Y. et al. (2022). Indium-promoted $ZnZrO_x$/nano-ZSM-5 for efficient conversion of CO_2 to aromatics with high selectivity. *J. Environ. Chem. Eng.* 10: 108032–108044.

47 Wang, T., Yang, C., Gao, P. et al. (2021). $ZnZrO_x$ integrated with chain-like nanocrystal HZSM-5 as efficient catalysts for aromatics synthesis from CO_2 hydrogenation. *Appl. Catal. B* 286: 119929–119941.

48 Du, H., Zhong, Z., Zhang, B. et al. (2020). Ex-situ catalytic upgrading of vapors from microwave-assisted pyrolysis of bamboo with chemical liquid deposition modified HZSM-5 to enhance aromatics production. *J. Anal. Appl. Pyrolysis* 149: 104857–104868.

49 Gao, P., Li, S., Bu, X. et al. (2017). Direct conversion of CO_2 into liquid fuels with high selectivity over a bifunctional catalyst. *Nat. Chem.* 9: 1019–1024.

50 Tian, H., Jiao, J., Zha, F. et al. (2022). Hydrogenation of CO_2 into aromatics over ZnZrO–Zn/HZSM-5 composite catalysts derived from ZIF-8. *Catal. Sci. Technol.* 12: 799–811.

51 Guo, L., Gao, X., Gao, W. et al. (2023). High-yield production of liquid fuels in CO_2 hydrogenation on a zeolite-free Fe-based catalyst. *Chem. Sci.* 14: 171–178.

52 Brübach, L., Hodonj, D., and Pfeifer, P. (2022). Kinetic analysis of CO_2 hydrogenation to long-chain hydrocarbons on a supported iron catalyst. *Ind. Eng. chem. Res.* 61: 1644–1654.

53 Shi, Z., Yang, H., Gao, P. et al. (2018). Direct conversion of CO_2 to long-chain hydrocarbon fuels over K–promoted CoCu/TiO_2 catalysts. *Catal. Today* 311: 65–73.

54 Song, G., Li, M., Yan, P. et al. (2020). High conversion to aromatics via CO_2-FT over a CO-reduced Cu-Fe_2O_3 catalyst integrated with HZSM-5. *ACS Catal.* 10: 11268–11279.

55 Wang, Y., Kazumi, S., Gao, W. et al. (2020). Direct conversion of CO_2 to aromatics with high yield via a modified Fischer-Tropsch synthesis pathway. *Appl. Catal. B* 269: 118792–118801.

56 Noreen, A., Li, M., Fu, Y. et al. (2020). One-pass hydrogenation of CO_2 to multibranched isoparaffins over bifunctional zeolite-based catalysts. *ACS Catal.* 10: 14186–14194.

57 Ramirez, A., Gong, X., Caglayan, M. et al. (2021). Selectivity descriptors for the direct hydrogenation of CO_2 to hydrocarbons during zeolite-mediated bifunctional catalysis. *Nat. Commun.* 12: 5914–5927.

58 Wei, J., Yao, R., Ge, Q. et al. (2021). Precisely regulating Brønsted acid sites to promote the synthesis of light aromatics via CO_2 hydrogenation. *Appl. Catal. B* 283: 119648–119659.

59 Moon, S., Chae, H.-J., and Park, M.B. (2018). Oligomerization of light olefins over ZSM-5 and beta zeolite catalysts by modifying textural properties. *Appl. Catal. A* 553: 15–23.

60 Ramirez, A., Dutta Chowdhury, A., Dokania, A. et al. (2019). Effect of zeolite topology and reactor configuration on the direct conversion of CO_2 to light olefins and aromatics. *ACS Catal.* 9: 6320–6334.

61 Zeng, X., Jin, F., Yao, G. et al. (2016). Theoretical study on highly efficient reduction of CO_2 by in-situ produced hydrogen using Al under mild hydrothermal conditions. *Int. J. Hydrogen Energy* 41: 9140–9144.

62 Jin, F., Zeng, X., Liu, J. et al. (2014). Highly efficient and autocatalytic H_2O dissociation for CO_2 reduction into formic acid with zinc. *Sci. Rep.* 4: 4503–4511.

63 Gomez, L.Q., Shehab, A.K., Al-Shathr, A. et al. (2020). H_2-free synthesis of aromatic, cyclic and linear oxygenates from CO_2. *ChemSusChem* 13: 647–658.

64 Tian, G., He, C., Chen, Y. et al. (2010). Hydrothermal reactions from carbon dioxide to phenol. *ChemSusChem* 3: 323–324.

11

CO_2 Hydrogenation to Alkenes over Zeolite-based Catalysts

Sen Wang[1], Qian Zhang[1,2], Shiying Li[1], Mei Dong[1], Jianguo Wang[1,2], Weibin Fan[1,]*

[1]*State Key Laboratory of Coal Conversion, Institute of Coal Chemistry, Chinese Academy of Sciences, Taiyuan, Shanxi, P.R. China*
[2]*University of Chinese Academy of Sciences, Beijing, PR China*

*Corresponding author: fanwb@sxicc.ac.cn

11.1 Introduction

With extensive use of carbon-containing resources, huge amounts of CO_2 were released into the atmosphere, and the emitted amount has exceeded 30 billion tons in 2020, which makes the global atmospheric CO_2 level reach 417 ppm [1, 2]. The significant increase in CO_2 concentration leads to a serious greenhouse effect, acidification of seawater, and sea level rise, as well as other environmental problems [3–6]. Reduction of carbon footprint and control of greenhouse effect have become a global consensus. Capture, utilization, and storage of CO_2 (CCUS) have been regarded as a feasible method to achieve carbon neutrality [7, 8]. However, CO_2 storage is difficult to propagate due to high cost and demand for specific geological conditions [9]. Although mineralization is the most traditional way for chemical carbon sequestration, the reaction rate is slow, along with the formation of useless or cheap products [10]. In comparison, hydrogenation of CO_2 into high-value fuels and chemicals is feasible from a point of economic view [11–14]. CO_2 is a nontoxic, abundant, and clean carbon resource, it can be selectively hydrogenated into C1 compounds, such as methane (CH_4) [15–18], methanol (CH_3OH) [19–23], carbon monoxide (CO) [24–27], and C_{2+} products (e.g., alkenes, aromatics, gasolines, diesels, liquefied petroleum gas (LPG), etc.) (see Figure 11.1) [28–32].

Among them, alkenes, especially $C_2^=–C_4^=$ light olefins, are significant platform molecules in the preparation of plastics, rubber, and pharmaceuticals [33–35]. Ethene and its derivatives enable the synthesis of ethylene oxide (EO) and polyvinyl chloride (PVC), while propene and butene are extensively used to produce

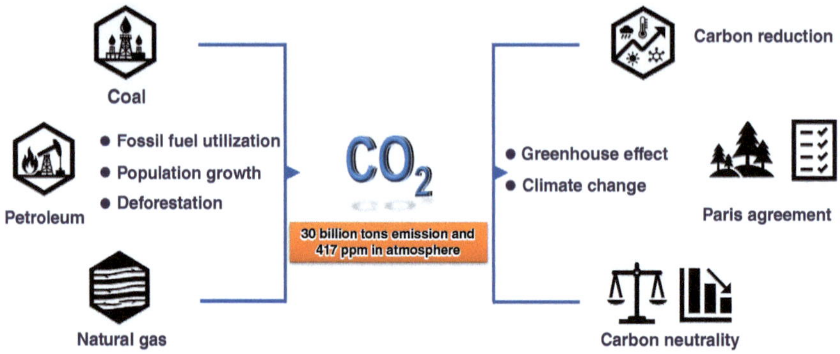

Figure 11.1 Schematic diagram of CO_2 generation and its impact on the global climate.

polypropylene, propylene oxide, tertiary butyl ether, ethyl tertiary butyl ether, etc. [36, 37]. Alkenes are generally obtained by steam cracking of naphtha or alkanes dehydrogenation, but this process requires high reaction temperature, resulting in a large energy consumption. Therefore, the transformation of CO_2 into alkenes cannot only significantly control CO_2 content released into the atmosphere, but also opens up new routes for the preparation of commodity chemicals.

Hydrogenation of CO_2 with renewable hydrogen (H_2), produced from photolysis and/or electrolysis of H_2O, can be achieved through thermal catalysis, electrocatalysis, or photocatalysis process [38–41]. Although the electro- and photo-catalytic CO_2 reduction reaction (CO_2 RR) has been extensively explored and important progress has been made, the obtained products are mainly C1 components, such as CH_4, CO, and CH_3OH [42–44]. The generation of C_{2+} alkenes with high yield through electrocatalysis or photocatalysis process remains challenging, due to the high free energy for the process of C—C coupling and low conversion efficiency. In contrast, thermal catalytic hydrogenation of CO_2 to C_{2+} oxygenates or hydrocarbons, is more feasible and also easy to scale up. Nowadays, two routes, including Fischer–Tropsch (FT) synthesis pathway and methanol-intermediate pathway (see Figure 11.2), were adopted for CO_2 hydrogenation to alkenes [45–48]. In the FT synthesis route, the production of alkenes is achieved through the combination of the reverse water-gas shift (RWGS) and CH_x–CH_x coupling reactions on iron (Fe) or cobalt (Co)-based catalysts [49, 50]. FT synthesis route generally gives high CO_2 activation ability; however, the complexity of CH_x–CH_x coupling reactions results in the generation of hydrocarbons ranging from C1 to C_{20+} [51–53]. The hydrocarbon product composition in FT synthesis is determined by the ASF rule, where the light alkenes and light alkanes ($C_2^=$–$C_4^=$ and C_2^0–C_4^0) selectivities are no higher than 58%, while that of undesired methane can reach 25% [54, 55]. Recently, some investigations indicated that use of sulfur (S) and sodium (Na)-modified Fe nanoparticles (NPs) or the cobalt carbide (CoC_x) nanoprisms exposed [101] and [020] facets can break the ASF model to some extent and the light olefins ($C_2^=$–$C_4^=$) selectivity gets to around 61% in syngas conversion [56, 57]. Despite these important advances, considerable improvements in the selectivity of target product (e.g., light olefins or even a single specific olefin) remain a challenge on a single Fe or Co-based FT catalyst. Introduction of additional

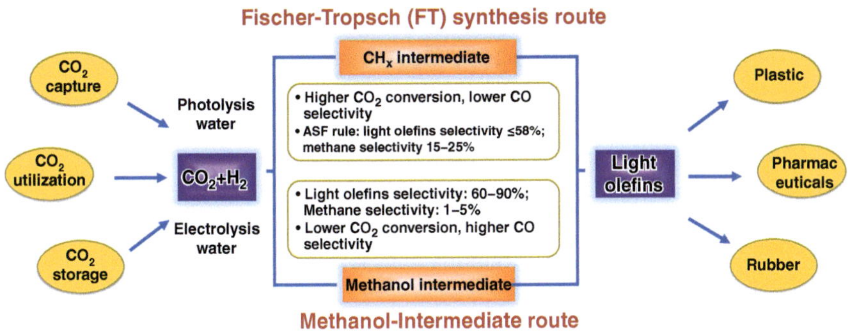

Figure 11.2 Schematic diagram of CO_2 conversion to olefins through the FT synthesis pathway and the methanol-intermediate pathway.

active components, such as acidic zeolites, into FT synthesis catalysts is regarded as an effective method to make the product spectrum tunable, as these acidic zeolites can catalyze the cracking of long-chain products [58–61].

In addition to the FT synthesis process, CO_2 can be hydrogenated into olefins via the methanol-intermediate route. This process is initially proposed by Bao's group and Wang's group and succeeded in the conversion of CO to alkenes [62, 63]. In this route, oxygen-containing species, including ketene and methanol, are first produced on metal oxides, such as $ZnCrO_x$ and $ZnZrO_x$ [64, 65]. The generated methanol quickly moves into acidic zeolites and is then converted to various hydrocarbons [66, 67]. The effective separation of CO/CO_2 activation and hydrocarbons formation steps makes the product distribution in this route do not obey the ASF rule. Consequently, the proportion of light olefins in hydrocarbons reached 74–80%, being higher than that of the traditional FT synthesis catalysts [62, 63].

Nowadays, zeolites have been widely used as an acid-catalytic moiety to regulate the distribution and composition of hydrocarbon products in FT synthesis and methanol-intermediate routes [68–70], due to their specific pore channel and tunable acidic property [71]. When the silicon (Si) atom was replaced by the aluminum (Al) atom, a negative charge was produced in the zeolite framework, and it was subsequently neutralized by hydrogen ions (H^+), forming a Brønsted acid site for methanol dehydration, alkylation, oligomerization, isomerization, and cracking reactions [72–75]. The reaction rate of these elemental steps is controlled by the content, strength, and position of acid sites in the zeolite framework [76, 77]. Moreover, the pore structure of zeolites also plays a vital role in determining the formation and diffusion of active intermediates [78–80].

Herein, this chapter summarizes the applications of various composite catalysts comprising metal/metal oxides and zeolites in CO_2 hydrogenation to alkenes. The relationship between the product distribution in CO_2 hydrogenation and the pore structure and acidic properties of zeolite is discussed in detail. Notably, this work emphasizes the generation of light olefins. Besides light olefins, higher olefins are also industrial-related raw materials [81–83]; the information about their formation in CO/CO_2 hydrogenation can be referred to other published works [84–86].

11.2 The FT Synthesis Route for Light Olefins Formation from CO_2 Hydrogenation

In CO_2 hydrogenation, the FT synthesis route generally combines the RWGS, and CO-FT synthesis reactions [87, 88]. CO_2 is initially transformed into CO via the process of the RWGS reaction (Eq. 11.1) and the generated CO is subsequently dissociated and hydrogenated into CH_x, which is further converted into hydrocarbons with different carbon-chain lengths via coupling of CH_x intermediates (Eq. 11.2).

$$CO_2 + H_2 \rightarrow CO + H_2O \quad (11.1)$$

$$nCO + 2nH_2 \rightarrow C_n H_{2n} + nH_2O \quad (11.2)$$

This makes the hydrocarbon distribution follow the ASF rule that results in the selectivity of light olefins hardly exceeding 58% [54, 55]. Although the S/Na-modified Fe NPs and CoC_x nanoprisms exposed [101] and [020] facets and have the potential to break the ASF rule to some extent, giving a light olefin selectivity of ~61%, large amounts of methane and/or C_{5+} long-chain hydrocarbons are still produced [56, 57]. Nowadays, the introduction of zeolite is considered to be an effective way to control product distribution, because the acidic zeolite can selectively tailor the C—C bond of heavier hydrocarbons, and its specific pore structure determines the formation and evolution of intermediate species.

11.2.1 Fe-based Oxide/Zeolite Bifunctional Catalysts

Fe-based metal/metal oxides have attracted increasing attention in the conversion of CO/CO_2 to olefins, because of the low methanation activity and strong resistance to contaminants [89–92]. Iron oxide (FeO_x) is first reduced to metallic iron (Fe^0) and then rapidly carbonized to form Fe carbides. Wu et al. analyzed the structural evaluation of Fe species in FT synthesis, and they found that an Fe_5C_2-Fe_3O_4 core-shell structure was quickly generated during FT synthesis (see Figure 11.3) [93]. The Fe_5C_2 exposing the (202) crystal facet mainly contributed to the CO dissociation with the assistance of H. When tested in CO conversion, this sample showed the light olefins ($C_2^=$–$C_4^=$) selectivity of around 30%. Ramirez et al. offered a highly effective potassium (K)-modified Fe_2O_3 catalyst, which can selectively convert CO_2 into olefins [94]. The ferric oxide (Fe_2O_3) and iron carbide (Fe_5C_2) were clearly identified in the spent catalyst, and they controlled the relative activity of RWGS and CO-FT synthesis reactions, respectively. Introduction of the K promoter not only elevated the alkalinity of the catalyst and improved the adsorption ability of CO_2, but also balanced the content of FeO_x and FeC_x species. A high catalytic activity (CO_2 conversion of ~44%) and alkene selectivity (~63% for $C_2^=$–$C_{10}^=$) were simultaneously achieved on $Fe_2O_3@K_2CO_3$ in CO_2 hydrogenation at 350°C and 3.0 MPa. Nevertheless, the hydrocarbon distribution is controlled by the ASF rule that leads to the light olefins ($C_2^=$–$C_4^=$) selectivity below 34%.

Wang et al. constructed the FeZnZr@H-ZSM-5 core-shell catalyst to effectively break the ASF model and regulate the product distribution in FT synthesis [95].

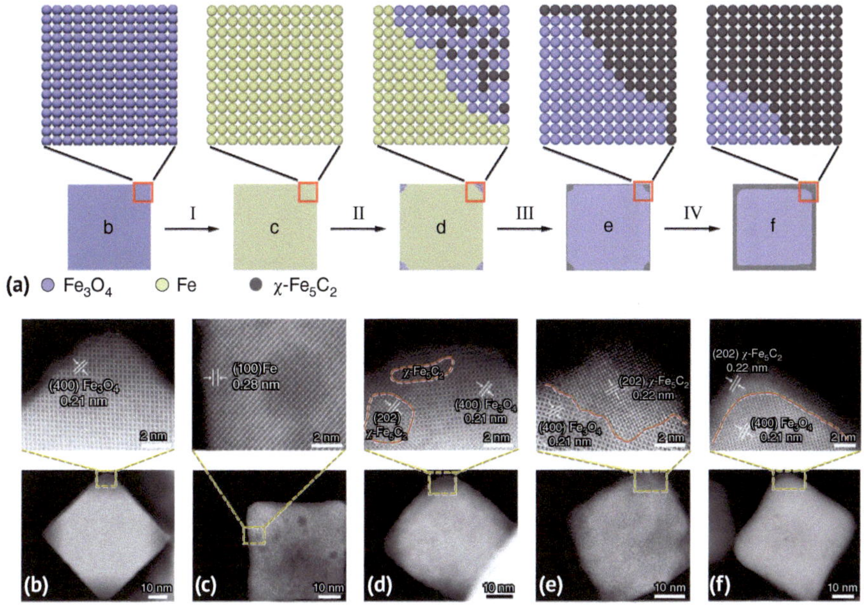

Figure 11.3 Structural evolution of Fe species during the CO-FT synthesis process: (a) schematic diagram of the reduction and carbonization of Fe species and (b–f) lattice fringes tested from HAADF-STEM images of various Fe particles. *Source*: [93]/Springer Nature/CC BY 4.0.

This sample cannot only suppress CO formation by increasing the CO* adsorption and dissociation activity, but also enhance the generation of C_{5+} isoalkane, because of the 10-membered ring (MR) medium pore structure and proper acidic properties of H-ZSM-5 (see Figure 11.4). At 340°C and 5.0 MPa, the proportion of C_{5+} isoalkane in gasoline reached above 93%.

Various zeolites were also used to fabricate the Fe_2O_3@K_2CO_3/zeolite bifunctional catalysts. It is found that the introduction of an acidic zeolite (i.e., H-ZSM-5) moiety promotes aromatics generation, whereas for Fe_2O_3@K_2CO_3/H-MOR, it delivers more ethene and propene at the expense of long-chain alkenes [96]. It should be noticed that the product's composition of the composite catalyst is controlled by the proximity of the Fe-based oxide and the acidic zeolite. When the spatial arrangement of the two groups was changed from dual-bed filling to powder mixing, the unexpected CO became the dominant product. This is due to the migration of part of K from Fe_2O_3 to zeolite, resulting in the poisoning of acidic zeolites. The reaction mechanism is investigated using advanced ^{1}H–^{13}C and ^{13}C–^{13}C MAS NMR. CO_2 and H_2 were confirmed to be activated on Fe_2O_3@K_2CO_3 to produce CO, which was further dissociated and hydrogenated into hydrocarbons on Fe_5C_2. Some remaining CO diffused to the pore channel of the zeolite and then transformed into methoxy groups. These methoxy undergone methylation, oligomerization, and aromatization, yielding light olefins and aromatics.

The catalytic role of zeolite topologic structure in determining the generation of hydrocarbons from CO_2 hydrogenation was further investigated [97]. The results

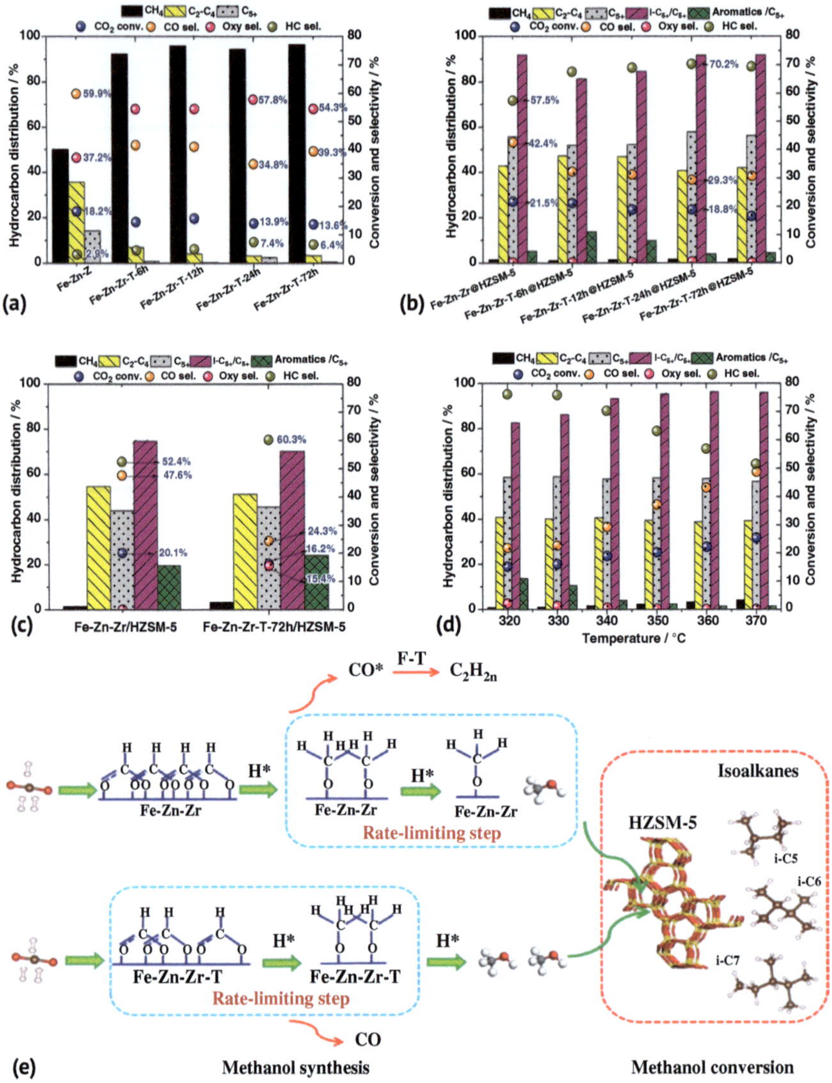

Figure 11.4 Catalytic performance of FeZnZr@H-ZSM-5 composite in CO_2 hydrogenation: (a–d) CO_2 conversion and products selectivity of FeZnZr@H-ZSM-5 at different conditions, and (e) schematic diagram for the formation of hydrocarbons in CO_2 hydrogenation. *Source*: Reproduced from [95]/with permission of American Chemical Society.

suggested that the MOR, SAPO-34, ZSM-58, Beta, and USY mainly produce light olefins, while more paraffins, long (olefinic) hydrocarbons, and aromatics are generated over FER, ZSM-22, and ZSM-5 zeolites in CO_2 hydrogenation. This phenomenon can be rationalized by the fact that 8-MR and 12-MR zeolites with smaller or larger pores cannot properly match the transition states that lead to the light olefins directly entering into the effluents, whereas for 10-MR zeolites (MFI, TON, and FER), they show the appropriate confinement effects, promoting the further growth of light olefins.

Although the addition of zeolite can inhibit the generation of heavy hydrocarbons by cleaving the C—C bonds on strong acid centers, the acid content of zeolite needs to be finely regulated, it will result in serious over-hydrogenation of olefins. Dokania and co-workers confirmed that introducing Ca additives on zeolite by the impregnation method not only decreases the acid content, but also induces the production of some Lewis acid centers inside the channel of H-ZSM-5 [98]. The diminished Brønsted acid content and the increased Lewis acid content inhibited H_2 transfer reactions, thus inhibiting further growth of light olefins. The prepared Fe-K/Ca-ZSM-5 sample offered high catalytic activity with CO_2 conversion higher than 50%; meanwhile, the selectivity of $C_2^=$–$C_4^=$ is ~38%. A combination of ^{13}C MAS NMR, FT-IR, and probe experiments demonstrated that the presence of Ca dopants also promotes the ketene and acetate species formation in acidic zeolites that are further converted to light olefins through hydrogenation. Liu and co-workers found that the $C_2^=$–$C_4^=$ selectivity is significantly increased from 11% on Fe/H-ZSM-5 to 29% on FeNa/H-ZSM-5 [99]. It can be rationalized that the introduction of proper amount of Na effectively neutralizes some acid sites, thereby suppressing the over-cracking and hydrogenation reactions.

Notably, the distribution of light olefins on these reported Fe-based oxide/zeolite bifunctional catalysts is still very low and generally below 40%. One possible reason is the use of medium-pore H-ZSM-5 (5.1 × 5.5 Å and 5.3 × 5.6 Å) as an acidic zeolite moiety that easily produces long-chain hydrocarbons. As a comparison, SAPO-34 with a narrower 8-MR pore opening (3.8 × 3.8 Å) can effectively catalyze the cracking of long-chain hydrocarbons into light olefins. Ramirez et al. claimed that Fe_2O_3@K_2O and ZrS/SAPO-34 composites delivered a high CO_2 conversion (~20%) and $C_2^=$–$C_4^=$ selectivity (~40–45%) [100]. ZrS/SAPO-34 was synthesized by directly impregnating commercial SAPO-34 in $Zr(SO_4)_2$ solution, followed by drying and calcination. The ZrS is uniformly distributed on the H-SAPO-34 surface, generating an oxide shell layer (~150 nm for layer thickness). The super acidic ZrS species promoted the cracking of long-chain hydrocarbons into light olefins. Recently, an SAPO-34 encapsulated FeC catalyst was prepared by Qiu and co-workers through adding Fe/C@Si into the synthesis gel of SAPO-34 [101]. The surface Si-OH of Fe/C@Si can strongly bond with metal ions (Al^{3+} and P^{5+}), which promotes the growth of SAPO-34 on the Fe/C@Si surface. The sample with core-shell structure showed a stronger confinement effect, making the $C_2^=$–$C_4^=$ selectivity reach 52.6%, being superior to Fe/C+SAPO-34 (44.6%) prepared by mechanical mixing. However, more methane was also produced on the Fe/C@Si-SAPO-34 sample, probably because of the faster diffusion rate of H_2 in the SAPO-34 shell, resulting in a larger H_2/CO ratio and promoting the methanation.

A significant breakthrough was achieved over the tandem catalyst comprising Fe-Cu-K (FCK) oxide moiety and acidic SAPO-34 moiety. The FCK/SAPO-34 composite presents a $C_2^=$–$C_4^=$ selectivity of around 63%, at CO_2 conversion of ~49% [102]. With the aid of CO_2-TPD, H_2-TPD, and DFT calculation, H_2 and CO_2 were confirmed to preferentially adsorb on the (100) crystal facet of Cu-Fe alloy to produce CO and some methanol intermediates, which are then converted into hydrocarbons over θ-Fe_3C species. Besides acting as an acid-catalytic unit, the H-SAPO-34 zeolite also enhances the reduction of FCK. Such a phenomenon originates from the fact that the introduction of H-SAPO-34 elevates the dispersion of FCK and promotes the H_2 spillover effect, and thus, enhances the reduction and carbonization of FeO_x to form θ- Fe_3C species. Additionally, increase in the contact distance between FCK and SAPO-34 considerably accelerates the rapid migration of C_{5+} heavy products from the FCK moiety to the acidic H-SAPO-34 moiety, thereby generating more light olefins.

11.2.2 Co-based Bifunctional Catalyst

In the low-temperature FT synthesis process (220–260 °C), Co-based catalysts showed superior catalytic activity in terms of the formation of long-chain hydrocarbons [103–105]. The product distribution of Co-based catalysts is controlled by many factors, such as Co particle size, morphology, exposed facets, etc. [106–109]. Ismail and co-workers prepared the cobalt-iron (Co-Fe) nanocrystals supported by carbon nanotubes, which show the C_2–C_4 olefins selectivity of 34.5%, but that of C_{5+} hydrocarbons and methane also reaches 26 and 39.5%, respectively [110]. Through controlling the morphology and exposed facets, Zhong et al. indicated that the generated CoC_x nanoprisms during reaction mainly exposed (101) and (020) crystal facets. These unique CoC_x exhibit light olefins selectivity as high as 60.8%, while that of methane is only 5.0%. The residual Na promoted carbonization of metal Co, while the Mn dopant contributed to the formation of nanoprisms morphology. Nevertheless, much more C_{5+} hydrocarbons with a selectivity of 31.4% are produced, and their selectivity is rapidly elevated with the decrease of reaction pressure. H-ZSM-5 and H-BEA zeolites were usually adopted to modulate the product selectivity [111]. Tsubaki's group designed the mesoporous Y zeolites supported by Co NPs, which showed different product distribution in syngas conversion through altering the texture properties and acidity of zeolites [112]. It was found that when the Co NPs were combined with mesoporous parent Y zeolite (Y_{meso}-Na), the steric restriction of mesoporous Y limited the production of C_{21+}, whereas the proportion of C_5–C_{20} hydrocarbons was improved. As expected, the traditional SiO_2 or Al_2O_3 supported Co catalysts produced much more C_{21+} products due to the ASF rule. Use of acidic Y_{meso}-H as a zeolite moiety gives rise to the generation of much more CH_4 and C_2^0–C_4^0 alkanes, because of the cracking of heavy hydrocarbons on strongly acidic Y_{meso}-H. When the strong acid sites were exchanged by Ce^{3+} and La^{3+} ions, the over-cracking of liquid fuels (C_5–C_{20}) was inhibited. The selectivity of gasoline (C_5–C_{11}) reached 74% over Co/Y_{meso}-Ce, while most of the components were jet fuel (C_8–C_{16}) (72%) over Co/Y_{meso}-La. All these observations consolidate that both the pore channel and the acidic property of zeolite play essential roles in syngas to hydrocarbons.

Although these Co-based samples exhibit exceptional activity in syngas conversion, when the feedstock is changed from CO to CO_2, they usually produce much more methane, due to their high activity toward CO_2 methanation [113, 114]. Use of alkali metal additives plays a positive role in enhancing the interaction of CO_2 with Co

NPs and elevates the surface C/H ratio that favors olefin production. Moreover, Liu et al. indicated that the decrease in the acid content of zeolite results in the elevation of light olefins selectivity from 17.5 to 35.4%, along with the decline of methane selectivity from 42.3 to 11.1% over K-CoO_x/H-ZSM-5 catalysts in CO_2 hydrogenation [115]. One possible reason is that an increase in the zeolite Si/Al ratio promotes the reduction of Co oxide, thus enhancing the C—C coupling ability.

As a whole, although the addition of acidic zeolite effectively inhibits the heavy product formation in the FT synthesis process, the selectivity of the aimed light olefin is still unsatisfactory, and it is hard to exceed 61%, due to the limitation of the ASF rule. Therefore, the design of highly efficient FT-based bifunctional or multifunctional catalysts that can significantly improve the light olefins selectivity has a long way to go.

11.3 Methanol-Intermediate Route for Light Olefins Formation from CO_2 Hydrogenation

For the methanol-intermediate route, CO_2 is adsorbed and activated on surface defects of metal oxides to produce carbonates and bicarbonates, which are subsequently reacted with active H to produce methanol via hydrogenation and dissociation reactions (Eq. 11.3) [62–64]. The formed methanol quickly moves into the acidic zeolite moiety and then converts into alkenes through the typical methanol-to-olefin (MTO) reaction (Eq. 11.4).

$$CO_2 + 3H_2 \rightarrow CH_3OH + H_2O \tag{11.3}$$

$$nCH_3OH \rightarrow C_nH_{2n} + nH_2O \tag{11.4}$$

Considering that CO_2 hydrogenation is an exothermic reaction, along with the reduction of volume, low temperature (200–280°C), and high pressure (3–5 MPa) are favorable for methanol synthesis [116, 117]. However, the subsequent methanol conversion is thermodynamically possible at high reaction temperature (400–470°C) and low pressure (0.1–1.0 MPa) [118]. When CO_2 to methanol and MTOs were integrated into one reactor, the cascade process was generally conducted at 300–400°C and 1–3 MPa to rationally match the CO_2 activation and olefin formation. This means that the traditional methanol synthesis catalysts (e.g., CuO-ZnO-Al_2O_3) are not suitable for coupling with zeolite to produce light olefins, as such high temperature easily induces the rapid sintering and agglomeration of copper species [119, 120]. Thus, it is urgent to design a new metal oxide that can be used for methanol synthesis at relatively high reaction temperatures. Additionally, the morphology, size, structure, and acidity of zeolite moieties should be controllably regulated, as they control the formation and evolution of active intermediate species as well as the contribution of the alkene cycle and aromatic cycle in the subsequent MTO process (see Figure 11.5) [121, 122].

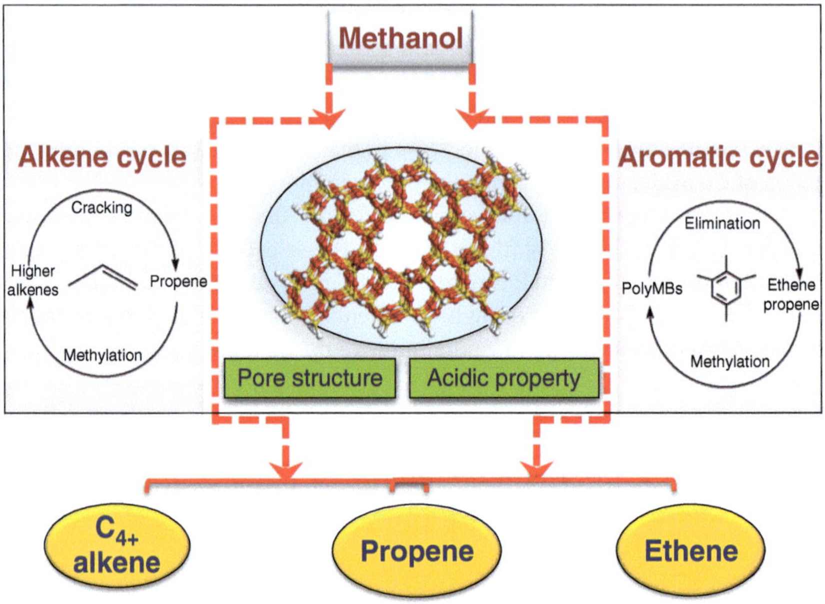

Figure 11.5 Reaction scheme of MTOs via dual-cycle mechanism.

11.3.1 Zn-based Oxides/Zeolite Bifunctional Catalysts

Wang and co-workers found that $ZnZrO_x$ binary metal oxide shows exceptional activity for methanol synthesis under industrial conditions [123]. The CO_2 conversion and methanol selectivity are higher than 10 and 91%, respectively, at 320°C and 5 MPa. The catalytic performance remains stable after a reaction of around 500 hours, and no evident deactivation is observed even doping SO_2 or H_2S impurities in reaction system. This confirms that $ZnZrO_x$ oxide exhibits superior stability than the traditional CuO-ZnO-Al_2O_3 catalyst at relatively high temperatures. Based on this, a $ZnZrO_x$/SAPO-34 tandem catalyst was then prepared [124]. According to the XRD patterns, the $ZnZrO_x$ oxide showed a typical tetragonal zirconium oxide phase, while the zinc oxide phase cannot be observed, indicating the formation of $ZnZrO_x$ solid solution structure. H-SAPO-34 has a typical CHA framework structure, and its mean size is around 2–5 μm. Upon powder mixing of the two groups, the $ZnZrO_x$ oxide was uniformly dispersed on the H-SAPO-34 external surface that is beneficial to the rapid diffusion of generated methanol intermediate onto the acidic zeolite moiety. In CO_2 hydrogenation, $ZnZrO_x$/SAPO-34 showed 80% selectivity of $C_2^=$–$C_4^=$ alkenes and 14% of C_2^0–C_4^0 alkanes, while that of methane and C_{5+} products were no higher than 3%, at 380 °C and 2 MPa (see Figure 11.6). The product composition over $ZnZrO_x$/SAPO-34 deviated significantly from the ASF model. The catalytic performance of this sample at different conditions was then evaluated. Although the CO_2 conversion can be further improved through raising reaction temperature, it also produced more CO by-product, because high reaction temperature considerably enhanced the RWGS reaction. The formation of CO was

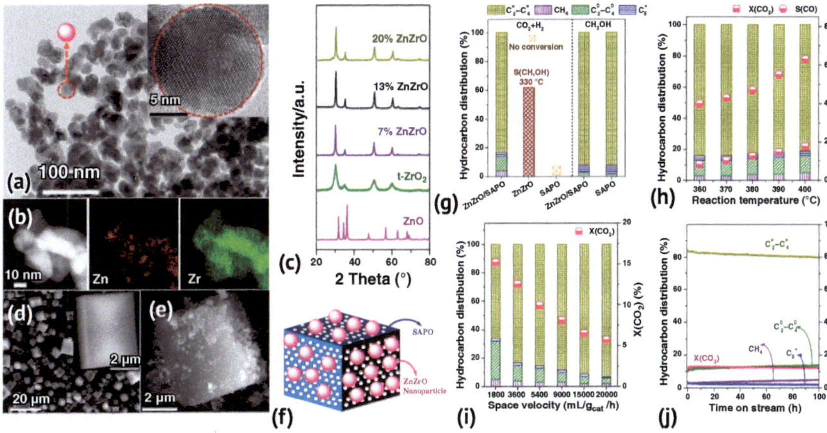

Figure 11.6 Catalyst characterization and reaction performance of $ZnZrO_x$/SAPO-34 in CO_2 hydrogenation: (a) morphology of $ZnZrO_x$, (b) element distribution on $ZnZrO_x$, (c) crystal structure of various samples, (d and e) morphology of SAPO-34 zeolite and $ZnZrO_x$/SAPO-34 composite, (f) schematic diagram of composite catalyst, and (g–j) reaction performance of $ZnZrO_x$/SAPO-34 for alkenes formation in CO_2 hydrogenation. *Source*: Reproduced from [124]/with permission of American Chemical Society.

slightly suppressed by raising the reaction pressure; however, high pressure resulted in the generation of more alkanes. In situ DRIFTS was performed to clarify the type and structure of intermediates. When CO_2 and H_2 mixture was introduced, the IR peaks associated with formate (2980, 2880, and 2835 cm^{-1}) and methoxy (2930 and 2824 cm^{-1}) quickly appeared on $ZnZrO_x$ oxide and gradually intensified with the reaction time, indicating that methanol was mainly produced via formate and methoxy intermediates.

The generation of methanol on $ZnZrO_x$ can be further enhanced through adding La dopants [125]. It is demonstrated that doping La additives into the lattice of $ZnZrO_x$ induces the generation of surface defects, and thus, promotes CO_2 activation. In situ DRIFTS illustrated that the IR peaks characteristic of HCOO* at around 1590, 1360, and 1310 cm^{-1}, and that of methoxy at around 1040–1070 cm^{-1} are visible only after a reaction of one minute, and their intensity is stronger on La-modified $ZnZrO_x$ than on $ZnZrO_x$.

After that, a series of bifunctional composite catalysts was prepared for light olefines formation via the methanol-intermediate route. A $ZnGaO_x$ binary metal oxide with the Zn/Ga molar ratio of 1/2 (Zn/Ga = 1/2) was prepared [126]. XRD patterns suggested the formation of a spinel structure of $ZnGa_2O_4$ with the particle size of around 7 nm. This was corroborated by HRTEM images that lattice fringes of 0.210, 0.252, and 0.295 nm, ascribed to (400), (311), and (220) crystal facets of $ZnGa_2O_4$ spinel, are clearly visible. Methanol and dimethyl ether (DME) are the dominant products over $ZnGaO_x$ in CO_2 hydrogenation even at a high temperature of 450 °C. When H-SAPO-34 zeolite was adopted as the reaction system, the unconverted methanol was sharply decreased to below 1%, while that of light olefins was increased to above 80% in hydrocarbons at 350–400 °C. Unlike the traditional FT synthesis route, the selectivity of undesired CH_4 was only 1% over $ZnGaO_x$/H-SAPO-34. Notably, $ZnGaO_x$/H-SAPO-34 bifunctional catalyst showed higher catalytic activity than that of standalone $ZnGaO_x$ oxide at the same conditions. This is due to the quick consumption of methanol on acidic H-SAPO-34 that drives the synthesis of methanol intermediates on $ZnGaO_x$ oxide. This thermodynamic driving force also significantly decreased the CO selectivity from 68% on $ZnGaO_x$ oxide to 46% on $ZnGaO_x$/H-SAPO-34 composite, because the production of CO and CH_3OH was competitive. Therefore, the generation of light olefins from CO_2 hydrogenation on metal oxide/zeolite composite catalysts is a thermodynamics and kinetics coupling process. Moreover, the $ZnGaO_x$/H-SAPO-34 composite can also catalyze CO hydrogenation to light olefins, although CO hydrogenation is slower than that of CO_2 hydrogenation at the same conditions.

$ZnAl_2O_4$ spinel offered a similar methanol (including DME) selectivity (99%), but a higher CO_2 conversion of 12%, compared to that of $ZnGa_2O_4$ (98 and 9.8%) in CO_2 hydrogenation [127]. This is because $ZnAl_2O_4$ has a larger concentration of surface oxygen defects, enhancing CO_2 adsorption and activation. In addition, Zn content in $ZnAlO_x$ strongly determines the production of methanol, as it influences the H_2 dissociation ability. It was confirmed that Al_2O_3 was almost inactive for methanol synthesis. Moreover, when Zn is replaced by Mg, the CO_2 conversion is considerably decreased, arising from the much lower activity of Mg in H_2 dissociation. As expected, the introduction of H-SAPO-34 zeolite cannot only promote the transformation of methanol and DME into alkenes, but also suppress the formation of CO. $ZnAl_2O_4$/H-SAPO-34 composites gave $C_2^=$–$C_4^=$ selectivity higher than 87%, while the formation of CH_4 was considerably inhibited to below 0.7% at 370 °C and 3.0 MPa. Although an increase of Zn content in $ZnAlO_x$ oxide can enhance CO_2 activation, when the molar ratio of Zn/(Zn+Al) is ≥ 0.3, much more C_2^0–C_4^0 alkanes were produced, due to over-hydrogenation of alkenes. It can be verified by the H_2–D_2 exchange experiment that the HD formation rate is elevated with the increase of Zn amount, and ZnO exhibits the highest H_2 dissociation activity. Moreover, the CO_2 conversion and product distribution of $ZnAlO_x$/H-SAPO-34 bifunctional catalyst are also controlled by the Brønsted acid content of H-SAPO-34 zeolite. Increasing the Brønsted acid content from 0 to 0.2 mmol g^{-1} results in a sharp decrease of unconverted methanol and DME, whereas more light olefins are produced. However, when the acid content is higher than 0.21 mmol g^{-1}, the proportion of C_2^0–C_4^0 is elevated, along with the decline of $C_2^=$–$C_4^=$. This result confirmed that an excessive number of acidic sites can also induce over-hydrogenation of alkenes.

Although these reported $ZnZrO_x$/H-SAPO-34, $ZnAlO_x$/H-SAPO-34, and $ZnGaO_x$/H-SAPO-34 composites offered a high $C_2^=$–$C_4^=$ selectivity (around 80% in hydrocarbons), the low CO_2 conversion (< 20%) limits their practical application. To overcome this issue, Mou et al. fabricated Mn_2O_3-ZnO binary oxides by the coprecipitation method [128]. Addition of Mn gives rise to the migration of the peaks characteristic of the (100), (002), and (101) facets of ZnO to higher angles, due to the replacement of larger Zn^{2+} (0.74 Å) by smaller Mn^{3+} (0.66 Å). The interplanar spacings of 2.48, 2.60, and 2.81 Å, assigned to (101), (002), and (100) facets of ZnO, and of 2.71 Å, assigned to the (103) facet of $ZnMn_2O_4$ spinel, were observed in HRTEM images. This indicated the simultaneous existence of $ZnMn_2O_4$ and ZnO phases. According to the O1s-XPS, doping a proper amount of Mn into ZnO can promote the generation of more defects that enhance the adsorption of CO_2. H_2 is mainly activated on ZnO, and the generated active H* quickly migrates to the oxygen vacancies, interacting with adsorbed CO_2 to form methanol. Therefore, the collaboration of $ZnMn_2O_4$ with ZnO strengthens the synthesis of methanol. After the introduction of H-SAPO-34 zeolite, 20%Mn_2O_3-ZnO/H-SAPO-34 showed the selectivity of $C_2^=-C_4^=$ higher than 80% in hydrocarbons, while the unexpected methane is considerably decreased with selectivity of only 3.3%. More importantly, the CO_2 conversion reached 29.8%, being much higher than that of ZnO/H-SAPO-34 (18.1%) and Mn_2O_3/H-SAPO-34 (8.8%). The contact distance of Mn_2O_3-ZnO oxide and H-SAPO-34 strongly controlled the product distribution. Dual-bed filling mode showed low CO_2 conversion and $C_2^=-C_4^=$ selectivity of only 20 and 49%, respectively, while the undesired CO got to 90.1%, suggesting that too long contact distance limits the tandem efficiency of metal oxide and zeolite. When Mn_2O_3-ZnO and H-SAPO-34 were assembled by particle packing mode, the CO_2 conversion and $C_2^=-C_4^=$ selectivity were slightly elevated to 22% and 50%, along with the decrease of CO selectivity to 80.9%. Much higher CO_2 conversion of ~30% and $C_2^=-C_4^=$ selectivity of ~80%, but lower CO selectivity of 60% were obtained over Mn_2O_3-ZnO/H-SAPO-34 constructed by ball milling method. It can be seen that the decrease in the contacting distance of Mn_2O_3-ZnO and H-SAPO-34 is more favorable for the hydrogenation of CO_2 to light olefins. Notably, too long ball milling time (> 60 min) destroyed the crystal structure and reduced the acid content of SAPO-34, thus, making $C_2^=-C_4^=$ selectivity decrease to 56.7%.

It can be seen that in Zn-based oxide, Zn plays a positive role in H_2 dissociation, while the surface defects determine CO_2 adsorption and activation ability. Therefore, controlling Zn content and increasing the surface defect concentration of metal oxide can effectively promote methanol synthesis. Moreover, the contacting distance between metal oxide and acidic zeolite is also a key factor affecting light olefins selectivity, as it determines the diffusion and transmission of methanol between the two groups.

11.3.2 In-based Metal Oxide/Zeolite Bifunctional Catalysts

Recently, some investigations indicated that compared with traditional CuO-ZnO-Al_2O_3 catalysts, indium (In)-based oxides exhibit superior catalytic performance in methanol synthesis at high temperatures of 280–330 °C [129–132]. Martin and co-workers reported that the In_2O_3 oxide showed the methanol selectivity of 100%, with catalytic stability as high as 1000 hours in CO_2 hydrogenation under industrially relevant conditions [133]. Ye et al. analyzed the important role of surface defects on In_2O_3(110) in methanol formation from CO_2 hydrogenation [134]. They examined the formation energy of various surface O vacancies and found that removing O from the D1 site of In_2O_3(110) is more favorable, whereas the generation of O defects on the D4 site requires higher energy. They also investigated the formation mechanism of some key intermediates and demonstrated that methanol is mainly produced through the HCOO-H_2CO-CH_3O route (see Figure 11.7).

Dang et al. calculated the reaction thermodynamics and kinetics of several elemental reactions for methanol synthesis and RWGS reactions on different crystal facets of In_2O_3 oxide [135]. They found that the cubic-In_2O_3(111) showed the lowest free energy for methanol synthesis (0.64 eV), while hexagonal-In_2O_3(012) (0.69 eV) was more active for the production of CO. It was found that both cubic-In_2O_3(111) and hexagonal-In_2O_3(104) surfaces favor the methanol formation. Under the guidance of theoretical calculations, several In_2O_3 oxides exposing different crystal facets were fabricated, and it was found that hexagonal-In_2O_3(104) with more defects shows superior catalytic activity to methanol formation, whereas the generation of CO has been restricted. The CO_2 conversion and methanol selectivity are ~17 and ~92% at 300°C, 5.0 MPa, and 9000 ml/g/h. However, cubic-In_2O_3 exhibits lower methanol formation activity, probably because of the lower adsorption energy of CO_2 on the cubic-In_2O_3(111) surface that suppresses the CO_2 activation to some extent. Notably, the methanol selectivity is still up to 80% over hexagonal-In_2O_3 even

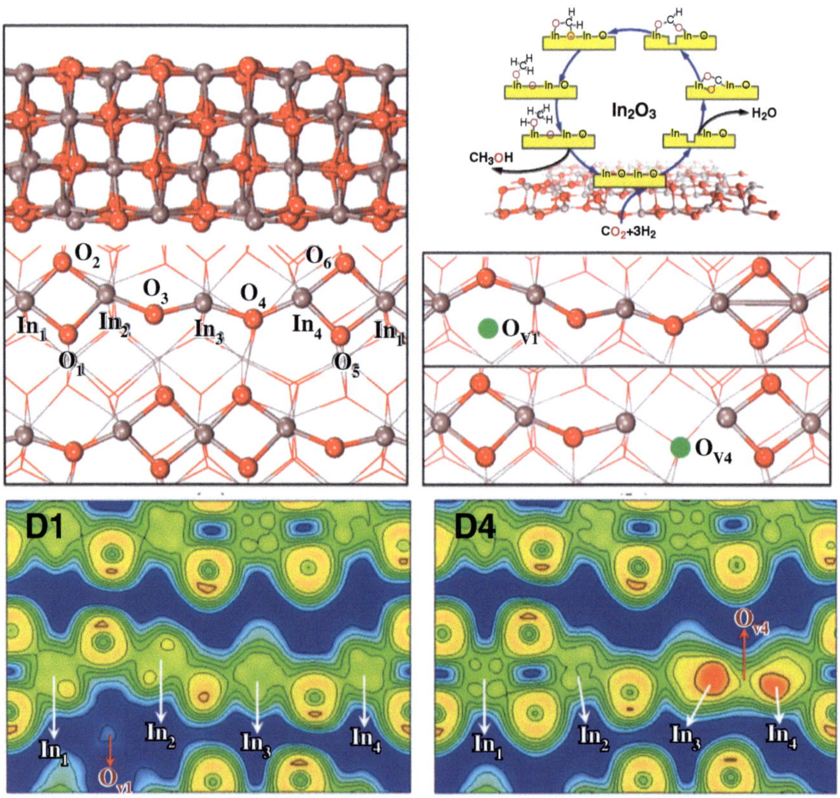

Figure 11.7 Optimized structure and the electronic state of the In_2O_3 model and the reaction mechanism for methanol synthesis from CO_2 hydrogenation. *Source:* Reproduced from [134]/with permission of American Chemical Society.

at a high temperature of 360 °C. It seems that In-based oxide has important potential for alkenes formation from CO_2 hydrogenation through the methanol-intermediate route, as it can make the optimal temperature of methanol synthesis and methanol conversion more match.

Subsequently, Numpilai et al. designed a hybrid sample involving In_2O_3 and H-SAPO-34, which gave the $C_2^=-C_4^=$ yield of 7.31% and the CO_2 conversion exceeded 34.6% at 360 °C and 2.5 MPa [136]. It is also found that H_2 pretreatment can generate more surface oxygen defects, thereby improving the adsorption ability of CO_2 on In_2O_3. Moreover, when the reduction temperature was higher than 223 °C, some metallic In was produced and they were regarded as the main sites for H_2 dissociation.

A $InZrO_x$/H-SAPO-34 bifunctional catalyst was fabricated by Gao and co-workers, and it exhibited CO_2 conversion of 35.5% and $C_2^=-C_4^=$ selectivity of 76.4% in hydrocarbons, while that of methane was only 4.3% at 400 °C and 3.0 MPa (see Figure 11.8) [137]. DFT calculation results indicated that Zr dopants prefer to substitute the In4 atoms of In_2O_3, which produces more oxygen vacancies. It was corroborated by NAP-XPS and CO_2-TPD results that a large number of surface defects and stronger desorption peaks

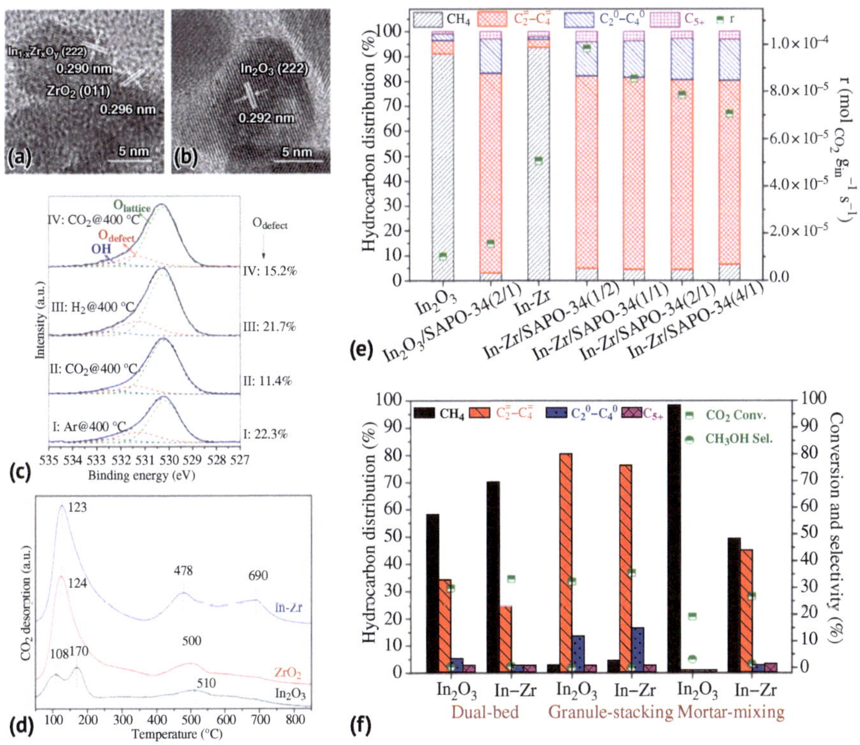

Figure 11.8 Characterizations and catalytic performance: (a, b) morphology of $InZrO_x$ (a) and In_2O_3 (b), (c) O(1s) XPS of $InZrO_x$ exposing different atmosphere, (d) CO_2-TPD profiles of various In-based oxides, and (e and f) catalytic performance of various composite catalysts in conversion of CO_2 to alkenes. *Source*: Reproduced from [137]/with permission of American Chemical Society.

of CO_2 were observed on $InZrO_x$ than on In_2O_3 and ZrO_2. This makes $InZrO_x$ show higher CO_2 conversion than the other two samples. In addition, the results of in situ XRD suggested that Zr dopants suppress the rapid aggregation of In_2O_3 NPs, and the mean size of $InZrO_x$ oxide only slightly grew from ~10 to ~12 nm as the reaction progressed. As a result, the CO_2 conversion remains stable over $InZrO_x$/H-SAPO-34 even after a reaction of 152 hours. It should be noticed that much more CH_4 is generated on $InZrO_x$/H-SAPO-34 prepared by powder mixing. Such a phenomenon was rationalized that overly tight contact of $InZrO_x$ and H-SAPO-34 induces the rapid migration of some In species from oxides to zeolites, thereby poisoning the acid sites and deactivating the composite catalyst.

Wang and co-workers also observed that some metal In species easily migrated from metal oxides into acidic zeolite, especially at high reaction temperature [138]. The catalytic activity and product selectivity of a composite consisting of H-ZSM-5 and different metal oxides in CO_2 hydrogenation were evaluated. Surprisingly, the dominant product over In_2O_3/H-ZSM-5 bifunctional catalyst was methane (~7%) and unconverted methanol and DME (92%), while the formation of C_{2+} products was considerably inhibited, with selectivity < 1%. However, when In_2O_3 was substituted by Cr_2O_3 or ZrO_2, they gave an opposite result that most of the products were C_{2+} hydrocarbons (> 90%), while the unconverted methanol and DME were undetectable. After reaction, some In NPs were defected on the H-ZSM-5 surface, confirming that In species easily migrated from oxides to zeolite. Therefore, the authors indicated that close intimacy (e.g., powder mixing or ball milling) was detrimental for the formation of C_{2+} hydrocarbons over composites comprising In_2O_3 and acidic zeolites, as the migration of In species will seriously poison the acidic zeolite.

Recently, Zhang et al. fabricated a composite catalyst comprising $InCeZrO_x$ and H-SAPO-34 and this sample gave light olefins selectivity > 93% in hydrocarbons, while those of methane and C_2^0–C_4^0 alkanes were only 1.9 and 2.4%, respectively [139]. Doping In additives into $CeZrO_x$ not only enhances the CO_2 adsorption by forming more surface oxygen defects, but also enhance the H_2 dissociation. Moreover, when the In/(Ce+Zr) molar ratio is ≤ 0.7, the rapid In migration is obviously suppressed and the catalytic activity of $InCeZrO_x$(0.7)/H-SAPO-34 composite catalyst remains stable even after reaction of 200 hours. The enhanced catalytic lifetime of $InCeZrO_x$(≤ 0.7) is due to introduction of In into the lattice of $CeZrO_x$, which effectively inhibits the In migration by inhibiting InO_x species reduction. In addition, increasing the acid amount of SAPO-34 or elevating the weight ratio of zeolite moiety can both improve the stability of composite catalyst. This is because all of these methods improve the non-poisoned acid content of H-SAPO-34 zeolite.

Although In species is easy to migrate, the high catalytic activity of In-based oxide made it attacked much attention in CO_2 hydrogenation. Dang and co-workers found that the surface defect content of $InZrO_x$ is determined by the In/Zr ratio [140]. With the decrease of In/Zr molar ratio, the diffraction peaks of In_2O_3 gradually disappeared, along with the enhancement of tetragonal ZrO_2. This is in line with the observation of HRTEM images that the lattice fringes of In_2O_3 are undetectable, whereas those of ZrO_2 become clearly visible, when In/Zr molar ratio is below 4/1. Doping Zr additives considerably promotes the formation of more defects. The amounts of surface defects showed a volcano-type curve with the variation of Zr content, and it reached the peak value of 24.1% on $In_4Zr_1O_x$ oxide. As expected, $InZrO_x$(4/1)/H-SAPO-34 composite exhibited more superior catalytic performance than the other samples. The C_2–C_4 alkene and alkane selectivity reached 96%, where light olefins account for 77.6%. Wang et al. demonstrated that the introduction of Cr additives into In_2O_3 lattice sites can enhance the interaction of formate and $InCrO_x$ oxide, and thus, lowering the energy barriers for formate and methanol formation [141]. The prepared $InCrO_x$/H-SAPO-34 composite showed CO_2 conversion and light olefins selectivity of 19.8 and 82.4%, respectively, which endows it with a higher light olefins yield (7.7%) than that of In_2O_3/H-SAPO-34 (4.7%).

11.3.3 Ga-based Metal Oxide/Zeolite Bifunctional Catalysts

Besides Zn-based or In-based catalysts, the gallium (Ga)-based catalysts are also promising samples for the synthesis of methanol, because of high structural stability. Liu et al. illustrated that gallium nitride (GaN) can selectively transform CO_2 into DME; the DME selectivity gets to ~80% (CO free) at high temperature of 360 °C [142]. This result suggests the significant potential of Ga-based catalysts for methanol and DME synthesis at relatively high temperatures. A series of $MZrO_x$ samples were prepared by Wang et al. using coprecipitation synthesis strategy, and their catalytic results in methanol synthesis were assessed [143]. It was found that $GaZrO_x$ showed highly comparable methanol formation activity to that of $ZnZrO_x$ oxide at 300 °C and 2.0 MPa. Through combining the XPS, solid-state NMR, EPR, and in situ DRIFTS, Feng et al. indicated that defective GaOZr is highly active for the formation of methanol [144]. H_2 is preferentially activated on Ga sites, forming gallium hydride species; the CO_2 adsorbed on the defect sites interacted with gallium hydride to generate formate, methoxy, and then methanol through successive hydrogenations. The prepared $GaZrO_x$ with an nGa/(nGa + nZr) molar ratio of 0.23 exhibited superior activity in CO_2 hydrogenation; CO_2 conversion and methanol selectivity reached 9.0 and 72.7%, at 330 °C and 3.0 MPa.

Zhang et al. fabricated $GaZrO_x$/H-SAPO-34 composite for light olefins formation from CO_2 hydrogenation [145]. Both CO_2 conversion and $C_2^=$–$C_4^=$ selectivity show a volcano-type curve with the increase of Ga content in $GaZrO_x$ oxide (see Figure 11.9). The maximum CO_2 conversion (~27%) and light olefins selectivity (~89%) are acquired on $GaZrO_x$(1/4)/H-SAPO-34 at 390 °C and 3.0 MPa. The olefin/paraffin ratio and light olefins yield reach 14% and 11.3%, respectively. $GaZrO_x$(1/4)/H-SAPO-34 also exhibited high stability; the CO_2 conversion and the light olefins selectivity are well kept at around 26 and 88%, while the CO selectivity is no higher than 52%, after a reaction of 100 hours.

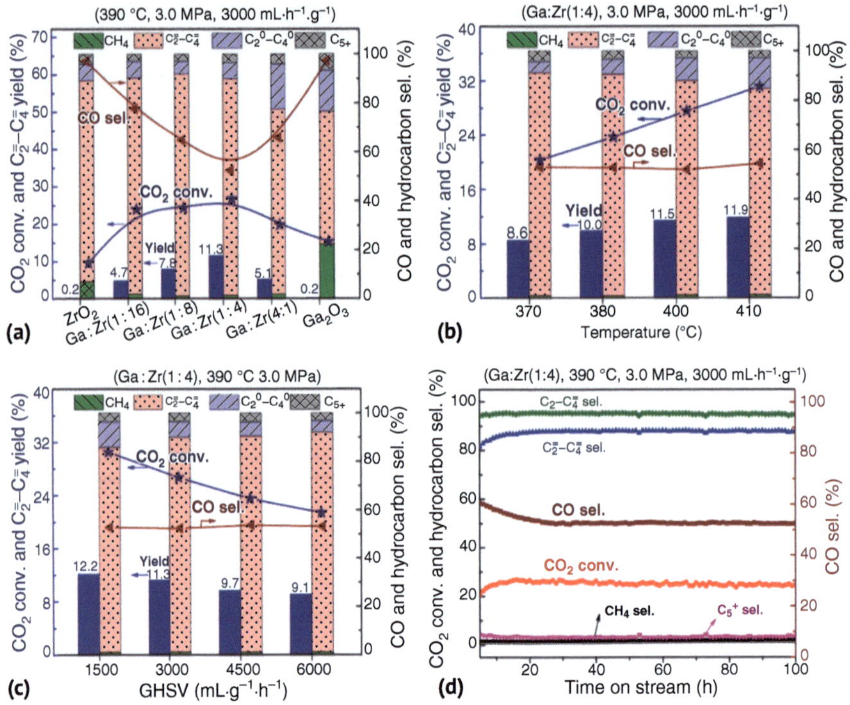

Figure 11.9 Catalytic activity and product distribution of $GaZrO_x$/SAPO-34 composite in CO_2 hydrogenation, (a) effect of Ga/Zr ratio, (b) effect of reaction temperature, (c) effect of GHSV, and (d) catalytic lifetime. *Source*: Reproduced from [145]/with permission of Elsevier.

11.3.4 Cu-based Metal Oxide/Zeolite Bifunctional Catalysts

Although traditional Cu-based samples (e.g., CuO-ZnO-Al_2O_3) are an important industrial catalyst for methanol synthesis from CO/CO_2 hydrogenation, direct coupling of CuO-ZnO-Al_2O_3 oxide with acidic zeolite generally produces large amounts of undesired CO or light alkanes, because of serious RWGS reaction and the over-hydrogenation of alkenes [146]. Recently, Chen and co-workers indicated that adjustment of the contacting distance between Cu-based oxides and acidic zeolites can restrict the over-hydrogenation of light olefins to some extent [147]. They prepared a core-shell CuZnZr@SAPO-34 bifunctional catalyst using silica sol as the binder. CuZnZr@SAPO-34 was found to be fairly uniform with a spherical structure, in which the cubic-type granules of SAPO-34 zeolites are closely distributed on the CuZnZr surface in a thickness of about 40 μm. In CO_2 hydrogenation, CuZnZr@H-SAPO-34 showed higher $C_2^=$–$C_4^=$ selectivity (42.5%), but lower to methane and C_2^0–C_4^0 alkane (54.5%), than those of the CuZnZr/SAPO-34(4/1) prepared by powder mixing (34.7 and 65.3%). The formation of light olefins is further enhanced, with the $C_2^=$–$C_4^=$ selectivity raised to 60.5% through neutralizing some acid sites of zeolite shell by Zn-modification. The reduction of acid content of zeolite inhibits secondary hydrogenation of generated alkenes. Sedighi et al. reported the $CuCeO_x$/SAPO-34 composite and this sample offered the $C_2^=$–$C_4^=$ selectivity

> 70% [148]. However, the unexpected CO selectivity also reached 57% with CO_2 conversion of ~13%, at 382 °C and 2.0 MPa. Therefore, although these Cu-based samples show some potential for light olefins formation, their strong hydrogenation ability and high RWGS reaction rate cause the yield of light olefins to be much lower than that of the Zn-based, In-based, or Ga-based catalysts.

11.4 Control of the Product Distributions in CO_2 Hydrogenation to Light Olefins

According to the above investigations, the methanol-intermediate route has greater potential for tailing the alkenes distribution in CO_2 hydrogenation, as this route breaks the restriction of ASF rule, though the catalytic activity of the methanol-intermediate route is generally inferior to that of the FT synthesis route. Several excellent composite catalysts have been prepared, and they exhibited the selectivity of light olefins of 60–90% in hydrocarbons, while the generation of CH_4 is considerably inhibited with the selectivity of < 5%. Although important progress has been made for the methanol-intermediate route, further elevation of light olefins selectivity beyond 90% is still a challenge. More importantly, the composition of obtained olefin products is very fixed (ethene: ~25–30%, propene: ~35–40%, and butene: ~10–15%), and it is difficult to selective conversion of CO_2 into a single specific olefin. One method to regulate the light olefin composition is to modulate the structure and electronic state of metal oxides to form ketene, acetate, or ethanol intermediates, which can be directly converted into ethene [149]. Jiao and co-workers demonstrated that much more ketenes are generated on $ZnCrO_x$ in syngas conversion, and it is preferred to interact with the acid sites sitting in 8-MR of H-MOR [150]. The strong steric restriction of 8-MR side pockets promoted hydrogenation of ketene to ethene. Moreover, the selectivity of olefin products also depends on the contribution of the dual cycle in the subsequent MTO process, which is controlled by the pore channel and acidic properties of zeolite [151–153]. Westgard Erichsen et al. investigated the MTO reaction on zeolite with isostructural, but different strengths of acid sites [154, 155]. The strongly acidic H-SSZ-24 easily induced the formation of polymethylbenzenes (polyMBs) intermediates, which enhanced the aromatic cycle and promoted the generation of more ethene, whereas the weakly acidic H-SAPO-5 was favorable for forming C_{4+} long-chain hydrocarbons. Wang et al. investigated the influence of the supercage structure of SAPO-n zeolites on product selectivity in syngas conversion [156]. Integration of $ZnAlO_x$ oxide with SAPO-17 zeolite (supercage size: 6.7 × 6.7 Å) gave the proportion of C_2, C_3, and C_4 of 38, 33, and 9%, respectively, whereas for $ZnAlO_x$/SAPO-34 (supercage size: 7.0 × 7.4 Å) and $ZnAlO_x$/SAPO-18 (supercage size: 7.0 × 8.5 Å), the C_3 and C_4 selectivities were increased to 55 and 13%, and 44 and 29%, along with the decrease of C_2 selectivity to 27 and 14%, respectively (see Figure 11.10). Clearly, the elevation of the supercage size of zeolites enhances the production of heavier hydrocarbons, because of the weakening of the steric-hindrance effect. Therefore, inducing

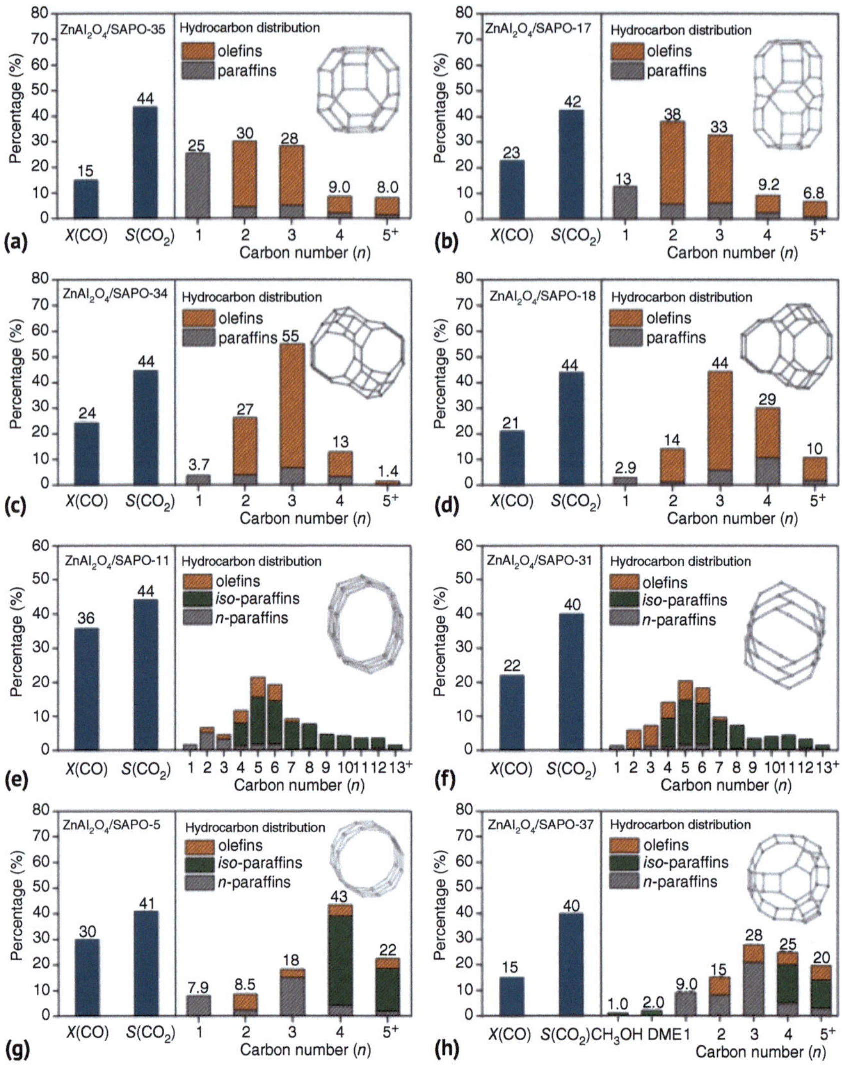

Figure 11.10 Catalytic performance of composite catalysts comprising $ZnAlO_x$ and SAPO-n zeolites in CO hydrogenation to hydrocarbons: (a) $ZnAlO_x$/SAPO-35, (b) $ZnAlO_x$/SAPO-17, (c) $ZnAlO_x$/SAPO-34, (d) $ZnAlO_x$/SAPO-18, (e) $ZnAlO_x$/SAPO-11, (f) $ZnAlO_x$/SAPO-31, (g) $ZnAlO_x$/SAPO-5, and (h) $ZnAlO_x$/SAPO-37. *Source*: Reproduced from [156]/with permission of Elsevier.

the generation of ketene or ethanol, and modulating the pore structure (supercage and window size) and acidity (acid sites content, strength, and location) of zeolites will be effective methods to control the distribution of olefin products in CO_2 hydrogenation.

11.4.1 Regulation of Light Olefins Distribution via Ethanol-Intermediate Route

A significant increase in the proportion of single ethene in products is very important to reduce separation cost, but is also rather challenging, as CO_2 hydrogenation to hydrocarbons involves many elementary reactions that lead to the generation of a wide product spectrum. In the FT synthesis route, the hydrocarbons are produced through CH_x–CH_x coupling that makes product compositions obey the ASF distribution rule, resulting in the selectivity of ethene and total light olefins no higher than 30 and 58%, respectively [157, 158]. In the methanol-intermediate route, the production of alkenes is determined by the dual-cycle mechanism, in which the aromatic cycle mainly produces ethene and propene, whereas the alkene cycle, generates more propene and butene [159, 160]. Coexistence of these two cycles results in the ethene selectivity below 40%.

Moreover, ethanol dehydration to ethene is very fast with few side reactions, and the selectivity of ethene can reach >90% [161]. There is no doubt that if ethanol intermediates are produced over metal oxides in CO_2 hydrogenation and are rapidly dehydrated on acidic zeolite, it is expected to greatly promote ethene formation. Recently, Fe-based metal oxides have been confirmed to have high ethanol production activity [162–164]. Yao and co-workers proposed a FeNaS catalyst, which exhibited higher alcohol ($C_{2+}OH$) space-time yield (STY) higher than 80 mg g^{-1} h^{-1} [165]. Among them, ethanol accounted for more than 70% of $C_{2+}OH$. Wang et al. designed KCuZnAl/Na-Fe@C composite [166]. It gave CO_2 conversion and ethanol selectivity of 40 and 35%, with at 320 °C and 5.0 MPa. Despite their excellent catalytic properties, these Fe-based catalysts used for ethanol synthesis are not suitable for direct coupling with zeolites to prepare ethene. This is because besides ethanol, they also simultaneously produce large amounts of methane and C_2–C_4 alkanes, due to their strong dissociation and hydrogenation activity [167, 168]. Therefore, elevating ethanol selectivity and decreasing the generation of unexpected methane and other alkanes play a vital role in CO_2 hydrogenation to ethene. Based on this, Cr_2O_3(SG)/H-SAPO-34 was developed by Wang et al. [169]. This sample exhibited $C_2^=$–$C_4^=$ selectivity in hydrocarbons > 95%. In addition, ethene accounts for ~74% of alkenes, with the E/P ratio higher than 3 (see Figure 11.11). The authors suggested that rapid transformation of ethanol into ethene is the key to improving ethene selectivity, as the ethanol selectivity in hydrocarbons + oxygenates over Cr_2O_3(SG) oxide reaches 30.5% at initial stage of CO_2 hydrogenation. It can be confirmed that the addition of a small amount of ethanol significantly improved ethene and total light olefin selectivities to 71.8 and 96.2%, respectively. Multiple characterizations evidence that ethanol is produced through the interaction of methanol with CO_2 and H* on Cr_2O_3(SG) oxide. As a result, ethene selectivity over Cr_2O_3(SG)/H-SAPO-34 is significantly increased, compared to that of traditional bifunctional catalysts via the methanol-intermediates route.

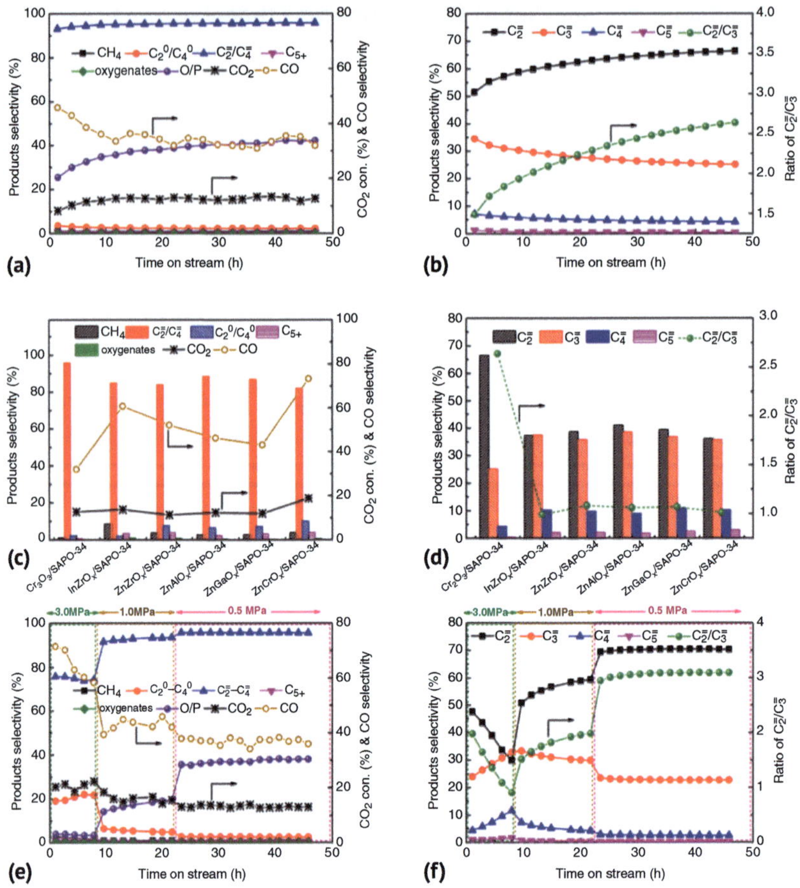

Figure 11.11 Catalytic result of Cr_2O_3(SG)/H-SAPO-34 in CO_2 hydrogenation to light olefins: (a and b) catalytic activity and products selectivity of Cr_2O_3(SG)/H-SAPO-34, (c and d) comparison of various samples at the same reaction conditions, and (e and f) catalytic performance of Cr_2O_3(SG)/H-SAPO-34 at different reaction pressures. *Source*: Reproduced from [169]/with permission of Elsevier.

11.4.2 Regulation of Light Olefins Distribution via Controlling the Pore Structure and Acidic Properties of Zeolite

It has been widely accepted that the products selectivity in methanol conversion is controlled by pore channel and acidic property of zeolites [170–173]. Wang and co-worker fabricated $InZrO_x$/SAPO-5 composite, showing the selectivity of light hydrocarbons higher than 83% [174]. Notably, butane was the main product of C_2–C_4 hydrocarbons. This may be due to the one-dimensional (1D) 12-MR pore structure of SAPO-5 zeolite, facilitating the formation and diffusion of C_4 hydrocarbon. Ticali and co-workers found that propene accounts for 50.6% of $C_2^=$–$C_4^=$ over $ZnZrO_x$/ZSM-5 in CO_2 hydrogenation. Nevertheless, the $C_2^=$–$C_4^=$ selectivity is only 27.7%, whereas much more C_{5+} (27.5%) were produced [175]. It can be seen that using zeolites with 10- or 12-MR pore channels easily produces C_{4+} hydrocarbons.

Recently, Li and co-workers gave an interesting result that the prepared $ZnZrO_x$/bio-ZSM-5 bifunctional catalyst shows $C_2^=-C_4^=$ selectivity of >64%, while that of aromatics is below 1% [176]. They used rice husks as a template agent for the preparation of ZSM-5 zeolite, which showed weaker acid strength and lower acid density than those of the commercial ZSM-5 sample. Reduction of Brønsted acid strength and density can inhibit aromatization reactions and decrease amounts of aromatics. Moreover, the external acid contents of this sample are selectively removed by means of TEOS deposition, which, hence, suppressing over-hydrogenation of light olefins. Despite the important progress, the $C_2^=-C_4^=$ selectivity is still lower than the $ZnZrO_x$/H-SAPO-34 under similar reaction conditions, as the narrow 8-MR pore opening of H-SAPO-34 can effectively limit generation of C_{5+} long-chain products. This is why H-SAPO-34 has been largely adopted as an acidic moiety for hydrogenation of CO/CO_2 to light olefins.

Recently, Zhang and co-workers confirmed that the alkenes distribution is strongly related to the acidic strength of zeolite [177]. Several H-SAPO-34(F) catalysts with different acid content and acid strength were synthesized through doping hydrofluoric acid during synthesis. Upon coupling with $GaCrO_x$ oxide, the $GaCrO_x$/H-SAPO-34(F) bifunctional catalysts displayed much higher ethene (~34%) and propene (~49%) selectivity, while that of butene is considerably decreased to ~5%, compared with that of $GaCrO_x$/H-SAPO-34 (33, 45, and 10% for ethene, propene, and butene, respectively). They pointed out that H-SAPO-34(F) having higher strong acid content and acid strength, promotes cracking of long-chain alkene into propene, but further transformation of propene to butene via methylation requires higher free energy. Therefore, the $GaCrO_x$/H-SAPO-34(F) composite produces more ethene and propene than $GaCrO_x$/H-SAPO-34.

A significant breakthrough in the regulation of product distribution was made by Wang, Guo, and co-workers [178, 179]. They designed a $ZnCeZrO_x$/H-RUB-13 composite, which exhibits $C_2^=-C_4^=$ selectivity of around 80% in hydrocarbons, similar to that of $ZnCeZrO_x$/H-SAPO-34. However, when H-RUB-13 zeolite was used as the zeolite moiety, it offered a much higher $(C_3^= + C_4^=)/C_2^=$ ratio of 6.1 than the H-SAPO-34 (1.8) (see Figure 11.12). The results of XPS, DFT calculation, and probe experiments reveal that methanol was formed on surface defects of $ZnCeZrO_x$ oxide via formate and methoxy intermediates. Subsequently, methanol was transformed into alkenes on acidic RUB-13 through a dual-cycle mechanism. It is shown that the contribution of alkene cycle is higher on RUB-13 than on SAPO-34, originated from the weaker acid strength of the former than the latter. As a result, the formation of polyMBs and methylcyclopentadienes was considerably inhibited. In addition, more Al atoms sitting in the non-distorted 8-MR pore channel of RUB-13 also promote rapid diffusion of propene and butene.

Very recently, $GaZrO_x$/H-SSZ-13 catalyst system was designed and it showed great potential for CO_2 hydrogenation to single propane, in which the propane and LPG (propane + butane) selectivities in hydrocarbons reach 79.5 and 89.4%, respectively, with CO_2 conversion of 43.4% [180]. The results of various in situ DRIFTS, ^{13}C MAS NMR, and DFT calculation confirmed that an increase in the acid content and acid

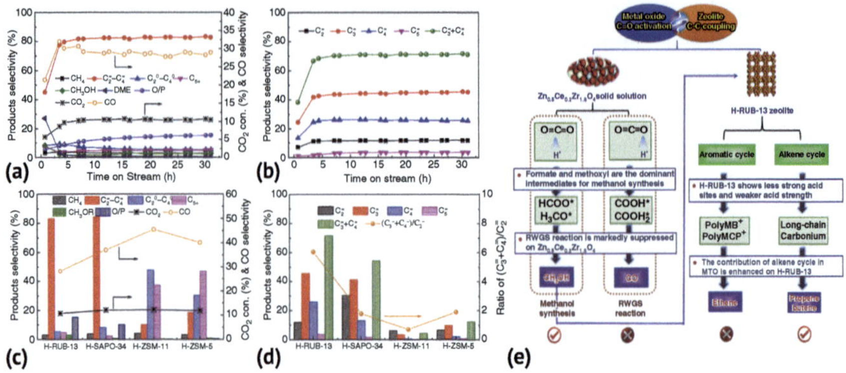

Figure 11.12 Catalytic performance of $ZnCeZrO_x$/H-RUB-13 sample for propene and butene formation in CO_2 hydrogenation: (a, b) catalytic stability of $ZnCeZrO_x$/H-RUB-13, (c, d) catalytic activity and production selectivity of various composite catalysts, and (e) schematic diagram for propene and butene formation on $ZnCeZrO_x$/H-RUB-13. *Source*: Reproduced from [178]/with permission of Elsevier.

strength of H-SSZ-13 zeolite seriously restricts the generation of aromatics at high H_2 pressure. This led to suppression of the aromatics cycle but enhancement of the alkene cycle, thereby producing more $C_3^=$ and $C_4^=$ than $C_2^=$, despite that these olefins are promptly transformed into alkanes on strongly acidic SSZ-13 zeolite. This work provides another piece of evidence that the distribution of hydrocarbon products could be controllably regulated through changing the structure and acidic properties of zeolite, which made it possible to synthesize a single hydrocarbon in CO_2 hydrogenation.

11.5 Conclusion and Outlook

Effective hydrogenation of CO_2 to high-value hydrocarbons represents a promising and feasible strategy to control the amount of CO_2 in the atmosphere and achieve carbon neutrality. Olefins, especially light olefins ($C_2^=$–$C_4^=$), are important platform molecules for the production of polymers, solvents, cosmetics, etc. Nowadays, FT synthesis and methanol-intermediate routes have been widely adopted for the preparation of light olefins from CO_2 hydrogenation. For CO_2-FT synthesis pathway, Fe-based catalysts show higher light olefins selectivity, while Co-based samples easily induce the formation of methane and heavy hydrocarbons (C_{5+}). Control of surface alkalinity, morphology, size, and exposed facets of these FT synthesis catalysts is regarded as an effective method to adjust the adsorption capacity of H_2 and CO_2, and inhibit the production of undesired methane. In addition, the formation of heavy hydrocarbons in FT synthesis can be suppressed to some extent by the introduction of additional acidic zeolites. Nevertheless, the uncontrollability of CH_x–CH_x coupling makes the hydrocarbon distribution generally follow the ASF model, which leads to $C_2^=$–$C_4^=$ selectivity no higher than 58%.

In the methanol-intermediate route, CO_2 is transformed into methanol on the surface defects of metal oxides via successive hydrogenation and dissociation reactions. The produced methanol rapidly moves onto acidic zeolites to form various hydrocarbon products via the hydrocarbon-pool mechanism. Various bifunctional catalysts composed of Zn-based, In-based, Ga-based, or Cu-based metal oxides and acidic zeolites have been fabricated, and they showed the $C_2^=–C_4^=$ selectivity higher than 60% in hydrocarbons. This makes the methanol-intermediate route superior to the FT synthesis route, because it effectively breaks the limitation of the ASF rule. Nevertheless, these reported bifunctional catalysts via the methanol-intermediate route always show low CO_2 conversion, but high CO by-product selectivity. This is because the tandem catalytic process was usually conducted at 350–400 °C to promote the rapid dehydration of methanol intermediates. High temperature inhibits methanol synthesis, but induces the formation of much more CO, because the RWGS is endothermic. Moreover, SAPO-34 zeolite is widely used as an acid catalytic moiety. Although it shows high $C_2^=–C_4^=$ selectivity, the distribution of ethene (25–35%), propene (35–40%), and butene (10–15%) is rather fixed, that makes preparation of target or even single olefin become challenging.

Regardless of the FT synthesis and methanol-intermediate routes, zeolites are essential for tailoring C—C bond and regulating the product selectivity. In the future, the following aspects of research may attract much more attention.

1. Evaluating the relationship between the acid sites distribution of zeolites and the products selectivity. Nowadays, many efforts have been devoted to unraveling the influence of pore structure, morphology, and acid content and strength of zeolites on CO_2 activation and alkenes formation. In addition, the change of the distribution of Al atoms in the zeolite framework also strongly determines the type and structure of reaction intermediates in CO/CO_2 hydrogenation. It was found that CO can be selectively converted into ethene, when more acid sites are sitting in 8-MR pore of H-MOR [150]. Moreover, an increase of acid sites located on the non-distorted 8-MR pores of H-RUB-13 promotes the formation of propene and butene, due to the enhancement of the alkene cycle [178]. The Al siting or acid distribution in the zeolite framework could be controllably adjusted through changing Si, Al, or P sources, using different template agents or introducing proper amount of alkali metal cations during synthesis [181–183]. Therefore, the design of new bifunctional catalysts through optimizing Al location in zeolite will be an effective method to regulate the alkene distribution in CO_2 hydrogenation.
2. Revealing the reaction mechanism over zeolites under real conditions. The adsorption and activation behavior of CO_2 and H_2 on Fe, Co and other metal/metal oxides have been widely investigated. In the FT synthesis route, CO_2 is usually transformed into CO via the RWGS process, and the generated CO is subsequently dissociated and hydrogenated to produce long-chain hydrocarbons via the coupling of CH_x intermediates, whereas for the methanol-intermediate route, CO_2 is initially transformed into methanol on surface defects of metal oxides via hydrogenation and dissociation reactions. The generated methanol

species are then converted into alkenes on acidic zeolites. Since CO_2 hydrogenation is preferentially conducted at high H_2 pressure, the type and structure of intermediates generated on zeolites are determined by detailed conditions. In contrast to the typical MTO reaction under an atmospheric N_2 atmosphere, few aromatic species were produced over strong acidic H-SSZ-13 zeolite in CO_2 hydrogenation, because of the low energy barrier of the hydrogenolysis reaction between aromatics and H_2 on the strong acid sites [180]. Xie and co-workers found that co-feeding CO can prevent the over-hydrogenation of light olefins, because the adsorption of alkenes and CO is highly competitive on zeolites [184]. Therefore, the reaction mechanism of methylation, oligomerization, cracking, hydrogenation, and hydrogenolysis on the acid sites of zeolites under real conditions (high pressure and $H_2/CO/CO_2$ coexistence) needs to be deeply explored.

3. Reducing the CO by-product selectivity. Although the methanol-intermediate route has greater potential in elevating the light olefins selectivity, the generation of large amounts of CO limits its practical application. Considering that the RWGS process is endothermic, decrease of the operating temperature could suppress CO formation [185]. This requires construction of highly active composite catalysts that have high activity for CO_2 hydrogenation at mild conditions. Moreover, adjustment of the surface hydrophilicity and hydrophobicity of zeolites may contribute to promoting conversion of generated methanol intermediates, which can relatively inhibit the RWGS reaction, as the methanol synthesis and the RWGS reaction are highly competitive in CO_2 hydrogenation.

Acknowledgments

The authors sincerely thanks for the financial supports of the National Key R&D Program of China (2023YFB4103102), the National Natural Science Foundation of China (21991090; 21991092; U22A20431; 22322208; 22272195), and the Fundamental Research Program of Shanxi Province (202203021224009).

Competing Interests

The authors declare no competing interests.

References

1 Global climate change, carbon dioxide. https://climate.nasa.gov/vital-signs/carbon-dioxide/ (accessed 1/12/2022).

2 Global climate change, global temperature. https://climate.nasa.gov/vital-signs/global-temperature/ (accessed 1/12/2022).

3 Olah, G.A., Goeppert, A., and Prakash, G.K.S. (2009). Chemical recycling of carbon dioxide to methanol and dimethyl ether: from greenhouse gas to renewable, environmentally carbon neutral fuels and synthetic hydrocarbons. *J. Org. Chem.* 74: 487–498.

4 Bushuyev, O.S., De Luna, P., Dinh, C.T. et al. (2018). What should we make with CO_2 and how can we make it? *Joule* 2: 825–832.

5 Centi, G., Quadrelli, E.A., and Perathoner, S. (2013). Catalysis for CO_2 conversion: a key technology for rapid introduction of renewable energy in the value chain of chemical industries. *Energy Environ. Sci.* 6: 1711–1731.

6 Peter, S.C. (2018). Reduction of CO_2 to chemicals and fuels: a solution to global warming and energy crisis. *ACS Energy Lett.* 3: 1557–1561.

7 Siefert, N.S. and Litster, S. (2013). Exergy and economic analyses of advanced IGCC-CCS and IGFC-CCS power plants. *Appl. Energy* 107: 315–328.

8 Samanta, A., Zhao, A., Shimizu, G.K.H. et al. (2012). Post-combustion CO_2 capture using solid sorbents: a review. *Ind. Eng. Chem. Res.* 51: 1438–1463.

9 Cuéllar-Franca, R.M. and Azapagic, A. (2015). Carbon capture, storage and utilization technologies: a critical analysis and comparison of their life cycle environmental impacts. *J. CO_2 Util.* 9: 82–102.

10 Lackner, K.S., Butt, D.P., and Wendt, C.H. (1997). Progress on binding CO_2 in mineral substrates. *Energy Convers. Manage.* 38: S259–S264.

11 Olah, G.A., Mathew, T., Goeppert, A. et al. (2018). Difference and significance of regenerative versus renewable carbon fuels and products. *Top. Catal.* 61: 522–529.

12 Bao, J., Yang, G.H., Yoneyama, Y. et al. (2019). Significant advances in C1 catalysis: highly efficient catalysts and catalytic reactions. *ACS Catal.* 9: 3026–3053.

13 Wang, W., Wang, S., Ma, X. et al. (2011). Recent advances in catalytic hydrogenation of carbon dioxide. *Chem. Soc. Rev.* 40: 3703–3727.

14 De, S., Dokania, A., Ramirez, A. et al. (2020). Advances in the design of heterogeneous catalysts and thermocatalytic processes for CO_2 utilization. *ACS Catal.* 10: 14147–14185.

15 Aziz, M.A.A., Jalil, A.A., Triwahyono, S. et al. (2015). CO_2 methanation over heterogeneous catalysts: recent progress and future prospects. *Green Chem.* 17: 2647–2663.

16 Graça, I., González, L.V., Bacariza, M.C. et al. (2014). CO_2 hydrogenation into CH_4 on NiHNaUSY zeolites. *Appl. Catal. B* 147: 101–110.

17 Shen, L., Xu, J., Zhu, M. et al. (2020). Essential role of the support for nickel-based CO_2 methanation catalysts. *ACS Catal.* 10: 14581–14591.

18 Wang, H., Xu, K., Yao, X. et al. (2018). Undercoordinated site-abundant and tensile-strained nickel for low temperature CO_x methanation. *ACS Catal.* 8: 1207–1211.

19 Álvarez, A., Bansode, A., Urakawa, A. et al. (2017). Challenges in the greener production of formates/formic acid, methanol, and DME by heterogeneously catalyzed CO_2 hydrogenation processes. *Chem. Rev.* 117: 9804–9838.

20 Studt, F., Sharafutdinov, I., Abild-Pedersen, F. et al. (2014). Discovery of a Ni-Ga catalyst for carbon dioxide reduction to methanol. *Nat. Chem.* 6: 320–324.

21 Gutterød, E.S., Lazzarini, A., Fjermestad, T. et al. (2020). Hydrogenation of CO_2 to methanol by Pt nanoparticles encapsulated in UiO-67: deciphering the role of the metal–organic framework. *J. Am. Chem. Soc.* 142: 999–1009.

22 Frei, M.S., Mondelli, C., Cesarini, A. et al. (2020). Role of zirconia in indium oxide-catalyzed CO_2 hydrogenation to methanol. *ACS Catal.* 10: 1133–1145.

23 An, B., Zhang, J., Cheng, K. et al. (2017). Confinement of ultrasmall Cu/ZnO_x nanoparticles in metal—organic frameworks for selective methanol synthesis from catalytic hydrogenation of CO_2. *J. Am. Chem. Soc.* 139: 3834—3840.

24 Zhang, X., Zhu, X., Lin, L. et al. (2017). Highly dispersed copper over β-Mo_2C as an efficient and stable catalyst for the reverse water gas shift (RWGS) reaction. *ACS Catal.* 7: 912—918.

25 Porosoff, M.D., Yang, X., Boscoboinik, J.A. et al. (2014). Molybdenum carbide as alternative catalysts to precious metals for highly selective reduction of CO_2 to CO. *Angew. Chem. Int. Ed.* 126: 6823—6827.

26 Zhou, W., Cheng, K., Kang, J.C. et al. (2019). New horizon in C1 chemistry: breaking the selectivity limitation in transformation of syngas and hydrogenation of CO_2 into hydrocarbon chemicals and fuels. *Chem. Soc. Rev.* 48: 3193–3228.

27 Zhao, B., Yan, B., Jiang, Z. et al. (2018). High selectivity of CO_2 hydrogenation to CO by controlling the valence state of nickel using perovskite. *Chem. Commun.* 54: 7354–7357.

28 Wei, J., Yao, R., Ge, Q. et al. (2018). Catalytic hydrogenation of CO_2 to isoparaffins over Fe based multifunctional catalysts. *ACS Catal.* 8: 9958—9967.

29 Wei, J., Ge, Q.J., Yao, R.W. et al. (2017). Directly converting CO_2 into a gasoline fuel. *Nat. Commun.* 8: 15174.

30 Gao, P., Li, S.G., Bu, X.N. et al. (2017). Direct conversion of CO_2 into liquid fuels with high selectivity over a bifunctional catalyst. *Nat. Chem.* 9: 1019–1024.

31 Ni, Y.M., Chen, Z.Y., Fu, Y. et al. (2018). Selective conversion of CO_2 and H_2 into aromatics. *Nat. Commun.* 9: 3457.

32 Wang, S.W., Wu, T.J., Lin, J. et al. (2019). FeK on 3D graphene—zeolite tandem catalyst with high efficiency and versatility in direct CO_2 conversion to aromatics. *ACS Sustain. Chem. Eng.* 7: 17825—17833.

33 Torres Galvis, H.M. and de Jong, K.P. (2013). Catalysts for production of lower olefins from synthesis gas: a review. *ACS Catal.* 3: 2130—2149.

34 Ye, R.P., Ding, J., Gong, W. et al. (2019). CO_2 hydrogenation to high-value products via heterogeneous catalysis. *Nat. Commun.* 10: 5698.

35 Liu, Y.T., Deng, D.H., and Bao, X.H. (2020). Catalysis for selected C1 chemistry. *Chem* 6: 2497–2514.

36 McClellan, P.P. (1950). Manufacture and uses of ethylene oxide and ethylene glycol. *Ind. Eng. Chem.* 42: 2402—2407.

37 Amberg, L.O. and Robinson, A.E. (1961). Ethylene-propylene rubber. *Ind. Eng. Chem.* 53: 368—370.

38 Xie, S.J., Ma, W.C., Wu, X.J. et al. (2021). Photocatalytic and electrocatalytic transformations of C1 molecules involving C–C coupling. *Energy Environ. Sci.* 14: 37—89.

39 Wu, J., Huang, Y., Ye, W. et al. (2017). CO_2 reduction: from the electrochemical to photochemical approach. *Adv. Sci.* 4: 1700194.
40 Prieto, G. (2017). Carbon dioxide hydrogenation into higher hydrocarbons and oxygenates: thermodynamic and kinetic bounds and progress with heterogeneous and homogeneous catalysis. *ChemSusChem* 10: 1056–1070.
41 Olah, G.A., Mathew, T., and Prakash, G.K.S. (2017). Chemical formation of methanol and hydrocarbon ("Organic") derivatives from CO_2 and H_2-carbon sources for subsequent biological cell evolution and life's origin. *J. Am. Chem. Soc.* 139: 566–570.
42 Zhang, Z., Xiao, J.P., Chen, X.J. et al. (2018). Reaction mechanisms of well-defined metal–N_4 sites in electrocatalytic CO_2 reduction. *Angew. Chem. Int. Ed.* 130: 16577–16580.
43 Jiang, K., Sandberg, R.B., Akey, A.J. et al. (2018). Metal ion cycling of Cu foil for selective C–C coupling in electrochemical CO_2 reduction. *Nat. Catal.* 1: 111–119.
44 Gao, D.F., Arán-Ais, R.M., Jeon, H.S. et al. (2019). Rational catalyst and electrolyte design for CO_2 electroreduction towards multicarbon products. *Nat. Catal.* 2: 198–210.
45 Jahangiri, H., Bennett, J., Mahjoubi, P. et al. (2014). A review of advanced catalyst development for Fischer—Tropsch synthesis of hydrocarbons from biomass derived syn-gas. *Catal. Sci. Technol.* 4: 2210—2229.
46 Zhang, Q., Kang, J., and Wang, Y. (2010). Development of novel catalysts for Fischer—Tropsch synthesis: tuning the product selectivity. *ChemCatChem* 2: 1030—1058.
47 Guo, L.S., Sun, J., Ge, Q.J. et al. (2018). Recent advances in direct catalytic hydrogenation of carbon dioxide to valuable C_{2+} hydrocarbons. *J. Mater. Chem. A* 6: 23244–23262.
48 Cheng, K., Kang, J., King, D.L. et al. (2017). Chapter 3: advances in catalysis for syngas conversion to hydrocarbons. *Adv. Catal.* 60: 125—208.
49 Wang, P., Chen, W., Chiang, F.K. et al. (2018). Synthesis of stable and low CO_2 selective ε-iron carbide Fischer–Tropsch catalysts. *Sci. Adv.* 4: eaau2947.
50 Ojeda, M., Nabar, R., Nilekar, A.U. et al. (2010). CO activation pathways and the mechanism of Fischer—Tropsch synthesis. *J. Catal.* 272: 287—297.
51 Cheng, Q.P., Tian, Y., Lyu, S.S. et al. (2018). Confined small-sized cobalt catalysts stimulate carbon-chain growth reversely by modifying ASF law of Fischer–Tropsch synthesis. *Nat Commun.* 9: 3250.
52 Lu, J., Yang, L., Xu, B. et al. (2014). Promotion effects of nitrogen doping into carbon nanotubes on supported iron Fischer–Tropsch catalysts for lower olefins. *ACS Catal.* 4: 613–621.
53 Torres Galvis, H.M., Koeken, A.C.J., Bitter, J.H. et al. (2013). Effects of sodium and sulfur on catalytic performance of supported iron catalysts for the Fischer–Tropsch synthesis of lower olefins. *J. Catal.* 303: 22—30.
54 Flory, P.J. (1936). Molecular size distribution in linear condensation polymers. *J. Am. Chem. Soc.* 58: 1877–1885.

55 Friedel, R.A. and Anderson, R.B. (1950). Composition of synthetic liquid fuels. I. Product distribution and analysis of C_5–C_8 paraffin isomers from cobalt catalyst. *J. Am. Chem. Soc.* 72: 1212–1215.

56 Torres Galvis, H.M., Bitter, J.H., Khare, C.B. et al. (2012). Supported iron nanoparticles as catalysts for sustainable production of lower olefins. *Science* 335: 835–838.

57 Zhong, L.S., Yu, F., An, Y.L. et al. (2016). Cobalt carbide nanoprisms for direct production of lower olefins from syngas. *Nature* 538: 84–87.

58 Sartipi, S., Parashar, K., Valero-Romero, M.J. et al. (2013). Hierarchical H-ZSM-5-supported cobalt for the direct synthesis of gasoline-range hydrocarbons from syngas: advantages, limitations, and mechanistic insight. *J. Catal.* 305: 179–190.

59 Bao, J., He, J., Zhang, Y. et al. (2008). A core/shell catalyst produces a spatially confined effect and shape selectivity in a consecutive reaction. *Angew. Chem. Int. Ed.* 47: 353–356.

60 Peng, X.B., Cheng, K., Kang, J.C. et al. (2015). Impact of hydrogenolysis on Fischer–Tropsch synthesis selectivity: diesel fuel production over mesoporous zeolite Y-supported cobalt nanoparticles. *Angew. Chem. Int. Ed.* 54: 4553–4556.

61 Sartipi, S., Alberts, M., Santos, V.P. et al. (2014). Insights into the catalytic performance of mesoporous H-ZSM-5-supported cobalt in Fischer–Tropsch synthesis. *ChemCatChem* 6: 142–151.

62 Jiao, F., Li, J.J., Pan, X.L. et al. (2016). Selective conversion of syngas to light olefins. *Science* 351: 1065–1068.

63 Cheng, K., Gu, B., Liu, X. et al. (2016). Direct and highly selective conversion of synthesis gas into lower olefins: design of a bifunctional catalyst combining methanol synthesis and carbon-carbon coupling. *Angew. Chem. Int. Ed.* 55: 4725–4728.

64 Cheng, K., Zhou, W., Kang, J.C. et al. (2017). Bifunctional catalysts for one-step conversion of syngas into aromatics with excellent selectivity and stability. *Chem* 3: 334–347.

65 Zhu, Y.F., Pan, X.L., Jiao, F. et al. (2017). Role of manganese oxide in syngas conversion to light olefins. *ACS Catal.* 7: 2800–2804.

66 Olsbye, U., Svelle, S., Bjørgen, M. et al. (2012). Conversion of methanol to hydrocarbons: how zeolite cavity and pore size controls product selectivity. *Angew. Chem. Int. Ed.* 51: 5810–5831.

67 Olsbye, U., Svelle, S., Lillerud, K.P. et al. (2015). The formation and degradation of active species during methanol conversion over protonated zeotype catalysts. *Chem. Soc. Rev.* 44: 7155–7176.

68 Ronda-Lloret, M., Rothenberg, G., and Raveendarn Shiju, N. (2019). A critical look at direct catalytic hydrogenation of carbon dioxide to olefins. *ChemSusChem* 12: 3896–3914.

69 Arslan, M.T., Qureshi, B.A., Ali Gilani, S.Z. et al. (2019). Single-step conversion of H_2-deficient syngas into high yield of tetramethylbenzene. *ACS Catal.* 9: 2203–2212.

70 Wang, Y., Tan, L., Tan, M.H. et al. (2019). Rationally designing bifunctional catalysts as an efficient strategy to boost CO_2 hydrogenation producing value-added aromatics. *ACS Catal.* 9: 895–901.

71 Corma, A. (1995). Inorganic solid acids and their use in acid-catalyzed hydrocarbon reactions. *Chem. Rev.* 95: 559–614.

72 Davis, M.E. (2002). Ordered porous materials for emerging applications. *Nature* 417: 813–821.

73 Dědeček, J., Sobalík, Z., and Wichterlová, B. (2012). Siting and distribution of framework aluminium atoms in silicon-rich zeolites and impact on catalysis. *Catal. Rev. Sci. Eng.* 54: 135–223.

74 Li, Y., Li, L., and Yu, J.H. (2017). Applications of zeolites in sustainable chemistry. *Chem* 3: 928–949.

75 Stöcker, M. (1999). Methanol to hydrocarbons: catalytic materials and their behavior. *Micropor. Mesopor. Mater.* 29: 3–48.

76 Gounder, R. and Iglesia, E. (2009). Catalytic consequences of spatial constraints and acid site location for monomolecular alkane activation on zeolites. *J. Am. Chem. Soc.* 131: 1958–1971.

77 Knott, B.C., Nimlos, C.T., Robichaud, D.J. et al. (2018). Consideration of the aluminum distribution in zeolites in theoretical and experimental catalysis research. *ACS Catal.* 8: 770–784.

78 Hemelsoet, K., Van der Mynsbrugge, J., De Wispelaere, K. et al. (2013). Unraveling the reaction mechanisms governing methanol to olefins catalysis by theory and experiment. *ChemPhysChem* 14: 1526–1545.

79 Yang, M., Fan, D., Wei, Y.X. et al. (2019). Recent progress in methanol to olefins (MTO) catalysts. *Adv. Mater.* 31: 1902181.

80 Wang, S., Qin, Z.F., Dong, M. et al. (2022). Recent progress on MTO reaction mechanisms and regulation of acid site distribution in the zeolite framework. *Chem Catal.* 2: 1657–1685.

81 Small, B.L. and Brookhart, M. (1998). Iron-based catalysts with exceptionally high activities and selectivities for oligomerization of ethylene to linear α-olefins. *J. Am. Chem. Soc.* 120: 7143–7144.

82 Lu, Y., Yan, Q., Han, J. et al. (2017). Fischer–Tropsch synthesis of olefin-rich liquid hydrocarbons from biomass derived syngas over carbon-encapsulated iron carbide/iron nanoparticles catalyst. *Fuel* 193: 369–384.

83 Zhang, C., Cao, C.X., Zhang, Y.L. et al. (2021). Unraveling the role of zinc on bimetallic Fe_5C_2-ZnO catalysts for highly selective carbon dioxide hydrogenation to high carbon α-olefins. *ACS Catal.* 11: 2121–2133.

84 Goud, D., Gupta, R., Maligal-Ganesh, R. et al. (2020). Review of catalyst design and mechanistic studies for the production of olefins from anthropogenic CO_2. *ACS Catal.* 10: 14258–14282.

85 Ma, Z.Q. and Porosoff, M.D. (2019). Development of tandem catalysts for CO_2 hydrogenation to olefins. *ACS Catal.* 9: 2639–2656.

86 Wei, J., Yao, R.W., Han, Y. et al. (2021). Towards the development of the emerging process of CO_2 heterogenous hydrogenation into high-value unsaturated heavy hydrocarbons. *Chem. Soc. Rev.* 50: 10764–10805.

87 Van Der Laan, G.P. and Beenackers, A.A.C.M. (1999). Kinetics and selectivity of the Fischer-Tropsch synthesis: a literature review. *Catal. Rev. Sci. Eng.* 41: 255–318.

88 Wei, J., Yao, R.W., Ge, Q.J. et al. (2021). Precisely regulating Brønsted acid sites to promote the synthesis of light aromatics via CO_2 hydrogenation. *Appl. Catal. B* 283: 119648.

89 Xu, K., Sun, B., Lin, J. et al. (2014). ε-Iron carbide as a low-temperature Fischer-Tropsch synthesis catalyst. *Nat. Commun.* 5: 5783–5790.

90 Liu, X., Liu, J., Yang, Y. et al. (2021). Theoretical perspectives on the modulation of carbon on transition-metal catalysts for conversion of carbon-containing resources. *ACS Catal.* 11: 2156–2181.

91 de Smit, E., Cinquini, F., Beale, A.M. et al. (2010). Stability and Reactivity of ϵ–χ–θ iron carbide catalyst phases in Fischer–Tropsch synthesis: controlling μ_c. *J. Am. Chem. Soc.* 132: 14928–14941.

92 Li, Z.L., Wu, W.L., Wang, M.L. et al. (2022). Ambient-pressure hydrogenation of CO_2 into long-chain olefins. *Nat. Commun.* 13: 2396.

93 Wu, W.L., Luo, J.H., Zhao, J.K. et al. (2024). Facet sensitivity of iron carbides in Fischer–Tropsch synthesis. *Nat. Commun.* 15: 6108.

94 Ramirez, A., Ould-Chikh, S., Gevers, L. et al. (2019). Tandem conversion of CO_2 to valuable hydrocarbons in highly concentrated potassium iron catalysts. *ChemCatChem* 11: 2879–2886.

95 Wang, X.X., Zeng, C.Y., Gong, N.N. et al. (2021). Effective suppression of CO selectivity for CO_2 hydrogenation to high-quality gasoline. *ACS Catal.* 11: 1528–1547.

96 Ramirez, A., Dutta Chowdhury, A., Dokania, A. et al. (2019). Effect of zeolite topology and reactor configuration on the direct conversion of CO_2 to light olefins and aromatics. *ACS Catal.* 9: 6320–6334.

97 Ramirez, A., Gong, X., Caglayan, M. et al. (2021). Selectivity descriptors for the direct hydrogenation of CO_2 to hydrocarbons during zeolite-mediated bifunctional catalysis. *Nat. Commun.* 12: 5914.

98 Dokania, A., Dutta Chowdhury, A., Ramirez, A. et al. (2020). Acidity modification of ZSM-5 for enhanced production of light olefins from CO_2. *J. Catal.* 381: 347–354.

99 Liu, R.J., Ma, Z.Q., Sears, J.D. et al. (2020). Identifying correlations in Fischer-Tropsch synthesis and CO_2 hydrogenation over Fe-based ZSM-5 catalysts. *J. CO_2 Util.* 41: 101290.

100 Ramirez, A., Dutta Chowdhury, A., Caglayan, M. et al. (2020). Coated sulfated zirconia/SAPO-34 for the direct conversion of CO_2 to light olefins. *Catal. Sci. Technol.* 10: 1507–1517.

101 Qiu, T., Wang, L., Lv, S. et al. (2017). SAPO-34 zeolite encapsulated Fe_3C nanoparticles as highly selective Fischer-Tropsch catalysts for the production of light olefins. *Fuel* 203: 811–816.

102 Ding, J., Liu, Q., Ye, R.P. et al. (2021). Metal–support interactions in Fe–Cu–K admixed with SAPO-34 catalysts for highly selective transformation of CO_2 and H_2 into lower olefins. *J. Mater. Chem. A* 9: 21877–21887.

103 Dorner, R.W., Hardy, D.R., Williams, F.W. et al. (2010). Willauer, heterogeneous catalytic CO_2 conversion to value-added hydrocarbons. *Energy Environ. Sci.* 3: 884–890.

104 Owen, R.E., Plucinski, P., Mattia, D. et al. (2016). Effect of support of Co-Na-Mo catalysts on the direct conversion of CO_2 to hydrocarbons. *J. CO_2 Util.* 16: 97–103.

105 Khodakov, A.Y., Chu, W., and Fongarland, P. (2007). Advances in the development of novel cobalt Fischer—Tropsch catalysts for synthesis of long-chain hydrocarbons and clean fuels. *Chem. Rev.* 107: 1692—1744.

106 Cheng, K., Zhang, L., Kang, J., et al. (2015). Selective transformation of syngas into gasoline-range hydrocarbons over mesoporous H-ZSM-5-supported cobalt nanoparticles. *Chem. Eur. J.* 21: 1928—1937.

107 den Breejen, J.P., Radstake, P.B., Bezemer, G.L. et al. (2009). On the origin of the cobalt particle size effects in Fischer—Tropsch catalysis. *J. Am. Chem. Soc.* 131: 7197—7203.

108 Khodakov, A.Y. (2009). Fischer—Tropsch synthesis: relation between structure of cobalt catalysts and their catalytic performance. *Catal. Today*. 144: 251—257.

109 Xie, J.X., Paalanen, P.P., van Deelen, T.W. et al. (2019). Promoted cobalt metal catalysts suitable for the production of lower olefins from natural gas. *Nat. Commun.* 10: 167.

110 Ismail, A.S.M., Casavola, M., Liu, B. et al. (2019). Atomic-scale investigation of the structural and electronic properties of cobalt—iron bimetallic Fischer—Tropsch catalysts. *ACS Catal.* 9: 7998—8011.

111 Sartipi, S., Alberts, M., Meijerink, M.J. et al. (2013). Towards liquid fuels from biosyngas: effect of zeolite structure in hierarchical-zeolite-supported cobalt catalysts. *ChemSusChem* 6: 1646–1650.

112 Li, J., He, Y.L., Tan, L., et al. (2018). Tsubaki, integrated tuneable synthesis of liquid fuels via Fischer–Tropsch technology. *Nat. Catal.* 1: 787–793.

113 Parastaev, A., Muravev, V., Osta, E.H. et al. (2022). Breaking structure sensitivity in CO_2 hydrogenation by tuning metal–oxide interfaces in supported cobalt nanoparticles. *Nat. Catal.* 5: 1051–1060.

114 Liu, Z.H., Gao, X.H., Liu, B. et al. (2022). Highly stable and selective layered Co-Al-O catalysts for low-temperature CO_2 methanation. *Appl. Catal. B* 310: 121303.

115 Liu, R.J., Leshchev, D., Stavitski, E. et al. (2021). Selective hydrogenation of CO_2 and CO over potassium promoted Co/ZSM-5. *Appl. Catal. B* 284: 19787.

116 Anicic, B., Trop, P., and Goricanec, D.J.E. (2014). Comparison between two methods of methanol production from carbon dioxide. *Energy* 77: 279–289.

117 Larmier, K., Liao, W.C., Tada, S. et al. (2017). CO_2-to-methanol hydrogenation on zirconia-supported copper nanoparticles: reaction intermediates and the role of the metal–support interface. *Angew. Chem. Int. Ed.* 56: 2318–2323.

118 Yarulina, I., De Wispelaere, K., Bailleul, S. et al. (2018). Structure–performance descriptors and the role of Lewis acidity in the methanol-to-propylene process. *Nat. Chem.* 10: 804–812.

119 Kattel, S., Ramírez, P. J., Chen, J. G. et al. (2017). Active sites for CO_2 hydrogenation to methanol on Cu/ZnO catalysts. *Science* 355: 1296–1299.

120 Zhang, L.X., Zhang, Y.C., and Chen, S.Y. (2012). Effect of promoter SiO_2, TiO_2 or SiO_2-TiO_2 on the performance of CuO-ZnO-Al_2O_3 catalyst for methanol synthesis from CO_2 hydrogenation. *Appl. Catal. A* 415–416: 118–123.
121 Li, T., Shoinkhorova, T., Gascon, J. et al. (2021). Aromatics production via methanol-mediated transformation routes. *ACS Catal.* 11: 7780–7819.
122 Yarulina, I., Chowdhury, A. D., Meirer, F. et al. (2018). Recent trends and fundamental insights in the methanol-to-hydrocarbons process. *Nat. Catal.* 1: 398–411.
123 Wang, J.J., Li, G.N., Li, Z.L. et al. (2017). A highly selective and stable ZnO-ZrO_2 solid solution catalyst for CO_2 hydrogenation to methanol. *Sci. Adv.* 3: e1701290.
124 Li, Z.L., Wang, J.J., Qu, Y.Z. et al. (2017). Highly selective conversion of carbon dioxide to lower olefins. *ACS Catal.* 7: 8544–8548.
125 Zhang, W.Y., Wang, S., Guo, S.J. et al. (2022). Effective conversion of CO_2 into light olefins over a bifunctional catalyst consisting of La-modified $ZnZrO_x$ oxide and acidic zeolite. *Catal. Sci. Technol.* 12: 2566–2577.
126 Liu, X.L., Wang, M.H., Zhou, C. et al. (2018). Selective transformation of carbon dioxide into lower olefins with a bifunctional catalyst composed of $ZnGa_2O_4$ and SAPO-34. *Chem. Commun.* 54: 140–143.
127 Liu, X.L., Wang, M.H., Yin, H.R. et al. (2020). Tandem catalysis for hydrogenation of CO and CO_2 to lower olefins with bifunctional catalysts composed of spinel oxide and SAPO-34. *ACS Catal.* 10: 8303–8314.
128 Mou, J., Fan, X.Q., Liu, F. et al. (2021). CO_2 hydrogenation to lower olefins over Mn_2O_3-ZnO/SAPO-34 tandem catalysts. *Chem. Eng. J.* 421: 129978.
129 Frei, M.S., Capdevila-Cortada, M., García-Muelas, R. et al. (2018). Mechanism and microkinetics of methanol synthesis via CO_2 hydrogenation on indium oxide. *J. Catal.* 361: 313–321.
130 Rui, N., Wang, Z., Sun, K. et al. (2017). CO_2 hydrogenation to methanol over Pd/In_2O_3: effects of Pd and oxygen vacancy. *Appl. Catal. B* 218: 488–497.
131 Frei, M.S., Mondelli, C., García-Muelas, R. et al. (2019). Atomic-scale engineering of indium oxide promotion by palladium for methanol production via CO_2 hydrogenation. *Nat. Commun.* 10: 3377.
132 Wang, J.Y., Liu, C.Y., Senftle, T.P. et al. (2020). Variation in the In_2O_3 crystal phase alters catalytic performance toward the reverse water gas shift reaction. *ACS Catal.* 10: 3264–3273.
133 Martin, O., Martín, A.J., Mondelli, C. et al. (2016). Indium oxide as a superior catalyst for methanol synthesis by CO_2 hydrogenation. *Angew. Chem. Int. Ed.* 55: 6261–6265.
134 Ye, J.Y., Liu, C.J., Mei, D.H. et al. (2013). Active oxygen vacancy site for methanol synthesis from CO_2 hydrogenation on In_2O_3(110): a DFT study. *ACS Catal.* 3: 1296–1306.
135 Dang, S.S., Qin, B., Yang, Y. et al. (2020). Rationally designed indium oxide catalysts for CO_2 hydrogenation to methanol with high activity and selectivity. *Sci. Adv.* 6: eaaz2060.

136 Numpilai, T., Wattanakit, C., Chareonpanich, M. et al. (2019). Optimization of synthesis condition for CO_2 hydrogenation to light olefins over In_2O_3 admixed with SAPO-34. *Energ. Convers. Manage.* 180: 511–523.

137 Gao, P., Dang, S.S., Li, S.G. et al. (2018). Direct production of lower olefins from CO_2 conversion via bifunctional catalysis. *ACS Catal.* 8: 571–578.

138 Wang, Y.H., Wang, G.Y., van der Wal, L.I. et al. (2021). Visualizing element migration over bifunctional metal-zeolite catalysts and its impact on catalysis. *Angew. Chem. Int. Ed.* 60: 17735–17743.

139 Zhang, W.Y., Wang, S., Guo, S.J. et al. (2022). Effective conversion of CO_2 into light olefins along with generation of low amounts of CO. *J. Catal.* 413: 923–933.

140 Dang, S.S., Gao, P., Liu, Z.Y. et al. (2018). Role of zirconium in direct CO_2 hydrogenation to lower olefins on oxide/zeolite bifunctional catalysts. *J. Catal.* 364: 382–393.

141 Wang, S., Wang, P.F., Qin, Z.F. et al. (2020). Enhancement of light olefin production in CO_2 hydrogenation over In_2O_3-based oxide and SAPO-34 composite. *J. Catal.* 391: 459–470.

142 Liu, C., Kang, J.C., Huang, Z.Q. et al. (2021). Gallium nitride catalyzed the direct hydrogenation of carbon dioxide to dimethyl ether as primary product. *Nat Commun.* 12: 2305.

143 Wang, J.J., Tang, C.Z., Li, G.N. et al. (2019). High-performance M_aZrO_x (Ma = Cd, Ga) solidsolution catalysts for CO_2 hydrogenation to methanol. *ACS Catal.* 9: 10253–10259.

144 Feng, W.H., Yu, M.M., Wang, L.J. et al. (2021). Insights into bimetallic oxide synergy during carbon dioxide hydrogenation to methanol and dimethyl ether over $GaZrO_x$ oxide catalysts. *ACS Catal.* 11: 4704–4711.

145 Zhang, P., Ma, L.X., Meng, F.H. et al. (2022). Boosting CO_2 hydrogenation performance for light olefin synthesis over $GaZrO_x$ combined with SAPO-34. *Appl. Catal. B* 305: 121042.

146 Prieto, G., Zečević, J., Friedrich, H. et al. (2013). Towards stable catalysts by controlling collective properties of supported metal nanoparticles. *Nat. Mater.* 12: 34–39.

147 Chen, J.Y., Wang, X., Wu, D.K. et al. (2019). Hydrogenation of CO_2 to light olefins on CuZnZr@(Zn-)SAPO-34 catalysts: strategy for product distribution. *Fuel* 239: 44–52.

148 Sedighi, M. and Mohammadi, M. (2020). CO_2 hydrogenation to light olefins over Cu-CeO_2/SAPO-34 catalysts: product distribution and optimization. *J. CO_2 Util.* 35: 236–244.

149 Wang, C.M., Wang, Y.D., and Xie, Z.K. (2016). Methylation of olefins with ketene in zeotypes and its implications for the direct conversion of syngas to light olefins: a periodic DFT study. *Catal. Sci. Technol.* 6: 6644–6649.

150 Jiao, F., Pan, X.L., Gong, K. et al. (2018). Shape-selective zeolites promote ethylene formation from syngas via a ketene intermediate. *Angew. Chem. Int. Ed.* 57: 4692–4696.

151 Bjørgen, M., Svelle, S., Joensen, F. et al. (2007). Conversion of methanol to hydrocarbons over zeolite H-ZSM-5: on the origin of the olefinic species. *J. Catal.* 249: 195–207.

152 Wang, S., Li, Z.K., Qin, Z.F. et al. (2021). Catalytic roles of the acid sites in different pore channels of H-ZSM-5 zeolite for methanol-to-olefins conversion. *Chin. J. Catal.* 42: 1126–1136.

153 Wang, S., Wang, P.F., Qin, Z.F. et al. (2018). Relation of catalytic performance to the aluminum siting of acidic zeolites in the conversion of methanol to olefins, viewed via a comparison between ZSM-5 and ZSM-11. *ACS Catal.* 8: 5485–5505.

154 Westgård Erichsen, M., Svelle, S., and Olsbye, U. (2013). H-SAPO-5 as methanol-to-olefins (MTO) model catalyst: towards elucidating the effects of acid strength. *J. Catal.* 298: 94–101.

155 Westgård Erichsen, M., Svelle, S., and Olsbye, U. (2013). The influence of catalyst acid strength on the methanol to hydrocarbons (MTH) reaction. *Catal. Today.* 215: 216–223.

156 Wang, M.H., Kang, J.C., Xiong, X.W. et al. (2021). Effect of zeolite topology on the hydrocarbon distribution over bifunctional ZnAlO/SAPO catalysts in syngas conversion. *Catal. Today.* 371: 85–92.

157 Xu, Y., Zhai, P., Deng, Y.C. et al. (2020). Highly selective olefin production from CO_2 hydrogenation on iron catalysts: a subtle synergy between manganese and sodium additives. *Angew. Chem. Int. Ed.* 59: 21736–21744.

158 Wang, X., Wu, D.K., Zhang, J.L. et al. (2019). Highly selective conversion of CO_2 to light olefins via Fischer-Tropsch synthesis over stable layered K–Fe–Ti catalysts. *Appl. Catal. A* 573: 32–40.

159 Svelle, S., Olsbye, U., Joensen, F. et al. (2007). Conversion of methanol to alkenes over medium- and large-pore acidic zeolites: steric manipulation of the reaction intermediates governs the ethene/propene product selectivity. *J. Phys. Chem. C* 111: 17981–17984.

160 Ilias, S. and Bhan, A. (2013). Mechanism of the catalytic conversion of methanol to hydrocarbons. *ACS Catal.* 3: 18–31.

161 Chen, B.H., Lu, J.Z., Wu, L.P. et al. (2016). Dehydration of bio-ethanol to ethylene over iron exchanged HZSM-5. *Chin. J. Catal.* 37: 1941–1948.

162 Liu, G.B., Yang, G.H., Peng, X.B. et al. (2022). Recent advances in the routes and catalysts for ethanol synthesis from syngas. *Chem. Soc. Rev.* 51: 5606—5659.

163 Xu, D., Ding, M.Y., Hong, X.L. et al. (2020). Mechanistic aspects of the role of K promotion on Cu—Fe-based catalysts for higher alcohol synthesis from CO_2 hydrogenation. *ACS Catal.* 10: 14516—14526.

164 Pahija, E., Panaritis, C., Gusarov, S. et al. (2022). Experimental and computational synergistic design of Cu and Fe catalysts for the reverse water-gas shift: a review. *ACS Catal.* 12: 6887—6905.

165 Yao, R.W., Wei, J., Ge, Q.J. et al. (2021). Monometallic iron catalysts with synergistic Na and S for higher alcohols synthesis via CO_2 hydrogenation. *Appl. Catal. B* 298: 120556.

166 Wang, Y., Wang, K.Z., Zhang, B.Z. et al. (2021). Direct conversion of CO_2 to ethanol boosted by intimacy-sensitive multifunctional catalysts. *ACS Catal.* 11: 11742–11753.

167 Li, Y.W., Gao, W., Peng, M. et al. (2020). Interfacial Fe_5C_2-Cu catalysts toward low-pressure syngas conversion to long-chain alcohols. *Nat. Commun.* 11: 61.

168 Xu, D., Wang, Y.Q., Ding, M.Y. et al. (2021). Advances in higher alcohol synthesis from CO_2 hydrogenation. *Chem* 7: 849–881.

169 Wang, S., Zhang, L., Wang, P.E. et al. (2022). Highly effective conversion of CO_2 into light olefins abundant in ethene. *Chem* 8: 1376–1394.

170 Galadima, A. and Muraza, O. (2015). Recent developments on silicoaluminates and silicoaluminophosphates in the methanol-topropylene reaction: a mini review. *Ind. Eng. Chem. Res.* 54: 4891–4905.

171 Wu, X.Q., Chen, W., Xu, S.T. et al. (2021). Dynamic activation of C1 molecules evoked by zeolite catalysis. *ACS Cent. Sci.* 7: 681–687.

172 Chowdhury, A.D., Paioni, A.L., Houben, K. et al. (2018). Bridging the gap between the direct and hydrocarbon pool mechanisms of the methanol-to-hydrocarbons process. *Angew. Chem. Int. Ed.* 57: 8095–8099.

173 Liu, Y., Kirchberger, F. M., Müller, S. et al. (2019). Critical role of formaldehyde during methanol conversion to hydrocarbons. *Nat. Commun.* 10: 1462.

174 Wang, J.Y., Zhang, A.F., Jiang, X. et al. (2018). Highly selective conversion of CO_2 to lower hydrocarbons (C_2-C_4) over bifunctional catalysts composed of In_2O_3-ZrO_2 and zeolite. *J. CO_2 Util.* 27: 81–88.

175 Ticali, P., Salusso, D., Ahmad, R. et al. (2021). CO_2 hydrogenation to methanol and hydrocarbons over bifunctional Zn-doped ZrO_2/zeolite catalysts. *Catal. Sci. Technol.* 11: 1249–1268.

176 Li, W., Wang, K.C., Zhan, G.W. et al. (2021). Design and synthesis of bioinspired ZnZrOx&Bio-ZSM-5 integrated nanocatalysts to boost CO_2 hydrogenation to light olefins. *ACS Sustain. Chem. Eng.* 9: 6446–6458.

177 Zhang, W.Y., Wang, S., Guo, S.J. et al. (2022). GamCrOx/H-SAPO-34(F), a highly efficient bifunctional catalyst for the direct conversion of CO_2 into ethene and propene. *Fuel* 329: 125475.

178 Wang, S., Zhang, L., Zhang, W. et al. (2020). Selective conversion of CO_2 into propene and butene. *Chem* 6: 3344–3363.

179 Guo, S.J., Wang, S., Zhang, W.Y. et al. (2022). Catalytic performance of various zinc-based binary metal oxides/H-RUB-13 for hydrogenation of CO_2. *Ind. Eng. Chem. Res.* 61: 10409–10418.

180 Wang, S., Zhang, L., Wang, P.F. et al. (2022). Highly selective hydrogenation of CO_2 to propane over GaZrOx/H-SSZ-13 composite. *Nat. Catal.* 5: 1038–1050.

181 Liang, T.Y., Chen, J.L., Qin, Z.F. et al. (2016). Conversion of methanol to olefins over H-ZSM-5 zeolite: reaction pathway is related to the framework aluminum siting. *ACS Catal.* 6: 7311–7325.

182 Dedecek, J., Balgová, V., Pashkova, V. et al. (2012). Synthesis of ZSM-5 zeolites with defined distribution of Al atoms in the framework and multinuclear MAS NMR analysis of the control of Al distribution. *Chem. Mater.* 24: 3231–3239.

183 Wang, S., He, Y., Jiao, W.Y. et al. (2019). Recent experimental and theoretical studies on Al siting/acid site distribution in zeolite framework. *Curr. Opin. Chem. Eng.* 23: 146–154.

184 Xie, J.X., Firth, D.S., Cordero-Lanzac, T. et al. (2022). MAPO-18 catalysts for the methanol to olefins process: influence of catalyst acidity in a highpressure syngas (CO and H_2) environment. *ACS Catal.* 12: 1520–1531.

185 Yang, H., Zhang, C., Gao, P. et al. (2017). A review of the catalytic hydrogenation of carbon dioxide into value-added hydrocarbons. *Catal. Sci. Technol.* 7: 4580–4598.

12

Application of Zeolites in CO_2 to Oxygenates

Limin Zhang[1], Miao Zhang[1], Wenhui Li[1], Guanghui Zhang[1,], Chunshan Song[1,2,*], Xinwen Guo[1,*]*

[1]*State Key Laboratory of Fine Chemicals, Frontier Science Center for Smart Materials, PSU-DUT Joint Center for Energy Research, School of Chemical Engineering, Dalian University of Technology, Dalian, Liaoning, China*
[2]*Department of Chemistry, Faculty of Science, The Chinese University of Hong Kong, Hong Kong, China*

*Corresponding authors: gzhang@dlut.edu.cn; chunshansong@cuhk.edu.hk; guoxw@dlut.edu.cn

12.1 Introduction

The global development and utilization of large amounts of fossil fuels have led to a significant increase in the atmospheric concentration of CO_2, causing climate problems. From the Keeling Curve, as of August 2024, CO_2 concentration in the atmosphere exceeded 421 ppm, about 53% higher than that before the industrial revolution [1]. The first global stocktake proposed in the Conference of the Parties (COP28) as the meeting of the Parties for the Paris Agreement in December 2023 made the appeals of accelerating efforts globally toward net-zero emission energy systems and utilizing zero- and low-carbon fuels around mid-century [2]. Storing clean energy in functional platform molecules by CO_2 conversion technology is an important strategy for achieving CO_2 reduction and sustainable utilization of carbon resources. Therefore, thermocatalytic CO_2 conversion with "green H_2" obtained with renewable energy to organic oxygenates, including methanol, ethanol, dimethyl ether (DME), and formic acid as platform molecules for energy storage, is crucial for the transformation of energy structure. These oxygenates can be used not only as important chemical raw materials, but also as clean fuel for households, cars, ships, etc. They emit no dust, sulfur, or nitrogen oxides, and have fewer carbon emissions compared with traditional fuels [3, 4]. Many countries and regions around the world, including the EU, China, and the United States, are speeding up the layout of green fuels markets, such as methanol, ethanol, and DME, reducing reliance on fossil fuels. Since 2023, the growth rate of methanol-powered ship orders has exceeded that of LNG-powered ships. Methanol has become an important option for the future fuel of ships. Some countries, such as Brazil and the United States, have implemented a strategy of using bioethanol as an alternative energy source, making ethanol fully competitive with gasoline [5]. However, the industrial synthesis of oxygenates still relies on unsustainable fossil fuels or biomass as raw materials, which is in contrast to the original intention of sustainable development. Therefore, it is imperative to vigorously develop CO_2 conversion technology to produce oxygenates to promote carbon reduction and energy transition. Below, we will provide an overview of the role of zeolite in four conversion reactions of CO_2 hydrogenation to methanol, ethanol, DME, and formic acid.

12.2 CO_2 Hydrogenation to Methanol

Methanol is an important platform molecule that can be used to produce value-added chemicals or fuels through chemical processes such as methanol-to-olefins (MTO), methanol-to-aromatics (MTA), etc. Methanol also has the advantages of high energy density and easy storage and transportation, making it an excellent clean fuel (see Figure 12.1a) [6, 7]. The market demand for methanol-derived fuels has significantly increased from 6% in 2011 to 22% in 2016 [8]. With the depletion of fossil energy, it is highly meaningful to use methanol as a substitute for fossil fuels and as a raw material for various chemicals. As a part of CCUS (CO_2 capture, utilization and storage), thermocatalytic CO_2 hydrogenation with renewable H_2 to methanol is an important pathway for the effective circulation of global carbon resources and the transformation of energy strategies. "Methanol economy" proposed by Olah points to methanol as an indispensable chemical raw material for the future society, and CO_2 hydrogenation is one of the most feasible ways to produce renewable methanol [9].

$$CO_2 + 3H_2 \Longleftrightarrow CH_3OH + H_2O\ \Delta H_{298K} = -49.5\ \text{kJ}\cdot\text{mol}^{-1} \tag{12.1}$$

$$CO_2 + H_2 \Longleftrightarrow CO + H_2O\ \Delta H_{298K} = +41.1\ \text{kJ}\cdot\text{mol}^{-1} \tag{12.2}$$

$$CO + 2H_2 \Longleftrightarrow CH_3OH\ \Delta H_{298K} = -90.4\ \text{kJ}\cdot\text{mol}^{-1} \tag{12.3}$$

CO_2 hydrogenation to methanol is shown in Eq. (12.1). The main side reaction is the reverse water-gas shift (RWGS), as shown in Eq. (12.2). The by-product CO from RWGS can be further hydrogenated to produce methanol, as shown in Eq. (12.3). Thermodynamic analysis on CO_2/CO hydrogenation to methanol is shown (see Figure 12.1b) [10, 11]. CO_2/CO hydrogenation to methanol is an exothermic reaction, while relatively low temperatures, high pressures, and high H_2/CO_2 ratios favor the forward reaction. For understanding the reaction mechanism of CO_2 hydrogenation to methanol, researchers mainly carry out FT-IR, isotopic labeling, density functional theory (DFT) calculations, etc. to clarify the intermediates in the reaction process. However, CO_2 hydrogenation to methanol is a structure-sensitive reaction, the catalytic performance is closely related to the particle size, surface structure, interface composition, and electronic properties of the active phase, which also affect the types of the reaction intermediates [22]. Therefore, researchers

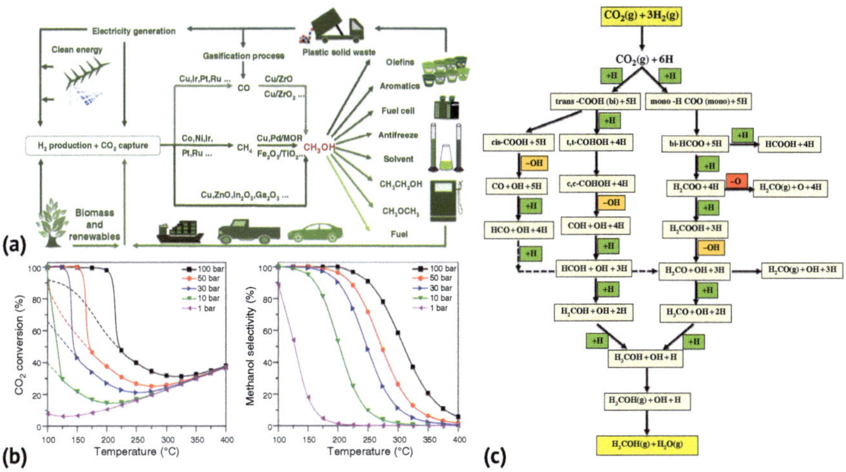

Figure 12.1 Fundamental properties for methanol synthesis from CO_2 hydrogenation. (a) The simplified carbon cycle with methanol as a platform molecule in the future. *Source*: Reproduced from [8] / Frontiers Media SA / CC BY 4.0. (b) CO_2 conversion and methanol selectivity at phase and chemical equilibrium. Dashed lines in panel b represent the chemical equilibrium predicted by gas-phase thermodynamics. *Source*: Reproduced from [11] / with permission of American Chemical Society. (c) Reaction mechanism network of methanol synthesis on Cu(111). *Source*: Reproduced from [17] / with permission of Elsevier.

have observed different intermediates in different catalytic reactions and proposed different methanol formation pathways. Currently, the widely known mechanisms include: (1) the HCOO* mechanism, (2) the revised HCOO* mechanism, (3) the CO* mechanism, and (4) the trans-COOH mechanism (see Figure 12.1c) [10, 17]. For instance, on the ZnO-ZrO_2 solid solution catalyst [18], Wang et al. inferred that CO_2 hydrogenation to methanol follows the HCOO* pathway. H_2 is adsorbed and dissociated on the Zn site. CO_2 is adsorbed on the coordinatively unsaturated Zr site. The adsorbed CO_2^* undergoes hydrogenation to generate stable HCOO* intermediate species. HCOO* undergoes hydrogenation to form H_2COO^*, H_2CO^*, H_3CO^*, and ultimately methanol. It is worth noting that even for the same reaction pathway, there are still significant differences in the rate-determining step, intermediate adsorption states, etc., with different catalytic systems. At present, the catalysts for CO_2 hydrogenation to methanol mainly include: (1) Cu-based catalysts, (2) noble metal catalysts (Pd, Pt, etc.), (3) metal oxide catalysts (ZnO, In_2O_3, etc.), and (4) other catalysts (MoS_2, Ni-Ga intermetallic compounds, etc.).

The catalysts for CO_2 hydrogenation to methanol should have high activity, selectivity, and stability for industrial applications. Despite significant efforts and progress made by researchers in improving catalytic performance for CO_2 hydrogenation to methanol, the low methanol yields due to thermodynamic limitations and the deactivation caused by harsh reaction conditions limit further development. From Eqs. (12.1) and (12.2), it can be seen that H_2O is inevitably generated along with methanol and has a relatively high content. Therefore, according to the Le Chatelier's principle, in situ removal of H_2O during the reaction process can favor the forward reaction and suppress the reverse reaction of methanol steam reforming, which provides an opportunity to break through the thermodynamic limitations of methanol yield. Zachopoulos et al. [19] conducted a detailed thermodynamic analysis and process simulation of CO_2 hydrogenation to methanol with enhanced adsorption. Compared with conventional CO_2 hydrogenation to methanol, the CH_3OH yield in the water-sorption-enhanced (SE) process can be increased by up to 130% in the 100–350 °C range. Due to a greater impact on the equilibrium of RWGS than that of CO_2 hydrogenation to methanol, in-situ water sorption reduces the selectivity of methanol and promotes the generation of CO. But the yield of methanol can be greatly improved in the presence of adsorbents, with an increased per-pass CO_2 conversion from 27 to ~100% and a corresponding increase of the CH_3OH yield from 20 to ~58%. Meanwhile, the in-situ removal of water will greatly simplify the downstream separation of liquid products and may improve energy efficiency in practical applications.

On the other hand, H_2O can also be the initiator of catalyst deactivation. Frei et al. [20] observed that water by-product led to In_2O_3 catalyst deactivation due to sintering when concentrations were four times higher than the amounts produced by the reaction under standard reaction conditions. In addition, Prašnikar et al. [21] observed that the Al_2O_3 surface area decreases with increasing molar fraction of H_2O while the Cu particle size increases on the industrial catalyst $Cu/Zn/Al_2O_3$ (see Figure 12.2). The ZnO phase can cover Cu particles with the induction of the reaction atmosphere. The sintering of Cu particles or the destruction of the Cu/ZnO interface will ultimately lead to the deactivation of the catalysts. It is worth noting that not only H_2O, but also the reaction conditions of high temperature and pressure can cause catalyst deactivation. Therefore, in-situ removal of water and stabilizing the active metal species are important directions for researchers to address challenges of improving the activity and stability of methanol synthesis from CO_2 hydrogenation.

Zeolite is a silicon-based porous material with unique pore properties, and some zeolites with specific pore sizes have excellent characteristics in separating water. And these can also be prepared into the zeolite membrane, which combines with

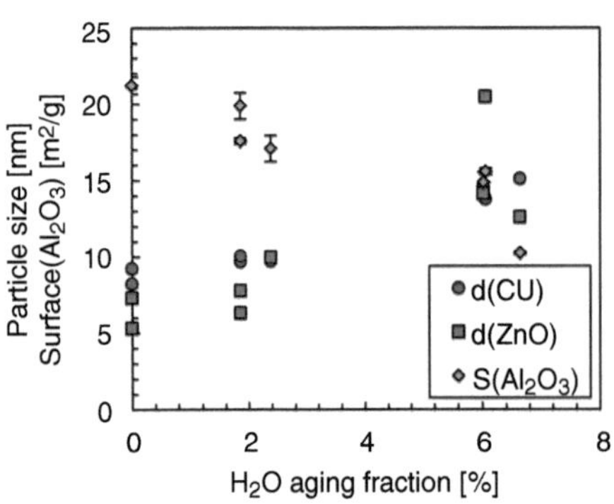

Figure 12.2 H_2O impact on the Cu and ZnO particle size and on the Al_2O_3 surface. *Source*: Reproduced from [21] / American Chemical Society / CC BY 4.0.

the catalysts for CO_2 hydrogenation to methanol to form the membrane reactor [23–25]. The catalytic membrane reactor (CMR) integrates catalytic reaction with separation process into one unit, which is more convenient and energy-efficient for process intensification since the by-products can be continuously removed from the reaction. It is difficult to obtain an ideal membrane to effectively and stably separate H_2O from other small gas molecules under the CO_2 hydrogenation conditions of high temperature and pressure. Therefore, the zeolite membranes with good thermal and mechanical stability are suitable for the harsh reaction conditions of CO_2 hydrogenation to methanol. Furthermore, porous supports can disperse and anchor metal species due to their large surface area and the metal-support interaction to stabilize the active center and inhibit sintering, ensuring the stability of the catalyst. Metal nanoparticles can be further confined within the pores of porous materials, especially zeolites, realizing high metal dispersion and small particle size [26]. Yang et al. [27] prepared core-shell catalysts Cu/ZnO@m-SiO_2, in which Cu/ZnO nanoparticles are encapsulated within mesoporous silica shells. The metal particle size remained below 5 nm at a metal loading of 17%. The catalytic performance showed no obvious decline within 160 hours, while the metal particle did not undergo sintering due to the isolation and confinement of the silica shell. These properties of zeolites mentioned above enable them to be particularly favored by researchers. Below, we will discuss the impact of zeolites on H_2O removal and the dispersion and stability of the active center.

12.2.1 The Function of Separating Water

First, zeolites can serve as the adsorbent for water to enhance the synthesis of methanol from CO_2 hydrogenation (e.g., Zeolite A, 13X, etc.) with regular pores that have good adsorption capacity for H_2O, even under high temperature conditions, and good thermal and regeneration ability. So, they are widely used as an adsorbent for water or made into a zeolite membrane to promote methanol generation.

Among the three types of A-type zeolites, zeolite 3A exhibited the best performance due to lower Henry's adsorption constants and higher reciprocal diffusion time constants of water and methanol, which facilitates the regeneration of the adsorbent. Under the optimized process conditions, the performance of CO_2 hydrogenation to methanol has been greatly improved, achieving a CO_2 conversion of 99.6% in one step, a methanol selectivity of 83–96%, and productivities of 0.023–0.13 mol $m^3 \cdot s^{-1}$ [28]. Maksimov et al. [29] studied the promotion effect of zeolite 3A SE CO_2 hydrogenation to methanol with selective in situ adsorption of water through experiments and process simulations. Methanol outlet concentration during the SE stage is ca. 150–290% of the steady-state values recorded after the complete adsorbent saturation, while 220–510% for CO. The increase in reactor pressure has the most significant impact on improving the methanol production rate. Especially over 4.0 MPa, the improvement of methanol formation and the reduction of CO production were achieved (see Figure 12.3c,d).

Although the SE CO_2 hydrogenation process has the advantages of breaking the thermodynamic limits and significantly improving methanol yield, its disadvantages

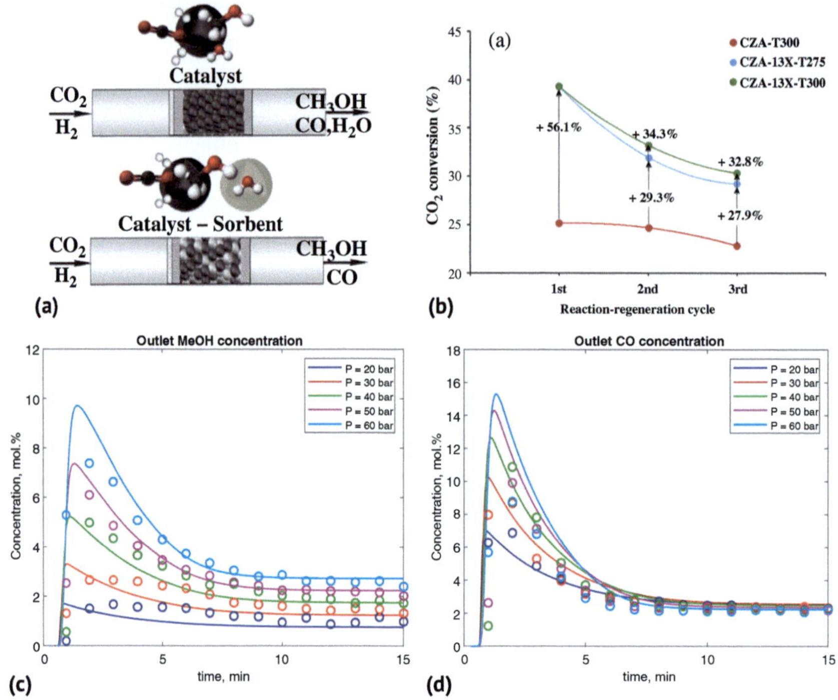

Figure 12.3 SE process for the methanol synthesis from CO_2 hydrogenation. (a) Schematics of the SE process. *Source:* [29] / with permission of Elsevier / CC BY-NC-ND 4.0. (b) Effect of consecutive reaction-regeneration cycles on integrated CO_2 conversion. *Source*: Reproduced from [30] / with permission of American Chemical Society. Outlet concentrations during the SE process with varied reactor pressure. (c) CH_3OH concentration. (d) CO concentration. *Source*: Reproduced from [29] / with permission of Elsevier.

of poor continuity and regeneration cannot be ignored. Heracleous et al. [30] studied the use of zeolite 13X experimentally as an efficient water sorbent in the SE CO_2 hydrogenation over $Cu/ZnO/Al_2O_3$ under multiple reaction-regeneration cycles. The greatest improvement of CO_2 hydrogenation to methanol by an adsorption-enhanced process occurred at 225 °C, with the methanol yield increasing by 115% compared with the conventional process. But, as the reaction progresses, the sorbent zeolite 13X gradually becomes saturated, the enhancement effect weakens, and the system eventually reaches the same steady state as using only the $Cu/ZnO/Al_2O_3$ catalyst (see Figure 12.3a,b). The enhancement period decreased from 120 minutes at 225°C to 90 minutes at 275°C. Furthermore, the adsorption enhancement effect significantly decreases as the cycle progresses, which is due to the blockage of pores by incompletely desorbed adsorbates, resulting in a significant reduction in the surface area of the adsorbent. In addition, the crystallinity of the adsorbent decreased after undergoing regeneration cycles, which can also lead to a decrease in adsorption performance.

The CMR can achieve continuous and effective separation of by-products without regeneration cycles compared with the adsorption-enhanced process, which greatly expands its applications. Although some zeolites have small pores and exhibit excellent small molecule adsorption and separation performance, larger zeolite crystal pores can significantly reduce performance under high-temperature and high-pressure conditions of CO_2 hydrogenation to methanol. Harsh reaction conditions can also accelerate the dewatering of zeolites, leading to the contraction of zeolite unit cells and the expansion of crystal pores. These challenges limit the application of zeolite membranes in methanol synthesis from CO_2 hydrogenation, and the key to the breakthrough is to prepare a defect-free zeolite membrane [31, 32].

Li et al. [25] designed a defect-free membrane assembled with NaA zeolitic nanochannels, with a thickness of 3–4 mm, using a ceramic hollow fiber as support, achieving a breakthrough in the application of zeolite membrane in CO_2 hydrogenation to methanol. In the simulation experiment of separating mixed gases under reaction conditions, NaA zeolite membrane exhibited excellent H_2O/gas separation performance of two to three orders of magnitude lower gas permeance but comparable water permeance compared with previously reported separation results. Even under higher pressure conditions, it still possessed good performance and stability. From theoretical simulations, excellent separation performance originates from a gating effect of Na^+ located in the 8-oxygen ring apertures that regulated their effective size. In the reaction of CO_2 hydrogenation to methanol, the CO_2 conversion with a membrane reactor is increased by 2.6–3.0 times compared with a traditional reactor, while the selectivity of methanol did not change significantly, overcoming the thermodynamic equilibrium limitation. Correspondingly, the space-time yield (STY) of methanol was greatly enhanced from 339 $mg \cdot g_{cat}^{-1} \cdot h^{-1}$ to 809 $mg \cdot g_{cat}^{-1} \cdot h^{-1}$ with GHSV of 10 500 $ml \cdot g_{cat}^{-1} \cdot h^{-1}$. Yue et al. [24] prepared a hydrophilic zeolite LTA membrane on the porous α-Al_2O_3 and combined it with Cu-ZnO-Al_2O_3-ZrO_2 catalytic layer to form a bifunctional zeolite CMR. Thanks to the in-situ removal of byproduct H_2O vapor, the CO_2 conversion was greatly improved, and catalyst deactivation was significantly suppressed, showing excellent performance of 36.1% CO_2 conversion, 100% CH_3OH selectivity, and over 80 hours of stability. Similarly, Deng et al. [23] prepared a defect-free NaA zeolite membrane on the alumina hollow fiber support without any surface modification by a template-free method. This zeolite membrane possessed a water flux up to 4.8×10^{-7} $mol \cdot m^{-2} \cdot s^{-1} \cdot Pa^{-1}$ and a superior separation factor for H_2O/H_2 (>10000), H_2O/CO_2 (>10000), and H_2O/CH_3OH (>300) under high temperature and high pressure ($T \geq 200\ °C$ and $p \geq 2.0$ MPa), showing its potential for CO_2 hydrogenation to methanol.

12.2.2 The Function of Confining and Stabilizing Metal Species

Uniform pore structure and large specific surface area are characteristics of zeolites as excellent supports. A simple loading method could achieve high dispersion of metal species. Xiang et al. [33] prepared hierarchical-structure SAPO-56 with Cu loading for CO_2 hydrogenation. Highly dispersed Cu nanoparticles in the pore channel and the specific Lewis acid sites derived from the extra-framework Al species located in the six-membered rings underwent H_2 adsorption (dissociation) and CO_2 adsorption, respectively, achieving 16.4% of CO_2 conversion, 80.9% of methanol selectivity, and over 100 hours of stability. There are some unique sites, such as silanol groups, which can anchor and stabilize metal species. Ma et al. [34] synthesized S-1-supported CuZn catalyst 10CuO-2ZnO/S-1 via the wet impregnation method (see Figure 12.4a). The metal particle size is as small as 2 nm. Due to the anchoring effect of silanol nests on S-1, only Cu^+ species exist on the surfaces. There is a synergistic effect between Cu^+ species and ZnO, with ZnO responsible for H_2 dissociation and Cu^+ species at the ZnO-Cu_2O interface responsible for CO_2 adsorption. No deactivation was observed and Cu particles were still well dispersed without sintering after running for 100 hours. The adsorption of metal cations by the zeolite framework with negative charge can achieve the dispersion and stability of metal species. Li et al. [35] prepared a 13X encapsulated Cu catalyst via ion exchange and hydrogen reduction at a tiny positive pressure. The catalyst exhibited high methanol selectivity of $\geq$95% and stability of 120 hours. And the introduction of Zn further improved stability to 180 hours.

Encapsulating active metal species inside zeolite pore channels is an effective strategy to enhance the dispersion and stability of metal nanoparticles [38, 39]. The in-situ growth method or ligand-protected in-situ hydrothermal route is widely used by researchers to prepare the encapsulated catalysts. During the synthesis process, metal species are easily embedded into the zeolite frameworks, which can achieve greater dispersion of metal species, enhance interaction with the zeolite, and reduce particle sintering in high-temperature reactions. Those metal species that interact with zeolite frameworks can further improve catalytic activity. Xiang et al. [40] synthesized

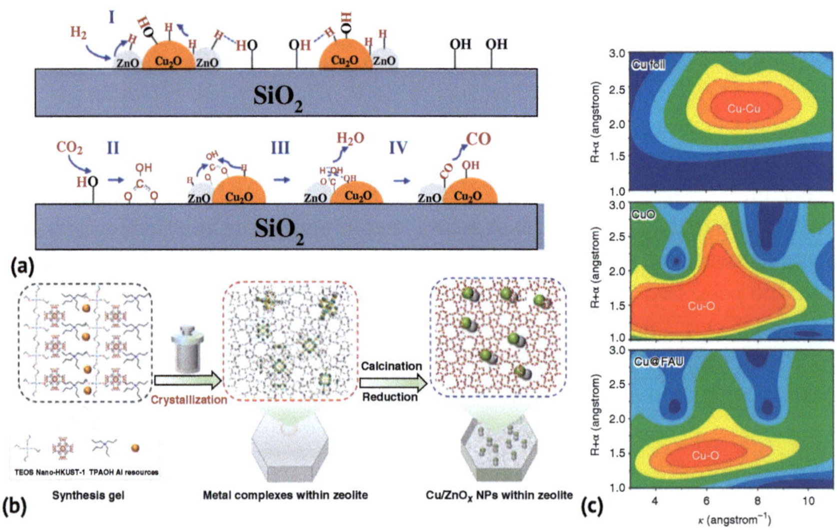

Figure 12.4 Zeolite-based catalysts with highly dispersed and confined metal species for methanol synthesis from CO_2 hydrogenation. (a) Diagram of the reaction mechanism on the silanol nests anchored to the 10CuO-2ZnO/S-1 catalyst. *Source*: Reproduced from [34] / The Royal Society of Chemistry. (b) Schematic diagram of in-situ hydrothermal synthesis of zeolite with ligand protection. *Source*: [36] / with permission of American Chemical Society. (c) WT EXAFS oscillations of Cu foil, CuO, and zeolite-encaged mononuclear copper centers, catalyst Cu@FAU. *Source*: Reproduced with permission from [37] / Oxford University Press / CC BY 4.0.

the low-silica zeolite X with a hierarchical structure that enables the promotion of zeolite hydrothermal stability and avoids the blockage of pores and channels from in-situ introduced Cu and Zn metal species. The electron-deficient Lewis acid centers and their interaction with ZnO species and the well-dispersed Cu nanoparticles enhance the adsorption of CO_2 and the dissociation of H_2, thus promoting the generation of methanol with a high STY of 51.1 $g_{CH_3OH} \cdot g_{Cu}^{-1} \cdot h^{-1}$. Similarly, Chai et al. [37] obtained zeolite-encaged mononuclear Cu catalyst Cu@FAU via in situ hydrothermal route for CO_2 hydrogenation to methanol (see Figure 12.4c). The results of H_2-TPR, X-ray absorption spectroscopy, and DFT calculations confirmed that Cu only exists in the frameworks of FAU zeolite as uniform and well-defined mononuclear sites Cu^{2+}. All the reaction steps that take place on $Cu^{\delta+}$—O^{2+} Lewis pairs confined in zeolites are thermodynamically and kinetically favorable from DFT calculations. Cu@FAU exhibited a high CH_3OH STY of 12.8 mmol $\cdot g_{cat}^{-1} \cdot h^{-1}$ at a selectivity of 89.5% and a high stability of 200 hours. MOFs, constructed by the coordination of metal ions/clusters with organic ligands, can also be added into the synthetic gel of zeolites to prepare the highly dispersed zeolite-encapsulated metal catalysts. Cui et al. [36] introduce ultrasmall Cu/ZnO_x NPs (~2 nm) into the crystalline particles of the Na-ZSM-5 zeolite via an in-situ hydrothermal route using CuZn-HKUST-1 as self-sacrificial metal stabilizers (see Figure 12.4b). The interaction of metal NPs and the zeolite framework protects Cu/ZnO_x NPs from sintering and aggregation and thus contributes to the excellent stability during the calcination, reduction, and methanol synthesis process. Cu/ZnO_x@ZSM-5 exhibited good catalytic performance of 44.88 $g_{MeOH} \cdot g_{Cu}^{-1} \cdot h^{-1}$ and stability of over 100 hours.

Finally, we have summarized the performance of zeolite-based catalysts for methanol synthesis from CO_2 hydrogenation as follows (see Table 12.1).

12.3 CO_2 Hydrogenation to Ethanol

Ethanol is a common alcohol widely used in various fields, including food, medicine, and in organic synthesis for the production of acetaldehyde, ethylene, ethylamine, ethyl acetate, acetic acid, and chloroethane in the chemical industry. In addition, ethanol is a high-quality and environmentally friendly liquid fuel. As a fuel additive, it can increase the oxygen content of gasoline and promote full combustion of the fuel [43–45]. Compared with methanol synthesis, CO_2 hydrogenation to ethanol is more challenging due to the activation of C=O/C—O and its matching with C—C coupling [46].

$$2CO_2 + 6H_2 \Longleftrightarrow CH_3CH_2OH + 3H_2O \ \Delta H_{298K} = -173.7 \ kJ \cdot mol^{-1} \tag{12.4}$$

$$CO_2 + 3H_2 \Longleftrightarrow CH_4 + H_2O \ \Delta H_{298K} = -165.0 \ kJ \cdot mol^{-1} \tag{12.5}$$

As shown in Eq. (12.4), CO_2 hydrogenation to ethanol is also an exothermic and volume-decreasing reaction, while relatively low temperatures and high pressures favor the forward reaction for ethanol synthesis. The side reactions are shown in Eqs. (12.1–12.3) and (12.5). From the thermodynamic calculations

Table 12.1 Catalytic performance of zeolite-based catalysts for methanol synthesis from CO_2 hydrogenation.

	Reaction conditions					Methanol			
Catalyst	*T* (°C)	*p* (MPa)	H_2/CO_2 ratio	CO_2 conv. (%)	CO sel. (%)	Sel. (%)	STY[a]	Stability (h)	References
Cu/ZnO$_x$(1.38)@Na-ZSM-5	250	3.0	3	~3	~27	~72	44.9[b]	100	[36]
Cu@FAU	240	3.0	3	11.5	10.5	89.5	409.6	200	[37]
CX-5	260	3.0	3	9.1	4	96.0	74.9	120	[41]
3Cu-ZnO$_x$/13X	290	3.0	3	9.7	2.4	97.6	81.1	200	[35]
5%Cu/SAPO-56	280	3.0	3	16.4	19.1	80.9	568.6	90	[33]
ZnO/Cu@S-1	250	2.0	3	4.7	44.7	55.3	1.3[b]	—	[42]
CuZn-HMLSX[c]	22.7	3.5	3	22.7	0	100	51.1[b]	—	[40]
10CuO-2ZnO/S-1	250	3.0	3	7.5	88	10	0.64	100	[34]
Cu-ZnO-Al_2O_3 + 3A[d]	250	5.0	—	99.62	—	82.7	—	—	[28]
CZA + 13X[d]	250	7.0	3	~40	~30	~70	—	—	[30]
CZA + NaA[e]	220	3.5	3	55.7	~30	~70	685	100	[25]
Cu-ZnO-Al_2O_3-ZrO_2 + LTA[e]	260	3.0	3	36.1	0	100	—	80	[24]

a. The unit is $mg_{MeOH} \cdot g_{cat}^{-1} \cdot h^{-1}$.
b. The unit is $g_{MeOH} \cdot g_{Cu}^{-1} \cdot h^{-1}$.
c. The CuZn-HMLSX catalyst was evaluated in the tank reactor.
d. SE process.
e. Catalytic membrane reactor.

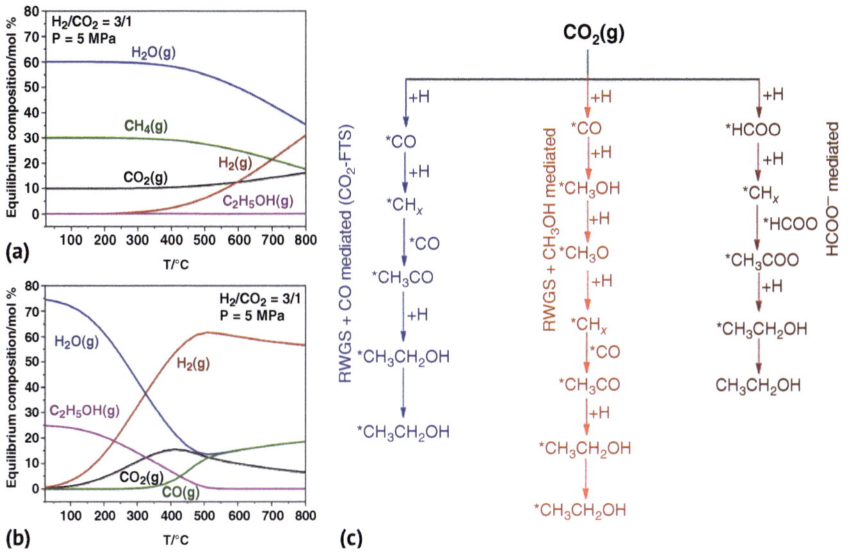

Figure 12.5 Fundamental properties of ethanol synthesis from CO_2 hydrogenation. Equilibrium composition of CO_2 hydrogenation to (a) methane and ethanol and (b) CO and ethanol at 5 MPa (H_2/CO_2 = 3/1). *Source*: Reproduced from [47] / with permission of Elsevier. (c) Three reaction pathways of ethanol synthesis from CO_2 hydrogenation. *Source*: Reproduced from [48] / with permission of Elsevier.

(see Figure 12.5a,b) [47], when the methanation reaction exists, the generation of ethanol tends to zero because hydrocarbon products are thermodynamically more favored products. Therefore, CO_2 methanation is a main reaction that researchers strive to suppress when CO and CH_3CH_2OH are the target products. The generation of ethanol has an optimal temperature between 250–320 °C. Compared with methanol synthesis, ethanol synthesis from CO_2 hydrogenation involves carbon chain growth, making the reaction mechanism more complex and more difficult to explore. Researchers have proposed multiple possible reaction mechanisms, including the CO-mediated mechanism, the formate/methoxy-mediated mechanism, the alkyl-formate coupling, and the formyl-methanol coupling mechanism (see Figure 12.5c) [48]. The insertion of non-dissociatively adsorbed CO* species is a widely recognized key intermediate for the formation of oxygen-containing compounds. For instance, on the Rh-based catalysts, CO_2 undergoes the RWGS to generate CO, which is then hydrogenated to form $*CH_x$. Subsequently, non-dissociatively adsorbed CO* is inserted into $*CH_3$ to form $*CH_3CO$ and then undergoes multiple steps of hydrogenation to form the target product CH_3CH_2OH [49]. From the mechanism of ethanol synthesis provided, it can be inferred that $*CH_x$ intermediate species play an important role in the carbon chain growth of oxygen-containing intermediates. Therefore, it requires the catalyst to suppress the hydrogenation or coupling process of $*CH_x$ species to prevent the generation of hydrocarbon products and to facilitate $*CH_x$ bonding with oxygen-containing intermediate species to only generate ethanol [50]. The current catalysts for ethanol synthesis from CO_2 hydrogenation can be mainly divided into the following three categories: (1) noble metal catalysts (Rh,

Pd, Ir, Pt, and Au); (2) transition metal carbide catalysts (Fe_5C_2, Co_2C, and Mo_2C); (3) modified Cu-based catalysts. The catalytic performance of zeolite-based catalysts for ethanol synthesis from CO_2 hydrogenation is summarized in the following table (see Table 12.2).

The unique pore and framework microenvironment of zeolites is often utilized in ethanol synthesis by highly dispersing metal species and constructing a unique catalytically active centers. Iltsiou et al. [51] prepared the mesoporous encapsulated zeolite catalyst Cu@m-Beta by the recrystallization method for CO_2 hydrogenation to ethanol in tank reactors. They compared the differences of CO_2 conversion to ethanol in different solvents (including water, cyclohexane, toluene, tetrahydrofuran, and N, N-dimethylformamide), finding that H_2O yielded the highest ethanol yield reaching a productivity of 50.3 $mmol{\cdot}g^{-1}{\cdot}h^{-1}$. The improvement in water performance is due to its characteristic of high polarity, which enables it to effectively solvate reactants and promote efficient mass transfer, resulting in higher reaction rates. In addition, under the condition of high H_2 pressure (4–5 MPa), a high productivity of up to 20.51 $mmol{\cdot}g^{-1}{\cdot}h^{-1}$ of isopropanol was detected. Although the tank reactors have a good advantage in ethanol synthesis, the fixed-bed reactors are suitable for practical applications of continuous production. Ding et al. [52] reported Na-Beta confined Cu nanoparticle catalysts for CO_2 conversion to ethanol in fixed-bed reactors (see Figure 12.6a,b). The unique copper nanoparticles in irregular shapes with step surface and the intimate interaction between Cu nanoparticles and Na-Beta are the key factors for ethanol synthesis from CO_2 hydrogenation. The CO_2 molecule is first activated and hydrogenated to $^*CH_3$ species on the Cu nanoparticle surface, and then $^*CH_3$ species and the adsorbed $^*CO_2$ bond together to form key intermediate species CH_3COO^* on the step surface with a Cu vacancy, which is further hydrogenated to form ethanol. As a result, Cu@Na-Beta exhibited CO_2 conversion of 7.9% and high ethanol selectivity of ~100% (CO-free) and stability of over 100 hours, corresponding to ethanol STY of ~398 $mg \cdot g_{cat}^{-1} \cdot h^{-1}$.

On the basis of zeolite dispersion and confinement, the introduction of other promoters can further construct the active interface sites to enhance the coupling of key intermediate species to form ethanol. Wang et al. [53] reported a catalyst with RhMn nanoparticles confined within the zeolite crystals for ethanol synthesis from syngas conversion. Dispersed and stable Mn—O—$Rh^{\delta+}$ structure was formed, promoting moderate adsorption of CO and facilitating the basic C—C coupling reaction of C_2 oxygen-containing groups formation. The zeolite sheath hindered the Rh sintering and stabilized the desired Mn—O—$Rh^{\delta+}$ structure. Wang et al. [54] utilized the dispersion and confinement effect of mesoporous MCM-41 on metals to prepare the VO_x-promoted Rh-based catalysts for CO_2 hydrogenation to ethanol. Highly dispersed VO_x-Rh interface sites were formed, which were more conducive to the adsorption of tilted CO. It can easily dissociate into $^*CH_x$, which is beneficial for CO insertion to further generate ethanol. The dissociative adsorbed CO is almost equal to that of non-dissociative adsorbed CO in an appropriate amount VO_x-promoted Rh-0.3VO_x/MCM-41, thereby exhibiting the highest CO_2 conversion (12.1%) and ethanol selectivity (24.1%). Zhang et al. [55] prepared Na-modified Rh nanoparticles embedded in zeolite S-1 (Na-Rh@S-1) (see Figure 12.7). The introduction of

Table 12.2 Catalytic performance of zeolite-based catalysts for ethanol synthesis from CO_2 hydrogenation.

	Reaction conditions					Ethanol			
Catalyst	T (°C)	p (MPa)	H_2/CO_2 ratio	CO_2 conv. (%)	CO sel. (%)	Sel.[a] (%)	STY[b]	Stability (h)	References
Cu@Na-Beta	300	1.3	3	7.9	30.5	~69.5	258	100	[52]
Rh-0.3VO_x/MCM-41	250	3.0	3	12.1	20.1	24.1	47.9	—	[54]
Na-Rh@S-1	250	5.0	3	10.0	32.0	24.0	24.2	100	[55]
Cu@m-Beta[c]	200	6.0	5	20.0	10.0	63.0	35.8	—	[51]
$Zn_{12.5}$-$Ce_{12.5}$/ZSM-5[d]	150	0.1	—	—	—	—	7.4	—	[56]

a. The ethanol selectivity including CO.
b. The unit is $mg \cdot g_{cat}^{-1} \cdot h^{-1}$.
c. The Cu@m-Beta catalyst was evaluated in the tank reactor.
d. The $Zn_{12.5}$-$Ce_{12.5}$/ZSM-5 was evaluated for chemical looping processes.

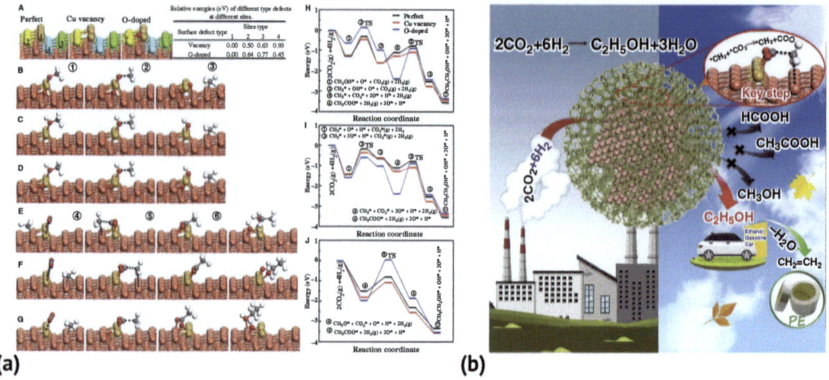

Figure 12.6 Ethanol synthesis from CO_2 hydrogenation over the Cu@Na-Beta. (a) Reaction pathway of CO_2 hydrogenation to CH_3CH_2OH and reaction energetics calculated by DFT. (b) Summary of the CO_2 hydrogenation to ethanol over Cu@Na-Beta. *Source*: Reproduced from [52] / with permission of Elsevier.

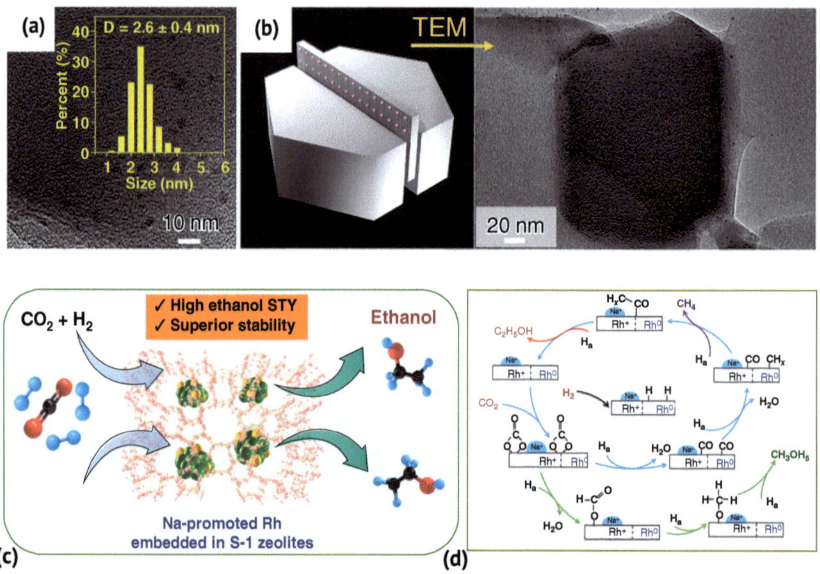

Figure 12.7 Ethanol synthesis from CO_2 hydrogenation over the modified Rh-based Na-Rh@S-1 catalysts. (a) TEM image and size distributions of Rh nanoparticles. (b) Tomographic TEM image. *Source*: Reproduced with permission from [55] / American Chemical Society. (c) Summary of the CO_2 hydrogenation to ethanol over Na-Rh@S-1. (d) Proposed reaction mechanism for CO_2 hydrogenation over Na-Rh@S-1. *Source*: Reproduced from [55] / with permission of American Chemical Society.

Na promoted the coexistence of oxidic Rh^+ species with metallic Rh^0, and formed the Na^+-Rh^+/Rh^0 interface site, which is the key factor for the transformation of the main products from CH_4 to oxygenates (mainly including CH_3OH and CH_3CH_2OH). On the surface of Na-Rh@S-1, CO_2 is activated to carbonate and then hydrogenated to form *CO species on the oxidic Rh^+ site. Some *CO migrate to the metallic Rh^0 site

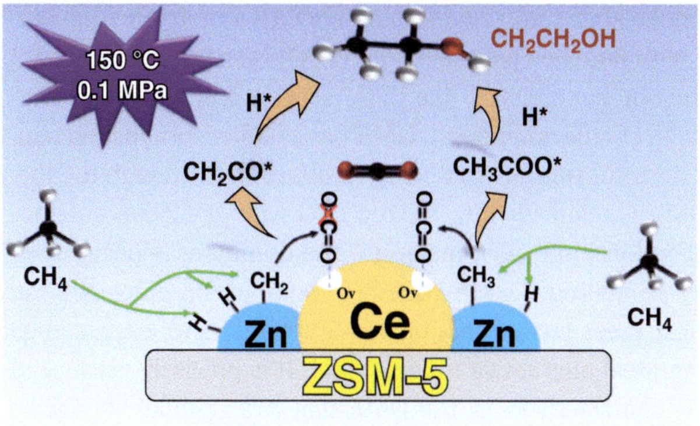

Figure 12.8 Proposed reaction mechanism of ethanol synthesis from CH_4 and CO_2 coupling over Zn–Ce/ZSM-5. *Source*: Reproduced from [56] / with permission of American Chemical Society.

and undergo a hydrogenation process to form $^*CH_x$ species. *CO on the oxidic Rh^+ site and $^*CH_x$ on the metallic Rh^0 site insert and bond to form $^*CO—CH_x$ species on the interface and then are further hydrogenated to form ethanol (see Figure 12.7d). 0.19Na-Rh@S-1 exhibited excellent catalytic performance of ethanol synthesis from CO_2 hydrogenation with 72 $mmol \cdot g_{Rh}^{-1} \cdot h^{-1}$. Owing to the dispersal and protective effects of the zeolite framework on metal species, it possessed stability of over 100 hours.

In addition, except for ethanol synthesis from CO_2 hydrogenation, researchers have also developed a new approach for coupling CO_2 and alkanes to produce ethanol. Liu et al. [56] reported the direct coupling of CH_4 and CO_2 to ethanol as the sole liquid product on the Zn–Ce/ZSM-5 by a two-step technique (see Figure 12.8). The $ZnOH^+$ sites are created and active in CH_4 dissociation to form CH_x^* intermediates, and the presence of oxygen vacancies in Ce sites preferably adsorbed CO_2 to form HCOO* and CO* intermediates. The C—C coupling is achieved through the coupling of CH_3^* and CO_2^*, as well as the insertion of CO_2^* into CH_2^*, to form CH_3COO^* and CH_2CO^* intermediates, which are ultimately hydrogenated to generate ethanol. As a result, the activation of CO_2 and CH_4 to produce ethanol at low temperature and atmospheric pressure was achieved, providing an attractive route for ethanol synthesis from greenhouse gases under mild conditions.

Finally, we have summarized the performance of zeolite-based catalysts for ethanol synthesis from CO_2 hydrogenation as follows (see Table 12.2).

12.4 CO_2 Hydrogenation to DME

DME is the simplest ether, and the features of being noncarcinogenic, noncorrosive, and almost nontoxic features make DME a green chemical with good application prospects. Due to the characteristics of easy storage, low pollution, high combustion efficiency, and high cetane number, DME can be used as a substitute for fuel. Meanwhile, DME only contains the elements C, H, and O, which prevents the production of harmful gases such as sulfur oxides and smoke after combustion, making DME a more environmentally friendly fuel. Blending with liquefied petroleum gas (LPG) is one of the important applications of DME. On the one hand, DME, like LPG, is a convenient gas fuel with similar physical properties [57]. On the other hand, DME has a higher combustion temperature and does not emit harmful substances during combustion. Although the theoretical calorific value of DME is relatively low, mixing LPG with DME can enhance combustion and reduce dependence on LPG, making it the dominant application of DME in civilian fuels today. In addition, DME with high combustion efficiency can achieve smokeless emissions, meet environmental requirements, and have a high cetane number, making it an ideal alternative to diesel fuel. It is precisely because of the environmentally friendly characteristics of the DME that it is gradually replacing other aerosol propellants such as chlorofluorocarbons, which have serious damage to the atmospheric ozone layer, thereby affecting the Earth's ecological environment and human health [58]. Furthermore, DME is an important chemical raw material that can be used to synthesize many chemical products, such as methyl acetate, dimethyl sulfate, olefins, isoparaffinic gasoline, or participate in the synthesis of chemical products.

DME can be usually produced by the dehydrogenation of methanol, and it can also be produced by syngas ($CO + H_2$). In recent years, people have paid more attention to the conversion of CO_2, so the direct production of DME by CO_2 hydrogenation has become a research hotspot. CO_2 hydrogenation to DME can be performed in three ways: (1) an indirect way consisting of CO_2 hydrogenation to methanol and then methanol dehydration to DME, which needs two reactor sections and two catalysts. The reactions are shown in Eqs. (12.1) and (12.6), (2) direct way with direct CO_2 hydrogenation to DME, and the most obvious difference is that the direct way only needs one reactor and one bifunctional catalyst. The reaction is shown in Eq. (12.7), and (3) RWGS way. This way consists of three steps—RWGS reaction and CO hydrogenation to methanol, then methanol dehydration as mentioned earlier, which is shown in Eqs. (12.2), (12.3), and (12.7). In the reactions mentioned above, almost all the reactions are exothermic except the RWGS reaction, so high temperature is not conducive to the reactions. Besides, methanol dehydration to DME can shift the thermodynamic equilibrium for CO_2 hydrogenation, resulting in the improvement of CO_2 conversion.

Indirect CO_2 hydrogenation to DME (CO_2 hydrogenation to methanol) (Eq. 12.1) and methanol dehydration to DME (Eq. 12.6):

$$2CH_3OH \Longleftrightarrow CH_3OCH_3 + H_2O\ \Delta H_{298K} = -23.4\,kJ\cdot mol^{-1} \quad (12.6)$$

Direct CO_2 hydrogenation to DME (Eq. 12.7):

$$2CO_2 + 6H_2 \Longleftrightarrow CH_3OCH_3 + 3H_2O\ \Delta H_{298K} = -122.4\,kJ\cdot mol^{-1} \quad (12.7)$$

The catalyst used in direct CO_2 hydrogenation to DME should be a bifunctional or hybrid catalyst according to the reaction process of CO_2 hydrogenation to methanol and methanol dehydration to DME. In general, the catalyst includes two components: one is usually metal active sites for CO_2 hydrogenation to methanol; the other is usually solid acid catalysts like γ-Al_2O_3 and zeolites [59]. For the part of methanol generation, the most commonly used catalysts are Cu-based catalysts, such as Cu-ZnO, Cu-ZnO-Al_2O_3, and Cu-ZnO-ZrO_2, and other metals like In, Ti, and noble metals. Pt, Pd are also used as additives to promote the dispersion of active sites and enhance reaction performance. For the part of methanol dehydration, solid acid catalysts like γ-Al_2O_3 and zeolites are used. Among them, zeolites are widely used due to their large specific surface area, adjustable acidity, and uniform pore structure. The commonly used types of zeolites are ZSM-5, SAPO-based zeolites, FER, MOR, as well as small-pore zeolites like LTA, which is often used because of its hydrophilicity. In direct CO_2 hydrogenation to DME, the combination of the above two components is required to form a hybrid catalyst or a bifunctional catalyst. There are several ways to mix two active components, which can be divided into two categories: (1) bifunctional catalysts where the metal active sites and the acid sites coexist in a single catalyst, so it can achieve one-step direct CO_2 hydrogenation to produce DME using just one catalyst, and (2) physically mixed catalysts for which there are three mixing methods frequently used: double-bed-layer filled catalysts, granule-mixed catalysts, and powder-mixed catalysts (see Figure 12.9) [60]. To sum up, the catalyst system of CO_2 hydrogenation to DME is complicated, and it is necessary and challenging to design efficient catalysts.

As mentioned above, zeolites can play an important role in CO_2 hydrogenation to DME. The catalytic performance of zeolite-based catalysts for DME synthesis from CO_2 hydrogenation is summarized in the following table (see Table 12.3). With the large specific surface area, the tunable acidity, and the ordered channel structure, zeolites perform better than other solid acid catalysts. The main function of zeolites is to serve as the methanol dehydration catalyst, providing the acid active sites. Besides, the large specific surface area and microporous structure of zeolites are conducive to the dispersion of active metals and the diffusion of reactant molecules, which facilitates the contact between reactant molecules and active sites [62, 63]. Furthermore, some zeolites with ultra-small microporous structures like LTA zeolites, are beneficial to CO_2 hydrogenation to DME due to the hydrophilicity, which can promptly separate the water produced by CO_2 hydrogenation and methanol

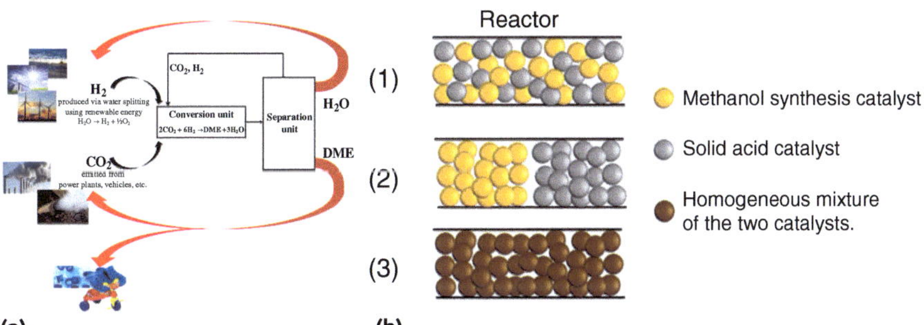

Figure 12.9 (a) Proposed carbon cycle loop involving CO_2 as energy vector. *Source:* Reproduced from [61] / MDPI / CC BY 4.0. (b) Different mixing methods of physical mixed catalysts in a fixed-bed reactor. *Source*: Reproduced from [60] / with permission of American Chemical Society.

Table 12.3 Catalytic performance of zeolite-based catalysts for DME synthesis from CO_2 hydrogenation.

	Reaction conditions					DME			
Catalyst	*T* (°C)	*p* (MPa)	H_2/CO_2 ratio	CO_2 conv. (%)	CO sel. (%)	Sel. (%)	STY[a]	Stability (h)	References
5%PdZn/TiO_2 + ZSM-5	270	2.0	3	11	61.7	32.3	30.7	—	[80]
Pd/ZnO/bio-ZSM-5	300	3.0	3	10.8	42.1	30.2	—	60	[67]
50CIZO-50SAPO	250	3.0	3	4.3	—	65.1	80.5	—	[65]
CZZr-10/DeSi-ZSM-5	260	3.0	3	14.2	20.6	60.4	52.8	—	[81]
CZA + FER	250	3.0	3	19.8	51.9	37.8	230.5	>30	[82]
CZA + Al-MCM-41	250	3.0	3	18.6	63.8	10.5	30.1	>30	[82]
CZA + HZSM-5	240	2.8	3	25.6	—	55	55.1	100	[83]
2PdCuAl//Z	260	2.0	3	11.3	78.0	11.8	15.3	—	[84]
CZZA/HZSM-5	240	2.8	3	26.2	—	69.8	—	100	[66]
ZZ-M	240	3.0	3	16.1	54.3	33.9	129.3	—	[64]
CZ-CNT/HZ	250	5.0	3	18.9	33.9	51.8	578	200	[85]
CZZ(C)/FER	250	2.0	3	17.5	58.3	28.4	23.0	—	[86]
CZZH-300	250	3.0	3	22.2	23.8	67.6	138.6	100	[87]
CuO-TiO_2-ZrO_2/HZSM-5	250	3.0	3	15.6	39.2	47.5	—	—	[88]
CZZ/HZSM-5	250	5.0	3	34.1	24.3	67.7	142.2	—	[89]
CZZ-FER8	260	3.0	3	22.0	47.5	38.5	160.6	—	[90]
OX-FER/2	280	5.0	3	~30	~23	~62	752	—	[69]

OX-MOR/2	280	5.0	3	~27	~38	~50	559	—	[69]
UNIT-25/9-OX	240	5.0	3	23.6	24.5	49.3	68.9	—	[91]
CuO-ZnO/ZSM-5	260	2.0	3	19.2	—	65.0	—	—	[92]
CZZ-MOR	260	5.0	3	23.2	38.1	50.8	245.7	—	[70]
CZZ-FER	260	5.0	3	26.0	31.5	55.7	302.0	—	[70]
CZZ-MFI	260	5.0	3	21.3	46.6	40.5	179.9	—	[70]
ZZ-NC	240	3.0	3	~14	~56	~32	225	—	[68]
CZA/FER(12)	250	5.0	3	29.7	21.9	55.2	208.4	—	[93]
CZZH-300	250	3.0	3	22.2	23.8	67.6	138.7	100	[87]
CZZr@S-11[b]	275	3.0	3	~11	—	~75	—	120	[72]
PdZn/ZSM-5 (S)	270	2.0	3	16.6	63.1	33.6	39.1	—	[94]
CZA-S (3D)	240	3.0	3	38.9	23.0	77.0	276.8	—	[95]
CuO-ZnO-Al_2O_3@HZSM-5	270	3.0	3	48.3	~20	48.5	108.2	—	[71]
CZZA/HZSM-5[c]	300	3.8	3	22.8	23.0	64.5	317.1	—	[79]
Cu-ZnO-Pt@HZSM-5[c]	280	3.25	3	41.1	0	~100	—	60	[78]

a. The unit is $mg_{DME} \cdot g_{cat}^{-1} \cdot h^{-1}$.
b. The reaction gas is $H_2 + CO + CO_2$ with $CO_2/CO_x = 0.5$ and $H_2/CO_x = 3$ molar ratios.
c. The catalysts were evaluated in the membrane reactor.

dehydration, thereby promoting the conversion of CO_2. Types of zeolites, preparation methods, mixing methods, and ratios of two components can all affect the reaction performance of CO_2 hydrogenation to DME, so it is important to regulate the catalysts to achieve efficient reaction performance.

12.4.1 Physically Mixed Catalysts

The physical mixing of catalysts is the simplest way to combine two functional components into one catalytic system, achieving dual functions (one is metal sites for CO_2 hydrogenation to methanol, and the other is acid sites for methanol dehydration to DME) in one catalytic system simultaneously. There are usually three ways to achieve physical mixing of two functional components: double-bed-layer filled catalysts, granule-mixed catalysts, and powder-mixed catalysts. Different mixing methods bring different proximity between metal and acid sites, as well as different distributions of active sites, resulting in different reaction performance and product selectivity.

Adjusting the mixing degree of dual functions by changing the mixing method is beneficial for studying the proximity of metal and acid sites. The impact of proximity between two functional components was studied by Frusteri et al. as early as 2013. They compared the reaction performance of three catalysts with different proximity: homogeneous physical mixture (ZZ-M), dual-bed composed of ZCZ methanol catalyst in first bed and HZ30 catalyst in the second bed (ZZ-D), and mono-bed constituted by a homogeneous solid mixture prepared by grinding of powdered catalysts (ZZ-G). They found that the homogeneous mixed (ZZ-M) catalyst showed a superior specific productivity of 430 $g_{total\ MeOH} \cdot kg_{cat}^{-1} \cdot h^{-1}$ at 513 K. By exploring the zeolite content, a higher CO_2 conversion rate is obtained with an optimal mass ratio of 1:1 between Cu-ZnO-ZrO_2 and H-ZSM-5 zeolite, which is benefited by a balanced concentration of metal/oxide and weak-medium acid sites [64].

Yao et al. [65] also demonstrated the importance of proximity between Cu-In-Zr-O (CIZO) catalyst and SAPO-34 zeolite, and explained the differences in reaction performance caused by different proximity from the perspective of reaction mechanism. They compared three catalysts (powder mixing, pellet mixing, and dual bed) and found that catalytic reactivity exhibited a decreasing trend with increasing proximity between CIZO and SAPO-34 zeolite (see Figure 12.10). By means of in-situ DRIFTS (see Figure 12.10), the CO_2 hydrogenation to DME reaction pathway was investigated, and it demonstrated that closer proximity induces the shortcut Methoxy → DME pathway rather than a typical Methoxy → Methanol → DME route, resulting in more effective reaction performance (see Figure 12.10). Fan et al. [66] found that the proximity can also affect the catalyst deactivation. The deactivation rate of the separated mode of CuO-ZnO-ZrO_2-Al_2O_3 (CZZA) and ZSM-5 was faster than the mixed one, and the loss of copper active sites caused by water and the increase in carbon deposition on ZSM-5 zeolite are the main reasons for deactivation, according to thermogravimetric analysis. In a word, an optimal proximity between two active components in the catalyst is crucial for the one-step direct CO_2 hydrogenation to DME reaction.

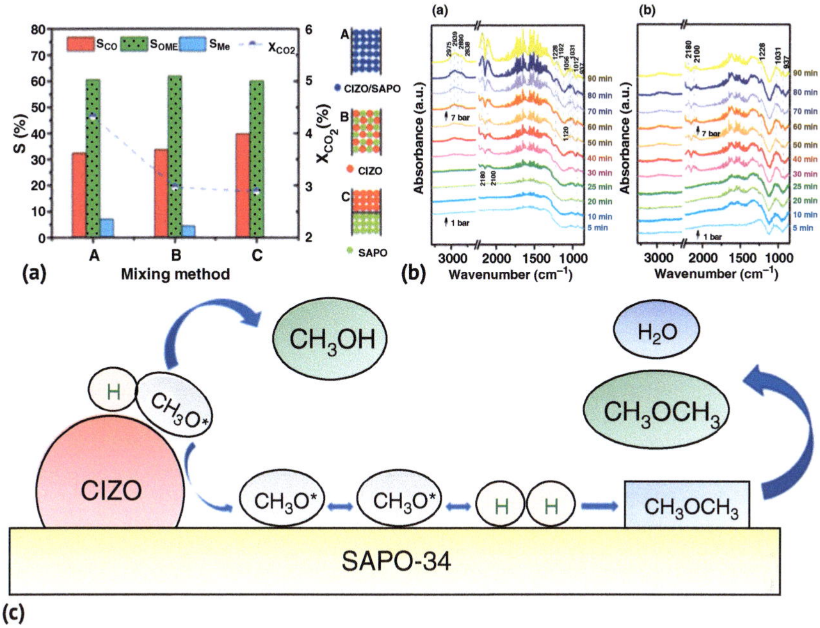

Figure 12.10 Reaction performance and mechanism of CIZO–SAPO bifunctional catalyst. (a) Influence of mixing method (proximity) on the catalytic performance of the CIZO–SAPO bifunctional catalyst. Reaction conditions: $T = 250\ °C$, $p = 30$ bar, and GHSV = 6000 mL · g_{cat}^{-1} · h^{-1}. (b) In-situ DRIFTS spectra with (a) mortar mixing and (b) direct mixing bifunctional catalysts. (c) Graphic representation of the synergy between CIZO–SAPO bifunctional active sites in DME generation. *Source*: Reproduced from [65] / with permission of American Chemical Society.

Due to the complexity of the catalyst system, the requirements for proximity may vary for different catalysts. Li et al. [67] synthesized hierarchically porous bio-ZSM-5 with high pore connectivity using bio-templates like rice husk and luffa sponge (see Figure 12.11), and the physically mixed catalysts of Pd/ZnO component and bio-ZSM-5 showed excellent reaction performance in CO_2 hydrogenation to DME. They then adjusted the spatial proximity between bio-ZSM-5 zeolite and Pd/ZnO components by controlling the size of Pd/ZnO particles. The results indicated that the closest proximity between the two components is not conducive to the generation of DME, because zinc cations may migrate and replace acidic sites on bio-ZSM-5, resulting in low DME selectivity. The suitable proximity in the powder-mixed catalyst was the best combination of bio-ZSM-5 and Pd/ZnO, achieving a CO_2 conversion of 10.8% and a DME selectivity of 31% under 300°C (see Figure 12.11).

12.4.2 Integrated Bifunctional Catalysts

Unlike the physically mixed catalysts, the integrated bifunctional catalysts are synthesized through one-pot preparation route, making the combination between metal and acid sites tighter. There are two most commonly used methods for synthesizing

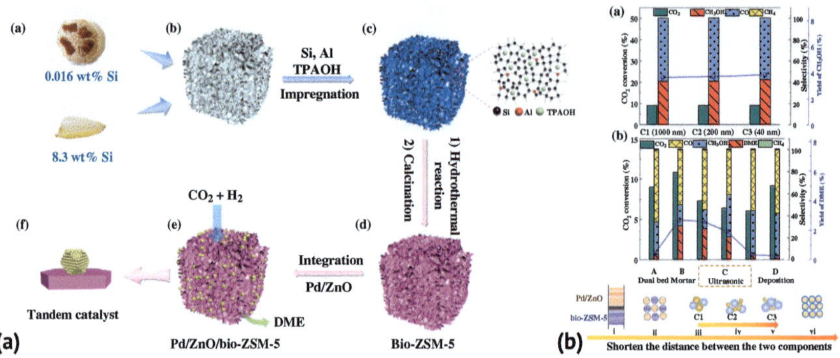

Figure 12.11 Preparation and performance of Pd/ZnO/bio-ZSM-5 catalyst. (a) Schematic illustrations for the synthesis of bio-ZSM-5. *Source*: Reproduced with permission from [67] / American Chemical Society. (b) Effect of particle size of Pd/ZnO and the integration manner on catalytic behaviors of Pd/ZnO/bio-ZSM-5. *Source*: Reproduced from [67] / with permission of American Chemical Society.

integrated bifunctional catalysts—one is coprecipitating the metal sites in a solution containing zeolite dispersed, and the other is constructing a unique core-shell structure. For the former method, the specific synthesis steps are as follows: first, solubilize the metal precursors in ethanol and coprecipitate with oxalic acid under stirring in a solution containing the zeolite finely dispersed, then filter, dry, and calcine after stirring for several hours. The metal sites can be better dispersed through this method compared with physical mixing. For the latter method, the unique core-shell structure is composed of the metal component as a core and the acid zeolite component as a shell. Due to the better efficiency of bifunctional catalysts, researchers have explored the application of bifunctional catalysts in CO_2 hydrogenation to DME.

The most obvious characteristic of bifunctional catalysts is the type of synthesis process, and any synthesis detail may affect the structure and physicochemical properties of the catalyst, thereby altering its catalytic performance. Frusteri et al. [68] found that the preparation method can significantly affect the catalyst's properties thereby affecting the activity. A series of bifunctional catalysts (Cu-ZnO-ZrO_2/H-ZSM-5) were synthesized through different kinds of precipitants, including sodium bicarbonate (ZZ-CC), ammonium carbonate (ZZ-NC), oxalic acid (ZZ-OC), and urea (ZZ-UH) (oxalic acid seems to be the most commonly used precipitant among them). The bifunctional catalyst using ammonium carbonate as precipitant showed an excellent reaction performance with high CO_2 conversion and DME selectivity, exhibiting a maximum STY of 0.225 $kg_{DME} \cdot kg_{cat}^{-1} \cdot h^{-1}$ (see Figure 12.12a). As acid sites in bifunctional catalysts, the type of zeolites also has a significant impact on the catalyst properties and reaction performance. The 2D ferrierite (FER) and 1D mordenite (MOR) zeolites were compared, and the author considered that the lamellar zeolite framework of FER zeolite is beneficial for the anchorage of metal oxide clusters, thus reducing the mass-transfer constraints in the reaction process, thereby achieving better catalytic activity in CO_2 to DME. The FER-based bifunctional catalyst showed excellent activity and selectivity, exhibiting an interesting

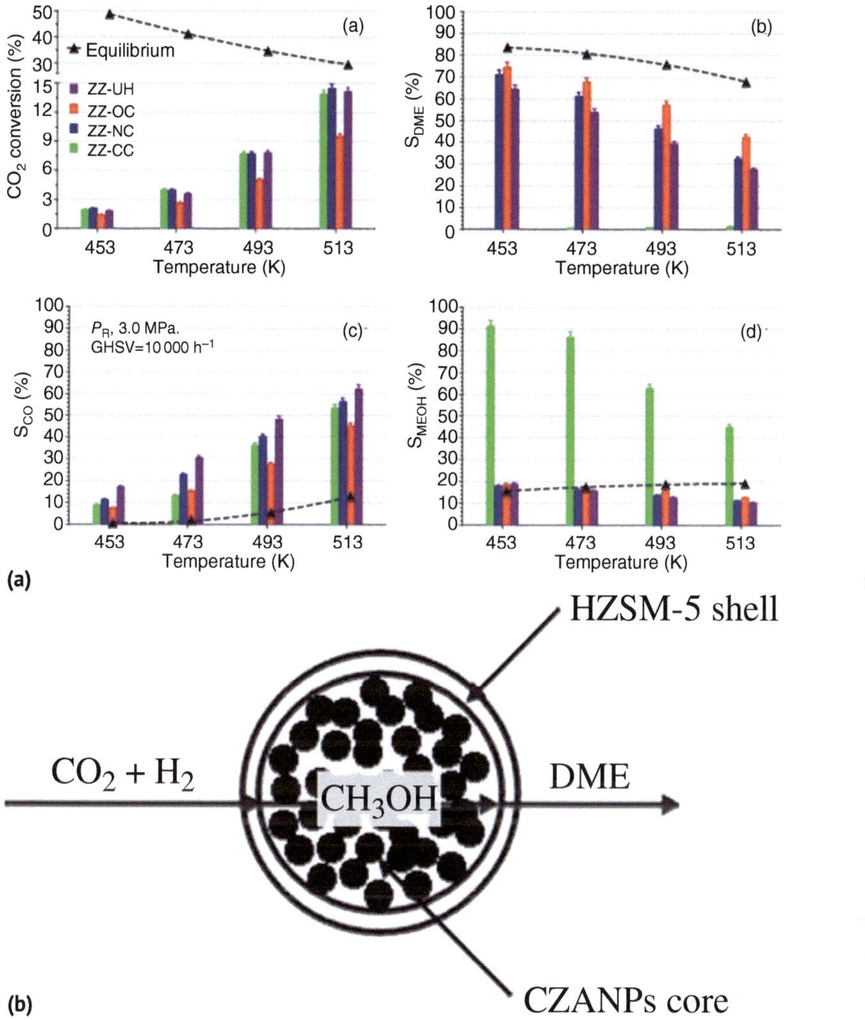

Figure 12.12 DME synthesis from CO_2 hydrogenation over integrated bifunctional catalysts. (a) Reaction performance of different precipitating agents. *Source*: Reproduced from [68] / with permission of Elsevier. (b) Schematic image of the capsule catalyst in the CO_2 hydrogenation. *Source*: Reproduced from [71] / with permission of Elsevier.

DME productivity (STY) of 752 $g_{DME} \cdot kg_{cat}^{-1} \cdot h^{-1}$ [69]. They later found that FER zeolite can not only ensure better dispersion of metal oxide clusters, but also facilitate the generation of Lewis basic sites for CO_2 activation, thereby achieving higher CO_2 activation efficiency and DME selectivity [70].

The core-shell bifunctional catalyst is a unique structure with a metal site as a core and zeolite as a shell. Benefited by this special structure, the reactant molecules CO_2 and H_2 can enter the core through the zeolite channels and generate methanol, then the methanol has to diffuse outward through the zeolite, and the restricted micro-environment is conducive to the dehydration of methanol to produce DME. The core-shell catalysts can increase the collision probability between methanol and acid sites of zeolites, achieving higher DME selectivity than traditional hybrid catalysts and suppressing the further dehydration of DME to form alkane and alkene. Liu et al. [71] constructed a core-shell catalyst with H-ZSM-5 packed CuO-ZnO-Al_2O_3 nanoparticles (see Figure 12.12b), exhibiting good performance of CO_2 conversion of 48.3% and high DME selectivity of 48.5% at 270°C. In addition to ZSM-5, Ateka et al. has conducted extensive research on SAPO-type zeolites [72–74]. They came to a similar conclusion in the core-shell catalyst (CuO-ZnO-ZrO_2@SAPO-11) that the unique structure was beneficial to the generation of DME. They later revealed the influence of the operating conditions on the deactivation of the core-shell catalyst CuO-ZnO-ZrO_2@SAPO-11, and they found that coke deposition was the main reason for catalyst deactivation.

12.4.3 Zeolite Catalytic Membrane Reactor

In the process of CO_2 hydrogenation to DME, a large amount of water will be generated, which may expedite the sintering of active metal sites and inhibit the production of DME [26, 75]. As mentioned earlier, zeolites play multiple roles in CO_2 hydrogenation to DME, and zeolites have also been used as a membrane to in situ separate the water steam generated during the CO_2 hydrogenation to DME for the reactor. According to Le Chatelier's principle, in situ removal of water can promote the positive shift of equilibrium to achieve higher CO_2 conversion and DME generation. Membrane reactor integrates both reaction and separation in just one unit. The most common types of membrane reactors are packed bed membrane reactors (PBMRs) and CMRs.

As for PBMR, the main function of the membrane is separation, so the diffusion and transport performance of different components is very important, measured by mass transport flux. Diban et al. [76] used a bifunctional catalyst (CuO-ZnO-Al_2O_3/γ-Al_2O_3) in a packed bed reactor (PBR) in CO_2 hydrogenation co-fed with syngas for the direct synthesis of DME, and used a mathematical model to theoretically evaluate the performance of the zeolite membrane for the in-situ H_2O removal in a PBMR. The diagram of the PBMR and the details are shown (see Figure 12.13a), in which the mass transport rate through the zeolite membrane of all the components

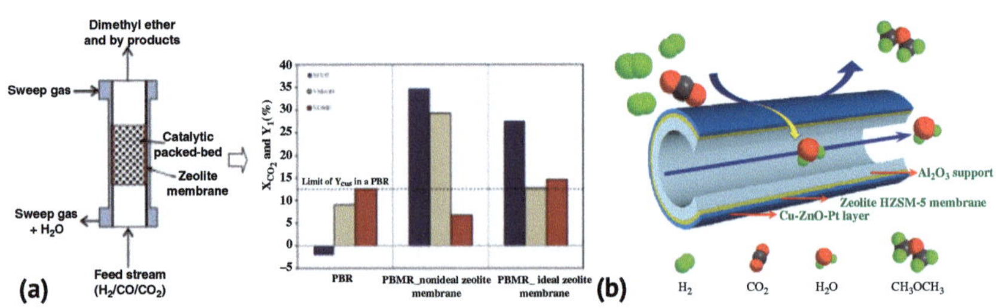

Figure 12.13 Application of zeolite membrane in CO_2 hydrogenation to DME. (a) PBMR with in-situ H_2O removal for DME production and the reaction performance of different membranes. *Source*: Reproduced from [76] / with permission of Elsevier. (b) Scheme of CO_2 hydrogenation to DME in a Cu-ZnO-Pt@HZSM-5 CMR. *Source*: Reproduced from [78] / with permission of Elsevier.

present in the system can be described. They illustrated that the application of almost all zeolite membranes can promote the conversion of CO_2 hydrogenation to DME, but the DME yield is tightly related to the membrane's H_2O permselectivity. The ideal H_2O selective membranes achieved DME yields up to approximately 16.7%, much higher than low permselective zeolite membranes like ZSM-5 and MOR. Furthermore, Ateka et al. used LTA zeolite membrane in a reaction equipment due to its hydrophilicity, and validated the model for simulating the direct CO_2 hydrogenation to DME in a PBMR. The experimental results and mathematical model indicated that PBMR achieved high yields of DME at higher temperatures and space-time than membrane-free PBRs [77].

Most studies were conducted on a laboratory scale, and improving the maturity of this technology remains a huge challenge. Dong et al. [79] developed a novel prototype-scale CMR for CO_2 hydrogenation to DME, which was equipped with a bifunctional Cu-ZnO-ZrO_2-Al_2O_3/HZSM-5 catalyst as the DME production catalyst and a hydrophilic NaA zeolite membrane for a separate section. Compared with normal fixed-bed reactor without zeolite membrane, this CMR with NaA membrane showed a significantly higher CO_2 conversion and DME yield, increasing from 21.4% and 8.71% to 33.7% and 22.8%, respectively, and realized a superior high DME production rate of 1.31 kg per day at 300 °C and GHSV of 8400 $mL \cdot g^{-1} \cdot h^{-1}$. The hydrophilic ZSM-5 zeolite is a well-known acid catalyst used in methanol dehydration to DME because of its strong acidity, as well as separation functions due to its hydrophilicity. Yue et al. [78] fabricated a Cu-ZnO-Pt@HZSM-5 CMR for CO_2 hydrogenation to DME (see Figure 12.13b). They claimed that the HZSM-5 membrane can in situ separate byproduct water, effectively breaking the limitations of thermodynamic equilibrium and inhibiting the deactivation of the catalyst induced by water, thereby maintaining high catalytic activity. Compared with catalytic fixed-bed reactor (CFBR), CMR exhibited better activity, achieving higher CO_2 conversion (24.9% in the CFBR and 41.1% in the CMR) and DME selectivity (53.7% in the CFBR and 100% in the CMR). Therefore, zeolite membranes are expected to achieve better reaction performance by in-situ removal of unwanted byproduct molecules, and how to achieve higher selectivity of zeolite membranes may be a future research hotspot.

Finally, we have summarized the performance of zeolite-based catalysts for DME synthesis from CO_2 hydrogenation as follows (see Table 12.3).

12.5 CO_2 Hydrogenation to Formic Acid

Formic acid is an important organic acid with the simplest structure, and it is an important C1 oxygenate which is widely used in industries such as chemical, leather, dyes, pharmaceuticals, and rubber. It can be used as a metal surface treatment agent and industrial solvent. In addition, formic acid is usually used for efficient hydrogen storage with a significantly high hydrogen content of 53 g H_2 per L [96–98]. The process of hydrogen storage can actually be simplified as CO_2 hydrogenation to formic acid ($CO_2 + H_2 \Longleftrightarrow HCOOH$), and the release process of hydrogen is actually the dehydrogenation of formic acid ($HCOOH \Longleftrightarrow CO_2 + H_2$). During CO_2 hydrogenation to formic acid, the flammable H_2 can be stored in the safe and nontoxic formic acid. Overall, using formic acid as a hydrogen storage material can simultaneously solve the problems of difficult storage of hydrogen and the environmental disadvantage of carbon dioxide, killing two birds with one stone, making this reaction a research hotspot.

As for thermodynamics of CO_2 hydrogenation to formic acid, the conversion of gas-phase CO_2 and H_2 to liquid-phase formic acid is thermodynamically unfavorable as shown in Eq. (12.8). Therefore, the most effective method is to drive the reaction in liquid phase because the transfer of liquid phase is more favorable in thermodynamics as shown in Eq. (12.9). In liquid-phase reaction, the reaction medium can significantly influence the reaction, and researchers usually introduce the base additives to further alleviate thermodynamic limitations, as shown in Eq. (12.10) [99, 100]. Although the liquid-phase reaction is more favorable in thermodynamics, it is more complicated to convert gas system into liquid and the mechanism is difficult to reveal. So, it is still necessary to further study the hydrogenation of gas-phase CO_2 to formic acid. No matter what kind of phase reaction it is, the catalytic performance can be significantly influenced by the reaction conditions and the catalytic active sites.

$$CO_2(g) + H_2(g) \Longleftrightarrow HCOOH(l)\Delta G_{298K} = +32.9\ kJ \cdot mol^{-1} \tag{12.8}$$

$$CO_2(aq) + H_2(g) + H_2O \Longleftrightarrow HCOO^{-}(aq) + H_2O^{+}\Delta G_{298K} = -4\ kJ \cdot mol^{-1} \tag{12.9}$$

$$CO_2(aq) + H_2(g) + NH_3(aq) \Longleftrightarrow HCOO^{-}(aq) + NH_4^{+}(aq)\Delta G_{298K} = -9.5\ kJ \cdot mol^{-1} \tag{12.10}$$

For CO_2 hydrogenation to formic acid in the liquid phase, common reaction solvents include water, alcohols, and ionic liquids. Furthermore, base additives like amines, hydroxides, and bicarbonates are added to further convert the formic acid to shift the reaction equilibrium towards the product. In the gas-phase reaction of CO_2 hydrogenation to formic acid, the only two reactants result in pure product. But in the liquid phase, the reaction becomes more complex and variable. For instance, the presence of water and some inorganic base additives can lead to the generation of carbonic acid or further dehydrogenation to form bicarbonate and carbonate ions. The presence of organic base additives makes the reaction a complex system containing amines, resulting in more variable products. Due to the complexity of the reaction system, the catalytic performance can be affected by the solution pH, CO_2 pressure, reaction medium, temperature, and so on [100, 101]. Therefore, suitable catalysts should be selected for the above reaction conditions. The commonly used catalysts are noble metals such as Pd, Ru, Au, and Ag due to its high activity in CO_2 hydrogenation to formic acid. The two most considered key factors in catalyst design are the number of active sites and the electronic structure of metal sites. And with the development of catalyst history, the research has focused more on nanoscale catalysts and single-atom catalysts in order to improve the catalytic efficiency [100]. For instance, Lee et al. [102] used mesoporous graphitic carbon nitride to disperse Pd metal (named as Pd/mpg-C_3N_4) and realized a small size of ~1.7 nm, resulting in reversible formic acid hydrogen storage process. To get a better efficiency of CO_2 hydrogenation to formic acid, more effective metal sites are needed, and a unique active site structure needs to be identified. For this purpose, it is important to improve the metal dispersion and optimize the active sites for the process to move forward.

The reaction system of CO_2 hydrogenation to formic acid/formate is complicated due to the multiple phases of gas-liquid-solid, where the CO_2 and H_2 gas need to enter the solution and further diffuse to the active sites on the surface of the solid catalyst for reaction, so the mechanism of the whole process is difficult to clarify. The commonly recognized reaction pathways are roughly as follows: (1) H_2 undergoes heterolysis to form protons and hydrides. H_2 may also undergo homolytic cleavage in some catalyst system. (2) The hydride species approach the electrophilic carbon atoms in CO_2/HCO_3^- to form formate intermediates on the metal. In the above-mentioned reaction routes, CO_2 insertion into metal-hydride bonds (the second step, the process of forming formate intermediates containing metal) is identified to be the rate-determining step. Thongnuam et al. [103] reported that CO_2 hydrogenation to formic acid on Lewis acid zeolites proceeds in two steps by DFT with the M06-L functional. First, CO_2 abstracts the hydrogen atom to form a formate intermediate. Then, intermediate and hydrogen species further react to form formic acid. They identified that the first step of hydrogenation is the rate-determining step in the catalyst system of Sn-ZSM-5. After comparing the perfect zeolite with defective one, they considered that the defect structure of zeolite is important for CO_2 hydrogenation to formic acid. It was supported by a free energy profile that the activation barrier for the defect structure is 33.0 kcal·mol^{-1}, which is lower than that for the perfect one (38.2 kcal·mol^{-1}) (see Figure 12.14a). The catalytic activity of different zeolite types and tetravalent metal centers were also considered, and they obtained the activation energy order (see Figure 12.14b): Sn-ZSM-5 (33.0 kcal·mol^{-1}) $<$ Sn-BEA (37.9 kcal·mol^{-1}) $<$ Sn-FAU (38.5 kcal·mol^{-1}), and the different tetravalent metal centers have different catalytic activity in the order of Ge $<$ Sn $<$ Zr $<$ Hf (see Figure 12.14c). Due to the complex reaction system of gas-liquid-solid phases, the particular reaction pathway would be influenced by multiple factors, such as catalysts, solution, base additives, and so on.

Szyja et al. [104] considered that the H_2O in the reaction system has a significant impact, especially in the catalyst system of zeolites. Take Ir_4/FAU for example, when there is no H_2O in the system, the barrier of the pathway to form the Ir-OOCH formate species is substantially lower than the Ir-bound COOH species, resulting in more formic acid. However, the existence of H_2O could promote the production of CO by reducing the activation barrier, which is competitive to the pathway to formic acid. In a way, the removal of water in the catalyst system is a necessary condition for ensuring the production of formic acid.

Zeolites, with unique pore structure and outstanding hydrothermal stability, have been widely used in many industrial reactions. Up to 250 topological structures and diverse characteristics endow zeolites with a wide range of applications. Zeolites are commonly used as excellent support materials in hydrogenation and dehydrogenation reactions [63, 105, 106]. In CO_2 hydrogenation to formic acid, zeolites itself are usually inactive, and they can promote the reaction performance in two ways: (1) dispersing

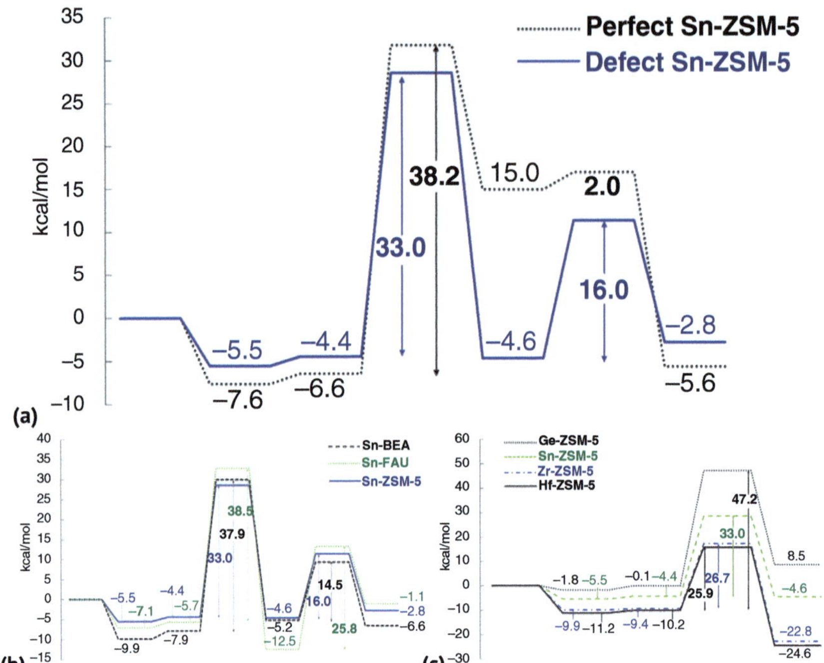

Figure 12.14 Free energy profile for CO_2 hydrogenation to formic acid (energies in kcal·mol^{-1}) over MFI zeolite. (a) Contrast of the perfect Sn-ZSM-5 zeolite and the defect Sn-ZSM-5 zeolite. (b) Contrast on the defect Sn-zeolite of Sn-ZSM-5, Sn-FAU, and Sn-BEA. (c) Contrast on defect frameworks of Ge-ZSM-5, Sn-ZSM-5, Zr-ZSM-5, and Hf-ZSM-5. *Source*: Reproduced from [103] / The Royal Society of Chemistry.

the metal active sites due to the channel structure and large specific surface area of zeolites. Better dispersion means more accessible active sites, which is beneficial for CO_2 hydrogenation and (2) providing microenvironmental conditions for building and confining unique active centers. Zeolites are suitable for many guests' molecules, accommodating them to form active centers within the zeolites' matrix [26, 39]. The detailed content will be illustrated through the following examples.

In CO_2 hydrogenation to formic acid, the most commonly used catalysts are noble metals, so it is extremely important to improve the utilization of metals. Unlike other disordered supports like SiO_2 and Al_2O_3, zeolites with an ordered pore structure are more conducive to metal dispersion. SBA-15 is a material with an ordered mesoporous structure, which is usually used to disperse the metal sites. Mori et al. [107] prepared a catalyst with PdAg nanoparticles supported on phenylamine-grafted mesoporous silica SBA-15, which was used for formic acid and CO_2-mediated hydrogen delivery and storage. The mesoporous channel structure and the weakly basic phenylamine functional group jointly assist in anchoring metal sites and achieving high dispersion of active sites (see Figure 12.15), resulting in high activity for the hydrogenation of CO_2 to formic acid.

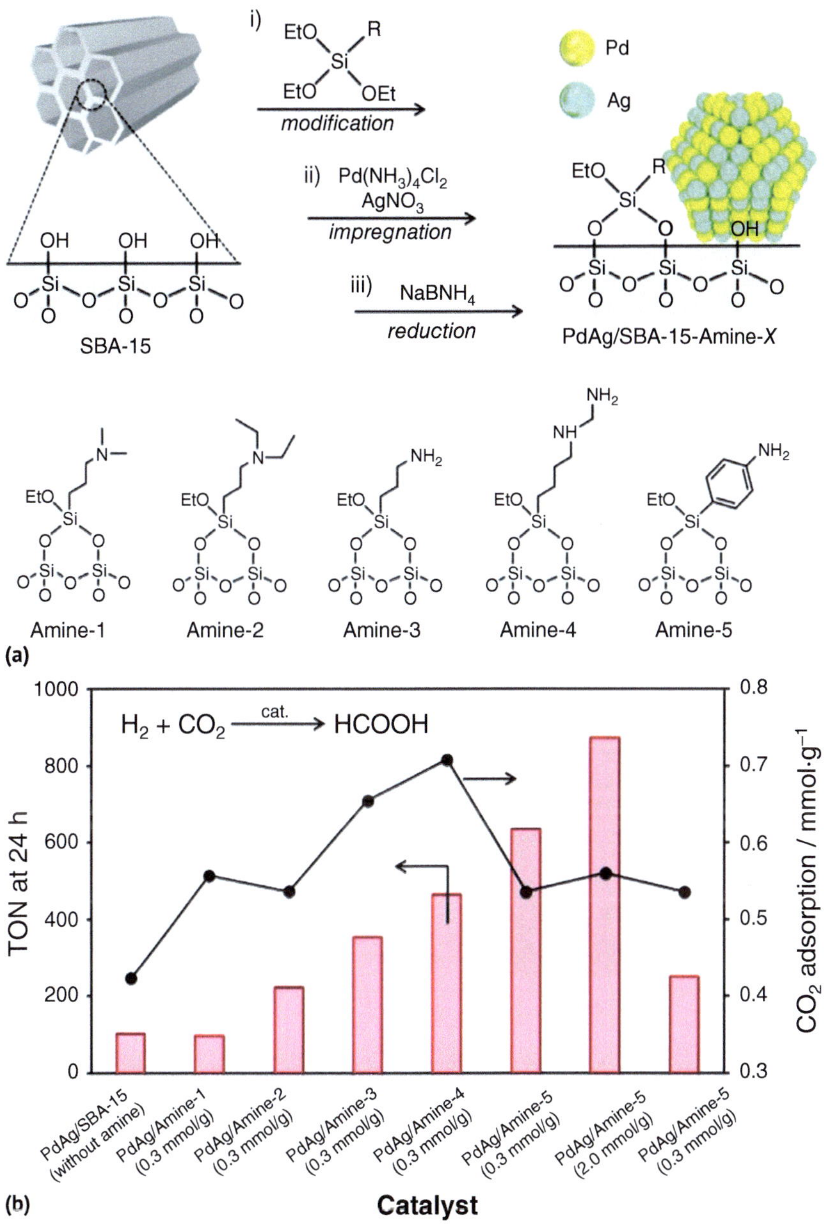

Figure 12.15 Formic acid/CO_2-mediated chemical hydrogen delivery/storage over PdAg/SBA-15. (a) Synthesis schematic illustration of PdAg nanoparticles supported on phenylamine-grafted mesoporous silica SBA-15, named as PdAg/Amine-x (x = 1–5) *Source*: Reproduced with permission from [107] / Royal Society of Chemistry. (b) Effect of grafted amine-functional groups on CO_2 hydrogenation to formic acid. Reaction conditions: 0.05 g catalyst, 1 mol/L $NaHCO_3$ aqueous solution (10 mL), total pressure 2.0 MPa (CO_2/H_2 = 1) at 373 K. *Source*: Reproduced from [107] / The Royal Society of Chemistry.

Zeolites are suitable for most guest molecules, accommodating them to form active centers within zeolites' matrix. The confinement of the zeolite is usually beneficial for preventing metal sintering and loss, resulting in higher reactivity and better stability. Sun et al. [108] proposed a ligand-protected method where ethylenediamine (en) and tetraethylenepentamine (TEPA) were used to facilitate the entry of subnanometer bimetallic Pd–Mn clusters into the zeolite matrix (see Figure 12.16a). The confinement of S-1 zeolite and the synergistic effect between Pd and Mn metal jointly formed an electron-enriched Pd surface, which was proved by XAS (see Figure 12.16b). Benefiting from the subnanometer metal clusters and the unique active metal sites, the $PdMn_{0.6}$@S-1 catalyst exhibited an excellent formate generation rate of 526 $mol_{formate} \cdot mol_{Pd}^{-1} \cdot h^{-1}$ at room temperature (25 °C) with NaOH as the base additive. The formate generation rate can achieve a high level of 2151 $mol_{formate} \cdot mol_{Pd}^{-1} \cdot h^{-1}$ at 80 °C (see Figure 12.16c). In addition, this catalyst also did well in the reverse process (the dehydrogenation of formic acid/formate) without any additive. Y-type zeolite has also been found to be useful in CO_2 hydrogenation to formic acid later by Jin et al. [109]. After comparing Ru-Beta zeolite, Ru-deAl-Beta, and $RuCl_3$ solution without zeolites with Ru-deAl-REUSY, they found that ruthenium (Ru)-modified REUSY exhibited a more significant promoting effect on the reaction of CO_2 hydrogenation to formic acid.

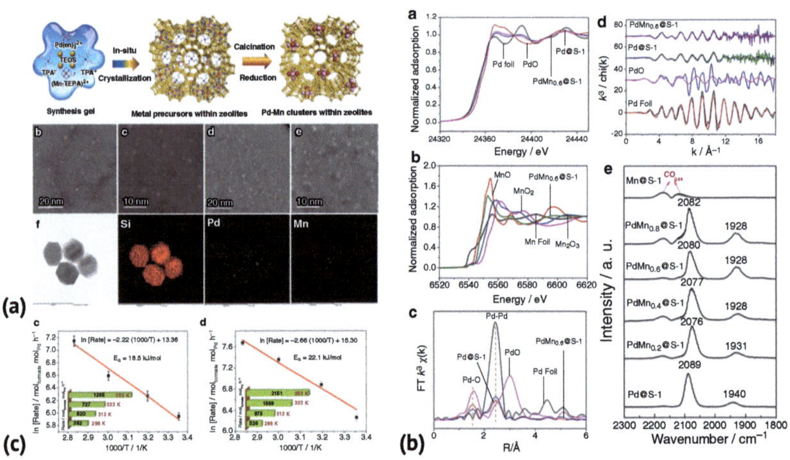

Figure 12.16 Zeolite-encaged Pd–Mn nanocatalysts for CO_2 hydrogenation and formic acid dehydrogenation. (a) Schematic of synthetic procedure and aberration-corrected STEM-HAADF images of $PdMn_x$@S-1 catalysts. *Source*: Reproduced with permission from [108] / John Wiley & Sons. (b) XANES spectra and in-situ CO-DRIFTS measurements of Pd@S-1, Mn@S-1, and $PdMn_x$@S-1 catalysts. (c) CO_2 hydrogenation performance over $PdMn_{0.6}$@S-1 catalyst in the NEt_3 solution and the NaOH solution at different temperatures. Reaction conditions: 0.005g catalyst, 1.5 mol/L aqueous NEt_3/NaOH solution (2 mL), H_2/CO_2 = 20 bar/20 bar. *Source*: Reproduced from [108] / John Wiley & Sons.

12.6 Conclusion and Outlook

Reacting the captured CO_2 with "green hydrogen" to produce oxygenates such as methanol, ethanol, DME, and formic acid for use in green fuels and energy storage is an important strategy for achieving carbon reduction and alleviating energy crises. Developing efficient and stable catalysts is crucial for CO_2 conversion, while the application of zeolite-based materials is promising. In the present chapter, we reviewed the application of zeolites in various processes of CO_2 hydrogenation to methanol, ethanol, DME, formic acid and covered the following effects: (1) the framework and pore microenvironment of zeolite can disperse, stabilize, and confine active metal species, thereby constructing unique catalytic active sites to improve catalytic performance and stability; (2) the tunable acid sites of zeolites for the synthesis of DME are indispensable for constructing the metal-zeolite bifunctional catalyst. Additionally, the easily controllable properties of zeolite morphology and structure also leave ample room for imagination in the design of the bifunctional catalysts; (3) the application of zeolite membranes can achieve in-situ removal of by-product H_2O during the reaction process, breaking the thermodynamic limitations and significantly improving the conversion efficiency of CO_2 to the target product.

With the increasing demand for chemicals and the development of industrial technology, current catalysts also need further improvement to enhance their performance and match the development of catalytic process technology. The design and application of zeolite-based catalysts should also be further developed to promote the practical application of CO_2 conversion technology. We consider that the development of zeolites in the field of CO_2 utilization in the future should focus on the following aspects: (1) the simple and convenient methods for preparation of stable and defect-free zeolite membranes for continuous-flow operation in harsh conditions are of great significance for achieving high reaction efficiency. (2) It would be very meaningful to achieve suitable metal loading with high dispersion on the zeolites, not only for CO_2 conversion but also for related reactions such as light alkane conversion. (3) Achieving the controllable preparation of high-performance integral zeolite catalytic materials through the combination of high-throughput CO_2 conversion processes, 3D printing technology, and artificial intelligence (e.g., via machine learning) is of great significance for the future development of the advanced zeolite catalysts.

Acknowledgments

This work was financially supported by the Special Project for Key Research and Development Program of Xinjiang Autonomous Region (No. 2022B01033-3), the Fundamental Research Funds for the Central Universities (DUT22LAB602), and the Liaoning Revitalization Talent Program (XLYC2203126).

References

1 The Keeling Curve. The Keeling Curve is a daily record of global atmospheric carbon dioxide concentration maintained by Scripps Institution of Oceanography at UC San Diego. https://keelingcurve.ucsd.edu/ (accessed August 2024).

2 United Nations Climate Change. Framework convention on climate change. https://unfccc.int/.

3 Boretti, A. (2024). Carbon dioxide hydrogenation for sustainable energy storage. *Int. J. Hydrogen Energy* 58: 1386–1395.

4 Dieterich, V., Buttler, A., Hanel, A. et al. (2020). Power-to-liquid via synthesis of methanol, DME or Fischer–Tropsch-fuels: a review. *Energy Environ. Sci.* 13: 3207–3252.

5 Silvio Silvério da Silva and Chandel, A.K. (2014). *Biofuels in Brazil. Fundamental Aspects, Recent Developments, and Future Perspectives*. Switzerland: Springer International Publishing.

6 He, M.Y., Sun, Y.H., and Han, B.X. (2013). Green carbon science: scientific basis for integrating carbon resource processing, utilization, and recycling. *Angew. Chem. Int. Ed.* 52: 9620–9633.

7 Simon Araya, S., Liso, V., Cui, X.T. et al. (2020). A review of the methanol economy: the fuel cell route. *Energies* 13: 596.

8 Zhang, X.B., Zhang, G.H., Song, C.S. et al. (2021). Catalytic conversion of carbon dioxide to methanol: current status and future perspective. *Front. Energy Res.* 8: 621119.

9 Olah, G.A. (2005). Beyond oil and gas: the methanol economy. *Angew. Chem. Int. Ed.* 44: 2636–2639.

10 Saeidi, S., Najari, S., Hessel, V. et al. (2021). Recent advances in CO_2 hydrogenation to value-added products-current challenges and future directions. *Prog. Energy Combust. Sci.* 85: 100905.

11 Stangeland, K., Li, H.L., and Yu, Z.X. (2018). Thermodynamic analysis of chemical and phase equilibria in CO_2 hydrogenation to methanol, dimethyl ether, and higher alcohols. *Ind. Eng. Chem. Res.* 57: 4081–4094.

12 Grabow, L.C. and Mavrikakis, M. (2011). Mechanism of methanol synthesis on Cu through CO_2 and CO hydrogenation. *ACS Catal.* 1: 365–384.

13 Palomino, R.M., Ramírez, P.J., Liu, Z.Y. et al. (2018). Hydrogenation of CO_2 on ZnO/Cu(100) and ZnO/Cu(111) catalysts: role of copper structure and metal–oxide interface in methanol synthesis. *J. Phys. Chem. B* 122: 794–800.

14 Le Valant, A., Comminges, C., Tisseraud, C. et al. (2015). The Cu-ZnO synergy in methanol synthesis from CO_2. Part 1. Origin of active site explained by experimental studies and a sphere contact quantification model on Cu+ZnO mechanical mixtures. *J. Catal.* 324: 41–49.

15 Behrens, M., Studt, F., Kasatkin, I. et al. (2012). The active site of methanol synthesis over $Cu/ZnO/Al_2O_3$ industrial catalysts. *Science* 336: 893–897.

16 Samson, K., Śliwa, M., Socha, R.P. et al. (2014). Influence of ZrO_2 structure and copper electronic state on activity of Cu/ZrO_2 catalysts in methanol synthesis from CO_2. *ACS Catal.* 4: 3730–3741.

17 Zhao, Y.F., Yang, Y., Mims, C. et al. (2011). Insight into methanol synthesis from CO_2 hydrogenation on Cu(111): complex reaction network and the effects of H_2O. *J. Catal.* 281: 199–211.

18 Wang, J.J., Li, G.N., Li, Z.L. et al. (2017). A highly selective and stable ZnO-ZrO_2 solid solution catalyst for CO_2 hydrogenation to methanol. *Sci. Adv.* 3: e1701290.

19 Zachopoulos, A. and Heracleous, E. (2017). Overcoming the equilibrium barriers of CO_2 hydrogenation to methanol via water sorption: a thermodynamic analysis. *J. CO_2 Util.* 21: 360–367.

20 Frei, M.S., Capdevila-Cortada, M., García-Muelas, R. et al. (2018). Mechanism and microkinetics of methanol synthesis via CO_2 hydrogenation on indium oxide. *J. Catal.* 361: 313–321.

21 Prašnikar, A., Pavlišič, A., Ruiz-Zepeda, F. et al. (2019). Mechanisms of copper-based catalyst deactivation during CO_2 reduction to methanol. *Ind. Eng. Chem. Res.* 58: 13021–13029.

22 Bahruji, H., Bowker, M., Hutchings, G. et al. (2016). Pd/ZnO catalysts for direct CO_2 hydrogenation to methanol. *J. Catal.* 343: 133–146.

23 Deng, Y.Z., Li, Z., Chen, T. et al. (2022). Low-cost and facile fabrication of defect-free water permeable membrane for CO_2 hydrogenation to methanol. *Chem. Eng. J.* 435: 133554.

24 Yue, W.Z., Li, Y.H., Wei, W. et al. (2021). Highly selective CO_2 conversion to methanol in a bifunctional zeolite catalytic membrane reactor. *Angew. Chem. Int. Ed.* 60: 18289–18294.

25 Li, H., Qiu, C., Ren, S. et al. (2020) Na^+-gated water-conducting nanochannels for boosting CO_2 conversion to liquid fuels. *Science* 367: 667-671.

26 Velty, A. and Corma, A. (2023). Advanced zeolite and ordered mesoporous silica-based catalysts for the conversion of CO_2 to chemicals and fuels. *Chem. Soc. Rev.* 52: 1773–1946.

27 Yang, H.Y., Gao, P., Zhang, C. et al. (2016). Core-shell structured Cu@m-SiO_2 and Cu/ZnO@m-SiO_2 catalysts for methanol synthesis from CO_2 hydrogenation. *Catal. Commun.* 84: 56–60.

28 Dobladez, J.A.D., Maté, V.I.Á., Torrellas, S.Á. et al. (2021). Comparative simulation study of methanol production by CO_2 hydrogenation with 3A, 4A and 5A zeolites as adsorbents in a PSA reactor. *Sep. Purif. Technol.* 262: 118292.

29 Maksimov, P., Laari, A., Ruuskanen, V. et al. (2021). Methanol synthesis through sorption enhanced carbon dioxide hydrogenation. *Chem. Eng. J.* 418: 129290.

30 Heracleous, E., Koidi, V., and Lappas, A.A. (2023). Experimental investigation of sorption-enhanced CO_2 hydrogenation to methanol. *ACS Sustain. Chem. Eng.* 11: 9684–9695.

31 Zhang, Y.T., Du, P., Shi, R. et al. (2020). Blocking defects of zeolite membranes with WS_2 nanosheets for vapor permeation dehydration of low water content isopropanol. *J. Membr. Sci.* 597: 117625.

32 Sorenson, S.G., Payzant, E.A., Gibbons, W.T. et al. (2011). Influence of zeolite crystal expansion/contraction on NaA zeolite membrane separations. *J. Membr. Sci.* 366: 413–420.

33 Xiang, M., Shi, Z.S., Zhang, X.P. et al. (2024). Facile synthesis of hierarchical SAPO-56 zeolite as a highly efficient catalyst for CO_2 hydrogenation to methanol. *Fuel* 361: 130663.

34 Ma, B.R., Pan, H.X., Yang, F. et al. (2022). Efficient CO_2 catalytic hydrogenation over CuO_x-ZnO/silicalite-1 with stable Cu^+ species. *Catal. Sci. Technol.* 12: 5850–5860.

35 Li, Y.N., Du, T., Chen, C. et al. (2024). Highly dispersed Cu-ZnOx regulated in zeolite for promoted performance in CO_2 hydrogenation to methanol. *Micropor. Mesopor. Mater.* 366: 112937.

36 Cui, W.G., Li, Y.T., Yu, L. et al. (2021). Zeolite-encapsulated ultrasmall Cu/ZnO_x nanoparticles for the hydrogenation of CO_2 to methanol. *ACS Appl. Mater. Interfaces* 13: 18693–18703.

37 Chai, Y.C., Qin, B., Li, B.N. et al. (2023). Zeolite-encaged mononuclear copper centers catalyze CO_2 selective hydrogenation to methanol. *Natl. Sci. Rev.* 10: nwad043.

38 Liu, L.C. and Corma, A. (2020). Evolution of isolated atoms and clusters in catalysis. *Trends Chem.* 2: 383–400.

39 Shamzhy, M., Opanasenko, M., Concepción, P. et al. (2019). New trends in tailoring active sites in zeolite-based catalysts. *Chem. Soc. Rev.* 48: 1095–1149.

40 Xiang, M., Shi, Z.S., Tong, L.C. et al. (2023). Fast synthesis of highly crystallized LSX zeolite with boosted carbon dioxide hydrogenation activity. *Appl. Catal. B: Environ.* 329: 122576.

41 Li, Y.N., Du, T., Chen, C. et al. (2024). Copper nanoparticles encapsulated in zeolite 13X for highly selective hydrogenation of CO_2 to methanol. *J. Environ. Chem. Eng.* 12: 111856.

42 Kanomata, R., Awano, K., Fujitsuka, H. et al. (2024). Development of silicalite-1 encapsulated Cu-ZnO catalysts for methanol synthesis by CO_2 hydrogenation. *Chem. Eng. J.* 485: 149896.

43 Pang, J.F., Zheng, M.Y., and Zhang, T. (2019). Chapter 2: synthesis of ethanol and its catalytic conversion. *Adv. Catal.* 64: 89–191.

44 Surisetty, V.R., Dalai, A.K., and Kozinski, J. (2011). Alcohols as alternative fuels: an overview. *Appl. Catal. A: Gen.* 404: 1–11.

45 Eagan, N.M., Kumbhalkar, M.D., Buchanan, J.S. et al. (2019). Chemistries and processes for the conversion of ethanol into middle-distillate fuels. *Nat. Rev. Chem.* 3: 223–249.

46 Ali, S.S., Ali, S.S., and Tabassum, N. (2022). A review on CO_2 hydrogenation to ethanol: reaction mechanism and experimental studies. *J. Environ. Chem. Eng.* 10: 106962.

47 Xu, D., Wang, Y.Q., Ding, M.Y. et al. (2021). Advances in higher alcohol synthesis from CO_2 hydrogenation. *Chem* 7: 849–881.

48 Li, X.P., Ke, J.C., Li, R. et al. (2023). Research progress of hydrogenation of carbon dioxide to ethanol. *Chem. Eng. Sci.* 282: 119226.

49 Kusama, H., Okabe, K., Sayama, K. et al. (1996). CO_2 hydrogenation to ethanol over promoted Rh/SiO_2 catalysts. *Catal. Today* 28: 261–266.
50 Asare Bediako, B.B., Qian, Q.L., and Han, B.X. (2021). Synthesis of C^{2+} chemicals from CO_2 and H_2 via C—C bond formation. *Acc. Chem. Res.* 54: 2467–2476.
51 Iltsiou, D., Mielby, J., and Kegnæs, S. (2023). Direct conversion of CO_2 into alcohols using Cu-based zeolite catalysts. *ChemPlusChem* 89: e202300313.
52 Ding, L.P., Shi, T.T., Gu, J. et al. (2020). CO_2 hydrogenation to ethanol over Cu@ Na-Beta. *Chem* 6: 2673–2689.
53 Wang, C.T., Zhang, J., Qin, G.Q. et al. (2020). Direct conversion of syngas to ethanol within zeolite crystals. *Chem* 6: 646–657.
54 Wang, G.S., Luo, R., Yang, C.S. et al. (2019). Active sites in CO_2 hydrogenation over confined VO_x-Rh catalysts. *Sci. China Chem.* 62: 1710–1719.
55 Zhang, F.Y., Zhou, W., Xiong, X.W. et al. (2021). Selective hydrogenation of CO_2 to ethanol over sodium-modified rhodium nanoparticles embedded in zeolite Silicalite-1. *J. Phys. Chem. C* 125: 24429–24439.
56 Liu, Y.J., Zhang, C.Y., Wang, R.J. et al. (2024). Low-temperature thermocatalytic coupling of CH_4 and CO_2 to ethanol over Zn-Ce/ZSM-5 by a stepwise technique. *ACS Catal.* 14: 8823–8835.
57 De Falco, M., Capocelli, M., and Centi, G. (2016). Dimethyl ether production from CO_2 rich feedstocks in a one-step process: thermodynamic evaluation and reactor simulation. *Chem. Eng. J.* 294: 400–409.
58 Mondal, U. and Yadav, G.D. (2019). Perspective of dimethyl ether as fuel: part I. Catalysis. *J. CO_2 Util.* 32: 299–320.
59 Ghosh, A., Nag, D., Chatterjee, R. et al. (2024). CO_2 to dimethyl ether (DME): structural and functional insights of hybrid catalysts. *Catal. Sci. Technol.* 14: 1387–1427.
60 Álvarez, A., Bansode, A., Urakawa, A. et al. (2017). Challenges in the greener production of formates/formic acid, methanol, and DME by heterogeneously catalyzed CO_2 hydrogenation processes. *Chem. Rev.* 117: 9804–9838.
61 Catizzone, E., Bonura, G., Migliori, M. et al. (2018). CO_2 recycling to dimethyl ether: state-of-the-art and perspectives. *Molecules* 23: 31.
62 Weckhuysen, B.M. and Yu, J.H. (2015). Recent advances in zeolite chemistry and catalysis. *Chem. Soc. Rev.* 44: 7022–7024.
63 Li, Y., Li, L., and Yu, J.H. (2017). Applications of zeolites in sustainable chemistry. *Chem* 3: 928–949.
64 Bonura, G., Cordaro, M., Spadaro, L. et al. (2013). Hybrid Cu-ZnO-ZrO_2/H-ZSM5 system for the direct synthesis of DME by CO_2 hydrogenation. *Appl. Catal. B: Environ.* 140–141: 16–24.
65 Yao, L.B., Shen, X.C., Pan, Y.B. et al. (2020). Unravelling proximity-driven synergetic effect within CIZO–SAPO bifunctional catalyst for CO_2 hydrogenation to DME. *Energy Fuels* 34: 8635–8643.
66 Fan, X., Jin, B.T., Ren, S.J. et al. (2021). Roles of interaction between components in CZZA/HZSM-5 catalyst for dimethyl ether synthesis via CO_2 hydrogenation. *AIChE J.* 67: e17353.

67 Li, W., Wang, K.C., Zhan, G.W. et al. (2020). Hydrogenation of CO_2 to dimethyl ether over tandem catalysts based on biotemplated hierarchical ZSM-5 and Pd/ZnO. *ACS Sustain. Chem. Eng.* 8: 14058–14070.

68 Frusteri, F., Cordaro, M., Cannilla, C. et al. (2015). Multifunctionality of Cu-ZnO-ZrO_2/H-ZSM5 catalysts for the one-step CO_2-to-DME hydrogenation reaction. *Appl. Catal. B: Environ.* 162: 57–65.

69 Bonura, G., Frusteri, F., Cannilla, C. et al. (2016). Catalytic features of CuZnZr-zeolite hybrid systems for the direct CO_2-to-DME hydrogenation reaction. *Catal. Today* 277: 48–54.

70 Frusteri, F., Migliori, M., Cannilla, C. et al. (2017). Direct CO_2-to-DME hydrogenation reaction: new evidences of a superior behaviour of FER-based hybrid systems to obtain high DME yield. *J. CO_2 Util.* 18: 353–361.

71 Liu, R., Tian, H.F., Yang, A.M. et al. (2015). Preparation of HZSM-5 membrane packed CuO-ZnO-Al_2O_3 nanoparticles for catalysing carbon dioxide hydrogenation to dimethyl ether. *Appl. Surf. Sci.* 345: 1–9.

72 Sánchez-Contador, M., Ateka, A., Aguayo, A.T. et al. (2018). Direct synthesis of dimethyl ether from CO and CO_2 over a core-shell structured CuO-ZnO-ZrO_2@SAPO-11 catalyst. *Fuel Process. Technol.* 179: 258–268.

73 Sánchez-Contador, M., Ateka, A., Ibáñez, M. et al. (2019). Influence of the operating conditions on the behavior and deactivation of a CuO-ZnO-ZrO_2@SAPO-11 core-shell-like catalyst in the direct synthesis of DME. *Renew. Energy* 138: 585–597.

74 Ateka, A., Sánchez-Contador, M., Portillo, A. et al. (2020). Kinetic modeling of CO_2+CO hydrogenation to DME over a CuO-ZnO-ZrO_2@SAPO-11 core-shell catalyst. *Fuel Process. Technol.* 206: 106434.

75 Vu, T.T.N., Desgagnés, A., and Iliuta, M.C. (2021). Efficient approaches to overcome challenges in material development for conventional and intensified CO_2 catalytic hydrogenation to CO, methanol, and DME. *Appl. Catal. A: Gen.* 617: 118119.

76 Diban, N., Urtiaga, A.M., Ortiz, I. et al. (2013). Influence of the membrane properties on the catalytic production of dimethyl ether with in situ water removal for the successful capture of CO_2. *Chem. Eng. J.* 234: 140–148.

77 Ateka, A., Rodriguez-Vega, P., Cordero-Lanzac, T. et al. (2021). Model validation of a packed bed LTA membrane reactor for the direct synthesis of DME from CO/CO_2. *Chem. Eng. J.* 408: 127356.

78 Yue, W.Z., Wan, Z., Li, Y.H. et al. (2022). Synthesis of Cu-ZnO-Pt@HZSM-5 catalytic membrane reactor for CO_2 hydrogenation to dimethyl ether. *J. Membr. Sci.* 660: 120845.

79 Dong, Q.B., Xu, W.L., Fan, X. et al. (2022). Prototype catalytic membrane reactor for dimethyl ether synthesis via CO_2 hydrogenation. *Ind. Eng. Chem. Res.* 61: 14656–14663.

80 Bahruji, H., Armstrong, R.D., Ruiz Esquius, J. et al. (2018). Hydrogenation of CO_2 to dimethyl ether over brønsted acidic PdZn catalysts. *Ind. Eng. Chem. Res.* 57: 6821–6829.

81 Singh, R., Tripathi, K., Pant, K.K. et al. (2022). Unravelling synergetic interaction over tandem Cu-ZnO-ZrO_2/hierarchical ZSM5 catalyst for CO_2 hydrogenation to methanol and DME. *Fuel* 318: 123641.

82 Cara, C., Secci, F., Lai, S. et al. (2022). On the design of mesostructured acidic catalysts for the one-pot dimethyl ether production from CO_2. *J. CO_2 Util.* 62: 102066.

83 Ren, S.J., Shoemaker, W.R., Wang, X.F. et al. (2019). Highly active and selective Cu-ZnO based catalyst for methanol and dimethyl ether synthesis via CO_2 hydrogenation. *Fuel* 239: 1125–1133.

84 Navarro-Jaén, S., Virginie, M., Morin, J.C. et al. (2022). Hybrid monometallic and bimetallic copper-palladium zeolite catalysts for direct synthesis of dimethyl ether from CO_2. *New J. Chem.* 46: 3889–3900.

85 Zhang, M.H., Liu, Z.M., Lin, G.D. et al. (2013). Pd/CNT-promoted $CuZrO_2$/HZSM-5 hybrid catalysts for direct synthesis of DME from CO_2/H_2. *Appl. Catal. A: Gen.* 451: 28–35.

86 Sheng, Q.T., Ye, R.P., Gong, W.B. et al. (2020). Mechanism and catalytic performance for direct dimethyl ether synthesis by CO_2 hydrogenation over CuZnZr/ferrierite hybrid catalyst. *J. Environ. Sci.* 92: 106–117.

87 Li, L.Y., Mao, D.S., Xiao, J. et al. (2016). Facile preparation of highly efficient CuO-ZnO-ZrO_2/HZSM-5 bifunctional catalyst for one-step CO_2 hydrogenation to dimethyl ether: influence of calcination temperature. *Chem. Eng. Res. Des.* 111: 100–108.

88 Wang, S., Mao, D.S., Guo, X.M. et al. (2009). Dimethyl ether synthesis via CO_2 hydrogenation over CuO-TiO_2-ZrO_2/HZSM-5 bifunctional catalysts. *Catal. Commun.* 10: 1367–1370.

89 Fang, X., Jia, H., Zhang, B. et al. (2021). A novel in situ grown Cu-ZnO-ZrO_2/HZSM-5 hybrid catalyst for CO_2 hydrogenation to liquid fuels of methanol and DME. *J. Environ. Chem. Eng.* 9: 105299.

90 Bonura, G., Migliori, M., Frusteri, L. et al. (2018). Acidity control of zeolite functionality on activity and stability of hybrid catalysts during DME production via CO_2 hydrogenation. *J. CO_2 Util.* 24: 398–406.

91 Frusteri, F., Bonura, G., Cannilla, C. et al. (2015). Stepwise tuning of metal-oxide and acid sites of CuZnZr-MFI hybrid catalysts for the direct DME synthesis by CO_2 hydrogenation. *Appl. Catal. B: Environ.* 176-177: 522–531.

92 Godini, H.R., Kumar, S.R., Tadikamalla, N. et al. (2022). Performance analysis of hybrid catalytic conversion of CO_2 to DiMethyl ether. *Int. J. Hydrogen Energy* 47: 11341–11358.

93 Ham, H., Xuan, N.T., Jung, H.S. et al. (2021). Crucial factors to maximize DME productivity on hydrophobic bifunctional Cu-ZnO-Al_2O_3/ferrierite by direct CO_2 hydrogenation. *Catal. Today* 369: 112–122.

94 Bahruji, H., Abdul Razak, S., Mahadi, A.H. et al. (2022). PdZn on ZSM-5 nanoparticles for CO_2 hydrogenation to dimethyl ether: comparative in situ analysis with Pd/TiO_2 and PdZn/TiO_2. *React. Kinet. Mech. Catal.* 135: 2973–2991.

95 Yang, X.Y., Sun, S., Ding, J.J. et al. (2012). Preparation, structure and performance of [CuO-ZnO-Al_2O_3]/[HZSM-5] core-shell bifunctional catalysts for one-step synthesis of dimethyl ether from CO_2+H_2. *Acta Phys. Chim. Sin.* 28: 1957–1963.

96 Eppinger, J. and Huang, K.W. (2017). Formic acid as a hydrogen energy carrier. *ACS Energy Lett.* 2: 188–195.

97 Supronowicz, W., Ignatyev, I.A., Lolli, G. et al. (2015). Formic acid: a future bridge between the power and chemical industries. *Green Chem.* 17: 2904–2911.

98 Enthaler, S., von Langermann, J., and Schmidt, T. (2010). Carbon dioxide and formic acid-the couple for environmental-friendly hydrogen storage? *Energy Environ. Sci.* 3: 1207–1217.

99 Jessop, P.G., Ikariya, T., and Noyori, R. (1995). Homogeneous hydrogenation of carbon dioxide. *Chem. Rev.* 95: 259–272.

100 Sun, R.Y., Liao, Y.H., Bai, S.T. et al. (2021). Heterogeneous catalysts for CO_2 hydrogenation to formic acid/formate: from nanoscale to single atom. *Energy Environ. Sci.* 14: 1247–1285.

101 Jessop, P.G., Joó, F., and Tai, C.C. (2004). Recent advances in the homogeneous hydrogenation of carbon dioxide. *Coord. Chem. Rev.* 248: 2425–2442.

102 Lee, J.H., Ryu, J., Kim, J.Y. et al. (2014). Carbon dioxide mediated, reversible chemical hydrogen storage using a Pd nanocatalyst supported on mesoporous graphitic carbon nitride. *J. Mater. Chem. A* 2: 9490–9495.

103 Thongnuam, W., Maihom, T., Choomwattana, S. et al. (2018). Theoretical study of CO_2 hydrogenation into formic acid on Lewis acid zeolites. *Phys. Chem. Chem. Phys.* 20: 25179–25185.

104 Szyja, B.M., Smykowski, D., Szczygieł, J. et al. (2016). A DFT study of CO_2 hydrogenation on faujasite-supported Ir4 clusters: on the role of water for selectivity control. *ChemCatChem* 8: 2500–2507.

105 Zhang, Q., Yu, J.H., and Corma, A. (2020). Applications of zeolites to C1 chemistry: recent advances, challenges, and opportunities. *Adv. Mater.* 32: 2002927.

106 Martínez, C. and Corma, A. (2011). Inorganic molecular sieves: preparation, modification and industrial application in catalytic processes. *Coord. Chem. Rev.* 255: 1558–1580.

107 Mori, K., Masuda, S., Tanaka, H. et al. (2017). Phenylamine-functionalized mesoporous silica supported PdAg nanoparticles: a dual heterogeneous catalyst for formic acid/CO_2-mediated chemical hydrogen delivery/storage. *Chem. Commun.* 53: 4677–4680.

108 Sun, Q.M., Chen, B.W.J., Wang, N. et al. (2020). Zeolite-encaged Pd-Mn nanocatalysts for CO_2 hydrogenation and formic acid dehydrogenation. *Angew. Chem. Int. Ed.* 59: 20183–20191.

109 Jin, M.Y., Song, Y.B., Wang, C.L. et al. (2023). Catalytic transfer hydrogenation of CO_2 to formic acid with glycerol as the hydrogen donor over Ru-REUSY catalysts. *Mol. Catal.* 551: 113655.

Part V

Methane Conversion

13

Application of Zeolites in Methane to Value-added Chemicals

Jin Liu[1,2], *Bo Wu*[1], *Liangshu Zhong*[1,2,*]

[1]*Center for Low-Carbon Conversion Science and Engineering, Shanghai Advanced Research Institute, Chinese Academy of Sciences, Shanghai, PR China*
[2]*School of Physical Science and Technology, ShanghaiTech University, Shanghai, PR China*

*Corresponding author: zhongls@sari.ac.cn

13.1 Introduction

As an important organic compound, methane (CH_4) is attracting increasing attention owing to its natural abundance and competitive production costs [1–3]. Methane originates from several fossil and renewable sources, notably shale gas [4], coalbed methane [5], combustible ice [3], as well as biogas [6]. Presently, the predominant utilization of methane involves combustion for energy generation, and only a tiny fraction is used as feedstock for chemical production. However, methane storage and long-distance transportation are economically unviable due to their extremely low boiling point (−164 °C). Notably, methane represents the second-largest contributor to greenhouse gas emissions after CO_2, and the emission of methane will affect the global climate [7–9]. Therefore, there is significant interest in developing methods for the on-site conversion of methane to condensable value-added chemicals such as methanol, olefins, and aromatics [10]. Methane conversion to value-added chemicals rather than direct combustion is a promising strategy to maximize the value of methane and reduce carbon emissions.

Methane is the most stable and inert hydrocarbon. Key physicochemical properties of methane include its small polarizability (2.84×10^{-40} C^2 m^2 J^{-1}), elevated ionization energy (1206.1 kJ mol^{-1}), modest proton affinity (543.9 kJ mol^{-1}), extremely weak acidity (pKa = 48), and the strongest known C—H bond (439.3 kJ mol^{-1}) [11, 12]. Therefore, the activation of C—H bonds and subsequent conversion of methane into value-added chemicals present formidable challenges. Methane can be broadly classified into indirect and direct routes (see Figure 13.1). For the indirect approach, methane undergoes initial conversion to synthesis gas

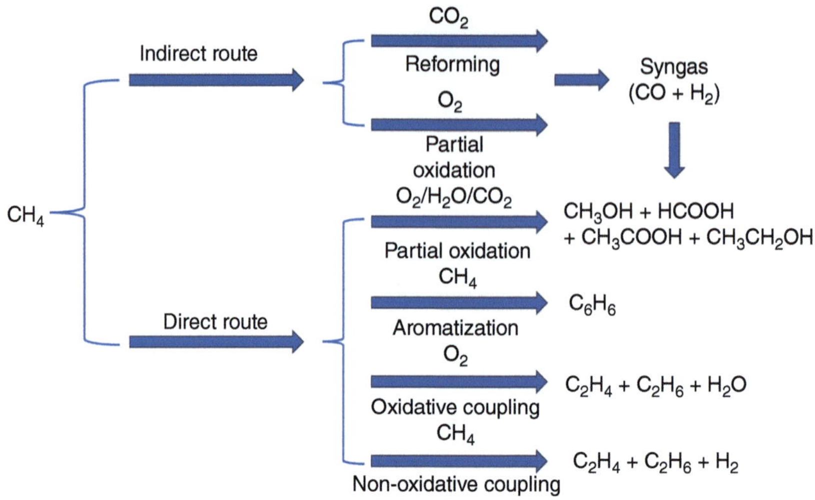

Figure 13.1 Approaches for methane valorization into various chemical feedstocks.

(a CO and H_2 mixture) via steam reforming, dry reforming, or partial oxidation, then the as-obtained syngas is converted to a variety of valuable products, such as alcohols, olefins, and heavy hydrocarbons [13–16]. At present, the indirect route for methane conversion has been successfully applied in industry. However, the current industrial pathway is highly energy-intensive, as the conversion typically requires high temperatures. Additionally, coke deposition and catalyst deactivation inevitably occur under severe reaction conditions. Furthermore, the intermediates or target products exhibit higher reactivity than methane, as their C—H bonds are comparatively weaker, facilitating activation once methane's strong C—H bond is cleaved. As a result, enabling direct methane conversion under mild reaction conditions is essential to reduce energy demands and enhance methane utilization efficiently.

The direct catalytic conversion of methane proceeds exclusively through either oxidative or non-oxidative pathways. The oxidative process of CH_4 includes steam reforming [14, 17], dry reforming [18, 19], and partial oxidation (yielding syngas) [20], oxidative coupling (producing ethylene) [13, 21, 22], and complete oxidation (forming CO_2) [23, 24]. Notably, the formation of CO_2 is inevitable for the incomplete oxidation process due to complete oxidation is more thermodynamically favorable. Unlike oxidative transformation, a non-oxidative process for methane conversion is endothermic and unfavorable in thermodynamics. Converting methane through a non-oxidative dehydrogenation reaction can efficiently produce ethylene, benzene, hydrogen, and carbon-based materials [25–27].

Many catalysts have been developed for methane conversion to various desired chemicals, complemented by in-depth review studies addressing methane activation strategies [2, 26–28]. Typically, Horn et al. [2] reviewed the challenges in methane activation and comprehensively summarized the status and existing bottleneck problems in the catalytic methane functionalization for chemical synthesis.

Recently, Schwach et al. [28] summarize recent advances in high-temperature methane conversion and elucidate the reaction mechanisms underlying direct conversion of methane to chemicals via oxidative coupling of methane (OCM), methane dehydroaromatization (MDA) and methane to olefins, aromatics, and hydrogen (MTOAH). Tang et al. [29] also provided their insights on the low-energy catalytic methane conversion processes.

Zeolites are typical porous crystalline materials with tunable acidity, high thermal, and hydrothermal stability [30, 31]. Zeolites have demonstrated exceptional catalytic performance and found widespread application in heterogeneous processes. The past few years have witnessed growing applications of metal-functionalized zeolite catalysts in alkane transformation reactions, including methane conversion. This chapter presents a comprehensive overview of recent advancements in zeolite-based catalysts for methane conversion to value-added chemicals. This work categorizes the applications of zeolite-based catalysts in methane conversion, highlighting their role in producing syngas, oxygenated compounds (such as methanol and acetic acid), olefins, and aromatics. The challenges and perspectives for future methane conversion with higher activity and selectivity will also be discussed.

13.2 Direct Methane Conversion to Chemicals

13.2.1 Methane Conversion to Syngas

Syngas production from methane occurs via high-temperature thermochemical processes, including steam reforming, dry reforming, and partial oxidation. Syngas production is the most developed technology for methane conversion.

13.2.2 Steam Reforming of Methane

The steam reforming of methane (SRM) process involves the endothermic reaction of CH_4 with H_2O (steam) to form H_2 and CO $(CH_4 + H_2O \longrightarrow CO + 3H_2,\ \Delta H_{298K} = 206\ \text{KJ mol}^{-1})$, coupled with the exothermic water-gas shift (WGS) reaction $(CO + H_2O \longrightarrow CO_2 + H_2,\ \Delta H_{298K} = -41\ \text{KJ mol}^{-1})$ [32].

To achieve methane steam reforming, temperatures exceeding 750 °C are necessary to surmount thermodynamic limitations. However, high-temperature processes accompanied by high energy loss and harsh process conditions inevitably lead to metal sintering, rapid carbon deposition, and eventual catalyst deactivation. Thus, improving catalyst stability by reducing catalyst sintering and carbon deposition is of great significance for the design of SRM catalysts. Commonly, the comparatively low cost of Ni-based catalysts, as opposed to noble metals (Rh, Ru), drives their widespread use in industrial SRM applications [33–35], despite the higher catalytic activity and lower coking tendency of Rh/Ru-based catalysts [36–38]. Al-Ubaid and Wolf [39] found that zeolite-based Ni/Na-Y SRM catalysts exhibited promising performance for SRM. The designed core-shell Ni/Al_2O_3@Sil-1 [40] showed

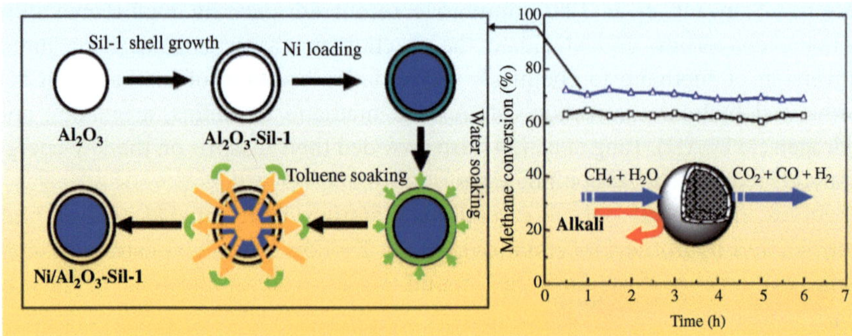

Figure 13.2 Structural engineering and catalytic properties of Ni/Al_2O_3-Sil-1 core-shell architecture. *Source*: [40] / with permission of Elsevier.

promising coke resistance and hydrothermal stability (see Figure 13.2), making it suitable for direct internal reforming-molten carbonate fuel cell (DIR-MCFC). Other cases for core-shell SRM catalysts include NiO/M-Sil-1 [41] and core-shell structured Ce-Zr/Ni-Mg with H-β zeolite encapsulation [42]. Although core-shell structures effectively prevent sintering and coking, their catalytic performance is invariably compromised by diffusion barriers.

13.2.3 Dry Reforming of Methane

Dry reforming of methane (DRM) enables the simultaneous conversion of methane and carbon dioxide—two potent greenhouse gases—into syngas (a mixture of CO and H_2), yielding a product with a significantly reduced H_2/CO ratio (typically 0.7~1.0) compared to the conventional steam reforming process (~ 3.0). DRM shares SRM's highly endothermic nature but faces greater catalyst deactivation risks from sintering and carbon deposition [43]. Noble metals, including Pt, Ru, and Rh, are highly active DRM catalysts with superior resistance to carbon deposition [36, 44, 45]. While noble metals resist coking, Ni catalysts quickly deactivate due to carbon deposition from the Boudouard reaction $\left(2CO \longrightarrow C + CO_2,\ \Delta H_{298K} = -172\ \text{KJ mol}^{-1}\right)$ and methane pyrolysis $\left(CH_4 \longrightarrow C + 2H_2,\ \Delta H_{298K} = 75\ \text{KJ mol}^{-1}\right)$ [46, 47]. The application of zeolites as supports for Ni-based DRM catalysts exhibits excellent catalytic performance, attributed to their high surface areas, well-defined structures, and high CO_2 adsorption as well as activation capacity. It was demonstrated that the catalytic performance depends on zeolite acidity, reducibility, and morphology [48].

A suitable preparation method and dopants of DRM catalysts are vital to regulate metal dispersion and particle size [49–51]. For instance, ultrasound-assisted synthesis of Ni/ZSM-5 yielded smaller Ni nanoparticles and superior activity [52]. Enhancement of the dispersion and stabilization of Ni-based DRM catalysts can be achieved by adding a second metal as a promoter. Estephane et al. [51] prepared bimetallic Ni-Co/ZSM-5 by the wet impregnation method and found that higher

DRM catalytic activity and lower carbon deposition could be obtained after adding the metal cobalt. Moreover, the coke resistance and the lifetime of Ni-based catalysts could also be improved when Mn was introduced [53]. The doping of rare-earth metals (La, Y, and Ce) [54–56] and noble metals (Ru, Pt) [57, 58] has also been investigated. The incorporation of Ce accelerates gasification rates, effectively suppressing carbon deposition [59], and a suitable amount of Ce was greatly beneficial to the dispersion of Ni nanoparticles and coking resistance [60]. Remarkably, Ni-Ce/ZSM-5 (2 wt.% Ce) exhibited high CH_4 (99%) and CO_2 (94%) conversions while maintaining robust stability (see Figure 13.3) [61]. It was suggested that the addition of Ce could promote Ni high dispersion and favor CH_4 dissociation and H_2 production as well as the gasification of coke due to the electric transformation and enhancement of CO_2 absorption and carbon oxidation. In addition, Ni-Ru and Ni-Pt bimetallic catalysts were explored to enhance catalytic activity and coke resistance. It was hypothesized that Ru or Pt enhances NiO reducibility [57, 58].

Confinement or stabilization of Ni species in porous structures was developed to mitigate the deactivation of Ni-based catalysts [62, 63]. with metal Ni encapsulated of Ni in silicalite-1 yielded a Ni/silicalite-1 catalyst with enhanced stability in the DRM process (see Figure 13.4) [64]. It was suggested that the high stability of Ni/silicalite-1 was attributed to the formation of Ni—O—Si species which can minimize active metal sintering and coking. Using a seed-directed synthesis approach, Xu et al. [65] successfully prepared Ni@SiO_2-S1 with fully encapsulated Ni nanoparticles. The migration and aggregation of Ni nanoparticles, as well as coke deposition, were effectively suppressed. Furthermore, the stability of Ni-based catalysts can be improved by precisely controlling both the encapsulation degree and Ni nanoparticle dispersion.

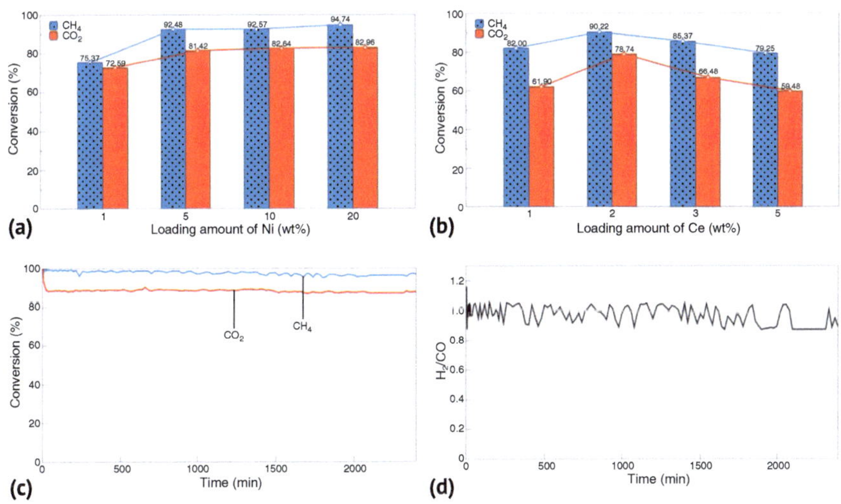

Figure 13.3 (a, b) Influence of Ni and Ce loadings on CH_4 and CO_2 conversions. (c, d) Time-dependent stability of conversion rates and H_2/CO ratio. *Source*: [61] / with permission of Elsevier.

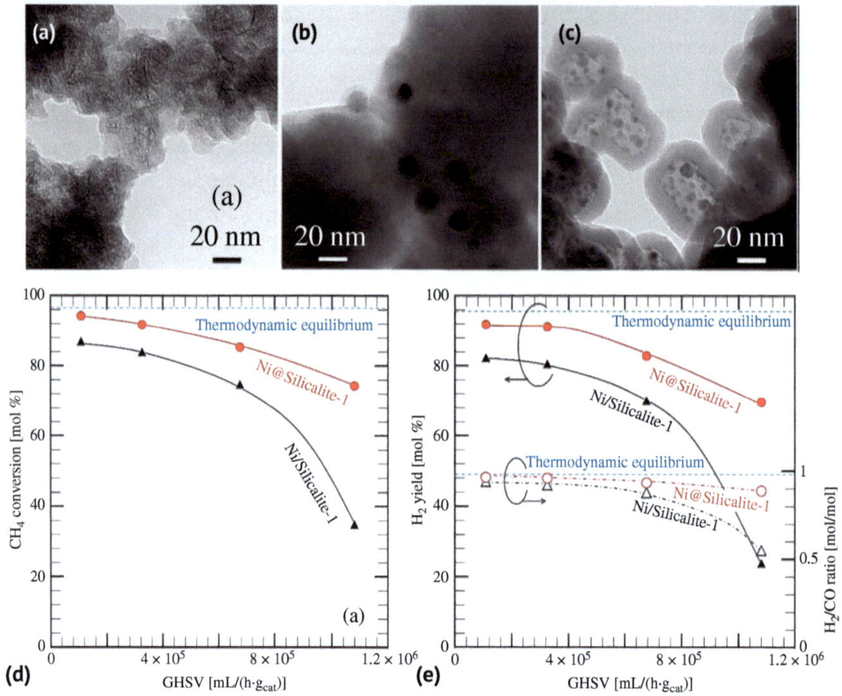

Figure 13.4 (a–c) TEM images of the prepared Ni-loaded catalysts before H_2 reduction. (d, e) Conversions of CH_4 and CO_2 for DRM over Ni@Silicalite-1 and Ni/silicalite-1 at various GHSV (reaction temperature: 850 °C; time on stream: 1 h; inlet gas composition: $CH_4/CO_2/Ar/He$ = 20/20/40/10 mL/min). *Source*: Reproduced from [64] / with permission of Elsevier.

13.2.4 Direct Conversion of Methane to Oxygenates

The most common oxygenate derived from methane direct conversion is methanol. This crucial C_1 intermediate is mainly synthesized through an indirect process: first, methane steam reforming generates syngas (CO/H_2 mixture), which then undergoes Fischer–Tropsch synthesis [29]. However, high temperature and pressure are indispensable for the indirect pathway. Alternative methods for converting methane to methanol under mild conditions are urgently needed. The selective oxidation of methane to methanol represents a promising pathway for CH_4 valorization. In nature, methane oxidation to methanol is primarily facilitated by two classes of enzymes: iron-dependent soluble methane monooxygenases (sMMOs) and copper-containing particulate MMO. For these monooxygenases, the active sites are well defined and isolated, enabling feasible access to study the electronic and structural properties via multiple advanced techniques. Bioinspired zeolite catalysts with precisely positioned iron and copper sites facilitate the selective partial oxidation of methane to methanol (see Figure 13.5a,b) [66].

A major step forward was realized by Sushkevich and coworkers [67]. They achieved a breakthrough by showing that dimeric Cu sites in Cu-MOR catalysts

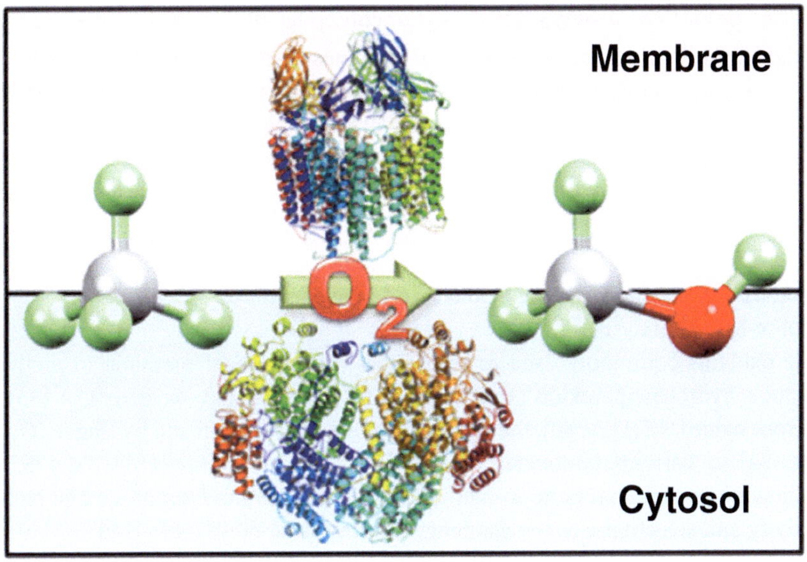

(a)

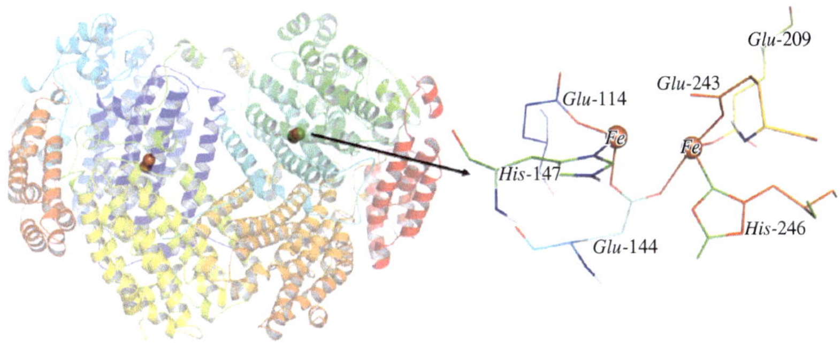

(b)

Figure 13.5 (a) Proposed mechanism for enzymatic methane-to-methanol conversion using molecular oxygen. (b) Three-dimensional structure of the sMMO hydroxylase from Methylococcus capsulatus (Bath). *Source*: [66] / with permission of American Chemical Society.

enable methane-to-methanol conversion using water as the oxidant in a stepwise manner. This strategy contains multistep processes, including catalyst activation in helium (He) in 400 °C, following sequential exposure of the catalyst to methane (7 bar) and water at 200 °C, the system yielded a consistent methanol production of 0.204 mol_{MeOH} mol_{Cu}^{-1} h^{-1} in the zeolite framework. Isotope labeling experiments, advanced characterization techniques, and density functional theory (DFT) calculations revealed that water is the source of oxygen to regenerate the zeolite active centers and renders methanol desorption energetically favorable. A mechanism involving methane oxidation at Cu^{II} oxide active centers, followed by Cu^{I} reoxidation by water with concurrent formation of hydrogen, was also proposed.

Fe-zeolites represent another well-established system for selective methane oxidation to methanol, where a mononuclear high-spin (S=2) Fe(IV)=O species [α-Fe(IV)=O] acts as the catalytic center [68–70]. The zeolite lattice enforces a rigid coordination environment that promotes H-atom abstraction. Synder and co-workers found that while Fe-BEA (wide pores) loses all activity after one turnover, Cu-CHA (narrow pores) regenerates 40% of its active sites, sustaining catalytic cycling [71]. They demonstrated that the small limited diffusion imposed by small pores suppresses methyl radicals from the active site following C—H activation, thus enhancing radical recombination to methanol.

13.2.4.1 Direct Conversion of Methane to C_1 Oxygenates over Zeolite-based Catalysts

The oxidants commonly used for the direct conversion of methane to methanol include hydrogen peroxide (H_2O_2) [68, 72–74], molecular oxygen (O_2) [75–78], nitrous oxide (N_2O) [79, 80], the mixture of molecular oxygen and hydrogen (O_2/H_2) [81, 82] or carbon monoxide (O_2/CO) [83, 84], and water (O_2/H_2O) [75, 85]. Achieving efficient methane-to-methanol oxidation demands catalysts that balance activity and selectivity. A key challenge is preventing methanol's over-oxidation—due to its higher reactivity—into formic acid and CO_2. Advanced catalytic strategies are urgently needed to control methanol over-oxidation. Recent studies demonstrate that Cu-modified zeolite catalysts can effectively enhance methanol selectivity by blocking the overoxidation of methane reaction pathways, leading to formic acid and CO_2 formation.

13.2.4.1.1 Selective Oxidation of Methane to Methanol With Hydrogen Peroxide as Oxidant

As a green oxidant, hydrogen peroxide is extensively used in chemical transformations, generating only water and oxygen as byproducts. Hammond and co-workers demonstrated that methane could be selectively converted to formic acid over an Fe/S-1 (Fe/ZSM-5) catalyst under mild conditions (50 °C) using hydrogen peroxide as an oxidant [68]. Formic acid was found to be the main product in the liquid products (87%), and the selectivity of methanol was only 12%. Through combined extended X-ray absorption spectroscopy (EXAFS) and DFT calculation, they identified dinuclear iron species of $\left[Fe_2(\mu_2 - OH)_2(OH)_2(H_2O)_2\right]^{2+}$ as the active sites (see Figure 13.6a) [68]. Upon doping Fe/ZSM-5 with Cu, methanol became the dominant liquid product (85% selectivity), revealing that Cu plays a crucial role in suppressing the overoxidation of methanol to formic acid (see Figure 13.6b) [86]. A continuous operation with hydrogen peroxide as an oxidant was also investigated, and around 0.5% methane conversion can be obtained with a methanol selectivity of 92.2% [87]. Recent work by Fang and co-workers revealed that MOR-anchored dinuclear Fe sites, when enhanced with Cu cations, serve as active catalysts for the H_2O_2-driven oxidation of methane to methanol [88].

Yu and co-workers [89] recently engineered Cu-enhanced ZSM-5 catalysts featuring mononuclear Fe species for methane hydroxylation with hydrogen peroxide. Through a multi-technique approach—ultraviolet visible spectroscopy (UV-vis), Mössbauer spectroscopy, and EXAFS results—they elucidated the mononuclear

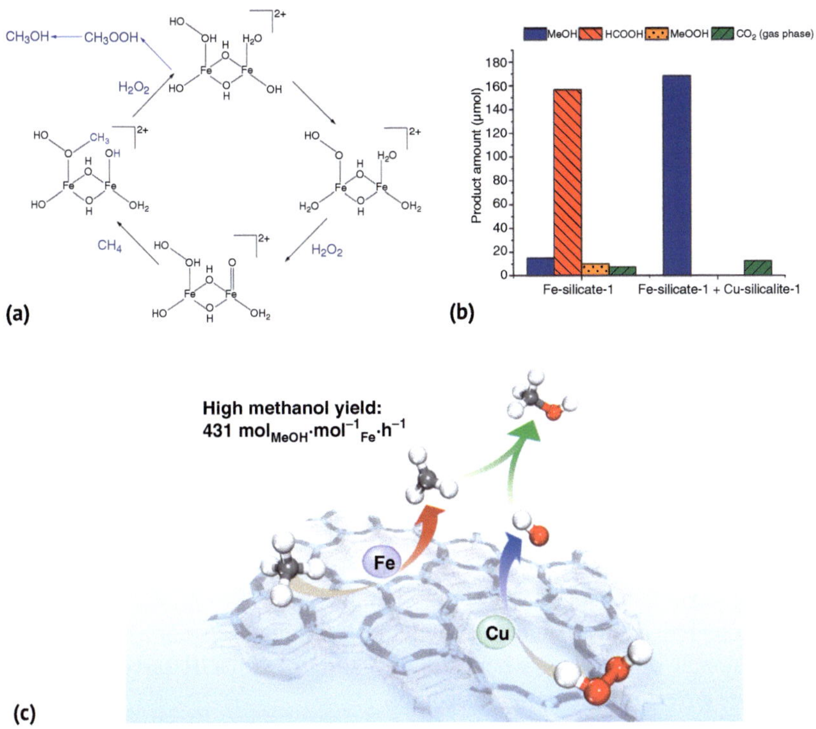

Figure 13.6 (a) A binuclear Fe^{2+} extra-framework cation in ZSM-5 catalyzes the oxidation of methane to CH_3OOH using H_2O_2, with the active center maintaining a formal +2 charge throughout the catalytic cycle. (b) The co-presence of Cu-silicalite-1 with Fe-silicalite-1 switches methane oxidation selectivity from formic acid to methanol. *Source*: [86]/with permission of American Chemical Society. (c) Cu-promoted Fe/ZSM-5 catalyzes the selective oxidation of methane to methanol with enhanced efficiency. *Source*: [89] / with permission of American Chemical Society.

$[(HO)_2\text{-}Fe(III)\text{-}(H_2O)_2]^+$ structure as the key active sites. The optimized Cu-Fe(2/0.1)/ZSM-5 demonstrated exceptional catalytic performance, achieving a remarkable turnover number of 431 mol_{MeOH} mol_{Fe}^{-1} h^{-1}, and the methanol selectivity was up to 80%. The addition of Cu species enabled the homolytic cleavage of H_2O_2 into hydroxyl radical (•OH), the resulting intermediate subsequently abstracted a hydrogen atom from methane to generate methyl radical (•CH_3), followed by their fast recombination with hydroxyl radicals to selectively produce methanol (see Figure 13.6c). Expanding on these results, Huang et al. [73] synthesized a Pd single-atom catalyst (SAC) (0.01 wt. % Pd/ZSM-5) on ZSM-5, demonstrating high selectivity in methane oxidation with hydrogen peroxide. The major product under different reaction temperatures was formic acid. With the impregnation of CuO to 0.01 wt. % Pd/ZSM-5, around 78~86% methanol selectivity was observed in liquid products. Detailed structure characterization verified the presence of Pd single atoms located in the internal surface of micropores of ZSM-5 with a moiety denoted as Pd_1O_4.

13.2.4.1.2 Selective Oxidation of Methane to Methanol With Molecular Oxygen as Oxidant

Molecular oxygen is a more economical and sustainable alternative to H_2O_2 for oxidation, owing to its abundance and low cost. Narsimhan et al. [75] first reported a continuous flow selective oxidation of methane to methanol over a series of Cu-exchanged zeolites with varying topologies at low temperature (210~225 °C). Notably, the Cu-Na-ZSM-5 exhibited a methanol yield of 1.81 $\mu mol_{gcat}^{-1}\ h^{-1}$ with 70.6% selectivity among all products. Building on previous work, Sun et al. [85] developed a Cu-CHA catalyst that enables simultaneous optimization of methane conversion and methanol selectivity in the direct catalytic oxidation process (see Figure 13.7a). Under optimized reaction conditions (300 °C, O_2/CH_4 <1:2000) that promoted the Cu^{2+}-Cu^{+}-Cu^{2+} redox cycle, the Cu-CHA catalyst exhibited outstanding catalytic performance, reaching a methanol space-time yield (STY) of 543 mmol $mol_{Cu}^{-1}\ h^{-1}$ with 91% product selectivity. Through isotopic labeling studies and temperature-programmed surface reactions, it was unequivocally demonstrated that water provides both the oxygen and hydrogen atoms for methanol hydroxyl formation. In continuous operation, both methanol selectivity and STY showed no degradation over 500 hours, demonstrating exceptional catalyst stability. Theoretical analysis demonstrated that methane activation proceeds via CuOH monomer/dimer sites, where the formation of CuOOH serves as the key intermediate governing methanol selectivity.

To ensure operational safety in continuous methane-to-methanol conversion, the use of water as the sole oxidant—eliminating the need for additional oxidants such as O_2—represents a highly desirable strategy. Jeong conducted the methane steam with water (33 mol% CH_4 + 66.7 mol% H_2O) over Cu-H-MOR catalysts (see Figure 13.7b) [90]. 100% of methanol selectivity can be obtained without the formation of carbon dioxide under 350 °C. The introduction of an 85 ppm concentration of molecular

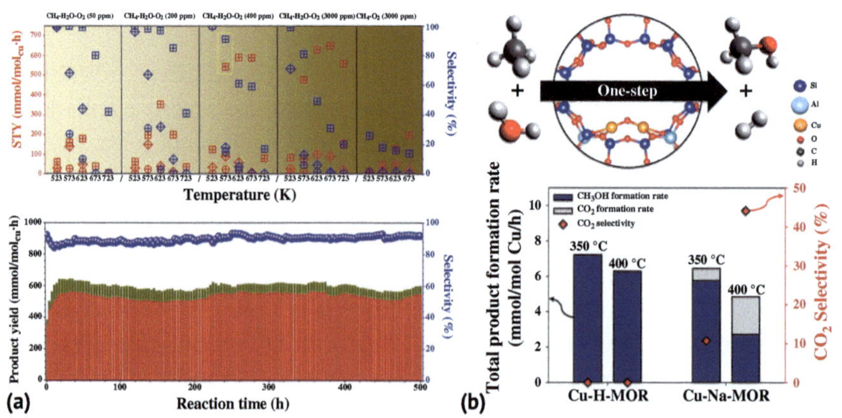

Figure 13.7 (a) Steady-state methanol production via selective methane catalytic oxidation. *Source*: [85] with permission of Elsevier. (b) Methane oxidation over Cu-H-MOR and Cu-Na-MOR with water as the oxidant. *Source*: [90] / with permission of American Chemical Society.

oxygen can improve the methanol yield from 7.2 to 98 mmol mol_{Cu}^{-1} h^{-1}, while keeping the high methanol selectivity of 99%.

The pursuit of selective methane-to-methanol conversion at temperatures below 200 °C continues to drive fundamental and applied catalysis research. However, under oxidation conditions, the over-oxidation of intermediates will lead to low yield and selectivity of methanol. In a breakthrough approach, Qi and co-workers [76] designed a closed catalytic cycle by developing ZSM-5-supported Au nanoparticles to catalyze methane to methanol under a temperature range from 120 to 240 °C using a batch reactor. The oxygen selectivity ranged from 60 to 100%. Remarkably, even after just one hour of reaction, the system exhibited a productivity of 1.9 mol mol_{Au}^{-1} h^{-1} (9.5 mmol g_{Au}^{-1} h^{-1}), surpassing the performance of reported Cu-based catalysts under comparable conditions.

Despite molecular oxygen being the oxidant of choice for selective methane oxidation to methanol, improving its activation kinetics is essential to achieve superior catalytic performance at low temperatures. As reported by Ab Rahim and co-workers [81], benchmark TiO_2-supported PdAu alloys are regarded as the highly active sites for selective oxidation of methane to methanol with molecular oxygen in the presence of molecular hydrogen by coupling with the *in-situ* generation of hydrogen peroxide. Experimental and spectroscopic evidence indicates that hydroxyl radicals, generated by the reaction of molecular oxygen and hydrogen, preferentially promote methanol formation compared to H_2O_2 as the initial oxidant.

The above system offers a novel strategy to enhance selective methane-to-methanol conversion through accelerated molecular activation. However, the low productivity of methanol and the abundant unused H_2O_2 need to be further addressed. To solve these problems, Jin et al. [82] developed an organosilane-decorated ZSM-5 encapsulated AuPd alloys (AuPd@ZSM-5-C_{16}) to catalyze methane to methanol under low temperatures. Organosilane functionalization facilitates the diffusion of H_2, O_2, and CH_4 to active sites while spatially confining in situ generated hydroperoxides, thereby enhancing catalytic performance. The high enrichment efficiency for AuPd@ZSM-5-C_{16} was around 76–86%, much higher than the pristine AuPd@ZSM-5 (8%). The confined high concentration of local peroxide around AuPd alloys nanoparticles significantly enhances methane oxidation activity, with AuPd@ZSM-5-C_{16} achieving a remarkable STY of 92 mmol g_{AuPd}^{-1} h^{-1} and methanol selectivity up to 92%. More importantly, around 17% of methane conversion was obtained in this system. However, the high usage of noble metals (Au and Pd) in this system needs to be further optimized. Wu et al. [91] engineered a ZSM-5-supported PdCu bimetallic catalyst system for the selective methane oxidation to oxygenates, revealing a unique active structure comprising PdO nanoparticles coexisting with atomically dispersed Cu species. PdO nanoparticles promote H_2O_2 synthesis, whereas single-atom Cu centers catalyze the selective decomposition of H_2O_2 into •OH radicals, driving the selective oxidation of methane. The hydroxyl radicals can subsequently react with methane to form methyl, and finally lead to the formation of methanol (see Figure 13.8a). The PdCu/ZSM-5 catalyst showed

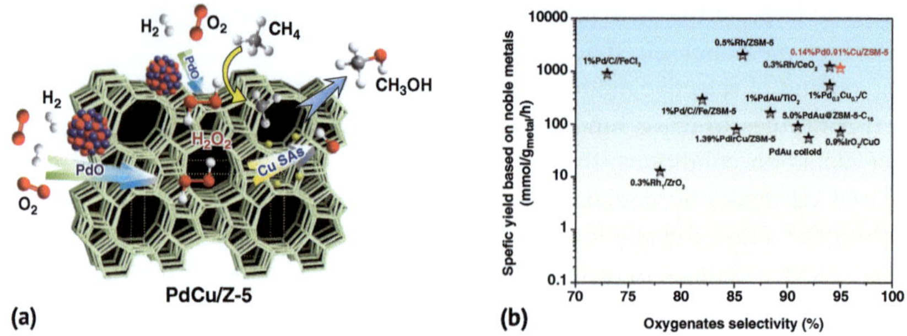

Figure 13.8 (a) Reaction pathway for the selective catalytic oxidation of methane with H_2/O_2 on PdCu/Z-5. (b) Comparative evaluation of catalytic performance and oxygenate selectivity in methane oxidation. *Source*: [91] / with permission of John Wiley & Sons.

exceptional catalytic activity of 1178 mmol mol_{Pd}^{-1} h^{-1} at 0.18 wt. % Pd loadings, demonstrating efficient Pd utilization (see Figure 13.8b).

Remias et al. [92] previously reported a comparable catalytic system employing Pd/C and copper chloride in a trifluoroacetic acid/water mixture, which demonstrated selective conversion of methane, ethane, and butane using oxygen and carbon monoxide as reactants. First, Pd facilitates the WGS reaction, converting CO and H_2O into H_2. The resulting H_2 then reacts with O_2 to produce H_2O_2. Finally, the *in situ* generation of H_2O_2 activates alkane to form oxygenates (see Figure 13.9a).

Gu and co-workers [93] demonstrated that TiO_2-supported single-site Rh catalysts on TiO_2 convert methane into methanol with O_2 and CO ($CH_4 + O_2 + CO$). Notably, when promoted by copper cations, the Rh/TiO_2 catalyst exhibited remarkable performance, producing methanol at millimole levels with >99% selectivity. H_2O_2 was formed as a reactive intermediate in the reaction, and O_2 was activated and converted to H_2O_2 using H_2 generated from the WGS reaction. In addition, Yu et al. [94] reported that ZSM-5-supported Rh_2O_3 nanoparticles can convert methane to oxygenates at 150 °C in the $CH_4 + O_2 + CO$ system. Spectroscopic and control experiments indicated that CO drives the WGS reaction to produce in-situ H species, which subsequently react with O_2 to form OOH*, H_2O_2, and OH* as active intermediates. The dominant OH* species will thus render high activity for methane oxidation (see Figure 13.9b). Beyond Pd and Rh species, ZSM-5-supported Ir species are capable of converting methane to methanol under $CH_4 + O_2 + CO$ system [95]. The IrCuPd/ZSM-5 system delivered an 80% methanol selectivity while reaching an STY of 1200 μmol g_{cat}^{-1}.

13.2.4.2 Direct Conversion of Methane to Acetic Acid over Zeolite-based Catalysts

While selective methane oxidation to C_1 oxygenates is well-studied, achieving high selectivity toward C_{2+} oxygenates under mild conditions presents both economic advantages and catalytic challenges. The industrial production of acetic acid primarily relies on methanol carbonylation catalyzed by Rh/Ir complexes [96]. Shan and co-workers [97] developed ZSM-5-supported rhodium (0.5% Rh/ZSM-5) to

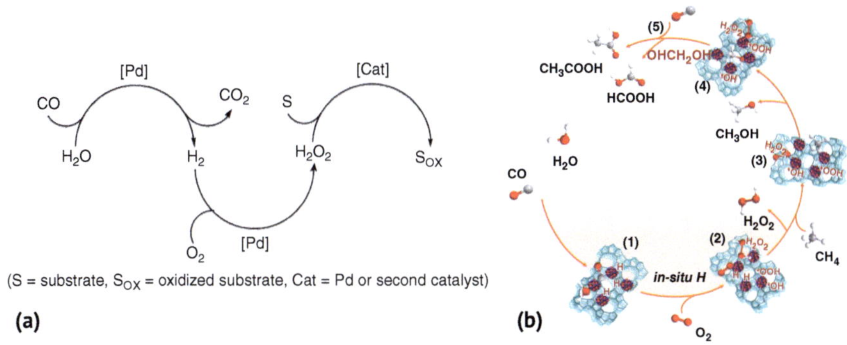

Figure 13.9 (a) Lower alkanes are selectively oxidized via a catalytic pathway utilizing in-situ generated H_2O_2 derived from the co-reaction of CO, O_2, and H_2O. *Source*: [92] with permission of Elsevier. (b) Mechanistic pathway for the CO-assisted direct oxidation of methane with molecular oxygen catalyzed by Rh/ZSM-5. *Source*: [94] / with permission of John Wiley & Sons.

catalyze methane to acetic acid with molecular oxygen as an oxidant, with the presence of carbon monoxide. Combined HAADF-STEM, CO-DRIFTS, and EXAFS analysis confirmed the atomic dispersion of mononuclear Rh species in the micropores of ZSM-5. The as-obtained 0.5% Rh/ZSM-5 catalyst can provide a high STY of 21 295 μmol g_{cat}^{-1} toward acetic acid within three hours at 150 °C. More importantly, the acetic acid selectivity was up to 60%. In contrast, when Rh species were supported on Na-form ZSM-5 (0.5 Rh-Na-ZSM-5), the major product turned out to be methanol, and less acetic acid was obtained, indicating that the Brønsted acid sites of the zeolite were necessary for the formation of acetic acid.

Tang and co-workers [98] demonstrated that single-site Rh_1O_5 species anchored on ZSM-5 (0.1 wt. % Rh loading) efficiently convert methane into acetic acid. DFT calculations suggested a Rh_1O_5 configuration in ZSM-5, with the Rh_1 center bonded to three lattice oxygen atoms (Si—O) and two oxygen atoms from adsorbed reactants (see Figure 13.10a). At 150 °C, the single-site Rh_1O_5 catalyst achieved a turnover frequency (TOF) of 0.10 acetic acid molecules per Rh site per second with ~70% selectivity, outperforming free Rh^+ cations by more than three orders of magnitude (see Figure 13.10b). However, under standard reaction conditions, no acetic acid was observed when methanol was used as an initial reactant. Step-forward isotopic experiments and DFT suggested that acetic acid was formed from the direct coupling of Rh-CH_3 and CO to an acetyl group and then reacted with the hydroxyl group to form acetic acid (see Figure 13.10c). To further improve the catalytic performance in the $CH_4 + O_2 + CO$ system, the strategy to enrich the density of mononuclear species in the channel of zeolites is highly desirable. Oda and co-workers [99] prepared a Rh-incorporated zeolite by in situ high-temperature treatment of H^+-MFI with $RhCl_3$. The molten $RhCl_3$ phase facilitated Rh migration into the zeolite channels, where confinement and ion exchange with H^+ led to stable Rh incorporation. Under the $CH_4 + O_2 + CO$ system, the as-synthesized $RhCl_3$@H^+-MFI catalyst demonstrated exceptional performance, achieving a record TOF of 592 mol mol_{Rh}^{-1} h^{-1} and a yield of 14.6 mmol g_{cat}^{-1} h^{-1} with 76.9% selectivity.

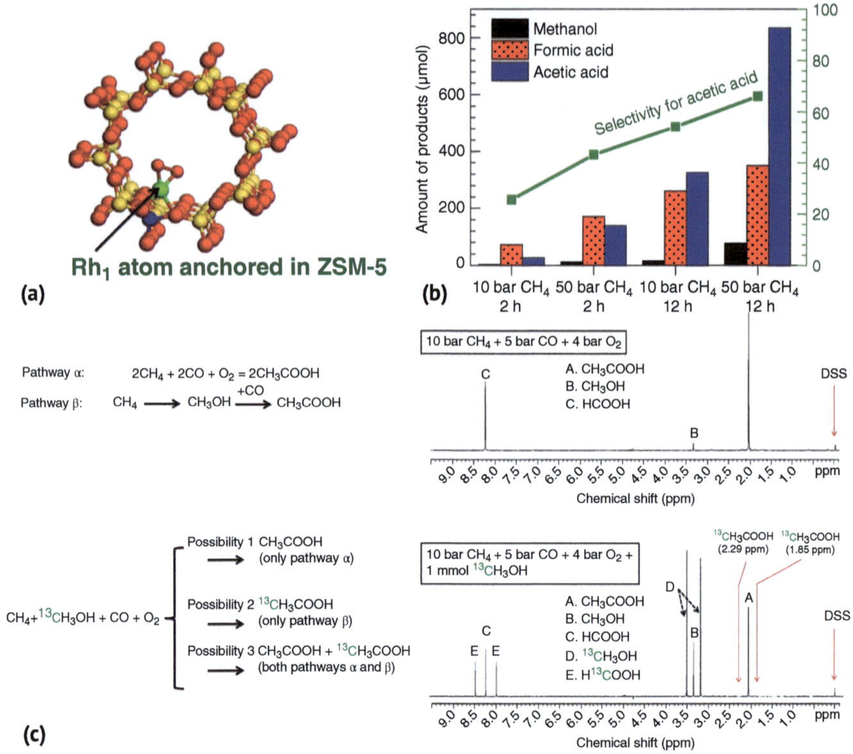

Figure 13.10 (a) Schematic of atomically dispersed Rh sites in ZSM-5. (b) Single rhodium atoms anchored in micropores for efficient transformation of methane under mild conditions. (c) Isotopic labeling probes methanol-CO coupling in acetic acid formation. *Source*: [100] / with permission of Springer Nature / CC BY 4.0.

While zeolite-based catalysts enable selective methane-to-acetic acid oxidation under mild conditions, challenges remain in noble metal costs and acetic acid selectivity. Fe/ZSM-5 emerges as the benchmark catalyst for selective methane oxidation to oxygenates using H_2O_2 as the oxidant. Wu and co-workers [101] developed a series of ZSM-5-supported Fe species (Fe/ZSM-5) to convert methane to acetic acid. The as-prepared binuclear Fe catalysts exhibited a high acetic acid content of 66% in liquid products. The exceptional CH_3COOH selectivity stems from a unique $\left[Fe(III) - (\mu O)_2 - Fe(III) - (OH)_2\right]$ binuclear site, which lowers the energy barrier for $\bullet CH_3/CO^*/OH^*$ coupling over CH_4 oxidation to CH_3OH, as revealed by spectroscopy and DFT. (see Figure 13.11a,b). Precise identification of iron active sites in Fe/ZSM-5 is critical for rationally designing next-generation catalysts to enhance both activity and acetic acid selectivity in methane oxidation. Wang et al. developed a series of Fe/ZSM-5 samples via the vacuum impregnation method [100]. The catalyst achieved an acetic acid STY of 12.01 mmol g_{cat}^{-1} h^{-1} with 63.2% selectivity at 50°C, setting a new benchmark for low-temperature methane oxidation (see Figure 13.11c). The high yield was attributed to mononuclear Fe^{3+} species, with their structural identity and catalytic role confirmed

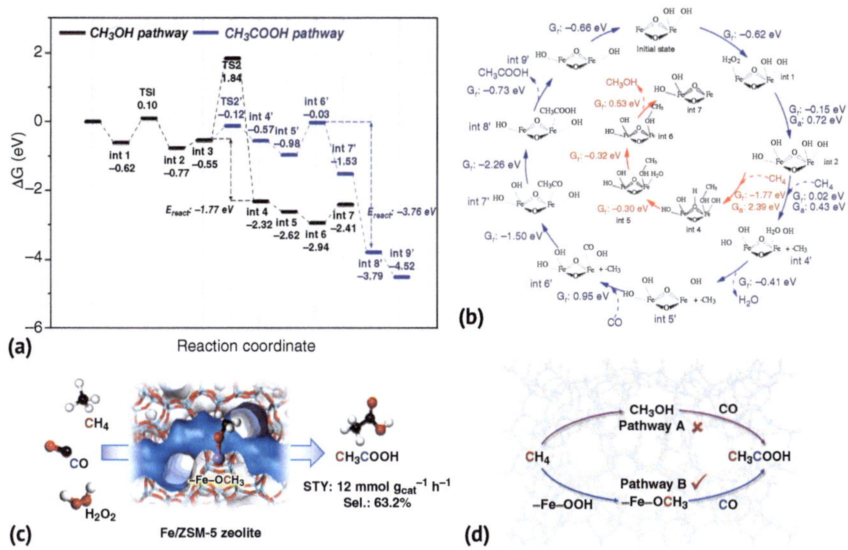

Figure 13.11 (a) DFT calculated free-energy diagram for CH_4 direct conversion with CO and H_2O_2 to form CH_3OH and CH_3COOH under the temperature of 30 °C. (b) DFT calculations reveal binuclear Fe sites facilitate methane C—H activation and selective oxidation via a low-energy pathway. (c) Direct conversion of CH_4 and CO to acetic acid is achieved over Fe/ZSM-5. (d) Two possible reaction pathways for the formation of acetic acid from methane and carbon monoxide. *Source*: [100] / with permission of Elsevier.

by UV-vis spectroscopy and structure-activity correlations. In-situ Raman and DRIFTS studies reveal a plausible reaction pathway for the direct conversion of methane to acetic acid using H_2O_2 and CO, providing key mechanistic insights into C—C coupling (see Figure 13.11d).

13.2.5 OCM to C_{2+} Hydrocarbons

Ethylene, a cornerstone chemical for plastics and petrochemicals, remains pivotal in modern industrial processes [103, 104]. While ethylene is predominantly produced via steam cracking of naphtha or ethane, OCM offers a direct route to convert methane—an abundant feedstock—into higher hydrocarbons, including ethylene. The process was first reported and achieved on alumina-supported metal oxides in the early 1980s by Keller and Bhasin [102]. With oxygen as the oxidant, the reaction includes the formation of ethane $\left(2CH_4 + 1/2\ O_2 \longrightarrow C_2H_6 + H_2O\ \Delta H_{298k} = -177\ \text{KJ mol}^{-1}\right)$ and followed by the oxidative dehydrogenation $\left(C_2H_6 + 1/2\ O_2 \longrightarrow C_2H_4 + H_2O\ \Delta H_{298K} = -105\ \text{kJ mol}^{-1}\right)$ [105]. The catalytic mechanism for OCM has been proposed and commonly suggested that the OCM processes involve three major steps [106]: (1) the formation of active oxygen atoms on the catalyst surface; (2) the formation of methyl radical by the interaction of methane molecules with the active oxygen atoms; and (3) the reaction proceeds via methyl radical coupling to form ethane, which subsequently dehydrogenates to ethylene—a key mechanistic step in methane valorization [107]. Many kinds of OCM catalysts have been developed to enhance the yield of C_2 hydrocarbons (C_2H_4 and C_2H_6) and the ability to suppress overoxidation. Lithium-doped magnesia (Li/MgO), featuring [Li^+O^-] active sites, was among the first catalysts shown to efficiently cleave C—H bonds in methane, generating methyl radicals for oxidative coupling to ethylene [108]. Lanthanide oxide catalysts have emerged as promising alternatives for OCM, where surface oxygen vacancies facilitate the generation of reaction oxygen species for methane activation [109, 110]. The Mn_2O_3-Na_2WO_4/SiO_2 system, first reported by Fang et al. [111], remains one of the most promising OCM catalysts, achieving 20~30% methane conversion with 60~80% C_2 selectivity—a benchmark performance in the field [112].

Zeolites offer a versatile platform for OCM catalyst design, with Mo/zeolite systems emerging as particularly promising due to their tunable acidity and confinement effects that enhance methane activation [113]. Single-site Mo oxide species, anchored either on framework aluminum or external silicon sites of zeolites, were identified as key active centers. Remarkably, these isolated species can be regenerated under oxygen treatment, highlighting their dynamic redox behavior. Metal-exchanged zeolites, including Li or Cs-ion-exchanged NaX [114] or exchanged $AlPO_4$-5 [115], and MgO-modified ZSM-5 [116] were also investigated for OCM reaction. The basicity of zeolites would affect the OCM performance, and higher basicity may lead to higher activity but lower selectivity towards the desired products. Ce-incorporated MFI zeolites, synthesized via a mechanochemical approach, provide a tunable platform for subsequent metal deposition (see Figure 13.12), offering new opportunities to tailor active sites for catalytic applications [117]. The Pd/Ce-MFIHT catalysts exhibited superior OCM performance, driven by synergistic interactions between Pd species and highly dispersed Ce sites, highlighting the importance of metal-oxide cooperativity in methane activation. Wang et al. [21]

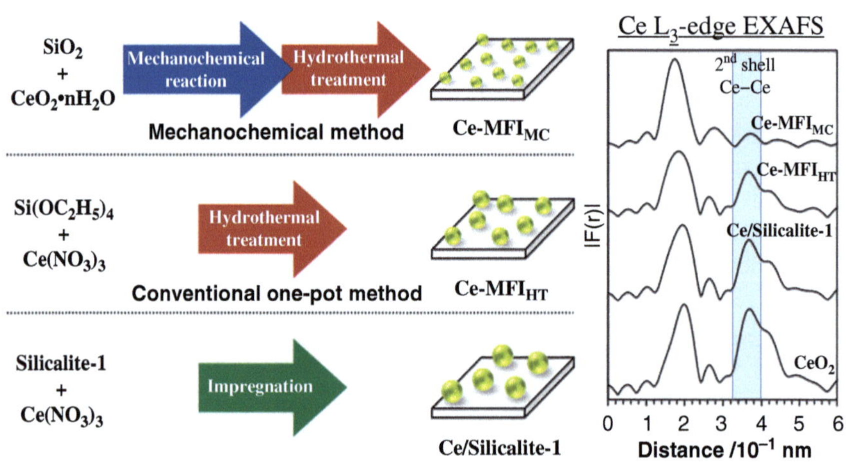

Figure 13.12 The preparation method and Ce L_3-edge EXAFS for Ce-containing samples. *Source*: [117]/with permission of American Chemical Society.

reported a TiO_2-doped Mn_2O_3-Na_2WO_4/SiO_2 catalyst using Ti-MWW zeolite as both a TiO_2 source and structural support, demonstrating enhanced OCM performance through tailored oxide-zeolite interactions, which exhibited 26% methane and 76% C_2-C_3 selectivity. While the microporous framework degrades under reaction conditions, the in situ formation of $MnTiO_x$ enhances oxygen activation at Mn sites, while $NaWO_x$ facilitates methane C—H bond scission at remarkably low temperatures (650 °C), revealing dynamic active-site evolution in OCM catalysis. Despite significant advances in catalyst design, the industrial implementation of OCM remains elusive, hindered by insufficient catalyst stability and unsatisfactory C_2 yields under practical high-temperature conditions.

13.2.6 Non-oxidative Conversion of Methane to C_{2+} Hydrocarbons

Direct methane conversion to C_{2+} hydrocarbons can be achieved even under non-oxidative conditions, opening new pathways for selective alkane activation. The non-oxidative process includes the non-OCM (NOCM) to ethylene ($2CH_4 \longrightarrow C_2H_6 + H_2$, $\Delta H_{298K} = 64.4$ kJ mol^{-1}, $2CH_4 \longrightarrow C_2H_4 + 2H_2$, $\Delta H_{298K} = 201.5$ kJ mol^{-1}) [118–120], MDA ($6CH_4 \longrightarrow C_6H_6 + 9H_2$, $\Delta H_{298K} = 532$ kJ mol^{-1}) [121, 122], and deep MDA to solid and hydrogen ($CH_4 \longrightarrow C + 2H_2$, $\Delta H_{298K} = 64.4$ kJ mol^{-1}) [123, 124]. In comparison, non-oxidative pathways avoid the drawbacks of the over-oxidation of products (the formation of CO_x) in the OCM process, and can obtain high-quality hydrogen besides the hydrocarbons with high selectivity. However, the NOCM remains thermodynamically constrained at low temperatures (< 450 °C), necessitating elevated reaction conditions to drive C—H activation [125, 126].

Zeolites have been widely applied for non-oxidative methane conversion. In 1993, Wang et al. [127] first reported an efficient zeolite catalyst Mo/ZSM-5. The combination of Mo and ZSM-5 creates a bifunctional catalyst that enhances MDA reactions. Mechanistic studies suggest that methane activation to CH_x intermediates occurs via carbonization of Mo-O-Al species, while Brønsted acid sites facilitate the critical step of CH_x aromatization to higher hydrocarbons [27, 128]. Bifunctional catalysts combining transition metals (Fe, V, Cu, W, and Ga) with zeolite frameworks (MWW, Y, etc.) have emerged as promising systems for MDA, leveraging synergistic metal-acid interactions [129–132]. Chen et al. compared conventional bifunctional zeolite catalysts and novel catalysts for NOCM (see Figure 13.13).

Guo et al. [121] demonstrated that a single-site Fe@SiO_2 catalyst enables efficient and stable non-oxidative methane conversion, selectively producing hydrogen and C_{2+} hydrocarbons—a critical step toward sustainable methane valorization. However, a mixture of products was formed in the non-shape-selective catalysts. With the aid of zeolites with particular acidity and pore architecture, high selectivity to a special target product can be achieved. For example, Mo_2C/[B]ZSM-5 was reported to show a high selectivity towards ethylene with few aromatics [119]. It was suggested that the substitution of boron led to a lower acidity of zeolite support and suppressed the oligomerization reaction of ethylene. Pt-Bi/ZSM-5 also provided a

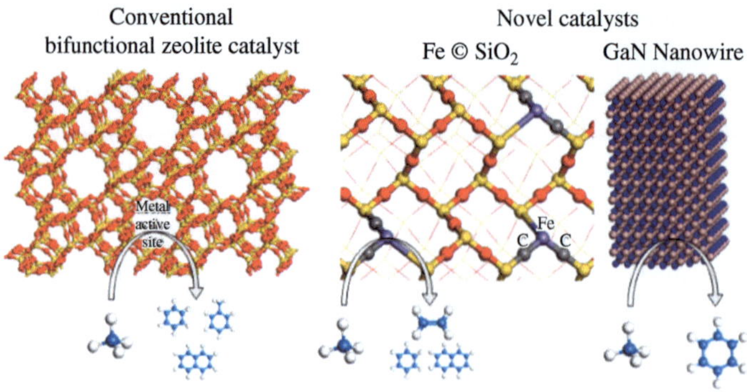

Figure 13.13 Conventional bifunctional zeolite catalysts and novel catalysts for NOCM. *Source*: [132]/with permission of American Chemical Society.

high C_2 selectivity (> 90%) for non-oxidative methane conversion [120]. However, limited methane conversion still existed as the main disadvantage.

The influence of zeolite topology on Mo-based catalysts for non-oxidative methane conversion was systematically investigated by Zhang et al. [130]. Notably, H-type silica-alumina zeolites exhibited channel dimensions closely matching the kinetic diameter of benzene (~6Å), suggesting a potential structure–function relationship for selective aromatic formation. Zeolites with 10-membered ring (MR) pores exhibited higher benzene selectivity and lower coke formation than smaller 8-MR pores or larger 12-MR pores. High activity and high benzene selectivity were obtained on MoO_3/H-ZSM-11. In contrast, MoO_3 supported on large-pore, low-acidity H-SAPO-5 and H-SAPO-11 showed no detectable hydrocarbon formation, underscoring the critical interplay between zeolite confinement and acid strength in methane activation. Compared with Mo/ZSM-5, higher benzene selectivity and stability can be achieved over Mo/MCM-22 due to the special framework structure of MCM-22, which was favorable for the diffusion of products and dispersion of Mo species [133, 134].

13.3 Summary and Prospects

As a clean energy carrier and cost-effective hydrocarbon resource, methane holds great potential for fuel and chemical synthesis. Yet, the kinetic and thermodynamic barriers associated with C—H bond activation in methane continue to hinder the development of practical direct conversion processes. In this chapter, we mainly summarize the progress of different reaction routes of direct methane conversion over zeolite-based thermocatalysis systems. Significant achievements have been made, including the construction of novel catalytic systems and concepts, identification of truly active sites, explanation of the role of zeolite, and investigation of the reaction mechanisms. Although various hydrocarbons and oxygenates, including olefins, aromatics, methanol, and acetic acid, could be produced from direct methane conversion, the large-scale industrial available technologies still rely on the indirect route through the production of syngas. Many challenges still exist in terms of pursuing highly active and selective catalytic materials for the direct methane conversion process.

High-temperature methane conversion technology includes methane reforming to syngas (DRM and SRM), OCM or non-OCM to hydrocarbons. Among which, methane reforming is an industrially mature technology. However, the harsh reaction conditions under high temperature indicate the feature of high energy cost and CO_2 emission. High-temperature operation inevitably induces metal sintering and coke deposition, leading to rapid catalyst deactivation. The high-temperature catalyst stability needs special attention. Compared with DRM or SRM, OCM or non-OCM (MDA) routes are more competitive since the C_{2+} chemicals could be directly obtained via one-step methane conversion. Despite recent advances, the yield of C_{2+} products remain severely constrained. Innovative approaches to enhance the activity and selectivity of heterogeneous catalysts are urgently needed. Emerging evidence indicates that optimizing metal nanoparticle size, dispersion, and reducibility can significantly improve catalyst stability. In addition, it is also indicated that the regulation of zeolite type and properties (i.e., Si/Al ratio, topology, acidity, mesoporosity, and pore structure), and the strategic incorporation of a second metal offer a powerful approach to modulate metal-support interactions, thereby fine-tuning catalytic performance. Moreover, solar energy-driven photothermal catalysis for high-temperature methane conversion would be a promising research direction since the solar energy could not only provide the required high energy input, but also may decrease activation energy for C—H bond and inhibit coke formation.

Direct and selective methane oxidation under mild conditions offers a sustainable route to oxygenates such as methanol and acetic acid, circumventing the energy-intensive processes and catalyst degradation associated with high-temperature operation. However, most of the reported reactions are carried out in the liquid phase, and it is difficult to balance the C—H activation and product overoxidation. Without suitable product protection, high methane conversion is typically accompanied by a low yield of oxygenated products and high CO_2 by-products. It would be a promising strategy to functionalize target products and decrease the residence time of target products on the catalyst surface. A successful example of methanol protection is to form the methanol derivatives. Another strategy is to construct the hydrophobic interface that could drive the hydrophilic-methanol away from the catalyst surface. Precise control over product selectivity is critical, as unregulated hydroxyl radical formation leads to a complex mixture of oxygenates rather than a single desired product. While various oxidants have been explored, H_2O_2 exhibits superior C—H bond activation efficiency. However, its high cost remains a key barrier to commercializing H_2O_2-based catalytic processes. Recent studies suggest that tandem functional catalysts could couple the in-situ formation of H_2O_2 with C—H activation to prepare specific oxygenates with high selectivity. Another solution to achieve selective oxidation is to develop flow-type reaction systems. For example, microfluidic technology might be an effective strategy to strengthen mass diffusion, improve catalytic efficiency, and hinder product overoxidation. Moreover, it is currently a grand challenge to catalyze CH_4 oxidation to form value-added C_{2+} products. Significant advances are still required to achieve efficient C—H activation and selective C—C coupling, which remain key challenges in catalytic chemistry.

SACs show great potential in the selective activation of CH_4. The rational choices of metal atom, zeolite support, and the design of the desired coordination environment should be carefully considered. A deeper understanding of the atomic-scale structure of active sites and their role in governing C—H activation and product selectivity is essential for advancing catalytic science. Monitoring the dynamic evolution of the chemical environments surrounding single-atom metal centers throughout the catalytic cycle is crucial for elucidating the fundamental nature of SACs and establishing precise structure–performance relationships. Furthermore, in-depth insight into the reaction mechanism and reaction intermediates also requires a synergistic approach combining in-situ/operando spectroscopy, kinetic studies, and theoretical simulations, which is essential to guide the development of advanced catalysts for selective methane valorization.

References

1 Lunsford, J.H. (1990). The catalytic conversion of methane to higher hydrocarbons. *Catal. Today* 6: 235–259.
2 Horn, R. and Schlögl, R. (2015). Methane activation by heterogeneous catalysis. *Catal. Lett.* 145: 23–39.
3 Konno, Y., Fujii, T., Sato, A. et al. (2017). Key findings of the world's first offshore methane hydrate production test off the coast of Japan: toward future commercial production. *Energy Fuels* 31: 2607–2616.
4 Kerr, R.A. (2010). Natural gas from shale bursts onto the scene. *Science* 328: 1624–1626.
5 Moore, T.A. (2012). Coalbed methane: a review. *Int. J. Coal Geol.* 101: 36–81.
6 Ryckebosch, E., Drouillon, M., and Vervaeren, H. (2011). Techniques for transformation of biogas to biomethane. *Biomass Bioenergy* 35: 1633–1645.
7 Fletcher, S.E.M. and Schaefer, H. (2019). Rising methane: a new climate challenge. *Science* 364: 932–933.
8 McFarland, E. (2012). Unconventional chemistry for unconventional natural gas. *Science* 338: 340–342.
9 Hammitt, J.K. (2021). The future costs of methane emissions. *Nature* 592: 514–515.
10 Franz, R., Uslamin, E.A., and Pidko, E.A. (2021). Challenges for the utilization of methane as a chemical feedstock. *Mendeleev Commun.* 31: 584–592.
11 Shilov, A.E. and Shulpin, G.B. (1997). Activation of C-H bonds by metal complexes. *Chem. Rev.* 97: 2879–2932.
12 Geerts, J.W.M.H., Hoebink, J.H.B.J., and van der Wiele, K. (1990). Methanol from natural gas. Proven and new technologies. *Catal. Today* 6: 613–620.

13 Zhou, W., Cheng, K., Kang, J. et al. (2019). New horizon in C_1 chemistry: breaking the selectivity limitation in transformation of syngas and hydrogenation of CO_2 into hydrocarbon chemicals and fuels. *Chem. Soc. Rev.* 48: 3193–3228.

14 Jiao, F., Li, J., Pan, X. et al. (2016). Selective conversion of syngas to light olefins. *Science* 351: 1065–1068.

15 Zhong, L., Yu, F., An, Y. et al. (2016). Cobalt carbide nanoprisms for direct production of lower olefins from syngas. *Nature* 538: 84–87.

16 Yu, H., Wang, C., Lin, T. et al. (2022). Direct production of olefins from syngas with ultrahigh carbon efficiency. *Nature Commun.* 13: 5987.

17 Ligthart, D.A.J.M., Van Santen, R.A., and Hensen, E.J.M. (2011). Influence of particle size on the activity and stability in steam methane reforming of supported Rh nanoparticles. *J. Catal.* 280: 206–220.

18 Buelens, L.C., Galvita, V.V., Poelman, H. et al. (2016). Super-dry reforming of methane intensifies CO_2 utilization via Le Chatelier's principle. *Science* 354: 449–452.

19 Song, Y., Ozdemir, E., Ramesh, S. et al. (2020). Dry reforming of methane by stable Ni-Mo nanocatalysts on single-crystalline MgO. *Science* 367: 777–781.

20 Elbadawi, A.H., Ge, L., Li, Z. et al. (2021). Catalytic partial oxidation of methane to syngas: review of perovskite catalysts and membrane reactors. *Catal. Rev.* 63: 1–67.

21 Wang, P., Zhao, G., Wang, Y. et al. (2017). $MnTiO_3$-driven low-temperature oxidative coupling of methane over TiO_2-doped Mn_2O_3-Na_2WO_4/SiO_2 catalyst. *Sci. Adv.* 3: e1603180.

22 Takanabe, K., Khan, A.M., Tang, Y. et al. (2017). Integrated in situ characterization of a molten salt catalyst surface: evidence of sodium peroxide and hydroxyl radical formation. *Angew. Chem. Int. Ed.* 56: 10403–10407.

23 Cargnello, M., Jaén, J.J.D., Garrido, J.C.H. et al. (2012). Exceptional activity for methane combustion over modular Pd@CeO_2 subunits on functionalized Al_2O_3. *Science* 337: 713–717.

24 Tao, F.F., Shan, J.J., Nguyen, L. et al. (2015). Understanding complete oxidation of methane on spinel oxides at a molecular level. *Nat. Commun.* 6: 7798.

25 Baltrusaitis, J., Jansen, I., and Christus, J.S. (2014). Renewable energy based catalytic CH_4 conversion to fuels. *Catal. Sci. Technol.* 4: 2397–2411.

26 Kolesnichenko, N.V., Ezhova, N.N., and Snatenkova, Y.M. (2020). Lower olefins from methane: recent advances. *Russ. Chem. Rev.* 89: 191.

27 Kosinov, N. and Hensen, E.J. (2020). Reactivity, selectivity, and stability of zeolite-based catalysts for methane dehydroaromatization. *Adv. Mater.* 32: 2002565.

28 Schwach, P., Pan, X., and Bao, X. (2017). Direct conversion of methane to value-added chemicals over heterogeneous catalysts: challenges and prospects. *Chem. Rev.* 117: 8497–8520.

29 Tang, Y., Li, Y., and Tao, F.F. (2022). Activation and catalytic transformation of methane under mild conditions. *Chem. Soc. Rev.* 51: 376–423.

30 Masoumifard, N., Guillet-Nicolas, R., and Kleitz, F. (2018). Synthesis of engineered zeolitic materials: from classical zeolites to hierarchical core-shell materials. *Adv. Mater.* 30: 1704439.

31 Liang, J., Liang, Z., Zou, R. et al. (2017). Heterogeneous catalysis in zeolites, mesoporous silica, and metal–organic frameworks. *Adv. Mater.* 29: 1701139.
32 Van Hook, J.P. (1980). Methane-steam reforming. *Catal. Rev.* 21: 1–51.
33 Sousa-Aguiar, E.F., Noronha, F.B., and Faro Jr, A. (2011). The main catalytic challenges in GTL (gas-to-liquids) processes. *Catal. Sci. Technol.* 1: 698–713.
34 Chen, L., Qi, Z., Zhang, S. et al. (2020). Catalytic hydrogen production from methane: a review on recent progress and prospect. *Catalysts* 10: 858.
35 Mosinska, M., Szynkowska, M.I., and Mierczynski, P. (2020). Oxy-steam reforming of natural gas on Ni catalysts—a mini review. *Catalysts* 10: 896.
36 Jones, G., Jakobsen, J., Shim, S. et al. (2008). First principles calculations and experimental insight into methane steam reforming over transition metal catalysts. *J. Catal.* 259: 147–160.
37 Zhou, L., Martirez, J.M.P., Finzel, J. et al. (2020). Light-driven methane dry reforming with single atomic site antenna-reactor plasmonic photocatalysts. *Nat. Energy* 5: 61–70.
38 Tang, Y., Wei, Y., Wang, Z. et al. (2019). Synergy of single-atom Ni_1 and Ru_1 sites on CeO_2 for dry reforming of CH_4. *J. Am. Chem. Soc.* 141: 7283–7293.
39 Al-Ubaid, A. and Wolf, E.E. (1987). Activity, FTIR studies and catalyst characterization during methane steam reforming on Ni/Y-zeolite catalysts. *Appl. Catal.* 34: 119–134.
40 Zhang, J., Zhang, T., Zhang, X. et al. (2014). New synthesis strategies for Ni/Al_2O_3-Sil-1 core-shell catalysts for steam reforming of methane. *Catal. Today* 236: 34–40.
41 Wei, N., Zhang, J., Zhong, H. et al. (2019). NiO/M-Sil-1 catalysts by *in-situ* coupled assembly of metal oxides and silicalite-1 zeolite with improved stability for methane steam reforming reaction. *J. Nanosci. Nanotechnol.* 19: 5831–5837.
42 Cimenler, U., Joseph, B., and Kuhn, J.N. (2015). Molecular-size selective H-β zeolite-encapsulated Ce-Zr/Ni-Mg catalysts for steam reforming. *Appl. Catal. A* 505: 494–500.
43 Pakhare, D. and Spivey, J. (2014). A review of dry (CO_2) reforming of methane over noble metal catalysts. *Chem. Soc. Rev.* 43: 7813–7837.
44 Carrara, C., Múnera, J., Lombardo, E.A. et al. (2008). Kinetic and stability studies of Ru/La_2O_3 used in the dry reforming of methane. *Topics Catal.* 51: 98–106.
45 Kehres, J., Jakobsen, J.G., Andreasen, J.W. et al. (2012). Dynamical properties of a Ru/$MgAl_2O_4$ catalyst during reduction and dry methane reforming. *J. Phys. Chem. C* 116: 21407–21415.
46 Gadalla, A.M. and Bower, B. (1988). The role of catalyst support on the activity of nickel for reforming methane with CO_2. *Chem. Eng. Sci.* 43: 3049–3062.
47 Wang, S., Lu, G.Q., and Millar, G.J. (1996). Carbon dioxide reforming of methane to produce synthesis gas over metal-supported catalysts: state of the art. *Energy Fuels* 10: 896–904.
48 Fakeeha, A.H., Khan, W.U., Al-Fatesh, A.S. et al. (2013). Stabilities of zeolite-supported Ni catalysts for dry reforming of methane. *Chinese J. Catal.* 34: 764–768.
49 Xu, H., Zeiger, B.W., and Suslick, K.S. (2013). Sonochemical synthesis of nanomaterials. *Chem. Soc. Rev.* 42: 2555–2567.

50 Li, Z., Zhuang, T., Dong, J. et al. (2021). Sonochemical fabrication of inorganic nanoparticles for applications in catalysis. *Ultrason. Sonochem.* 71: 105384.

51 Estephane, J., Aouad, S., Hany, S. et al. (2015). CO_2 reforming of methane over Ni-Co/ZSM-5 catalysts. Aging and carbon deposition study. *Int. J. Hydro. Energy* 40: 9201–9208.

52 Vafaeian, Y., Haghighi, M., and Aghamohammadi, S. (2013). Ultrasound assisted dispersion of different amount of Ni over ZSM-5 used as nanostructured catalyst for hydrogen production via CO_2 reforming of methane. *Energy Convers. Manag.* 76: 1093–1103.

53 Najfach, A.J., Almquist, C.B., and Edelmann, R.E. (2021). Effect of Manganese and zeolite composition on zeolite-supported Ni-catalysts for dry reforming of methane. *Catal. Today* 369: 31–47.

54 Zhang, W.D., Liu, B.S., Zhu, C. et al. (2005). Preparation of La_2NiO_4/ZSM-5 catalyst and catalytic performance in CO_2/CH_4 reforming to syngas. *Appl. Catal. A* 292: 138–143.

55 Zhang, X., Cao, E., Tian, Y. et al. (2022). Synthesis of coal tar pitch-derived heteroatom-doped porous carbon materials for aqueous zinc-ion hybrid supercapacitors. *Carbon Resour. Convers.* 5: 193–199.

56 Wang, J.B., Hsiao, S.Z., and Huang, T.J. (2003). Study of carbon dioxide reforming of methane over Ni/yttria-doped ceria and effect of thermal treatments of support on the activity behaviors. *Appl. Catal. A* 246: 197–211.

57 Crisafulli, C., Scirè, S., Minicò, S. et al. (2002). Ni-Ru bimetallic catalysts for the CO_2 reforming of methane. *Appl. Catal. A* 225: 1–9.

58 Pawelec, B., Damyanova, S., Arishtirova, K. et al. (2007). Structural and surface features of PtNi catalysts for reforming of methane with CO_2. *Appl. Catal. A* 323: 188–201.

59 Inaba, M., Murata, K., Saito, M. et al. (2006). Hydrogen production by gasification of cellulose over Ni catalysts supported on zeolites. *Energy Fuels* 20: 432–438.

60 Tang, P., Zhu, Q., Wu, Z. et al. (2014). Methane activation: the past and future. *Energy Environ. Sci.* 7: 2580–2591.

61 Gao, N., Cheng, M., Quan, C. et al. (2020). Syngas production via combined dry and steam reforming of methane over Ni-Ce/ZSM-5 catalyst. *Fuel* 273: 117702.

62 Jiang, H.T., Hua, W., Li, H.Q. et al. (2013). Recent progresses on some coke resistant Ni-Based catalysts for carbon dioxide reforming of methane. *Adv. Mater. Res.* 650: 85–91.

63 Drobná, H., Kout, M., Sołtysek, A. et al. (2017). Analysis of Ni species formed on zeolites, mesoporous silica and alumina supports and their catalytic behavior in the dry reforming of methane. *React. Kin. Mech. Catal.* 121: 255–274.

64 Fujitsuka, H., Kobayashi, T., and Tago, T. (2021). Development of silicalite-1-encapsulated Ni nanoparticle catalyst from amorphous silica-coated Ni for dry reforming of methane: achieving coke formation suppression and high thermal stability. *J. CO_2 Util.* 53: 101707.

65 Xu, S., Slater, T.J.A., Huang, H. et al. (2022). Developing silicalite-1 encapsulated Ni nanoparticles as sintering-/coking-resistant catalysts for dry reforming of methane. *Chem. Eng. J.* 446: 137439.

66 Wang, V.C.C., Maji, S., Chen, P.P.Y. et al. (2017). Alkane oxidation: methane monooxygenases, related enzymes, and their biomimetics. *Chem. Rev.* 117: 8574–8621.

67 Sushkevich, V.L., Palagin, D., Ranocchiari, M. et al. (2017). Selective anaerobic oxidation of methane enables direct synthesis of methanol. *Science* 356: 523–527.

68 Hammond, C., Forde, M.M., Ab Rahim, M.H. et al. (2012). Direct catalytic conversion of methane to methanol in an aqueous medium by using copper-promoted Fe-ZSM-5. *Angew. Chem. Inter. Ed.* 51: 5129–5133.

69 Bols, M.L., Hallaert, S.D., Snyder, B.E. et al. (2018). Spectroscopic identification of the α-Fe/α-O active site in Fe-CHA zeolite for the low-temperature activation of the methane C-H bond. *J. Am. Chem. Soc.* 140: 12021–12032.

70 Snyder, B.E.R., Vanelderen, P., Bols, M.L. et al. (2016). The active site of low-temperature methane hydroxylation in iron-containing zeolites. *Nature* 536: 317–321.

71 Snyder, B.E.R., Bols, M.L., Rhoda, H.M. et al. (2021). Cage effects control the mechanism of methane hydroxylation in zeolites. *Science* 373: 327–331.

72 Vitillo, J.G., Bhan, A., Cramer, C.J. et al. (2019). Quantum chemical characterization of structural single Fe (II) sites in MIL-type metal-organic frameworks for the oxidation of methane to methanol and ethane to ethanol. *ACS Catal.* 9: 2870–2879.

73 Huang, W., Zhang, S., Tang, Y. et al. (2016). Low-temperature transformation of methane to methanol on Pd_1O_4 single sites anchored on the internal surface of microporous silicate. *Angew. Chem. Inter. Ed.* 55: 13441–13445.

74 Williams, C., Carter, J.H., Dummer, N.F. et al. (2018). Selective oxidation of methane to methanol using supported AuPd catalysts prepared by stabilizer-free sol-immobilization. *ACS Catal.* 8: 2567–2576.

75 Narsimhan, K., Iyoki, K., Dinh, K. et al. (2016). Catalytic oxidation of methane into methanol over copper-exchanged zeolites with oxygen at low temperature. *ACS Cent. Sci.* 2: 424–429.

76 Qi, G., Davies, T.E., Nasrallah, A. et al. (2022). Au-ZSM-5 catalyses the selective oxidation of CH_4 to CH_3OH and CH_3COOH using O_2. *Nat. Catal.* 5: 45–54.

77 Cui, W.G., Zhang, G.Y., Hu, T.L. et al. (2019). Metal-organic framework-based heterogeneous catalysts for the conversion of C_1 chemistry: CO, CO_2 and CH_4. *Coord. Chem. Rev.* 387: 79–120.

78 Tian, J., Tan, J., Zhang, Z. et al. (2020). Direct conversion of methane to formaldehyde and CO on B_2O_3 catalysts. *Nat. Commun.* 11: 5693.

79 Baek, J., Rungtaweevoranit, B., Pei, X. et al. (2018). Bioinspired metal-organic framework catalysts for selective methane oxidation to methanol. *J. Am. Chem. Soc.* 140: 18208–18216.

80 Fan, L., Cheng, D., Song, L. et al. (2019). Direct conversion of CH_4 to oxy-organics by N_2O using freeze-drying FeZSM-5. *Chem. Eng. J.* 369: 522–528.

81 Ab Rahim, M.H., Forde, M.M., Jenkins, R.L. et al. (2013). Oxidation of methane to methanol with hydrogen peroxide using supported gold-palladium alloy nanoparticles. *Angew. Chem. Inter. Ed.* 52: 1280–1284.

82 Jin, Z., Wang, L., Zuidema, E. et al. (2020). Hydrophobic zeolite modification for in situ peroxide formation in methane oxidation to methanol. *Science* 367: 193–197.
83 Li, B., Song, X., Feng, S. et al. (2021). Direct conversion of methane to oxygenates on porous organic polymers supported Rh mononuclear complex catalyst under mild conditions. *Appl. Catal. B* 293: 120208.
84 Fang, Z., Huang, M., Liu, B. et al. (2022). Identifying the crucial role of water and chloride for efficient mild oxidation of methane to methanol over a $[Cu_2(\mu\text{-}O)]^{2+}$-ZSM-5 catalyst. *J. Catal.* 405: 1–14.
85 Sun, L., Wang, Y., Wang, C. et al. (2021). Water-involved methane-selective catalytic oxidation by dioxygen over copper zeolites. *Chem* 7: 1557–1568.
86 Freakley, S.J., Dimitratos, N., Willock, D.J. et al. (2021). Methane oxidation to Methanol in water. *Acc. Chem. Res.* 54: 2614–2623.
87 Xu, J., Armstrong, R.D., Shaw, G. et al. (2016). Continuous selective oxidation of methane to methanol over Cu- and Fe-modified ZSM-5 catalysts in a flow reactor. *Catal. Today* 270: 93–100.
88 Fang, Z., Murayama, H., Zhao, Q. et al. (2019). Selective mild oxidation of methane to methanol or formic acid on Fe–MOR catalysts. *Catal. Sci. Technol.* 9: 6946–6956.
89 Yu, T., Li, Z., Lin, L. et al. (2021). Highly selective oxidation of methane into methanol over Cu-promoted monomeric Fe/ZSM-5. *ACS Catal.* 11: 6684–6691.
90 Jeong, Y.R., Jung, H., Kang, J. et al. (2021). Continuous synthesis of methanol from methane and steam over copper-mordenite. *ACS Catal.* 11: 1065–1070.
91 Wu, B., Lin, T., Huang, M. et al. (2022). Tandem catalysis for selective oxidation of methane to oxygenates using oxygen over PdCu/Zeolite. *Angew. Chem. Int. Ed.* 61: e202204116.
92 Remias, J.E. and Sen, A. (2002). Palladium-mediated aerobic oxidation of organic substrates: the role of metal versus hydrogen peroxide. *J. Mol. Catal. A* 189: 33–38.
93 Gu, F., Qin, X., Li, M. et al. (2022). Selective catalytic oxidation of methane to methanol in aqueous medium over copper cations promoted by atomically dispersed Rhodium on TiO_2. *Angew. Chem. Inter. Ed.* 61: e202201540.
94 Yu, X., Mao, J., Wu, B. et al. (2023). Boosting direct oxidation of methane with molecular oxygen at low temperature over Rh/ZSM-5 catalyst. *ChemCatChem* 15: e202300077.
95 Li, M., Shan, J., Giannakakis, G. et al. (2021). Single-step selective oxidation of methane to methanol in the aqueous phase on iridium-based catalysts. *Appl. Catal. B* 292: 120124.
96 Kalck, P., Le Berre, C., and Serp, P. (2020). Recent advances in the methanol carbonylation reaction into acetic acid. *Coord. Chem. Rev.* 402: 213078.
97 Shan, J., Li, M., Allard, L.F. et al. (2017). Mild oxidation of methane to methanol or acetic acid on supported isolated rhodium catalysts. *Nature* 551: 605–608.
98 Tang, Y., Li, Y., Fung, V. et al. (2018). Single rhodium atoms anchored in micropores for efficient transformation of methane under mild conditions. *Nat. Commun.* 9: 1231.

99 Oda, A., Horie, M., Murata, N. et al. (2022). Highly efficient CO-assisted conversion of methane to acetic acid over Rh-encapsulated MFI zeolite prepared using $RhCl_3$ molten salt. *Catal. Sci. Technol.* 12: 5488–5494.

100 Wang, C.W., Sun, Y., Wang, L.J. et al. (2023). Oxidative carbonylation of methane to acetic acid on an Fe-modified ZSM-5 zeolite. *Appl. Catal. B* 329: 122549.

101 Wu, B., Lin, T., Lu, Z. et al. (2022). Fe binuclear sites convert methane to acetic acid with ultrahigh selectivity. *Chem* 8: 1658–1672.

102 Keller, G.E. and Bhasin, M.M. (1982). Synthesis of ethylene via oxidative coupling of methane: I. Determination of active catalysts. *J. Catal.* 73: 9–19.

103 Gielen, D.J., Moriguchi, Y., and Yagita, H. (2002). CO_2 emission reduction for Japanese petrochemicals. *J. Clean. Prod.* 10: 589–604.

104 Penteado, A., Esche, E., Salerno, D. et al. (2016). Design and assessment of a membrane and absorption based carbon dioxide removal process for oxidative coupling of methane. *Ind. Eng. Chem. Res.* 55: 7473–7483.

105 Galadima, A. and Muraza, O. (2016). Revisiting the oxidative coupling of methane to ethylene in the golden period of shale gas: a review. *J. Ind. Eng. Chem.* 37: 1–13.

106 Lunsford, J.H. (1995). The catalytic oxidative coupling of methane. *Angew. Chem. Int. Ed.* 34: 970–980.

107 Simon, Y., Baronnet, F., Côme, G.M. et al. (2004). Detailed mechanism of the oxidative coupling of methane. *Stud. Surf. Sci. Catal.* 147: 571–576.

108 Ito, T. and Lunsford, J.H. (1985). Synthesis of ethylene and ethane by partial oxidation of methane over lithium-doped magnesium oxide. *Nature* 314: 721–722.

109 Sugiyama, S., Matsumura, Y., and Moffat, J.B. (1993). A comparative study of the oxides of lanthanum, cerium, praseodymium, and samarium as catalysts for the oxidative dehydrogenation of methane in the presence and absence of carbon tetrachloride. *J. Catal.* 139: 338–350.

110 Le Van, T., Che, M., Kermarec, M. et al. (1990). Structure sensitivity of the catalytic oxidative coupling of methane on lanthanum oxide. *Catal. Lett.* 6: 395–400.

111 Fang, X., Li, S., Gu, J. et al. (1992). Preparation and characterization of W-Mn catalyst for oxidative coupling of methane. *J. Mol. Catal. A* 6: 255–261.

112 Arndt, S., Otremba, T., Simon, U. et al. (2012). Mn-Na_2WO_4/SiO_2 as catalyst for the oxidative coupling of methane. What is really known? *Appl. Catal. A* 425: 53–61.

113 Gao, J., Zheng, Y., Jehng, J.-M. et al. (2015). Identification of molybdenum oxide nanostructures on zeolites for natural gas conversion. *Science* 348: 686–690.

114 Kovacheva, P., Arishtirova, K., and Davidova, N. (1997). Characterization of lithium and cesium modified zeolite catalysts in the oxidative conversion of methane. *Appl. Catal. A* 149: 277–287.

115 Kovacheva, P., Arishtirova, K., and Davidova, N. (1999). Oxidative methylation of toluene with methane catalyzed by cesium modified molecular sieves. *Appl. Catal. A* 178: 111–115.

116 Kovacheva, P., Davidova, N., and Weiss, A.H. (1994). Oxidative conversion of methane over MgO/ZSM-5 catalysts. *Stud. Surf. Sci. Catal.* 82: 403–409.

117 Yabushita, M., Yoshida, M., Osuga, R. et al. (2021). Mechanochemical route for preparation of MFI-type zeolites containing highly dispersed and small Ce species and catalytic application to low-temperature oxidative coupling of methane. *Ind. Eng. Chem. Res.* 60: 10101–10111.

118 Soulivong, D., Norsic, S., Taoufik, M. et al. (2008). Non-oxidative coupling reaction of methane to ethane and hydrogen catalyzed by the silica-supported tantalum hydride: $(\equiv SiO)_2$ Ta-H. *J Am. Chem. Soc.* 130: 5044–5045.

119 Sheng, H., Schreiner, E.P., Zheng, W. et al. (2018). Non-oxidative coupling of methane to ethylene using Mo_2C/[B]ZSM-5. *ChemPhysChem* 19: 504–511.

120 Xiao, Y. and Varma, A. (2018). Highly selective nonoxidative coupling of methane over Pt-Bi bimetallic catalysts. *ACS Catal.* 8: 2735–2740.

121 Guo, X., Fang, G., Li, G. et al. (2014). Direct, nonoxidative conversion of methane to ethylene, aromatics, and hydrogen. *Science* 344: 616–619.

122 Hao, J., Schwach, P., Fang, G. et al. (2019). Enhanced methane conversion to olefins and aromatics by H-donor molecules under nonoxidative condition. *ACS Catal.* 9: 9045–9050.

123 Palmer, C., Tarazkar, M., Kristoffersen, H.H. et al. (2019). Methane pyrolysis with a molten Cu-Bi alloy catalyst. *ACS Catal.* 9: 8337–8345.

124 Agarwal, N., Freakley, S.J., McVicker, R.U. et al. (2017). Aqueous Au-Pd colloids catalyze selective CH_4 oxidation to CH_3OH with O_2 under mild conditions. *Science* 358: 223–227.

125 Scurrell, M.S. (1987). Prospects for the direct conversion of light alkanes to petrochemicalfeedstocks and liquid fuels - a review. *Appl. Catal.* 2: 1–22.

126 Scurrell, M.S. (1988). Factors affecting the selectivity of the aromatization of light alkanes on modified ZSM-5 catalysts. *Appl. Catal.* 41: 89–98.

127 Wang, L., Tao, L., Xie, M. et al. (1993). Dehydrogenation and aromatization of methane under non-oxidizing conditions. *Catal. Lett.* 21: 35–41.

128 Xu, Y., Bao, X., and Lin, L. (2003). Direct conversion of methane under nonoxidative conditions. *J. Catal.* 216: 386–395.

129 Weckhuysen, B.M., Wang, D., Rosynek, M.P. et al. (1998). Conversion of methane to benzene over transition metal ion ZSM-5 zeolites: I. Catalytic characterization. *J. Catal.* 175: 338–346.

130 Zhang, C.L., Li, S., Yuan, Y. et al. (1998). Aromatization of methane in the absence of oxygen over Mo-based catalysts supported on different types of zeolites. *Catal. Lett.* 56: 207–213.

131 Vollmer, I., Yarulina, I., Kapteijn, F. et al. (2019). Progress in developing a structure-activity relationship for the direct aromatization of methane. *ChemCatChem* 11: 39–52.

132 Sun, K., Ginosar, D.M., He, T. et al. (2018). Progress in nonoxidative dehydroaromatization of methane in the last 6 years. *Ind. Eng. Chem. Res.* 57: 1768–1789.

133 Ma, D., Zhu, Q., Wu, Z. et al. (2005). The synergic effect between Mo species and acid sites in Mo/HMCM-22 catalysts for methane aromatization. *Phys. Chem. Chem. Phys.* 7: 3102–3109.

134 Shu, Y., Ohnishi, R., and Ichikawa, M. (2003). Improved methane dehydrocondensation reaction on HMCM-22 and HZSM-5 supported rhenium and molybdenum catalysts. *Appl. Catal. A* 252: 315–329.

Part VI

Methanol Conversion

14

Application of Zeolites in Methanol/Dimethyl Ether to Fuels

*Faisal Zafar, Mansoor Ali, Jong Wook Bae**

School of Chemical Engineering, Sungkyunkwan University (SKKU), Suwon, Gyeonggi-do, Republic of Korea

*Corresponding author: finejw@skku.edu

14.1 Introduction

Since fossil fuels have been regarded as vital sources of energy for driving robust economic developments, almost all sectors of industry have significantly relied on the utilization of these resources [1]. About 80% of the world's energy needs are met by fossil fuels [2]; it is also an important feedstock in the chemical industry for the production of different organic and inorganic chemicals [3]. According to the International Energy Agency (IEA) in its Oil Market Report (2022), global oil supply has quickly returned to pre-pandemic levels as economies emerged from the lockdowns, and it is forecasted that oil demand in 2023 will rise further to 101.8 million barrels per day from 99.7 million barrels per day in 2022 [4]. The recovery will drive an unprecedented increase in CO_2 emissions (> 30 billion tons) into the environment and an increase of atmospheric CO_2 concentration up to > 420 ppm [3, 5]. CO_2, as one of the greenhouse gases, is not only a cause of ocean acidification but also traps solar radiation, which has been believed to be a main contributor to current climate change phenomena. Although different strategies have been proposed to convert anthropogenic CO_2 back to useful chemical products, the most versatile, simple, and easy to obtain liquid products from CO_2 are methanol, which can be produced by electrochemical reduction, thermal catalytic hydrogenation, and other methods [6–9]. The generated renewable and sustainable methanol can be subsequently converted into formaldehyde, acetic acid (HOAc), and olefins for further transformation into plastics, adhesives, and many other products [10, 11]. Methanol industry spans the entire world, with a rise in demand from 75 million metric tons in 2015 to more than 110 million metric tons in 2022, of which 200,000 tons of methanol are used directly as a transportation fuel as well [11]. When methanol and its derivative fuels are burned, they release CO_2, which can be recycled to close the carbon cycle [12, 13]. Moreover, as the natural reserves of fossil fuels are limited and depleting over time, it becomes imperative to explore alternative technologies to produce synthetic fuels. One promising technology is the conversion of methanol into fuels that can serve as substitutes for traditional petroleum-based fuels. Compared to traditional fuels, methanol-derived fuels offer numerous advantages by reducing the emissions of environmentally harmful pollutants, such as sulfur oxides (SO*x*), nitrogen oxides (NO*x*), and particulate matter (PM). Additionally, if produced from renewable sources, they have the potential to significantly reduce greenhouse gas emissions [14].

Dimethyl ether (DME) is a colorless gas with a sweet, ether-like smell, and it is a simple ether compound with the chemical formula of CH_3OCH_3. DME can be produced through dehydration of methanol or catalytic conversion of synthesis gas, a mixture of carbon monoxide and hydrogen. It is widely used as a solvent, refrigerant, and clean-burning alternative to fossil fuels [15]. Both methanol and DME are attractive chemical feedstocks due to their abundance, low cost, and potential for low-carbon emission. However, they have drawbacks such as relatively low energy content and susceptibility to oxidation and other chemical reactions. Therefore, many ongoing research projects aim to optimize their use and to develop more efficient and sustainable methods for their production and utilization. Methanol and DME can be further converted into hydrocarbons through the process called "methanol-to-hydrocarbon" (MTH) or "DME-to-hydrocarbon" (DTH). Those processes involve the use of heterogeneous catalysts that facilitate the conversion of these molecules into larger hydrocarbon molecules [16, 17]. The catalyst materials used for MTH and DTH conversion are typically zeolites, which are crystalline materials with a three-dimensional network of pores and channels [18]. Zeolite catalytic materials with their tunable structural and chemical properties have a wide range of industrial and commercial applications. They are composed of a three-dimensional network of silicon and oxygen atoms, with aluminum and other metal ions occupying the remaining voids in the structures [19, 20]. Zeolites have high surface areas and well-defined pore size distributions, which are responsible for unique adsorption and catalytic properties. Zeolites are commonly used as catalysts in the chemical industry to facilitate various reactions, including alkylation, isomerization, and aromatization. They are used as adsorbents in applications, such as water treatment, air purification, and natural gas storage as well. In addition, zeolites have potential applications in other fields such as medicine, agriculture, and environmental protection [21].

During the MTH and DTH reactions, methanol and DME molecules can be diffused into zeolite pores and undergo a series of reactions such as dehydration, cracking, and oligomerization to form larger hydrocarbons. These reactions involve the breaking and forming of chemical bonds, resulting in the rearrangements of carbon and hydrogen atoms [22]. These processes have the potential to reduce greenhouse gas emissions and to increase the sustainability of the transportation sector, as they offer a pathway to produce liquid fuels from renewable sources, as depicted in Figure 14.1. However, the efficiency of these processes continues to be

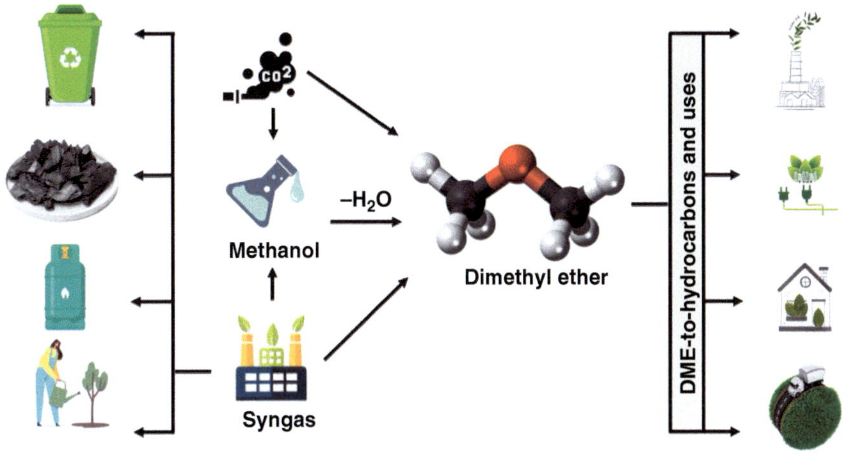

Figure 14.1 Production and utilization of different hydrocarbons produced from renewable sources.

a critical subject of active research, and there is a need for advancements in catalyst designs and process optimizations to enhance the selectivity and yield of desired hydrocarbons. Choosing the proper types of zeolites with their desired qualities can increase significant opportunities to enhance the transformations of methanol/DME, as well as optimizing the yield of particular products with the specific sets of hydrocarbons, can continue to pose challenging tasks from both scientific and technical points of view. Therefore, a wide variety of zeolites has been investigated, where the zeolites have their distinct properties and abilities for the conversion of methanol or DTH. The goals of the present review chapter are to present recent advancements and active research areas in designing effective zeolites for the MTH and DTH reactions by focusing on the effects of different physiochemical properties of the zeolites, reaction conditions, and so on. In addition, the chapter also summarized brief future challenges and opportunities for further advancing the technologies of the MTH and DTH reactions by using proper zeolite applications.

14.2 Zeolites for Methanol/DME Conversion to Fuels

The MTH reaction is the process by which methanol, a simple alcohol compound, is converted into hydrocarbons, which are various organic compounds. The MTH process has garnered significant attention in recent years due to the abundance of methanol as a feedstock and the increasing demand for hydrocarbons as clean fuels and chemical intermediates. The first stage in MTH conversion involves methanol dehydration to produce the mixtures containing methanol, DME, and water in their equilibrium compositions. Subsequently, the mixture of oxygenates undergoes a series of reactions involving hydrogen transfer (HT), isomerization, alkylation, and polycondensation, which leads to the production of lower olefins, alkanes, naphthenes, and higher olefins on zeolite catalysts [16] as presented in Figures 14.2 and 14.3. In this

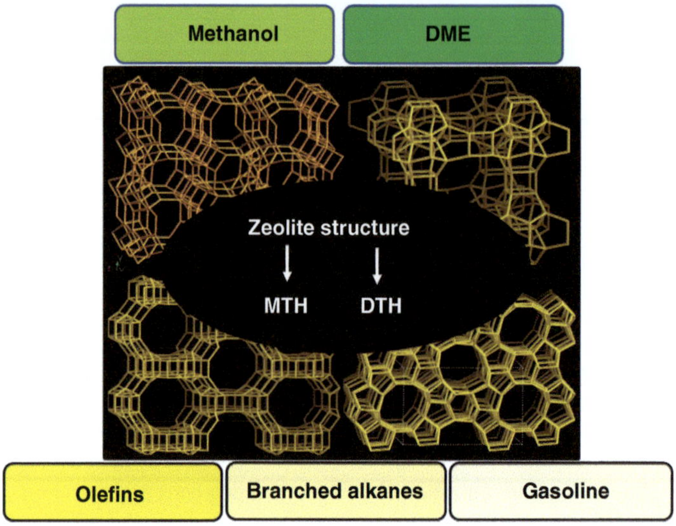

Figure 14.2 Methanol/DME conversion to hydrocarbons by the MTH and DTH processes using zeolite catalysts.

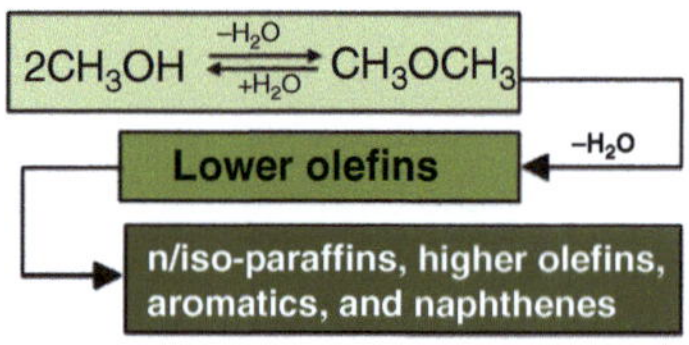

Figure 14.3 A summarily description of methanol/DME conversion to hydrocarbons.

section, we discuss various methods for converting MTH, their advantages and disadvantages, and the current state of research in this field.

One commercial technology for converting MTH is a methanol-to-olefins (MTO) process, where the methanol is first converted into olefins such as ethylene or propylene, via catalyst-assisted reactions and the olefins are then further converted into hydrocarbons through a series of reaction steps such as oligomerization or polymerization. The MTO process has the advantages of producing high-value products such as gasoline and polyethylene, but it also has a relatively low yield and requires expensive SAPO-31 catalysts. Since the Mobil Corporation proposed MTO in 1977, there have been considerable advancements in the catalyst synthesis, reaction mechanisms, process research, reactor designs, and commercialization developments [23]. In 2010, Dalian Institute of Chemical Physics (DICP) achieved a critical milestone by launching the world's first 600 000-ton/year MTO process based on SAPO-34 catalyst, which marked a significant step towards lower olefins (ethylene and propylene) production from coal [24]. Subsequently, DICP improved olefin yield and production capacity of the MTO reaction with their DMTO-II and DMTO-III technologies. Meanwhile, UOP and Norsk Hydro utilized the SAPO-34 zeolite to operate a fluidized bed reactor for a low-pressure MTO process.

Additionally, Shanghai Research Institute of Petrochemical Technology developed a fluidized bed SMTO technology, while Lurgi Company built the first methanol-to-propylene (MTP) plant based on ZSM-5 zeolite in 2011 by using their fixed-bed MTP process. The usage of zeolite catalysts was pivotal in the industrialization of MTO and MTP processes, and these large-scale efforts significantly drove the development of high-performance catalyst preparation methods with mechanistic knowledge, which enhanced production efficiency [25, 26]. Several MTO catalysts showed promising activity, their effectiveness is hindered by low selectivity to desired products and deactivation resulting from the changes in oxidation states, catalyst sintering, and coke depositions [27].

Another technology for converting MTH is the methanol-to-gasoline (MTG) process, where the methanol is directly converted into gasoline-range hydrocarbons through a series of reaction steps, including vapor-phase reforming and deoxygenation. The fuels resulting from the MTG process are mainly composed of iso-paraffins, methyl-substituted aromatics, naphthenes, and olefins. The MTG process has the advantage of direct production of fuels which is readily usable as fuels, but it also has a relatively lower yield and requires expensive catalysts. In 1985, Mobil partnered with the New Zealand government to establish an MTG plant with 14 500 barrels per day based on natural gas, which was previously converted into methanol through synthesis gas [28, 29]. The MTG process involves the dehydration of crude methanol over solid acid catalysts (i.e., H-ZSM-5), to produce an equilibrium mixture of methanol, DME, and water. This mixture is then fed into the gasoline synthesis loop, where it is converted into C_1–C_{11} hydrocarbons with 80% selectivity of C_{5+} benzene fraction under the stronger exothermic reaction conditions. To regulate the temperature in an adiabatic reactor, light hydrocarbons, hydrogen, and carbon oxides (which can be produced in small amounts by cracking reactions) are recycled and added to the feed. Although the H-ZSM-5 has unique properties, coke formations gradually occur, and the catalyst's activity needs to be renewed by the controlled coke burn-off step. As a result, the MTG section comprises the parallel reactors that enable intermittent regeneration [30]. As the DME is an intermediate for the MTG, MTO, or methanol-to-aromatics (MTA) process, the various final products can also be produced from the DME as a primary reactant. Since the DME is a clean-burning, low-carbon alternative to traditional fossil fuels, the production of DME from syngas in a single-step reaction has aroused significant interest in the conversion of DME-to-gasoline (DTG) and other valuable products [20, 31]. DME synthesis is more thermodynamically favorable than methanol synthesis, making it possible to utilize syngas from diverse sources [32]. The DTG process is simpler than the MTG process, because it does not involve an exothermic reaction of methanol dehydration to DME. Furthermore, the high reactivity of DME offers numerous advantages to the DTG process over the MTG process. In conclusion, the conversions of methanol/DTH are active research areas with significant potential to address a growing demand for hydrocarbons as fuels and chemical feedstock. Although several processes have been developed for methanol/DME conversion to hydrocarbons, there are still considerable amounts of work needed to optimize the processes for economic feasibility.

14.3 Effects of Reaction Variables for Methanol/DME Conversions

Reaction conditions for MTH and DTH reactions have a great influence on catalytic activity, product distribution, and catalyst lifetime as well [33]. This section includes valuable information with respect to the influences of various reaction conditions.

14.3.1 Effects of Reaction Temperature and Pressure

Since MTH reaction is a highly exothermic reaction, and catalyst deactivation rates strongly depend on the operating temperatures, which increases with an increase of maximum process temperature [34], where the conversion rate generally increases, and product selectivity decreases with an increase of reaction temperatures. It is a general trend of MTO reactions that propylene and ethylene selectivity increase with an increase in the reaction temperatures; however, propylene selectivity decreases at higher temperatures above 450 °C with significant coke formations [35]. The polyethylene glycol (PEG)-treated SAPO-18 catalyst used for MTO reaction at 425 °C showed an improved selectivity to light olefins with 100% methanol conversion and higher structural stability of the zeolite [36]. Furthermore, the SAPO-34 catalyst operated at the temperature range of 300–500 °C for MTO reaction displayed an enhanced propylene selectivity as an increase in reaction temperature at 300–400 °C, however, propylene selectivity was decreased with further temperature increase above 400–500 °C [37]. The Mn-modified HZSM-5 catalyst used for the MTO reaction in the temperature range of 400–550 °C exhibited an increase in propylene selectivity with an increase in reaction temperature [38]. When the reaction temperature increased from 380 to 480 °C on zinc (Zn) and phosphate (P)-modified ZSM-5 for MTH reaction [39], the catalyst lifetime decreased with an increase in reaction temperature. When the temperature was raised from 380 to 420 °C, the yield of olefins remarkably declined with an accelerated methanol conversion and aromatics formation rate. Conversely, the olefin yield was drastically increased from 420 to 480 °C, and paraffin yield continually decreased in the temperature range from 380 to 480 °C. In addition, the HZSM-5 catalyst at temperatures of 300, 350, 375, 400, and 450 °C for the MTG reaction as shown in Figure 14.4a1, a2 [33]. At the reaction temperature of $\leq$350 °C, the catalyst was rapidly deactivated prior to complete conversion of methanol, while the temperature increases above 375 °C leads to complete methanol conversion with a higher gasoline yield and improved catalyst stability. The reason is the formation of volatile hydrocarbons at low temperature (300 °C), which seem to be caused by the strongly interacting and bonded intermediates on the active sites in the micropores of ZSM-5 zeolite, which leads to an internal coke formation and deactivation even at lower temperatures [40]. However, a further rise of temperature above 400 °C facilitates the C_1–C_4 hydrocarbon formation, resulting in the suppressed gasoline-range hydrocarbon yield. It is well reported that the HT rate decreases at higher temperatures due to the weaker adsorption of the formed olefins at an elevated temperature. Generally, the aromatics yield consistently increases as the temperatures increase,

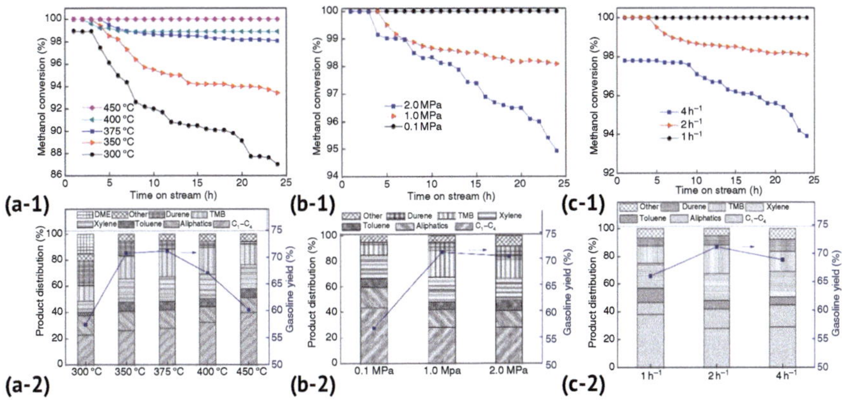

Figure 14.4 Effect of (a) temperature (experimental conditions: T = 300–450 °C, P = 1.0 MPa, and WHSV = 2 h^{-1}), (b) pressure (experimental conditions: T = 375 °C, WHSV 2 h^{-1}, P = 0.1–2 MPa), and (c) WHSV (experimental conditions: T = 375 °C, P = 1.0 MPa, and WHSV 1–4 h^{-1}) on performance of the nanocrystal ZSM-5 catalyst in MTG: methanol conversion, product distribution, and gasoline yield. *Source*: [33] / with permission of Elsevier.

which facilitates the yield of smaller molecular size components such as benzene and toluene, as well as reduces the yield of larger molecular size trimethylbenzene [41].

In addition, for the MTH process, its operation under an optimal reaction pressure is another key factor for attaining the maximized selectivity and stability as well. At low pressure, lighter hydrocarbons are mainly formed, while higher pressure increases the content of C_{5+} hydrocarbons, including aromatics [42–45]. The methylation and cracking steps associated with a relative rate of propagation of olefins and aromatics are greatly influenced by a methanol feed pressure, and an increase of propylene to ethylene molar ratio from 2 to 18 was observed on the H-ZSM-5 when the methanol feed pressure decreased from 52.5 to 0.6 kPa [46]. The 0.1 wt. % Ge-modified SAPO-34 catalyst operated at P = 0.1 MPa and T = 450 °C for the MTO reaction showed an enhanced ethylene and propylene selectivity with an excellent stability for 180 hours [47]. The ZSM-5 supported on SiC foams also revealed 99.5% methanol conversion with a propylene selectivity of 45.1% at the reaction condition of P = 0.25 MPa, T = 500 °C by using a microreactor [48]. Wan et al. tested the HZSM-5 catalyst at various reaction pressures of 0.1, 1.0, and 2.0 MPa for the MTG process, as shown in Figure 14.4b1, b2. The increase in reaction pressures decreases methanol conversion and C_1–C_4 selectivity, while aromatics selectivity was increased with an increase in pressure. Although the catalyst tested at 0.1 MPa displayed excellent catalytic stability and higher C_1–C_4 selectivity, the catalyst activity evaluated at 1.0 MPa exhibited a higher gasoline yield due to a higher methanol conversion. Meanwhile, the higher reaction pressures lead to the formation of coke precursors, resulting in fast catalyst deactivation [33], and it is well accepted that the higher pressure hinders the secondary cracking reaction activity to suppress the formation of light gaseous molecules.

14.3.2 Effects of Space Velocity

For the MTH reaction, weight hourly space velocity (SV) is one of the crucial factors, which simultaneously alters hydrocarbon distribution. Although methanol or DME conversion rate to hydrocarbons is very slow at a lower space time, the conversions can be rapidly increased under a narrow specific SV range, and the observations imply an autocatalytic nature of the MTH reaction as well [49]. The increase of SVs may decrease selectivity to desired products and conversion as well. Chang and Silvestri [23] studied the impacts of SV of DME/methanol for the MTH conversion on ZSM-5, and they observed that as DME/methanol SV decreases, selectivity towards C_2–C_5 olefins decreases while the selectivity towards paraffins, C_{6+} olefins, and aromatics increases. Similarly, Chang et al. [50] observed that after achieving complete conversion of DME/methanol to 100% over H-ZSM-5 at 500 °C, the selectivity to C_2–C_3 olefins was decreased while the selectivity towards paraffins and aromatics increased with a decrease of methanol SV. Hadi et al. [51] performed the MTO reaction on Ce-impregnated Mn/HZSM-5 at different SV from 2.51 to 8.37 h^{-1}, and they reported that a higher propylene selectivity was observed at SV of 2.5 h^{-1}. Golubev et al. [52] examined the Mg-HZSM-5/Al_2O_3 catalyst at various SV from 0.8 to 6.3 h^{-1} for the MTO reaction, and an optimum lighter olefin selectivity and methanol conversion were observed at SV = 0.8 h^{-1}, however, an increased SV causes a lower methanol conversion because of the shortened reactant contact time on the catalyst surface. Similarly, Heriyanto et al. [53] reported that the light olefin selectivity was significantly decreased with an increase of SV from 4.76 to 14.28 h^{-1} due to the shortened contact time of methanol. Lacarriere et al. [54] verified that the catalyst lifetime of two different topology zeolites, such as MCM-22 and MCM-36 was abruptly dropped with an increase of SV from 1 to 3 h^{-1} for the MTA reaction as well. Wan et al. [33] investigated the HZSM-5 for the MTG reaction at different SVs of 1, 2, and 4 h^{-1} as shown in Figure 14.4c1, c2, and methanol conversion were decreased with a rise of SV and higher gasoline yield was obtained at SV = 2 h^{-1}, whereas an incomplete methanol conversion at SV = 4 h^{-1} implies that the reactant molecules don't make enough contact time on the catalyst surfaces. The decrease of C_1–C_4 selectivity at a higher SV leads to the increased selectivity of higher aromatics, which also induced a rapid coke formation.

14.3.3 Effects of Co-reactant Feeding for Product Distributions

Since the MTH reaction is an exothermic reaction, which can be hybridized with an endothermic reaction to optimize an energy balance with an enhanced productivity of desired products, the lower energy-consuming and thermoneutral catalytic process can be achieved by coupling a highly endothermic hydrocarbon cracking reaction with the MTH reaction. Therefore, the coupled methanol hydrocarbon cracking (CMHC) process has been recently suggested [55] with a methanol coupling reaction with various liquids and gaseous hydrocarbons [56–59]. For example, the energy-neutral process with the composite of methanol and n-butane over two different Ni-doped ZSM-5 and P-doped ZSM-5 exhibited an improved selectivity towards C_2–C_4 and propylene compared to the sole 1-butylene cracking or MTO reaction [56, 60]. In addition, n-butanol co-feeding with methanol for the MTA reaction over Ga-modified H-ZSM-5 [61] revealed that n-butanol addition affected the dual cycle mechanisms as shown in Figure 14.5a. The simple MTA reaction is initiated with the olefin-based cycles and the formed intermediates are promptly converted to larger intermediates, which subsequently convert to the aromatics for the MTA reaction. Due to those reaction mechanisms, the polycyclic aromatics can be accumulated in the pores and active sites, which can be easily converted to the coke precursors. Wherein, with n-butanol co-feeding for the MTA reaction, the n-butanol is preferentially dehydrated into n-butene, which contributes to an enhanced olefin-based catalytic cycles, meanwhile, the aromatics-based cyclic mechanisms and accumulations of coke precursors are remarkably suppressed as well. The methanol conversion mixed with liquid C_5–C_6 hydrocarbons into aromatic hydrocarbons by using a two-stage fluidized bed reactor exhibited a superior activity compared to the sole methanol or C_5–C_6 hydrocarbon conversion [59], where the C_5–C_6 hydrocarbons coupling with methanol suppresses the self-decomposition of methanol without affecting methanol conversion, and aromatics and hydrogen yield are increased with decreased coke formation. Besides, the additional effects

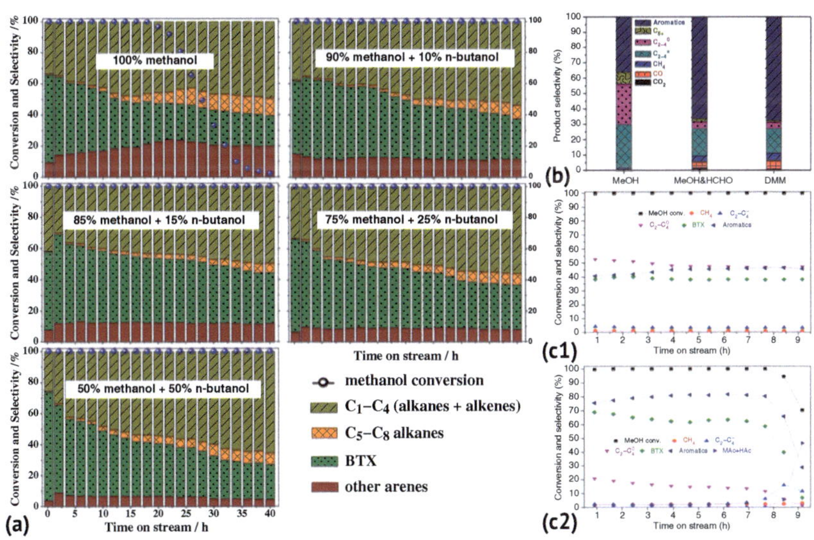

Figure 14.5 (a) Methanol conversion and hydrocarbon product distribution over 4% Ga-HZ-5 catalysts at 450 °C and with a TOS of 40 hours using methanol or a mixture of methanol and n-butanol as reagents. *Source*: [61] / with permission of American Chemical Society. (b) Catalytic results for the conversion of various C1 feedstocks over H-ZSM-5. Comparison for the conversion of MeOH, MeOH and HCHO (a mixture of MeOH and formalin), and DMM; 0.6 g HZSM-5, P = 0.1 MPa, T = 450 °C, He flow = 50 mL/min, feed rate = 0.26 mmol min^{-1}, MeOH/HCHO/H_2O (molar ratio) = 2/1/2.7; catalytic data were collected at 1.5 hours on stream, and the feedstocks were completely converted. *Source*: [63] / with permission of American Chemical Society. (c) MTA and CMTA (CO + methanol-to-aromatics) performance over Z-25. Reaction conditions: T = 400 °C; WHSV = 0.15 g/(g·h); N_2 or CO/CH_3OH = 113; P = 4 MPa. *Source*: [64] / with permission of John Wiley & Sons.

of supercritical fluid (SCF) of iso-octane as a co-reactant for MTH reaction over H-ZSM-5 zeolite [58] revealed that the SCF-assisted MTH reaction exhibited an enhanced hydrocarbon yield and improved catalyst lifetime than a gas-phase MTH reaction. Similarly, co-feeding of naphthalene during the MTH reaction over HZSM-5 shortens the induction period by accelerating the formation of methylbenzene species [62], where the co-fed naphthalene serves as a primary aromatic species and promotes aromatic-based cycles to increase the C_2H_4/C_3H_6 ratio at a smaller methanol conversion. Despite the larger molecular size of naphthalene, it can act as an intermediate for hydrocarbon pool (HCP) mechanisms in the internal acidic sites of the HZSM-5, where the mechanisms are proposed by coupling methanol with formaldehyde (HCHO) to form aromatics over HZSM-5 at 450 °C [63]. As displayed in Figure 14.5b, a lower aromatics selectivity with dominant C_2–C_4 hydrocarbons was observed for methanol conversion due to the possible HT reaction. Interestingly, the aromatics selectivity was greatly improved with the addition of HCHO to methanol ($CH_3OH/HCHO/H_2O = 2/1/2.7$), which implies that the introduction of HCHO strongly inhibits the HT reaction. Since the dimethoxymethane (DMM) can be formed by the reaction of $2CH_3OH + HCHO = CH_3OCH_2OCH_3 + H_2O$, pure DMM conversion was also performed over the HZSM-5 for comparison and it is obvious from the results in Figure 14.5b, which revealed that DMM conversion provides a higher aromatic selectivity and product distribution similar to that of ($CH_3OH + HCHO$) conversion. This observation also suggests that H_2O in the system has little effect on aromatization by HCHO feeding. In addition, a new aromatization mechanism by coupling CO with methanol to form aromatics over HZSM-5 was proposed and the results are displayed in Figure 14.5c1, c2, where HZSM-5 catalyzes carbonylation of methanol to form methyl acetate (MeOAc) and HOAc with the presence of CO by the Koch reaction [64, 65]. Interestingly, when the methanol was combined with CO co-feed, an initial aromatic selectivity increased significantly to 80% with BTX selectivity of 65%, whereas CH_4 and C_2–C_4 paraffin selectivity were found to be only 3 and 20%, respectively.

14.4 Effects of Zeolite Morphology and Pore Size Distribution

Besides the optimizations of Brønsted acidity and zeolite topology, the zeolite morphology and channel distributions play vital roles in regulating the aromatic distribution by controlling the diffusion of reactant and product molecules according to their pore structures and properties [66]. The incorporation of secondary pores and the decrease of zeolite particle sizes significantly affected the MTH reaction performance. The catalyst deactivation and formation of undesired secondary reactions can be inhibited by using hierarchical and nano-sized zeolites by shortening the path lengths and by efficiently facilitating mass transfer of molecules in those micropores [19, 67–71]. These highly characteristic hierarchical and nano-structured zeolites with diverse pore structures, such as single crystalline, it's very small size and interconnected mesopores have been extensively applied to improve the MTH activity and to increase the potential applications with the production of industrially desired products [72–77]. Various procedures have been introduced in the literature to selectively prepare hierarchically structured ZSM-5, such as soft template-based synthesis methods including dual templating and amphiphilic organosilanes [78–82], hard template-based synthesis methods [83, 84], and posttreatment methods like acid-treated dealumination [85] or alkaline-treated desilication [86] by partially extracting Si or Al atoms to generate mesopores in the main zeolite structures [87]. For example, the different morphology of ZSM-5 comprised of nanosphere or nanorod with micro-sized structures and similar acidic properties was synthesized and it was investigated to verify the morphology effects of the ZSM-5 particles for the MTH process [88]. Eventually, it was found that the ZSM-5 with aggregated nanosphere morphology showed a significantly larger external surface area (0.75 m^2g^{-1}) and mesopore volume (0.36 cm^3g^{-1}) compared to the micro-sized ZSM-5. The nano-sphere crystallite ZSM-5 showed a higher selectivity towards aromatics and iso-paraffins with a slow deactivation, since the larger external surface area and mesopore volume are benefits to improve diffusion property and to inhibit coke formation as well. In addition, the MTA reaction activity on the Zn-impregnated ZSM-5 with sphere, rod, coffin, and hollow-structured morphology [89] was verified and the results are displayed in Figure 14.6a1, a2, where the zeolite morphology has substantial effects on the number of acid sites, distribution, and state of Zn species. The order of catalytic activity of the Zn-impregnated ZSM-5 was found to be in the order of Hollow > Sphere > Rod > Coffin. The excellent catalytic activity of the hollow morphology was attributed to its mesopore shells and hollow capsule structures, which facilitate the generation of higher and well-dispersed active $(ZnOH)^+$ sites. In addition, the templates such as the types of organic structure-directing agents (OSDA) are a crucial variable for zeolite structure synthesis [90, 91]. For example, the high-silica ZSM-5 zeolite can be prepared by the conventional hydrothermal method in the presence of various OSDAs such as tetrapropyl ammonium hydroxide (TPAOH), tetrapropyl ammonium bromide (TPABr), n-butylamine (NBA), and morpholine (MOR) [92], and they were utilized for MTO reaction. It was demonstrated that the ZSM-5 synthesized by using TPABr exhibited a higher crystallinity, larger acidic sites, and reasonable surface area, along with enhanced methanol conversion and propylene selectivity. Similarly, the synthesized ZSM-23 with various OSDAs was verified to elucidate the effects of particle size and morphology for MTH reaction at 400 °C. The ZSM-23 catalysts showed an improved production rate of hydrocarbons in the range of C_3–C_{6+} with negligible amounts of aromatics (<3%) [93]. Recently, the metal-organic framework (MOF)-assisted Zr-modified SAPO-34 zeolite with hierarchical structures was applied for the MTO reaction [94], and the Zr-modified SAPO-34 zeolites with hierarchical porous structures exhibit an excellent MTO catalytic performance with a longer lifetime for more than 1500 minutes with a remarkably improved selectivity to light olefins of 98.66% compared to those of conventional SAPO-34. Those slight variations in crystal size and morphology have significant effects on methanol conversion and catalyst stability because of the preferential crystal growth along with the microchannel

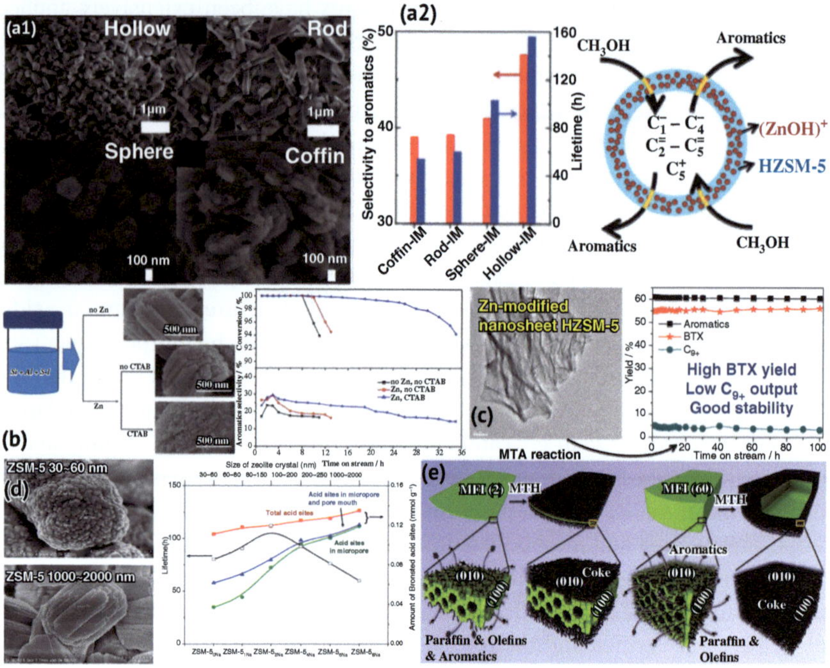

Figure 14.6 (a) SEM images of HZSM-5 zeolite surface in Hollow-ZSM-5, Rod-ZSM-5, Sphere-ZSM-5, and Coffin-ZSM-5, selectivity for aromatics of the MTA process and catalyst lifetime Zn/HZSM-5 zeolites (reaction conditions: T = 390 °C, P = 0.5 MPa, $WHSV_{Methanol}$ = 3.18 h^{-1}) and schematics of the catalytic reaction processes over Hollow-IM. *Source*: Adapted with permission [89] / The Royal Society of Chemistry. (b) Effect of Zn species and CTAB during synthesis on the aggregate morphology and catalytic performance. *Source*: Reproduced with permission from [96] / Elsevier. (c) TEM image of Zn-modified nanosheet HZSM-5 and total output of aromatics, the outputs of BTX, and C_{9+} aromatics. *Source*: Reproduced with permission from [97] / American Chemical Society. (d) SEM images of aggregated ZSM-5 zeolite crystals with different crystal sizes and lifetime and BAS amount versus crystal size of ZSM-5 zeolites. *Source*: Reproduced with permission from [98] / American Chemical Society. (e) Schematic diagrams show smaller coke depositions on the MFI(2) and MFI(60). *Source*: [99] / with permission of Elsevier.

direction. Besides, the improved liquid hydrocarbon selectivity for the MTG reaction on Zn/ZSM-5 catalyst synthesized with second soft template, dimethyldiallyl ammonium chloride acrylamide copolymer (PDDA) was also observed [95], and the introduction of hierarchy to Zn/HZSM-5 catalyst by varying the amount of CTAB in the seed solution was also enlarged the catalyst lifetime by reducing particle size and by increasing surface area and pore volume as well as enhancing diffusion properties [96] as shown in Figure 14.6b. In addition, Zn-modified nanosheet ZSM-5 zeolite was also investigated for the MTA reaction [90] and the results are shown in Figure 14.6c, where the catalyst showed good stability for 100 hours on-stream with a BTX yield of >54.5% and C_{9+} aromatic byproducts lower than 5.1%.

In addition, the crystal sizes of zeolites are also another important factor to control their overall catalytic activity, selectivity, and lifetime. Several researchers revealed the effects of crystal sizes for the MTA [100, 101], MTG [102], and MTO [103] reactions over the ZSM-5 zeolite. In general, the larger crystallites promoted the cracking reactions of nonaromatic C_5–C_9 components, dealkylation of polymethyl benzenes and HT rate, which eventually declined catalyst lifetime. However, the smaller crystallite sizes of ZSM-5 showed a much longer lifetime due to the enlarged surface area and shortened residence time of the products formed in zeolite pores [101]. In addition, the aggregated nano-sized ZSM-5 crystallites with different size distributions from 30–50 to 1000–2000 nm were elucidated for the MTP reaction [98], where the smaller crystallites increased the surface ratio of weak to strong acid sites as well as external Brønsted acid sites (BAS). As shown in Figure 14.6d, the ratio of Brønsted to Lewis (B/L) acid sites and the amount of internal BAS were decreased with decreasing crystallite sizes, and the smaller crystallite sizes and larger internal BAS clearly revealed an excellent catalytic lifetime and activity for MTH. The decrease in crystallite sizes and mesopores in zeolites enhanced propylene selectivity; however, all products were diffused out only through the larger straight channels of MFI-type nanosheets, which revealed non-uniform coke distributions on the (010) plane. The product distributions and extent of coke precursor formations seem to be controlled by anisotropic diffusion behaviors on different crystal planes, and the coke depositions on different planes, such as (010) and (100) of anisotropic MFI zeolites were found to be significant in the distinct *b*-axis lengths for the MTG reaction [99]. On the conventional MFI-type zeolite, the aromatics themselves were only diffused out through the straight channels with random diffusions of smaller molecules from both channels, which were responsible for significant coke depositions on (010) and (100) planes selectively formed from different precursors, as shown in Figure 14.6e.

14.5 Effects of Zeolite Topology and Pore Size Distribution

The MTH and DTH reactions mainly take place on zeolite crystallites, and product distributions strongly depend on the dimensionality of various zeolite channel structures. Typically, the channel dimensions with various 8-, 10-, and 12-membered ring (MR) windows (with their openings of 4, 5.5, and 7 Å, respectively) can be formed by varying the number of tetrahedral site atoms. Therefore, for the MTH or DTH reaction, the appropriate topology seems to be crucial for increasing the yield of desired products [104–106]. The more detailed descriptions of those zeolite topologies can be obtained from the International Zeolite Association (IZA) database of zeolite systems [107], and topologies of some zeolites are displayed in Figure 14.7a. Small pore zeolites with 8-MR channels, including SAPO-34 (CHA), RUB-13 (RTH), SAPO-18 (AEI), and SAPO-35 (LEV) showed an excellent catalytic activity for the production of light olefins for MTO reaction [68, 104, 108, 109].

The narrow pore openings of 8-MR channels prevent the migration of large intermediates, which increases light alkenes (mainly ethylene and propylene) [110].

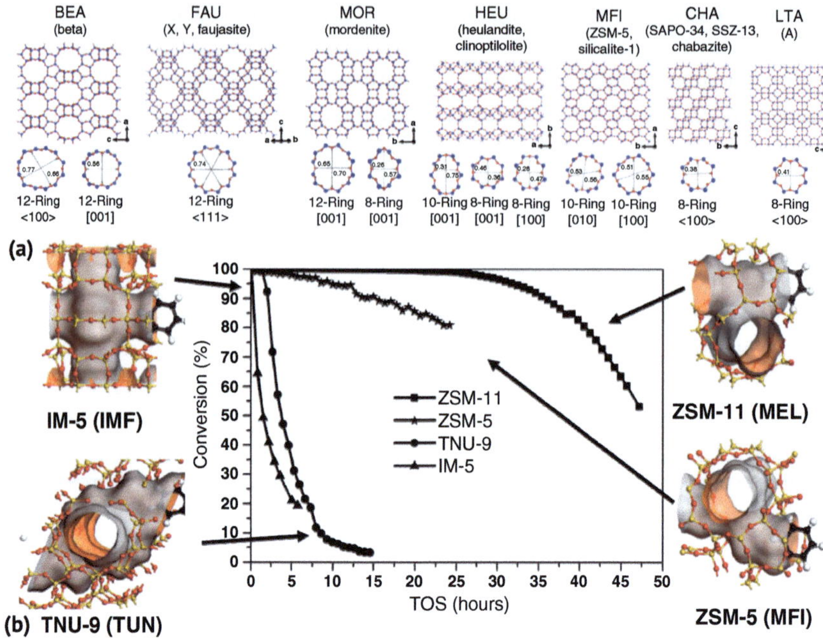

Figure 14.7 (a) Types of widely used zeolite frameworks, and three-letter codes assigned by the IZA (with bold character) according to each framework types, where the names of zeolites are belong to the corresponding framework type such as ball-stick atomic model with different pore windows and blue spheres denote T atoms and red spheres represent oxygen atoms according to the size of rings (nm) and the pore directions. *Source*: [21] / with permission of Elsevier. (b) Channel intersections for four topologies, where benzene is placed close to the channel mouth for easier size comparisons and conversion of methanol over IM-5, TNU-9, ZSM-11, and ZSM-5 at atmospheric pressure, 350 °C, and SV = 9 h^{-1}. *Source*: Adapted with permission [120] / The Royal Society of Chemistry.

Based on the recent study, the small-pore SAPO-14 zeolite with main 8-MR channels exhibited the highest propylene selectivity of 77.8% for MTO reaction [111] compared to SAPO-34 and SAPO-18 zeolites, where the SAPO-14 zeolite inhibits the aromatic-based mechanisms, which were known for the formation route of heavy hydrocarbons. In addition, an alternate zeolite of SSZ-13 [106], which is similar to the SAPO-34 with a CHA topology for MTO reaction [112, 113], showed the higher acid strengths to enhance MTO reaction activity at relatively lower temperatures with a rapid deactivation compared to SAPO-34 [113, 114]. In contrast, the medium pore-sized zeolites with 10-MR channels, such as ZSM-5 (MFI) and ZSM-48 (MRE) are excellent for improving propylene selectivity with moderate space limitation [99, 115]. In addition, H-MCM-22 zeolite with 10-MR channels and two independent pore openings is also a good candidate for MTH reaction, with its activity for catalytic cracking, isomerization, alkylation, and disproportionation [116]. Typically, zeolites with large 12-MR channels such as BEA zeolite facilitate the higher selectivity to heavy hydrocarbons through aromatic-mediated mechanisms due to the lack of spatial confinement effects within 12-MR channels [117].

Zeolites possessing one-dimensional 10-MR channels [118], such as EU-1 (EUO), ZSM-48 (MRE), ZSM-23 (MTT), and ZSM-22 (TON) with their slight variations of channel structures showed significant impacts on product distributions for MTH reaction. In addition, conversion of methyl mercaptan (CH_3SH) to H_2S and hydrocarbons over various zeolite topologies revealed the decreasing order of catalytic activity, such as H-ZSM-5 > H-Y > H-FER > H-BEA > H-MOR > H-SAPO-34 [119], and the H-ZSM-5 displayed the highest activity and stability. Interestingly, when methanol conversion was also compared with CH_3SH over the HZSM-5, significant variations for product distributions were observed as well. Similarly, zeolites with three-dimensional 10-MR channels, such as IM-5, TNU-9, ZSM-11, and ZSM-5 with their Si/Al ratio = 14–24 and crystal size less than 2 μm, were investigated for MTH reaction [120] at T = 350 °C, P = 0.1 MPa, and WHSV = 9 h^{-1}. All catalysts exhibited almost 100% initial methanol conversion and similar product distributions. However, the catalyst lifetime was significantly changed according to their topology with the decreasing order of ZSM-11 > ZSM-5 >> TNU-9 > IM-5 as displayed in Figure 14.7b. The larger cavities in the TNU-9 and IM-5 were responsible for more rapid deactivation due to the preferential formations of polyaromatics. These stability variations also suggested that the polymethyl benzene intermediates were more favorable in the IM-5 and TNU-9 zeolite than in the ZSM-5 and ZSM-11, where the methylation and cracking reactions are dominant. Different zeolite topologies through HCP intermediates and reaction pathways showed different constraints on product distributions. Among all above discussed topologies, the ZSM-5 with medium pores (10-MR channels) and three-dimensional structures of 5.4 × 5.6 × 5.5 Å as well as SAPO-34 with small pores (8-MR channels) and three-dimensional cage structures with pore openings of 3.8 × 3.8 Å can be industrially used for an enhanced selectivity towards C_{5+} hydrocarbons (aromatics and alkanes) as well as C_2–C_4 hydrocarbons (olefin), respectively.

14.6 Effects of Si/Al Ratio and Al Distribution in Zeolite Frameworks

It has been well reported that Si/Al ratios of zeolites affect the ratios of B/L acid sites, which directly change the strengths and numbers of acidic sites, resulting in altering product distributions [121]. The abundant strong acidic sites generally cause the overreactions of surface intermediates, which eventually block the zeolite micropore channels by the deposition of coke precursors [122]. The relationships between the Si/Al ratios of ZSM-5 and hydrocarbon distributions for MTG reaction were investigated [123] on the ZSM-5 zeolites having different SiO_2/Al_2O_3 ratios such as 23, 47, 107, 217, and 411 with their similar crystal sizes and structural properties at the reaction conditions of T = 375 °C, P = 1 MPa, and WHSV = 2 h^{-1}. Conversions of methanol were kept almost constant with an increase of SiO_2/Al_2O_3 ratio and gradually decreased above the SiO_2/Al_2O_3 ratio of 107, and a lower selectivity of C_2–C_4 alkenes and higher selectivity of C_1–C_4 alkanes were observed on the ZSM-5 having lower SiO_2/Al_2O_3 ratios. In addition, the decrease of SiO_2/Al_2O_3 ratios facilitated aromatization reactions with a higher selectivity to durene and the formed coke precursors accelerated catalyst deactivation as well. As shown in Figure 14.8a, the ZSM-5 with an appropriate SiO_2/Al_2O_3 ratio of 217 showed complete methanol conversion with higher gasoline production rate and smaller coke formations. Similarly, with an increase in the SiO_2/Al_2O_3 ratio of ZSM-5, the selectivity to aromatics gradually decreased and the selectivity to paraffin, olefins, and naphthalene was inversely increased [124] due to the decreased acidic sites, as shown in Figure 14.8b. In addition, the stepwise catalytic conversion of MTA on the physically mixed ZSM-5 (high Si/Al ratio) with Zn-doped ZSM-5 (low Si/Al ratio) [125] as displayed in Figure 14.8c revealed that methanol reactant was first catalytically converted to light hydrocarbons on the ZSM-5 having high Si/Al ratio, which was subsequently converted to aromatics over Zn-doped ZSM-5 having low Si/Al ratio. With an increase in the mixing ratio of ZSM-5 and Zn-doped /ZSM-5 (9:1), an improved aromatics selectivity and catalyst stability for 219 hours on stream were observed compared to the Zn-doped/ZSM-5 itself for 22 hours. The additional study for methanol conversion to aromatics was reported over the quadruple-packed ZSM-5 beds with a gradient increasing acidity [126], where each ZSM-5 bed has a different Si/Al ratio with the sequence of Z480/Z240/Z120/ZnZ60 from top to bottom bed as displayed in Figure 14.8d. The top side-bed of the weak acidic Z480 with a higher

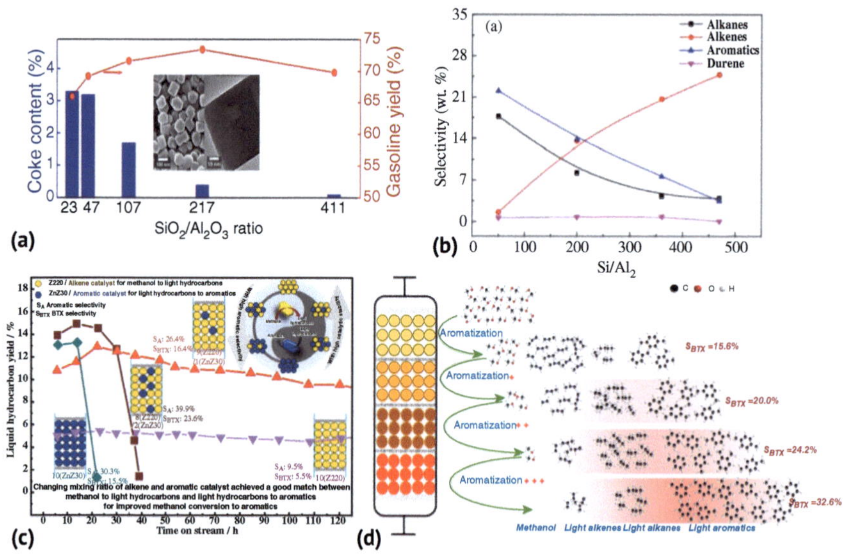

Figure 14.8 (a) Effect of SiO_2/Al_2O_3 ratio on the gasoline yield and generated coke species over nanocrystal ZSM-5 zeolites for methanol conversion to gasoline-range hydrocarbons. *Source*: Adapted with permission [123] / The Royal Society of Chemistry. (b) Effect of SiO_2/Al_2O_3 ratio on product selectivity over ZSM-5. *Source*: [124] / with permission of American Chemical Society. (c) Catalytic synergy between the low Si/Al ratio of Zn/ZSM-5 and the high Si/Al ratio of HZSM-5 for a higher performance of methanol conversion to aromatics. *Source*: [125] / with permission of American Chemical Society, and (d) Conversion of MTA over quadruple ZSM-5 sequence beds (Z480/Z240/Z120/ZnZ60) with gradient-increasing acidity. *Source*: [126] / with permission of Elsevier.

SiO_2/Al_2O_3 ratio of 480 converted methanol into light alkanes and alkenes intermediates with small amount of BTX aromatics, and the light alkanes and alkenes intermediates with small amounts of unreacted methanol were slowly transported on the subsequent two layers of ZSM-5 with a gradually increased acidity (lower SiO_2/Al_2O_3 ratios of 240:120), which were converted into aromatics. Finally, the light hydrocarbons from upper layers, especially light alkenes, were further aromatized on the bottom-layer Zn-doped ZSM-5 with a lower SiO_2/Al_2O_3 ratio of 60. Eventually, the aromatics selectivity steadily increased from 15.6 to 32.6% using the sequential ZSM-5 beds configuration, which was also found to be much higher than that of 17.6% for the widely recognized two-step MTA reaction.

Furthermore, the distribution and location of main framework aluminum species (Al_F) in the zeolite frameworks were also regarded as key factors for both redox and acid-catalyzed reactions. In general, besides the weak BAS assigned to Si-OH groups, all other acidic sites are closely related with the framework Al species, and the strong acidic sites are selectively formed by the Al substitutions for Si atoms [127]. Therefore, the catalytic performances of zeolites are strongly related to Al locations because the reaction routes are directly influenced by the presence of differently structured pores (or their sizes) in the zeolite frameworks [73, 128, 129]. The Al_F locations in the zeolite frameworks can be controlled either by using direct functionalization processes, such as the incorporation of special heteroatoms, as well as the utilization of different structure-directing agents (SDAs)) [130–133] or by using post-synthesis methods such as dealumination treatments [85, 134]. The locations of Al_F can effectively adjust the reaction pathways of aromatic- and alkene-based reaction cycles for MTO reaction based on the energetics between ZSM-5 and ZSM-11 zeolite [121]. Many researchers also discovered that the positions of Al_F in zeolites, such as H-ITQ, H-MCM-22, and H-ZSM-5 can be adjusted by incorporating basic atoms such as Ge and B species or by altering silica sources as well [131, 135–137], where an excellent linear correlation between $B_{sin}/(B_{poc} + B_{sup})$ (the amount of BAS in the sinusoidal channels divided by sum in the surface pockets and supercages) or $Al_{(56)}/\left(Al_{(50)} + Al_{(61)}\right)$ ratio for B-incorporated H-MCM-22 zeolites and their lifetimes for MTH reaction were reported [131], and the results are displayed in Figure 14.9a as confirmed by Py-IR measurement and ^{27}Al MAS NMR analysis. Those results confirmed that the framework $Al_{(56)}$ species was responsible for BAS in the sinusoidal channels, whereas the framework $Al_{(50)}$ and $Al_{(61)}$ species were also responsible for BAS in the supercages and surface pockets, respectively. The enhancement of the $Al_{(56)}$ species on the H-MCM-22 frameworks or BAS in the sinusoidal channels can effectively enhance long-term catalytic stability for the MTH reaction. Consequently, the appropriate regulation of reaction routes for MTO reaction largely improved the propylene selectivity and catalyst stability. In addition, the distributions of Al_F and BAS on the H-ZSM-5 can be significantly tuned by utilizing suitable Si sources and crystallization temperature, which significantly changes reaction pathways for dual cycle mechanisms of methanol conversion [73, 136] as displayed in Figure 14.9b. The Al_F atoms located at the channel intersections of the H-ZSM-5 were more suitable for the propagation of aromatics-based cycles, while

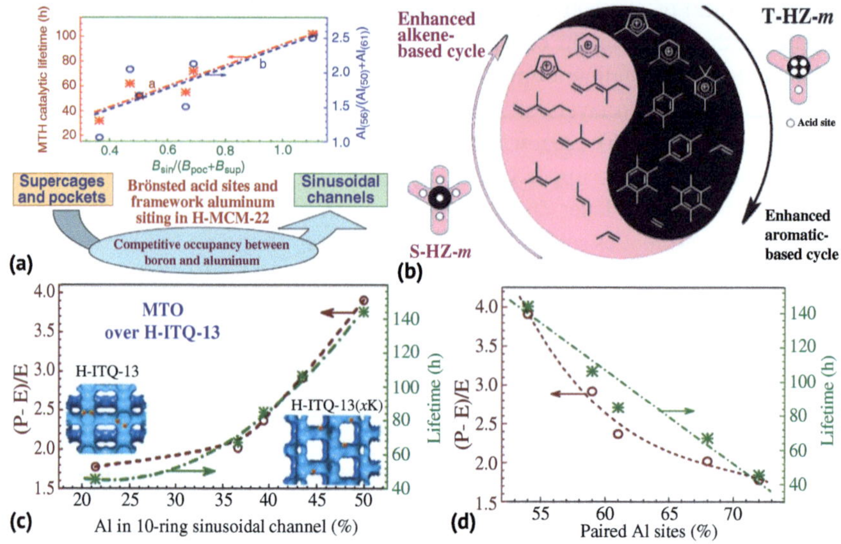

Figure 14.9 (a) Relationships between $B_{sin}/(B_{poc} + B_{sup})$ ratio measured by Py-IR on the boron-incorporated H-MCM-22 zeolites and their catalytic lifetime for MTH reaction, and the $Al_{(56)}/(Al_{(50)}+Al_{(61)})$ ratio determined by ^{27}Al MAS NMR. *Source*: [131] / with permission of American Chemical Society. (b) Conversion of MTO over H-ZSM-5 zeolite: reaction pathway is related to the framework aluminum siting is licensed under CC BY 4.0. *Source*: [136] / with permission of American Chemical Society. (c) Relation of the catalytic lifetime and (P−E)/E ratio for MTO (at 450 °C and SV of methanol with 1 h^{-1} with complete methanol conversion over various H-ITQ-13(xK) zeolites to the proportion of the paired Al atoms (measured using Co^{2+}-exchange). (d) Fraction of Al atoms sited in the 1-MR sinusoidal channels (T9 sites assigned to the resonance peak at 55.5 ppm from ^{27}Al NMR spectra). *Source*: [138] / with permission of Elsevier.

the Al_F atoms positioned in the sinusoidal or straight channels of the H-ZSM-5 were more beneficial for alkene-based cycles. The catalytic performances of H-ITQ-13 zeolite for MTO reaction were also improved by controlling the distributions of aluminum sites in the frameworks [138]. With an addition of appropriate content of K^+ ions to the synthesis gel can significantly reduce the amount of paired Al as well as distribute more Al atoms in the 10-MR sinusoidal channels out from the channel intersections. As displayed in Figure 14.9c, d, the (P–E)/E ratio (where, E and P represent the selectivity to ethene and propene, respectively) and the catalyst lifetime were enhanced significantly with an increase of the fraction of Al atoms located in 10-MR sinusoidal channels on the H-ITQ-13(K), while the (P–E)/E ratio and stability were decreased linearly with an increase of the proportion of the paired Al atoms. There have been many efforts to properly control the quantities and chemical environments of the Si and Al atoms in the silicoaluminophosphate and aluminosilicate zeolites. However, it seems to be difficult to precisely control the locations of those atoms within the zeolite main frameworks. Despite these challenges, many researchers continue to work on improving our understanding and control of these factors, since those properties can significantly change the catalytic performance.

14.7 Effects of B/L and External/Internal Acidic Sites on the Zeolites

The carbon efficiency and product distribution of the MTH or DTH reaction can be influenced by the type, density, and strength of acidic sites of various zeolites. The stronger acidic sites seem to quickly deactivate due to their higher surface energy to over-activate chemical reactions, which can cause the formation of unwanted byproducts such as coke precursors by accumulating on the catalyst surfaces to reduce the catalytic activity irreversibly. On the other hand, the milder acidic sites may be more stable with lower deactivation rates, resulting in improved catalytic performance with much longer lifetimes [54]. Therefore, it is important to properly control the strengths of acid sites in the zeolites to optimize their catalytic performances and lifetimes. The Brønsted acid density and strength of zeolites can be modified by changing the Si/Al ratio in the zeolite main frameworks as well as by introducing other acidic metal atoms into the zeolite structures. These modifications can largely affect the catalytic activity of the zeolites, which can also be optimized according to the different types of reactions as well. The acid strengths of MFI zeolite (H-[Al]-ZSM-5, H-[Fe]-ZSM-5, and H-[Al, Fe]-ZSM-5) were controlled by the incorporation of Al^{3+} and/or Fe^{3+} into the main frameworks of the zeolites to maximize propylene and propylene/ethylene ratio for the MTH reaction [18]. In the study, the acidic strengths of three different types of MFI zeolites followed the increasing order of H-[Fe]-ZSM-5 < H-[Al, Fe]-ZSM-5 < H-[Al]-ZSM-5, where the density of BAS on the H-[Al, Fe]-ZSM-5 and H-[Fe]-ZSM-5 was decreased more easily with an increase of temperature compared to the H-[Al]-ZSM-5 with its largest decrease on the H-[Fe]-ZSM-5. At a low temperature of 400 °C, the H-[Fe]-ZSM-5 showed a higher propylene selectivity and propylene/ethylene ratio with a lower methanol conversion and a higher rate of DME formation. However, at high temperatures above 450 °C, the propylene selectivity on the H-[Fe]-ZSM-5 was significantly lower than the other two zeolites. Therefore, the best overall catalytic performance was observed on the H-[Al, Fe]-ZSM-5, which was attributed to the combination of the stronger Al-based and weaker Fe-based BAS. Similarly, the smaller BAS and larger weak to medium acid sites on the ZSM-5/ZSM-11 inter-growing composites were responsible for a higher propylene/ethylene ratio during MTH reaction [35].

In addition, the strengths of BAS in isomorphously framework-substituted ZSM-5 zeolites were increased in the order of B(OH)Si < Fe(OH)Si < Ga(OH)Si < Al(OH)Si, where the B-substituted ZSM-5 had very lower acid strengths, which was found to be less active than the H-[Al]-ZSM-5 zeolite for MTO reaction. Similarly, Ce-modified ITQ-13 zeolite reduced the content of BAS as well as formed new Lewis acid sites, which inhibited the formation of aromatics and accelerated coke formations by simultaneously producing more propene and butene product [139]. The comparisons of the MCM-22 (Si/Al ratio = 18) and MCM-36 (Si/Al ratio = 54) revealed that the B/L acid sites ratio in the MCM-36 was higher than that of the MCM-22. However, it showed three times lower acid site density as compared to the MCM-22, where larger aromatics selectivity and coke formations on the MCM-22 were observed at 450 °C and SV = 1–2 h^{-1}, whereas the MCM-22 selectively produced aliphatic hydrocarbons due to higher density of acid sites [54]. It is also believed that the MTH reaction in the sinusoidal channels is preferential by methylation-cracking reactions, which involve the cleavages of chemical bonds through the addition of methyl groups. The catalytic actions in the supercages can control HCP mechanisms, which are related to the accumulations and releases of hydrocarbons formed within the supercages. It has been also reported that the acidic sites located on the external pockets of zeolitic materials can be detrimental to the product distribution and stability for the MTH reaction, which has negative effects to decrease the overall efficiency of the MTH reaction [140]. The B/L acid sites of zeolites are recognized as the catalytic active sites for methanol conversion through HCP mechanisms, and the larger concentrations of strong BAS can selectively cause catalyst deactivation due to a rapid coke deposition with a higher yield of C_2–C_4 hydrocarbons due to the cracking reactions. However, the weak BAS along with Lewis acid sites can lead to moderate hydrocracking activity of heavier hydrocarbons with a higher selectivity to C_5–C_{11} hydrocarbons [126]. On the ZSM-5 and Beta zeolite, both aromatic and alkene-based mechanisms are possible at low concentrations of BAS, and the alkene-based mechanisms predominate by showing higher selectivity to light olefins and lower selectivity of aromatics [124, 141, 142]. However, the aromatic-based mechanisms are efficiently responsible for a higher aromatic selectivity when the concentration of BAS is higher. A similar trend was observed in the SAPO-18 zeolite, where a higher acidic density favors more aromatic-based cycles than olefin-based mechanisms [143], where an improved MTO catalytic performance with higher olefin selectivity and enhanced stability was obtained over SAPO-based zeolites containing BAS with moderate acidic density.

The aromatics yield strongly depends on the amount of BAS, where the yield was monotonically increased with an increase of BAS, while the yields of paraffin, olefin, and naphthalene showed an opposite trend [124] as displayed in Figure 14.10a. It suggests that the larger content of BAS is more favorable to form aromatics rather than olefins, where the HZ-5(50) showed six times higher aromatic yield than HZ-5(470). Besides, the acid density of the ZSM-5 with weak base ($NaHCO_3$, Na_2CO_3, and $NH_3 \cdot H_2O$) modifiers were applied for MTA reaction [144], where the ratio of B/L acid sites was changed significantly from 0.107 to 7.35 after the modification with weak base chemicals. The $NH_3 \cdot H_2O$-treated ZSM-5 with the highest B/L ratio exhibited a largely enhanced aromatics yield of 56.1% with a higher BTX selectivity of 40.3%. Similarly, the ZSM-5 treated with various amounts of NaOH was applied for the MTA reaction, and the moderate alkaline treatment with 0.2–0.3 ml (NaOH)/L generated mesoporous structures in the ZSM-5 frameworks without significant acidity changes. However, the ZSM-5 treated with 0.4–0.5 ml (NaOH)/L showed a decreased number of acidic sites via dealumination along with desilication, which enhanced catalyst lifetime at a little lower conversion of methanol [145].

The conversion of MTP and MTG by using H-ZSM-5 has been successfully implemented for industrial applications; however, main byproducts such as ethane and butanes from MTP reaction and 1,2,4,5-tetramethylbenzene from MTG reaction are not sufficiently suppressed [30]. Generally, light aromatics are formed in the

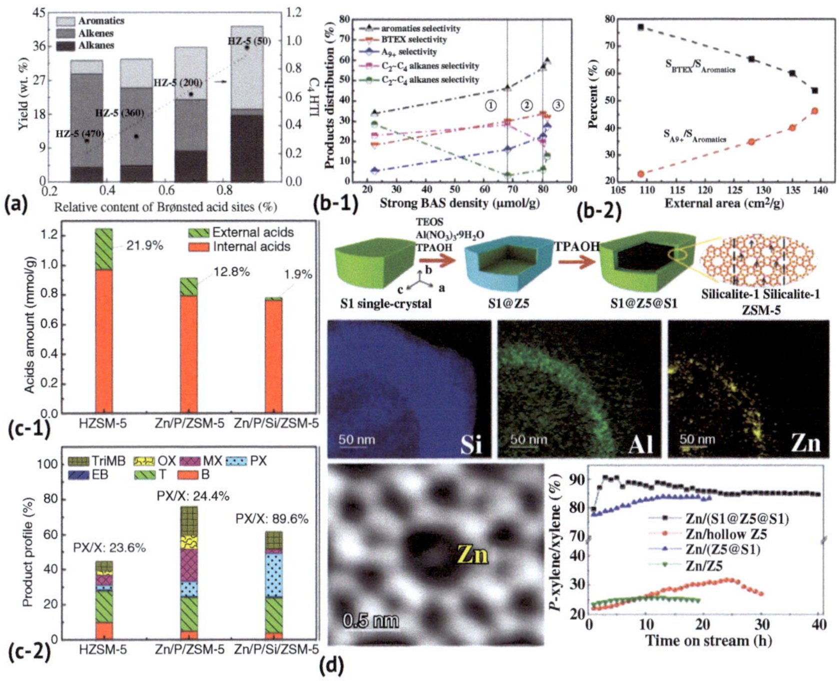

Figure 14.10 (a) C_4 HT index (C_4 HTI) and product yield over HZSM-5 catalysts according to strong BAS density. *Source*: Adapted with permission [124]/Springer Nature. (b) Product selectivity according to (b-1) BAS density and (b-2) external area for MTG reaction at TOS = 4 hours over Z5-NS-t catalysts. *Source*: [148] / with permission of Elsevier. (c) External and internal acid amounts of the three catalysts and product profiles of MTA from the different catalysts, where TriMB is trimethylbenzene. *Source*: [149] / with permission of American Chemical Society. (d) Rational design of zinc/zeolite catalyst: selective formation of PX from MTA reaction. *Source*: Reproduced from [153] / with permission of John Wiley & Sons.

internal BAS of the ZSM-5 through aromatization reaction, whereas C_{9+} aromatics can be formed on the external BAS by alkylation of the formed light aromatics. It implies that the product distribution can be significantly changed by the different locations of BAS, such as internal and external surfaces of the ZSM-5 zeolite. Therefore, the larger external acid sites of the H-ZSM-5 under considerable concentrations of reactants are responsible for the secondary reactions by forming many undesired byproducts, which can be properly suppressed by various posttreatments of zeolite surfaces [146, 147]. For example, the rapid and low-cost synthetic method of ZSM-5 zeolite with nanosheet-like morphology was proposed for MTG reaction to confirm the effects of different locations and densities of BAS on the product distribution [148]. As displayed in Figure 14.10b, the BTEX selectivity was found to be higher than A_{9+} aromatics, which was increased with an increase of BAS due to the introduction of Al species into zeolite frameworks rather than outer surfaces by controlling HT reactions on the BAS located in zeolite micropores. However, the selectivity to C_2–C_4 alkenes is greatly reduced due to an increased strength of BAS, which favors the aromatic-based cycles. Furthermore, the ratio of A_{9+} aromatics in total aromatics ($S_{A9+}/S_{Aromatics}$) increased significantly with an increase in external surface areas, while the ratio of BTEX in total aromatics ($S_{BTEX}/S_{Aromatics}$) was decreased as displayed in Figure 14.10b1, b2. Therefore, it is obvious that the larger content of strong BAS on the external surfaces of nanosheet-structured ZSM-5 enhances the further methylation of BTEX to A_{9+} aromatics.

Recently, the Zn/P/Si/ZSM-5 consisting Zn/P/ZSM-5 core having stronger acid sites (high acid density) coated with outer SiO_2 shells having weak acid sites (low acid density) were investigated for MTA reaction [149, 150], which showed an increased para-xylene (PX) selectivity of 89.6% among xylenes and total aromatic selectivity of 61.7% as displayed in Figure 14.10c1, c2. Similarly, one-step direct formation of PX from methanol over Mg-Zn-Si-ZSM-5 catalyst was studied by tuning pore openings and surface acidic properties [151], where PX selectivity was significantly enhanced up to 98.9% among xylenes, and the formation of the much larger C_{9+} aromatics was suppressed by eliminating external acidic sites and by narrowing pore openings with silica-coating passivation. In addition, the core-shell structured Zn/ZSM-5@ Silicalite-1 catalyst prepared by an environmentally benign route exhibited a higher BTX selectivity with PX selectivity of 84.1% among all xylenes. The usages of inactive silicalite-1 zeolite shells prevented the alkylation of toluene with methanol by significantly suppressing external surface acidic sites on the Zn/ZSM-5@Silicalite-1 [152]. The zeolite synthesis strategy to achieve higher PX selectivity and higher catalyst stability was also proposed by passivating external acidic sites and by shortening the diffusion length of intermediates [153], where the triple-shell MFI single crystal catalyst, which is quite different from the core-shell catalyst for MTA reaction. The catalyst showed almost complete methanol conversion and 90% PX selectivity by eliminating external acidic sites as well as by simultaneously improving diffusion natures as displayed in Figure 14.10d.

14.8 Effects of Metal Modifications to Catalytic Activity and Stability

Solid acid zeolites with their different pore structures and topologies, acidity according to Si/Al ratio and bridging hydroxyl groups are responsible for different B/L acid sites, which makes them widely used for many organic synthesis reactions [154]. In addition, the surface modifications of those zeolites with different metal cations can provide functionalities to zeolites by generating new acidic sites [20, 155]. One way to modify the zeolite surfaces is to carry out an ion exchange (IE), in which metal ions are introduced into the zeolite structures by replacing existing metal ions, which can be done by using a solution containing the desired metal ions for the adsorption onto the zeolite surfaces. The metal ions can be incorporated into zeolite structures in different amounts and locations, depending on the conditions of IE processes [94, 156]. Another way to modify zeolite surfaces is done by the impregnation (IM) method, in which metal ions are introduced into zeolite structures by depositing onto the zeolite surfaces, which can be done by using a solution containing the desired metal ions and the metal ions can be incorporated into the zeolite structures in different amounts and locations according to IM conditions [153].

Various metal-doped ZSM-5 have been investigated for MTH reaction, such as Ag-ZSM-5 [157], Zn-ZSM-5, P-ZSM-5 [158, 159], Mn-ZSM-5 [38], Mg-ZSM-5 [160], B-ZSM-5 [161], Ga-ZSM-5 [162], and Cd-HZSM-5 [156]. The metal modifications reduced the number of BAS on the catalyst surfaces, thereby improving the catalyst activity and stability as well as the desired products [163]. The intergrown ZSM-5/ZSM-11 with phosphorus species was applied for MTH reaction, and the modification of the intergrowth zeolites revealed the decreased BAS with significantly improved propylene selectivity and a higher propylene/ethylene ratio of ~8 at 400 °C, where the phosphorus modification also suppressed the secondary reactions by significantly suppressing the formations of aromatics and alkanes [35]. In addition, the catalytic performances can be effectively enhanced by the presence of secondary metals such as Ni-Mn/HZSM-5 [155] and Ce-Mn/HZSM-5 [51]. Various metals such as potassium (K), cerium (Ce), molybdenum (Mo), silver (Ag), and cesium (Cs) species on SAPO-34 zeolite were investigated for MTO reaction. The Zn-doped ZSM-11 generated a new BAS for MTO reaction [164], and a bimetallic catalyst with 1 wt. % Zn and 1 wt. % Sn doped on ZSM-5 also exhibited the highest BTX selectivity of 64.1% with 100% methanol conversion at MTA reaction condition of T = 450 °C, P= 0.1 MPa, WHSV = 0.8 h^{-1} [163].

The bimetal-modified zeolite of H-[Fe, Al]-ZSM-5 with Fe content in the range of 0–1.2 wt. % and various crystallization duration were also investigated for the MTG and DTG reactions at T = 390 °C and P = 0.1 MPa. For the MTG reaction, the H-[Fe, Al]-ZSM-5 modified with 0.8 wt. % Fe showed aromatics yield of 30%, while the catalytic performance and aromatics selectivity were decreased with an increase of Fe content [165]. For the DTG reaction, the H-[Fe, Al]-ZSM-5 crystallized for 24 hours exhibited the best catalytic performance with DME conversion of 100% and gasoline yield of 71.5% with stability for 125 hours due to its suitable acid properties [166]. In addition, the bimetallic Ca and Ga-modified ZSM-5 was investigated for MTA reaction at T = 450 °C, P = 0.1 MPa, and SV = 5.3 h^{-1} [167], where the small amount of Ca (0.02 wt. %) extends the catalyst lifetime by maintaining a higher aromatic selectivity. The integral yield of BTEX was increased from 3 to 4 g_{carbon}/g_{cat} and the selectivity to light olefins was also increased from 8 to 12 g_{carbon}/g_{cat} on the Ga/HZSM-5 and Ca(0.02)Ga/HZSM-5, respectively, as displayed in Figure 14.11a. The interactions between Ca and Ga on the ZSM-5 enhanced moderated dehydrogenation rates as confirmed by the smaller hydrogen formation over the Ca(0.02)Ga/HZSM-5 than the Ga/HZSM-5 for MTA reaction. The Ga-doped ZSM-5 (Ga/Al ratio of 0.25, 0.5, 0.75, and 1) was investigated for MTA reaction at T = 470 °C, P = 0.1 MPa, and SV = 4 h^{-1}, and the catalyst with Ga/Al ratio of 1 showed an improved catalyst stability for methanol conversion up to 14 days [162]. On the CuO/NH_4-ZSM-5 (3, 5, 7, and 9 wt. % CuO) for the MTA reaction, the catalyst with 7%CuO/NH_4-ZSM-5 exhibited methanol conversion of 100% with aromatic selectivity of 49% at T = 400 °C and P = 0.1 MPa [168]. The HZSM-5-modified with gadolinium (Gd) metal oxide by IM or IE method was investigated for the MTA reaction at T = 400 °C, P = 0.1 MPa, and SV = 4.75 h^{-1}, where the catalyst prepared by

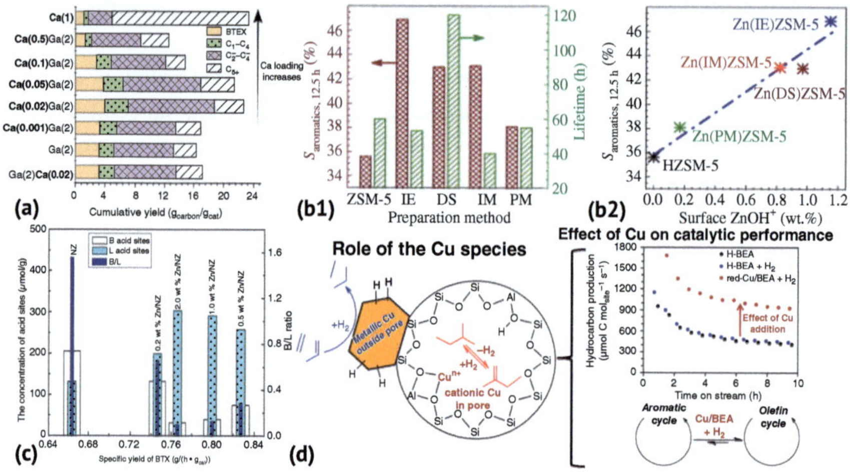

Figure 14.11 (a) Integral yields of the main groups in MTA products before MeOH conversion drops below 20% is licensed under CC BY 4.0. *Source*: [167] / with permission of American Chemical Society / CC BY 4.0. (b) Correlation between the generated coke species with selectivity to aromatics and the surface amount of $ZnOH^+$ species in the Zn-containing HZSM-5 catalysts prepared by various methods for methanol conversion to gasoline-range hydrocarbons. *Source*: [171] / with permission of Elsevier. (c) Relationship of acidity and specific yield of BTX for MTA reaction over Zn/NZ catalysts. *Source*: [173] / with permission of American Chemical Society. (d) Hydrocarbon production rates, which were based on the number of acid sites on H-BEA zeolite. *Source*: [174] / with permission of American Chemical Society.

IE method (GdHZ-IE) possessed higher amount Lewis acid sites, which facilitated a higher aromatics selectivity of 35.2% and enhanced stability compared to other catalysts [169]. On the Zn-doped ZSM-5 catalysts tested for MTA reaction at T = 450 °C and SV = 4.74 h^{-1}, BTX selectivity of 32.5% was attributed to the well-dispersed ZnO, which also induced a rapid catalyst deactivation due to the presence of ZnO clusters in the ZSM-5 channels [22]. Similarly, the Zn-impregnated on nano-sized ZSM-5 showed an enhanced aromatic selectivity of 62% and an excellent stability for MTA reaction at T = 470°C, P = 0.1 MPa, and SV = 0.75 h^{-1} [170]. In addition, the Zn-modified ZSM-5 prepared by four different methods, such as IM, IE, physical mixing (PM), and direct synthesis (DS) was investigated for MTA reaction [171] as displayed in Figure 14.11b1, b2. The distribution of acidic sites, nature of Zn species, and catalytic activity were affected by the Zn incorporation method. The Zn(IE)/ZSM-5 containing a higher fraction of surface $ZnOH^+$ species showed the highest selectivity to aromatics, whereas the Zn(DS)/ZSM-5 with a smaller particle size and larger mesoporous volume exhibits the longest catalytic lifetime. In another study, the ZSM-5 modified with different Zn salts such as zinc sulfate (S), zinc acetate (Ac), zinc nitrate (N), and zinc chloride (Cl) was studied for MTA reaction [172], where the natures of Zn species on the ZSM-5 were greatly influenced with the types of Zn salts by largely altering acidic strengths and distributions of acidic sites. The Zn(S)/HZSM-5 with moderate BAS effectively improved the MTA performance by showing a total aromatics selectivity of 77.9%. In addition, among Zn (0.2, 0.5, 1.0, and 2.0 wt. %) impregnated on the nano HZSM-5 for MTA reaction at T = 400°C and

P = 0.1 MPa as displayed in Figure 14.11c, the 0.5 wt. % Zn/NZSM-5 showed a 100% methanol conversion with an aromatics yield of 51.7% [173]. Additionally, the DME conversion to 2,2,3-trimethylbutane over Cu-modified large-pore BEA zeolite (Cu/BEA) with metallic Cu sites, Lewis acidic cationic sites, and BAS was enhanced to enlarge C_4 (e.g., isobutene) and C_7 (e.g., 2,2,3-trimethylbenzene) selectivity by suppressing the aromatic selectivity [174]. As displayed in Figure 14.11d, metallic Cu sites facilitated H_2 activation and ethylene/propylene hydrogenation, while the cationic Cu sites were responsible for a HT reaction and hydrogenation of C_4–C_7 olefins produced via the olefin catalytic cycle.

14.9 Reaction Mechanisms for Methanol/DME Conversion and Deactivations

Methanol can be converted into hydrocarbons on the solid acid zeolites by indirect mechanisms known as "HCP" mechanisms, where the formed olefins and aromatics act as platform chemicals [175–179]. For the possible mechanisms of the MTH reaction, two aspects have been still under debate in terms of (1) origins of first C—C bond formation and (2) ways to MTH propagation. It has been reported that two kinds of catalytic cycles are involved: (1) olefins can be periodically methylated to form branched hydrocarbons which are susceptible to cracking and (2) aromatics are also periodically methylated and dealkylated to form light olefins [180, 181]

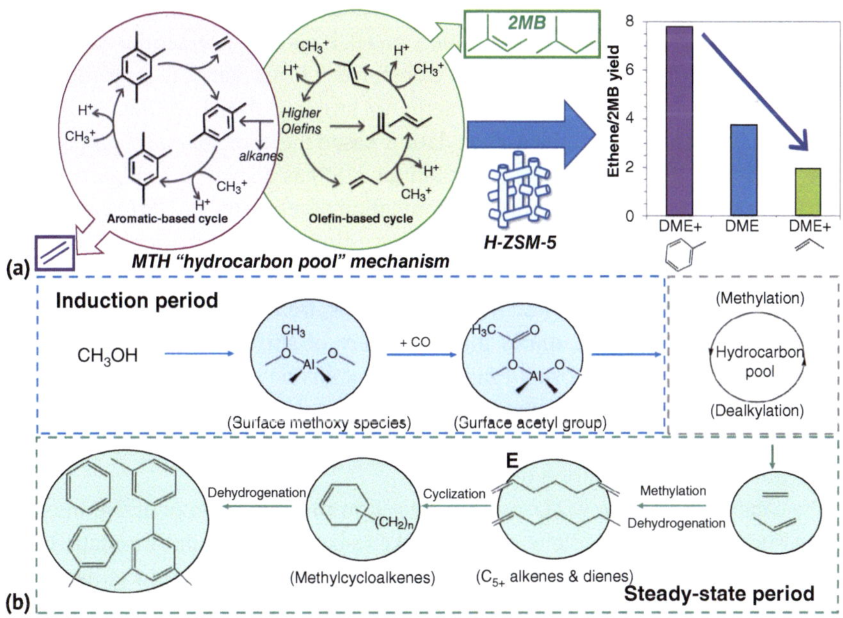

Figure 14.12 (a) Dual olefin and aromatic methylation/cracking cycles for MTHs on H-ZSM-5. *Source*: [183] / with permission of Elsevier. (b) Reaction pathway from MTA via MTO and olefins-to-aromatics route over metal-modified ZSM-5 zeolites is licensed under CC BY 4.0. *Source*: [186] / with permission of American Chemical Society / CC BY 4.0.

as displayed in Figure 14.12a. It was observed that the yield of ethane/2MB (2-methylbutane + 2-methyl-2-butene) was declined by a factor of 1.9 when propene was co-fed with DME compared to that from DME conversion itself due to an increased propagation reaction of the olefin-based cycle. In contrast, the ratio of ethane/2MB was increased by a factor of 2.1 when toluene was co-fed with DME due to an increased propagation reaction of the aromatic-based cycle. Those dual cycle mechanisms were greatly contributed to understanding the HCP mechanisms. Besides, the participation of the main six-step chemical reactions, such as olefin methylation, olefin cracking, HT, cyclization, aromatic methylation, and aromatic dealkylation have been well reported for dual catalytic cycles for the HCP mechanism during MTH reaction [182–184]. In addition, the possible induction periods for MTH reaction at the temperature <255 °C were reported [185], where three steps of induction period were divided as an initiation step of C—C bond formation, HCP species formation, and autocatalysis. The HCP species formation step was reported to be rate-determining step (RDS) of the MTH reaction with the highest energy barrier, and a critical concentration of HCP species at which the autocatalysis reaction initiates (at the third step) can proceed by driving the HCP species on the catalyst surfaces, which lowers the activation energy barrier for second and third steps and eventually reduces the induction period [62]. Similar reaction pathways for MTA by multistep reaction routes of MTA and olefins to aromatics route over metal-modified ZSM-5 zeolites were proposed [186] as displayed in Figure 14.12b. Since the MTA reaction pathways are similar with that of MTO and MTG reaction, an induction period to build up the HCP species and steady-state period for the formation of heavier hydrocarbons by an autocatalysis were also applied to explain the reaction mechanisms. The generation of HCP species is essential to produce light olefins, which can be subsequently transformed to aromatics through dehydrogenation and cyclization steps for the methanol aromatization over metal-modified ZSM-5 zeolites. In addition, the carbonylation-based mechanisms were proposed by experimental data and theoretical calculations [187] as displayed in Figure 14.13a, which also suggests that the possible carbonylation of methanol or DME was responsible for the formation of first C—C bonds resulting in the formation of HOAc or MeOAc intermediates in the presence of water. A series of acid-catalyzed reactions such as acetylation, decarboxylation, aldol condensation, and cracking reaction can convert the surface intermediates into a mixture of surface-bound hydrocarbons and HCP, resulting in the formation of initial olefins on the catalyst surfaces. The carbonylation-based mechanisms have been reported to have much lower energy barriers for the formation of first C—C bonds than the previously proposed mechanisms due to an easy initiation of MTH [188].

The catalytic cycles for MTO reaction based on the formation of active intermediates such as cyclopentadiene through alkene-based cycles and aromatic-based cycles are well-related to the gradual methylation with methanol to enhance the propagations of C—C bond, and the extended entities like alkyl chain carrier can be split off to form light olefins [189] as displayed in Figure 14.13b. For the conversion of methanol or DME on the SAPO-34, the surface methoxymethyl cations ($CH_3OCH_2^+$) were formed through the reaction of methoxy and DME [190],

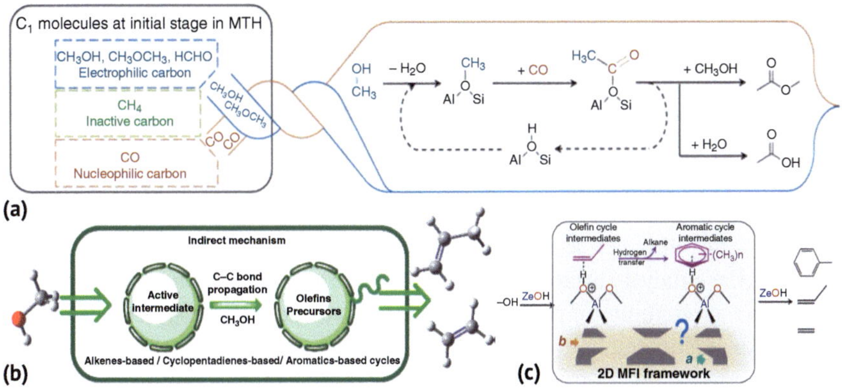

Figure 14.13 (a) First C—C bond formation for MTH reaction by coupling between nucleophilic and electrophilic carbon atoms, and the adsorption of MeOH (or DME) on the BAS forms surface methoxy groups, which proceed nucleophilic attack with CO by forming surface acetyl groups, which also contain the first C—C bonds and the acetyl groups dissociate into MeOAc and HOAc in a gas phase. *Source*: [187] / with permission of John Wiley & Sons. (b) Common natures of the indirect mechanisms for methanol conversion. *Source*: [189] / with permission of American Chemical Society. and (c) Schematic illustrations of MTHs reaction catalyzed by ZSM-5 zeolite with MFI structures, which consist of sinusoidal and straight channels, and the reaction follows the HCP mechanisms with olefinic and aromatic cycles simultaneously is licensed under CC BY 4.0. *Source*: [192] / with permission of John Wiley & Sons / CC BY 4.0.

where the intermediates further reacted with methanol or DME to form intermediates containing C—C bonds such as 1,2-DMM or 2-methoxyethanol. Subsequent formation of oxonium cations and ether groups caused the formation of propene as the major product, which follows the HCP mechanism. For the methanol or DME conversion to olefins on SAPO-34 at 400–460 °C [191], methanol can be a more competent feeding reactant for olefin production than DME. The distribution of products for the DTO reaction differed from the MTO reaction, particularly in terms of the rates at which methane and light olefins were formed. When the DME was used as a feed, the catalyst's lifespan was increased and the relatively lower selectivity to light olefins for the DTO reaction compared to the MTO reaction can be attributed to the increased rates of methanol and methane formation. Similarly, on ZSM-5, the DTO reaction exhibits a lower olefin selectivity with excellent catalyst stability compared to that of the MTO reaction, which was attributed to the effects of the channel geometry of ZSM-5 frameworks on the reaction intermediates of the MTH reaction by combining the anisotropic zeolite crystallites [192, 193]. The results clearly revealed that the subtle differences in the geometry of zeolite channels can alter the compositions of reaction intermediates as displayed in Figure 14.13c. For example, the narrower sinusoidal channels in the α-oriented ZSM-5 crystallites more favored olefin-based cycles with an enlarged homologation activity of alkanes. The more extended straight channels in the b-oriented ZSM-5 crystallites favor the aromatic-based cycles by generating a higher degree of alkylation to form ethylene.

The more understanding of deactivation mechanisms of solid acid zeolites during MTH (or DME) are still under vigorous investigation, and the formation of coke precursors on the active sites is one of the major reasons, which refers to the accumulation of solid carbon particles or coke precursors during the cracking of heavy hydrocarbons to produce lighter and more valuable products. The catalyst deactivation by coke depositions is affected by different factors such as dimensions of channels and cavities of zeolites, acid sites and locations, Si/Al ratio, crystallite size, reaction conditions, surface intermediates, and final product [194]. The coke precursors can be preferentially formed when heavy hydrocarbons are partially condensed and oligomerized, which can increase the accumulation of carbon species on the catalyst surfaces during the MTH reaction, which eventually suppressing the accessibility of reactants on the active sites by being decorated with surface coke precursors.

14.10 Summary and Perspectives

Methanol can be produced from different renewable sources such as biomass, natural gas, and coal by initially converting them into syngas through gasification and reforming reactions. Syngas can be transformed into methanol by COx hydrogenation, and methanol can be further processed into useful products such as gasoline, olefins (ethylene and propylene), and aromatics depending on the catalyst types and processes (Figure 14.2). If renewable feedstocks are acquired and processed sustainably, the production of methanol from renewable sources can be an eco-friendly alternative method to replace conventional fossil fuel-based productions, which reduce the emission of CO_2 up to 95% and NOx up to 80%, as well as SOx and PM emission suppression [195]. A biochemical company Enerkem, partnered with the Shell company to build a new advanced methanol production facility in the Port of Rotterdam utilizing solid waste, which is expected to be operated by 2024 [196]. The facility will also employ carbon capture technology to reduce carbon emissions and utilize green hydrogen and sustainable biomass as feedstocks to produce methanol, which will be used as a low-carbon fuel in the transportation sector and as a feedstock for the chemical industry. Significant advances have been made in catalyst synthesis, reaction mechanisms, process developments, reactor designs, and commercialization developments in the MTO field since the Mobil Corporation proposed it in 1977. In 2010, the DICP accomplished major breakthroughs by launching the world's first 600000-ton/year MTO facility by using SAPO-34 zeolite, which represented a significant milestone in the production of lower olefins (ethylene and propylene) from coal. The DICP has improved the olefin yield and production capacity of MTO with their DMTO-II and DMTO-III technologies. In addition, the UOP and Norsk Hydro employed the SAPO-34 zeolite to develop a fast fluidized bed reactor for a low-pressure MTO process. Moreover, Shanghai Research Institute of Petrochemical Technology developed a fluidized bed SMTO technology, while Lurgi Company constructed the first MTP plant by using ZSM-5 zeolite in 2011 with their fixed-bed MTP process. As the various MTH processes become more widely commercialized, it is expected that the research fields will focus on efficient plant operations and optimal catalysts.

The need for cleaner energy resources in response to climate change and the limitations of fossil fuels have led to an increased interest in using methanol directly as a fuel or converting it into useful hydrocarbons for different chemical applications. Methanol is the simplest molecule for energy storage and used to produce a diverse array of products, where the conversions of MTH are widely studied through zeolite-catalyzed processes. In addition, conversions of oxygenates to hydrocarbons has been successfully employed in various commercial applications, resulting in the establishment of new plants. This has led to a growing interest in this process within both academic and industrial circles. The present review chapter critically addressed the literature concerning the use of methanol/DME for the production of different hydrocarbons to elaborate the roles of different structural and chemical properties of zeolites, reaction mechanisms to selectively synthesize desired products and technological challenges in converting methanol/DTH. The zeolites have higher selectivity when it comes to producing hydrocarbons from methanol/DME, and the properties of the catalysts are adjusted to increase the conversion rate, hydrocarbon yield, and selectivity. This study examines the effects of reaction conditions and properties of zeolites that directly influence methanol/DME conversion to hydrocarbons, such as co-reactant injection, acid site types and strengths, morphology and pore size distribution, zeolite topology, Si/Al ratio, hydrophobicity and hydrophilicity, and metal modifications of those zeolites.

The product distributions of the MTH and DTH reactions can be affected by both operating conditions and zeolite properties, which favors the propagation of the aromatic- or olefin-based cycles. Commercially, the ZSM-5 and SAPO-34 are the most common zeolites used to produce olefins. Studies for the MTH/DTH reactions have indicated that most of the zeolites showed an increased conversion rate of reactants and decreased selectivity of olefins with an increase in reaction temperature. High reaction temperatures showed a lower HT rate due to weaker adsorption of the adsorbed olefins, by producing more aromatics which influences the coke formation in the internal micropores, resulting in a rapid deactivation of the zeolites. Similarly, the higher reaction pressures lead to C_{5+}/aromatics hydrocarbons while lower pressures tend to generate lighter hydrocarbons. In addition, the SV not only influenced the conversion of methanol/DME but also altered product selectivity, when the SV was elevated, the catalyst showed less contact time which resulted in their lower conversions.

The concentration, location, and strength of acid sites on the zeolites are crucial in determining their overall acidity and catalytic activity. The larger concentration of BAS from a higher Al content in the zeolites generally leads to the formation of aromatics and ethylene. The Si/Al ratio is linearly correlated with the propylene selectivity, while decrease in aluminum content resulted in a decrease of aromatics selectivity. The incorporation of metals in zeolites can reduce the BAS resulted in the suppressed HT and cyclization reaction to form smaller paraffins and aromatics by simultaneously enhancing the selectivity of propylene. Most of the recent efforts have been directed towards the selective formation of propylene; however, it is also one of the main products of the aromatic-based cycles, and undesired coke formation also follows this route. Given the volatility of the olefin market, it would be advantageous to have the ability to readily modify zeolite structure and catalytic reaction conditions in order to adjust product selectivity. The most promising solutions appear to involve the co-feeding of the necessary products to facilitate the desired cycles, which highlights the need for further work on the reactor designs. The quest for novel, high-performance zeolites with specific framework topology and morphology, acid strengths and high stability are continuously required, where optimal zeolites with desired topology (such as pore opening and cavity) and acidity can create significant opportunities to enhance the production of desired products.

Acknowledgments

This work was supported by the Carbon Neutral Industrial Strategic Technology Development Program funded by the Ministry of Trade, Industry & Energy (MOTIE) of the Republic of Korea (RS-2023-00261088). This work was also supported by the National Research Foundation of Korea (NRF) grant funded by the Korean government (RS-2024-00405818).

References

1 Fredric, B., Viktoras, K., Christopher, O. et al. (2022). *Petrochemicals and Climate Change: Tracing Globally Growing Emissions and Key Blind Spots in a Fossil-Based Industry. IMES/EESS Report*. Vol. 126. Lund: Lund University.
2 Holechek, J.L., Geli, H.M., Sawalhah, M.N. et al. (2022). A global assessment: can renewable energy replace fossil fuels by 2050? *Sustainability* 14 (8): 4792–4813.
3 Goeppert, A., Czaun, M., Jones, J.P. et al. (2014). Recycling of carbon dioxide to methanol and derived products-closing the loop. *Chem. Soc. Rev.* 43: 7995–8048.
4 Oil Market Report (2022). International Energy Agency (IEA). https://www.iea.org/reports/oil-market-report-december-2022.
5 Schwiderowski, P., Ruland, H., and Muhler, M. (2022). Current developments in CO_2 hydrogenation towards methanol – a review related to industrial application. *Curr. Opin. Green Sustain. Chem.* 38: 100688–100695.
6 Jadhav, S.G., Vaidya, P.D., Bhanage, B.M. et al. (2014). Catalytic carbon dioxide hydrogenation to methanol: a review of recent studies. *Chem. Eng. Res. Design.* 92: 2557–2567.
7 Nielsen, D.U., Hu, X.M., Daasbjerg, K. et al. (2018). Chemically and electrochemically catalysed conversion of CO_2 to CO with follow-up utilization to value-added chemicals. *Nat. Catal.* 1: 244–254.
8 Zafar, F., Zhao, R., Ali, M. et al. (2022). Unprecedented contributions of In_2O_3 promoter on ordered mesoporous Cu/Al_2O_3 for CO_2 hydrogenation to oxygenates. *Chem. Eng. J.* 439: 135649–135661.

9 Wang, X., Woo, H.Y., Shen, D. et al. (2025). Cu–ZnO nanoparticles encapsulated in ZSM-5 for selective conversion of carbon dioxide into oxygenates. *EES Catal.* 3: 420–434.

10 Natte, K., Neumann, H., Beller, M. et al. (2017). Transition-metal-catalyzed utilization of methanol as a C1 source in organic synthesis. *Angew. Chem. Int. Ed.* 56: 6384–6394.

11 Moellenbruck, F., Kempken, T., Dierks, M. et al. (2018). Cogeneration of power and methanol based on a conventional power plant in Germany. *J. Energy Storage* 19: 393–401.

12 Shen, D., Han, S.B., Wang, X. et al. (2022). Effects of Sn promoter on the ordered mesoporous Co_3O_4-Al_2O_3 mixed metal oxide for Fischer-Tropsch synthesis reaction. *Catalysts.* 12: 1447–1458.

13 Ali, M., Koo, H.M., Kasipandi, S. et al. (2020). Direct synthesis of liquid fuels and aromatics from syngas over mesoporous FeZrOx catalyst mixed with Mo/ferrierite. *Fuel.* 264: 116851–116861.

14 Dang, S., Yang, H., Gao, P. et al. (2019). A review of research progress on heterogeneous catalysts for methanol synthesis from carbon dioxide hydrogenation. *Catal. Today* 330: 61–75.

15 Mondal, U. and Yadav, G.D. (2019). Perspective of dimethyl ether as fuel: part I. Catalysis. *J. CO_2 Util.* 32: 299–320.

16 Khadzhiev, S., Magomedova, M., and Peresypkina, E. (2014). Mechanism of olefin synthesis from methanol and dimethyl ether over zeolite catalysts: a review. *Pet Chem.* 54: 245–269.

17 Lee, D., Kim, J.J., Ali, M. et al. (2022). Mechanistic kinetic modeling for catalytic conversion of DME to gasoline-range hydrocarbons over nanostructured ZSM-5. *Catal. Sci. Technol.* 12: 4798–4810.

18 Lee, K.Y., Lee, S.W., and Ihm, S.K. (2014). Acid strength control in MFI zeolite for the methanol-to-hydrocarbons (MTH) reaction. *Ind. Eng. Chem. Res.* 53: 10072–10079.

19 Li, S., Yang, H., Wang, S. et al. (2022). Improvement of adsorption and catalytic properties of zeolites by precisely controlling their particle morphology. *Chem. Commun.* 58: 2041–2054.

20 Ali, M., Kim, J.J., Zafar, F. et al. (2023). Effects of ZSM-5 morphology and Fe promoter for dimethyl ether conversion to gasoline-range hydrocarbons. *Catalysts.* 13: 910–923.

21 Li, Y., Li, L., and Yu, J. (2017). Applications of zeolites in sustainable chemistry. *Chem* 3: 928–949.

22 Ruiz, M.G., Casados, D.A.S., Pliego, J.A. et al. (2020). ZSM-5 zeolites modified with Zn and their effect on the crystal size in the conversion of methanol to light aromatics (MTA). *React. Kinet. Mech. Catal.* 129: 471–490.

23 Chang, C.D. and Silvestri, A.J. (1977). The conversion of methanol and other O-compounds to hydrocarbons over zeolite catalysts. *J. Catal.* 47: 249–259.

24 Bao, J., Yang, G., Yoneyama, Y. et al. (2019). Significant advances in C1 catalysis: highly efficient catalysts and catalytic reactions. *ACS Catal.* 9: 3026–3053.

25 Sun, Q., Xie, Z., and Yu, J. (2018). The state-of-the-art synthetic strategies for SAPO-34 zeolite catalysts in methanol-to-olefin conversion. *Nat. Sci. Rev.* 5: 542–558.

26 Yang, M., Fan, D., Wei, Y. et al. (2019). Recent progress in methanol-to-olefins (MTO) catalysts. *Adv. Mater.* 31: 1902181–1902195.

27 Yarulina, I., Chowdhury, A.D., Meirer, F. et al. (2018). Recent trends and fundamental insights in the methanol-to-hydrocarbons process. *Nat. Catal.* 1: 398–411.

28 Yurchak, S. (1988). Development of Mobil's fixed-bed methanul-to-gasoline (MTG) process. *Stud. Surf. Sci. Catal.* 36: 251–272.

29 Chang, C.D. (1992). The New Zealand gas-to-gasoline plant: an engineering tour de force. *Catal. Today* 13: 103–111.

30 Olsbye, U., Svelle, S., Bjørgen, M. et al. (2012). Conversion of methanol to hydrocarbons: how zeolite cavity and pore size controls product selectivity. *Angew. Chem. Int. Ed.* 51: 5810–5831.

31 Kasipandi, S. and Bae, J.W. (2019). Recent advances in direct synthesis of value-added aromatic chemicals from syngas by cascade reactions over bifunctional catalysts. *Adv. Mater.* 31: 1803390–1803407.

32 Ateka, A., Pérez-Uriarte, P., Gamero, M. et al. (2017). A comparative thermodynamic study on the CO_2 conversion in the synthesis of methanol and of DME. *Energy*. 120: 796–804.

33 Wan, Z., Li, G., Wang, C. et al. (2018). Effect of reaction conditions on methanol to gasoline conversion over nanocrystal ZSM-5 zeolite. *Catal. Today*. 314: 107–113.

34 Echevskii, G., Ione, K., Nosyreva, G. et al. (1988). Effect of the temperature regime of methanol conversion to hydrocarbons on coking of zeolite catalysts and their regeneration. *Appl. Catal.* 43: 85–89.

35 Li, P., Zhang, W., Han, X. et al. (2010). Conversion of methanol to hydrocarbons over phosphorus-modified ZSM-5/ZSM-11 intergrowth zeolites. *Catal. Lett.* 134: 124–130.

36 Nazari, M., Behbahani, R.M., Moradi, G. et al. (2016). A facile synthesis route for modifying the catalytic performance of SAPO-18 in MTO process. *J. Porous Mater.* 23: 1037–1046.

37 Borodina, E., Sharbini Harun Kamaluddin, H., Meirer, F. et al. (2017). Influence of the reaction temperature on the nature of the active and deactivating species during methanol-to-olefins conversion over H-SAPO-34. *ACS Catal.* 7: 5268–5281.

38 Hadi, N., Niaei, A., Nabavi, S. et al. (2014). Development of a new kinetic model for methanol to propylene process on Mn/H-ZSM-5 catalyst. *Chem. Biochem. Eng. Q.* 28: 53–63.

39 Li, H., Li, X.G., and Xiao, W.D. (2019). Deactivation kinetics of individual C_6–C_9 aromatics' generation from methanol over Zn and P co-modified HZSM-5. *RSC Adv.* 9: 22327–22335.

40 Wan, Z., Li, G.K., Wang, C. et al. (2018). Relating coke formation and characteristics to deactivation of ZSM-5 zeolite in methanol to gasoline conversion. *Appl. Catal. A: Gen.* 549: 141–151.

41 Huang, X., Li, H., Xiao, W.D. et al. (2016). Insight into the side reactions in methanol-to-olefin process over HZSM-5: a kinetic study. *Chem. Eng. J.* 299: 263–275.

42 Shareh, F.B., Kazemeini, M., Asadi, M. et al. (2014). Metal promoted mordenite catalyst for methanol conversion into light olefins. *Pet. Sci. Technol.* 32: 1349–1356.

43 Wan, Z., Wu, W., Chen, W. et al. (2014). Direct synthesis of hierarchical ZSM-5 zeolite and its performance in catalyzing methanol to gasoline conversion. *Ind. Eng. Chem. Res.* 53: 19471–19478.

44 Ma, Z., Fu, T., Wang, Y. et al. (2019). Silicalite-1 derivational desilication-recrystallization to prepare hollow nano-ZSM-5 and highly mesoporous micro-ZSM-5 catalyst for methanol to hydrocarbons. *Ind. Eng. Chem. Res.* 58: 2146–2158.

45 Bleken, F.L., Barbera, K., Bonino, F. et al. (2013). Catalyst deactivation by coke formation in microporous and desilicated zeolite H-ZSM-5 during the conversion of methanol to hydrocarbons. *J. Catal.* 307: 62–73.

46 Arora, S.S. and Bhan, A. (2017). The critical role of methanol pressure in controlling its transfer dehydrogenation and the corresponding effect on propylene-to-ethylene ratio during methanol-to-hydrocarbons catalysis on H-ZSM-5. *J. Catal.* 356: 300–306.

47 Hu, X., Yuan, L., Cheng, S. et al. (2019). GeAPSO-34 molecular sieves: synthesis, characterization and methanol-to-olefins performance. *Catal. Commun.* 123: 38–43.

48 Jiao, Y., Jiang, C., Yang, Z. et al. (2013). Synthesis of highly accessible ZSM-5 coatings on SiC foam support for MTP reaction. *Micropor. Mesopor. Mater.* 181: 201–207.

49 Chen, N.Y. and Reagan, W.J. (1979). Evidence of autocatalysis in methanol to hydrocarbon reactions over zeolite catalysts. *J. Catal.* 59: 123–129.

50 Chang, C.D., Chu, C.T., and Socha, R.F. (1984). Methanol conversion to olefins over ZSM-5: I. Effect of temperature and zeolite $SiO_2Al_2O_3$. *J. Catal.* 86: 289–296.

51 Hadi, N., Niaei, A., Nabavi, S. et al. (2015). Effect of second metal on the selectivity of Mn/H-ZSM-5 catalyst in methanol to propylene process. *J. Ind. Eng. Chem.* 29: 52–62.

52 Golubev, K.B., Batova, T.I., Kolesnichenko, N.V. et al. (2019). Synthesis of C_2–C_4 olefins from methanol as a product of methane partial oxidation over zeolite catalyst. *Catal. Commun.* 129: 105744–105748.

53 Heriyanto, H., Muraza, O., Nasser, G.A. et al. (2020). Development of new kinetic models for methanol to hydrocarbons over a Ca-ZSM-5 Catalyst. *Energy Fuels* 34: 6245–6260.

54 Lacarriere, A., Luck, F., Świerczyński, D. et al. (2011). Methanol to hydrocarbons over zeolites with MWW topology: effect of zeolite texture and acidity. *Appl. Catal. A: Gen.* 402: 208–217.

55 Martin, A., Nowak, S., Lu, B. et al. (1989). Coupled conversion of methanol and C4 hydrocarbons to lower olefins. *Appl. Catal.* 50: 149–155.

56 Mier, D., Aguayo, A.T., Gayubo, A.G. et al. (2010). Synergies in the production of olefins by combined cracking of n-butane and methanol on a HZSM-5 zeolite catalyst. *Chem. Eng. J.* 160: 760–769.

57 Gong, T., Zhang, X., Bai, T. et al. (2012). Coupling conversion of methanol and C_4 hydrocarbon to propylene on La-modified HZSM-5 zeolite catalysts. *Ind. Eng. Chem. Res.* 51: 13589–13598.

58 Stewart, C. and Roberts, C.B. (2015). Investigation of supercritical isooctane as a reaction medium in the conversion of methanol to hydrocarbons over H-ZSM-5. *Ind. Eng. Chem. Res.* 54: 7811–7821.

59 Su, C., Qian, W., Xie, Q. et al. (2016). Conversion of methanol with C_5–C_6 hydrocarbons into aromatics in a two-stage fluidized bed reactor. *Catal. Today.* 264: 63–69.

60 Mier, D., Aguayo, A.T., Gayubo, A.G. et al. (2010). Catalyst discrimination for olefin production by coupled methanol/n-butane cracking. *Appl. Catal. A: Gen.* 383: 202–210.

61 Dai, W., Yang, L., Wang, C. et al. (2018). Effect of n-butanol cofeeding on the methanol to aromatics conversion over Ga-modified nano H-ZSM-5 and its mechanistic interpretation. *ACS Catal.* 8: 1352–1362.

62 Qi, L., Li, J., Wei, Y. et al. (2016). Role of naphthalene during the induction period of methanol conversion on HZSM-5 zeolite. *Catal. Sci. Technol.* 6: 3737–3744.

63 Ni, Y., Zhu, W., and Liu, Z. (2019). H-ZSM-5-catalyzed hydroacylation involved in the coupling of methanol and formaldehyde to aromatics. *ACS Catal.* 9: 11398–11403.

64 Chen, Z., Ni, Y., Zhi, Y. et al. (2018). Coupling of methanol and carbon monoxide over H-ZSM-5 to form aromatics. *Angew. Chem. Int. Ed.* 57: 12549–12553.

65 Cheung, P., Bhan, A., Sunley, G.J. et al. (2006). Selective carbonylation of dimethyl ether to methyl acetate catalyzed by acidic zeolites. *Angew. Chem. Int. Ed.* 118: 1647–1650.

66 Zhang, Q., Yu, J., and Corma, A. (2020). Applications of zeolites to C1 chemistry: recent advances, challenges, and opportunities. *Adv. Mater.* 32: 2002927–2002957.

67 Zhang, Q., Mayoral, A., Terasaki, O. et al. (2019). Amino acid-assisted construction of single-crystalline hierarchical nanozeolites via oriented-aggregation and intraparticle ripening. *J. Am. Chem. Soc.* 141: 3772–3776.

68 Sun, C., Wang, Y., Zhao, A. et al. (2020). Synthesis of nano-sized SAPO-34 with morpholine-treated micrometer-seeds and their catalytic performance in methanol-to-olefin reactions. *Appl. Catal. A: Gen.* 589: 117314–117324.

69 Zhang, L., Liu, H., Yue, Y. et al. (2019). Design and in situ synthesis of hierarchical SAPO-34@ kaolin composites as catalysts for methanol to olefins. *Catal. Sci. Technol.* 9: 6438–6451.

70 Xu, Z., Li, J., Huang, Y. et al. (2019). Size control of SSZ-13 crystals with APAM and its influence on the coking behaviour during MTO reaction. *Catal. Sci. Technol.* 9: 2888–2897.

71 Shao, J., Fu, T., Ma, Z. et al. (2019). Facile creation of hierarchical nano-sized ZSM-5 with a large external surface area via desilication–recrystallization of silicalite-1 for conversion of methanol to hydrocarbons. *Catal. Sci. Technol.* 9: 6647–6658.
72 Wang, N., Hou, Y., Sun, W. et al. (2019). Modulation of b-axis thickness within MFI zeolite: correlation with variation of product diffusion and coke distribution in the methanol-to-hydrocarbons conversion. *Appl. Catal. B: Environ.* 243: 721–733.
73 Kim, S., Park, G., Woo, M.H. et al. (2019). Control of hierarchical structure and framework-Al distribution of ZSM-5 via adjusting crystallization temperature and their effects on methanol conversion. *ACS Catal.* 9: 2880–2892.
74 Gallego, E.M., Li, C., Paris, C. et al. (2018). Making nanosized CHA zeolites with controlled Al distribution for optimizing methanol-to-olefin performance. *Chem. A Europ. J.* 24: 14631–14635.
75 Martín, N., Li, Z., Martínez-Triguero, J. et al. (2016). Nanocrystalline SSZ-39 zeolite as an efficient catalyst for the methanol-to-olefin (MTO) process. *Chem. Commun.* 52: 6072–6075.
76 Martínez-Franco, R., Li, Z., Martínez-Triguero, J. et al. (2016). Improving the catalytic performance of SAPO-18 for the methanol-to-olefins (MTO) reaction by controlling the Si distribution and crystal size. *Catal. Sci. Technol.* 6: 2796–2806.
77 Javdani, A., Ahmadpour, J., and Yaripour, F. (2019). Nano-sized ZSM-5 zeolite synthesized via seeding technique for methanol conversions: a review. *Micropor. Mesopor. Mater.* 284: 443–458.
78 Xiao, F.S., Wang, L., Yin, C. et al. (2006). Catalytic properties of hierarchical mesoporous zeolites templated with a mixture of small organic ammonium salts and mesoscale cationic polymers. *Angew. Chem. Int. Ed.* 118: 3162–3165.
79 Liu, F., Willhammar, T., Wang, L. et al. (2012). ZSM-5 zeolite single crystals with b-axis-aligned mesoporous channels as an efficient catalyst for conversion of bulky organic molecules. *J. Am. Chem. Soc.* 134: 4557–4560.
80 Choi, M., Na, K., Kim, J. et al. (2009). Stable single-unit-cell nanosheets of zeolite MFI as active and long-lived catalysts. *Nature*. 461: 246–249.
81 Na, K., Jo, C., Kim, J. et al. (2011). Directing zeolite structures into hierarchically nanoporous architectures. *Science*. 333: 328–332.
82 Ali, M., Zafar, F., Shen, D. et al. (2024). Contributions of ZSM-5 morphology over hybridized ZnO-ZrO_2/ZSM-5 for direct CO_2 hydrogenation activity to aromatics. *Fuel*. 378: 132925–132941.
83 Tao, Y., Kanoh, H., and Kaneko, K. (2003). ZSM-5 monolith of uniform mesoporous channels. *J. Am. Chem. Soc.* 125: 6044–6045.
84 Jacobsen, C.J., Madsen, C., Houzvicka, J. et al. (2000). Mesoporous zeolite single crystals. *J. Am. Chem. Soc.* 122: 7116–7117.
85 Meng, F., Wang, Y., and Wang, S. (2016). Methanol to gasoline over zeolite ZSM-5: improved catalyst performance by treatment with HF. *RSC Adv.* 6: 58586–58593.
86 Qi, R., Fu, T., Wan, W. et al. (2017). Pore fabrication of nano-ZSM-5 zeolite by internal desilication and its influence on the methanol to hydrocarbon reaction. *Fuel Process. Technol.* 155: 191–199.

87 Liu, C., Su, J., Liu, S. et al. (2020). Insights into the key factor of zeolite morphology on the selective conversion of syngas to light aromatics over a Cr_2O_3/ZSM-5 catalyst. *ACS Catal.* 10: 15227–15237.
88 Fu, T., Zhou, H., and Li, Z. (2016). Effect of particle morphology for ZSM-5 zeolite on the catalytic conversion of methanol to gasoline-range hydrocarbons. *Catal. Lett.* 146: 1973–1983.
89 Wang, K., Dong, M., Niu, X. et al. (2018). Highly active and stable Zn/ZSM-5 zeolite catalyst for the conversion of methanol to aromatics: effect of support morphology. *Catal. Sci. Technol.* 8: 5646–5656.
90 Álvaro-Muñoz, T., Márquez-Álvarez, C., and Sastre, E. (2012). Use of different templates on SAPO-34 synthesis: effect on the acidity and catalytic activity in the MTO reaction. *Catal. Today.* 179: 27–34.
91 Yin, C., Wei, Y., Wang, F. et al. (2013). Introduction of mesopority in zeolite ZSM-5 using resin as templates. *Mater. Lett.* 98: 194–196.
92 Yaripour, F., Shariatinia, Z., Sahebdelfar, S. et al. (2015). Conventional hydrothermal synthesis of nanostructured H-ZSM-5 catalysts using various templates for light olefins production from methanol. *J. Nat. Gas Sci. Eng.* 22: 260–269.
93 Molino, A., Holzinger, J., Łukaszuk, K. et al. (2019). Synthesis of ZSM-23 (MTT) zeolites with different crystal morphology and intergrowths: effects on the catalytic performance in the conversion of methanol to hydrocarbons. *Catal. Sci. Technol.* 9: 6782–6792.
94 Yao, J., Rong, Y., Gao, Z. et al. (2022). Metal-organic framework-assisted synthesis of Zr-modified SAPO-34 zeolites with hierarchical porous structure for the catalytic transformation of methanol to olefins. *Catal. Sci. Technol.* 12: 894–905.
95 Wang, X., Gao, X., Dong, M. et al. (2015). Production of gasoline range hydrocarbons from methanol on hierarchical ZSM-5 and Zn/ZSM-5 catalyst prepared with soft second template. *J. Energy. Chem.* 24: 490–496.
96 Sun, L., Wang, Y., Chen, H. et al. (2018). Direct synthesis of hierarchical Zn-ZSM-5 with addition of CTAB in a seeding method and improved catalytic performance in methanol to aromatics reaction. *Catal. Today.* 316: 91–98.
97 Gong, Q., Fang, T., Xie, Y. et al. (2021). High-efficiency conversion of methanol to BTX aromatics over a Zn-modified nanosheet-HZSM-5 zeolite. *Ind. Eng. Chem. Res.* 60: 1633–1641.
98 Wu, Z., Zhao, K., Zhang, Y. et al. (2019). Synthesis and consequence of aggregated nanosized ZSM-5 zeolite crystals for methanol to propylene reaction. *Ind. Eng. Chem. Res.* 58: 10737–10749.
99 Wang, N., Sun, W., Hou, Y. et al. (2018). Crystal-plane effects of MFI zeolite in catalytic conversion of methanol to hydrocarbons. *J. Catal.* 360: 89–96.
100 Niu, X., Gao, J., Wang, K. et al. (2017). Influence of crystal size on the catalytic performance of H-ZSM-5 and Zn/H-ZSM-5 in the conversion of methanol to aromatics. *Fuel Process. Technol.* 157: 99–107.
101 Zhang, Y., Qu, Y., and Wang, J. (2018). Effect of crystal size on the catalytic performance of HZSM-5 Zeolite in the methanol to aromatics reaction. *Pet. Sci. Technol.* 36: 898–903.

102 Shao, J., Fu, T.J., Chang, J.W. et al. (2017). Effect of ZSM-5 crystal size on its catalytic properties for conversion of methanol to gasoline. *J. Fuel Chem. Technol.* 45: 75–83.

103 Bleken, F.L., Chavan, S., Olsbye, U. et al. (2012). Conversion of methanol into light olefins over ZSM-5 zeolite: strategy to enhance propene selectivity. *Appl. Catal. A: Gen.* 447–448: 178–185.

104 Zhang, L., Wang, S., Shi, D. et al. (2020). Methanol to olefins over H-RUB-13 zeolite: regulation of framework aluminum siting and acid density and their relationship to the catalytic performance. *Catal. Sci. Technol.* 10: 1835–1847.

105 Yarulina, I., Wispelaere, K.D., Bailleul, S. et al. (2018). Structure-performance descriptors and the role of Lewis acidity in the methanol-to-propylene process. *Nature. Chem.* 10: 804–812.

106 Li, Z., Navarro, M.T., Martínez-Triguero, J. et al. (2016). Synthesis of nano-SSZ-13 and its application in the reaction of methanol to olefins. *Catal. Sci. Technol.* 6: 5856–5863.

107 Zeolite (2017). Welcome to the database of zeolite structures. http://www.iza-structure.org/databases.

108 Konnov, S.V., Pavlov, V.S., and Ivanova, I.I. (2020). Effect of coating with silica on acidic and catalytic properties of SAPO-18 in MTO conversion. *Micropor. Mesopor. Mater.* 300: 110158–110164.

109 Pinilla-Herrero, I., Márquez-Álvarez, C., and Sastre, E. (2016). Methanol-to-olefin reaction on SAPO-35 catalysts synthesized with controlled crystal size and using mesoporogen additives. *Catal. Today*. 277: 29–36.

110 Dusselier, M. and Davis, M.E. (2018). Small-pore zeolites: synthesis and catalysis. *Chem. Rev.* 118: 5265–5329.

111 Yang, M., Li, B., Gao, M. et al. (2020). High propylene selectivity in methanol conversion over a small-pore SAPO molecular sieve with ultra-small cage. *ACS Catal.* 10: 3741–3749.

112 Yuen, L.T., Zones, S.I., Harris, T.V. et al. (1994). Product selectivity in methanol to hydrocarbon conversion for isostructural compositions of AFI and CHA molecular sieves. *Micropor. Mesopor. Mater.* 2: 105–117.

113 Bleken, F., Bjørgen, M., Palumbo, L. et al. (2009). The effect of acid strength on the conversion of methanol to olefins over acidic microporous catalysts with the CHA topology. *Top. Catal.* 52: 218–228.

114 Bordiga, S., Regli, L., Cocina, D. et al. (2005). Assessing the acidity of high silica chabazite H-SSZ-13 by FTIR using CO as molecular probe: comparison with H-SAPO-34. *J. Phys. Chem. B.* 109: 2779–2784.

115 Zhang, J., Huang, Z., Xu, L. et al. (2019). Verifying the olefin formation mechanism of the methanol-to-hydrocarbons reaction over H-ZSM-48. *Catal. Sci. Technol.* 9: 2132–2143.

116 Wang, P., Huang, L., Li, J. et al. (2015). Catalytic properties and deactivation behavior of H-MCM-22 in the conversion of methanol to hydrocarbons. *RSC Adv.* 5: 28794–28802.

117 Liu, Z., Dong, X., Zhu, Y. et al. (2015). Investigating the influence of mesoporosity in zeolite Beta on its catalytic performance for the conversion of methanol to hydrocarbons. *ACS Catal.* 5: 5837–5845.

118 Teketel, S., Skistad, W., Benard, S. et al. (2012). Shape selectivity in the conversion of methanol to hydrocarbons: the catalytic performance of one-dimensional 10-ring zeolites: ZSM-22, ZSM-23, ZSM-48, and EU-1. *ACS Catal.* 2: 26–37.

119 Hulea, V., Huguet, E., Cammarano, C. et al. (2014). Conversion of methyl mercaptan and methanol to hydrocarbons over solid acid catalysts – a comparative study. *Appl. Catal. B: Environ.* 144: 547–553.

120 Bleken, F., Skistad, W., Barbera, K. et al. (2011). Conversion of methanol over 10-ring zeolites with differing volumes at channel intersections: comparison of TNU-9, IM-5, ZSM-11 and ZSM-5. *Phys. Chem. Chem. Phy.* 13: 2539–2549.

121 Wang, S., Wang, P., Qin, Z. et al. (2018). Relation of catalytic performance to the aluminum siting of acidic zeolites in the conversion of methanol to olefins, viewed via a comparison between ZSM-5 and ZSM-11. *ACS Catal.* 8: 5485–5505.

122 Li, M., Zhou, Y., Ju, C. et al. (2016). Remarkable increasing of ZSM-5 lifetime in methanol to hydrocarbon reaction by post engineering in fluoride media. *Appl. Catal. A: Gen.* 512: 1–8.

123 Wan, Z., Wu, W., Li, G.K. et al. (2016). Effect of SiO_2/Al_2O_3 ratio on the performance of nanocrystal ZSM-5 zeolite catalysts in methanol to gasoline conversion. *Appl. Catal. A: Gen.* 523: 312–320.

124 Huang, H., Zhu, H., Zhang, Q. et al. (2019). Effect of acidic properties of hierarchical HZSM-5 on the product distribution in methanol conversion to gasoline. *Kor. J. Chem. Eng.* 36: 210–216.

125 Fu, T., Shao, J., and Li, Z. (2021). Catalytic synergy between the low Si/Al ratio Zn/ZSM-5 and high Si/Al ratio HZSM-5 for high-performance methanol conversion to aromatics. *Appl. Catal. B: Environ.* 291: 120098–120113.

126 Fu, T., Guo, Y., Li, Z. et al. (2022). Selective conversion of methanol to aromatics with superior catalytic stability by relay catalysis over quadruple ZSM-5 sequence beds with gradient-increasing acidity. *Fuel.* 315: 123241–123254.

127 Dědeček, J., Tabor, E., and Sklenak, S. (2019). Tuning the aluminum distribution in zeolites to increase their performance in acid-catalyzed reactions. *ChemSusChem.* 12: 556–576.

128 Gounder, R. and Iglesia, E. (2009). Catalytic consequences of spatial constraints and acid site location for monomolecular alkane activation on zeolites. *J. Am. Chem. Soc.* 131: 1958–1971.

129 Dědeček, J., Sobalík, Z., and Wichterlová, B. (2012). Siting and distribution of framework aluminium atoms in silicon-rich zeolites and impact on catalysis. *Catal. Rev.* 54: 135–223.

130 Liu, M., Yokoi, T., Yoshioka, M. et al. (2014). Differences in Al distribution and acidic properties between RTH-type zeolites synthesized with OSDAs and without OSDAs. *Phys. Chem. Chem. Phys.* 16: 4155–4164.

131 Chen, J., Liang, T., Li, J. et al. (2016). Regulation of framework aluminum siting and acid distribution in H-MCM-22 by boron incorporation and its effect on the catalytic performance in methanol to hydrocarbons. *ACS Catal.* 6: 2299–2313.

132 Di Iorio, J.R. and Gounder, R. (2016). Controlling the isolation and pairing of aluminum in chabazite zeolites using mixtures of organic and inorganic structure-directing agents. *Chem. Mater.* 28: 2236–2247.

133 Xue, Y., Li, J., Wang, P. et al. (2021). Regulating Al distribution of ZSM-5 by Sn incorporation for improving catalytic properties in methanol to olefins. *Appl. Catal. B: Environ.* 280: 119391–119403.

134 Wang, Y., Yokoi, T., Namba, S. et al. (2016). Improvement of catalytic performance of MCM-22 in the cracking of n-hexane by controlling the acidic property. *J. Catal.* 333: 17–28.

135 Wang, S., Wei, Z., Chen, Y. et al. (2015). Methanol to olefins over H-MCM-22 zeolite: theoretical study on the catalytic roles of various pores. *ACS Catal.* 5: 1131–1144.

136 Liang, T., Chen, J., Qin, Z. et al. (2016). Conversion of methanol to olefins over H-ZSM-5 zeolite: reaction pathway is related to the framework aluminum siting. *ACS Catal.* 6: 7311–7325.

137 Li, L., Chen, Y., Xu, S. et al. (2016). Oriented control of Al locations in the framework of Al-Ge-ITQ-13 for catalyzing methanol conversion to propene. *J. Catal.* 344: 242–251.

138 He, S., Fan, S., Yuan, K. et al. (2022). Enhancing the catalytic performance of H-ITQ-13 zeolite in the conversion of methanol to olefins through regulating the aluminum distribution in its framework. *Appl. Catal. A: Gen.* 637: 118604–118615.

139 He, S., Wang, S., Fan, S. et al. (2023). Improvement of the catalytic performance of ITQ-13 zeolite in methanol to olefins via Ce modification. *Catal. Today.* 410: 184–192.

140 Min, H.K., Park, M.B., and Hong, S.B. (2010). Methanol-to-olefin conversion over H-MCM-22 and H-ITQ-2 zeolites. *J. Catal.* 271: 186–194.

141 Guo, H., Ge, T., Lv, J. et al. (2019). Mesoporogen-free synthesis of high-silica hierarchically structured ZSM-5 zeolites and their superior performance for the methanol-to-propylene reaction. *Eur. J. Inorg. Chem.* 2019: 51–58.

142 Zhang, J., Xu, L., Zhang, Y. et al. (2018). Hydrogen transfer versus olefins methylation: On the formation trend of propene in the methanol-to-hydrocarbons reaction over Beta zeolites. *J. Catal.* 368: 248–260.

143 Liu, Y., Kirchberger, F.M., Müller, S. et al. (2019). Critical role of formaldehyde during methanol conversion to hydrocarbons. *Nat. Commun.* 10: 1462–1470.

144 Feng, W., Gao, X., Ding, C. et al. (2017). Effect of weak base modification on ZSM-5 catalyst for methanol to aromatics. *Appl. Organomet. Chem.* 31: 3625–3631.

145 Wei, Z., Xia, T., Liu, M. et al. (2015). Alkaline modification of ZSM-5 catalysts for methanol aromatization: the effect of the alkaline concentration. *Front. Chem. Sci. Eng.* 9: 450–460.

146 Alabi, W., Atanda, L., Jermy, R. et al. (2012). Kinetics of toluene alkylation with methanol catalyzed by pure and hybridized HZSM-5 catalysts. *Chem. Eng. J.* 195: 276–288.

147 Li, H., Dong, P., Ji, D. et al. (2020). Effect of the post-treatment of HZSM-5 on catalytic performance for methanol to aromatics. *ChemistrySelect.* 5: 3413–3419.

148 Su, X., Liu, B., Feng, C. et al. (2022). Rapid and low-cost synthesis of ZSM-5 zeolite nanosheet assemblies for conversion of methanol to gasoline. *Micropor. Mesopor. Mater.* 344: 112215–112228.

149 Zhang, J., Qian, W., Kong, C. et al. (2015). Increasing para-xylene selectivity in making aromatics from methanol with a surface-modified Zn/P/ZSM-5 catalyst. *ACS Catal.* 5: 2982–2988.

150 Qiao, J., Wang, J., Frenkel, A.I. et al. (2020). Methanol to aromatics: isolated zinc phosphate groups on HZSM-5 zeolite enhance BTX selectivity and catalytic stability. *RSC Adv.* 10: 5961–5971.

151 Li, J., Tong, K., Xi, Z. et al. (2016). Highly-efficient conversion of methanol to p-xylene over shape-selective Mg–Zn–Si-HZSM-5 catalyst with fine modification of pore-opening and acidic properties. *Catal. Sci. Technol.* 6: 4802–4813.

152 Zhang, J., Wang, L., Wu, Z. et al. (2019). Solvent-free synthesis of core–shell Zn/ZSM-5@ silicalite-1 catalyst for selective conversion of methanol to BTX aromatics. *Ind. Eng. Chem. Res.* 58: 15453–15458.

153 Wang, N., Li, J., Sun, W. et al. (2022). Rational design of zinc/zeolite catalyst: selective formation of p-xylene from methanol to aromatics reaction. *Angew. Chem. Int. Ed.* 61: 202114786–202114792.

154 Martínez, C. and Corma, A. (2011). Inorganic molecular sieves: preparation, modification and industrial application in catalytic processes. *Coord. Chem. Rev.* 255: 1558–1580.

155 Hadi, N., Niaei, A., Nabavi, S.R. et al. (2016). An intelligent approach to design and optimization of M-Mn/H-ZSM-5 (M: Ce, Cr, Fe, Ni) catalysts in conversion of methanol to propylene. *J. Taiwan Inst. Chem. Eng.* 59: 173–185.

156 Tian, H., Shi, D., Yu, L. et al. (2022). Transformation of methanol to trimethylbenzene catalyzed by cadmium modified HZSM-5 zeolites. *React. Kinet. Mech. Catal.* 135: 333–348.

157 Zeng, D., Yang, J., Wang, J. et al. (2007). Solid-state NMR studies of methanol-to-aromatics reaction over silver exchanged HZSM-5 zeolite. *Micropor. Mesopor. Mater.* 98: 214–219.

158 Jia, Y., Wang, J., Zhang, K. et al. (2020). Highly shape-selective Zn-P/HZSM-5 zeolite catalyst for methanol conversion to light aromatics. *Appl. Organomet. Chem.* 34: 5932–5946.

159 Li, H., Li, X.G., Xiao, W.D. (2021). Collaborative effect of zinc and phosphorus on the modified HZSM-5 zeolites in the conversion of methanol to aromatics. *Catal. Lett.* 151: 955–965.

160 Chen, C., Zhang, Q., Meng, Z. et al. (2015). Effect of magnesium modification over H-ZSM-5 in methanol to propylene reaction. *Appl. Pet. Res.* 5: 277–284.

161 Yaripour, F., Shariatinia, Z., Sahebdelfar, S. et al. (2015). Effect of boron incorporation on the structure, products selectivities and lifetime of H-ZSM-5 nanocatalyst designed for application in methanol-to-olefins (MTO) reaction. *Micropor. Mesopor. Mater.* 203: 41–53.

162 Li, M., Zhou, Y., Oduro, I.N. et al. (2016). Comparative study on the catalytic conversion of methanol and propanal over Ga/ZSM-5. *Fuel.* 168: 68–75.

163 Xin, Y., Qi, P., Duan, X. et al. (2013). Enhanced performance of Zn–Sn/HZSM-5 catalyst for the conversion of methanol to aromatics. *Catal. Lett.* 143: 798–806.
164 Meng, X., Chen, C., Liu, J. et al. (2016). Effects of zinc incorporation on hierarchical ZSM-11 catalyst for methanol conversion. *Appl. Pet. Res.* 6: 41–47.
165 Li, J., Miao, P., Li, Z. et al. (2015). Hydrothermal synthesis of nanocrystalline H Fe, Al. ZSM-5 zeolites for conversion of methanol to gasoline. *Energy Convers. Manage.* 93: 259–266.
166 Li, J., Han, D., Zi, Z. et al. (2022). The synthesis of H Fe, Al. ZSM-5 zeolites with uniform nanocrystals for dimethyl ether to gasoline reaction. *Fuel.* 313: 122643–122654.
167 Liu, C., Uslamin, E.A., Khramenkova, E. et al. (2022). High stability of methanol to aromatic conversion over bimetallic Ca, Ga-modified ZSM-5. *ACS Catal.* 12: 3189–3200.
168 Kianfar, E., Salimi, M., Pirouzfar, V. et al. (2018). Synthesis of modified catalyst and stabilization of CuO/NH_4-ZSM-5 for conversion of methanol to gasoline. *Int. J. Appl. Ceramic Technol.* 15: 734–741.
169 Kim, S., Park, G., Kim, S.K. et al. (2018). Gd/HZSM-5 catalyst for conversion of methanol to hydrocarbons: effects of amounts of the Gd loading and catalyst preparation method. *Appl. Catal. B: Environ.* 220: 191–201.
170 Shen, K., Wang, N., Qian, W. et al. (2014). Atmospheric pressure synthesis of nanosized ZSM-5 with enhanced catalytic performance for methanol to aromatics reaction. *Catal. Sci. Technol.* 4: 3840–3844.
171 Niu, X., Gao, J., Miao, Q. et al. (2014). Influence of preparation method on the performance of Zn-containing HZSM-5 catalysts in methanol-to-aromatics. *Micropor. Mesopor. Mater.* 197: 252–261.
172 Bi, Y., Wang, Y., Chen, X. et al. (2014). Methanol aromatization over HZSM-5 catalysts modified with different zinc salts. *Chin. J. Catal.* 35: 1740–1751.
173 Zhang, G.Q., Bai, T., Chen, T.F. et al. (2014). Conversion of methanol to light aromatics on Zn-modified nano-HZSM-5 zeolite catalysts. *Ind. Eng. Chem. Res.* 53: 14932–14940.
174 Schaidle, J.A., Ruddy, D.A., Habas, S.E. et al. (2015). Conversion of dimethyl ether to 2, 2, 3-trimethylbutane over a Cu/BEA catalyst: role of Cu sites in hydrogen incorporation. *ACS Catal.* 5: 1794–1803.
175 Dahl, I.M. and Kolboe, S. (1993). On the reaction mechanism for propene formation in the MTO reaction over SAPO-34. *Catal. Lett.* 20: 329–336.
176 Dahl, I.M. and Kolboe, S. (1994). On the reaction mechanism for hydrocarbon formation from methanol over SAPO-34: I. Isotopic labeling studies of the co-reaction of ethene and methanol. *J. Catal.* 149: 458–464.
177 Dahl, I.M. and Kolboe, S. (1996). On the reaction mechanism for hydrocarbon formation from methanol over SAPO-34: 2. Isotopic labeling studies of the co-reaction of propene and methanol. *J. Catal.* 161: 304–309.
178 Haw, J.F., Song, W., Marcus, D.M. et al. (2003). The mechanism of methanol to hydrocarbon catalysis. *Acc. Chem. Res.* 36: 317–326.

179 Wang, S., Chen, Y., Qin, Z. et al. (2019). Origin and evolution of the initial hydrocarbon pool intermediates in the transition period for the conversion of methanol to olefins over H-ZSM-5 zeolite. *J. Catal.* 369: 382–395.

180 Svelle, S., Joensen, F., Nerlov, J. et al. (2006). Conversion of methanol into hydrocarbons over zeolite H-ZSM-5: ethene formation is mechanistically separated from the formation of higher alkenes. *J. Am. Chem. Soc.* 128: 14770–14771.

181 Bjørgen, M., Svelle, S., Joensen, F. et al. (2007). Conversion of methanol to hydrocarbons over zeolite H-ZSM-5: on the origin of the olefinic species. *J. Catal.* 249: 195–207.

182 Ilias, S. and Bhan, A. (2013). Mechanism of the catalytic conversion of methanol to hydrocarbons. *ACS Catal.* 3: 18–31.

183 Ilias, S., Khare, R., Malek, A. et al. (2013). A descriptor for the relative propagation of the aromatic-and olefin-based cycles in methanol-to-hydrocarbons conversion on H-ZSM-5. *J. Catal.* 303: 135–140.

184 Ilias, S. and Bhan, A. (2014). The mechanism of aromatic dealkylation in methanol-to-hydrocarbons conversion on H-ZSM-5: what are the aromatic precursors to light olefins? *J. Catal.* 311: 6–16.

185 Qi, L., Wei, Y., Xu, L. et al. (2015). Reaction behaviors and kinetics during induction period of methanol conversion on HZSM-5 zeolite. *ACS Catal.* 5: 3973–3982.

186 Li, T., Shoinkhorova, T., Gascon, J. et al. (2021). Aromatics production via methanol-mediated transformation routes. *ACS Catal.* 11: 7780–7819.

187 Liu, Y., Müller, S., Berger, D. et al. (2016). Formation mechanism of the first carbon–carbon bond and the first olefin in the methanol conversion into hydrocarbons. *Angew. Chem. Int. Ed.* 55: 5723–5726.

188 Chowdhury, A.D., Houben, K., Whiting, G.T. et al. (2016). Initial carbon-carbon bond formation during the early stages of the methanol-to-olefin process proven by zeolite-trapped acetate and methyl acetate. *Angew. Chem. Int. Ed.* 128:16072–16077.

189 Zhang, W., Zhi, Y., Huang, J. et al. (2019). Methanol to olefins reaction route based on methylcyclopentadienes as critical intermediates. *ACS Catal.* 9: 7373–7379.

190 Li, J., Wei, Z., Chen, Y. et al. (2014). A route to form initial hydrocarbon pool species in methanol conversion to olefins over zeolites. *J. Catal.* 317: 277–283.

191 Ghavipour, M., Behbahani, R.M., Rostami, R.B. et al. (2014). Methanol/dimethyl ether to light olefins over SAPO-34: comprehensive comparison of the products distribution and catalyst performance. *J. Nat. Gas Sci. Eng.* 21: 532–539.

192 Fu, D., Paioni, A.L., Lian, C. et al. (2020). Elucidating zeolite channel geometry–reaction intermediate relationships for the methanol-to-hydrocarbon process. *Angew. Chem. Int. Ed.* 59: 20024–20030.

193 Fu, D., van der Heijden, O., Stanciakova, K. et al. (2020). Disentangling reaction processes of zeolites within single-oriented channels. *Angew. Chem. Int. Ed.* 59: 15502–15506.

194 Schulz, H. (2010). "Coking" of zeolites during methanol conversion: basic reactions of the MTO-, MTP-and MTG processes. *Catal. Today*. 154: 183–194.

195 Deka, T.J., Osman, A.I., Baruah, D.C. et al. (2022). Methanol fuel production, utilization, and techno-economy: a review. *Environ. Chem. Lett.* 20: 3525–3554.

196 Hobson, C. and Márquez, C. (2018). Renewable methanol report. ATA Markets Intelligence. https://methanol.org/wp-content/uploads/2019/01/MethanolReport.pdf

15

Application of Zeolites in Methanol-to-olefins

*Yong Wang, Peipei Xiao, Toshiyuki Yokoi**

Institute of Integrated Research, Institute of Science Tokyo, Yokohama, Japan

*Corresponding author: yokoi@cat.res.titech.ac.jp

15.1 Introduction

Light olefins, particularly ethylene and propylene, serve as crucial feedstocks in the petrochemical industry. Currently, these compounds are primarily produced through steam (thermal) cracking of a variety of hydrocarbon sources (e.g., gaseous raw materials, naphtha, and heavier fractions). Unlike traditional steam cracking, catalytic cracking using catalysts operate at reduced temperatures, thereby conserving energy and decreasing CO_2 emissions. Moreover, this method allows for better regulation of the propylene-to-ethylene ratio. Another significant method for selectively producing light olefins is the MTO process, since methanol can be sourced from plentiful alternatives like coal, natural gas, and biomass through non-petrochemical routes, helping to alleviate the scarcity of petroleum resources. The MTO process, when catalyzed by appropriate catalysts, provides another route for the large-scale production of basic chemicals using non-petroleum resources. In recent decades, the MTO process using zeolite catalysts has attracted substantial attention. It has proven to be a successful technology for bypassing crude oil while producing highly demanded light olefins. Notably, MTO technologies, such as those developed by Hydro/UOP (MTO), Lurgi (MTP), and DICP (DMTO), have successfully achieved commercial operation [1].

Zeolites have demonstrated excellent MTO catalytic performance, attributed to their appropriate acidic properties, excellent thermal and hydrothermal stability, as well as significant tolerance to deactivation. Currently, silicoaluminophosphate SAPO-34 (CHA) and aluminosilicate ZSM-5 (MFI) are widely used in the MTO process, typically implemented through fluidized-bed and fixed-bed technologies, respectively. SAPO-34 demonstrates remarkably high selectivity toward light olefins (ethylene, propylene, and butenes), reaching up to 90% or higher; however, it is prone to rapid deactivation. ZSM-5 exhibits excellent catalytic stability and an extended lifetime, with propylene selectivity ranging from *ca.* 40 to 50%. However, its effectiveness in selectively producing ethylene is relatively limited [2, 3].

To achieve the highest yield of light olefins and extend the lifetime of the catalyst in the MTO process, it is essential to design zeolite catalysts with enhanced performance by selecting suitable structural configurations and adjusting their physicochemical properties. It is widely recognized that the topology of zeolites significantly impacts MTO performance, primarily due to shape-selectivity effects, which include selectivity toward reactants, transition states, and final products. Specifically, the characteristics of intermediate species formed during the reaction process are significantly influenced by channel dimensionality, pore size, and cavity dimensions or geometry. These structural features directly impact the catalytic activity, product selectivity, and overall catalyst durability. Since the MTO reaction is a classic acid-catalyzed reaction, the acidic properties of zeolites, including acid density, strength, type, and spatial distribution, are key factors in determining their catalytic effectiveness. Besides, other factors such as particle size, hierarchical pore structure, and morphology also significantly influence the diffusion properties of products, thereby affecting product distribution and catalyst lifetime. In addition, the catalytic performance is considerably affected by parameters including reaction temperature, methanol weight hourly space velocity (WHSV), and co-feeding approaches.

In recent years, considerable attention has been given to the MTO process, especially in the selective production of light olefins, with extensive studies conducted on both reaction mechanisms and zeolite catalysts [1–14]. However, it remains important to review recent developments, particularly since 2010, in this field, focusing on the relationships between structure, properties, and performance of different zeolite catalysts. In this research, we first introduce the development of the MTO reaction mechanism. Then, we focus on discussing the impact of various factors, such as topology, acidity, particle size, hierarchical pore structure, and morphology of zeolite catalysts, on MTO catalytic performance, particularly emphasizing their effects on light olefins selectivity and catalyst lifetime. In addition, the effect of reaction conditions will also be discussed. Finally, we provide conclusions and an outlook.

15.2 MTO Reaction Mechanism on Zeolite

Studies on the reaction mechanism of MTO conversion are crucial for advancing both fundamental scientific understanding and practical industrial applications. After decades of intensive research, the MTO reaction mechanism is now well elucidated and primarily comprises the so-called "direct mechanism" during the induction stage, and the hydrocarbon pool (HCP)-based "dual-cycle mechanism" during the autocatalytic period.

15.2.1 First C—C Bond Formation

To explain the formation of the initial C—C bond in the MTO reaction, more than 20 direct mechanisms have been proposed in previous studies. These mechanisms encompass numerous reactive intermediates, such as oxonium ylides, carbocations, carbenes, and radical species, which were suggested in earlier research. However, these direct mechanisms have been challenged due to the lack of experimental support and their inability to adequately explain certain experimental observations. Moreover, theoretical calculations indicate that the direct coupling of methanol or dimethyl ether (DME) involves overcoming a significant energy barrier within these reaction pathways [3].

Until 2003, Wang et al. [15] discovered the significant catalytic activity of surface methoxy species (SMS), which are generated when methanol adsorbs onto the Brønsted acid site (BAS) of zeolites, in hydrocarbon formation. It was suggested that SMS plays a critical role in forming the first C—C bond. Due to its high reactivity, SMS is considered a plausible intermediate in the initial stage of the MTO reaction; however, the exact mechanism by which it contributes to the initial C—C bond formation is still not fully understood and requires further investigation. In recent years, thanks to advancements in spectroscopic techniques and theoretical studies, considerable progress has been achieved in elucidating the mechanism of initial C—C bond formation. These studies have provided robust experimental and theoretical evidence for the direct mechanism during the initial stages of the MTO process [16–28]. Among all proposed direct mechanisms, the SMS formed upon methanol adsorption onto BAS is undoubtedly the most experimentally verified intermediate. Recently, Yarulina et al. [6] systematically summarized the critical steps in the initial stages of the MTO reaction on zeolites (see Figure 15.1).

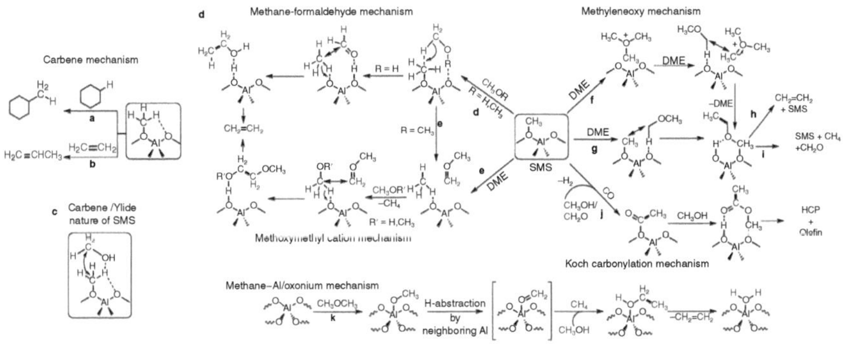

Figure 15.1 Several direct mechanisms proposed to explain the initial stages of the MTO reaction on zeolites. (a, b) Carbene mechanism. (c) Carbene/Ylide nature of SMS. (d, e) Traditional methane–formaldehyde mechanism and modified methoxymethyl cation mechanism. (f–i) Methyleneoxy mechanism. (j) Koch-type carbonylation mechanism. (k) Methane–aluminium/oxonium mechanism. *Source*: Reproduced from [6] / with permission of Springer Nature.

15.2.1.1 Carbene Mechanism

The concept of a carbene-insertion mechanism was first proposed by Wang et al., based on NMR spectroscopy analysis and carbene-trapping studies [15]. The carbene/ylide character of the SMS was confirmed through a carbene-trapping experiment, in which cyclohexane was used as a probe molecule. In this process, methylcyclohexane was formed as a result of the carbene/ylide species inserting into the sp^3 C—H bond of cyclohexane (see Figure 15.1a–c). In 2011, Yamazaki et al. [16, 17] proposed that the hydrocarbon species containing C—C bonds are formed through a coupling reaction involving a carbene-like SMS and methanol or DME, based on IR spectroscopic evidence. Furthermore, Chowdhury et al. [18] presented findings that substantiate the carbene-mediated mechanism in C—C bond formation, using two-dimensional (2D) magic-angle spinning (MAS) solid-state NMR analysis.

15.2.1.2 Methoxymethyl Cation Mechanism

In 2014, Li et al. [19] proposed the methoxymethyl cation mechanism based on experimental and theoretical studies of methanol conversion over SAPO-34. They suggested that the methoxymethyl carbocation ($CH_3OCH_2^+$), rather than formaldehyde, forms as a crucial intermediate through the interaction between SMS and DME. Subsequently, the $CH_3OCH_2^+$ species reacts with another methanol or DME molecule, leading to the formation of either 1,2-dimethoxyethane or 2-methoxyethanol, thereby establishing the initial C—C bond (see Figure 15.1d, e). Density functional theory (DFT) calculations indicate that the formation of methoxymethyl cations from methoxy groups and DME (the initial step) involves an energy barrier about 14.5 kJ mol^{-1} lower than that required for the formation of methane and formaldehyde, resulting in a *ca.* four-fold increase in the reaction rate constant. Moreover, the energy barrier required to produce 1,2-dimethoxyethane from methoxymethyl cations and DME (the second step) is merely 94.6 kJ mol^{-1}, which is about 30 kJ mol^{-1} lower than the energy barrier for the formation of ethanol from methane and formaldehyde. Notably, the rate constant increases by about 10^5 times during this stage. These results indicate that the methoxymethyl cation-based mechanism is thermodynamically more favorable than the methane/formaldehyde pathway. According to theoretical calculations, the formation of $CH_3OCH_2^+$ is more preferred through the reaction between SMS and formaldehyde [20]. However, although the formation of $CH_3OCH_2^+$ cations has been confirmed by *in situ* IR spectroscopy, the direct formation of 1,2-dimethoxyethane or 2-methoxyethanol from methoxymethyl cations and DME or methanol still lacks sufficient experimental evidence.

15.2.1.3 Methoxy-mediated DME/Methanol Mechanism

Recently, Wu et al. [21] investigated methanol conversion on ZSM-5 zeolite under realistic MTO reaction conditions and proposed the methyleneoxy mechanism (see Figure 15.1f–i). Solid-state nuclear magnetic resonance (ssNMR) spectroscopy revealed the presence of SMS and trimethyloxonium (TMO) in the effluent, which are formed from the reaction between DME and SMS/methanol, coinciding with the initial production of ethylene. More significantly, specific surface methyleneoxy analogue species (R—O—CH_2—H—Zeo; R=H, CH_3), formed from activated DME, were for the first time directly detected using *in situ* ssNMR spectroscopy and recognized as the key intermediates involved in the formation of the initial C—C bond. The activated DME/methanol can undergo methylation through either the SMS or TMO pathways, resulting in the formation of surface-adsorbed ethanol/methylethyl ether species that contain the initial C—C bond and subsequently release ethylene. However, the specific contributions of methane and formaldehyde to the direct formation of the C—C bond are not fully understood, even though they have been suggested to participate in the catalytic process. Very recently, Wu and Sun et al. [22, 23] observed the generation of surface ethoxy species during the initial C—C bond formation through *in situ* ^{13}C MAS NMR analysis. Combining theoretical calculations with spectroscopic evidence, two plausible reaction pathways, namely SMS-mediated and TMO-mediated methanol/DME conversion, were proposed.

15.2.1.4 Carbonylation-based Mechanism

In 2016, Liu et al. [24] first proposed that acetic acid and methyl acetate (MA) ($CH_3CO_2CH_3$), formed through the carbonylation of methanol or DME, serve as the initial intermediates containing a C—C bond in the MTH reaction over ZSM-5 catalysts. It was suggested that the decomposition of methanol into formaldehyde and CO occurs readily even on chemically inert surfaces, such as silicate, Na-ZSM-5, or γ-Al_2O_3. The carbon atom in methanol, DME, and formaldehyde exhibits electrophilic character, whereas the carbon atom in CO displays nucleophilic behavior. This promotes the formation of a C—C bond between electrophilic and nucleophilic carbon atoms (see Figure 15.1j). The carbonylation-based mechanism features an energy barrier of 80 kJ mol^{-1} for the initial C—C bond formation, which is considered lower compared to previously suggested mechanisms. By combining ssNMR spectroscopy with UV–Vis diffuse reflectance spectroscopy and mass spectrometry on an active SAPO-34 catalyst, Chowdhury et al. [18] obtained spectroscopic evidence indicating the presence of surface-bound acetate species, MA, and dimethoxymethane during the MTO reaction. This finding offers substantial support for the carbonylation-based mechanism. Interestingly, spectroscopic analysis also confirmed the existence of the carbene mechanism, indicating that several direct C—C bond formations occur simultaneously. In addition, Plessow and Studt have provided theoretical support for the Koch carbonylation mechanism in the MTO reaction [25].

15.2.1.5 Methane–aluminum/Oxonium Mechanism

Although it is widely accepted that the SMS formed by the adsorption of methanol on BAS are intermediates for the initial C—C bond formation in the early stage of the MTO reaction, LAS arising from extra-framework Al (EFAl) species also play a crucial role in this process. Comas-Vives et al. [26] utilized spectroscopic studies combined with computational approaches to reveal that the LAS present on the 110 facet of γ-Al_2O_3 effectively facilitate the activation of DME, resulting in the generation of methane, alkenes, and formate species on the surface. A transient oxonium ion intermediate is formed through hydrogen transfer between SMS and methanol coordinated at neighboring Al sites. This intermediate can subsequently interact with methane to produce the initial C—C bond or combine with an additional DME molecule to form a surface-bound formate species (see Figure 15.1k). Using advanced $^{13}C - \{^{27}Al\}$ double-resonance ssNMR techniques, Wang et al. [27] observed that SMS interacts with an EFAL species during the methanol conversion on ZSM-5 zeolite. When ^{13}C–formaldehyde was co-fed, ^{13}C MAS NMR analysis verified the formation of surface ethoxy species, MA, surface acetate species, and acetaldehyde. Consequently, the high reactivity of the SMS–EFAL complex promotes the formation of surface ethoxy species and ethanol, which serve as key intermediates in ethylene generation during the early reaction stage. Additionally, theoretical calculations reveal that the reactions between SMS–BAS and both formaldehyde and $CH_2OCH_3^+$ species are associated with higher activation barriers. Therefore, it has been confirmed that SMS–EFAL, not SMS–BAS, serves as the critical surface species initiating the C—C bond formation. Additionally, Chu et al. [28] theoretically revealed the mechanism of initial C—C bond formation during the early stage of the MTO process, proposing a novel methane–formaldehyde reaction pathway on LAS/BAS in zeolites. A formaldehyde-analogue intermediate, specifically an Al—OCH_2^+ species formed via hydride abstraction from a surface Al—OCH_3 group, was proposed for the first time as a crucial intermediate in the initial C—C bond generation. Theoretical calculations demonstrate that the highly electrophilic nature of this intermediate significantly promotes its reaction with methane, making the entire process thermodynamically favorable.

15.2.2 HCP Mechanism

In the 1990s, Dahl and Kolboe [29] proposed the widely accepted "HCP mechanism" for methanol conversion over the SAPO-34 zeolite. In this initial conceptual model, $(CH_2)_n$ represents hydrocarbon species adsorbed within the pores of the catalyst, which act as key intermediates and cocatalysts in the MTO process. These species undergo sequential methylation reactions, followed by side-chain elimination reactions that produce light olefins, thereby regenerating the hydrocarbon species and completing the catalytic cycle.

15.2.2.1 Nature of HCP Species

Although polymethylbenzenes (polyMBs) and their corresponding carbenium ions are considered potential active HCP species [30], the direct observation of these species and a clear understanding of their roles in the MTO reaction on zeolite catalysts under realistic reaction conditions remain significant challenges. Currently, through the application of advanced techniques such as ^{13}C isotope labeling, co-feeding experiments, GC-MS analysis, ssNMR, and IR spectroscopy, various carbocations and their associated neutral forms have been successively identified as reactive intermediates over different zeolites.

For example, by employing GC-MS and ^{13}C MAS NMR techniques, Li et al. [31] first observed the formation of the heptamethylbenzenium cation ($heptaMB^+$) and the deprotonated form of heptamethylmethylenecyclohexadiene (HMMC) within the DNL-6 (RHO) zeolite during the MTO reaction. Subsequently, they also observed the participation of heptaMB$^+$ and pentamethylcyclopentenyl cation (pentaMCP$^+$) in the MTO reaction over the CHA-type zeolites [32]. These findings suggest that the formation of dominant HCP species is significantly influenced by the structural and acidic properties of the zeolite. To fully grasp the significance of zeolite cavities and the distinct roles of confined intermediate species, it is essential to investigate the interaction between the zeolite framework and these confined species. Therefore, Li et al. [33] investigated the MTO reactions over SAPO-35 (LEV), SAPO-34 (CHA), and DNL-6 (RHO) to explore how the distinct cavity structures influence the confinement effects on intermediates. Various polyMB$^+$ cations and polyMCP$^+$ cations were identified in SAPO-35 and SAPO-34, with molecular sizes and reactivities differing from those observed in DNL-6. Specifically, SAPO-35 primarily produced 1,2,3-triMB, 1,2,3,5-tetraMB, and pentaMB, whereas SAPO-34 mainly generated 1,2,4,5-tetraMB and pentaMB. In contrast, DNL-6 was found to produce hexaMB. These results, further validated through theoretical analysis, indicate that cavity size governs the reactions occurring within confined spaces, thereby determining the nature and reactivity of the intermediates involved.

15.2.2.2 Dual-cycle Mechanism

In 2012, Olsbye et al. [3] proposed a refined HCP mechanism (see Figure 15.2), referred to as the dual-cycle mechanism, which consists of an olefin-based cycle and an aromatic-based cycle over ZSM-5 zeolite. This proposal marked a significant advancement in understanding the HCP mechanism. In the olefin cycle, the methylation of light olefins produces higher olefins, which then mainly undergo cracking to produce propylene and butene. Meanwhile, in the aromatic cycle, MBs and their protonated counterparts serve as active intermediates and further produce light olefins. According to the dual-cycle mechanism, ethylene and propylene are derived from the aromatic-based cycle, while propylene and higher olefins are produced from the olefin-based cycle.

It is noted that the olefin- and aromatic-based cycles are not completely separated. The olefins generated in the aromatic cycle can participate in the olefin cycle to form higher olefins, while the higher olefins generated in the olefin cycle can be

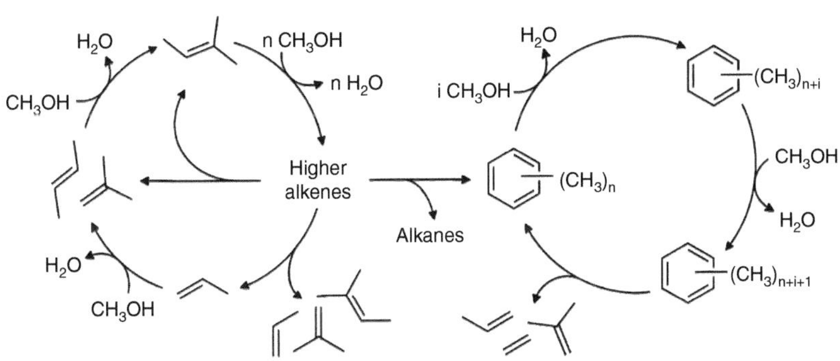

Figure 15.2 Proposed dual-cycle approach for methanol conversion over ZSM-5 zeolite. *Source*: Reproduced from [3] / with permission of John Wiley & Sons.

converted into aromatic compounds, thereby reinforcing the aromatic cycle. Dai et al. [34] provided direct spectroscopic evidence supporting the dual-cycle mechanism and elucidated the connection between the two cycles. Several key primary intermediates, such as alkylcyclopropanes, dienes, polyMCP$^+$, polyMCH$^+$, and polyMB$^+$, were directly observed using ssNMR spectroscopy. Through sequential deprotonation and hydride transfer steps, polyMCP$^+$ and polyMCH$^+$ can be transformed into polyMBs, which are identified as the active HCP species involved in the aromatic cycle.

Recently, Zhang et al. [35, 36] investigated the involvement of MCP and MCP$^+$ in the MTO reaction over SAPO-34 through *in situ* ^{13}C MAS NMR spectroscopy combined with theoretical calculations. It has been demonstrated that MCP$^+$ species are first formed in the early stage and then evolve into polyMBs, initiating the aromatic cycle. Moreover, cyclopentadiene intermediates are directly involved in the catalytic cycle for olefin formation. These findings led to the hypothesis that MCP$^+$ species serve as a bridge between the early and steady stages of the MTO reaction, giving rise to a newly proposed catalytic cycle known as the cyclopentadiene-based cycle. This theoretical progress extends the classical dual-cycle mechanism by clarifying the transformation pathways between the two cycles. Likewise, Hu et al. [37] demonstrated the essential contribution of cycloalkenes in facilitating olefin aromatization during the MTO reaction on ZSM-5 zeolite through ssNMR studies and DFT calculations. They proposed that the aromatization process of cycloalkenes serves as a bridge connecting the dual-cycle mechanism.

Given that olefins and aromatics can simultaneously reside within the channels and cavities of zeolites, a competitive interaction emerges in the dual-cycle mechanism. The relative contribution of this dual-cycle mechanism is determined by several factors, including zeolite topology, acidic properties, and reaction conditions. For example, cavities, crossing intersections, or large pores are essential for accommodating polyMBs, thereby promoting the dominance of the aromatic cycle; in contrast, narrow pores favor the olefin cycle. To describe the relative propagation of the olefin and aromatic cycles during methanol conversion on ZSM-5, Iliasa et al. [38] proposed the ratio of ethylene to 2-methylbutane + 2-methyl-2-butene ($C_2^=$/2-mb) yield as a descriptor. A higher $C_2^=$/2-mb yield suggests enhanced propagation of the aromatic cycle, whereas a lower $C_2^=$/2-mb yield indicates greater involvement of the olefin cycle.

15.2.2.3 Side-chain and Paring Route

In the aromatic-based mechanism, two reaction routes have been proposed: the side-chain route and the paring route (see Figure 15.3). Both routes involve carbenium ions as key intermediates in olefin production. Specifically, the paring route involves a contraction of 6-membered ring (MR) cations and an expansion of 5-MR cations. On the other hand, the side-chain route proceeds via methanol methylation on polyMB$^+$ cations, followed by the elimination of side-chain groups to generate olefins [39]. The *in situ* detection of MB$^+$ and MCP$^+$ cations, along with numerous theoretical and isotopic labeling studies, confirms the viability of both routes during the MTO process. These two routes consistently coexist and compete, with their

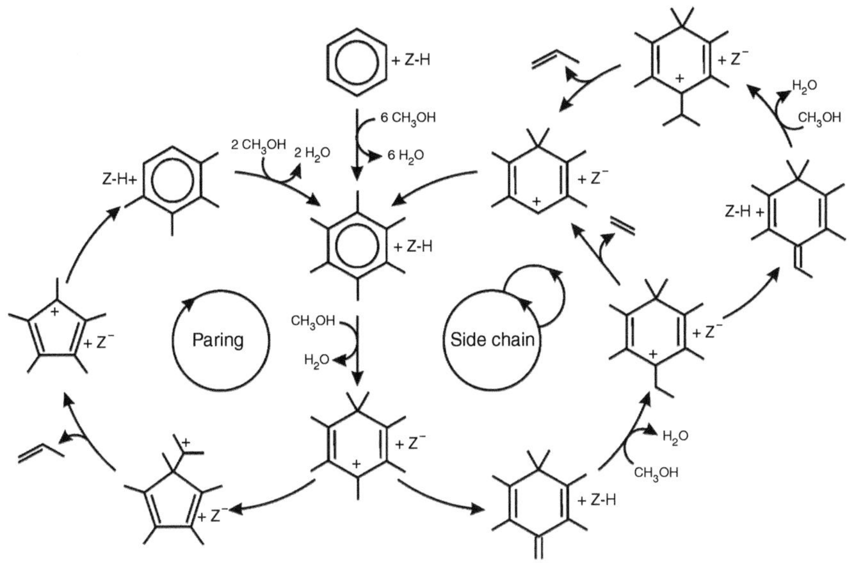

Figure 15.3 Schematic illustration of the paring and side-chain routes in MTO reaction over zeolite catalysts. *Source*: Reproduced from [39] / with permission of John Wiley & Sons.

relative dominance strongly dependent on the zeolite structure. For example, Li and Xu et al. [31, 32] conducted ^{12}C-/^{13}C-methanol transient switch experiments, which demonstrated that the side-chain route is the dominant mechanism for olefin production over DNL-6 zeolite. Although evaluations of the two catalytic routes with the paring and side-chain methylation mechanisms indicate that both are energetically feasible reaction routes, the side-chain methylation mechanism is more favorable, as it exhibits a lower energy barrier.

Recently, Ferri et al. [40, 41] proposed that the distribution of light olefins is closely related to the methylation degree of the entrapped MB^+, as supported by DFT calculations, reaction experiments, and NMR results. For example, the cavities in ITE and RTH zeolites stabilize fully methylated HCP species, thereby enhancing the paring route for propylene and butene production. In contrast, the CHA cavity stabilizes partially methylated aromatic intermediates, favoring the side-chain route for ethylene formation. Additionally, a quantitative factor known as the interaction energy ratio $E_{int(7/5)}$ ($E_{int(7/5)} = E_{int(g7MB+)}/E_{int(g5MB+)}$) was proposed to assess the relative contributions of the paring route and its effect on the distribution of light olefins.

15.2.3 Deactivation Mechanism

In the MTO reaction, zeolite catalysts typically experience rapid deactivation. The continuous accumulation of active cyclic HCP species leads to their progressive transformation into larger PAHs. These PAHs then accumulate on acidic active sites, obstruct pore channels, and ultimately lead to catalyst deactivation.

15.2.3.1 Deactivation Mode

As two industrially relevant MTO catalysts, SAPO-34 and ZSM-5 zeolites exhibit distinct deactivation mechanisms during methanol conversion. For SAPO-34, confined polyMB species progressively transform into PAHs, such as polymethylnaphthalene, phenanthrene, and pyrene, within the CHA cavity as the reaction proceeds. Given the cavity size, pyrene, with its four-ring structure, and its low-methyl homologs represent the largest aromatic molecules that can be accommodated. However, due to steric constraints, these bulky pyrene molecules are less likely to undergo further methylation reactions, thereby diminishing their effectiveness as co-catalysts. The formation and accumulation of PAHs inside the cavity not only occupy the catalytic active sites, hindering the access of methanol molecules to the active centers, but also block the diffusion pathways of guest molecules due to their large size. These combined effects lead to the deactivation of the SAPO-34 catalyst [42]. Different from SAPO-34, ZSM-5 lacks sufficient space to facilitate the generation of large aromatic compounds. This channel structure can only accommodate MB compounds with a limited number of methyl substitution groups, most of which are able to diffuse out. The structural characteristics of ZSM-5 indicate that there is minimal deposition of bulky coke species within the channel. Therefore, the deactivation of ZSM-5 is not attributed to coke accumulation on the internal surface of the zeolite, but rather to coke deposition on its outer surface [43].

15.2.3.2 Formation and Evolution of Coke Species

Previous research suggested that PAHs might be produced through the oligomerization of MB and olefins, followed by cyclization and hydrogen transfer processes. However, this proposed mechanism necessitates overcoming an extremely high energy barrier, which contradicts the experimentally observed rapid generation of PAHs. Wang et al. [44] emphasized the significant influence of intermolecular π -interactions on catalyst deactivation in the MTO reaction on SSZ-13 and ZSM-5 zeolites. 2D ssNMR spectroscopy demonstrated that π-interactions lead to spatial proximity between cyclopentenyl cations and aromatic compounds within the restricted environments of zeolite channels and/or cages. The generation of naphthalene, serving as a precursor to coke formation, is enhanced through reactions between aromatics and adjacent cyclopentenyl cations. This process is closely related to both the acid density and the structural topology of the zeolite.

To gain deeper insights into the comprehensive pathway of aromatic hydrocarbon evolution during the MTO reaction over SAPO-34, Wang et al. [45] introduced an innovative methodology that combines high-resolution MALDIFT-ICR mass spectrometry with isotope labeling techniques. Significantly, this research identified a novel and previously unknown mechanistic process, cage-passing growth, which leads to the generation of cross-linked multicore PAHs with structural features resembling those of graphene. A growth mechanism involving cage-passing was proposed to explain the formation of heavy PAHs during the MTO process (see Figure 15.4a). In this model, smaller PAHs, such as naphthalene, fluorene, phenanthrene, and pyrene, are initially formed within the isolated cages of zeolites. These

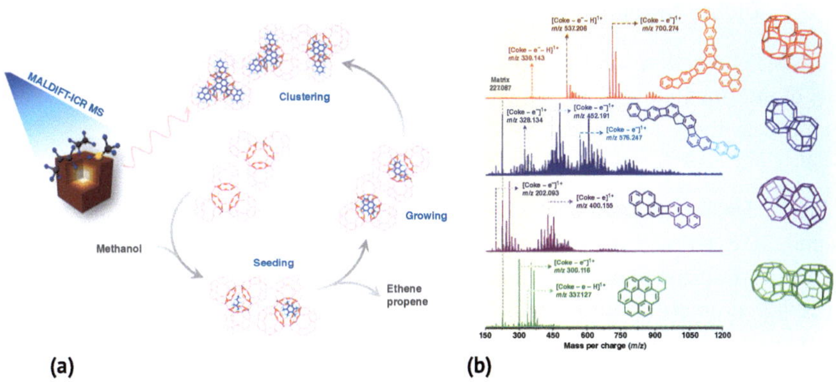

Figure 15.4 (a) Molecular evolution pathways of PAHs. (b) MALDIFT-ICR mass spectra of extracts obtained from various cage-structured zeolites and the proposed possible molecular structure of PAHs. *Source*: Reproduced from [45] / with permission of Springer Nature / CC BY 4.0.

compounds act as precursors or clustering seeds for subsequent coke formation. Subsequently, these species migrate into adjacent cages through the 8-MR pore openings, where they undergo condensation reactions, leading to the formation of graphite or graphite-like carbon structures via inter-cage cross-linking. The proposed mechanistic concept has been demonstrated to apply to other cage-based zeolites, such as SAPO-35, SAPO-18, and DNL-6. However, the possible molecular structures of PAHs within these zeolites are striking different, which is attributed to their distinct cage-connecting ways (see Figure 15.4b).

Recently, Fan et al. [46] compared the deactivation behavior of SSZ-13, beta, and ZSM-5 zeolites during the MTO process. Their results indicate that the alkylation of cyclic intermediates, such as MBs and cyclohexene, by cyclic carbocations, including cyclopentadienyl and cyclohexadienyl cations, plays a crucial role in the formation of PAHs. Additionally, these cross-linked PAHs serve as essential precursors in the formation of coke. The development of coke precursors is significantly influenced by the zeolite structure. In particular, the more spacious intracrystalline environment in SSZ-13 and Beta provides sufficient space for the alkylation and cross-linking of cyclic intermediates. In contrast, the narrow intracrystalline space in ZSM-5 hinders these processes because it is energetically less favorable.

15.3 Factors Affecting MTO Performance

Based on the reaction mechanism, the characteristics of HCP species play a crucial role in determining the reaction pathway, as they incorporate essential intermediates into the complex reaction network. Consequently, the nature and evolution of HCP species during the reaction have a pronounced effect on the MTO performance, especially the light olefin selectivity and catalyst lifetime. These characteristics are closely associated with a range of factors, such as zeolite topology, acidic properties, particle size and morphology, and reaction conditions.

15.3.1 Zeolite Topology

In the MTO reaction, zeolite topology primarily determines the nature and reactivity of intermediate species, which are crucial in influencing the catalytic pathway, product selectivity, and catalytic stability.

15.3.1.1 Small-pore Zeolite

Overall, 3D 8-MR zeolites with large cavities and small openings, such as SAPO-34, restrict the diffusion of bulky HCP intermediate species within the channels, thereby selectively promoting the formation of light olefins as the primary products. In addition to CHA-type zeolite, other cage-containing zeolites, including AEI [47], DDR [48], RTH [49], LEV [50], and ERI [51], also exhibit good MTO activity and light olefin selectivity. However, their performance varies in terms of activity, light olefin selectivity, and catalyst durability, which can be attributed to differences in cavity configuration. Conversely, 1D 8MR zeolites such as SAPO-39 (ATN), MCM-35 (MTF), and ERS-7 (ESV), and 2D 8MR RUB-37 (CDO), are inactive in the MTO reaction. This is attributed to the insufficient void space in these zeolite structures, which constrains the formation of intermediate polyMBs species [52]. These results indicate the critical role of cavity structure in small-pore zeolites and further support the importance of HCP species for realizing catalytic activity in the MTO reaction.

To illustrate the correlation between the distribution of light olefins and the configuration of zeolite cavities, Bhawe et al. [53] compared LEV, CHA, and AFX zeolites that exhibited comparable Si/Al ratios and crystal sizes. They found that the LEV zeolite, which has smaller cavities, favors ethylene formation. Ethylene selectivity increases with the reduction in cage size. The CHA zeolite exhibited the highest propylene selectivity. Li et al. [54] investigated the MTO process in *cha*, *lta*, and *lev* cavities using DFT calculations under the assumption of an identical side-chain mechanism. Their findings revealed that the *cha* cavity exhibits the lowest intrinsic free-energy barriers across all methylation and olefin-production steps, suggesting its superior MTO activity, which may be due to the appropriate confinement effect provided by the CHA structure.

Recently, Kang et al. [55, 56] carried out an extensive study on the MTO performance across more than 30 different silicoaluminate and SAPO zeolites, including 14 types of small-pore cage-like zeolitic structures. The findings indicated that the light olefins selectivity is primarily influenced by the framework topology, rather than by its elemental composition. A novel geometric concept, known as the cage-defining ring, was proposed based on a straightforward assumption regarding the alignment of HCP intermediates inside the cages (see Figure 15.5a, b). Studies indicate that there is a close relationship between the size of the cage-defining ring, which constrains the dimensions of intermediate species, and the distribution of light olefin products. Accordingly, four categories were identified based on differences in the light olefins' distribution (see Figure 15.5c): (1) LEV and ERI, which have 12-membered cage-defining rings with diameters below 7.45 Å, give higher ethylene selectivity than propylene; (2) CHA, AFX, and SFW, which have 12-membered cage-defining rings with a size of 7.45 Å, exhibit similar ethylene

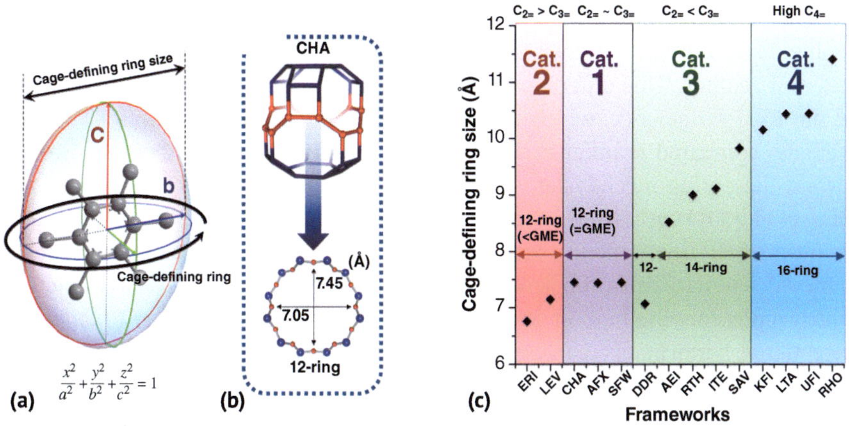

Figure 15.5 (a) Ellipsoidal representation of the cage-defining ring and its dimensions. (b) Identification of the cage-defining ring within a CHA-type cage structure. (c) Relation between cage-defining ring size, cage structure, and distribution of light olefins. *Source*: Reproduced from [56] / with permission of American Chemical Society.

and propylene selectivity along with low butene selectivity; (3) DDR, AEI, RTH, ITE, and SAV, which have wider 14-membered cage-defining rings, produce higher propylene selectivity than ethylene; (4) KFI, LTA, UFI, and RHO, which have the widest 16-membered cage-defining rings (with diameters larger than 10 Å), show high butene selectivity.

The influence of cavity configuration on the selectivity of light olefins can be elucidated through the HCP mechanism. Zhang et al. [57] investigated all elementary steps of the entire catalytic cycle at the molecular level, thereby clarifying the relationship between product selectivity and cavity structure. It was revealed that the preferential production of olefins was closely associated with the generation of HCP species, the development of olefin precursors, and the reaction pathways influenced by different types of cavity-containing zeolites. By employing an *in situ* ^{13}C NMR technique, the existence of tri- and tetra-methylated HCP intermediates inside the *lev* cages was verified. Theoretical predictions suggest that the side-chain methylation mechanism is energetically favorable, resulting in higher ethylene selectivity. Based on DFT calculations, reaction data, and NMR results, Ferri et al. [40, 41] suggested that the distribution of light olefins is influenced by the extent of methylation of the confined MB cations. Specifically, the ITE and RTH cavities stabilize the fully methylated HCP species, thereby facilitating the formation of propylene and butene via the paring mechanism. In contrast, the CHA cavity supports the stabilization of partially methylated HCP species, favoring the side-chain route that leads to ethylene production.

Recently, Li et al. [58] developed a novel method for synthesizing highly effective MTO zeolite catalysts by employing specifically designed organic structure-directing agents (OSDAs). The key to OSDA design lies in mimicking the transition states of known chemical reactions that occur during the MTO process. Based on this methodology, cage-based small-pore zeolites, including SSZ-13 (CHA), SAPO-18 (AEI), and RUB-13 (RTH), were successfully synthesized. In the MTO reaction, the RTH zeolite demonstrated a higher propylene-to-ethylene ratio compared to the CHA and AEI zeolites. This difference can be attributed to the unique cavity structure of the RTH framework, which offers superior stabilization of pentaMCP$^+$ intermediates compared to other zeolites. As a result, the reaction pathway is directed toward the paring mechanism, which favors propylene formation. Very recently, Yang et al. [59] synthesized a small-pore SAPO-14 (AFN) zeolite characterized by an ultra-small cage structure and a confined 8-MR channel system. This catalyst achieved a remarkable propylene selectivity of up to 77.3% in the MTO reaction, which represents the highest one-pass propylene selectivity reported to date. The exceptional performance can be attributed to its extremely small cage structure, moderate acidity, and low acid density, which effectively reduce side reactions such as hydrogen transfer and cyclization of olefins and inhibit the aromatic cycle. Therefore, the reaction pathway favors the olefin methylation/cracking pathway, leading to a notable enhancement in propylene selectivity. Moreover, the relatively narrow 8-MR pore openings limit the diffusion of C_4^+ hydrocarbons, further promoting the selective production of propylene.

15.3.1.2 Medium-pore Zeolite

Generally, 3D 10-MR medium-pore zeolites, such as ZSM-5, exhibit high selectivity for C_5^+ hydrocarbons, including alkanes and aromatic compounds. However, under optimized reaction conditions, such as high temperatures, the selectivity is shifted predominantly toward propylene rather than C_5^+ when a recycle process is utilized. Although ZSM-5 exhibits relatively low ethylene selectivity, it achieves a propylene selectivity of *ca.* 35%. Furthermore, by reducing the acid density and altering the particle size and morphology, the selectivity toward propylene can be enhanced to *ca.* 50%. It has been demonstrated that aromatic- and olefin-based cycles coexist during the MTO reaction process on ZSM-5 zeolite with an intersecting channel structure.

However, 1D 10-MR zeolites without intersections, such as ZSM-22 (TON), ZSM-23 (MTT), ZSM-48 (*MRE), and EU-1 (EUO), tend to generate lower amounts of light olefins and aromatic compounds and exhibit rapid deactivation. This is attributed to the limited width of the 10-MR channels, which restricts the formation of aromatic molecules, and thereby favoring the olefin-based cycle [60]. It was noted that ZSM-22 and ZSM-23 exhibit relatively low selectivity toward light olefins, which may be related to the acid sites on the external surface and pore mouth [61]. Compared to 1D 10-MR zeolites, 2D 10-MR zeolites such as MCM-22 (MWW) [62] exhibit higher light olefin selectivity, attributed to the channel intersections that provide larger space for the formation of aromatic intermediate species. Additionally, their relatively long lifetime can be attributed to the improved diffusion of reaction products.

By contrast, 3D 10-MR zeolites such as ZSM-11 (MEL) zeolite, which consists of two types of channel intersections (one with a size roughly equivalent to that of MFI and another approximately 30% larger), are considered excellent MTO catalysts. Shen et al. [63] compared the MTO performance of ZSM-5 and ZSM-11 with similar crystal sizes and Si/Al ratios (see Figure 15.6). ZSM-11 displayed a double catalyst lifetime compared to ZSM-5, attributed to its straight micropore channels that enhance molecular diffusion. Moreover, reducing the crystal size of ZSM-11 from 750 to 150 nm led to an eight-fold improvement in catalyst lifetime. Operando UV–vis microspectroscopy analyses indicated that both crystal size and pore structure distinctly influence the formation of HCP species and coke speciation during the initial (transient) reaction stage.

Besides, the ITQ-13 (ITH) zeolite, which consists of a 3D pore structure with 10 × 10 × 9-MR channels, is also considered an excellent MTO catalyst because it exhibits very high propylene selectivity [64]. Compared to ZSM-5, ITQ-13 has a relatively smaller channel size; however, the maximum diameter of the accessible sphere within ITQ-13 is slightly larger, resulting in variations in the HCP species generated during the MTO process between these two zeolites. DFT calculations indicated that both aromatic- and olefin-based cycles are feasible on ITQ-13 [65]. However, the olefin cycle is more favorable because the transition states involved in the olefin cycle are effectively stabilized within the ITQ-13 framework, resulting in very high propylene selectivity. It was demonstrated that the [Al, Ge]-ITQ-13

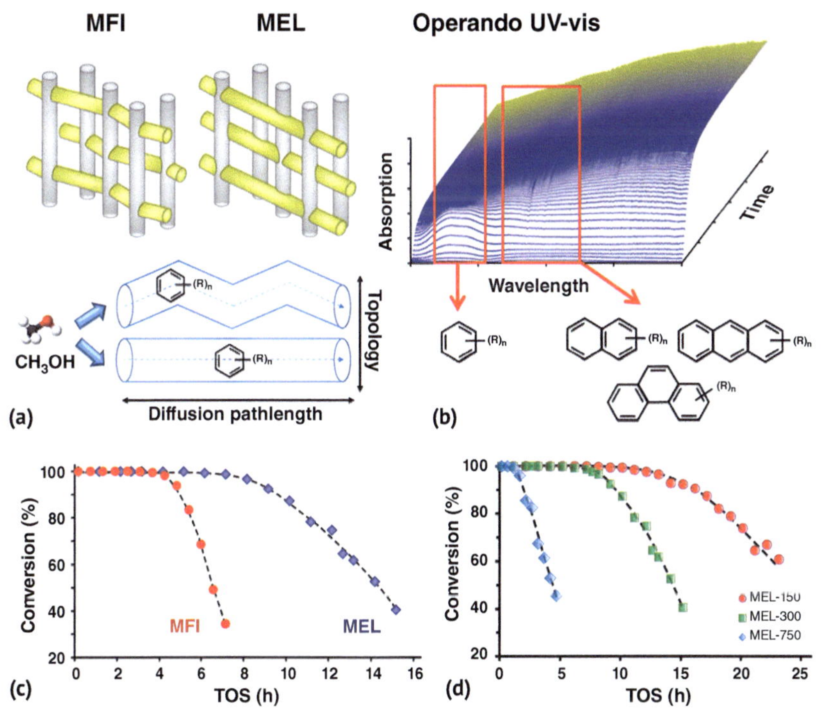

Figure 15.6 (a) Diagram illustration of the microporous structure of MFI and MEL zeolites. (b) Scheme representation of operando UV–vis diffuse reflectance spectra collected during the MTO reaction. (c, d) Change of methanol conversion with TOS for MEL-300 and MFI-300 catalysts (c) and MEL-150, MEL-300, and MEL-750 catalyst (d). Reactions conditions: T = 350 °C and WHSV = 9 h^{-1}. *Source*: Reproduced from [63] / with permission of American Chemical Society.

(Si/Al = 155) zeolite exhibited a propylene selectivity of 55.5% and a total light olefins (ethylene, propylene, and butene) selectivity of 94% [66], which was attributed to its low amount of acid sites and weak acidity, suppressing hydrogen transfer and aromatization reactions. However, the synthesis of this catalyst requires expensive Ge species, which significantly reduce the hydrothermal and thermal stabilities of the zeolite. Recently, Lei et al. [67] directly synthesized a Ge-free aluminosilicate ITH zeolite using a cationic oligomer as an OSDA. The aluminosilicate ITH (Si/Al = 94) zeolite, with a nanosheet-like crystal morphology, displayed significantly enhanced propylene selectivity and a prolonged catalyst lifetime compared to commercial ZSM-5. More recently, Ma et al. [68] found that Al-COE-12 (ITR) zeolite, which is a structural twin of the ITH zeolite, also exhibited superior propylene selectivity along with a prolonged catalyst lifetime compared to ZSM-5.

15.3.1.3 Large-pore Zeolite

Generally, 12-MR large-pore zeolites facilitate the formation of heavier hydrocarbons in the MTO reaction. This effect is mainly due to the dominance of the aromatic-mediated mechanism. Additionally, the 12-MR channels provide sufficient space for the accommodation of aromatic hydrocarbons. Therefore, the aromatic-based cycle can proceed efficiently within these channels, producing propylene and butene.

Among the various large-pore zeolites, beta (*BEA) is a promising candidate due to its comparatively mild acidity and 3D pore structure. Otomo et al. [69] reported that a template-free, dealuminated beta zeolite (Si/Al = 112) demonstrated exceptional selectivity to propylene (44%) and butene (26%) while maintaining over 90% methanol conversion. It was suggested that the high propylene selectivity arises from both the olefin- and aromatic-based cycles, particularly those involving higher MBs, whereas ethylene production remains limited in these mechanisms. Additionally, the long lifetime of high-silica beta was attributed to its large pore openings, which enhance the diffusion of bulky molecules and mitigate coke deposition. This work suggests that large-pore zeolites can achieve high selectivity to propylene and higher olefins when their acid properties, morphology, and particle size are carefully modulated.

The CON-type zeolite, characterized by its 3D pore structure featuring $12 \times 10 \times 10$-MR channels, exhibited excellent performance. Yoshioka et al. [70] found that the [B, Al]-CON (Si/Al = 100–200) zeolites exhibited extremely high propylene selectivity (60%) and butene selectivity (20%) due to their weak acidity. Moreover, sequential reactions such as oligomerization and subsequent cracking were effectively suppressed, resulting in low ethylene selectivity (<5%). It is noteworthy that the performance of CON zeolites depends on the synthesis approaches and conditions, including direct or post-synthesis methods, as well as static or tumbling conditions, which could be related to the different distributions of Al atoms (see Figure 15.7).

Besides, Hong et al. [71] found that the 3D $12 \times 10 \times 10$-MR aluminosilicate COE-6 (IWR) zeolite (Si/Al = 135) exhibited enhanced propylene selectivity (47%) and a significantly higher propylene-to-ethylene ratio (10.2). Ren et al. [72] reported that the 2D 12×8-MR mordenite (MOR) (Si/Al > 150) exhibited excellent propylene

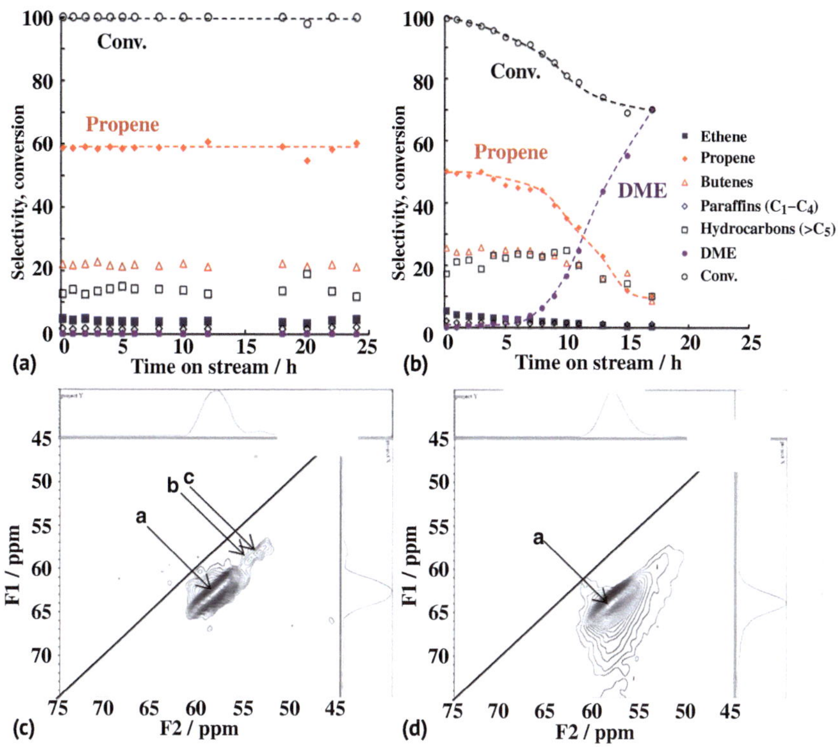

Figure 15.7 MTO catalytic performance and ^{27}Al MQMAS NMR spectra of [Al, B]-CON-D (Si/Al = 134, tumbling) (a, c) and [Al]-CON-P (Si/Al = 87) (b, d). Reaction conditions: T = 500 °C and WHSV = 6.6 g·hmol^{-1}. *Source*: Reproduced from [70] / with permission of American Chemical Society.

selectivity (63%) and a high propylene-to-ethylene ratio (> 10). In addition, two new types of 3D multipore zeolites containing interconnected 12- and 10- or 8-MR channels, i.e., MCM-68 (MSE) and YNU-5 (YFI), were also investigated by Park et al. [73]. It was found that MCM-68 showed high initial propylene selectivity; however, both of them exhibited inferior stability.

15.3.2 Acidic Property

The MTO process involves a series of reaction steps: the generation and stabilization of carbenium ions, subsequent alkylation, and rearrangement reactions, and finally the cracking process that yields light olefins. Concurrently, while olefin products are being generated and desorbed, retained olefins within the zeolite channels or cages may undergo additional condensation reactions, resulting in the formation of heavier coke deposits and subsequent catalyst deactivation. Due to these consecutive reactions occurring at the acid sites of zeolites, which progressively evolve into a complex reaction system, the acidic properties of the zeolite play a critical role in determining the methanol conversion pathway and, consequently, the MTO performance.

15.3.2.1 Acid Density

The BAS density in zeolite catalysts significantly influences MTO performance, as BAS are primarily responsible for catalyzing key steps in methanol conversion and olefin formation. The BAS density is strongly correlated with the content of Al and Si in aluminosilicate and SAPO zeolites, respectively, which can be adjusted by changing the composition of the gel or through dealumination. Overall, a moderate BAS density is ideal for balancing high methanol conversion, good olefin selectivity, and a long catalyst lifetime. Low BAS density may limit overall activity, but can help minimize coke formation and maintain catalyst stability over time. Although high BAS density can accelerate methanol conversion, it may also enhance side reactions, facilitate coke deposition, and hasten catalyst deactivation. Therefore, controlling BAS density is crucial for high MTO performance.

It has been demonstrated that light olefin selectivity and catalyst lifetime are closely related to the Si/Al ratio in zeolites, which affects the nature of HCP species and their evolution during the MTO reaction. For SSZ-13 with high Si/Al ratios, polyMBs are the primary HCP species, demonstrating high light olefin selectivity and prolonged catalyst lifetimes [74]. However, as the Si/Al ratio decreases, the dominant HCP species transition from MBs to methylnaphthalenes. A low Si/Al ratio (i.e., high acid density) facilitates side reactions of olefins, which reduce the light olefin selectivity and accelerate catalyst deactivation through coke formation. The Si/Al ratio significantly affects the preference between olefin- and aromatic-based cycles, thereby influencing product selectivity. As the Si/Al ratio in ZSM-5 increases, propylene selectivity increases while aromatic selectivity decreases. This suggests that a lower Al content promotes the olefin cycle for propylene production and inhibits secondary reactions leading to aromatic compounds. Conversely, a higher concentration of BAS enhances the preference for the aromatic cycle that favors aromatic formation [6].

To improve the MTO performance, phosphatation, alkaline earth metal modification (e.g., Ca, Sr, Ba), and transition metal modification (e.g., Zn) are applied, leading to a significant reduction in BAS density. Yarulina et al. [75] reported that incorporating calcium into ZSM-5 leads to a *ca.* 10-fold decrease in BAS density. This reduction effectively suppresses hydride transfer and cyclization reactions, resulting in minimal production of paraffins and aromatics and concurrently enhancing propylene yield to as high as 53%. Meanwhile, the introduction of calcium promotes the formation of $CaOCaOH^+$ species, which notably reduces the acid strength and facilitates the transformation of BAS into LAS, thereby influencing the MTO performance. Furthermore, the modification of pore geometry induced by the introduction of alkaline metal cations affects the formation of HCP species, thus favoring the olefin-based cycle [76].

15.3.2.2 Acid Strength

The acid strength of zeolites also significantly affects the MTO performance. Compared to aluminosilicate zeolites, SAPO zeolites demonstrate higher propylene selectivity and extended lifetimes, which can be attributed to their lower acid strength. This effectively suppresses hydrogen transfer reactions and aromatic cycles in the dual-cycle mechanism. Erichsen et al. [77] investigated the influence of zeolitic acid strength in a 1D 12-MR AFI zeolite on olefin- and aromatic-based cycles. The highly acidic SSZ-24 was more selective toward aromatic compounds and C_2–C_3 hydrocarbons, in contrast to the mildly acidic SAPO-5, which favored the formation of non-aromatic C_4^+ hydrocarbons. Co-reaction experiments combined with theoretical calculations demonstrate that acid strength greatly influences the reactivity of the HCP species. In the case of SSZ-24, benzene methylation is considerably more pronounced than propylene methylation, thereby making the aromatic-based cycle the predominant pathway. Conversely, SAPO-5 exhibits comparable reaction rates for both olefin and aromatic methylation, suggesting that a reduction in acid strength can suppress the aromatic-based cycle while favoring the olefin-based cycle.

To reduce the acid strength of zeolites, a commonly adopted approach involves incorporating alternative heteroatoms such as B and Ga to substitute framework Al. Studies have shown that gallosilicate CHA zeolite exhibits high light olefin selectivity while producing minimal amounts of ethane and propane. This is attributed to the reduced acid strength, which suppresses the hydrogen transfer reaction of olefins [78]. Similarly, [Ga, Al, B]-CON achieves a propylene yield of 60%, comparable to that of [Al, B]-CON. However, the reduced strength of BAS might suppress ethylene formation, resulting in an increased propylene-to-ethylene ratio [79]. Additionally, the acid strength can be significantly reduced through phosphatation and alkaline-earth metal modification, which poison the strong acid sites and introduce weak acid sites, thereby effectively suppressing the aromatic cycle. Consequently, both the propylene selectivity and catalyst lifetime are drastically enhanced [6].

15.3.2.3 Acid Type

It is well known that BAS plays a key role in the MTO reaction; however, the role of LAS was unclear in previous research. In zeolite catalysts, LAS often arise from EFAl species or defects in the framework structure, which have been proven to negatively affect MTO performance. Generally, LAS are less effective than BAS in methanol activation and olefin production, and they are more prone to promoting side reactions, thereby reducing olefin selectivity and increasing coke formation.

Recently, Zhao et al. [80] demonstrated that a certain amount of LAS in ZSM-5 enhances ethylene selectivity by promoting the conversion reactions that lead to an increased formation of MB species. More significantly, due to the steric constraints caused by the EFAl, MBs can be more effectively converted into ethylene. Moreover, Yarulina et al. [81] revealed that the active site is considerably complex: the simultaneous presence of both BAS and LAS can substantially improve catalyst performance (see Figure 15.8). The light olefins selectivity and lifetime of ZSM-5 were improved after Ca modification, which was attributed to a notable reduction and isolation of BAS, along with the formation of LAS. An appropriate amount of a specific LAS, namely $\left[Ca(\mu\text{-}OH)_2Ca\right]^{2+}$, can enhance the MTO reaction lifetime, which is attributed to the suppression of aromatics growth through methylation in the aromatic cycle. These results suggest that LAS play a significant role in

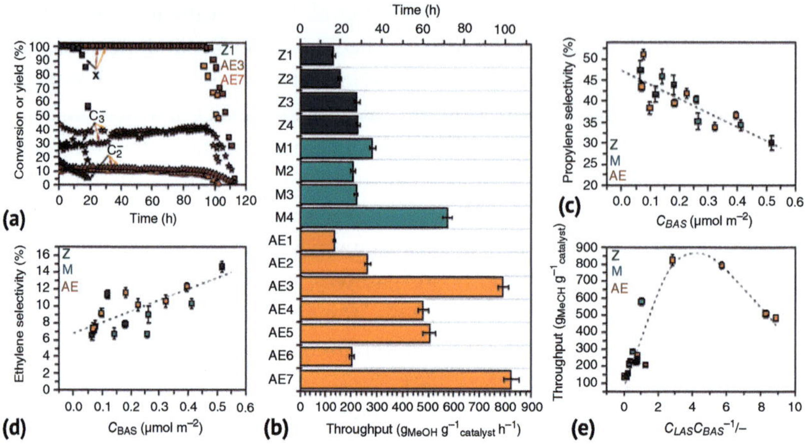

Figure 15.8 (a, b) MTO performance on different ZSM-5 zeolite catalysts. Reaction conditions: T = 500 °C and WHSV = 8 g_{MeOH} $g_{catalyst}^{-1}h^{-1}$. (c–e) Correlation between acidity and catalytic behavior. Z (black), ZSM-5 with varying Si/Al ratios modified pre-synthetically; M (green), post-synthetically demetallized ZSM-5; AE (orange), zeolites modified via post-synthetic introduction of alkaline-earth metals. *Source*: Reproduced from [81] / with permission of Springer Nature.

influencing the reactivity of intermediates throughout the reaction. Based on the dynamic nuclear polarization surface-enhanced NMR spectroscopy (DNP SENS) results, it has been demonstrated that the presence of LAS inhibits the formation of carbene/ylide species, which affects the balance between the aromatic and olefin cycles in the dual-cycle mechanism and thus regulates the olefin selectivity and extends the catalyst lifetime [82].

15.3.2.4 Al Distribution

The distribution of framework Al in zeolites encompasses both the spatial arrangement of Al atoms (e.g., isolated vs. paired Al) and their specific locations within the framework structure (e.g., channels, intersections, or defined tetrahedral (T) sites). In recent years, increasing attention has been focused on the effect of Al distribution in zeolites on the MTO performance. Generally, isolated Al sites lead to the formation of isolated BAS, which reduces secondary reactions (e.g., hydrogen transfer, oligomerization) of light olefins, thereby enhancing olefin selectivity and prolonging catalyst lifetime. On the contrary, paired Al sites result in their proximity and high BAS density, which promote secondary reactions of light olefins to produce PAHs (precursors to coke), thus decreasing olefin selectivity and leading to rapid deactivation.

For example, SSZ-13 zeolites with high Si/Al ratios exhibit higher propylene selectivity and a longer catalytic lifetime compared to those with low Si/Al ratios. This is attributed to the reduced amount of paired Al, which effectively suppresses the aromatic-based cycle in the dual-cycle mechanism [83]. Based on NMR analysis and steady-state isotopic switching experiment results, it has been demonstrated that paired Al enhances the reactivity of crucial HCP species, such as MBs, cyclopentenyl cations, and pentaMB ions, promoting the aromatic cycle in the MTO reaction over ZSM-5 [84]. Furthermore, theoretical calculations support these findings, showing that paired Al (i.e., BAS) markedly lowers the activation energy required for the production of both ethylene and propylene through a paring route, making the aromatic cycle more thermodynamically favorable. To decrease the amount of paired Al, numerous strategies, including the regulation of Si and Al sources, the type or ratio of organic/inorganic SDAs, etc., have been employed. For example, Nishitoba et al. [85] found that the paired Al became dominant in SSZ-13 when the aluminum source was changed from amorphous $Al(OH)_3$ to crystalline FAU zeolite (see Figure 15.9a). Gallego et al. [86] investigated the Al distribution in nanosized SSZ-13 zeolites synthesized from high-silica FAU zeolites, both in the presence and absence of Na^+. The study revealed that the CHA zeolite synthesized without Na^+ exhibited entirely isolated Al and demonstrated a longer catalyst lifetime compared to the sample synthesized with Na^+ in the MTO reaction (see Figure 15.9b, c).

Besides the spatial arrangement of Al atoms, the positions of these atoms also influence the nature of HCP species. When Al atoms (or acid sites) are located in more spacious regions, such as channel intersections, they are more likely to facilitate the development of large aromatic HCP species. Conversely, the development of such bulky HCP species is suppressed in narrow regions, such as straight and sinusoidal channels. Moreover, theoretical calculations suggest that the formation of olefin intermediates may be less sensitive to spatial limitations compared to aromatic intermediates [87]. Consequently, if more acid sites are located in confined spaces, the olefin-based cycle is favored, thereby increasing propylene selectivity. For example, Liu et al. [88] and Zhang et al. [89] found that RUB-13 (RTH) zeolite, which contains a higher number of Al atoms preferentially located at the T1 and T2

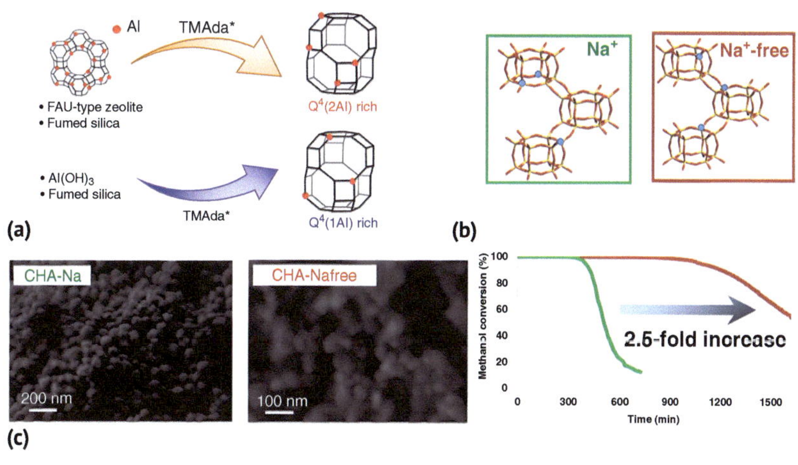

Figure 15.9 (a) Approach to regulating Al distribution in CHA zeolite through the selection of aluminum sources. *Source*: Reproduced from [85] / with permission of American Chemical Society. (b) Method for modulating Al distribution in CHA zeolite in the presence or absence of Na^+. (c) SEM images of CHA zeolite synthesized with or without Na^+, along with a comparison of catalytic stability in the MTO reaction. Reaction conditions: T = 350° and WHSV = 0.8 h^{-1}. *Source:* Reproduced from [86] / with permission of John Wiley & Sons.

sites within a non-distorted 8-MR channel, exhibited excellent MTO activity and high light olefin selectivity. Fan et al. [90] reported that the ITH zeolite, with Al atoms predominantly located in the straight and sinusoidal channels rather than at the channel intersections, exhibited an enhanced catalyst lifetime and higher propylene selectivity. Zhao et al. [91] found that a high-Si beta zeolite enriched with Al atoms located at the T9 site demonstrated superior propylene selectivity and the longest catalyst lifetime.

The location of Al in zeolites at specific positions can be regulated by changing the OSDAs or pore-filling agents, with or without Na^+, and by varying the starting silicon and aluminum sources during zeolite synthesis. For example, Yokoi et al. [92] found that in the ZSM-5 zeolite synthesized using the tetrapropylammonium cation (TPA^+) in the absence of Na^+, Al atoms are mainly located at the channel intersections. However, when both TPA^+ and Na^+ are present during synthesis, Al atoms are distributed not only at the intersections but also within the narrow straight and sinusoidal channels. This is because TPA^+ tends to occupy the intersections, while Na^+ resides within the channels. Furthermore, Yokoi and Biligetu et al. [93, 94] found that when large, branched-chain alcohols without charge, such as pentaerythritol (PET) and trimethylolethane (TME), are employed as pore-filling agents, Al atoms tend to occupy the narrow channels rather than the channel intersections. The ZSM-5 prepared with Al located in the narrow channels reduces coke formation in the confined spaces, thereby exhibiting a longer catalyst lifetime in the MTO reaction. In contrast, ZSM-5 with Al atoms located at the wide channel intersections promotes the aromatic-based cycle in the dual-cycle mechanism, which results in a shorter lifetime but enhances ethylene selectivity.

Liang et al. [95] effectively controlled the siting of Al atoms in ZSM-5 by adjusting the silicon sources used during the synthesis process. They further explored how the distribution of Al influences the MTO performance and reaction mechanism (see Figure 15.10). By combining ^{27}Al MAS NMR spectroscopy with UV–vis–DRS analysis of Co(II) cations, it was found that Al atoms were predominantly located in the sinusoidal and straight channels of S-HZ-m synthesized with silica sol as the silicon source. However, Al atoms were mainly located at the channel intersections of T-HZ-m synthesized with tetraethyl orthosilicate as the silicon source. Further catalytic tests combined with ^{13}C/^{12}C-methanol-switching experiments demonstrated that T-HZ-m tends to follow an aromatic cycle, resulting in increased aromatic selectivity. In contrast, S-HZ-m favors an olefin cycle, leading to enhanced selectivity toward propylene and C_4–C_5 olefins, as well as a prolonged catalyst lifetime.

Moreover, introducing a heteroatom source into the synthesis gel enables the precise placement of Al atoms at desired positions. For example, when an appropriate quantity of boron is incorporated into the MCM-22 framework, Al atoms are predominantly found within the sinusoidal channels, rather than in the surface pockets or supercages. Because boron and aluminum compete for occupancy at various framework T sites, the textural properties as well as the acid density, strength, and type are not significantly affected by the incorporation of boron. Therefore, the enhanced catalytic stability and improved selectivity for propylene and butene in the MTO reaction on [B, Al]-MCM-22 are attributed to the controlled distribution of Al

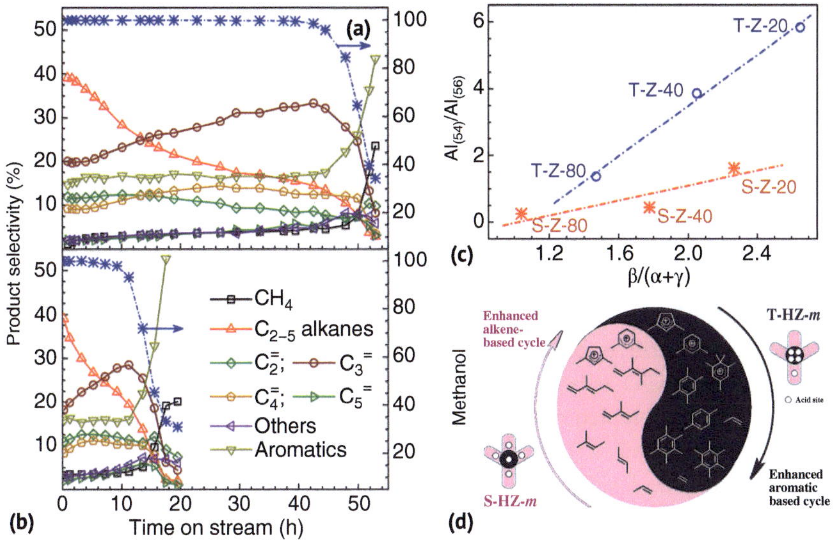

Figure 15.10 (a, b) MTO performance of different ZSM-5 catalysts, S-HZ-40 (a) and T-HZ-40 (b). Reaction conditions: T = 450 °C and WHSV = 4.23 h^{-1}. (c) Correlation between the $\beta/(\alpha + \gamma)$ ratio and the Al(54)/Al(56) ratio. (d) Connection between Al site positions within the ZSM-5 framework and the dual-cycle reaction mechanism. *Source*: Reproduced from [95] / with permission of American Chemical Society.

atoms at specific positions [62]. This strategy has also been applied to other zeolites. The incorporation of boron compels some Al atoms to relocate from the straight channels to the intersections in ZSM-11 [96]. This is supported by the enhanced production of aromatics on [B, Al]-ZSM-11 compared to ZSM-11 in the MTO reaction, which accelerates the aromatic cycle and ultimately results in increased ethylene production. Li et al. [97] reported that introducing boron into ZSM-5 led to a greater concentration of Al atoms within the straight and/or sinusoidal channels. This occurs because boron atoms preferentially occupy the T6 and T10 sites, which are located near or at the channel intersections. Consequently, ZSM-5 synthesized via the boron-assisted method exhibits a prolonged catalyst lifetime and improved propylene yield in the MTO reaction.

15.3.3 Effect of Zeolite Particle Size, Hierarchical Pore, and Morphology

Based on the HCP mechanism in the MTO reaction, polyMB$^+$ cations may undergo further conversion into large organic compounds, such as coke precursors, which accumulate within the channels or cavities, resulting in the coverage of active sites, diffusion limitations of products, and ultimately rapid catalyst deactivation. It has been demonstrated that reducing particle size, introducing hierarchical pore structures, and controlling zeolite morphology are effective strategies to suppress side reactions and reduce diffusion limitations. These approaches not only prolong catalyst lifetime but also improve light olefins selectivity.

15.3.3.1 Particle Size

Although cavity-containing small-pore zeolites show excellent light olefin selectivity in the MTO reaction, they deactivate quickly due to the accumulation of organic residues and coke deposition within their pores and cavities. Compared to micro-sized zeolites, nanosized zeolites exhibit enhanced diffusivity. Several approaches to synthesizing nanosized SAPO-34 have been proposed and summarized in recent reviews [98, 99]. To prepare nanosized zeolites, it is crucial to promote rapid and abundant nucleation during the initial pre-crystallization stage and/or to suppress crystal growth throughout the crystallization process. For example, Yang et al. [100] prepared small, cube-shaped SAPO-34 crystals (50–350 nm) through a post-synthesis method that combines milling and recrystallization. By using the milled SAPO-34 precursor as initial nuclei, the subsequent crystallization was significantly promoted. The nanosized SAPO-34 thus prepared demonstrated significantly enhanced MTO performance compared to the unprocessed sample. Similarly, nanoscale SAPO-5 and SAPO-35 were successfully prepared using this post-synthesis approach, indicating the broad applicability of the aforementioned method to SAPO zeolites.

Besides the post-synthesis method, nanosized SSZ-13 (50–200 nm) was prepared via a one-pot synthesis method by Li et al., with the addition of cetyltrimethylammonium bromide (CTAB) [101]. The nanosized SSZ-13 exhibited a reduced crystal size and abundant intracrystalline mesopores, which facilitate the mass transport of reactants and products, thereby leading to a significantly longer lifetime in the MTO reaction compared to its microsized counterpart. The nanosized SSZ-13 exhibited a higher C_2/C_3 ratio, suggesting a more prominent role of the aromatic cycle in the dual-cycle mechanism. Moreover, for nanosized SSZ-13, the hydrogen transfer reactions of olefins were effectively inhibited, as indicated by a low hydrogen transfer index (HTI, $C3/C_3^=$ or $C2/C_2^=$), thereby leading to a slower deactivation. Using this one-pot synthesis approach, a small-sized CON zeolite (Si/Al = 400) was prepared by Sawada et al. [102], achieving both a high propylene selectivity of up to 50% and an exceptionally long catalyst lifetime.

15.3.3.2 Hierarchical Pore

Hierarchically structured zeolites containing mesopores or macropores can significantly enhance the diffusion efficiency of reactants and products within the crystals, reduce the rate of catalyst deactivation, and thus prolong the catalyst lifetime. Several strategies have been provided for the synthesis of hierarchical SAPO-34, such as the "top-down" method (involving fluoride etching, basic or acidic etching, etc.) and the "bottom-up" method (using hard or soft templates, etc.) [98, 99]. For example, Ren et al. [103] synthesized hierarchical SAPO-34 through a simple tetraethylammonium hydroxide (TEAOH) etching post-treatment. The diffusion efficiency of reagents and products into and out of the crystal was greatly enhanced due to the presence of abundant micro-, meso-, and macroporous channels in the hierarchical SAPO-34, thereby reducing coke formation. As a result, the catalyst exhibited a notably extended lifetime in the MTO reaction, while maintaining high methanol conversion and light olefin selectivity. Park et al. [104] prepared a mesopore-containing [B, Al]-CON zeolite through alkaline treatment using an NH_4OH solution containing CTAB. The CON zeolite with mesopores exhibited a prolonged catalyst lifetime compared to the parent zeolite, while maintaining high propylene selectivity. The improved performance can be attributed to the combined influence of the alkaline treatment on both the pore structure and acidity adjustment.

15.3.3.3 Morphology

In recent years, nanosheet zeolites have gained increasing attention, as they offer easier separation compared to traditional nanoparticles. Their reduced diffusion path lengths lead to enhanced mass transfer performance, which can help suppress the secondary reactions of the formed olefins and minimize coke deposition during the MTO process. For example, Wu et al. [105] prepared sheet-like SAPO-34 (*ca.* 130 nm in thickness) under microwave conditions at elevated crystallization temperatures and in a short time. Compared with the hydrothermally synthesized cubic SAPO-34 (1.5–2.5 μm), the sheet-like SAPO-34 demonstrated an extended lifetime, higher ethylene selectivity (51.8%), and higher total light olefin selectivity (90.2%). Lei et al. [67] prepared aluminosilicate ITH nanosheet zeolite (30 nm in thickness) using a cationic oligomer as an OSDA, which exhibited enhanced propylene selectivity, extended lifetime, and similar hydrothermal stability in comparison to conventional ZSM-5.

Recently, Lu et al. [106] reported the successful synthesis of MOR nanosheets (*ca.* 11 nm in thickness) using an amphiphilic surfactant as the OSDA. This catalyst displayed a remarkably enhanced ethylene selectivity and extended lifetime compared to conventional bulk MOR (see Figure 15.11). This can be attributed to the prominent exposure of the (010) crystal planes resulting from thinning along the *b*-axis, as well as the existence of narrow 8-MR pores, both of which facilitate the mass transfer of ethylene. Additionally, the distribution of light olefins was found to be closely dependent on the morphology of zeolites. The MOR nanosheets exhibited an unexpectedly high selectivity to ethylene, whereas the conventional dealuminated

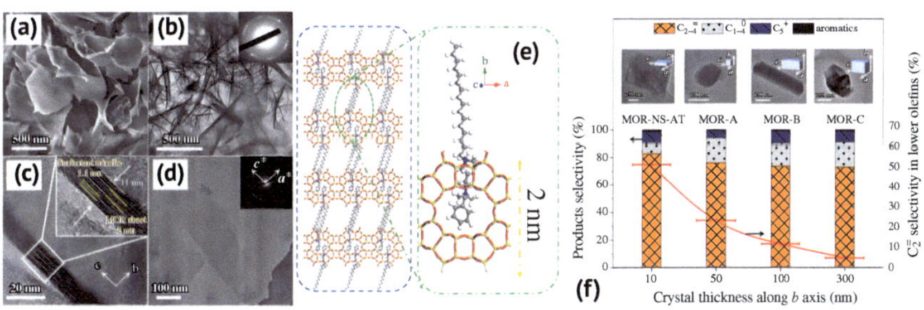

Figure 15.11 (a–d) SEM and TEM images of MOR-NS-P zeolite. (e) Schematic illustration of the MOR NS-P structure viewed along the *c*-axis. (f) Effect of the crystal thickness along the *b*-axis of MOR zeolites with different crystal morphologies on product distribution and ethylene selectivity in light olefins. Reaction conditions: T = 400 °C and WHSV = 1.0 h^{-1}. *Source*: Reproduced with permission from [106] / John Wiley & Sons.

MOR predominantly produced propylene and butene. These findings highlight that tailoring crystal morphology is a promising approach not only for prolonging the catalyst lifetime but also for adjusting the selectivity to desired light olefins.

15.3.4 Reaction Conditions

Apart from the physicochemical properties of zeolite catalysts, reaction conditions, including temperature, methanol WHSV, and co-feeding strategy, also significantly influence the MTO performance.

15.3.4.1 Temperature

Borodina et al. [107, 108] investigated the influence of reaction temperature on the nature of the active and deactivating species during the MTO reaction over SSZ-13 and SAPO-34, using a combination of operando UV–vis spectroscopy and online gas chromatography. For SAPO-34, the selectivity for propylene increases within the temperature range of 573–623 K but decreases when the temperature is increased from 623 to 773 K. The changes in olefin selectivity are related to the nature of HCP species at different temperatures. When the temperature ranges from 573 to 598 K, the active HCP species are predominantly composed of highly methylated benzene carbocations, which exhibit higher selectivity for propylene formation. In contrast, at higher temperatures ranging from 673 to 773 K, the formation of less methylated benzene carbocations and methylated naphthalene carbocations becomes more prevalent, promoting the production of ethylene. It has been observed that methylated naphthalene carbocations function as deactivating agents at lower temperatures by blocking micropores. However, at higher temperatures, they serve as active species, despite promoting the formation of extensive hydrocarbon deposits. Although SAPO-34 and SSZ-13 have similar active and deactivating species, the hydrocarbon species responsible for activity loss in SAPO-34 are more stable at higher temperatures. This difference can likely be attributed to the stronger acidity of SSZ-13, which facilitates the generation of more conjugated and bulky hydrocarbon species.

15.3.4.2 WHSV

Zhang et al. [109] investigated the impact of methanol WHSV on the MTO performance of ZSM-5 zeolite at 300 °C. The findings indicated that the extent of hydrogen transfer reactions is influenced by the WHSV (i.e., contact time), thereby affecting the balance between olefin and aromatic cycles. Specifically, lower WHSV and longer contact time facilitated hydrogen transfer reactions of olefins, promoting the formation and accumulation of MCPs retained in the catalyst. These MCPs contributed to the enhancement of the aromatic cycle, thereby increasing ethylene production. In contrast, higher WHSV and shorter contact time reduced the extent of hydrogen transfer reactions, suppressing the formation of MCPs. Consequently, the aromatic cycle weakened, and the olefin cycle became dominant, leading to increased production of C_3–C_7 olefins.

Shi et al. [110] demonstrated that, under comparable turnover numbers, reducing the partial pressure and WHSV of methanol during the MTO process on cage-type zeolites, such as SAPO-34, SSZ-13, and SSZ-39, enhances the cumulative ethylene-to-propylene ratio by *ca.* 2.2 to 2.4 times. The extent of MBs homologation among the trapped MBs was reduced due to the lower methanol concentration within the catalyst bed, thereby promoting the accumulation of aromatic precursors that favor ethylene production. These findings imply that the selectivity of light olefins is influenced not solely by the characteristics of the zeolite framework, but also significantly modulated by changes in the local methanol pressure, which may exert an equal or even greater effect.

15.3.4.3 Co-feeding

It is commonly observed that the introduction of water improves the stability of the catalyst during the MTO reaction. Recent studies indicate that water molecules may partially occupy acidic sites, thereby reducing the strong adsorption of methanol and olefins. This competitive adsorption reduces the reactivity of methanol and olefins, thereby inhibiting the rapid generation of large amounts of coke and enhancing catalyst stability. Moreover, for ZSM-58 (DDR) [111], the addition of water significantly enhances the selectivity toward light olefins, approaching *ca.* 100% with high methanol conversion. Recently, DeWispelaere et al. [112] clarified the role of water in the MTO reaction over SAPO-34 through a combination of advanced molecular dynamics simulations and *in situ* microspectroscopy analyses. Theoretical calculations indicate that the activation of methanol and propylene becomes less efficient, as these molecules compete for direct access to the BAS, resulting in reduced intrinsic methanol reactivity toward methoxide formation and thereby prolonging the induction period. This phenomenon is supported by *in situ* UV–vis microspectroscopy results. In addition, confocal fluorescence microscopy observations indicate that co-feeding water leads to a more uniform discoloration of zeolite during the reaction period. Water molecules occupy the acid sites, allowing methanol and olefins to diffuse more deeply into the crystal structure. Consequently, the coke species are distributed more evenly within the zeolite crystal, thereby enhancing the overall catalytic efficiency of the catalyst (see Figure 15.12).

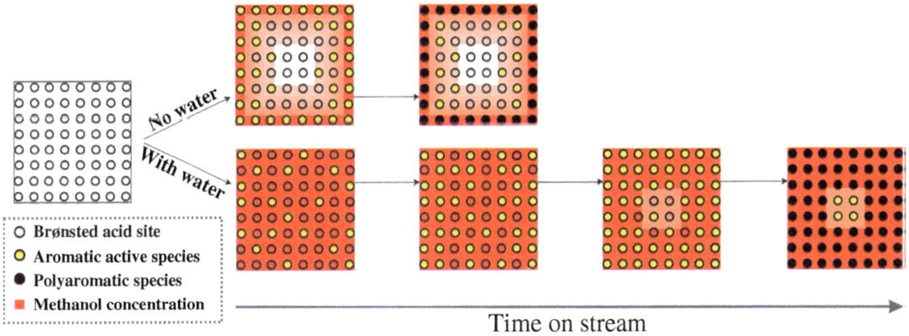

Figure 15.12 Schematic diagram of the influence of water addition on the MTO reaction and the formation of various aromatic and polyaromatic species. *Source:* Reproduced from Ref. [112] / with permission of American Chemical Society.

Very recently, Zhang et al. [113] found that in the MTO reaction over SAPO-34, water co-feeding achieved a remarkable ethylene selectivity of *ca.* 58%, while also extending the catalyst lifetime. *In situ* spectroscopic observations combined with isotope switching experiments demonstrated that water co-feeding promotes the accumulation of naphthyl species within SAPO-34 and enhances their reactivity. DFT calculations indicated that the water-mediated microenvironment plays a dual role: it not only enhances the activation of methanol but also facilitates proton transfer between the zeolite framework and organic intermediates, thereby reducing the energy barrier for ethylene production. Based on these findings, when naphthyl species were pre-accommodated in SAPO-34, water co-feeding resulted in an improved ethylene selectivity of 67.5%. However, the presence of water molecules induced irreversible hydrolysis of the P—O—Al bonds, resulting in partial degradation of the SAPO-34 structure and consequently reducing its catalytic durability after multiple regeneration processes.

Considerable advancements have recently been made in modulating the evolution of coke species during the MTO reaction, leading to a substantially extended catalyst lifetime. Arora et al. [114] first reported that the stability of the MTH reaction catalyzed by SAPO-34 could be remarkably enhanced (by a factor of 70) by co-feeding high-pressure H_2. They proposed that H_2 functioned as a hydrogen transfer reagent and intercepted the formation pathways of PAHs. A further mechanistic study [115] revealed that the higher proclivity of 1,3-butadiene to hydrogenation with H_2 hindered its further transformation and mitigated the formation of polycyclic species. Furthermore, Zhao et al. [116] demonstrated that when H_2 and H_2O were co-fed, a 200-fold increase in methanol processing capacity was achieved in the MTO reaction over SAPO-34. This notable enhancement can be attributed to the hydrogenation capacity of the BASs and the consequent retarded formation of heavy aromatic deposits.

15.3.4.4 Pre-coking

Although the single-pass lifetime of the catalyst can be significantly prolonged by water co-feeding, the zeolite framework partially collapses after regeneration cycles due to the irreversible hydrolysis of T—O—T bonds. However, Yang et al. [117] found that the hydrolysis of T—O—T bonds in SAPO-34 can be strongly inhibited by catalyst pre-coking, which was supported by 1D ^{27}Al, ^{31}P MAS NMR, and 2D ^{31}P-^{27}Al HETCOR NMR spectroscopy analysis. Therefore, combining catalyst pre-coking and water co-feeding represents an efficient strategy for enhancing both the catalytic efficiency and long-term stability of SAPO-34 catalysts in commercial MTO processes. However, rapid catalyst deactivation is insurmountable due to the coke formation. During the practical regeneration cycles of spent catalysts, coke is typically removed through air combustion or steam gasification to restore catalytic activity, which inevitably leads to the elimination of active HCPs (i.e., naphthalenic cations). Recently, Zhou et al. [118] proposed a novel regeneration approach that enables the directional transformation of coke into active HCPs within the SAPO-34 zeolite via steam cracking (see Figure 15.13). Under industrially relevant,

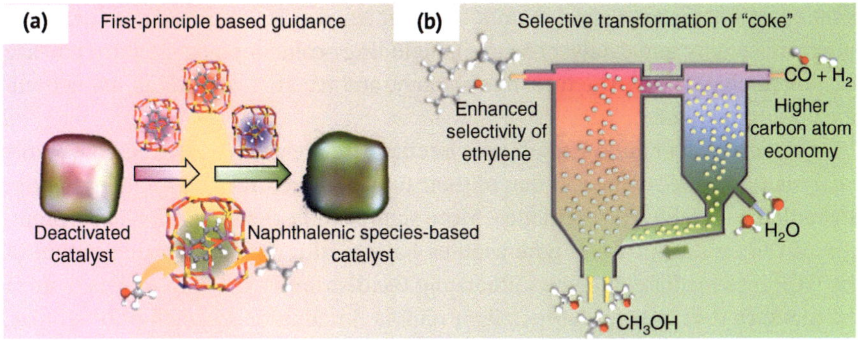

Figure 15.13 (a) Simulations based on first principles offer insights into the stability and functional behavior of organic intermediates encapsulated within nanoscale cavities. (b) Conversion of coke into specific naphthalenic-enriched catalytic species, along with enhanced MTO performance and atom utilization, through a circulating fluidized-bed reactor-regenerator system. *Source*: Reproduced from Ref. [118] / with permission of Springer Nature / CC BY 4.0.

continuous operating conditions, pilot-scale experiments conducted in a fluidized bed reactor-regenerator system demonstrated an unexpectedly high light olefins selectivity (85%). Additionally, the process yields 88% valuable synthesis gas components, namely CO and H_2, with minimal CO_2 produced as a regeneration byproduct. This approach thereby significantly enhances both the economic viability and environmental sustainability of the MTO process.

Moreover, a strategy integrating pre-coking and steaming was developed to directionally construct active naphthalenic species within the SAPO-34 catalyst [119]. This approach not only increases the selectivity of ethylene and propylene to *ca.* 89%, but also prolongs the catalyst lifetime by *ca.* 3.7-fold in the MTO reaction. The spatiotemporal distribution and evolution of carbonaceous species formed during the pre-coking and steaming processes were investigated using SIM, *in situ* UV–vis, online MS, and GC–MS. Pre-coking led to the formation and concentration of naphthalenes at both the rim and the center of the crystal, which promotes ethylene formation. Upon steaming, naphthalenes located at the crystal rim become more accessible to water molecules and are more prone to cracking at elevated temperatures, whereas those in the crystal center remain highly stable and are retained effectively. This method effectively inhibits the accumulation of coke compounds at the crystal rim, thereby prolonging the catalyst lifetime. Consequently, this feasible strategy can be readily applied to commercial MTO catalysts containing binders, showcasing significant potential for application in the current MTO industry.

15.4 Conclusions and Outlook

In summary, this study offers a comprehensive and up-to-date review of recent advances in the MTO reaction over zeolite catalysts, with a focus on both the reaction mechanism and the properties of zeolite catalysts. Based on the well-established reaction mechanism, various factors that influence MTO performance (i.e., light olefins selectivity and catalyst lifetime), including zeolite topology, acidity, particle size, hierarchical pore structure, morphology, and reaction conditions, are systematically summarized.

Firstly, cavity-containing small-pore zeolites are considered the most favorable catalysts for the MTO process due to their unique structural and acidic properties that enhance catalytic performance. More significantly, the dimensions and geometry of the cavities in small-pore zeolites have a critical impact on the nature of the HCP intermediate, thereby influencing the distribution of light olefins to meet the market's demands for various light olefins. In addition to small-pore zeolites, medium- or large-pore zeolites are also regarded as promising candidates for application in the MTO process, provided that their acidity, particle size, and morphology are appropriately tailored. Secondly, an appropriate balance between medium and strong BAS, together with specific LAS, as well as a higher concentration of isolated Al atoms, can significantly enhance the selectivity toward light olefins. Moreover, the catalytic lifetime can be extended by employing nanosized zeolites and/or nanosheet zeolites, which help reduce diffusion limitations. Finally, by fine-tuning reaction conditions, such as temperature, space velocity, co-feeding, and pre-coking strategies, the selectivity toward light olefins and the catalyst lifetime can be effectively optimized. It is noted that although significant progress has been made in improving the light olefins selectivity and prolonging the catalyst lifetime, achieving higher light olefin selectivity without compromising catalyst stability remains highly desirable.

References

1 Tian, P., Wei, Y.X., Ye, M. et al. (2015). Methanol to olefins (MTO): from fundamentals to commercialization. *ACS Catal.* 5: 1922–1938.

2 Stöcker, M. (1999). Methanol-to-hydrocarbons: catalytic materials and their behavior. *Micropor. Mesopor. Mater.* 29: 3–48.

3 Olsbye, U., Svelle, S., Bjørgen, M. et al. (2012). Conversion of methanol to hydrocarbons: how zeolite cavity and pore size controls product selectivity. *Angew. Chem. Int. Ed.* 51:5810–5831.

4 Ye, M., Li, H., Zhao, Y.F. et al. (2015). MTO processes development: the key of mesoscale studies. *Adv. Chem. Eng.* 47: 279–335.

5 Xu, S.T., Zhi, Y.C., Han, J.F. et al. (2017). Advances in catalysis for methanol-to-oefins conversion. *Adv. Catal.* 61: 37–122.

6 Yarulina, I., Chowdhury, Meirer, A.D.F. et al. (2018). Recent trends and fundamental insights in the methanol-to-hydrocarbons process. *Nat. Catal.* 1: 398–411.

7 Zhang, Q., Yu, J.H., and Corma., A. (2000). Applications of zeolites to C1 chemistry: recent advances, challenges, and opportunities. *Adv. Mater.* 32: e2002927.

8 Zhong, J.W., Han, J.F., Wei, Y.X. et al. (2021). Catalysts and shape selective catalysis in the methanol-to-olefin (MTO) reaction. *J. Catal.* 396: 23–31.

9 Wang, S., Qin, Z.F., Dong, M. et al. (2022). Recent progress on MTO reaction mechanisms and regulation of acid site distribution in the zeolite framework. *Chem. Catal.* 2: 1657–1685.

10 Lin, S.F., Zhi, Y.C., Liu, Z.Q. et al. (2022). Multiscale dynamical cross-talk in zeolite-catalyzed methanol and dimethyl ether conversions. *Natl. Sci. Rev.* 9: nwac151.

11 Li, J.L., Gao, M.K., Yan, W.F. et al. (2023). Regulation of the Si/Al ratios and Al distributions of zeolites and their impact on properties. *Chem. Sci.* 14: 1935–1959.

12 Wang, Y., Yokoi, T., and Tatsumi, T. (2023). Selective production of light olefins over zeolite catalysts: impacts of topology, acidity, and particle size. *Microp. Mesop. Mater.* 358: 112353.

13 Zhang, W.N., Lin, S.F., Wei, Y.X. et al. (2023). Cavity-controlled methanol conversion over zeolite catalysts. *Natl. Sci. Rev.* 10: nwad120.

14 Wang N., Wei, Y.X., and Liu, Z.M. (2023). Methanol to olefins (MTO): a condensed matter chemistry. *Prog. Chem.* 35: 839–860.

15 Wang, W., Buchholz, A., Seiler, M. et al. (2003). Evidence for an initiation of the methanol-to-olefin process by reactive surface methoxy groups on acidic zeolite catalysts. *J. Am. Chem. Soc.* 125: 15260–15267.

16 Yamazaki, H., Shima H., Imai, H. et al. (2011). Evidence for a “carbene-like” intermediate during the reaction of methoxy species with light alkenes on H-ZSM-5. *Angew. Chem. Int. Ed.* 50: 1853–1856.

17 Yamazaki, H., Shima, H., Imai, H. et al. (2012). Direct production of propene from methoxy species and dimethyl ether over H-ZSM-5. *J. Phys. Chem. C* 116: 24091–24097.

18 Chowdhury, A.D., Houben, K., Whiting, G.T. et al. (2016). Initial carbon–carbon bond formation during the early stages of the methanol-to-olefin process proven by zeolite-trapped acetate and methyl acetate. *Angew. Chem. Int. Ed.* 55: 15840–15845.

19 Li, J.F., Wei, Z.H., Chen, Y.Y. et al. (2014). A route to form initial hydrocarbon pool species in methanol conversion to olefins over zeolites. *J Catal.* 317: 277–283.

20 Peng, C., Wang, H., and Hu, P. (2016). Theoretical insights into how the first C–C bond forms in the methanol-to-olefin process catalysed by HSAPO-34. *Phys. Chem. Chem. Phys.* 18: 14495–14502.

21 Wu, X.Q., Xu, S.T., Zhang, W.N. et al. (2017). Direct mechanism of the first carbon–carbon bond formation in the methanol-to-hydrocarbons process. *Angew. Chem. Int. Ed.* 56: 9039–9043.

22 Wu, X., Xu, S., Wei, Y. et al. (2018). Evolution of C–C bond formation in the methanol-to-olefins process: from direct coupling to autocatalysis, *ACS Catal.* 8: 7356–7361.

23 Sun, T., Chen, W., Xu, S. et al. (2021). The first carbon-carbon bond formation mechanism in methanol-to-hydrocarbons process over chabazite zeolite. *Chem.* 7: 2415–2428.

24 Liu, Y., Muller, S., Berger, D. et al. (2016). Formation mechanism of the first carbon–carbon bond and the first olefin in the methanol conversion into hydrocarbons. *Angew. Chem. Int. Ed.* 55: 5723–5726.

25 Plessow, P.N. and Studt, F. (2017). Unraveling the mechanism of the initiation reaction of the methanol to olefins process using *ab Initio* and DFT calculations. *ACS Catal.* 7: 7987–7994.

26 Comas-Vives, A., Valla, M., Copéret, C. et al. (2015). Cooperativity between Al sites promotes hydrogen transfer and carbon-carbon bond formation upon dimethyl ether activation on alumina. *ACS Cent. Sci.* 1: 313–319.

27 Wang, C., Chu, Y., Xu, J. et al. (2018). Extra-framework aluminum-assisted initial C—C bond formation in methanol-to-olefins conversion on zeolite H-ZSM-5. *Angew. Chem. Int. Ed.* 57: 10197–10201.

28 Chu, Y., Yi, X., Li, C. et al. (2018). Brønsted/Lewis acid sites synergistically promote the initial C—C bond formation in the MTO reaction. *Chem. Sci.* 9: 6470–6479.

29 Dahl, I.M. and Kolboe, S. (1996). On the reaction mechanism for hydrocarbon formation from methanol over SAPO-34: 2. Isotopic labeling studies of the co-reaction of propene and methanol. *J. Catal.* 161: 304–309.

30 Wang, C., Chu, Y., Zheng, A. et al. (2014). New insight into the hydrocarbon-pool chemistry of the methanol-to-olefins conversion over zeolite H-ZSM-5 from GC-MS, solid-state NMR spectroscopy, and DFT calculations. *Chem. Eur. J.* 20: 12432–12443.

31 Li, J., Wei, Y., Chen, J. et al. (2012). Observation of heptamethylbenzenium cation over SAPO-type molecular sieve DNL-6 under real MTO conversion conditions. *J. Am. Chem. Soc.* 134: 836–839.

32 Xu, S.T., Zheng, A.M., Wei, Y.X. et al. (2013). Direct observation of cyclic carbenium ions and their role in the catalytic cycle of the methanol-to-olefin reaction over chabazite zeolites. *Angew. Chem. Int. Ed.* 52: 11564–11568.

33 Li, J.Z., Wei, Y.X., Chen, J.R. et al. (2015). Cavity controls the selectivity: insights of confinement effects on MTO reaction. *ACS Catal.* 5: 661–665.

34 Dai, W.L., Wang, C.M., Dyballa, M. et al. (2015). Understanding the early stages of the methanol-to-olefin conversion on H-SAPO-34. *ACS Catal.* 5: 317–326.

35 Zhang, W., Zhi, Y., Huang, J. et al. (2019). Methanol to olefins reaction route based on methylcyclopentadienes as critical intermediates. *ACS Catal.* 9: 7373–7379.

36 Zhang, W., Zhang, M., Xu, S. et al. (2020). Methylcyclopentenyl cations linking initial stage and highly efficient stage in methanol-to-hydrocarbon process. *ACS Catal.* 10: 4510–4516.

37 Hu, M., Wang, C., Gao, X. et al. (2020). Establishing a link between the dual cycles in methanol-to-olefins conversion on H-ZSM-5: aromatization of cycloalkenes. *ACS Catal.* 10: 4299–4305.

38 Iliasa, S., Khare, R., Malek, A. et al. (2013). A descriptor for the relative propagation of the aromatic-and olefin-based cycles in methanol-to-hydrocarbons conversion on H-ZSM-5. *J. Catal.* 303: 135—140.

39 Lesthaeghe, D., Horre, A., Waroquier, M. et al. (2009). Theoretical insights on methylbenzene side-chain growth in ZSM-5 zeolites for methanol-to-olefin conversion. *Chem. Eur. J.* 15: 10803—10808.

40 Ferri, P., Li, C.G., Paris, C. et al. (2019). Chemical and structural parameter connecting cavity architecture, confined hydrocarbon pool species, and MTO product selectivity in small-pore cage-based zeolites. *ACS Catal.* 9: 11542–11551.

41 Ferri, P., Li, C.G., Mill'an, R. et al. (2020). Impact of zeolite framework composition and flexibility on methanol-to-olefins selectivity: confinement or diffusion? *Angew. Chem. Int. Ed.* 29: 19708–19715.

42 Gao, S., Xu, S., Wei, Y. et al. (2018). Insight into the deactivation mode of methanol-to-olefins conversion over SAPO-34: coke, diffusion, and acidic site accessibility. *J Catal.* 367: 306–314.

43 Wennmacher, J.T.C., Mahmoudi, S., Rzepka, P. et al. (2022). Electron diffraction enables the mapping of coke in ZSM-5 micropores formed during methanol-to-hydrocarbons conversion. *Angew. Chem. Int. Ed.* 61: e202205413.

44 Wang, C., Hu, M., Chu, Y. et al. (2020). π-interactions between cyclic carbocations and aromatics cause zeolite deactivation in methanol-to-hydrocarbon conversion. *Angew. Chem. Int. Ed.* 59: 7198–7202.

45 Wang, N., Zhi, Y., Wei, Y. et al. (2020). Molecular elucidating of an unusual growth mechanism for polycyclic aromatic hydrocarbons in confined space. *Nat. Commun.* 11: 1079–1091.

46 Fan, S., Wang, H., Wang, P. et al. (2024). Formation and evolution of the coke precursors on the zeolite catalyst in the conversion of methanol to olefins. *Chem. Catal.* 4: 100927.

47 Dusselier, M., Deimund, M.A., Schmidt, J.E. et al. (2015). Methanol-to-olefins catalysis with hydrothermally treated zeolite SSZ-39. *ACS Catal.* 5: 6078–6085

48 Yarulina, I., Goetze, J., Gücüyener, C. et al. (2016). Methanol-to-olefins process over zeolite catalysts with DDR topology: effect of composition and structural defects on catalytic performance. *Catal. Sci. Technol.* 6: 2663–2678.

49 Yokoi, T., Yoshioka, M., Imai, H. et al. (2009). Diversification of RTH-type zeolite and its catalytic application. *Angew. Chem. Int. Ed.* 48: 9884–9887.

50 Chen, J.R., Li, J.Z., Wei, Y.X. et al. (2014). Spatial confinement effects of cage-type SAPO molecular sieves on product distribution and coke formation. *Catal. Commun.* 46: 36–40.

51 Alshafei, F.H., Park, Y., Zones, S.I. et al. (2021). Methanol-to-olefins catalysis on ERI-type molecular sieves: towards enhancing ethylene selectivity. *J. Catal.* 404: 620–633.

52 Deimund, M.A., Schmidt, J.E., and Davis, M.E. (2015). Effect of pore and cage size on the formation of aromatic intermediates during the methanol-to-olefins reaction. *Top Catal.* 58: 416–423.

53 Bhawe, Y., Moliner, M., Lunn, J.D. et al. (2012). Effect of cage size on the selective conversion of methanol to light olefins. *ACS Catal.* 2: 2490–2495.

54 Li, X., Sun, Q.M., Li, Y. et al. (2014). Confinement effect of zeolite cavities on methanol-to-olefin conversion: a density functional theory study. *J. Phys. Chem. C* 118: 24935–24940.

55 Kang, J.H., Walter, R., Xie, D. et al. (2018). Further studies on how the nature of zeolite cavities that are bounded by small pores influences the conversion of methanol to light olefins. *ChemPhysChem* 19: 412–419.

56 Kang, J.H., Alshafei, F.H., Zones, S.I. et al. (2019). Cage-defining ring: a molecular sieve structural indicator for light olefin product distribution from the methanol-to-olefins reaction. *ACS Catal.* 9: 6012–6019.

57 Zhang, W.N., Chen, J.R, Xu, S.T. et al. (2018). Methanol to olefins reaction over cavity-type zeolite: cavity controls the critical intermediates and product selectivity. *ACS Catal.* 8: 10950–10963.

58 Li, C.G., Paris, C., Martínez-Triguero, J. et al. (2018). Synthesis of reaction-adapted zeolites as methanol-to-olefins catalysts with mimics of reaction intermediates as organic structure-directing agents. *Nat. Catal.* 1: 547–554.

59 Yang, M., Li, B., Gao, M.B. et al. (2020). High propylene selectivity in methanol conversion over a small pore SAPO molecular sieve with ultra-small cage. *ACS Catal.* 10: 3741–3749.

60 Teketel, S., Skistad, W., Benard, S. et al. (2012). Shape selectivity in the conversion of methanol to hydrocarbons: the catalytic performance of one-dimensional 10-ring zeolites: ZSM-22, ZSM-23, ZSM-48, and EU-1. *ACS Catal.* 2: 26–37.

61 Wei, F., Cui, Z.M., Meng, X.J. et al. (2014). Origin of the low olefin production over HZSM-22 and HZSM-23 zeolites: external acid sites and pore mouth catalysis. *ACS Catal.* 4: 529–534.

62 Chen, J.L., Liang, T.Y., Li, J.F. et al. (2016). Regulation of framework aluminum siting and acid distribution in H-MCM-22 by boron incorporation and its effect on the catalytic performance in methanol to hydrocarbons. *ACS Catal.* 6: 2299–2313.

63 Shen, Y.F., Le, T.T., Fu, D.L. et al. (2018). Deconvoluting the competing effects of zeolite framework topology and diffusion path length on methanol to hydrocarbons reaction. *ACS Catal.* 8: 11042–11053.

64 Skistad, W., Teketel, S., Bleken, F.L. et al. (2014). Methanol conversion to hydrocarbons (MTH) over H-ITQ-13 (ITH) zeolite. *Top Catal.* 57: 143–158.

65 Ma, H., Chen, Y.Y., Wang, S. et al. (2018). Reaction mechanism for the conversion of methanol to olefins over H-ITQ-13 zeolite: a density functional theory study. *Catal. Sci. Technol.* 8: 521–533.

66 Zeng, P.H., Zhang, L.L., Guo, X.Z. et al. (2017). Catalytic performances of Al-ITQ-13 zeolites with different $SiO_2/A1_2O_3$ ratios in the conversion of methanol to propene. *J. Fuel. Chem. Technol.* 45: 1349–1355.

67 Lei, C., Dong, Z.Y., Martnez, C. et al. (2020). A cationic oligomer as an organic template for direct synthesis of aluminosilicate ITH zeolite. *Angew. Chem. Int. Ed.* 59: 15649–15655.

68 Ma, Y., Hu, J.Y., Fan, K. et al. (2023). Design of an organic template for synthesizing ITR zeolites under Ge-free conditions. *J. Am. Chem. Soc.* 145: 17284–17291.

69 Otomo, R., Müller, U., Feyen, M. et al. (2016). Development of a post-synthetic method fortuning the Al content of OSDA-free beta as a catalyst for conversion of methanol to olefins. *Catal. Sci. Technol.* 6: 713–721.

70 Yoshioka, M., Yokoi, T., and Tatsumi, T. (2015). Development of the CON-type aluminosilicate zeolite and its catalytic application for the MTO reaction. *ACS Catal.* 5: 4268–4275.

71 Hong, X., Chen, W., Zhang, G.Q. et al. (2019). Direct synthesis of aluminosilicate IWR zeolite from a strong interaction between zeolite framework and organic template. *J. Am. Chem. Soc.* 141: 18318–18324.

72 Ren, L., Wang, B.W., Lu, K. et al. (2021). Selective conversion of methanol to propylene over highly dealuminated mordenite: Al location and crystal morphology effects. *Chin. J. Catal.* 42: 1147–1159.

73 Park, S., Sato, G., Osuga, R. et al. (2021). Methanol-to-olefins reaction over large-pore zeolites: impact of pore structure on catalytic performance, *Chem. Ing. Tech.* 93: 990–1000.

74 Zhu, Q.J., Kondo, J.N., Ohnuma, R. et al. (2008). The study of methanol-to-olefin over proton type aluminosilicate CHA zeolites. *Micropor. Mesopor. Mater.* 112: 153–161.

75 Yarulina, I., Bailleul, S., Pustovarenko, A. et al. (2016). Suppression of the aromatic cycle in MTO reaction over ZSM-5 by post-synthetic modification using calcium. *ChemCatChem* 8: 3057–3063.

76 Liutkova, A., Zhang, H., Simons, J.F.M. et al. (2023). Ca cations impact the local environment inside HZSM-5 pores during the methanol-to-hydrocarbons reaction. *ACS Catal.* 13: 3471–3484.

77 Erichsen, M.W., Svelle, S., and Olsbye, U. (2013). The influence of catalyst acid strength on the methanol to hydrocarbons (MTH) reaction. *Catal. Today* 215: 216–223.

78 Zhu, Q.J., Hinode, M., Yokoi, T. et al. (2008). Methanol-to- olefin over gallosilicate analogues of chabazite zeolite, *Micropor. Mesopor. Mater.* 116: 253–257.

79 Park, S., Sato, G., Nishitoba, T. et al. (2020). Synthesis of Ga-containing CON-type material and its catalytic performance in methanol-to-olefins reaction. *Catal. Today* 352: 175–182.

80 Zhao, S.F., Yao, X.T., Yan, B.H. et al. (2017). Flexible regulation of $C_3^=/C_2^=$ ratio in methanol-to-hydrocarbons by delicate control of acidity of ZSM-5 catalyst. *Chin. Chem. Lett.* 28: 1318–1323.

81 Yarulina, I., De Wispelaere, K., Bailleul, S. et al. (2018). Structure–performance descriptors and the role of Lewis acidity in the methanol-to-propylene process. *Nat. Chem.* 10: 804–812.

82 Chowdhury, A.D., Yarulina, I., Abou-Hamad, E. et al. (2019). Surface enhanced dynamic nuclear polarization solid-state NMR spectroscopy sheds light on Brønsted–Lewis acid synergy during the zeolite catalyzed MTH. *Chem. Sci.* 10: 8946–8954.

83 Deimund, M.A., Harrison, L., Lunn, J.D. et al. (2016). Effect of heteroatom concentration in SSZ-13 on the methanol-to-olefins reaction. *ACS Catal.* 6: 542–550.

84 Wang, X.X., Wang, C., Chu, Y.Y. et al. (2024). Deciphering the link between zeolite crystal size, Brønsted acid site distribution, and dual-cycle selectivity in methanol-to-olefins. *ACS Catal.* 14: 15609–15621.

85 Nishitoba, T., Yoshida, N., Kondo, J.N. et al. (2018). Control of Al distribution in the CHA-Type aluminosilicate zeolites and its impact on the hydrothermal stability and catalytic properties. *Ind. Eng. Chem. Res.* 57: 3914–3922.

86 Gallego, E.M., Li, C.G., Paris, C. et al. (2018). Making nanosized CHA zeolites with controlled Al distribution for optimizing methanol-to-olefin performance. *Chem. Eur. J.* 24: 14631–14635.

87 Wang, S., Wei, Z., Chen, Y. et al. (2015). Methanol to olefins over H-MCM-22 zeolite: theoretical study on the catalytic roles of various pores *ACS Catal.* 5: 1131–1144.

88 Liu, M., Yokoi, T., Yoshioka, M. et al. (2014). Differences in Al distribution and acidic properties between RTH-type zeolites synthesized with OSDAs and without OSDAs. *Phys. Chem. Chem. Phys.* 16: 4155–4164.

89 Zhang, L., Wang, S., Shi, D.Z. et al. (2020). Methanol to olefins over H-RUB-13 zeolite: regulation of framework aluminum siting and acid density and their relationship to the catalytic performance. *Catal. Sci. Technol.* 10: 1835–1847.

90 Fan, K., Wu, Q.M., Liu, S. et al. (2024). Adjusting Al location in the framework of ITH zeolites for catalytic conversion of methanol to olefins. *Chin. J. Catal.* 56: 114–121.

91 Zhao, X.B., Wang, L.Y., Li, J.Z. et al. (2017). Investigation of methanol conversion over high-Si beta zeolites and the reaction mechanism of their high propene selectivity. *Catal. Sci. Technol.* 7: 5882–5589.

92 Yokoi, T., Mochizuki, H., Namba, S. et al. (2015). Control of the Al distribution in the framework of ZSM-5 zeolite and its evaluation by solid-state NMR technique and catalytic properties. *J. Phys. Chem. C.* 119: 15303–15315.

93 Yokoi, T., Mochizuki, H., Biligetu, T. et al. (2017). Unique Al distribution in the MFI framework and its impact on catalytic properties. *Chem. Lett.* 46: 798–800.

94 Biligetu, T., Wang, Y., Nishitoba, T. et al. (2017). Al distribution and catalytic performance of ZSM-5 zeolites synthesized with various alcohols. *J. Catal.* 353: 1–10.

95 Liang, T.Y., Chen, J.L., Qin, Z.F. et al. (2016). Conversion of methanol to olefins over H-ZSM-5 Zeolite_Reaction pathway is related to the framework aluminum siting. *ACS Catal.* 6: 7311–7325.

96 Wang, S., Li, S.Y., Zhang, L. et al. (2017). Insight into the effect of incorporation of boron into ZSM-11 on its catalytic performance for conversion of methanol to olefins. *Catal. Sci. Technol.* 7: 4766–4779.

97 Li, C.G., Vidal-Moya, A., Miguel, P.J. et al. (2018). Selective introduction of acid sites in different confined positions in ZSM-5 and its catalytic implications. *ACS Catal.* 8: 7688–7697.

98 Zhong, J.W., Han, J.F., Wei, Y.X. et al. (2017). Recent advances of the nano-hierarchical SAPO-34 in the methanol-to-olefin (MTO) reaction and other applications. *Catal. Sci. Technol.* 7: 4905–4923.

99 Sun, Q.M., Xie, Z.K., Yu, J.H. (2018). The state-of-the-art synthetic strategies for SAPO-34 zeolite catalysts in methanol-to-olefin conversion. *Natl. Sci. Rev.* 5: 542–558.

100 Yang, M., Tian, P., Wang, C. et al. (2014). A top-down approach to prepare silicoaluminophosphate molecular sieve nanocrystals with improved catalytic activity. *Chem. Commun.* 50: 1845–1847.

101 Li, Z., Navarro, M.T., Martínez-Triguero, J. et al. (2016). Synthesis of nano-SSZ- 13 and its application in the reaction of methanol to olefins. *Catal. Sci. Technol.* 6: 5856–5863.

102 Sawada, M., Matsumoto, T., Osuga, R. et al. (2022). Surfactant-assisted direct crystallization of CON-type zeolites with particle size and acid-site location controlled. *Ind. Eng. Chem. Res.* 61: 1733–1747.

103 Ren, S., Liu, G.J., Wu, X. et al. (2017). Enhanced MTO performance over acid treated hierarchical SAPO-34. *Chin. J. Catal.* 38: 123–130.

104 Park, S., Sato, G., Onozuk, H. et al. (2020). Development of mesopore-containing CON-type zeolite with unique acidic and catalytic properties. *Catal. Sci. Technol.* 10: 4293–4304.

105 Wu, L., Liu, Z.Y., Xia, L. et al. (2013). Effect of SAPO-34 molecular sieve morphology on methanol to olefins performance. *Chin. J. Catal.* 34: 1348–1356.

106 Lu, K., Huang, J., Ren, L. et al. (2020). High ethylene selectivity in methanol-to-olefin (MTO) reaction over MOR-zeolite nanosheets. *Angew. Chem. Int. Ed.* 59: 6258–6262.

107 Borodina, E., Sharbini Harun Kamaluddin, H., Meirer, F. et al. (2015). Influence of the reaction temperature on the nature of the active and deactivating species during methanol to olefins conversion over H-SSZ-13. *ACS Catal.* 5: 992–1003.

108 Borodina, E., Sharbini Harun Kamaluddin, H., Meirer, F. et al. (2017). Influence of the reaction temperature on the nature of the active and deactivating species during methanol-to-olefins conversion over H-SAPO-34. *ACS Catal.* 7: 5268–5281.

109 Zhang, M.Z., Xu, S.T., Wei, Y.X. et al. (2016). Changing the balance of the MTO reaction dual-cycle mechanism Reactions over ZSM-5 with varying contact times. *Chin. J. Catal.* 37: 1413–1422.

110 Shi, Z.C. and Bhan, A. (2021). Tuning the ethylene-to-propylene ratio in methanol-to-olefins catalysis on window-cage type zeolites. *J. Catal.* 395: 266–272.

111 Kumita, Y., Gascon, J., Stavitski, E. et al. (2011). Shape selective methanol to olefins over highly thermostable DDR catalysts. *Appl. Catal. A.* 391: 234–243.

112 DeWispelaere, K., Wondergem, C.S., Ensing, B. et al. (2016). Insight into the effect of water on the methanol-to-olefins conversion in H-SAPO-34 from molecular simulations and in situ microspectroscopy. *ACS Catal.* 6: 1991–2002.

113 Zhang, C.W., Wu, X.Q., Zhang, Y. et al. (2024). Water-assisted shape-selective production of ethene in methanol-to-olefins reaction on SAPO-34. *Chem. Catal.* 4: 101025.

114 Arora, S.S., Nieskens, D.L.S., Malek, A. et al. (2018). Lifetime improvement in methanol-to-olefins catalysis over chabazite materials by high-pressure H_2 co-feeds. *Nat. Catal.* 1: 666–672.

115 Arora, S.S., Shi, Z.C., and Bhan, A. (2019). Mechanistic basis for effects of high-pressure H_2 cofeeds on methanol-to-hydrocarbons catalysis over zeolites. *ACS Catal.* 9: 6407–6414.

116 Zhao, X.B., Li, J.Z., Tian, P. et al. (2019). Achieving a superlong lifetime in the zeolite-catalyzed MTO reaction under high pressure: synergistic effect of hydrogen and water. *ACS Catal.* 9: 3017–3025.

117 Yang, L., Wang, C., Zhang, L. et al. (2021). Stabilizing the framework of SAPO-34 zeolite toward long-term methanol-to-olefins conversion. *Nat. Commun.* 12: 4661.

118 Zhou, J., Gao, M., Zhang, J. et al. (2021). Directed transforming of coke to active intermediates in methanol-to-olefins catalyst to boost light olefins selectivity. *Nat. Commun.* 12: 17.

119 Wang, C., Yang, L., Gao, M. et al. (2022). Directional construction of active naphthalenic species within SAPO-34 crystals toward more efficient methanol-to-olefin conversion. *J. Am. Chem. Soc.* 144: 21408–21416.

Part VII

Other Reactions

16

Application of Zeolite in Dehydrogenation and Hydrogenation of Hydrocarbons

Yupeng Tian[1], Xinmei Liu[2,]*

[1]*State Key Laboratory of Chemical Safety, SINOPEC Research Institute of Safety Engineering Co. Ltd., Qingdao, China.*
[2]*State Key Laboratory of Heavy Oil Processing, College of Chemistry and Chemical Engineering, China University of Petroleum (East China), Qingdao, China*

*Corresponding author: lxmei@upc.edu.cn

16.1 Introduction

Zeolite has been playing a pivotal role in the refining of crude oil and chemical production. Its primary applications in refining include fluid catalytic cracking (FCC), alkylation, hydrocracking, and isomerization processes. These applications help to improve the efficiency and selectivity of converting crude oil into valuable products such as gasoline, diesel, and other refined fuels. In the field of petrochemicals, zeolites are widely used as catalysts in various important reactions. Examples include hydroxylation reactions, such as phenol synthesis; alkylation reactions, such as the production of ethylbenzene and isopropylene; oximation reactions, such as cyclohexanone oxime synthesis; and epoxidation reactions, such as propylene oxide synthesis. These catalytic properties make zeolites essential materials in modern chemical manufacturing.

Precisely positioning acidic centers within well-defined porous frameworks in industrial catalytic systems is crucial for the control of molecular-level reaction processes. These engineered structures allow for accurate regulation of the diffusion of reactants, products, and intermediate transition species through their shape-selective features. Modern catalytic systems are increasingly utilizing multifunctional zeolite materials, in which metal species are deliberately incorporated as nanoscale particles within the crystalline framework, thereby introducing complementary catalytic functions such as hydrogenation and dehydrogenation. The combination of nanoscience with catalytic technology has driven significant advancements in material design and heterogeneous catalytic processes. Compared to traditional bulk metals, subnanoscale metal catalysts exhibit significantly enhanced surface area, unique geometric configuration, and novel electronic structure. These distinct physicochemical characteristics place them at the forefront of catalytic research, especially in the development of precisely engineered supports containing isolated metal atoms, well-defined clusters, and tailored nanoparticles. The field has witnessed exponential growth over the past few decades, driven by the continuous pursuit of improving catalytic efficiency and selectivity in heterogeneous reaction systems.

Due to their exceptional catalytic performance, these species have been widely employed to facilitate the activation of small molecules, like H_2, O_2, CH_4, CO_2, CO, and various organic functional groups. However, as the size of the metal species decreases, their surface energy increases significantly. This phenomenon leads to thermodynamic instability of traditional supported metal particles on open-structured supports during high-temperature reactions, which further results in aggregation of particles and accelerated catalyst deactivation. Such inherent instability poses a major limitation to their industrial application in chemical processes. To overcome these stability issues, new strategies have emerged which focus on encapsulating active metal sites within porous frameworks or cavities, achieving precise spatial confinement of catalytic centers. The migration and aggregation of metal particles can be effectively suppressed by controlling their size and spatial distributions, and the thermal stability during catalytic reactions is consequently significantly enhanced. This structural feature also improves the resistance to sintering, ensuring long-term catalytic stability. Therefore, zeolites have emerged as promising supports for encapsulating metal components. Particularly, zeolites with well-defined crystalline frameworks exhibit a remarkable ability to stabilize metal nanoparticles through spatial confinement within their unique pore architectures.

Zeolites with diverse structural frameworks exhibit distinct nanoscale confinement properties within their porous systems. These characteristics allow for precise control over metal localization, spatial orientation, and atomic coordination environments. The tailoring of catalytic performance for specific chemical transformations is enabled. The synergistic integration of catalytically active metal centers with the zeolite matrix combines the advantages of both components. The atomic-level dispersion of metal species enhances the efficiency of catalytic reactions, while the inherent thermal stability of the crystalline zeolite framework minimizes metal leaching and thermal agglomeration. Importantly, molecular shape selectivity and sequential reaction pathways can be regulated through deliberate design of the zeolite architectures, and framework chemical composition. This unique capability positions metal-doped zeolite catalysts as superior alternatives to conventional supported metal catalysts in heterogeneous catalysis.

Undoubtedly, recent advances in synthesis methods and characterization techniques have significantly promoted the development of these materials. The latest progress in metal-zeolite catalytic systems can be attributed to two key factors: advanced characterization tools now enable atomic-scale detection of metal species within zeolite frameworks, and innovative synthetic approaches allow for the precise engineering of well-defined active metal centers. This chapter reviews hydrocarbon conversion processes catalyzed by metal-zeolite catalysts, with a focus on hydrogen transfer reactions involving hydrogenation and dehydrogenation. Particular attention is given to the synergistic effects observed in encapsulated metal-zeolite materials during these reactions.

16.2 Dehydrogenation Process

Conventional methods for producing light olefins primarily rely on steam cracking and FCC processes. With the growing market demand for propylene and other light olefins, there has been a significant promotion towards advancing dedicated production technologies. The industrial catalytic dehydrogenation of alkanes has been in operation since the 1930s, although these reactions confront considerable thermodynamic barriers. The dehydrogenation of propane to propylene and isobutane to isobutene are the main applications. Due to the highly endothermic nature of these reactions, elevated temperatures are required to achieve favorable thermodynamic conversion.

16.2.1 Direct Dehydrogenation of Light Alkanes

Despite the limited industrial application of zeolite-based catalytic systems to date, these materials have demonstrated significant potential as alternatives to conventional non-microporous supported dehydrogenation catalysts. Catalyst supports typically consist of low-acid or neutral materials, such as silica matrices, alkali-modified alumina substrates, or thermally treated hydrotalcite derivatives. These materials effectively suppress undesirable acid-catalyzed side reactions, such as isomerization and coke formation. The catalytic activity and stability are significantly influenced by the dispersion of the metal and the size of the nanoparticles. The dehydrogenation reaction mechanism has also been intensively studied to provide theoretical guidance for the optimization of the dehydrogenation performance of zeolite-based catalysts. Ivanova and co-workers elucidated the propane activation mechanism in the bifunctional Ga/Zn-modified H-ZSM-5 system using ^{13}C and ^{1}H MAS NMR techniques, providing key insights into the initial stages of propane dehydrogenation (PDH) [1]. Notably, optimal olefin selectivity is achieved when Brønsted acid sites (BAS) are minimized, as these sites promote undesirable side reactions such as oligomerization and coke formation. Iglesia et al. have carried out extensive studies on the mechanism of alkane activation using H-ZSM-5 catalysts in dehydrogenation reactions [2]. Their experimental investigation into the ion exchange behavior of H-ZSM-5 with zinc, cobalt, and gallium has yielded valuable insights into the underlying catalytic mechanism. The study demonstrated that the rate-determining step does not involve C—H bond activation, but rather the desorption of the generated alkyl species. This conclusion is consistent with the conclusions of Meusinger on alkane cracking using acidic zeolite catalysts [3]. Moreover, structural and compositional analyses of the zeolite framework suggest that the stability of these alkyl intermediates is closely correlated with the electronic properties and three-dimensional configuration of BAS. The acidic zeolites can promote propylene production through a monomolecular dehydrogenation mechanism, which is facilitated by thermally driven desorption, in propane conversion reactions operating under conditions of low alkane partial pressure and high temperature.

Metal oxide components are crucial for C—H bond activation, and current research mainly focuses on two approaches: modification of zeolite support with zinc, cobalt, or gallium to enhance alkane activation, and preparation of platinum-functionalized zeolite composites. El-Malki et al. identified Zn and Ga cations as active centers through both experimental and computational analyses, demonstrating distinct activation pathways on Ga- and Zn-modified H-ZSM-5 frameworks [4]. Artsiusheuski et al.'s study on the catalytic effect of acidic zeolites in the conversion of propane suggests that metal cations at the exchange site may play a crucial role in the activation of the C—H bond, although their primary function appears to be facilitating the recombination of hydrogen atom adsorbates after bond cleavage [5].

The dissociation energy of the C—C bond (~246 kJ mol^{-1}) is relatively lower than that of the C—H bond (~363 kJ mol^{-1}), which suggests that zeolite-derived materials are more favorable for hydrocarbon cracking in acidic environments rather than for dehydrogenation reactions. Platinum-based catalysts are widely used in industrial dehydrogenation processes, such as UOP's Oleflex technology and Phillips Petroleum's STAR method, due to platinum's exceptional ability to activate C—H bonds while minimizing C—C bond cleavage. However, major challenges in platinum-catalyzed reactions still influence the dehydrogenation performance like competing hydrogenolysis pathways and the accumulation of carbonaceous deposits. These limitations can be mitigated by introducing secondary metal components into the catalytic system, among which tin has been recognized as the most effective alloying element for improving catalytic performance. The conventional platinum-based support systems are susceptible to thermal degradation, as nanoscale metal particles tend to undergo sintering and deactivation under high-temperature reaction conditions. As previously discussed, platinum-based catalysts have been widely adopted in industrial dehydrogenation applications due to platinum's strong interaction with alkane C—H bonds and minimal influence on C—C bond cleavage. Although Pt-based catalysts exhibit excellent initial propylene selectivity, they are prone to structural degradation under harsh reaction conditions, primarily due to sintering of platinum particles, which results in an irreversible loss of catalytic activity. Highly crystalline zeolite materials with large surface areas and precisely defined microporous structures have emerged as ideal supports for preventing aggregation of metal particles.

Extensive studies have demonstrated that L-type hydrated nonacidic zeolites, which possess a one-dimensional 12-membered ring (MR) pore structure, are ideal support materials for platinum-containing catalysts used in the dehydrogenation and aromatization of C_6–C_8 alkanes. The unique pore architecture of these zeolites enhances their selectivity toward aromatic compounds by stabilizing platinum clusters at the nanoscale and restricting the bimolecular reaction pathways that lead to coke formation. Recent work by Hill et al. further demonstrated the high catalytic efficiency of Pt–Sn systems supported on L-zeolite frameworks in the dehydrogenation of isobutane. Mössbauer spectroscopic analysis of Pt–Sn/K-L composites indicated that the zeolite structure enhances alkane dissociation via a preferential adsorption mechanism [6]. The synergistic interaction between tin and potassium additives effectively controls the size of Pt–Sn alloy particles, which suppresses undesirable side reactions such as hydrogenolysis and structural isomerization, and promotes the selective production of isobutylene. Nawaz and co-workers reported enhanced catalytic performance of the Pt–Sn/ZSM-5 system in n-butane dehydrogenation, achieving an olefin selectivity of 90 wt. % at a Si/Al ratio of 300 while maintaining strong hydrothermal stability [7].

In the early 1980s, researchers at Mobil Corporation successfully incorporated metal elements into zeolite frameworks by employing hydrothermal crystallization techniques with suitable precursors. The deliberate incorporation of platinum species within LTA-type crystal structures imparts molecular discrimination properties to these catalysts, along with enhanced resistance to deactivation. The experimental results revealed that ruthenium-based assemblies and platinum aggregates confined within narrow-pore zeolite matrices demonstrate improved stability in the presence of sulfur-containing compounds. In these catalytic materials, the metal components form nanoscale particles with diameters typically ranging from 1 to 2 nm.

Moliner and co-workers demonstrated that platinum species synthesized through thiol-coordinated platinum complexes exhibit enhanced thermal stability compared to conventional Pt/SiO_2 systems when subjected to reaction temperatures ranging from 650 to 750 °C [8]. The researchers confirmed the reversible transformation between platinum nanoparticles of approximately 1 nm in size and dispersed individual Pt atoms under alternating redox conditions according to X-ray absorption spectroscopy and high-angle annular dark-field scanning transmission electron microscopy (HAADF-STEM) characterization. The study further revealed that the microporous mordenite framework serves as an effective trapping medium for metal species initially deposited on support materials such as SiO_2 or Al_2O_3. Experimental observations showed that the active metal species migrate from the outer surface into the confined channels of the zeolite structure, which emphasizes the dynamic nature of platinum under reaction conditions.

Recent studies have demonstrated the significant advantages of anchoring metal species within zeolite frameworks for the selective and durable dehydrogenation of light alkanes. Researchers employed an innovative synthetic strategy to immobilize ultra-stable platinum nanoparticles within the pore structures of silicon-rich MCM-22 molecular sieves (see Figure 16.1). During the structural transformation from two-dimensional layered precursors to three-dimensional crystalline networks, Liu et al. successfully encapsulated platinum nanostructures, including single-atom sites and sub-nano clusters, within the MCM-22 matrix (denoted as Pt@MCM-22) [9]. HAADF-STEM revealed that the platinum species in Pt@MCM-22 are predominantly present in the form of atomic-scale to nanoscale particles. The Pt species are uniformly distributed across surface defects, internal channels, and supercages of the zeolite framework. The coordination number (CN) of platinum in Pt@MCM-22

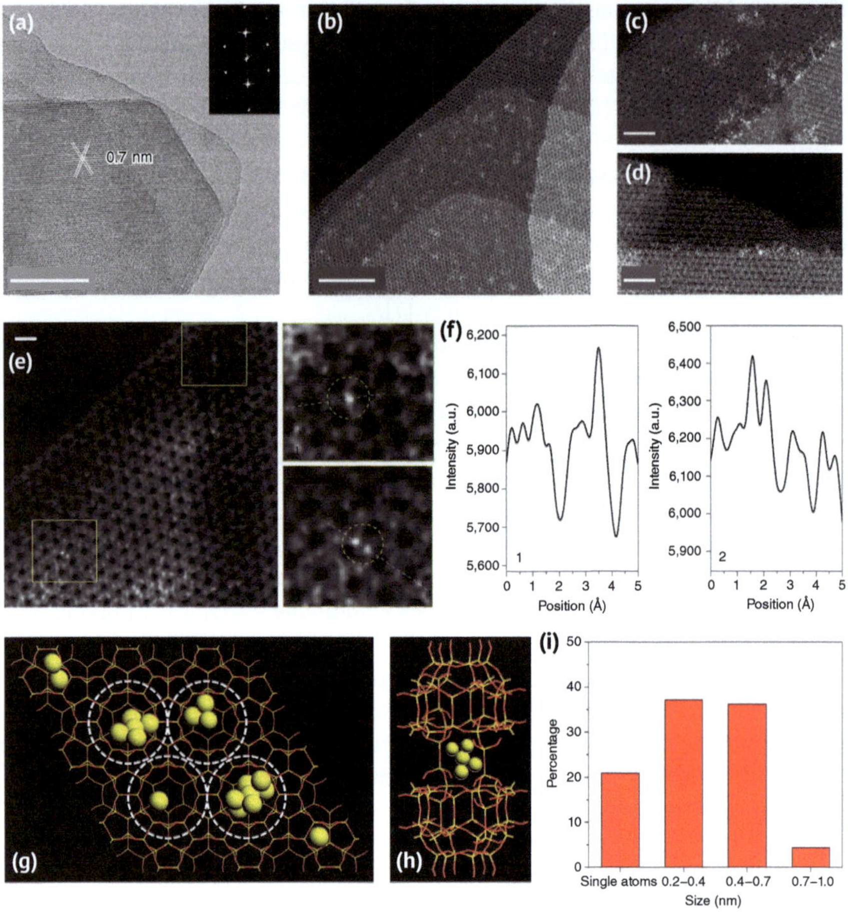

Figure 16.1 Atomic structures of Pt@MCM-22 for dehydrogenation of propane to propylene. (a) TEM image of Pt@MCM-22. (b) HAADF-STEM image of Pt@MCM-22. (c, d) Atomic structures of Pt clusters confined in MCM-22. (e) HAADF-HRSTEM image of Pt@MCM-22. (f) Corresponding intensity profiles obtained on the two zoomed areas in e. (g) Schematic illustration of Pt@MCM-22 in a 'top-down' view along the c axis. Pt clusters and individual Pt atoms are located in the surface cups, cavities and 12-MR supercages. (h) Schematic illustration of a Pt5 cluster in supercages of MCM-22. (i) Size distribution of subnanometric Pt species in a Pt@MCM-22 sample. *Source*: Reproduced from [9]/with permission of Springer Nature.

was determined to be approximately 4.7, indicating that the platinum clusters contain fewer than 13 atoms. These results suggest that the platinum nanoclusters likely adopt a cubic octahedral structure. The synthesized Pt@MCM-22 catalyst demonstrated high catalytic activity in dehydrogenation reactions. Compared to the Pt/MCM-22-imp catalyst prepared by conventional impregnation methods, the Pt@MCM-22 composite exhibited superior thermal stability and significantly enhanced catalytic activity. In the PDH reaction, the Pt@MCM-22 system retained 90% of its initial catalytic activity after five consecutive reaction-regeneration cycles at 550°C. In contrast, the impregnated catalyst showed a 40% decrease in activity under the same conditions. These findings demonstrate the effectiveness of structural confinement in preserving catalytic integrity under harsh reaction environments.

Studies have also shown that introducing secondary metal elements into the catalysts can effectively suppress platinum particle aggregation in bimetallic systems while simultaneously enhancing PDH efficiency, such as Sn, Fe, In, Zn, Mn, Co, and La. Notably, Liu and co-workers demonstrated the successful encapsulation of ultrafine PtSn clusters (0.5–0.6 nm) within the MFI-type zeolite framework through a simple one-pot hydrothermal synthesis method (see Figure 16.2) [10]. Advanced characterization techniques, such as HAADF-STEM and integrated differential phase contrast (iDPC) imaging, allowed for precise mapping of the spatial distribution of platinum-tin species within the sinusoidal 10-MR channels of the MFI zeolite. These confined PtSn nanoclusters exhibited significantly improved catalytic performance. The K-PtSn@MFI catalyst, in which platinum-tin species were confined within the sinusoidal 10-MR channels, exhibited superior catalytic performance in PDH compared to the conventional K-PtSn/MFI catalyst prepared via impregnation, along with excellent resistance to deactivation. In-situ EXAFS analysis, HAADF-STEM imaging, and CO adsorption infrared spectroscopy jointly

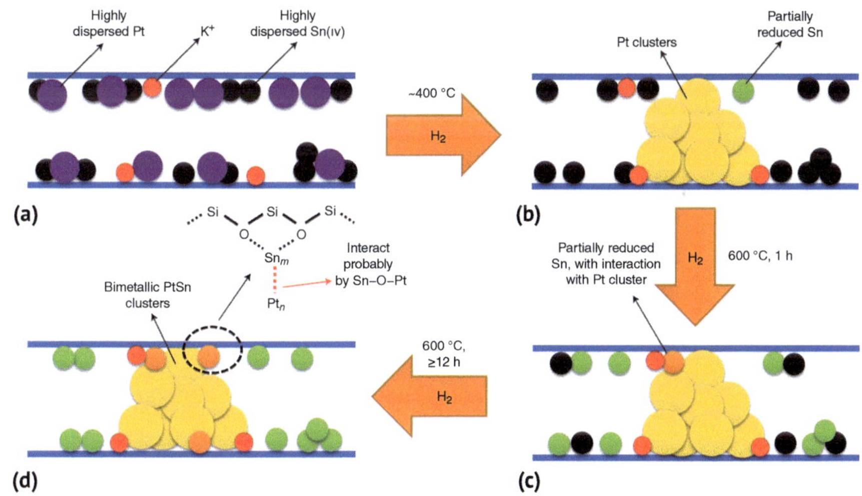

Figure 16.2 Schematic diagram for the evolution of K-PtSn@MFI during H_2 reduction and dehydrogenation. (a) In the pristine K-PtSn@MFI sample, both Pt and Sn(iv) exist as atomically dispersed species in the sinusoidal channels of the MFI zeolite. (b) After reduction by H_2 at ~400 °C, Pt single atoms will be reduced and form subnanometric Pt clusters. whereas only a few of the Sn(iv) species will be reduced. (c) When the temperature is increased to 600 °C, most of the Sn(iv) species are reduced to Sn(ii), but these reduced Sn species mainly remain separated from the Pt clusters. (d) After being kept in a flow of H_2 for a long time (≥12 h), some of the reduced Sn species migrate towards the Pt clusters and bimetallic PtSn clusters with reduced Sn species interacting with the external surface of Pt clusters are formed in the sinusoidal 10R channels of the MFI zeolite. It should be noted that, due to the complexity of the structures of the PtSn clusters, the models proposed in this Figure could be over simplified. *Source*: [10]/Springer Nature.

confirmed that the incorporation of partially reduced tin at the platinum-support interface reduced the deactivation rate of PDH by approximately tenfold. Under continuous reaction conditions at 550°C, the engineered catalyst demonstrated outstanding catalytic performance, characterized by enhanced propylene selectivity and prolonged operational stability.

In a complementary study, Sun et al. developed an innovative ligand-stabilized hydrogen reduction approach to immobilize ultrasmall platinum-zinc alloy clusters (denoted as $PtZn_4$@S-1-H) within siliceous MFI-type zeolites [11]. Advanced aberration-corrected STEM analysis confirmed that the platinum-zinc clusters in $PtZn_4$@S-1-H exhibit sub-nanometer dimensions, with sizes below 1 nm. Within the MFI-type zeolite framework, these nanoclusters are confined within channels measuring 5–6 Å, as confirmed by structural characterization. XANES and EXAFS analyses indicate that the Pt–Pt CN of $PtZn_4$@S-1-H is approximately 2, suggesting a triangular platinum configuration. In catalytic tests conducted at 550°C with a WHSV of 108 h^{-1}, the $PtZn_4$@S-1-H material exhibited excellent propylene production efficiency (65.5 mol $C_3H_6 \cdot gPt^{-1} \cdot h^{-1}$), maintaining 99.3% selectivity over a broad WHSV range (3.6–54 h^{-1}). Notably, the catalyst demonstrated negligible activity loss during prolonged operation (13000 minutes, WHSV = 3.6 h^{-1}, without H_2 supplementation), outperforming the monometallic Pt/S-1-H catalyst in PDH. This enhanced performance underscores the crucial role of zinc in inhibiting platinum nanoparticle aggregation and enhancing thermal stability. Experimental observations further demonstrate that the incorporation of cesium ions into the zeolite channels can effectively modulate catalytic performance. The regeneration durability of the $PtZn_4$@S-1-H-Cs catalyst was significantly enhanced, as evidenced by its consistent catalytic activity across four consecutive regeneration cycles. Extended operational stability tests conducted on both $PtZn_4$@S-1-H and Pt/S-1-H catalysts during PDH (WHSV = 3.6 h^{-1}, 550°C) further confirmed their sustained performance. Under controlled reaction conditions (0.3 g catalyst mixed with 1.0 g quartz sand, atmospheric pressure, feed ratio C_3H_8/N_2 = 10:30 mL · min^{-1}, WHSV = 3.6 h^{-1}), the $PtZn_4$@S-1-H-Cs catalyst maintained stable performance through multiple regeneration cycles. After the initial reaction, the catalyst was regenerated by heating in air (10 mL · min^{-1}) to 550°C, held at this temperature for one hour, and subsequently cooled down to 200°C under a nitrogen flow. This was followed by hydrogen reduction through gradient heating to 550°C, with a subsequent reactivation period of one hour. The PDH process was then resumed using fresh propane feedstock under identical reaction conditions.

Studies indicate that zeolite frameworks can promote the formation of catalytic metal phases and iron carbide species. Yun and Lobo examined β-zeolite and ZSM-5 zeolite as potential catalysts for PDH [12]. Their results show that while isolated iron sites within the framework act as the primary catalytic centers, extra-framework species also play an important role in dehydrogenation activity. The study suggests a redox-mediated reaction mechanism, whereby propane interacts with the active site via the formation of propane-derived radical cation intermediates.

As for the structural characteristics of zeolites, De Cola et al. investigated the influence of sodium in platinum-zinc-modified mesoporous and macroporous zeolites on the PDH process [13]. Their findings indicate that ZSM-5 and ZSM-11 frameworks containing 10-MRs with sodium are particularly active in promoting the formation of aromatic compounds, whereas macroporous molecular sieves such as β-zeolite, mordenite, L-type, and Y-type zeolites predominantly yield propylene. Notably, the Pt-Zn/β catalyst demonstrated exceptional stability under the test conditions, maintaining conversion levels close to equilibrium for more than 63 hours.

The catalytic dehydrogenation of ethane has also received widespread attention in both industrial applications and academic research. Ethylene can serve as a fundamental building block in the production of various polymer materials, like PE, PS, and PVC, which can also be used for the production of ethanol and ethylene oxide. Currently, industrial production of ethylene predominantly depends on energy-intensive processes, such as steam cracking and FCC, which utilize petroleum-derived feedstocks like naphtha and light diesel oil. These conventional methods face two major challenges. First, they require extremely high operating temperatures, typically above 1073 K, to maximize ethylene yield. Second, undesirable side reactions and coke deposition commonly occur, significantly reducing process selectivity.

The direct dehydrogenation of ethane (EDH) offers a promising alternative for the production of ethylene. Currently, platinum-based and chromium oxide-based catalysts, which operate through mechanisms similar to those in PDH, are the two main widely studied and industrially utilized catalytic systems for EDH. However, due to the high cost and limited availability of platinum, as well as the environmental hazards associated with chromium compounds, there is increasing research aimed at identifying more sustainable catalytic alternatives.

In recent years, microporous framework-stabilized base metal/oxide composites incorporating transition elements such as cobalt, nickel, zinc, molybdenum, gallium, and iron have garnered increasing attention in both academic and industrial research communities. Liu et al. employed an initial wet impregnation method to develop various zinc-modified SSZ-13 zeolite systems [14]. Advanced electron microscopy characterization demonstrated a uniform dispersion of zinc within the SSZ-13 matrix even at minimal metal loadings, with no discernible ZnO crystalline phase detected in the optimized $Zn_{0.83}$/Na-SSZ-13 samples. Furthermore, the researchers synthesized additional catalytic materials of zinc-exchanged ZSM-5, Zn-loaded MCM-22, and Zn-loaded NaY molecular sieves for comparative analysis. The Zn/Na-SSZ-13 catalyst exhibited over 90% selectivity toward ethylene in CO_2-enhanced ethane dehydrogenation, accompanied by significant improvements in CO_2 and ethane conversion efficiencies. The enhanced catalytic performance can be attributed to the unique microporous structure of SSZ-13, which effectively confines CO_2 within its channels, thereby promoting reactant activation. Density functional theory (DFT) calculations identified a bifurcated reaction pathway: (1) ethane dehydrogenation occurs at Zn^{2+} and dimeric $[Zn—O—Zn]^{2+}$ active sites, leading to the formation of ethylene; and (2) CO_2 participates in hydrogen removal from the catalytic surface. Importantly, the co-feeding of CO_2 reduced the dehydrogenation activation energy by 38%, thereby effectively suppressing methane generation through secondary hydrogenolysis.

Spectroscopic analyses provided evidence for the dual function of CO_2: stabilization of the transition state via hydrogen bonding and involvement in the formation of hydroxyl-assisted intermediates. Under different reaction conditions, the energy barriers for H_2 desorption from the $[Zn{-}O{-}Zn]^{2+}$ complex were calculated to be 26.13 and 13.71 kcal mol^{-1}, respectively. Comparative analysis revealed that these values were significantly lower than those in the CO_2-free system, where the corresponding barriers were approximately 38.57 and 35.73 kcal mol^{-1}.

Extensive research efforts have been devoted to developing zeolite-based catalysts for light alkane dehydrogenation using non-precious metals. This study presents a novel strategy for incorporating isolated Fe species into a silicon-rich MFI framework (Fe/S-1-EDTA) via a one-step hydrothermal crystallization process using iron-EDTA sodium complex precursors. In contrast to conventional preparation methods such as impregnation or ion exchange, structural characterization confirmed the absence of FeO clusters or oligomers in Fe/S-1-EDTA, with all iron atoms successfully incorporated into the zeolite framework, thereby demonstrating superior atomic-level control.

In addition to isolated iron species, other metal elements such as chromium, gallium, and indium also exhibit significant catalytic activity in ethane dehydrogenation. Maeno and co-workers synthesized the In-CHA catalyst through solid-state ion exchange, forming $[InH_2]^+$ hydride species within the zeolite framework, followed by high-temperature hydrogen treatment at 500°C (see Figure 16.3) [15]. During performance testing, the catalyst achieved 96.1% ethylene selectivity and 25.9% ethane conversion, maintaining stable catalytic activity over 90 hours of continuous operation with a carbon balance exceeding 98%. Notably, the In-CHA catalyst demonstrated excellent regenerability, fully recovering its catalytic activity after calcination in an oxygen atmosphere at 600°C and subsequent reduction in a hydrogen environment. Computational simulations suggest that $[InH_2]^+$ serves as the active dehydrogenation center, outperforming zinc-based systems. Furthermore, the formation of zeolite-anchored proton and carbon cation intermediates is significantly suppressed during hydrogenation. Zeolites of the CHA framework modified by gallium and indium ion exchange exhibit enhanced resistance to carbon deposition, improved selectivity, and extended catalytic stability. The $[InH_2]^+$ species demonstrate behavior analogous to Ga-H species in promoting alkane dehydrogenation, although their catalytic activity is modulated by the nature of the support.

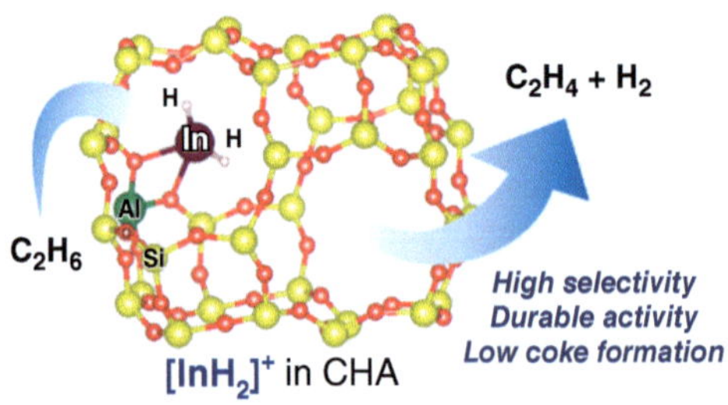

Figure 16.3 Isolated indium hydrides in CHA zeolites for dehydrogenation of ethane. *Source*: [15]/with permission of American Chemical Society.

Experimental studies by Mansoor et al. have shown distinct catalytic behaviors: ethane dehydrogenation is efficiently catalyzed by both $[GaH]_2^+$ and $[GaH]^+$ cations, while n-butane conversion is exclusively facilitated by the $[GaH]^+$ cation [16]. Moreover, various zeolite-supported metal oxide catalysts, such as GaO_x, MoO_x, VO_x, and CrO_x, have exhibited notable activity in the dehydrogenation of light alkanes.

16.2.2 Oxidative Dehydrogenation of Light Alkanes

Although microporous crystalline materials demonstrate high selectivity in direct dehydrogenation processes, their conversion efficiency is limited by thermodynamic constraints and the substantial energy consumption required at elevated reaction temperatures. Oxidative dehydrogenation (ODH) presents a promising alternative for the conversion of light alkanes to alkenes—especially ethane—by overcoming thermodynamic equilibrium limitations and allowing operation at considerably lower temperatures. Extensive research on the ODH of light hydrocarbons has investigated a variety of catalytic systems and oxidants. However, the transition to industrial-scale application has not yet been achieved. The safety concerns are one of the major challenges arising from the use of potentially explosive gas mixtures, particularly in O_2-based processes. Besides, the occurrence of competing overoxidation reactions may diminish olefin selectivity. Moreover, the process often results in the formation of carbon dioxide, with valuable hydrogen by-products being partially or completely consumed. Despite these drawbacks, molecular oxygen remains a widely investigated oxidant for dehydrogenation due to its compatibility with medium-temperature operations, cost-effectiveness, and broad availability.

The ODH of light alkanes generally proceeds via two distinct mechanisms. One approach employs a self-heating strategy, in which selective hydrogen oxidation generates thermal energy. This dual-function system not only overcomes equilibrium limitations through water formation but also compensates for the endothermic nature of the reaction by utilizing heat from combustion. These systems can utilize either bifunctional platinum-based monolithic catalysts or composite catalysts containing multiple active components. Alternative strategies for alkane activation have also been explored, including the direct involvement of oxygen through homolytic and heterolytic C—H bond cleavage pathways. Compared to direct dehydrogenation, ODH requires multifunctional catalysts that are capable of activating multiple reactants simultaneously. In particular, reducible metal oxides, especially vanadium-based systems, demonstrate oxygen activation capabilities at relatively low temperatures (300–400 °C), enabling efficient reaction conditions.

Among the various catalytic systems investigated, vanadium-containing catalysts have attracted considerable attention. The redox behavior of surface vanadium oxides is significantly influenced by their coordination environment and degree of aggregation, while the acid–base properties of the surrounding matrix play a crucial role in determining catalytic performance. Similar findings have been reported when vanadium is incorporated into structurally analogous frameworks, such as

$AlPO_4$-5 molecular sieves or zeolite structures. In 2001, Solsona et al. carried out a comprehensive study on the synthesis methods, oxidation states, and catalytic performance of vanadium species in microporous and mesoporous systems, which represented a major advancement in the field [17]. Although vanadium is commonly introduced into the precursor mixture prior to hydrothermal synthesis, only a small fraction is successfully incorporated into the framework. In vanadium-containing microporous silicates, both V^{5+} and V^{4+} species have been found to substitute for tetrahedrally coordinated silicon centers within the lattice structure.

Vanadium species typically coordinate with structural defects in zeolite frameworks, adopting tetrahedral or four-coordinate geometries. In crystalline aluminum vanadium phosphate systems, several binding mechanisms have been observed, such as the isomorphic substitution of phosphorus(V) or aluminum(III) ions within the framework structure, as well as stabilization via bonding interactions with phosphorus centers in the AlPO lattice. Comparative catalytic studies reveal that V-Si-1 and V-Si-2 catalysts synthesized via hydrothermal methods exhibit higher propylene selectivity in PDH than their impregnated counterparts, suggesting that vanadium is structurally incorporated into the zeolite matrix. This improved catalytic behavior is further supported by studies on vanadium-modified $AlPO_4$-5 microporous materials with enlarged channel dimensions, where transition metals selectively bind to specific crystallographic positions.

Conversely, non-reducible oxide materials employed in methane oxidative coupling require significantly higher temperatures (650–800 °C) to activate oxygen effectively. Since the 1990s, a wealth of comprehensive reviews has been published on oxygen-mediated ODH processes, including foundational contributions by Cavani and co-workers [18]. These reviews commonly contrast ODH with direct dehydrogenation, emphasizing catalytic systems that extend beyond zeolite-based structures. The introduction on ODH catalysts such as vanadium- and chromium-based oxides, nonstoichiometric nickel oxides, and lithium-doped alkaline earth metal oxides is discussed intensively.

Emerging studies indicate that boron-containing catalysts demonstrate remarkable selectivity in the ODH of alkanes for olefin production. Analogous to the aromatics cycle mechanism observed in methanol-to-olefins processes, Qiu et al. synthesized an MWW-structured borosilicate zeolite, in which the PDH activity was attributed to trivalent boron defects rather than tetrahedral boron configurations (see Figure 16.4) [19]. In contrast, a three-dimensional B-MCM-22 catalyst predominantly featuring framework-isolated $B(OSi)_3$ units was reported, which exhibited no PDH activity. Their comparative analysis indicated that catalytic activity depends on aggregated boron species rather than isolated boron units, suggesting that the boron clustering is necessary for effective dehydrogenation activity.

The one-dimensional pore structure of the AFI framework significantly enhances the catalytic efficiency of ODH of ethane and propane, achieving yields to alkenes of 15–17%. By substituting Al^{3+} with divalent Mg^{2+} or Co^{2+} in the $AlPO_4$-5 matrix, BAS are introduced, which exert differential effects on alkane conversion performance. In propane conversion, the nonacidic VAPO-5 catalyst exhibits excellent olefin selectivity, whereas in ethane dehydrogenation, the acidic variant demonstrates

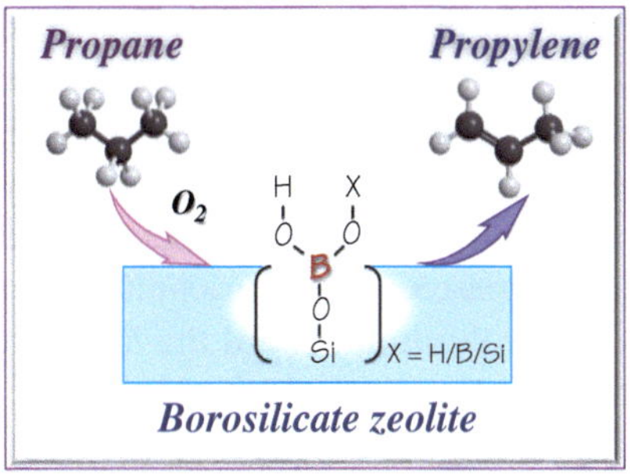

Figure 16.4 Illustration on the active sites of borosilicate zeolite for propane ODH. *Source*: [19]/with permission of Elsevier.

superior catalytic activity. Compared to VAPO-5, VMgAPO-5, and VCoAPO-5 display enhanced ethylene selectivity while preserving catalytic stability. A comprehensive study was conducted to investigate how the incorporation of vanadium and/or cobalt into aluminum phosphate microporous frameworks influences the catalytic performance in ethane ODH. Catalysts derived from the zero-dimensional $AlPO_4$-18 framework (VPO-18, CoAPO-18, and VCoAPO-18) demonstrated superior ethylene selectivity compared to analogous catalysts based on the one-dimensional 12-MR $AlPO_4$-5 structure, which is connected to a three-dimensional pore network through 8-MR channels. Notably, small-pore molecular sieves such as SAPO-34 and lanthanum-modified SAPO-34 exhibit both Brønsted and Lewis acidity, contributing to significant catalytic activity and product specificity in this dehydrogenation process. The enhanced ethylene yield suggests that the reaction predominantly occurs within the confined environment of the mordenite zeolite, underscoring the potential of SAPO-34-derived catalysts for this transformation. The high selectivity observed in narrow-pore zeolitic materials may result from the formation of hydrocarbon pool-like intermediates within their cage-like structures, followed by ethylene generation through a controlled reaction pathway.

The interplay between redox-active environments and acidic conditions plays a crucial role in determining the ODH performance of transition metal-impregnated zeolite catalysts in light alkane conversions. In a comparative study of nickel-, copper-, and iron-modified HY and KY zeolites, Lin et al. reported that HY-supported catalysts exhibited superior catalytic activity in ethane ODH and higher ethylene selectivity [20]. Mechanistic investigations suggest that this enhanced performance is attributed to the catalytic function of BAS, which promotes the extraction of β-H from the metal acetaldehyde intermediate formed during alkane activation. This key C—H bond cleavage step takes place at the metal center coordinated with the oxygen atom of the zeolite framework, enabling selective ethylene formation through a coordinated deprotonation mechanism. The elimination of β-hydrogen from the metal-ethoxy complex indicated the synergistic metal-support interaction that governs ethylene production. In contrast, the catalytic performance of V/Al/MCM-22 in propane ODH was enhanced by suppressing cracking side reactions through alkali metal ion substitution, which effectively modulates BAS concentration.

Lin et al. designed a bifunctional catalytic system that integrates VMgO and ZSM-5 zeolites for sequential ODH followed by cyclic aromatization, and the efficient propane-to-aromatics conversion was achieved subsequently [20]. However, the incorporation of gallium into the H-ZSM-5 framework was found to be detrimental, as it promotes hydrocarbon combustion. Experimental observations revealed that when these catalytic components coexist within a uniformly packed reactor bed, the product distribution significantly shifts toward the formation of high-value alkylated aromatics, rather than the benzene derivatives.

As previously emphasized, the use of oxygen as an ODH oxidant presents several limitations, such as the explosion hazard associated with flammable O_2/hydrocarbon mixtures, the high costs of nitrogen separation in air-based systems, and challenges in temperature control when using pure oxygen. Furthermore, olefin underproduction caused by competing oxidation side reactions remains insufficient to meet industrial feasibility demand. As a result, extensive research efforts have been directed toward alternative ODH pathways that employ less reactive oxidants, such as CO or nitrous oxide (N_2O). These approaches both involve the simultaneous introduction of hydrocarbon reactants and oxidizers into the reaction environment.

The utilization of carbon dioxide has demonstrated notable sustainability advantages and has become an environmentally preferable option across a range of industrial applications. This method efficiently mitigates the over-oxidation of propylene into gaseous by-products, thereby enhancing the efficiency of olefin production. Moreover, it promotes the removal of carbonaceous deposits formed during catalytic processes, thus prolonging the catalyst lifespan. Meanwhile, it is also important for its role in promoting the recycling of greenhouse gases while reducing environmental impact. Recent studies have conducted in-depth investigations into CO_2-mediated oxidative propane dehydrogenation (ODPC), offering a systematic evaluation of zeolite-based catalytic systems. Their research underscores the operational benefits of employing CO_2 as a tunable oxidant and outlines promising avenues for advancements in catalytic material development and process optimization.

Carbon dioxide plays multifunctional roles in catalytic systems beyond its participation in direct dehydrogenation reactions. Via the reverse Boudouard reaction, CO_2 interacts with coke deposits, contributing to enhanced catalyst stability and alleviating deactivation. However, under elevated temperatures, competing reaction pathways, such as dry reforming of alkanes, thermal cracking, and hydrogenolysis, can be activated, which collectively diminish olefin production efficiency. In the case of ODH of propane coupled with CO utilization (ODPC), an ideal catalytic material must concurrently support PDH and the reverse water-gas shift (RWGS) mechanism. To achieve this dual functionality, it requires the optimization of surface properties, which allows the simultaneous adsorption of H_2 and CO_2 with favorable reaction kinetics.

The catalytic system must possess basic sites for the weak acidic adsorption of CO, and acidic regions capable of binding alkanes and hydrogen molecules. Therefore, the catalytic behavior in reactions involving CO_2 and its role in the dehydrogenation mechanism are strongly influenced by the intrinsic properties of the catalyst. In the reactions of ODH using carbon dioxide (ODPC), various catalytic materials have been investigated, including zeolite frameworks, metal oxide-modified molecular sieves, and composite oxide systems. Although extensive studies have focused on mesoporous metal oxides supported on zeolites such as MCM-41 or SBA-15, chromium-impregnated zeolites have demonstrated significant performance in facilitating CO_2-assisted dehydrogenation of ethane and propane. In ethane conversion systems, experiments conducted at high temperatures (650–800 °C) under atmospheric pressure revealed stable catalytic performance and enhanced ethylene selectivity, which has been attributed to the decarbonization effect of CO. The primary zeolite material used was a silicon-rich ZSM-5 (SiO_2/Al_2O_3 ratio > 190), and spectral analysis confirmed the presence of high-valent chromium species (Cr=O and Cr—O types) at the active sites.

When light alkanes undergo dehydrogenation in a CO_2 environment, the RWGS reaction facilitates the conversion of generated hydrogen and CO_2 into carbon monoxide and water. This synergistic effect drives the dehydrogenation equilibrium toward product formation by consuming hydrogen. However, the coupled reaction system is strongly endothermic, imposing thermodynamic constraints that need higher temperatures and lower pressures for optimal performance, which becomes the key limitation of oxidative dehydrogenation of propane (ODP). As demonstrated by Du et al., the efficiency of ethane activation in a CO_2 medium at temperatures below 500 °C was significantly lower than that of propane and butane. Moreover, n-butane dehydrogenation products were more prone to adverse side reactions compared to propylene derivatives. These comparative analyses indicate that PDH in a CO_2 atmosphere represents the thermodynamically favorable pathway among light alkanes, which explains its prominence in current catalytic research.

During the ethane dehydrogenation reaction, chromium undergoes redox cycling in a CO atmosphere. Experimental evidence demonstrates that these catalytic components are reduced to metallic chromium upon exposure to ethane, yet the oxidizing capacity was recovered after CO_2 treatment. Comparative studies on CO_2-enhanced PDH have shown that in MFI-structured borosilicate-supported catalysts, the monomeric acidic form of hexavalent chromium exhibits higher catalytic performance compared to trivalent polyacidic chromium species. Notably, vapor-activated catalysts contain highly dispersed Cr^{3+} species, such as $[CrO]^+$ complexes, contributing to the significant catalytic activity. Similar CO-enhancing effects have also been observed on Ga_2O_3-based materials supported on high-silica H-ZSM-5 and H-ZSM-48 zeolites. Although the one-dimensional mesoporous structure of ZSM-48 provides improved propylene selectivity over ZSM-5, its stability is limited due to accelerated deactivation processes.

Al-Mamoori and co-workers recently proposed an innovative cyclic system that integrates ethane ODH with CO_2 capture technology, offering an environmentally sustainable strategy to reduce industrial carbon emissions while converting low-valued feedstocks into high-valued chemicals (see Figure 16.5) [21]. Their experimental setup employed a dual-function fixed-bed reactor comprising physically mixed K-Ca double-salt CO_2 adsorbents and chromium-enhanced H-ZSM-5 ODH catalysts. The operational sequence consisted of two distinct phases: an initial isothermal CO_2 adsorption phase using a 10% CO_2/Ar mixture at 600 °C, followed by a

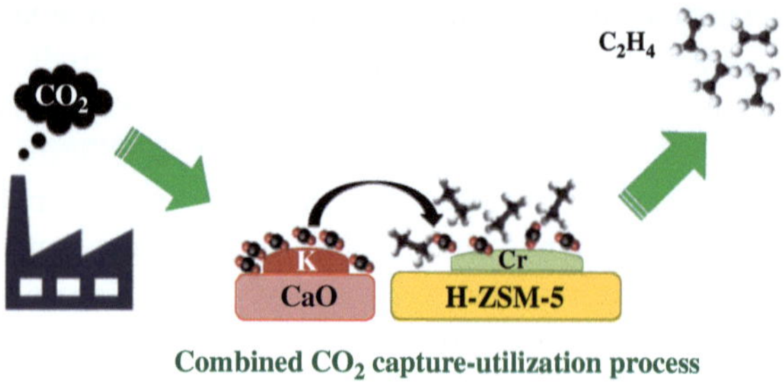

Figure 16.5 Scheme for ODH of ethane in an integrated CO_2 capture-utilization process. *Source*: [21]/with permission of Elsevier.

thermal transition phase in which the reactor temperature was increased to 700 °C. In the second stage, a 5% ethane/Ar flow was introduced into the catalyst bed, where CO thermodynamically released from the first stage was selectively utilized in the reaction. Although this conclusion demonstrated technical feasibility, both catalytic and adsorption efficiencies decreased after four consecutive adsorption-reaction cycles. This deactivation performance was attributed to particle agglomeration and a reduction in the activity of hexavalent chromium sites. Additionally, carbonaceous deposits were identified as the main factor in catalyst deactivation.

In N_2-driven dehydrogenation of light alkanes, iron-containing zeolite materials, such as mesoporous Fe-ZSM-5, have gathered considerable attention. Pannov's study demonstrated that the ZSM-5 framework can stabilize iron complexes, which are reduced to the ferrous state during activation, leading to the formation of unique α-sites [22]. Subsequent decomposition of N_2O enables reoxidation to the ferric state, resulting in the generation of highly reactive atomic oxygen species capable of oxidizing organic substrates under ambient conditions. Importantly, these active centers display methane activation capability even at ambient temperatures.

Preliminary studies in the early 1990s revealed that using N_2O as an oxidant in light alkane dehydrogenation exhibited excellent selectivity, although hydrocarbon conversion rates remained below 10%. Subsequently, Pérez-Ramírez et al. demonstrated that hydrothermally modified Fe-ZSM-5 catalysts significantly improved catalytic performance, achieving 24% propylene and 6% propanaldehyde yields at 525 °C [23]. This enhancement was attributed to two primary factors: the selective oxidation-promoting oxygen transfer capability of nitrogen oxides, and the presence of extra-framework iron species that stabilize the reactive α-oxygen intermediate generated during N_2O decomposition. Notably, tetrahedrally coordinated iron sites within the zeolite framework showed minimal activity in N_2O activation, which highlights the importance of controlled steam treatment to generate optimal dehydrogenation active sites. Further mechanistic investigations by the same research group provided detailed insights into the structural transformations of active iron centers during catalyst activation and reaction stages.

The interaction between nitrogen oxides and alkanes was examined using steam-activated Fe-ZSM-5 molecular zeolite. The results indicated that low-nuclear iron species, which are generated via high-temperature hydrothermal treatment, demonstrated enhanced reactivity toward atomic oxygen. These active sites serve dual roles: promoting both the adsorption and recombination of oxygen atoms, as well as enabling the efficient transfer of oxygen to hydrocarbon molecules. Kumar et al. demonstrated that the zeolite framework enhances the stability of active iron species [24]. Iron silicate zeolites exhibited higher catalytic efficiency in propane ODH compared to mesoporous Fe-SBA-15 materials. Despite comparable acidity and iron loading, the distinct pore structures of the two catalyst types led to differences in iron species distribution. The zeolite systems predominantly contained isolated Fe^{3+} ions. This study highlights how the microporous framework not only stabilizes active iron centers but also facilitates molecular interactions between reactants and catalytic sites. Nowińska et al. investigated the extent to which the acidic properties of zeolite supports affect the activity and selectivity of Fe-ZSM-5 catalysts [25]. To achieve this, they employed an experimental approach in which ZSM-5 substrates with different aluminum contents were modified via ion exchange using various transition metals (Fe^{3+}, Mn^{2+}, Co^{2+}). Comparative analysis of these metal-modified systems revealed that the iron-based catalyst displayed the highest catalytic activity, although this level of activity was achieved only after thermal activation at a calcination temperature of 900 °C. Furthermore, the distribution of products showed that carbon oxides were the primary products under oxygen-mediated dehydrogenation conditions, while oxidation with N_2O significantly enhanced the selectivity toward propylene. Bulánek et al. investigated the catalytic performance of Fe-ZSM-5 subjected to hydrothermal treatment in the ODH of propane under various oxidizing atmospheres, including O_2, N_2O, and their combinations [26]. Their findings revealed that catalyst durability reached its peak when a dual oxidant system was utilized in the absence of BAS. Under these optimized conditions, a propane conversion rate of 51.7% and propylene selectivity of 39.7% were attained.

Within the temperature range of 350–450 °C, N_2O in exhaust gases acts as a moderate oxidizing agent, facilitating the ODH of ethane over various iron-exchanged zeolite materials. Although mesoporous Fe-ZSM-5 demonstrated superior catalytic efficiency and ethylene selectivity compared to macroporous Fe-Y and Fe-mordenite catalysts, all three zeolite-supported catalysts maintained ethane dehydrogenation activity in the presence of N_2O. Notably, iron-modified silica catalysts and iron species supported on non-porous materials, such as silica gel or silica–alumina composites, displayed negligible ethane conversion under identical experimental conditions.

16.3 Hydrogenation of Unsaturated Hydrocarbons

Recent advancements have established zeolite-supported metal catalysts as highly efficient materials for the hydrogenation of unsaturated hydrocarbons, including aromatic compounds, alkenes, and alkynes. Their catalytic performance, determined by reactant conversion efficiency and product selectivity, is largely influenced by factors such as metal composition, particle size, and the structural properties of the zeolite, including pore diameter, surface acidity, and hydrophilicity. Moreover, these composite catalytic systems exhibit significant activity in sequential reaction processes, acting as bifunctional platforms that facilitate the efficient synthesis of target molecules through integrated catalytic pathways. Importantly, when metal species are selectively encapsulated within the zeolite framework, this structural arrangement markedly enhances the shape-selective characteristics of the hydrogenation process. This section reviews recent studies that demonstrate the exploration of zeolite-anchored metal catalysts in hydrogenation reactions of various unsaturated hydrocarbons, with an emphasis on novel applications and mechanistic insights.

16.3.1 Hydrogenation Reactions of Aromatic Compounds

The catalytic hydrogenation of aromatic species plays a pivotal role in petrochemical processing and pharmaceutical synthesis. Recent research advancements have demonstrated that zeolite-encapsulated metal catalysts exhibit superior performance in promoting the hydrogenation of various aromatic reactants, achieving both high conversion efficiency and excellent product selectivity.

Recent studies have highlighted the structure-dependent catalytic performance of microporous zeolite-encapsulated metal composites. Li and co-workers synthesized hollow ZSM-5 and S-1 zeolite matrices encapsulating transition metals (Co, Ni, Cu) through tetrapropyl-ammonium-mediated dissolution-recrystallization technology, achieving a metal loading of 8% with particle sizes ranging from 13 to 17 nm [27]. Experimental evaluations of hydrogenation for toluene and nitrotoluene over the Ni@S-1 catalyst, which possesses abundant surface structures, demonstrated that molecular size discrimination was strongly governed by zeolite pore size diameters. While toluene molecules could readily diffuse through the zeolite framework and undergo hydrogenation at the Ni active sites, the larger nitrotoluene molecules faced restricted access to the encapsulated metal particles, resulting in the inhibition of hydrogenation conversion. This distinct reactivity pattern underscores the essential role of the zeolite pore architecture in modulating molecular transport pathways for selective catalytic conversion.

Similarly, Li et al. utilized a mild desilication approach to synthesize platinum nanoparticles (2–3 nm in diameter) encapsulated within a ZSM-5 zeolite [28]. TEM microstructural analysis confirmed that these Pt nanocrystals were predominantly stable, either embedded within the zeolite framework structures or located at interfacial junctions (see Figure 16.6). The nanocapsular architecture effectively inhibits metal aggregation while preserving accessible active sites. Sub-nanometer platinum clusters are predominantly located near the zeolite framework. The superior encapsulation capability of the crystal shell effectively inhibits metal aggregation and maintains structural integrity even under extreme thermal reduction conditions in a hydrogen atmosphere at 750 °C. The engineered Pt@nanoshell system exhibited exceptional catalytic efficiency in aromatic hydrogenation, with a reaction rate approximately 100 times higher than that of conventional Pt/ZSM-5 composites. Comparative analysis of four platinum-based catalysts (Pt@nanoshell, Pt/SiO_2, Pt@hollow, and Pt@block) through measurements of activation energy and kinetic studies revealed that mass transfer limitations were the primary factor influencing

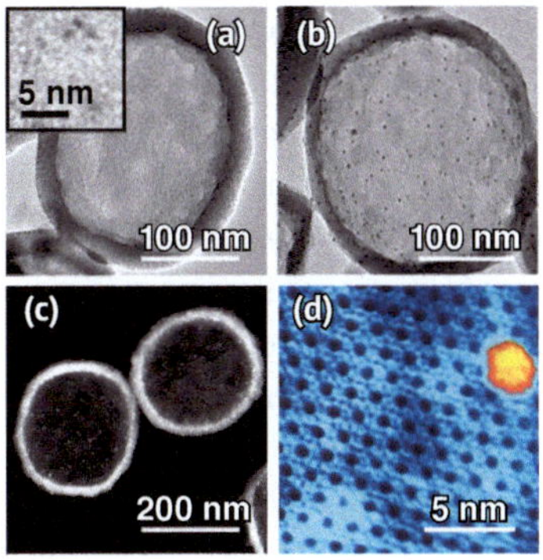

Figure 16.6 TEM images of Pt@nanoshell. Before (a) and after (b) reduction. STEM-HAADF (c) and HRTEM (d) of Pt particle on ZSM shell. *Source*: Reproduced from [28]/with permission of Elsevier.

catalytic efficiency. In the traditional Pt/ZSM-5 catalyst, the spatial confinement of platinum nanoparticles within the microporous structure induces molecular sieve effects that hinder the diffusion of larger methylene compounds (molecular size 0.87 nm), and the hydrogenation activity is reduced consequently. In contrast, the Pt@nanoshell catalyst features a distinct hierarchical structure with abundant mesoporous defects that facilitate full interaction between methylene species and platinum nanoparticles. This structural configuration enables an impressive hydrogenation rate of 45.3 μmol $\cdot$ $gPt^{-1} \cdot h^{-1}$ for trimethylbenzene at 100 °C. Furthermore, nickel nanoparticles confined within the zeolite matrix exhibit remarkable catalytic efficiency in aromatic hydrogenation, demonstrating higher selectivity for toluene formation compared with larger trimethyl derivatives.

In the recent work reported by Iida et al., a novel synthetic strategy was developed, which involved the impregnation of ammonium molybdate followed by thermal treatment [29]. Through solid-state ion exchange combined with subsequent carburization in a CH_4/H_2 atmosphere at 700 °C, molybdenum carbide nanoclusters were successfully confined within the FAU zeolite framework. Advanced characterization techniques, such as high-resolution transmission electron microscopy, revealed that 1 nm MoC_x clusters were uniformly dispersed within the microporous network of the zeolite. The resulting MoC_x/FAU catalyst exhibited exceptional catalytic performance in alkylation reactions, achieving an alkylation rate of 2.9 mmol $\cdot g^{-1} \cdot h^{-1}$. The gas-phase hydrogenation, deoxygenation, and alkylation of anisole exhibited notable catalytic activity. The Mo_2C_x/FAU composite demonstrated a methane selectivity as low as 0.2 mol%, which was approximately 60 times lower than that of the conventional MoC catalyst (12 mol%). Experimental findings indicate that the spatial proximity between BAS in the zeolite framework and the metal active centers plays a crucial role in enhancing catalytic performance. This deliberate design of catalytic components not only improves reaction selectivity but also enhances system stability. The catalytic activity in aromatic hydrogenation processes can also be enhanced by tailoring the chemical microenvironment of metal nanoparticles confined within the zeolite framework. Liu et al. accomplished the selective hydrogenation of phenol to cyclohexanone by in situ synthesizing a Pd@KL catalyst through the encapsulation of palladium nanoclusters within LTL-type zeolites [30]. HAADF-STEM characterization demonstrated that monodisperse palladium nanoparticles, with diameters ranging from 1.93 to 2.45 nm, were uniformly embedded within the crystalline zeolite framework. In contrast, Pd/KL catalysts prepared by conventional impregnation methods tend to aggregate into larger clusters during thermal treatment due to the lack of zeolite confinement, as palladium nanoparticles are predominantly deposited on the outer surface of the inert zeolite support. Under optimized reaction conditions, the Pd@KL catalyst achieved a phenol conversion of 99.1% and a cyclohexanone selectivity of 93.8%, which can be largely attributed to the enhanced basic microenvironment around the zeolite channel intersections. The combination of experimental data and mechanistic studies suggests that the kinetic confinement follows a dual activation mechanism. The hydrogen dissociation occurs at the Pd active center, while phenol molecules are activated through interactions with basic sites, particularly oxygen bridges within the zeolite framework or basic clusters on the outer surface. Subsequently, the dissociated hydrogen migrates to interact with the activated phenol intermediate, and the formation of cyclohexanone is promoted consequently. This behavior sharply contrasts with that observed in modified weakly acidic (Pd@NaKL) or strongly acidic (Pd@HKL), which favor alternative reaction pathways.

Beyond reactant shape selectivity, the microporous structure of zeolites allows for precise control over product distribution. Liu and co-workers employed $[Pt(NH_3)_4]^{2+}$ complexes as precursors to synthesize Y-zeolite-encapsulated platinum nanoparticle catalysts (Pt@Y) through a one-pot hydrothermal approach. In comparison with conventional Pt/Y catalysts, where platinum nanoparticles are deposited externally, the Pt@Y catalyst demonstrated markedly improved hydrogenation selectivity toward aromatic ketones [31]. Notably, the catalytic system preserved complete selectivity for aromatic alcohols even under full conversion of the reactant. The zeolite confinement effect plays a critical role in achieving superior selectivity for alcohol by modulating molecular interactions spatially. The encapsulated metal structure facilitates the preferential adsorption of the aromatic ketone's phenyl group within the micropores. Therefore, the concurrent hydrogenation of both the carbonyl and aromatic functionalities on the catalytic surface is prevented.

The chemical microenvironment surrounding metal nanoparticles within the zeolite framework plays a crucial role in directing reaction pathways, as acidic sites preferentially stabilize the dicyclic hexyl ether intermediate while suppressing phenol hydrogenation. This phenomenon highlights the fundamental importance of achieving precise spatial control over the catalytic environment in the design of selective reaction channels within zeolite-based catalytic systems.

16.3.2 Hydrogenation Reactions of Alkenes and Alkynes

The catalytic hydrogenation of unsaturated hydrocarbons, such as ethylene and propylene, is widely regarded as a fundamental reaction for evaluating the performance of zeolite-anchored metal catalysts and confirming the effective incorporation of metal species into frameworks of zeolites. Molineret al. introduced an innovative encapsulation approach that employs N,N,N-trimethyl-1-adamantylammonium template agents in combination with platinum-sulfur complexes as metal precursors, enabling the immobilization of sub-nano Pt clusters within high-silica CHA zeolite [8]. Importantly, these confined Pt clusters demonstrate exceptional thermal stability, and their initial size is maintained even after being treated at 650 °C in H_2O and H_2. In an oxygen-rich environment, platinum nanoparticles dissociate into individual Pt atoms, which subsequently reaggregate into nanoscale clusters under hydrogen-rich conditions, demonstrating full structural reversibility over multiple transformation cycles. Under the same reaction conditions, the Pt@CHA catalytic system achieved 80% conversion efficiency in the hydrogenation of ethylene to ethane, while showing negligible activity (only 2%) in the conversion of allene to propane. In comparison, silica-supported Pt nanocatalysts with an average particle size of 2 nm exhibited similar catalytic performance in hydrogenation reactions. Under identical experimental conditions, the hydrogenation efficiencies for ethylene and propylene were approximately 27%. When normalized to Pt/SiO_2, the Pt@CHA catalyst displayed enhanced catalytic activity in ethylene hydrogenation, suggesting that smaller platinum species contribute to higher reactivity in this reaction. The Pt@CHA catalyst demonstrated lower conversion efficiency in propylene hydrogenation, further supporting the successful encapsulation of platinum species within the zeolite microporous framework. In addition to light olefin hydrogenation, zeolite-supported metal catalysts can achieve selective saturation of n-hexene and cyclohexene through precise modifications of the zeolite structure and localization of metal sites. Importantly, the selective conversion of 1,4-hexadiene to 2-hexene can be realized by modifying the pore structure of crystalline zeolites.

Based on these findings, Corma's team proposed an innovative strategy that employs layered MCM-22 as a zeolite precursor to encapsulate sub-nanoscale platinum clusters within an MCM-22 matrix. The catalytic performance of Pt@MCM-22 was systematically evaluated through hydrogenation experiments on light olefins, with a focus on the dispersion of platinum within the porous framework. Compared to the surface deposited Pt/MCM-22-imp (1.3 mL propane $\cdot$ s^{-1} $\cdot$ gPt^{-1}), the Pt@MCM-22 catalyst exhibited a fivefold increase in propane conversion efficiency (approximately 6.4 mL propane $\cdot$ s^{-1} $\cdot$ gPt^{-1}). In the hydrogenation of isobutylene, both catalysts showed similar reaction activities, which can be attributed to the presence of platinum based cup shaped structures predominantly located on the external surface of the zeolite, rather than within the MCM-22 framework. These experimental findings confirm that the catalytically active platinum species are effectively confined within the three-dimensional zeolite structure during the hydrogenation reaction.

The catalytic partial hydrogenation of acetylene to ethylene is a crucial purification step in industrial organic synthesis. In practical petrochemical operations, ethylene streams often contain small quantities of acetylene impurities. Selective removal of these impurities via catalytic conversion is essential to ensure the purity of the ethylene feedstock prior to its application in downstream chemical processes.

The advances in polyethylene quality optimization are also anticipated to enhance the polymerization efficiency. Recent studies have demonstrated that nanoscale metal clusters encapsulated within zeolite frameworks exhibit superior catalytic performance in acetylene semi-hydrogenation, particularly with regard to olefin selectivity. Wang et al. developed the Pd@SOD catalytic system via a hydrothermal synthesis method, strategically utilizing sodium hydroxide and $[Pd(en)_2]Cl_2$ complexes as structural templates and metal precursors [32]. This strategy enabled the successful immobilization of palladium nanoclusters (0.8 ± 0.3 nm in diameter) within a sodalite-type zeolite matrix (see Figure 16.7). Experimental results indicated that at 150 °C, the Pd@SOD catalyst achieved 94.5% ethylene selectivity with a 99.8% acetylene conversion rate. In comparison, the surface-deposited Pd/SOD catalyst displayed a comparable conversion rate (99.9%) while significantly reducing ethylene selectivity (21.5%). This difference in catalytic performance can be primarily attributed to the spatial confinement effect imposed by the zeolite structure, wherein the restricted pore size (approximately 0.28 nm) of the SOD framework limits reactant diffusion and stabilizes reaction intermediates.

Compared to acetylene and ethylene molecules, larger molecules are unable to penetrate the SOD crystal framework. The Pd@SOD system facilitates acetylene transformation through a surface-mediated mechanism, in which surface hydroxyl groups interact with acetylene via hydrogen desorption at encapsulated palladium sites, contributing to high ethylene selectivity. In contrast, the Pd/SOD catalyst enables acetylene molecules to directly adsorb onto exposed Pd active sites, promoting rapid hydrogenation. This structural arrangement enhances intermediate-palladium interactions, which inhibit the desorption of ethylene and ultimately lead to ethane formation after over-hydrogenation.

Transition metal sulfides (TMS) can also be incorporated into the zeolite framework to promote hydrogenation reactions efficiently. A research team led by Lercher successfully encapsulated bimetallic (Mo_2S_2) and tetranuclear (Mo_4S_4) molybdenum disulfide clusters within the supercages of NaY zeolite (denoted as Mo_xS_y@NaY) through chemical vapor deposition followed by heat treatment of

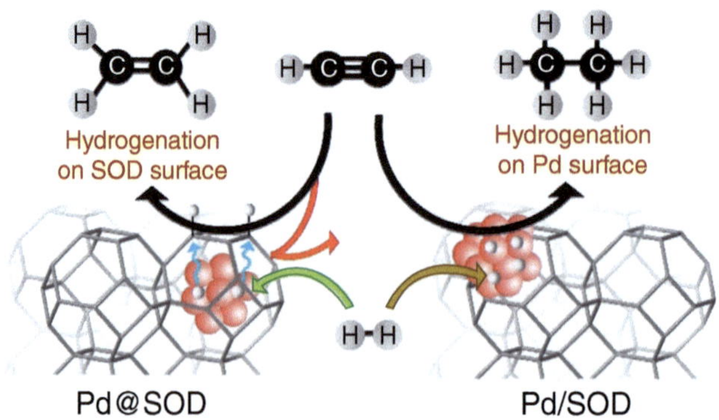

Figure 16.7 Synergism between metal catalysis and the spatial restriction effect of a small-pore zeolite in acetylene hydrogenation. *Source*: [32]/John Wiley & Sons.

reactive gas. Advanced characterization methods, such as HAADF-STEM imaging and EDS elemental mapping, verified the uniform dispersion of molybdenum species within the zeolite matrix, with sulfur distribution closely corresponding to that of molybdenum. This spatial correlation demonstrates that the cluster structures were effectively formed within the confined pore environment without compromising structural integrity. The material predominantly adopts binuclear and tetrahedral sulfide cluster configurations, with the binuclear structure being particularly well-defined. Structural characterization, supported by spectral analysis through electron paramagnetic resonance (EPR), reveals distinct paramagnetic signals consistent with the presence of a single unpaired electron per cluster. Furthermore, DFT computational modeling corroborates these experimental observations, indicating that the unpaired electron remains localized at the molybdenum center within each Mo_xS_y cluster. These findings provide strong evidence for the electronic structure and geometric arrangement of the active sites in the material. The synthesized MoS@NaY catalytic materials demonstrate not only high structural integrity but also robust functional stability under operational conditions, making them promising candidates for catalytic applications requiring long-term performance and resilience.

During the ethylene hydrogenation evaluation for seven days, the catalytic performance of the conventional MoS_2 catalyst deactivated significantly, primarily due to the structural degradation of the MoS_2 phase under reaction conditions. This observation underscores the enhanced stability conferred by spatially confining active metal sulfide sites within the zeolite framework. Zeolite-encapsulated metal catalysts demonstrate broad application across various olefin hydrogenation processes, such as the conversion of 1-hexene to hexane using S-1/Pd/ZSM-5 or Pd@FER catalysts, and the selective hydrogenation of 1,4-hexadiene to 2-hexene with Rh@CHA catalysts. However, their catalytic efficiency is somewhat diminished when applied to lighter olefins, such as ethylene, propylene, and isobutylene.

In addition to zeolite-encapsulated noble metal systems for alkyne hydrogenation, transition metal-based catalytic materials have also been proven to be active in the selective conversion of alkynes to alkenes. Zhang et al. developed a novel synthetic strategy employing nickel(II) diethylenetriamine complexes as both structural templates and metal sources to incorporate tetrahedrally coordinated nickel cations into the CHA molecular sieve framework [33]. This approach facilitates the formation of stable nickel-oxygen coordination structures through strong interactions between nickel ions and the oxygen atoms of the zeolite framework. The resulting Ni@CHA catalyst demonstrated superior performance in acetylene hydrogenation, exhibiting higher conversion efficiency and ethylene selectivity compared to conventional Ni/CHA materials prepared by ion exchange or impregnation methods. The catalytic activity was further improved through alkali metal modification, with sodium-modified Na-Ni@CHA achieving complete acetylene conversion at 180°C and 97% ethylene selectivity. DFT calculations revealed that hydrogen-activated Ni@CHA shows greater activity toward acetylene than ethylene. As for the

Ni@CHA catalyst, the rate-determining step of acetylene hydrogenation is the dissociation of molecular hydrogen. The tetrahedrally coordinated Ni(II) centers embedded within the zeolite framework act as intrinsic active sites for the selective hydrogenation of acetylene to ethylene.

References

1 Ivanova, I.I., Pomakhina, E.B., Rebrov, A.I. et al. (1998). ^{13}C MAS NMR mechanistic study of the initial stages of propane activation over H-ZSM-5 zeolite. *Top. Catal.* 6 (1): 49–59.

2 Iglesia, E., Baumgartner, J.E., and Price, G.L. (1992). Kinetic coupling and hydrogen surface fugacities in heterogeneous catalysis: I. Alkane reactions on Te/NaX, H-ZSM5, and Ga/H-ZSM5. *J. Catal.* 134 (2): 549–571.

3 Meusinger J. and Corma A.J. (1995). Activation of hydrogen on zeolites: kinetics and mechanism of n-heptane cracking on H-ZSM-5 zeolites under high hydrogen pressure. *J. Catal.* 152 (1): 189–197.

4 El-Malki, E.M., van Santen, R.A., and Sachtler, W.M. (1999). Introduction of Zn, Ga, and Fe into HZSM-5 cavities by sublimation: identification of acid sites. *J. Phys. Chem. B* 103 (22): 4611–4622.

5 Artsiusheuski, M.A., Verel, R., van Bokhoven, J.A. et al. (2023). Selective oxidative dehydrogenation of ethane and propane over copper-containing mordenite: insights into reaction mechanism and product protection. *Angew. Chem. Int. Ed.* 135 (44): e202309180.

6 Hill, J.M., Cortright, R., Dumesic, J.J. et al. (1998). Silica-and L-zeolite-supported Pt, Pt/Sn and Pt/Sn/K catalysts for isobutane dehydrogenation. *Appl. Catal. A* 168 (1): 9–21.

7 Nawaz, Z., Qing, S., Jixian, G. et al. (2010). Chemistry, effect of Si/Al ratio on performance of Pt-Sn-based catalyst supported on ZSM-5 zeolite for n-butane conversion to light olefins. *Ind. Eng. Chem.* 16 (1): 57–62.

8 Moliner, M., Gabay, J.E., Kliewer, C.E. et al. (2016). Reversible transformation of Pt nanoparticles into single atoms inside high-silica chabazite zeolite. *J. Am. Chem. Soc.* 138 (48): 15743–15750.

9 Liu L., Diaz U., Arenal R. et al. (2017). Generation of subnanometric platinum with high stability during transformation of a 2D zeolite into 3D. *Nat. Mater.* 16 (1): 132–138.

10 Liu, L., Lopez-Haro, M., Lopes, C.W. et al. (2020). Structural modulation and direct measurement of subnanometric bimetallic PtSn clusters confined in zeolites. *Nat. Catal.* 3 (8): 628–638.

11 Sun Q., Wang N., Fan Q. et al. (2020). Subnanometer bimetallic platinum–zinc clusters in zeolites for propane dehydrogenation. *Angew. Chem. Int. Ed.* 132 (44): 19618–19627.

12 Yun, J.H. and Lobo, R.F. (2014). Catalytic dehydrogenation of propane over iron-silicate zeolites. *J. Catal.* 312: 263–270.

13 De Cola, P.L., Gläser, R., Weitkamp, J.J. et al. (2006). Non-oxidative propane dehydrogenation over Pt–Zn-containing zeolites. *Chem. Eng. J.* 306: 85–97.

14 Liu J., He N., Zhang Z. et al. (2021). Highly-dispersed zinc species on zeolites for the continuous and selective dehydrogenation of ethane with CO_2 as a soft oxidant. *ACS Catal.* 11 (5): 2819–2830.

15 Maeno Z., Yasumura S., Wu X. et al. (2020). Isolated indium hydrides in CHA zeolites: speciation and catalysis for nonoxidative dehydrogenation of ethane. *J. Am. Chem. Soc.* 142 (10): 4820–4832.

16 Mansoor, E., Head-Gordon, M., Bell, A.T. (2018). Computational modeling of the nature and role of Ga species for light alkane dehydrogenation catalyzed by Ga/H-MFI. *ACS Catal.* 8 (7): 6146–6162.

17 Solsona, B., Blasco, T., Nieto, J.L. et al. (2001). Vanadium oxide supported on mesoporous MCM-41 as selective catalysts in the oxidative dehydrogenation of alkanes. *J. Catal.* 203 (2): 443–452.

18 Cavani, F., Ballarini, N., Cericola, A.J. et al. (2007). Oxidative dehydrogenation of ethane and propane: how far from commercial implementation? *Catal. Today* 127 (1–4): 113–131.

19 Qiu B., Jiang F., Lu, W.D. et al. (2020). Oxidative dehydrogenation of propane using layered borosilicate zeolite as the active and selective catalyst. *J. Catal.* 385: 176–182.

20 Lin, X., Hoel, C.A., Sachtler, W.M. et al. (2009). Oxidative dehydrogenation (ODH) of ethane with O_2 as oxidant on selected transition metal-loaded zeolites. *J. Catal.* 265 (1): 54–62.

21 Al-Mamoori, A., Lawson, S., Rownaghi, A.A. et al. (2020). Oxidative dehydrogenation of ethane to ethylene in an integrated CO_2 capture-utilization process. *Appl. Catal. B* 278: 119329.

22 Pannov, G., Sobolev, V., Kharitonov, A.J. et al. (1990). The role of iron in N_2O decomposition on ZSM-5 zeolite and reactivity of the surface oxygen formed. *J. Mol. Catal.* 61 (1): 85–97.

23 Pérez-Ramírez J. and Gallardo-Llamas A.J. (2005). Impact of the preparation method and iron impurities in Fe-ZSM-5 zeolites for propylene production via oxidative dehydrogenation of propane with N_2O. *Appl. Catal. A* 279 (1–2): 117–123.

24 Kumar, M.S., Pérez-Ramírez, J., Debbagh, M. et al. (2006). Evidence of the vital role of the pore network on various catalytic conversions of N_2O over Fe-silicalite and Fe-SBA-15 with the same iron constitution. *Appl. Catal. B* 62 (3–4): 244–254.

25 Nowińska, K., Wącław, A., and Izbińska, A.J. (2003). Propane oxydehydrogenation over transition metal modified zeolite ZSM-5. *Appl. Catal. A* 243 (2): 225–236.

26 Bulánek, R., Wichterlová, B., Novoveská, K. et al. (2004). Oxidation of propane with oxygen and/or nitrous oxide over Fe-ZSM-5 with low iron concentrations. *Appl. Catal. A* 264 (1): 13–22.

27 Li, S., Tuel, A., Laprune, D. et al. (2015). Transition-metal nanoparticles in hollow zeolite single crystals as bifunctional and size-selective hydrogenation catalysts. *Chem. Mater.* 27 (1): 276–282.

28 Li, S., Tuel, A., Meunier, F. et al. (2015). Platinum nanoparticles entrapped in zeolite nanoshells as active and sintering-resistant arene hydrogenation catalysts. *Catal. Today* 332: 25–30.

29 Iida, T., Shetty, M., Murugappan, K. et al. (2017). Encapsulation of molybdenum carbide nanoclusters inside zeolite micropores enables synergistic bifunctional catalysis for anisole hydrodeoxygenation. *ACS Catal.* 7 (12): 8147–8151.

30 Liu, C., Wang, J., Zhu, P. et al. (2022). Relating the performances of selective phenol hydrogenation with encapsulated palladium nanoparticles and surrounding distinct LTL-zeolite microenvironments. *Chem. Eng. J.* 430: 132589.

31 Chen, Q., Kang, H., Liu, X. et al. (2020). Selective hydrogenation of aromatic ketone over Pt@ Y zeolite through restricted adsorption conformation of reactants by zeolitic micropores. *ChemCatChem* 12 (7): 1948–1952.

32 Wang, S., Zhao, Z.J., Chang, X. et al. (2019). Activation and spillover of hydrogen on sub-1 nm palladium nanoclusters confined within sodalite zeolite for the semi-hydrogenation of alkynes. *Angew. Chem. Int. Ed.* 58 (23): 7668–7672.

33 Zhang, T., Chen, Z., Walsh, A.G. et al. (2020). Single-atom catalysts supported by crystalline porous materials: views from the inside. *Adv. Mater.* 32 (44): 2002910.

Index

Note: Page numbers in *italics* and **bold** refers to figures and tables respectively.

C

n

o

t

u